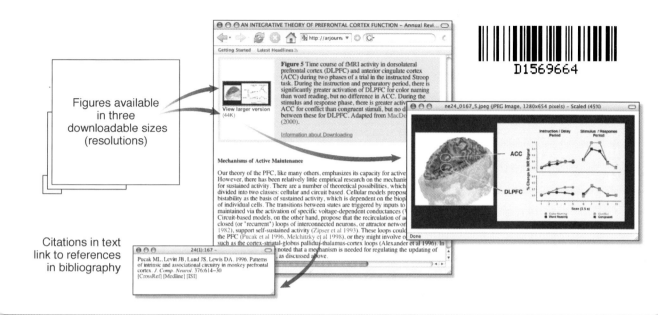

Figures available in three downloadable sizes (resolutions)

Citations in text link to references in bibliography

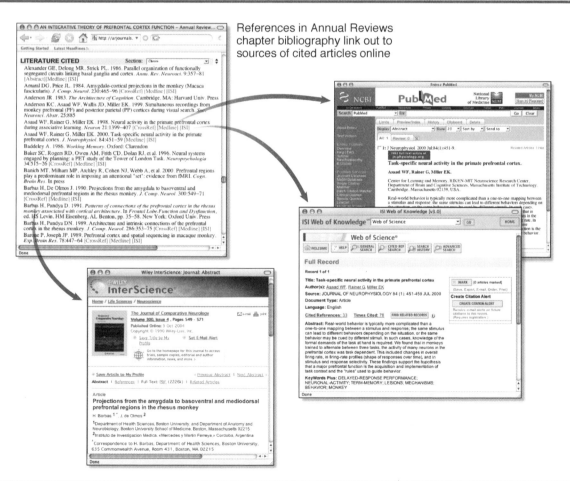

References in Annual Reviews chapter bibliography link out to sources of cited articles online

Annual Review of
Neuroscience

Editorial Committee (2005)

Cornelia I. Bargmann, Rockefeller University
Steven E. Hyman, Harvard University
Thomas M. Jessell, Columbia University
William T. Newsome, Stanford University
Steven E. Petersen, Washington University
Carla J. Shatz, Harvard Medical School
Charles F. Stevens, Salk Institute for Biological Studies
Marc Tessier-Lavigne, Genentech, Inc.
Huda Y. Zoghbi, Baylor College of Medicine

**Responsible for the Organization of Volume 28
(Editorial Committee, 2002)**

Steven E. Hyman
Thomas M. Jessell
William T. Newsome
Carla J. Shatz
Charles F. Stevens
Marc Tessier-Lavigne
Huda Y. Zoghbi
Steven E. Petersen (Guest)

Production Editor: Jennifer E. Mann
Bibliographic Quality Control: Mary A. Glass
Electronic Content Coordinator: Suzanne K. Moses
Subject Indexer: Suzanne Copenhagen

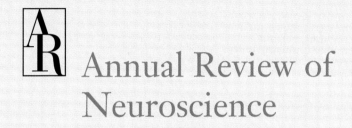

Annual Review of Neuroscience

Volume 28, 2005

Steven E. Hyman, *Editor*
Harvard University

Thomas M. Jessell, *Associate Editor*
Columbia University

Carla J. Shatz, *Associate Editor*
Harvard Medical School

Charles F. Stevens, *Associate Editor*
Salk Institute for Biological Studies

www.annualreviews.org • science@annualreviews.org • 650-493-4400

Annual Reviews
4139 El Camino Way • P.O. Box 10139 • Palo Alto, California 94303-0139

Annual Reviews
Palo Alto, California, USA

COPYRIGHT © 2005 BY ANNUAL REVIEWS, PALO ALTO, CALIFORNIA, USA. ALL RIGHTS RESERVED. The appearance of the code at the bottom of the first page of an article in this serial indicates the copyright owner's consent that copies of the article may be made for personal or internal use, or for the personal or internal use of specific clients. This consent is given on the condition that the copier pay the stated per-copy fee of $20.00 per article through the Copyright Clearance Center, Inc. (222 Rosewood Drive, Danvers, MA 01923) for copying beyond that permitted by Section 107 or 108 of the U.S. Copyright Law. The per-copy fee of $20.00 per article also applies to the copying, under the stated conditions, of articles published in any *Annual Review* serial before January 1, 1978. Individual readers, and nonprofit libraries acting for them, are permitted to make a single copy of an article without charge for use in research or teaching. This consent does not extend to other kinds of copying, such as copying for general distribution, for advertising or promotional purposes, for creating new collective works, or for resale. For such uses, written permission is required. Write to Permissions Dept., Annual Reviews, 4139 El Camino Way, P.O. Box 10139, Palo Alto, CA 94303-0139 USA.

International Standard Serial Number: 0147-006X
International Standard Book Number: 0-8243-2428-5

All Annual Reviews and publication titles are registered trademarks of Annual Reviews.

∞ The paper used in this publication meets the minimum requirements of American National Standards for Information Sciences—Permanence of Paper for Printed Library Materials, ANSI Z39.48-1992.

Annual Reviews and the Editors of its publications assume no responsibility for the statements expressed by the contributors to this *Annual Review*.

TYPESET BY TECHBOOKS, FAIRFAX, VA
PRINTED AND BOUND BY QUEBECOR WORLD, KINGSPORT, TN

Annual Review of
Neuroscience

Volume 28, 2005

Contents

Genetics of Brain Structure and Intelligence
    Arthur W. Toga and Paul M. Thompson 1

The Actin Cytoskeleton: Integrating Form and Function at the Synapse
    Christian Dillon and Yukiko Goda 25

Molecular Pathophysiology of Parkinson's Disease
    Darren J. Moore, Andrew B. West, Valina L. Dawson, and Ted M. Dawson 57

Large-Scale Genomic Approaches to Brain Development and Circuitry
    Mary E. Hatten and Nathaniel Heintz 89

Autism: A Window Onto the Development of the Social
and the Analytic Brain
    Simon Baron-Cohen and Matthew K. Belmonte 109

Axon Retraction and Degeneration in Development and Disease
    Liqun Luo and Dennis D.M. O'Leary 127

Structure and Function of Visual Area MT
    Richard T. Born and David C. Bradley 157

Growth and Survival Signals Controlling Sympathetic Nervous System
Development
    Natalia O. Glebova and David D. Ginty 191

Adult Neurogenesis in the Mammalian Central Nervous System
    Guo-li Ming and Hongjun Song 223

Mechanisms of Vertebrate Synaptogenesis
    Clarissa L. Waites, Ann Marie Craig, and Craig C. Garner 251

Olfactory Memory Formation in *Drosophila*: From Molecular
to Systems Neuroscience
    Ronald L. Davis 275

The Circuitry of V1 and V2: Integration of Color, Form, and Motion
    Lawrence C. Sincich and Jonathan C. Horton 303

Molecular Gradients and Development of Retinotopic Maps
Todd McLaughlin and Dennis D.M. O'Leary 327

Neural Network Dynamics
Tim P. Vogels, Kanaka Rajan, and L.F. Abbott 357

The Plastic Human Brain Cortex
Alvaro Pascual-Leone, Amir Amedi, Felipe Fregni, and Lotfi B. Merabet 377

An Integrative Theory of Locus Coeruleus-Norepinephrine Function:
Adaptive Gain and Optimal Performance
Gary Aston-Jones and Jonathan D. Cohen 403

Neuronal Substrates of Complex Behaviors in *C. elegans*
Mario de Bono and Andres Villu Maricq 451

Dendritic Computation
Michael London and Michael Häusser 503

Optical Imaging and Control of Genetically Designated Neurons
in Functioning Circuits
Gero Miesenböck and Ioannis G. Kevrekidis 533

INDEXES

Subject Index 565

Cumulative Index of Contributing Authors, Volumes 19–28 577

Cumulative Index of Chapter Titles, Volumes 19–28 582

ERRATA

An online log of corrections to *Annual Review of Neuroscience* chapters
may be found at http://neuro.annualreviews.org/

Related Articles

From the *Annual Review of Biochemistry,* Volume 74 (2005)

   The Biochemistry of Parkinson's Disease
     Mark R. Cookson

   Monitoring Energy Balance: Metabolites of Fatty Acid Synthesis as Hypothalamic Sensors
     Paul Dowell, Zhiyuan Hu, and M. Daniel Lane

   Copper-Zinc Superoxide Dismutase and Amyotrophic Lateral Sclerosis
     Joan Selverstone Valentine, Peter A. Doucette, and Soshanna Zittin Potter

From the *Annual Review of Biomedical Engineering,* Volume 7 (2005)

   Deterministic and Stochastic Elements of Axonal Guidance
     Susan Maskery and Troy Shinbrot

   Functional Electrical Stimulation for Neuromuscular Applications
     P. Hunter Peckham and Jayme S. Knutson

   Instrumentation Aspects of Animal PET
     Yuan-Chuan Tai and Richard Laforest

From the *Annual Review of Cell and Developmental Biology,* Volume 20 (2004)

   RNA Translation in Axons
     Michael Piper and Christine Holt

   Cortical Neuronal Migration Mutants Suggest Separate but Intersecting Pathways
     Stephanie Bielas, Holden Higginbotham, Hiroyuki Koizumi, Teruyuki Tanaka, and Joseph G. Gleeson

   Specification of Temporal Identity in the Developing Nervous System
     Bret J. Pearson and Chris Q. Doe

   The W$_{\text{NT}}$ Signaling Pathway in Development and Disease
     Catriona Y. Logan and Roel Nusse

From the *Annual Review of Genomics and Human Genetics,* Volume 5 (2004)

    Autism as a Paradigmatic Complex Genetic Disorder
        Jeremy Veenstra-VanderWeele, Susan L. Christian, and Edwin H. Cook, Jr.

    Mammalian Circadian Biology: Elucidating Genome-Wide Levels of Temporal Organization
        Phillip L. Lowrey and Joseph S. Takahashi

From the *Annual Review of Medicine,* Volume 56 (2005)

    The Silent Revolution: RNA Interference as Basic Biology, Research Tool, and Therapeutic
        Derek M. Dykxhoorn and Judy Lieberman

    Monogenic Obesity in Humans
        I. Sadaf Farooqi and Stephen O'Rahilly

From the *Annual Review of Physiology,* Volume 67 (2005)

    RNAi as an Experimental and Therapeutic Tool to Study and Regulate Physiological and Disease Processes
        Christopher P. Dillon, Peter Sandy, Alessio Nencioni, Stephan Kissler, Douglas A. Rubinson, and Luk Van Parijs

    Endocrinology of the Stress Response
        Evangelia Charmandari, Constantine Tsigos, and George Chrousos

    Lessons in Estrogen Biology from Knockout and Transgenic Animals
        Sylvia C. Hewitt, Joshua C. Harrell, and Kenneth S. Korach

    Regulation of Signal Transduction Pathways by Estrogen and Progesterone
        Dean P. Edwards

    Retinal Processing Near Absolute Threshold: From Behavior to Mechanism
        Greg D. Field, Alapakkam P. Sampath, and Fred Rieke

From the *Annual Review of Psychology,* Volume 55 (2004)

    On Building a Bridge Between Brain and Behavior
        Jeffrey D. Schall

    The Neurobiology of Consolidations, Or, How Stable is the Engram?
        Yadin Dudai

    Understanding Other Minds: Linking Developmental Psychology and Functional Neuroimaging
        R. Saxe, S. Carey, and N. Kanwisher

    Hypocretin (Orexin): Role in Normal Behavior and Neuropathology
        Jerome M. Siegel

Visual Mechanisms of Motion Analysis and Motion Perception
  Andrew M. Derrington, Harriet A. Allen, and Louise S. Delicato

Cumulative Progress in Formal Theories of Attention
  Gordon D. Logan

The Psychology and Neuroscience of Forgetting
  John T. Wixted

Schizophrenia: Etiology and Course
  Elaine Walker, Lisa Kestler, Annie Bollini, and Karen M. Hochman

Annual Reviews is a nonprofit scientific publisher established to promote the advancement of the sciences. Beginning in 1932 with the *Annual Review of Biochemistry*, the Company has pursued as its principal function the publication of high-quality, reasonably priced *Annual Review* volumes. The volumes are organized by Editors and Editorial Committees who invite qualified authors to contribute critical articles reviewing significant developments within each major discipline. The Editor-in-Chief invites those interested in serving as future Editorial Committee members to communicate directly with him. Annual Reviews is administered by a Board of Directors, whose members serve without compensation.

2005 Board of Directors, Annual Reviews

Richard N. Zare, *Chairman of Annual Reviews, Marguerite Blake Wilbur Professor of Chemistry, Stanford University*
John I. Brauman, *J.G. Jackson–C.J. Wood Professor of Chemistry, Stanford University*
Peter F. Carpenter, *Founder, Mission and Values Institute, Atherton, California*
Sandra M. Faber, *Professor of Astronomy and Astronomer at Lick Observatory, University of California at Santa Cruz*
Susan T. Fiske, *Professor of Psychology, Princeton University*
Eugene Garfield, *Publisher, The Scientist*
Samuel Gubins, *President and Editor-in-Chief, Annual Reviews*
Steven E. Hyman, *Provost, Harvard University*
Daniel E. Koshland Jr., *Professor of Biochemistry, University of California at Berkeley*
Joshua Lederberg, *University Professor, The Rockefeller University*
Sharon R. Long, *Professor of Biological Sciences, Stanford University*
J. Boyce Nute, *Palo Alto, California*
Michael E. Peskin, *Professor of Theoretical Physics, Stanford Linear Accelerator Center*
Harriet A. Zuckerman, *Vice President, The Andrew W. Mellon Foundation*

Management of Annual Reviews

Samuel Gubins, President and Editor-in-Chief
Richard L. Burke, Director for Production
Paul J. Calvi Jr., Director of Information Technology
Steven J. Castro, Chief Financial Officer and Director of Marketing & Sales

Annual Reviews of

Anthropology
Astronomy and Astrophysics
Biochemistry
Biomedical Engineering
Biophysics and Biomolecular Structure
Cell and Developmental Biology
Clinical Psychology
Earth and Planetary Sciences
Ecology, Evolution, and Systematics
Entomology
Environment and Resources

Fluid Mechanics
Genetics
Genomics and Human Genetics
Immunology
Law and Social Science
Materials Research
Medicine
Microbiology
Neuroscience
Nuclear and Particle Science
Nutrition
Pathology: Mechanisms of Disease

Pharmacology and Toxicology
Physical Chemistry
Physiology
Phytopathology
Plant Biology
Political Science
Psychology
Public Health
Sociology

SPECIAL PUBLICATIONS
Excitement and Fascination of Science, Vols. 1, 2, 3, and 4

Genetics of Brain Structure and Intelligence

Arthur W. Toga and Paul M. Thompson

Laboratory of Neuro Imaging, Department of Neurology, School of Medicine, University of California, Los Angeles, California 90095; email: toga@loni.ucla.edu, thompson@loni.ucla.edu

Key Words

heritability, IQ, brain mapping

Abstract

Genetic influences on brain morphology and IQ are well studied. A variety of sophisticated brain-mapping approaches relating genetic influences on brain structure and intelligence establishes a regional distribution for this relationship that is consistent with behavioral studies. We highlight those studies that illustrate the complex cortical patterns associated with measures of cognitive ability. A measure of cognitive ability, known as g, has been shown highly heritable across many studies. We argue that these genetic links are partly mediated by brain structure that is likewise under strong genetic control. Other factors, such as the environment, obviously play a role, but the predominant determinant appears to genetic.

Contents

INTRODUCTION 2
INTELLIGENCE 3
    Fluid and Crystallized Intelligence 5
BRAIN MAPPING 5
    Maps of Brain Structure 6
    Registration and Mapping 6
    Cortical Pattern Matching 6
    Statistical Maps 9
GENETIC INFLUENCES
  ON BRAIN STRUCTURE 9
    Related MRI Studies 10
    Candidate Genes and Brain
      Function 11
    Specific Genes and Brain Structure 11
HERITABILITY OF
  INTELLIGENCE 13
    Environmental Influences on
      Intelligence 14
    Gene x Environment Correlations . 15
BRAIN STRUCTURE AND
  INTELLIGENCE 16
    Environmental Influences on Brain
      Structure 16
CONCLUSION 17

INTRODUCTION

The relationship between genetics, brain structure, and intelligence is an age-old polemic evident in such diverse disciplines as phrenology, sociology, education, neuroscience, and politics. Measures of brain anatomy, inferred from cranial morphology (circa the 1800s) or made directly with imaging using magnetic resonance, have been correlated with a variety of cognitive assessments (see, e.g., Andreasen et al. 1993, McDaniel & Nguyen 2002). Numerous reviews of the literature (Herrnstein & Murray 1994, Jensen 1998)—in addition to personal experience—lead one to conclude, perhaps heretically, that we are not all created equal. But the question still deserves attention: What are the relative influences of nature (genetics) and nurture (environment) on the brain, and how do these affect intelligence?

Structural imaging of total brain gray and white matter volumes is perhaps the most obvious approach to correlate brain measures with general intelligence. Brain structure measured from MRI correlates with intelligence test scores as total brain volume (Gignac et al. 2003), as do the volumes of individual lobes and aggregate gray and white matter volumes (Posthuma et al. 2002). The quest for better specificity regarding regional correlations of brain structure with intelligence has required more sophisticated analytic techniques to achieve sufficient sensitivity. Voxel-based morphometry, where voxels belonging to an area in the brain are counted and analyzed (Ashburner & Friston 2000, Haier et al. 2004), and surface-based approaches, where three dimensional (3D) models of brain structures are compared across subjects (Thompson et al. 2001a), have each demonstrated regional differences in relationships to IQ.

That there is a clear relationship between intelligence and regional brain volumes does not shed light on why there are differences across individuals. Heritability of gray matter density (Thompson et al. 2001a) and familial contributions to brain morphology in general have been demonstrated repeatedly (Baaré et al. 2001a, Pfefferbaum et al. 2001). Studies of healthy twins (Posthuma et al. 2002) and cohorts of siblings discordant and concordant for a specific disease (Cannon et al. 2002, Narr et al. 2002) all provide evidence regarding the heritability of brain morphology. Finally, empirically we (Devlin et al. 1997) showed that monozygotic twins reared apart are more alike—for many cognitive measures including IQ—compared with fraternal twins raised together. This underscores the relevance of genetic factors in shaping intelligence and brain structure.

Interactions with the environment also contribute to differences in brain morphology. Several animal studies show that environmental stimulation can alter synaptic densities in the cortex of rodents reared in impoverished versus enriched environments (Greenough et al. 1970, Diamond 1988). Furthermore, animals maintained in enriched environments were better problem solvers (but not in all tests) than those not maintained in enriched environments (Forgays & Forgays 1952). Thus it is clear that several, interrelated factors influence cognitive function in general, and intelligence specifically.

Here we examine the recent application of sophisticated brain-mapping approaches relating genetic influences on brain structure and intelligence. We highlight those studies that illustrate the complex cortical patterns associated with measures of cognitive ability. Drawing on work with cohorts of subjects at risk for several genetically linked diseases, twins, and observations during brain maturation and degeneration with age, we characterize this interesting and important basis for human diversity.

INTELLIGENCE

Intelligence has several meanings, largely based on the context in which the term is used. Generally referring to competence and accomplishment, in neuroscience intelligence is typically referred to as general cognitive ability and quantified as Spearman's g—after its first proponent, Charles Spearman (1927), the statistician who developed factor analysis. Many psychometric and twin studies have used this cognitive measure to quantify intellectual function.

Intelligence testing began in 1897 with the work of the French psychologist Alfred Binet, who, together with Theodore Simon, developed tests to identify children who needed special remedial teaching. By developing norms for mental ability at each age, Binet could quantify whether a child was ahead of or behind his peers, and by how many years. German psychologist Wilhelm Stern noted that being a year ahead at age 5 was more significant than at age ten, so he multiplied the ratio of mental age to chronological age by 100 to obtain an intelligence quotient (or IQ—a term coined by American scientist Lewis Terman), with scores over 100 being above average. IQ tests, among them the Army Alpha and Beta Tests, were subsequently adopted by the U.S. army to help assign jobs to vast numbers of recruits; nearly two million American men had taken these tests by 1919. Lewis Terman at Stanford University subsequently adapted the Binet-Simon tests for the American school curriculum and published the Revised Stanford-Binet Intelligence Tests in 1937, 1960, and 1985. IQ tests began to be widely used in schools after World War I, largely to predict academic potential and to assign children to suitable classes according to intellectual ability. Traditional intelligence tests and scholastic aptitude tests (SAT) remain a key part of college admissions to this day. Among the tests still in use is the Wechsler Adult Intelligence Scale (WAIS). On the basis of work by psychologist David Wechsler in the 1930s, the WAIS (and its counterpart for assessing children—the WISC) provides separate scales for verbal, performance, and total IQ. These scales are often used to assist with psychiatric diagnosis.

In psychometric research, statistical analysis can distill from multiple tests a measure of mental ability that is independent—as far as possible—of the subject matter of the tests. In computing the g factor, for example, factor analysis isolates a component of intellectual function that is common to multiple cognitive tests, but not specific to the task being performed. IQ tests come in different forms, but they typically assess visuospatial, deductive, semantic, and symbolic reasoning ability. Specific subtests may evaluate a subject's ability to perform inferences, to detect similarities and differences in geometrical patterns or word patterns, and to process complex information quickly and accurately.

People differ substantially in their performance on these tests, but those who do well on one test tend to do well on others. The high correlations among scores on tests of spatial relations, logic, vocabulary, picture

Spearman's g: quantified general cognitive ability (intelligence); basic general factor of mental ability

completion, and even reaction times support the notion that there may be an overarching skill that underlies intellectual ability, rather than many distinct and independent abilities. Scores on a range of tests can be factor analyzed to give g, a single summary measure of cognitive ability. g is composed of a small number of (non-independent) subfactors representing more specific abilities (Carroll 1993, Deary 2001), but each of these correlates closely with g. One of the best tests for measuring "pure g" is thought to be Raven's Progressive Matrices, a nonverbal test of inductive reasoning.

The validity of g as a single, unitary measure of intelligence has been hotly debated by its advocates and detractors (Jensen 1969, Brand 2001, in favor; see Gould 1996, Kamin 1997 for contrary views). Most psychometric researchers agree that the g factor is sensitive to individual differences in abilities to learn, reason, and solve problems. It predicts scholastic achievement, employment, lifetime income, and even health-related parameters such as life expectancy (Gottfredson 1997). The ethics and validity of using IQ tests to predict educational potential, and in college admissions and recruitment decisions, are still somewhat controversial. In the 1960s, many boards of education rejected IQ testing because of concerns about possible cultural biases in test questions, and there was a general backlash against psychometric testing in admissions and hiring decisions, a political trend that has been reversed somewhat today.

From a scientific standpoint, some argue that the basic general factor of mental ability (g) can explain performance variations on individual mental tests (Spearman 1927, Jensen 1998). Most mental ability tests correlate with g, and the degree to which they do has been termed their g-loading [analogous to an octane rating for gasoline (Jensen 1980)]. Performance variations on different tasks may therefore depend on how much each task draws on a general cognitive process underlying mental ability (the unitary intelligence theory). Advocates of unitary intelligence have typically pointed to physiological parameters in the brain that are correlated with g, including reaction times, nerve conduction velocity, or cerebral glucose metabolism during problem solving (Haier et al. 1988). Other brain-based correlates of g have been observed in recent MRI studies showing that differences in frontal gray matter volumes correlate with g ($p < 0.0044$; $p < 0.0176$ after correction for multiple tests; Thompson et al. 2001a; see also Haier et al. 2004).

A more modular view, to some extent implicit in brain-mapping studies, interprets intelligence as reflecting multiple abilities that may have anatomically distinct biological substrates in the brain. Functional MRI, for example, can be used to build a more mechanistic model of intelligence because it can localize brain systems involved during cognitive tasks. The activation of specific neural systems in the frontal and parietal lobes correlates with g, which suggests that these regions interact to contribute to g (Prabhakaran et al. 1997, 2001; Duncan et al. 2000; Gray et al. 2002).

A contrary view of intelligence holds that important intellectual abilities are poorly assessed or entirely missed by standardized intelligence tests. Sternberg (1999) proposed a triarchic theory of intelligence, in which practical and creative intelligence are regarded on par with analytic skills. For Sternberg, analytic intelligence denotes one of three primary intellectual skills, namely one that is similar to the g factor—the ability to recognize and apply logical relations. Equally fundamental, however, are practical intelligence, which denotes pragmatic and social skills, and creative intelligence, or the ability to come up with imaginative solutions to problems rather than applying familiar logical rules or book knowledge. Social or emotion-related abilities have also been argued to be essential ingredients in mental function (Salovey et al. 2002).

A still broader view of intelligence has been popularized by Gardner (2000). Gardner posits at least seven types of intelligence (mathematical, spatial, musical, bodily-kinesthetic, intra-personal, and interpersonal). The case for multiple intelligences has been supported by studies of brain lesions that cause very specific neurological deficits but leave many cognitive abilities intact (e.g., speech or visuospatial skills). Gardner considers that proponents of the *g* factor confuse intelligence with a highly specific type of scholastic performance.

The most negative view of IQ testing is that inherent biases make cognitive tests a poor measure of individual competence. Detractors of IQ tests say that the ability to answer some questions may depend on a person's upbringing or cultural background, and that the questions assume a familiarity or agreement with certain cultural norms. Situational factors may also impair performance (Steele & Aronson 1995; Gould 1996, p. 166; Baumeister et al. 2002; Schmader & Johns 2004).

Fluid and Crystallized Intelligence

Even among psychometric researchers who agree that there is a general factor in cognitive ability, crystallized and fluid intelligence are often distinguished (Cattell 1971). Crystallized intelligence refers to the large body of factual knowledge that an individual accumulates over his/her lifespan, including, for example, vocabulary. This ability to apply knowledge to solve problems is largely determined by education and experience, and increases with age. Fluid intelligence, however, refers to analytical reasoning ability, as well as memory and information processing speed, and it declines somewhat with age. The fluid component of intelligence is thought to be largely genetically determined, however. In addition, fluid intelligence is strongly associated with working memory (Prabhakaran et al. 2001) and is correlated with activation during cognitively demanding tasks observed with functional MRI (Gray et al. 2003).

BRAIN MAPPING

Because of its promise in localizing brain function, functional brain imaging has been widely applied to map brain activation in a variety of psychiatric and neurological disorders. Brain activation can be examined noninvasively while subjects perform specific tasks or cognitive assessments [see Cabeza & Nyberg 2000, for a review of studies using positron emission tomography (PET) and functional MRI (fMRI)]. However, the cause of individual differences in hemodynamic-based functional measures—their heritability, for example—is largely unknown. Although it is clear that functional imaging provides the link between the anatomic maps and cognitive measures, the present paucity of data using fMRI may be due to the vagaries of neurovascular coupling, the variability of the response or the limitations of instrumentation, and protocols to date. Numerous efforts are underway to collect sufficient baseline data to attempt to improve sensitivity. The International Consortium for Brain Mapping has developed a battery of fMRI tests (Mazziotta et al. 2001) that exhibit stable baseline across subjects. These ultimately can be used to normalize other more cognitively challenging behavioral tasks in much the same way as structural scans are normalized for placement into an atlas. Similarly, the Bioinformatics Research Network has developed a series of tasks (http://www.nbirn.net) specifically designed to normalize across populations of schizophrenic patients and their normal matched controls. Thus, it is likely that in the near future we will see many more studies examining genetic influences on brain function using fMRI.

Structural brain mapping, in contrast, has already shown specific patterns related to intelligence (see above) and, as with other brain-mapping studies, can provide the anatomic framework to achieve improved sensitivity in functional studies. For this reason, we next review the steps required to create brain maps. These include maps of morphologic features, such as the 3D distribution of gray and white matter in the brain, and statistical maps that compile these maps from whole populations. To examine sources of morphological and

> **Brain map:** a relationship between points in a coordinate space and features or annotations describing brain structure and/or function

Cortical pattern matching: encoding both gyral patterning and gray matter variation

functional variability across subjects, we also review methods that combine imaging and genetic statistics to compute genetic influences on the brain (Thompson et al. 2001a, 2003). This combination as in correlation creates an important link between genetics, brain measures, and intelligence, shedding light on the systems involved in cognition and which factors affect their function.

Atlases to regionally chart the degree and extent of individual variations in brain structure require detailed descriptions of anatomy to achieve a morphometric comparison rather than the volumetric comparisons described above. To create atlases that contain detailed representations of anatomy, we have developed model-driven algorithms that use geometrical surfaces to represent structures in the brain, such as the cortex, hippocampus, ventricles, and corpus callosum (Thompson & Toga 1996, 2003). Anatomic models provide an explicit geometry for individual structures in each scan, such as landmark points, curves, or surfaces. These modeling approaches can also answer the following questions: How does anatomy differ across subjects and between groups? What is the degree of individual variability in anatomy, and how do these differences link with cognitive measures? What are the sources of these variations, and to what degree are they influenced by genes and environment? Brain mapping can provide answers to these and other questions; the answers are typically displayed in the form of a brain map.

Maps of Brain Structure

First we consider the analysis steps required to compute the patterns of genetic influences on brain structure, using a database of brain MRIs from twins (Thompson et al. 2001). The process can be conceived as shown in **Figure 1**, where a sequence of image analysis steps are applied to brain MRI scans from multiple subjects. The goal of such an analysis is typically to create color-coded maps of brain regions with structural differences between groups, or in this case to reveal where individual differences in brain structure depend on genetic factors.

Registration and Mapping

D MRI scans are first rotated and scaled to match a standardized brain template in stereotaxic space. This template may be either an average intensity brain dataset constructed from a population of young normal subjects (Mazziotta et al. 2001) or one specially constructed to reflect the average anatomy of a subgroup of defined homogeneous subjects (e.g., Mega et al. 2000, Thompson et al. 2000, Janke et al. 2001; see these papers for a discussion of disease-specific templates). Once aligned, a measure of the brain scaling imposed is retained as a covariate for statistical analysis. A tissue classification algorithm then segments the image data into regions representing gray matter, white matter, cerebrospinal fluid (CSF), and nonbrain tissues. Because the digital models reside in the same stereotaxic space as the atlas data, surface and volume models stored as lists of vector coordinates are amenable to digital transformation, as well as geometric and statistical measurement (Mega et al. 2000, Narr et al. 2003, Thompson et al. 2004, Zhou et al. 1999). The underlying 3D coordinate system is central to all atlas systems because it supports the linkage of structure models and associated image data with spatially indexed neuroanatomic labels.

Cortical Pattern Matching

MRI scans have sufficient resolution and tissue contrast, in principle, to track cortical gray and white matter differences in individual subjects. This affords the opportunity to measure regional degrees of heritability and establish structural and even gyral/sulcal relationships with specific cognitive measures. Even so, extreme variability in gyral patterns confounds efforts (*a*) to compare between groups and (*b*) to determine the average profile of patterns within a group. Cortical pattern matching methods (detailed further in **Figure 2**) address these challenges. They encode both gyral patterning and gray matter variation. This can substantially improve the statistical power to localize effects of genes and environmental factors on the brain. These cortical analyses can also be used to measure cortical asymmetries (Narr et al. 2001, Sowell et al. 2001).

Briefly, a 3D geometric model of the cortical surface is extracted from the MRI scan and flattened to a two-dimensional planar format (to avoid making cuts, a spherical topology can be retained; Fischl et al. 2001; Thompson et al. 1997, 2002). A complex deformation, or warping transform, is then applied that aligns the sulcal anatomy of each subject with an average sulcal pattern derived for the group. To improve feature alignment across subjects, all sulci that occur consistently can be digitized and used to constrain this transformation. As far as possible, this procedure adjusts for differences in cortical patterning and

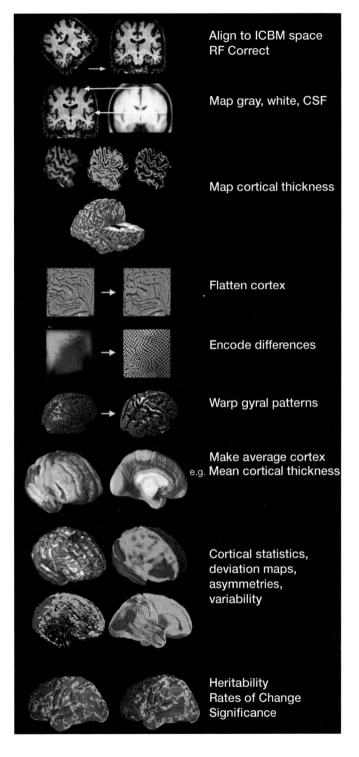

Figure 1

Analyzing cortical data. The schematic shows a sequence of image-processing steps that can be used to map how development and disease, or genetic factors, affect the cortex. Regions can also be identified where brain variation is linked with intelligence, specific cognitive measures, or clinical measures. The steps include aligning MRI data to a standard space, tissue classification, and cortical pattern matching, as well as averaging and comparing local measures of cortical gray matter volumes across subjects. (These procedures are detailed in the main text). To help compare cortical features of subjects whose anatomy differs, individual gyral patterns are flattened and aligned with a group average gyral pattern. Group variability and cortical asymmetry can also be computed. Correlations can be mapped between disease-related gray matter deficits and genetic risk factors. Maps may also be generated visualizing linkages between genes and morphology, cognitive scores, and other effects. The only steps here that are not currently automated are the tracing of sulci on the cortex. Some manual editing may also be required to assist algorithms that delete dura and scalp from images, especially if there is very little CSF in the subarachnoid space.

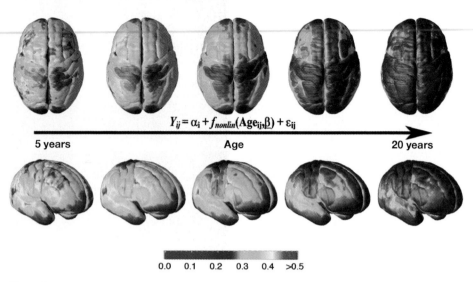

Figure 2

Right lateral and top views of the dynamic sequence of gray matter maturation over the cortical surface. The side bar shows a color representation in units of gray matter volume. Constructed from MRI scans of healthy children, these maps illustrate 15 years of brain development (ages 5–20; data from Gogtay et al. 2004). Red indicates more gray matter, blue less gray matter. Gray matter wanes in a back-to-front wave as the brain matures and neural connections are pruned. Areas performing more basic functions mature earlier; areas for higher-order functions (emotion, self-control) mature later. The prefrontal cortex, which handles reasoning and other executive functions, emerged late in evolution and is among the last to mature. Intriguingly, this sequence of brain changes is reversed in Alzheimer's disease (see **Figure 4**).

shape across subjects. Cortical measures can then be compared across subjects and groups.

Sulcal landmarks are used as anchors because homologous cortical regions are better aligned after matching sulci than by just averaging data at each point in stereotaxic space (see, e.g., fMRI studies by Rex et al. 2001; Zeineh et al. 2001, 2003; Rasser et al. 2004). Given that the deformation maps associate cortical locations with the same relation to the primary folding pattern across subjects, a local measurement of gray matter density is made in each subject and averaged across equivalent cortical locations. To quantify local gray matter, gray matter density can be measured to compare the spatial distribution of gray matter across subjects. This gray matter density measures the proportion of gray matter in a small region of fixed radius (15 mm) around each cortical point (Wright et al. 1995; Bullmore et al. 1999; Sowell et al. 1999, 2003; Ashburner & Friston 2000; Mummery et al. 2000; Rombouts et al. 2000; Baron et al. 2001; Good et al. 2001; Thompson et al. 2001a,b). Given the large anatomic variability in some cortical regions, high-dimensional elastic matching of cortical patterns (Thompson et al. 2000, 2001b) is used to associate measures of gray matter density from homologous cortical regions first across time and then also across subjects. One advantage of cortical matching is that it localizes differences relative to gyral landmarks; it also averages data from corresponding gyri, which would be impossible if data were only linearly mapped into stereotaxic space. The effects of age, gender, zygosity, g, and other measures on gray matter can be assessed at each cortical point (see **Figure 3** for an example in twins).

Statistical Maps

An algorithm then fits a statistical model, such as the general linear model (GLM) (Friston et al. 1995), to the data at each cortical location. This model results in a variety of parameters that characterize how gray matter variation is linked with other variables. The significance of these links can be plotted as a significance map. A color code can highlight brain regions where linkages are found, allowing us to visualize the strength of these linkages. In addition, estimated parameters can be plotted, such as (*a*) the local rates of gray matter loss with aging (see **Figure 2** for example) at each cortical location (e.g., as a percentage change per year), (*b*) regression parameters that identify heritability, and even (*c*) nonlinearities in the rates of brain change over time (e.g., quadratic regression coefficients; Sowell et al. 2003). In principle, any statistical model can be fitted, including genetic models that estimate genetic or allelic influences on brain structure (Thompson et al. 2003). Finally, permutation testing is typically used to ascribe an overall significance value for the observed map. This adjusts for the fact that multiple statistical tests are performed when a whole map of statistics is visualized. Subjects are randomly assigned to groups, often many millions of times on a supercomputer. A null distribution is built to estimate the probability that the observed effects could have occurred by chance, and the result is reported as a significance value for the overall map.

GENETIC INFLUENCES ON BRAIN STRUCTURE

Statistical maps of cortical anatomy can also be used to reveal genetic influences on brain morphology. **Figure 3** shows intersubject variations in cortical gray matter distribution and their heritability. In a study of genetic influences on brain structure (Thompson et al. 2001a), we began by computing

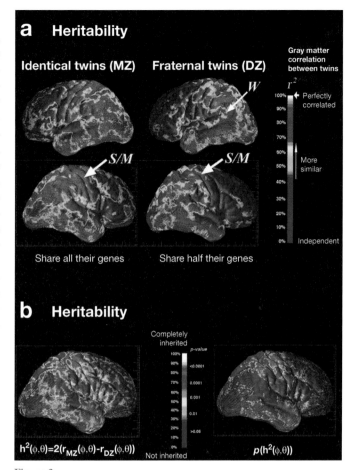

Figure 3

Heritability of gray matter. Intraclass correlation in gray matter density $g_{i,r}(\mathbf{x})$ for groups of identical and fraternal twins, after cortical pattern matching [giving maps $r_{MZ}(\phi, \theta)$ and $r_{DZ}(\phi, \theta)$, in **Figure 3*a***]. In behavioral genetics, a feature is heritable if r_{MZ} significantly exceeds r_{DZ}. An estimate of its heritability h^2 can be defined as $2(r_{MZ}-r_{DZ})$, with standard error $SE^2(h^2) = 4[((1 - r_{MZ}^2)^2/n_{MZ}) + ((1 - r_{DZ}^2)^2/n_{DZ})]$. **Figure 3*b*** shows a heritability map computed from the equation

$$h^2(\phi, \theta) = 2(r_{MZ}(\phi, \theta) - r_{DZ}(\phi, \theta)).$$

Regions in which significant genetic influences on brain structure are detected are shown in the significance map [**Figure 3*b*** (right)] $p[h^2(\phi, \theta)]$. Genetic influences on brain structure are pronounced in some frontal and temporal lobe regions, including the dorsolateral prefrontal cortex and temporal poles [denoted by *DLPFC* and *T* in **Figure 3*b*** (left)]. These effects were confirmed by assessing the significance of the effect size of h^2 by permutation (this involved repeated generation of null realizations of an h^2-distributed random field; for details of these permutation methods, see Thompson et al. 2004).

the intraclass correlations in gray matter (**Figure 3a,b**) in groups of monozygotic (MZ) and dizygotic (DZ) twins. Forty healthy normal subjects, consisting of 10 MZ and 10 age- (48.2 ± 3.4 years) and gender-matched DZ twin pairs were drawn from a twin cohort consisting of all the same-sex twins born in Finland between 1940 and 1957, inclusive, in which both members of each pair were alive and residing in Finland as of 1967 (Kaprio et al. 1990). Consistent with earlier studies reporting the high heritability of brain volume (Bartley et al. 1997), MZ within-pair gray matter differences were almost zero (intraclass $r \sim 0.9$ and higher, $p < 0.0001$ corrected; **Figure 3a**) in a broad anatomical band encompassing frontal, sensorimotor, and linguistic cortices, including Broca's speech and Wernicke's language comprehension areas. MZ twins are genetically identical, so any regional differences must be attributed to environmental effects or gene-environment interactions. Meanwhile, sensorimotor and parietal occipital, but not frontal, territories were significantly more similar in DZ twins than random pairs. Affinity was greatest in the MZ pairs, suggesting a genetic continuum in the determination of brain structure. In behavioral genetics, a feature is heritable if the identical twin correlation exceeds the fraternal twin correlation. Comparisons of MZ and DZ correlations suggested that frontal, sensorimotor, and anterior temporal cortices were under significant genetic control ($p < 0.05$, rejecting the hypothesis that $h^2 = 0$; one-tailed). Middle frontal regions, near Brodmann areas 9 and 46, displayed a 90%–95% genetic determination of structure (i.e., $h^2 \sim 0.90$–0.95). Many regions are under tight genetic control (bilateral frontal and sensorimotor regions, $p < 0.0001$; **Figure 3b**). Heritability estimates were comparable with twin-based estimates for the most highly genetically determined human traits, including fingerprint ridge count ($h^2 = 0.98$), height ($h^2 = 0.66$–0.92), and systolic blood pressure ($h^2 = 0.57$).

Related MRI Studies

The high heritability of gray matter volumes, visualized in **Figure 3**, corroborates earlier studies that revealed strong genetic influences on brain structure. Studies of healthy twins suggest that overall brain volume is highly genetically influenced (Bartley et al. 1997, Tramo et al. 1998). Volumes of some brain structures are also under strong genetic control, including the corpus callosum (Oppenheim et al. 1989, Pfefferbaum et al. 2000) and ventricles, whereas gyral patterns are much less heritable (Bartley et al. 1997, Biondi et al. 1998). Bartley et al. (1997) reported a 94% heritability for brain volume (identical twin correlation = 0.95, $p < 0.00,001$; fraternal twin correlation = 0.35, $p = 0.09$), on the basis of structural equation modeling in 10 MZ and 9 DZ pairs scanned with MRI. In elderly twins, Sullivan et al. (2001) found that the volume of the hippocampus was less heritable ($h^2 = 0.4$) than that of the adjacent temporal horns ($h^2 = 0.6$), corpus callosum ($h^2 = 0.8$), and intracranial volume ($h^2 = 0.8$). They suggested that environmental differences, perhaps interacting with genetic differences, may exert especially strong or prolonged influences on hippocampal size, consistent with its lifelong plasticity and fundamental role in learning and memory. A lower heritability figure for hippocampal size is consistent with its role in memory encoding, its vulnerability to plasma cortisol levels, and its plasticity in later life (Maguire et al. 2000; see also Lyons et al. 2001, for a related MRI study in monkeys). In a similar vein, Baaré and colleagues (2001b) found that individual differences in lateral ventricle volume were best explained by a structural equation model containing common (58%) and unique (42%) environmental factors, indicating genes to be of little or no influence. The same authors found that genetic factors almost entirely accounted for individual differences in whole brain (90%), gray (82%), and white (88%) matter volume, in a study based on a sizeable sample of 54 MZ and 58 DZ twin pairs and 34 of their full siblings. In their multivariate analysis of body height and volumes of gray matter, white matter, and the intracranial space, Baaré et al. (2001b) noted that a large part of the genetic influences were common to the three brain measures, and a smaller part was shared with height. Some genes may therefore have a general effect on the brain, whereas other genes may affect specific volumes.

More recently, Pfefferbaum et al. (2001) used diffusion imaging, which is sensitive to myelination levels and fiber orientation, to quantify the microstructure of the corpus callosum in 15 MZ and 18 DZ pairs. They found that anterior interhemispheric connecting pathways, in the callosal *genu*, were more susceptible than splenial pathways to environmental influences, perhaps reflecting the prolonged maturation of the frontal cortex well into adulthood (Sowell et al. 1999, Gogtay et al. 2004). Using bivariate genetic modeling, these authors also noted that intracranial volume and corpus callosum area were tightly correlated, a correlation due entirely to shared genetic effects between these two brain structures. Wright et al. (2002) extended this design to parcellate 92 regional gray matter volumes in 10 MZ and 9 DZ twin pairs scanned with MRI. Interregional relationships were summarized by principal component analysis of the resulting genetic correlation matrix. This analysis identified shared genetic effects on the frontal-parietal cortices and bilateral temporal cortex and insula. As the size and scope of these studies increases, decomposition of the genetic correlation matrix is likely to be a key exploratory tool to identify supraregional brain systems (Wright et al. 1999), which share common genetic influences and which may cut across conventional anatomic boundaries.

Candidate Genes and Brain Function

Heritable aspects of brain structure are important to identify because they provide endophenotypes to guide the search for specific genes whose variations are linked with brain structure and function. For example, recent functional imaging work has shown that a polymorphism in the human brain-derived neurotrophic factor (BDNF) gene is associated with poor memory performance and with working memory activation mapped with fMRI (Egan et al. 2001, 2003). Diamond et al. (2004) have recently shown striking specificity of COMT (catechol-*o*-methyltransferase) polymorphisms to some but not other prefrontal cortex–dependent tasks in children (Diamond et al. 2004).

Heritability of cognitive function is certainly complex and difficult to dissociate from environmental factors, among other influences. A recent review of candidate genes contributing to human cognition lists more than 70 suspects (Morley & Montgomery 2001). However, examining this list and relating them to spatial patterns of gene expression or segmenting those that are related to neuroanatomical regions involved in cognition results in a far more tractable problem. For example, the prefrontal cortex, an area highly involved in cognition (see Winterer & Goldman 2003, among others), links only 3 of the 70 identified by Morley & Montgomery (2001).

Specific Genes and Brain Structure

With current databases of structural brain images, there is now significant power to assess the effects of specific candidate genes on brain structure. The easiest context for evaluating these genetic influences is to examine alleles overtransmitted to individuals with specific diseases such as dementia or schizophrenia. Using statistical maps to visualize brain systems that are at genetic risk, brain images can also provide a quantitative index of disease liability in individuals at increased genetic risk for disease (Cannon et al. 2002, Narr et al. 2002, Thompson et al. 2003).

For example, the apolipoprotein E4 (ApoE4) allele is found in 38% of all Alzheimer's disease patients, but in only 15% of

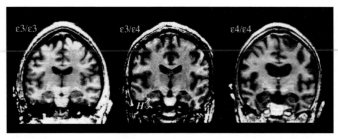

Figure 4

Patterns of brain structure associated with genetic risk for Alzheimer's disease. Brain structure is often significantly different from normal in subjects who are at genetic risk for a brain disorder but who are cognitively normal. Typical MRI scans are shown from healthy elderly subjects with zero, one, and two ε4 alleles of the ApoE gene, each of which confers increased risk for late-onset Alzheimer's disease (data courtesy of G. Small, UCLA Center on Aging). Note the hippocampal atrophy and ventricular enlargement in those at risk. The ε3 allele is most prevalent and is considered normal. Subjects at genetic risk may display metabolic and structural deficits before overt cognitive symptoms appear, which suggests that genetic and imaging information may identify candidates for early treatment in dementia (Small et al. 2000).

controls (Roses 1996). As shown in **Figure 4**, medial temporal brain structure shows profound atrophic changes in healthy ApoE4-positive individuals, and the ventricles expand, even before overt cognitive deficits can be detected. However, some brain regions are comparatively protected (e.g., frontal cortices in ApoE4 subjects with Alzheimer's disease; Geroldi et al. 1999, Hashimoto et al. 2001). Because neuroprotective drugs are effective in early dementia (Lehtovirta et al. 1995, 2000; Small et al. 2000) there is interest in associating patterns of brain change with specific genetic markers, especially if these patterns predict imminent disease onset among individuals at genetic risk.

Gray matter deficits are also found in healthy first-degree relatives of schizophrenia patients. Because these relatives are at increased genetic risk for developing the disorder themselves, there is great interest in understanding what factors promote or resist disease onset or progression (Weinberger et al. 1981, Suddath et al. 1990, Cannon et al. 2002, Thompson et al. 2003). **Figure 5** shows brain regions with significant reductions in gray matter density in healthy relatives of patients, relative to a population of normal controls who are not related to a schizophrenia patient (Cannon et al. 2002). This pattern of brain structure is intriguing because the observed deficit is associated with the degree of genetic risk (i.e., greater in MZ than DZ twins of a patient). There are also genetically mediated structural deficits in the hippocampus, and in the corpus callosum, in schizophrenic patients and in their healthy at-risk relatives (Narr et al. 2003). But, unlike the ApoE4 example, the specific genes involved in schizophrenia are currently unknown. By correlating alleles overtransmitted to patients with these structural variations, specific polymorphic genetic loci that mediate these deficits may be identified (Cannon et al. 2003). In this respect, brain mapping can assist in the search for genes involved in a disorder by generalizing genetic methods such as Haseman-Elston mapping to brain images.

Conversely, brain mapping may also help to establish the scope of brain abnormalities in diseases in which the genetic causes are already well understood. Williams syndrome, for example, results from a known genetic deletion in the 7q11.23 chromosomal region (Korenberg et al. 2000). The syndrome is associated with disrupted cortical development and mild-to-moderate mental retardation (Bellugi et al. 2000). By statistically averaging cortical thickness maps from 166 brain hemispheres and comparing Williams syndrome patients with healthy controls, we recently identified a sharply delimited region of language cortex with increased cortical thickness, revealing the cortical territory affected by the genetic deletion (Thompson et al. 2004). This selective augmentation of brain structure may underlie the relative strengths patients exhibit in language function. These maps also refine our knowledge of how the genetic deletion impacts the brain, providing new leads for molecular work on Williams syndrome and a link between genetic and behavioral findings in the disorder.

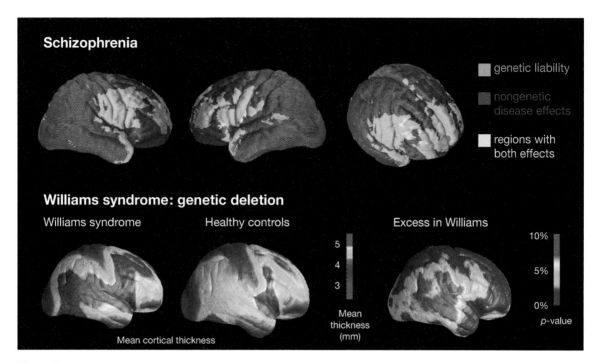

Figure 5
Patterns of brain structure associated with genetic risk for schizophrenia and with genetic deletion in Williams syndrome. Top row, last panel: Statistical combinations of brain scans from at-risk relatives of schizophrenia patients show that relatives have abnormally reduced gray matter density in the frontal cortex (*green*; data adapted from Cannon et al. 2002). In a twin design, statistical comparison of schizophrenia patients with their healthy identical twins reveals regions (*red*) in parietal and frontal cortices that have reduced gray matter density in disease. The source of these differences must be environmental in origin because the differences are based on averages of maps that subtract data from genetically identical twins. Bottom Row: Finally, group average maps of cortical thickness are shown for 43 subjects with Williams syndrome (*a*), and 40 matched healthy controls (*b*), which revealed that perisylvian language cortex is 10% thicker in the patients. Williams syndrome results from a known genetic deletion on chromosome 23q11. Composite brain maps such as these can help identify circumscribed cortical regions whose formation or maturation is influenced by the genetic lesion, perhaps during gyrogenesis.

HERITABILITY OF INTELLIGENCE

Before reviewing some of the brain substrates that correlate with intelligence, it is worth examining the evidence that there are genetic influences on intelligence; we argue that these genetic links are partly mediated by brain structure that is under strong genetic control. We review this literature only briefly because it has been examined thoroughly elsewhere (Herrnstein & Murray 1994, Gould 1996, Jensen 1998, Pinker 2002). In 1969, the debate regarding genetic influences on IQ became increasingly vitriolic after an article argued that there are racial differences in intelligence that may be genetic in origin (Jensen 1969). Most behavioral geneticists now agree that heredity plays a role in individual differences in intelligence, but some have argued that group differences in IQ include environmental influences or cultural biases in the tests (Lewontin 1975; see Jensen & Miele 2002, Gray & Thompson 2004, for reviews of arguments on this topic).

Correlations between related individuals show that both nature and nurture influence intelligence. Adopted MZ twins—raised apart—still correlate 0.72 for intelligence, i.e., one twin's intelligence strongly predicts the other's, despite their different rearing environments. This suggests an undeniable genetic component to intelligence. A popular line of attack against this argument states that several nongenetic factors could confound this association by making MZ twins more similar. For example, identical twins might be adopted into similar homes (selective placement). Sharing the same fetal environment might also make identical twins more alike cognitively or perhaps even less alike (via twin-twin competition for nutrition, transfusion effects, and so on). Also, fraternal twins may inadequately control for the effects of shared family environments (see Vogel & Motulsky 1997, Kamin & Goldberger 2002).

Nonetheless, adoption and family studies using sophisticated genetic model-fitting have shown g to be highly heritable across many studies, even more so than specific cognitive abilities [h^2 = 0.62, McClearn et al. (1997), Feldman & Otto (1997); h^2 = 0.48, Devlin et al. (1997); h^2 = 0.6–0.8, Finkel et al. (1998), Swan et al. (1990), Loehlin (1989), Chipuer et al. (1990), Plomin & Petrill (1997)]. The heritability of intelligence also increases with age: As we grow older, phenotype reflects genotype more closely. A strictly environmental theory would predict the opposite. Some IQ-related genes may not be switched on until adolescence or adulthood, but a more plausible explanation may be the existence of a gene by environment interaction (Boomsma et al. 1999, Rowe & Jacobson 1999). As individuals select or create environments that foster their genetic propensities throughout life, the genetic differences in cognition are greatly amplified (Plomin 1999). Jensen (1998) hypothesized that the more a mental test score correlates with general intelligence, or g, the higher its heritability is. If true, this hypothesis supports a biological rather than purely statistical basis for g.

Environmental Influences on Intelligence

Many environmental factors are known to influence intelligence favorably or adversely (Ceci & Williams 1997, Neisser 1998). By comparing identical twins reared apart and reared together, effects of different rearing environments can be established. Bouchard et al. (1990) found that growing up in the same family increased the IQ similarities for all types of relatives: Individual IQs correlated more highly with their MZ twins, siblings, and parents (0.86, 0.47, 0.42) if they grew up together than if they did not (0.72, 0.24, 0.22). Adopted children's IQs also correlate with their siblings (0.34) and adoptive parents (0.19), so 20%–35% of the observed population differences in IQ are thought to be due to differences in family environments. Intriguingly, these shared family environmental influences on IQ dissipate once young children leave home: As adults, adoptive relatives only correlated –0.01 for IQ (McGue et al. 1993), showing no lasting influence of shared upbringing on IQ. Those environmental influences on IQ that do last are thought to be experiences that an individual does not share with others, interpreted broadly to include the chemical environment in the womb and the multitude of random events in human experience that are hard to quantify or control.

Heritability does not imply inevitability because the environment can determine the relative impact of genetic variation (gene x environment interaction). For example, in a recent study of 320 pairs of twins who were born in the 1960s and given IQ tests at age 7, Turkheimer et al. (2003) found that environmental factors made a much bigger difference in the determination of childhood IQ in impoverished families relative to those with higher socioeconomic status. The heritability of IQ at the low end of the wealth spectrum was just 0.10 on a scale of zero to one, but it was 0.72 for families of high socioeconomic status. The importance of environmental influences on IQ was four times stronger in the poorest families than in the higher status families, which suggests that nature matters more on the high end of socioeconomic status and nurture matters more on the low end. The genetic contribution to intelligence therefore differs in different environments—a caveat against general inferences based on heritability data. The same could be said of certain physical attributes such as height, which is heritable when nutrition is not limiting.

Population-level increases in intelligence test scores have also been observed in recent decades. Dutch 18-year-old men tested in 1982 scored 20 IQ points (standard deviation = 15) higher than did 18-year-old men tested in 1952 (Dickens & Flynn 2001). This widely replicated population-level increase in intelligence is known as the Flynn Effect. Because genetic variation remained fairly stable over such a short time frame, these relatively rapid increases are attributed to nongenetic factors such as improved schooling and technology, better access to education, and improved nutrition. There has also been a reduction in some environmental toxins (such as lead) and hazards that are detrimental to IQ. Dickens & Flynn (2001) also proposed powerful gene-environment interactions to reconcile the paradox that IQ is highly heritable even though average scores have increased significantly in recent decades.

Positive environmental influences on intelligence are hard to identify, in part, because of the inevitable confounding of variables in large-scale epidemiological studies of cognition. For example, duration of breastfeeding during infancy has been associated with higher IQ in a group of more than 2000 children assessed at age 6 (Oddy et al. 2003). However, this association has been contested because it is confounded by maternal age, intelligence, and education, as well as smoking during pregnancy. After adjusting for these confounding factors, breastfeeding during infancy is still associated with enhanced childhood cognitive development (by 2–5 IQ points for full-term infants and 8 points for those with low birth weight; Drane & Logemann 2000).

Gene x Environment Correlations

The significant influence of heredity on IQ has been misinterpreted to imply that there is little point trying to educate or be educated, or that IQ is somehow impervious to change. This is a fallacy because many environmental factors, including family rearing environments, socioeconomic status, diet, and schooling, influence IQ. As noted elsewhere (Plomin & Kosslyn 2001), gray matter volume may be correlated with intelligence partly because more intelligent individuals seek out mentally challenging activities that increase the volume of their gray matter. Such strong gene x environment correlations may draw individuals with higher genetic potential into learning environments more conducive to intellectual advancement. Gifted individuals might either create or evoke situations that further promote their intellectual ability (termed active and reactive genotype-environment (GE) correlation, respectively; Plomin et al. 1977). These correlations make it impossible to conceptually differentiate effects of nature and nurture (Ridley 2003).

If environments are not randomly assigned to each individual but are, in part, individually selected on the basis of genetically influenced preferences (GE autocorrelation), it becomes impossible to discern which genetic effects act directly on intellectual function and which result from the action of environmental variation causally linked with genetic differences (Block 1995). One form of GE correlation can be estimated explicitly in adoption designs: the environment that parents provide their offspring (Neale 1997). Active and reactive correlations are more difficult to estimate, leading to suggestions that the notion of heritability conflicts with common sense (Sesardic 2002).

ApoE4:
apolipoprotein E4

BDNF:
brain-derived neurotrophic factor

COMT: catechol-*o*-methyltransferase

CSF: cerebrospinal fluid

DZ: dizygotic

fMRI: functional magnetic resonance imaging

g: a measure of cognitive ability

GLM: general linear model

MZ: monozygotic

PET: positron emission tomography

SAT: scholastic aptitude test

WAIS: Wechsler Adult Intelligence Scale

WISC: Wechsler Intelligence Scale for Children

BRAIN STRUCTURE AND INTELLIGENCE

If specific features of brain structure are under strong genetic control, investigators should determine whether any of these features are correlated with intelligence. If so, this correlation may not only reveal why IQ has repeatedly been found to be highly heritable, but also yield insight into possible neural mechanisms. To help understand this approach, we first review evidence that brain structure and intelligence are correlated before discussing evidence for the existence of genetic correlations between brain structure and intelligence (which means that the same sets of genes are implicated in determining both; Posthuma et al. 2002).

A recent meta-analysis (including a total of 1375 subjects) found that total brain volume and IQ were correlated significantly in all but 1 of 28 MRI studies, with an estimated correlation of 0.33 (McDaniel & Nguyen 2002). This finding implies that ∼10% of the population variability in IQ can be predicted from brain volume measures alone. Some studies have quoted slightly higher figures for these correlations (e.g., 0.41; Andreasen et al. 1993), and the exact value obtained will depend on the measurement error of the technique because measurement errors will tend to diminish any observed correlation (relative to the true correlation).

Linkages between brain structure and IQ also can be further localized by parcellating the brain into subregions or by creating maps of the correlations between gray matter and IQ. Recently, we found that intellectual function (*g*) was significantly linked with differences in frontal gray matter volumes, which were determined primarily by genetic factors (Thompson et al. 2001a). Posthuma et al. (2002) extended these findings using a cross-twin cross-trait (bivariate genetic) analysis to compute genetic correlations. They demonstrated that the linkage between gray matter volumes and *g* is mediated by a common set of genes. Haier et al. (2004) used voxel-based morphometry in two independent samples to identify substantial gray matter correlates of IQ. More gray matter was associated with higher IQ in all lobes, underscoring a distributed model of the neural basis of intelligence. Intriguingly, the strongest correlations are typically found between IQ and frontal gray matter volumes (Thompson et al. 2001a, Haier et al. 2004), the same brain regions that are under greatest genetic control. Frontal brain regions play a key role in working memory, executive function, and attentional processes, and their structure has rapidly expanded in recent primate evolution, consistent with their role in reasoning and intellectual function.

Environmental Influences on Brain Structure

Neural plasticity in humans may also lead to use-dependent structural adaptation in cerebral gray matter in response to environmental demands. At the gross level observable with MRI, there is already evidence that the human brain may adapt dynamically to reflect the cognitive demands of the environment. Neuroimaging studies have observed structural plasticity after training on difficult motor tasks such as juggling (Draganski et al. 2004). Increased hippocampal volumes have also been found in taxi drivers with enhanced spatial navigation skills (Maguire et al. 2000). Gaser & Schlaug (2003) also found gray matter increases in motor, auditory, and visual-spatial brain regions when comparing professional musicians (keyboard players) with a matched group of amateur musicians and nonmusicians. Brain structure is by no means unchanging even in health. Dynamic regional changes over the entire life span can be mapped (**Figure 2**), showing a progressive change in cortical volume. The heritability of brain structure, although certain, is neither final nor static. However, without genetic brain-mapping techniques (described in this review), strictly speaking it is not certain whether these brain differences are attributable to innate predisposition or due to adaptations in response to skill acquisition and repetitive rehearsal of those skills.

Intelligence therefore depends, to some extent, on structural differences in the brain that are under very strong genetic control. This indicates a partly neuroanatomical (structural) explanation for the high heritability of intelligence. These methods are currently being applied to large databases that assess the impact of candidate genes on brain structure, which allows causal pathways between genes, brain, and behavior to be pursued at the allelic level.

CONCLUSION

Currently, the most fruitful combination of genetics and imaging is perhaps their application to large patient populations. This shows great promise for seeking out genetic markers that are linked with brain structure, as well as intellectual function and cognition, more generally. Brain mapping can provide some of the hard data to establish a basis for why people vary in their general mental capacity. This review illustrates the bridge afforded by structural imaging between genetics and behavior.

Nature is not democratic. Individuals' IQs vary, but the data presented in this review and elsewhere do not lead us to conclude that our intelligence is dictated solely by genes. Instead genetic interactions with the environment suggest that enriched environments will help everyone achieve their potential, but not to equality. Our potential seems largely predetermined.

That our interpretation of intelligence, the brain, and heritability has succumbed to a variety of political and social pressures is undeniable. How the public chooses to use scientific findings in the establishment of policy, particularly in regards to education and law, however, is not the stuff of a chapter in *Annual Review of Neuroscience*. As our understanding of the complex relationships between genes, brain, and intelligence improves, what becomes of this knowledge remains to be seen.

ACKNOWLEDGMENTS

Funding support for this work was provided by grant P41 RR13642 from the National Center for Research Resources, the Center for Computational Biology, U54 RR21813, and a Human Brain Project grant and the International Consortium for Brain Mapping, funded by the National Institute of Mental Health and the National Institute on Drug Abuse (P20 MH/DA52176). Additional support was provided by grants from the National Institute for Biomedical Imaging and Bioengineering and the National Center for Research Resources (R21 EB01651, R21 RR019771; to P.T.).

LITERATURE CITED

Andreasen NC, Flaum M, Swayze V 2nd, O'Leary DS, Alliger R, et al. 1993. Intelligence and brain structure in normal individuals. *Am. J. Psychiatry* 150:130–34

Ashburner J, Friston KJ. 2000. Voxel-based morphometry—the methods. *NeuroImage* 11(6):805–21

Baare WF, Hulshoff Pol HE, Boomsma DI, Posthuma D, de Geus EJ, et al. 2001a. Quantitative genetic modeling of variation in human brain morphology. *Cereb. Cortex* 11:816–24

Baare WF, van Oel CJ, Hulshoff Pol HE, Schnack HG, Durston S, et al. 2001b. Volumes of brain structures in twins discordant for schizophrenia. *Arch. Gen. Psychiatry* 58(1):33–40

Baron JC, Chetelat G, Desgranges B, Perchey G, Landeau B, et al. 2001. In vivo mapping of gray matter loss with voxel-based morphometry in mild Alzheimer's disease. *NeuroImage* 14:298–309

Bartley AJ, Jones DW, Weinberger DR. 1997. Genetic variability of human brain size and cortical gyral patterns. *Brain* 120(2):257–69

Baumeister RF, Twenge J, Nuss CK. 2002. Effects of social exclusion on cognitive processes: Anticipated aloneness reduces intelligent thought. *J. Pers. Soc. Psychol.* 83:817–27

Bellugi U, Lichtenberger L, Jones W, Lai Z, St. George M. 2000. I. The neurocognitive profile of Williams Syndrome: a complex pattern of strengths and weakenesses. *J. Cogn. Neurosci.* 12(Suppl. 1):7–29

Biondi A, Nogueira H, Dormont D, Duyme M, Hasbon D, et al. 1998. Are the brains of monozygotic twins similar? A three-dimensional MR study. *Am. J. Neuroradiol.* 19(7):1361–67

Block N. 1995. How heritability misleads about race. *Cognition* 56:99–128

Boomsma DI, de Geus EJ, van Baal GC, Koopmans JR. 1999. A religious upbringing reduces the influence of genetic factors on disinhibition: evidence for interaction between genotype and environment on personality. *Twin Res.* 2:115–25

Bouchard TJ Jr, Lykken DT, McGue M, Segal NL, Tellegen A. 1990. Sources of human psychological differences: the Minnesota Study of Twins Reared Apart. *Science* 250:223–28

Brand C. 2001. *The G factor*. http://www.douance.org/qi/brandtgf.htm

Bullmore ET, Suckling J, Overmeyer S, Rabe-Hesketh S, Taylor E, Brammer MJ. 1999. Global, voxel, and cluster tests, by theory and permutation, for a difference between two groups of structural MR images of the brain. *IEEE Trans. Med. Imag.* 18:32–42

Cabeza R, Nyberg L. 2000. Imaging cognition II: an empirical review of 275 PET and fMRI studies. *J. Cogn. Neurosci.* 12(1):1–47

Cannon TD, Thompson PM, van Erp TGM, Toga AW, Poutanen V-P, et al. 2002. Cortex mapping reveals regionally specific patterns of genetic and disease-specific gray-matter deficits in twins discordant for schizophrenia. *Proc. Natl. Acad. Sci. USA* 99(5):3228–33

Cannon TD, van Erp TG, Bearden CE, Loewy R, Thompson P, et al. 2003. Early and late neurodevelopmental influences in the prodrome to schizophrenia: contributions of genes, environment, and their interactions. *Schizophr. Bull.* 29(4):653–69

Carroll J. 1993. *Human Cognitive Abilities: A Survey of Factor-Analytic Studies*. Cambridge, UK: Cambridge Univ. Press

Cattell RB. 1971. *Abilities: Their Structure, Growth, and Action*. Boston: Houghton Mifflin

Ceci SJ, Williams WM. 1997. Schooling, intelligence, and income. *Am. Psychol.* 52:1051–58

Chipuer HM, Rovine MJ, Plomin R. 1990. LISREL modeling: genetic and environmental influences on IQ revisited. *Intelligence* 14:11–29

Deary IJ. 2001. Human intelligence differences: a recent history. *Trends Cogn. Sci.* 5:127–30

Devlin B, Daniels M, Roeder K. 1997. The heritability of IQ. *Nature* 388(6641):468–71

Diamond A, Briand L, Fossella J, Gehlbach L. 2004. Genetic and neurochemical modulation of prefrontal cognitive functions in children. *Am. J. Psychiatry* 161:125–32

Diamond MC. 1988. *Enriching Heredity*. New York: Free Press

Dickens WT, Flynn JR. 2001. Heritability estimates versus large environmental effects: the IQ paradox resolved. *Psychol. Rev.* 108:346–69

Draganski B, Gaser C, Busch V, Schuierer G, Bogdahn U, May A. 2004. Neuroplasticity: changes in grey matter induced by training. *Nature* 427(6972):311–12

Drane DL, Logemann JA. 2000. A critical evaluation of the evidence on the association between type of infant feeding and cognitive development. *Paediatr. Perinat. Epidemiol.* 14:349–56

Duncan J, Seitz RJ, Kolodny J, Bor D, Herzog H, et al. 2000. A neural basis for general intelligence. *Science* 289(5478):457–60

Egan MF, Goldberg TE, Kolachana BS, Callicott JH, Mazzanti CM, et al. 2001. Effect of COMT Val 108/158 Met genotype on frontal lobe function and risk for schizophrenia. *Proc. Natl. Acad. Sci. USA* 98:6917–22

Egan MF, Kojima M, Callicott JH, Goldberg TE, Kolachana BS, et al. 2003. The BDNF val66met polymorphism affects activity-dependent secretion of BDNF and human memory and hippocampal function. *Cell* 112:257–69

Feldman MW, Otto SP. 1997. Twin studies, heritability, and intelligence. *Science* 278:1383–84

Finkel D, Pedersen NL, Plomin R, McClearn GE. 1998. Longitudinal and cross-sectional twin data on cognitive abilities in adulthood: the Swedish Adoption/Twin Study of Aging. *Dev. Psychol.* 34:1400–13

Fischl B, Liu A, Dale AM. 2001. Automated manifold surgery: constructing geometrically accurate and topologically correct models of the human cerebral cortex. *IEEE Trans Med. Imag.* 20(1):70–80

Forgays DG, Forgays JW. 1952. The nature of the effect of free-environmental experience in the rat. *J. Comp. Physiol. Psychol.* 45:322–28

Friston KJ, Frith CD, Frackowiack RS, Turner R. 1995. Characterizing dynamic brain responses with fMRI: a multivariate approach. *NeuroImage* 2(2):166–72

Gardner H. 2000. *Intelligence Reframed*. New York: Basic Books

Gaser C, Schlaug G. 2003. Gray matter differences between musicians and nonmusicians. *Ann. N. Y. Acad. Sci.* 999:514–17

Geroldi C, Pihlajamaki M, Laakso MP, DeCarli C, Beltramello A, et al. 1999. APOE-epislon4 is associated with less frontal and more medial temporal lobe atrophy in AD. *Neurology* 53(8):1825–32

Gignac G, Vernon PA, Wickett JC. 2003. Factors influencing the relationship between brain size and intelligence. In *The Scientific Study of General Intelligence*, ed. H Nyborg, pp. 93–106. Amsterdam: Pergamon

Gogtay N, Lusk L, Hayashi KM, Giedd JN, Greenstein D, et al. 2004. Dymanic mapping of human cortical development during childhood and adolescence. *Proc. Natl. Acad. Sci. USA* 101(21):8174–79

Good CD, Johnsrude IS, Ashburner J, Henson RN, Friston KJ, Frackowiak RS. 2001. A voxel-based morphometric study of ageing in 465 normal adult human brains. *NeuroImage* 14(1 Pt. 1):21–36

Gottfredson LS. 1997. Why g matters: the complexity of everyday life. *Intelligence* 24:79–132

Gould SJ. 1996 [1981]. *The Mismeasure of Man*. New York: Norton

Gray JR, Braver TS, Raichle ME. 2002. Integration of emotion and cognition in the lateral prefrontal cortex. *Proc. Natl. Acad. Sci. USA* 99:4115–20

Gray JR, Chabris CF, Braver TS. 2003. Neural mechanisms of general fluid intelligence. *Nat. Neurosci.* 6:316–22

Gray JR, Thompson PM. 2004. Neurobiology of intelligence: science and ethics. *Nat. Rev. Neurosci.* 5:471–82

Greenough WT, Fulcher JK, Yuwiler A, Geller E. 1970. Enriched rearing and chronic electroshock: effects on brain and behavior in mice. *Physiol. Behav.* 5(3):371–73

Haier RJ, Jung RE, Yeo RA, Head K, Alkire MT. 2004. Structural brain variation and general intelligence. *NeuroImage* 23(1):425–33

Haier RJ, Siegel BV, MacLachlan A, Soderling E, Lottenberg S, Buchsbaum MS. 1992. Regional glucose metabolic changes after learning a complex visuospatial/motor task: a positron emission tomography study. *Brain Res.* 570:134–43

Haier RJ, Siegel BV, Nuechterlein KH, Hazlett E, Wu JC, et al. 1998. Cortical glucose metabolic-rate correlates of abstract reasoning and attention studied with positron emission tomography. *Intelligence* 12:199–217

Haier RJ, Siegel B, Tang C, Abel L, Buchsbaum MS. 1992. Intelligence and changes in regional cerebral glucose metabolic-rate following learning. *Intelligence* 16:415–26

Haier RJ, White NS, Alkire MT. 2003. Individual differences in general intelligence correlate with brain function during nonreasoning tasks. *Intelligence* 31:429–41

Hashimoto M, Yasuda M, Tanimukai S, Matsui M, Hirono N, et al. 2001. Apolipoprotein E epsilon 4 and the pattern of regional brain atrophy in Alzheimer's disease. *Neurology* 57(8):1461–66

Herrnstein RJ, Murray C. 1994. *The Bell Curve*. New York: Free Press

Janke AL, de Zubicaray GI, Rose SE, Griffin M, Chalk JB, Galloway GJ. 2001. 4D deformation modeling of cortical disease progression in Alzheimer's dementia. *Magn. Reson. Med.* 46:661–66

Jensen AR. 1969. How much can we boost I.Q. and scholastic achievement? *Harvard Educ. Rev.* 39(1):1–123

Jensen AR. 1980. *Bias in Mental Testing*. New York: Free Press

Jensen AR. 1998. *The G Factor: The Science of Mental Ability*. Westport, CT: Praeger

Jensen AR, Miele F. 2002. *Intelligence, Race, and Genetics: Conversations with Arthur R. Jensen*. Boulder, CO: Westview

Kamin LJ. 1997. Twin studies, heritability, and intelligence. *Science* 278:1385

Kamin LJ, Goldberger AS. 2002. Twin studies in behavioral research: a skeptical view. *Theor. Popul. Biol.* 61(1):83–95

Kaprio J, Koskenvuo M, Rose RJ. 1990. Population-based twin registries: illustrative applications in genetic epidemiology and behavioral genetics from the Finnish Twin Cohort Study. *Acta Genet. Med. Gemellol.* 39(4):427–39

Korenberg JR, Chen XN, Hirota H, Lai Z, Bellugi U, et al. 2000. VI. Genome structure and cognitive map of Williams Syndrome. *J. Cogn. Neurosci.* 12(Suppl. 1):89–107

Lehtovirta M, Helisalmi S, Mannermaa A, Soininen H, Koivisto K, et al. 1995. Apolipoprotein E polymorphism and Alzheimer's disease in eastern Finland. *Neurosci Lett.* 185(1):13–15

Lehtovirta M, Partanen J, Konomen M, Hiltunen J, Helisalmi S, et al. 2000. A longitudinal quantitative EEG study of Alzheimer's disease: relation to apolioprotein E polymorphism. *Dement. Geriatr. Cogn. Disord.* 11(1):29–35

Lewontin RC. 1975. Genetic aspects of intelligence. *Annu. Rev. Genet.* 9:387–405

Loehlin JC. 1989. Partitioning environmental and genetic contributions to behavioral development. *Am. Psychol.* 44:1285–92

Lyons DM, Yang C, Sawyer-Glover AM, Moseley ME, Schatzberg AF. 2001. Early life stress and inherited variation in monkey hippocampal volumes. *Arch. Gen. Psychiatry* 58(12):1145–51

Maguire EA, Gadian DG, Johnsrude IS, Good CD, Asburner J, et al. 2000. Navigation-related structural change in the hippocampi of taxi drivers. *Proc. Natl. Acad. Sci. USA* 97(8):4398–403

Mazziotta J, Toga AW, Evans A, Fox P, Lancaster J, et al. 2001. A probabilistic atlas and reference system for the human brain: International Consortium for Brain Mapping (ICBM). *Phil. Trans. R. Soc. Lond. (R. Soc.)* 356:1293–322

McClearn GE, Johansson B, Berg S, Pedersen NL, Ahren F, et al. 1997. Substantial genetic influence on cognitive abilities in twins 80 or more years old. *Science* 276:1560–63

McDaniel MA, Nguyen NT. 2002. *A meta-analysis of the relationship between MRI-assessed brain volume and intelligence*. Presented at Proc. Int. Soc. Intel. Res., Nashville, TN

McGue M, Bouchard TJ Jr, Iacono WG, Lykken DT. 1993. Behavioral genetics of cognitive ability: a life-span perspective. In *Nature, Nurture and Psychology*, ed. R Plomin, GE McClearn, pp. 59–76. Washington, DC: Am. Psychol. Assoc.

Mega MS, Dinov ID, Lee L, Masterman DM, O'Connor SM, et al. 2000. Orbital and dorsolateral frontal perfusion defect associated with behavioral response to cholinesterase inhibitor therapy in Alzheimer's Disease. *J. Neuropsychiatr. Clin. Neurosci.* 12:209–18

Morley KI, Montgomery GW. 2001. The genetics of cognitive processes: candidate genes in humans and animals. *Behav. Genet.* 31(6):511–31

Mummery CJ, Patterson K, Price CJ, Asburner J, Frackowiack RS, Hodges JR. 2000. A voxel-based morphometry study of semantic dementia: relationship between temporal lobe atrophy and semantic memory. *Ann. Neurol.* 47(1):36–45

Narr KL, Cannon TD, Woods RP, Thompson PM, Kim S, et al. 2002. Genetic contributions to altered callosal morphology in schizophrenia. *J. Neurosci.* 22(9):3720–29

Narr KL, Thompson PM, Sharma T, Moussai J, Zoumalon C, et al. 2001. Three-dimensional mapping of gyral shape and cortical surface asymmetries in schizophrenia: gender effects. *Am. J. Psychiatry* 158(2):244–55

Narr KL, Thompson PM, Szeszko P, Robinson D, Jang S, et al. 2003. Regional specificity of hippocampal volume reductions in first-episode schizophrenia. *NeuroImage* 21(4):1563–75

Neale MC. 1997. *Mx: Statistical Modeling*. Richmond: Va. Commonwealth Univ., Dep. Psychiatry

Neisser U, ed. 1998. *The Rising Curve: Long-Term Gains in IQ and Related Measures*. Washington, DC: Am. Psychol. Assoc.

Oddy WH, Kendall GE, Blair E, De Klerk NH, Stanley FJ, et al. 2003. Breast feeding and cognitive development in childhood: a prospective birth cohort study. *Paediatr. Perinat. Epidemiol.* 17:81–90

Oppenheim JS, Skerry JE, Tramo MJ, Gazzaniga MS. 1989. Magnetic resonance imaging morphology of the corpus callosum in monozygotic twins. *Ann. Neurol.* 26:100–4

Pfefferbaum A, Sullivan EV, Carmelli D. 2001. Genetic regulation of regional microstructure of the corpus callosum in late life. *Neuroreport* 12(8):1677–81

Pfefferbaum A, Sullivan EV, Swan GE, Carmelli D. 2000. Brain structure in men remains highly heritable in the seventh and eighth decades of life. *Neurobiol. Aging* 21:63–74

Pinker S. 2002. *The Blank Slate*. New York: Viking

Plomin R. 1999. Genetics of childhood disorders. III. Genetics and intelligence. *J. Am. Acad. Child Adolesc. Psychiatry* 38(6):786–88

Plomin R, DeFries JC, Loehlin JC. 1977. Genotype-environment interaction and correlation in the analysis of human behavior. *Psychol. Bull.* 84:309–22

Plomin R, Kosslyn SM. 2001. Genes, brain and cognition. *Nat. Neurosci.* 4:1153–54

Plomin R, Petrill SA. 1997. Genetics and intelligence: What's new? *Intelligence* 24:53–78

Posthuma D, De Geus EJ, Baare WF, Hulshoff Pol HE, Kahn RS, Boomsma DI. 2002. The association between brain volume and intelligence is of genetic origin. *Nat. Neurosci.* 5:83–84

Prabhakaran V, Rypma B, Gabrieli JDE. 2001. Neural substrates of mathematical reasoning: a functional magnetic resonance imaging study of neocortical activation during performance of the Necessary Arithmetic Operations Test. *Neuropsychology* 15:115–27

Prabhakaran V, Smith JAL, Desmond JE, Glover GH, Gabrieli JDE. 1997. Neuronal substrates of fluid reasoning: an fMRI study of neocortical activation during performance of the Raven's Progressive Matrices Test. *Cogn. Psychol.* 33:43–63

Rasser PE, Johnston PJ, Lagopoulos J, Ward PB, Schall U, et al. 2005. Analysis of fMRI BOLD activation during the Tower of London Task using Gyral Patter and Intensity Averaging Models of Cerebral Cortex. *NeuroImage*. In press

Rex A, Pfeifer L, Fink H. 2001. Determination of NADH in frozen rat brain sections by laser-induced fluorescence. *Biol. Chem.* 382(12):1727–32

Ridley M. 2003. *Nature via Nurture: Genes, Experience, and What Makes Us Human*. New York: Harper Collins

Rombouts SA, Barkhof F, Witter MP, Schettens P. 2000. Unbiased whole-brain analysis of gray matter loss in Alzheimer's disease. *Neurosci. Lett.* 285(3):231–33

Roses AD. 1996. Apolipoprotein E and Alzheimer's disease. A rapidly expanding field with medical and epidemiological consequences. *Ann. N. Y. Acad. Sci.* 802:50–57

Rowe DC, Jacobson KC. 1999. Mainstream: research in behavioral genetics. In *Behavioral Genetics: The Clash of Culture and Biology*, ed. RA Carson, MA Rothstein, pp. 12–34. Baltimore/London: Johns Hopkins Univ. Press

Salovey P, Mayer JD, Caruso D. 2002. The positive psychology of emotional intelligence. In *The Handbook of Positive Psychology*, ed. CR Snyder, SJ Lopez, pp. 159–71. New York: Oxford Univ. Press

Schmader T, Johns M. 2004. Converging evidence that stereotype threat reduces working memory capacity. *J. Pers. Soc. Psychol.* 85:440–52

Sesardic N. 2002. Heritability and indirect causation. *Proc. Philos. Sci. Assoc.* 70:1002–14

Small GW, Ercoli LM, Silverman DH, Huang SC, Komo S, et al. 2000. Cerebral metabolic and cognitive decline in person at genetic risk for Alzheimer's disease. *Proc. Natl. Acad. Sci. USA* 97(11):6037–42

Sowell ER, Mattson SN, Thompson PM, Jernigan TL, Riley EP, Toga AW. 2001. Mapping callosal morphology and cognitive correlates: effects of heavy prenatal alcohol exposure. *Neurology* 57:235–44

Sowell ER, Peterson BS, Thompson PM, Welcome SE, Henkenius AL, Toga AW. 2003. Mapping cortical change across the human life span. *Nat. Neurosci.* 6(3):309–15

Sowell ER, Thompson PM, Holmes CJ, Batth R, Travner DA, et al. 1999. Localizing age-related changes in brain structure between childhood and adolescence using statistical parametric mapping. *NeuroImage* 9:587–97

Spearman C. 1927. *The Nature of "Intelligence" and the Principles of Cognition*. London, UK: Macmillan

Steele CM, Aronson J. 1995. Stereotype threat and the intellectual test performance of African Americans. *J. Pers. Soc. Psychol.* 69:797–811

Sternberg RJ. 1999. The theory of successful intelligence. *Rev. Gen. Psychol.* 3:292–16

Suddarth RI, Christison GW, Torrey EF, Casanova MF, Weinberger DR. 1990. Anatomical abnormalities in the brains of monozygotic twins discordant for schizophrenia. *N. Engl. J. Med.* 322(12):789–94

Sullivan EV, Pfefferbaum A, Swan GE, Carmelli D. 2001. Heritability of hippocampal size in elderly twin men: equivalent influence from genes and environment. *Hippocampus* 11(6):754–62

Swan GE, Carmelli D, Reed T, Harshfield GA, Fabsitz RR, Eslinger PJ. 1990. Heritability of cognitive performance in ageing twins. The National Heart, Lung, and Blood Institute Twin Study. *Arch. Neurol.* 47(3):259–62

Thompson P, Hayashi KM, de Zubicaray G, Janke AL, Rose SE, et al. 2003. Dynamics of gray matter loss in Alzheimer's disease. *J. Neurosci.* 23(3):994–1005

Thompson P, Hayashi KM, de Zubicaray G, Janke AL, Rose SE, et al. 2004. Mapping hippocampal and ventricular change in Alzheimer's disease. *NeuroImage* 22(4):1754–66

Thompson P, Toga AW. 1996. A surface-based technique for warping three-dimensional images of the brain. *IEEE Trans. Med. Imag.* 15(4):402–17

Thompson PM, Cannon TD, Narr KL, van Erp T, Poutanen V-P, et al. 2001a. Genetic influences on brain structure. *Nat. Neurosci.* 4(12):1253–58

Thompson PM, Cannon TD, Toga AW. 2002. Mapping genetic influences on human brain structure. *Ann. Med.* 24:523–36

Thompson PM, MacDonald D, Mega MS, Holmes CJ, Evans AC, Toga AW. 1997. Detection and mapping of abnormal brain structure with a probabilistic atlas of cortical surfaces. *J. Comp. Assist. Tomogr.* 21(4):567–81

Thompson PM, Toga AW. 2005. Cortical disease and cortical localization. *Nat. Encycl. Life Sci.* In press

Thompson PM, Vidal C, Giedd JN, Gochman P, Blumenthal J, et al. 2001b. Mapping adolescent brain change reveals dynamic wave of accelerated gray matter loss in very early-onset schizophrenia. *Proc. Natl. Acad. Sci. USA* 98(20):11650–55

Thompson PM, Woods RP, Mega MS, Toga AW. 2000. Mathematical/computational challenges in creating deformable and probabilistic atlases of the human brain. *Hum. Brain Mapp.* 9:81–92

Tramo MJ, Loftus WC, Stukel TA, Green RL, Weaver JB, Gazzaniga MS. 1998. Brain size, head size, and intelligence quotient in monozygotic twins. *Neurology* 50(5):1246–52

Turkheimer E, Haley A, Waldron M, D'Onofrio B, Gottesman II. 2003. Socioeconomic status modifies heritability of IQ in young children. *Psychol. Sci.* 14:623–28

Vogel F, Motulsky AG. 1997. *Human Genetics*. New York: Springer-Verlag

Weinberger DR, Delisi LE, Neophytides AN, Wyatt RJ. 1981. Familial aspects of CT scan abnormalities in chronic schizophrenic patients. *Psychiatry Res.* 4:65–71

Winterer G, Goldman D. 2003. Genetics of human prefrontal function. *Brain Res. Brain Res. Rev.* 43(1):134–63

Wright IC, McGuire PK, Poline JB, Travere JM, Murray RM, et al. 1995. A voxel-based method for the statistical analysis of gray and white matter density applied to schizophrenia. *NeuroImage* 2(4):244–52

Wright IC, Sham P, Murray RM, Weinberger DR, Bullmore ET. 2002. Genetic contributions to regional human brain variability estimated in a twin study by path analysis of parcellated grey matter maps. *NeuroImage* 17:256–71

Wright IC, Sharma T, Ellison ZR, McGuire PK, Fiston KJ, et al. 1999. Supra-regional brain systems and the neuropathology of schizophrenia. *Cereb. Cortex* 9(4):366–78

Zeineh MM, Engel SA, Thompson PM, Bookheimer SY. 2001. Unfolding the human hippocampus with high resolution structural and functional MRI. *Anat. Rec.* 265(2):111–20

Zeineh MM, Engel SA, Thompson PM, Bookheimer SY. 2003. Dynamics of the hippocampus during encoding and retrieval of face-name pairs. *Science* 299(5606):577–80

Zhou Y, Cheshire A, Howell CA, Ryan DH, Harris RB. 1999. Neuroautoantibody immunoreactivity in relation to aging and stress in apolipoprotein E-deficient mice. *Brain Res. Bull.* 49(3):173–79

The Actin Cytoskeleton: Integrating Form and Function at the Synapse

Christian Dillon and Yukiko Goda

MRC Cell Biology Unit and Laboratory for Molecular Cell Biology, University College London, London WC1E 6BT, United Kingdom; email: christian.dillon@ucl.ac.uk, y.goda@ucl.ac.uk

Key Words

LTP, LTD, synaptic plasticity, synaptic vesicle cycle, morphological plasticity, dendritic spine

Abstract

Synapses are highly specialized intercellular junctions that mediate the transmission of information between axons and target cells. A fundamental property of synapses is their ability to modify the efficacy of synaptic communication through various forms of synaptic plasticity. Recent developments in imaging techniques have revealed that synapses exhibit a high degree of morphological plasticity under basal conditions and also in response to neuronal activity that induces alterations in synaptic strength. The underlying molecular basis for this morphological plasticity has attracted much attention, yet its functional significance to the mechanisms of synaptic transmission and synaptic plasticity remains elusive. These morphological changes ultimately require the dynamic actin cytoskeleton, which is the major structural component of synapses. Delineating the physiological roles of the actin cytoskeleton in supporting synaptic transmission and synaptic plasticity, therefore, paves the way for gaining molecular insights into when and how synaptic machineries couple synapse form and function.

Contents

INTRODUCTION	26
PROPERTIES OF THE SYNAPTIC CYTOSKELETON	27
Organization of the Presynaptic Cytoskeleton	27
Organization of the Postsynaptic Cytoskeleton	29
Molecular and Pharmacological Control of Actin Dynamics	29
ACTIN REQUIREMENT FOR SYNAPSE FORMATION AND MAINTENANCE	30
Actin's Involvement in Synaptogenesis	31
Does Actin Help Maintain Synapse Structure?	32
A ROLE FOR ACTIN IN THE SYNAPTIC VESICLE CYCLE	35
Clustering of the Reserve Synaptic Vesicle Pool	36
Delivering Synaptic Vesicles to the Active Zone	37
Synaptic Vesicle Exocytosis	37
Synaptic Vesicle Endocytosis	40
ACTIVITY-DEPENDENT SYNAPTIC PLASTICITY	41
Actin Dynamics and Short-Term Plasticity	41
Stable Actin Remodeling and Long-Term Plasticity	43
Learning and Memory	46
CONCLUDING REMARKS	47

INTRODUCTION

The cytoskeleton plays a major role in virtually every cell biological process in eukaryotes, from cell division and cell motility to the intracellular trafficking of organelles. This remarkable cytoskeletal network is composed principally of three types of protein filaments—actin, microtubules, and intermediate filaments—each of which possesses unique biophysical and biochemical properties. The remodeling of these protein filaments by multiple intrinsic and extrinsic cues, which act through conserved signaling pathways, enables the cytoskeleton to control the amazing diversity of eukaryotic cell shapes, and moreover, to modify dynamic cellular behaviors.

In the nervous system, the cytoskeleton plays an important part in axon and dendrite formation, which allows neurons to establish their exquisite and complex morphology. The cytoskeleton also helps in the wiring of neural circuitry by driving both the guidance of neuronal processes and the formation of synapses, which are the sites of interneuronal communication. At the synapse, the cytoskeleton is likely to control dynamic synaptic functions, and in so doing, might induce morphological changes that are apparent from an extracellular vantage point. How is the functional capacity of a synapse integrated into the architecture of a synapse, and what sorts of functional changes are represented by the alterations in synapse morphology? In this chapter we review our current understanding of the synaptic cytoskeleton in an attempt to unravel its intimate relationship with the incredible adaptive properties of the synapse, and to decipher the connection between synapse form and function. We limit our discussion primarily to the actin cytoskeleton, which is the main structural component of both pre- and postsynaptic terminals. Also, we focus on excitatory synapses, which are the most extensively studied, and where the function of the actin cytoskeleton is best understood.

To examine the multifaceted involvement of actin at the synapse, we have divided this review into four main sections. In the first section we review the organization of the underlying synaptic architecture by focusing on the spatial distribution of actin and other cytoskeletal components that contribute to synapse morphology. We also look at some of the biophysical properties and regulation of the actin cytoskeleton at the synapse. In the second section we consider how the actin cytoskeleton contributes to synapse formation and maintenance. The third section addresses the roles proposed for actin in the presynaptic terminal. In particular, we discuss the involvement of actin during basal synaptic transmission by examining each stage of the synaptic vesicle cycle. Finally, we end the review by exploring the contribution of the actin cytoskeleton to activity-dependent structural and functional modifications at the synapse. In this section we also highlight recent data implicating actin remodeling in higher brain function and consider the impact of synaptic actin deregulation in human disease.

PROPERTIES OF THE SYNAPTIC CYTOSKELETON

Despite the considerable diversity of synapse types, every chemical synapse, which relies on intercellular communication via neurotransmitters, displays prototypical features of a synaptic junction in electron micrographs: i.e., a cluster of synaptic vesicles, a presynaptic active zone that is lined by a small number of synaptic vesicles, and a postsynaptic dense matrix (see **Figure 1**). The molecular determinants that confer these characteristic structural properties are not known, although components of the cytoskeleton are potential candidates. We begin the review with a quick tour of the cytoskeletal features of synaptic terminals.

Organization of the Presynaptic Cytoskeleton

Our first glimpse at the detailed arrangements of the presynaptic cytoskeleton was provided by early ultrastructural studies of the rat cerebellum, the frog neuromuscular junction (NMJ), and the electric organ from rays (Landis et al. 1988, Hirokawa et al. 1989). In these pioneering experiments, quick freeze-etched electron microscopy revealed a dense and varied meshwork of cytoskeletal filaments that were present throughout the terminal cytoplasm and were interconnected with the synaptic vesicle cluster. The main cytoskeletal features were essentially similar among synapses from divergent preparations, consistent with a high conservation of the basic appearance of the presynaptic architecture.

Two types of cytoskeletal filaments are readily discernable. First, connected to most synaptic vesicles are fine, short filaments of about 30 nm in length composed of the phosphoprotein synapsin I, which modulates neurotransmitter release (Landis & Reese 1983, Hirokawa et al. 1989, Greengard et al. 1994, Hilfiker et al. 1999). Although not a classical cytoskeletal element, synapsin filaments help form a lattice of interlinked synaptic vesicles (Li et al. 1995, Pieribone et al. 1995). Second, F-actin (filaments of actin) is abundantly distributed throughout the presynaptic terminal and is often associated with the synapsin filaments (Landis et al. 1988). Biochemical (Phillips et al. 2001), ultrastructural (Hirokawa et al. 1989, Phillips et al. 2001, Bloom et al. 2003), and functional (Morales et al. 2000) studies indicate that actin also constitutes an important component of the active zone (see **Figure 1**).

In addition to synapsin filaments and F-actin, two other structural components, microtubules and fodrin (a brain isoform of spectrin), are present in the presynaptic terminal. Microtubules (filaments of tubulin), when identified, are present at low levels and are found in the center of the terminal (Gotow et al. 1991, but see Gray 1983). Microtubules do not directly regulate synapse morphology or function; nevertheless, at *Drosophila* NMJ, microtubule-associated proteins have been implicated in synapse assembly and disassembly (Roos et al. 2000, Eaton et al. 2002). This requirement for microtubules may be specific to particular synapse types. In electron micrographs, a class of presynaptic filament with dimensions similar to fodrin extends out from the synaptic junction and may link to synaptic vesicles via synapsin I (Walker et al. 1985, Landis et al. 1988, Hirokawa et al. 1989, Phillips et al. 2001). Fodrin-like filaments and F-actin show an overlapping distribution within the synapse (Dunaevsky &

Active zone: the specialized region of the presynaptic terminal where neurotransmitters are released.

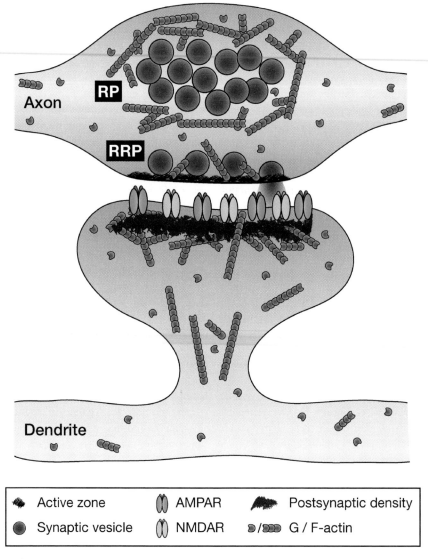

Figure 1

Schematic model showing the distribution of the actin cytoskeleton in a prototypical glutamatergic synapse. In the presynaptic bouton, synaptic vesicles containing neurotransmitters are associated with the actin cytoskeleton. Actin filaments surround the reserve pool (RP) of vesicles, where they may act as a corral or a scaffold for the synaptic vesicle cluster. Actin is also present at the active zone where vesicles belonging to the readily releasable pool (RRP) undergo exocytosis. Actin filaments in the postsynaptic compartment are present throughout the spine head and contact the postsynaptic density. Actin monomers (G-actin) are distributed homogeneously throughout the axon and dendrite.

Connor 2000), but the functional significance of this colocalization remains to be examined.

At the light microscopy level, imaging of fluorescently labeled, actin-specific probes has further clarified the subcellular localization of synaptic actin (Drenckhahn & Kaiser 1983, Drenckhahn et al. 1984, Fischer et al. 1998, Dunaevsky & Connor 2000, Matus 2000, Morales et al. 2000, Colicos et al. 2001, Zhang & Benson 2001, Zhang & Benson

2002). At the frog NMJ, F-actin is present in a ladder-shaped pattern along the linear array of apparent release sites and is excluded from the synaptic vesicle cluster in the center of the boutons (Dunaevsky & Connor 2000, Richards et al. 2004). A comparable pattern of actin localization has been reported at central synapses. In the lamprey reticulospinal giant synapse, F-actin forms a doughnut-shaped structure that surrounds a population of synaptic vesicles (Shupliakov et al. 2002) and is also found at lower levels in the core of this vesicle cluster (Bloom et al. 2003). In cultured hippocampal neurons, synaptic actin localization has been characterized by the pattern of exogenous expression of actin tagged with the green fluorescent protein (GFP-actin). Whereas GFP-actin per se is monomeric (globular or G-actin, see below), GFP-actin becomes incorporated into endogenous F-actin. The GFP signal, therefore, is contributed by both F- and G-actin. Although the relative abundance of F- and G-actin cannot be inferred, GFP-actin is enriched in the region surrounding the synaptic vesicle cluster (Sankaranarayanan et al. 2003), and a significant amount of GFP-actin colocalizes with the synaptic vesicle cluster (Morales et al. 2000, Colicos et al. 2001, Sankaranarayanan et al. 2003). It is tempting to speculate that this GFP-actin represents a network of vesicle-associated F-actin that provides a scaffold to the vesicle cluster. The distribution of G-actin has also been investigated by immunocytochemistry. Unlike F-actin, G-actin is distributed homogeneously throughout axonal and dendritic processes in hippocampal neurons, which indicates that its main role at the synapse is to maintain the synaptic F-actin pool (see below; Zhang & Benson 2002).

Organization of the Postsynaptic Cytoskeleton

The postsynaptic cytoskeleton has been best characterized in dendritic spines, which receive the majority of excitatory synaptic inputs in the brain. Dendritic spines are the small protrusions formed on the main dendrite shaft, and they take on various shapes—from thin or stubby to mushroom or cup-like. Moreover, the spines can move and alter their morphology dynamically. These properties have provoked much interest in identifying how the morphological plasticity of spines is coupled with the mechanism by which they receive and process incoming synaptic signals. Ultrastructural examination of dendritic spines from the rodent brain has revealed three prevalent forms of cytoskeletal filaments, all of which appear to consist of F-actin (see **Figure 1**; Fifkova & Delay 1982, Matus et al. 1982, Landis & Reese 1983, Cohen et al. 1985). First, a network of short, straight filaments of 4–6 nm in diameter and about 20 nm in length extends into the spine head from the synaptic junction. This filamentous mesh probably represents part of the postsynaptic density (PSD; see Sheng & Kim 2002, McGee & Bredt 2003 for reviews). Second, an additional network of F-actin is present throughout the spine head, some of which contacts the PSD. Finally, a class of filaments 5–7 nm in diameter, which may be a form of F-actin, is found in the neck of the spine, distal from the synaptic junction. The enrichment of F-actin in dendritic spines has been confirmed by histochemical studies of guinea-pig hippocampus using actin-specific probes (Drenckhahn et al. 1984), and by recent fluorescence resonance energy transfer experiments using coexpressed spectral variants of GFP-actin (Okamoto et al. 2004).

Molecular and Pharmacological Control of Actin Dynamics

As mentioned above, actin exists in two states within the cell: as polymerized two-stranded helical filaments (F-actin) or as monomers (G-actin), which provide the building blocks for filament assembly. F-actin is asymmetric, and actin monomers are added at a faster rate at the so-called barbed end compared with the pointed end of the filament (the nomenclature of barbed versus pointed ends originates from

> **Postsynaptic density:** an electron-dense structure that lies opposite the presynaptic active zone and provides a scaffold for molecules involved in neurotransmitter reception and signal transduction

> **Filopodia:** highly motile cellular protusions that are rich in F-actin

the arrowhead appearance of myosin heads that decorate F-actin). A couple of fundamental properties of F-actin are significant for its dynamic behavior. First, F-actin is assembled from weak, noncovalent interactions of G-actin. This property allows for rapid filament assembly and disassembly, which is required for quick remodeling of the actin cytoskeleton in response to changing stimuli. Second, the balance between F- and G-actin in subcellular regions is in a state of flux. As filaments are formed, the concentration of available monomers drops until a critical concentration of G-actin is reached, that is, when the net rates of filament assembly and disassembly are at steady-state. Owing to the difference in the polymerization rates at the two ends, the critical concentration of G-actin at the pointed end is higher than that at the barbed end. This difference, at a given cellular G-actin concentration, leads to a net loss of actin molecules at the pointed end and a simultaneous gain of monomers at the barbed end. Importantly, this phenomenon, known as actin treadmilling, creates a net flow of newly incorporated actin monomers through the filament and allows for a dynamic turnover of actin filaments.

A cohort of actin binding proteins (ABPs, reviewed in dos Remedios et al. 2003, Winder 2003) influences actin remodeling in different ways, resulting in a variety of actin filament structures. For instance, some ABPs act as F-actin bundling or cross-linking proteins (e.g., Arp2/3) that guide the formation of tight aggregates or two- and three-dimensional web-like networks of F-actin. Other ABPs affect F-actin turnover, for example, by (*a*) severing filaments (e.g., gelsolin and actin depolymerizing factor (ADF)/cofilin); (*b*) sequestering actin monomers (e.g., profilin); (*c*) binding to and capping the barbed end (e.g., CapZ) or pointed end (e.g., tropomodulin) of F-actin, thereby preventing the incorporation of G-actin; or (*d*) binding to F-actin and accelerating filament disassembly (e.g., ADF/cofilin). Far from being structural molecules that assist in organizing the cytoskeleton, ABPs are targets for a host of signal transduction cascades of intra- and extracellular origin. For a comprehensive overview of the molecular regulation of actin assembly/disassembly, we refer the readers to the following excellent recent reviews: dos Remedios et al. (2003), Pollard & Borisy (2003), Revenu et al. (2004).

In developing neurons, actin is highly enriched in axonal and dendritic growth cones, and an extensive body of work over the past two decades has identified a fundamental role for actin in the formation, growth, and guidance of neuronal processes (reviewed in da Silva & Dotti 2002, Luo 2002, Dent & Gertler 2003). Some of the same signaling pathways are also likely to regulate the actin cytoskeleton at the synapse (Meyer & Feldman 2002). The shared mechanisms, however, have made it difficult to probe the function of actin at synapses independently of its role in process outgrowth that precedes synaptogenesis. To date, therefore, investigations into the functional role of the synaptic actin cytoskeleton have largely relied on acute pharmacological manipulations in mature synaptic networks. Such studies have made use of a number of naturally occurring toxins that target actin. These agents are produced by certain species of plants, fungi, sponges, and bacteria and are used as part of a self-defense or infection mechanism. The toxins perturb actin dynamics by diverse modes of action, either (*a*) directly by binding to F- or G-actin or (*b*) indirectly by interacting with upstream actin-modulating proteins. Many of these compounds are membrane permeable, a favorable property suited for assessing the involvement of actin in mature neurons with established functional synaptic connections. Some of the more commonly used toxins are listed in **Table 1**.

ACTIN REQUIREMENT FOR SYNAPSE FORMATION AND MAINTENANCE

The enrichment of actin at the synapse implies that it has a major structural influence

Table 1 Pharmacological agents commonly used to manipulate the actin cytoskeleton

Compound	Source	Action	Cytoskeletal effects
Toxins that directly target actin (Cooper 1987, Saito et al. 1998, Spector et al. 1999, Terry et al. 1997, Yarmola et al. 2000)			
Cytochalasin B, Cytochalasin D	Fungal metabolites from *Metarrhizium anisopliae* and *Drechslera dematioideum*, respectively	Binds to barbed end of actin filaments	Inhibits subunit assembly and disassembly at barbed end, leads to actin depolymerization
Jasplakinolide (Jaspamide)	Indo-Pacific sponge, *Jaspis johnstoni*	Binds to F-actin	Promotes nucleation, stabilization, and polymerization of actin
Latrunculin A, Latrunculin B	Red Sea sponge, *Latrunculia magnifica*	Sequesters G-actin by binding in a 1:1 complex	Promotes actin depolymerization
Misakinolide A	Marine sponge, *Theonella* sp.	Binds to two-actin monomers	Prevents actin filament nucleation and elongation, and caps barbed end of F-actin
Phalloidin (Phallotoxin)	Fungal toxin from *Amanita phalloides* (death cap mushroom)	Binds to F-actin	Stabilizes F-actin
Swinholide A	Red Sea sponge, *Theonella swinhoei*	Binds to two-actin monomers	Prevents actin filament nucleation and elongation, and severs F-actin
Toxins that indirectly target actin (Barbieri et al. 2002, Boquet & Lemichez 2003, Busch & Aktories 2000, Lerm et al. 2000)			
C2 toxins	Various bacteria from *Clostridium* genus	ADP-ribosylation of actin	Depolymerization of F-actin
C3 exoenzymes	*Clostridium botulinum*	ADP-ribosylation of RhoA	Inhibits RhoA-induced actin assembly
Cytotoxic-necrotizing factor (CNF1)	*Escherichia coli*	Deamidation of RhoA, leading to its constitutive activation	Induction of RhoA-induced actin polymerization (e.g., formation of actin stress fibers)
Dermonecrotic toxin	*Bordetella* sp.	Deamidation of RhoA, leading to its constitutive activation	Induction of RhoA-induced actin polymerization (e.g., formation of actin stress fibers)
Lethal toxin	*Clostridium sordelli*	Glucosylation of Rac1 (also glucosylates Ras family members)	Inhibits actin polymerization induced by Rac1 activation
Toxin A and toxin B	*Clostridium difficile*	Glucosylation of Rho family members—RhoA, Rac1, and Cdc42—preventing their activation. Uncouples Rho from its effectors	Inhibits actin polymerization induced by RhoA, Rac1, or Cdc42 activation
Other compounds (Uehata et al. 1997)			
Y-27632	Synthetic compound	Inhibits Rho-associated protein kinase (ROCK), a downstream effector of RhoA	Inhibits RhoA-induced actin assembly

on both the pre- and postsynaptic compartments. But does actin play an instructive or permissive role in the formation and maintenance of the synapse? And, what influence, if any, does it have in the mechanisms that destabilize a synapse?

Actin's Involvement in Synaptogenesis

The process of synaptogenesis has recently been reviewed extensively (Goda & Davis 2003, Li & Sheng 2003, Ziv & Garner 2004), and therefore, we cover only a few salient points regarding the involvement of the actin cytoskeleton. In a popular model, synaptogenesis is driven by dynamic axo-dendritic interactions involving filopodia originating from dendrites (Fiala et al. 1998, Jontes & Smith 2000) and/or axons (Fiala et al. 1998, De Paola et al. 2003, Muller & Nikonenko 2003, Tashiro et al. 2003). That filopodia are most prevalent in young neurons undergoing active synaptogenesis is suggestive of their major role in synapse formation. A number of experimental observations support the view that dynamic actin filaments, in various forms, are a prerequisite for synapse generation: (*a*) actin-perturbing drugs inhibit both filopodial motility and synaptogenesis (Zhang & Benson 2001, Bonhoeffer & Yuste 2002); (*b*) clusters of presynaptic actin appear on the axon before functional synapses form, as if acting as a spatial tag or seed for synaptogenesis (Colicos et al. 2001); (*c*) activity-dependent synaptogenesis, as discussed below, is blocked by actin depolymerizing agents (Antonova et al. 2001); (*d*) disruption of signaling pathways implicated in synaptic actin reorganization results in synaptogenesis defects (e.g., Meng et al. 2002, Schenck et al. 2003, Takai et al. 2003, Zhang et al. 2003, Coyle et al. 2004, Ruiz-Canada et al. 2004); and (*e*) actin plays a part in the development of dendritic spines (spinogenesis; reviewed in Yuste & Bonhoeffer 2004). In particular, the molecular control of spinogenesis has been actively investigated. The emerging theme points to a central role for signal transduction pathways triggered at the cell surface that act through the Rho family of small GTPases (binary molecular switches that induce actin remodeling: see Hall 1998, Etienne-Manneville & Hall 2002 for reviews) in the formation of spines (Nakayama et al. 2000, Irie & Yamaguchi 2002, Penzes et al. 2003; **Figure 2**). Altogether, an ever-increasing number of signaling pathways and modulators of actin dynamics are implicated in regulating synapse formation at mammalian glutamatergic central synapses (see **Table 2** for summary), and the list is likely to expand further.

Does Actin Help Maintain Synapse Structure?

Several recent reports have addressed the contribution of actin in organizing and maintaining synapse structure (e.g., Allison et al. 1998, 2000; Zhang & Benson 2001). These studies have used latrunculin A, a toxin that sequesters available G-actin via formation of a nonpolymerizable 1:1 complex, to promote actin filament disassembly (see **Table 1**). Interestingly, in cultured hippocampal neurons, the requirement for actin filaments to maintain the localization of synaptic proteins and the overall structure of the synapse is developmentally regulated. For instance, immature synapses formed in young neurons (5–6 days in vitro) are rapidly dismantled by latrunculin A treatment, as determined by (*a*) a near complete loss of axonal clusters of synaptophysin (a synaptic vesicle-associated protein) and bassoon (a major structural component of the active zone), and (*b*) an absence of morphologically identifiable synapses at the ultrastructural level (Zhang & Benson 2001). As the synapse matures, it becomes increasingly resistant to latrunculin A treatment, such that by 18–20 days in vitro the distribution of presynaptic synaptophysin or bassoon is unaltered by actin depolymerization (Zhang & Benson 2001). This change in the stability of synaptic components correlates well with the formation of dendritic spines, which signifies the maturation of excitatory postsynaptic compartments. Additionally, the resistance to latrunculin A coincides with an increased expression of synaptic proteins that contain multiple protein-protein interaction domains, such as piccolo and PSD-95, which act as structural and scaffolding proteins at the pre- and postsynaptic terminals, respectively (Li & Sheng 2003). The development of a dense network of protein-protein interactions, therefore, is positioned to mediate an important synaptic transformation from a

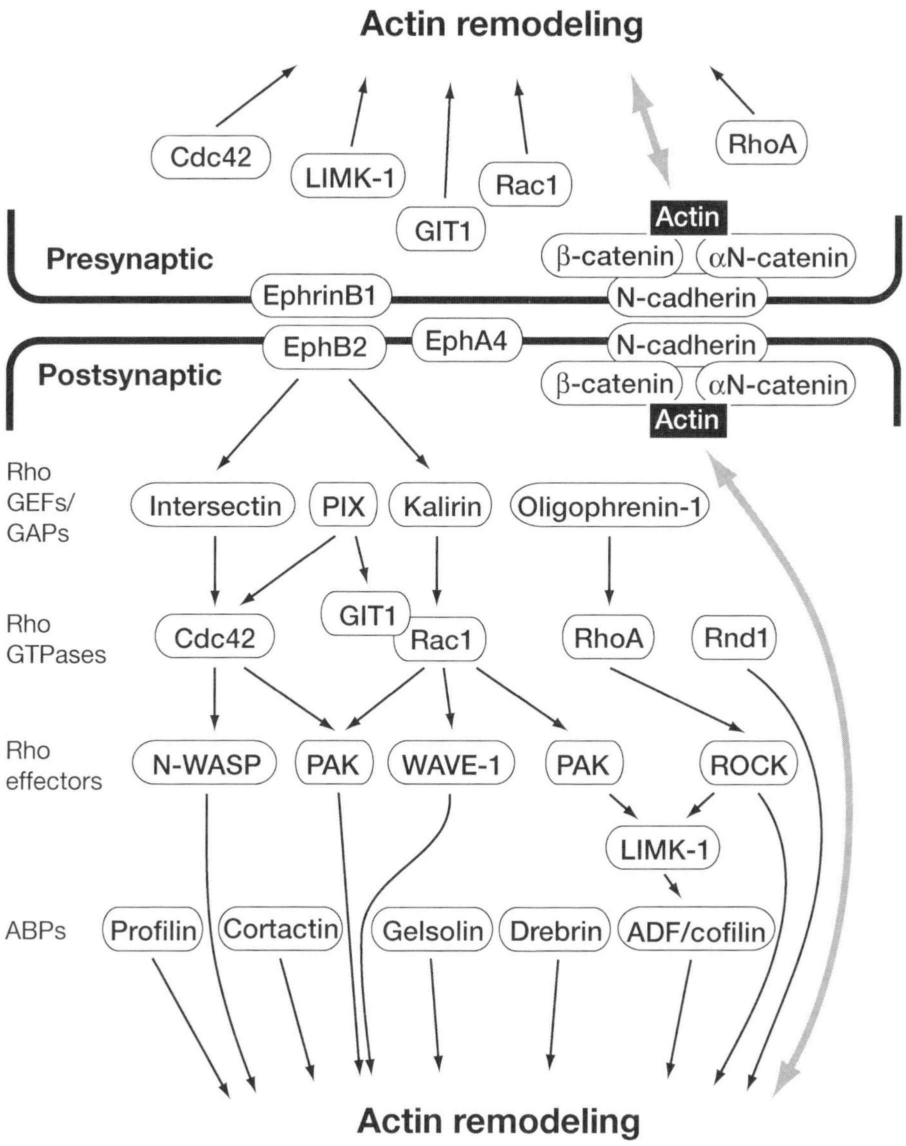

Figure 2
Signaling pathways involved in regulating actin at the synapse. Actin remodeling in the postsynaptic compartment can be initiated, for example, by trans-synaptic signaling through the EphB2 receptor tyrosine kinase via the small Rho GTPases, Rac1 and Cdc42. A number of other signaling molecules including actin binding proteins (ABPs) and proteins associated with Rho GTPase signaling are also important for maintaining spine integrity. In addition, adhesion molecules such as N-cadherin influence synaptic structure via indirect interactions with the actin cytoskeleton. The pathways involved in remodeling presynaptic actin remain to be defined in detail. See text and **Table 2** for details.

labile, actin-dependent structure to a stable form that is largely independent of dynamic actin filaments. Moreover, the interactions involving the postsynaptic scaffold are likely to facilitate the formation of the dendritic spines.

Note that the reported effects of perturbing actin dynamics on synapse morphology are not always uniform between different experimental conditions and/or preparations. For instance, latrunculin treatment in general does not produce discernable morphological changes in mature central synapses (Kim & Lisman 1999, Zhang & Benson 2001, but see Allison et al. 2000) or at the frog NMJ (Richards et al. 2004). Nevertheless, the close link between the actin cytoskeleton and spine morphology persists in mature synapses, as changes in dendritic spine morphology are induced when signaling molecules known to modulate the actin cytoskeleton are perturbed (see **Table 2** and **Figure 2**; e.g., Irie & Yamaguchi 2002, Penzes et al. 2003, Zhang

TABLE 2 Signaling molecules affecting synapse form and function through remodeling of the actin cytoskeleton. Synaptic phenotypes are given for knockout, activation, or inhibition studies where appropriate[1]

Protein	Functional type	Synaptic phenotype	Reference
αN-catenin	Links cadherin to cytoskeleton	Overexpression: increased spine density, stabilization of spines Knockout: increased dendritic spine length and decreased spine stability	Togashi et al. 2002, Abe et al. 2004
ADF/cofilin	ABP	Inhibition: defective maintenance of late phase of LTP	Fukazawa et al. 2003
β-catenin	Links cadherins to α-catenin and the cytoskeleton	Knockout: increased spine number and perforated synapses, reduced vesicle pool size and a faster decay of synaptic response to prolonged stimulation Phosphorylation mutant that increases spine β-catenin levels: increased PSD size, mEPSC frequency, and vesicle pool size	Murase et al. 2002, Bamji et al. 2003
Cdc42	Rho GTPase	Inhibition: blockade of spine formation, increased filopodia number	Irie & Yamaguchi 2002
Cortactin	ABP	Overexpression: increased spine length Inhibition: decreased spine number	Hering & Sheng 2003
Drebrin	ABP	Inhibition: defective postsynaptic clustering of PSD-95	Takahashi et al. 2003
EphA4	RTK	Activation: enhanced spine retraction Knockout: increased spine density and elongated structures	Murai et al. 2003
EphB1, 2, 3	RTK	Activation: induced spine formation Inhibition/knockout: blockade of spine formation, increased filopodia number	Ethell et al. 2001, Irie & Yamaguchi 2002, Henkemeyer et al. 2003, Penzes et al. 2003
Gelsolin	ABP	Knockout: altered LTD response, increased NMDAR currents	Furukawa et al. 1997, Star et al. 2002
GIT-1	Adaptor protein: targets PIX to synapse and binds PAK	Inhibition: decreased mushroom-shaped spines, increased filopodial protrusions	Zhang et al. 2003
Intersectin	Cdc42 GEF	Inhibition: disrupted spine formation, increased filopodia number	Irie & Yamaguchi 2002
Kalirin-7	Rac1 GEF	Overexpression: increased spine density Inhibition: decreased spine density	Penzes et al. 2001, Ma et al. 2003, Penzes et al. 2003
LIMK-1	Regulates activity of ADF/cofilin	Knockout: thin spine necks and enlarged heads, enhanced LTP, increased synaptic depression and frequency of mEPSCs	Meng et al. 2002
N-Cadherin	Adhesion molecule	Inhibition: longer dendritic protrusions, smaller spine head width, decreased synapsin puncta and FM dye uptake, inhibition of late phase of LTP	Bozdagi et al. 2000, Togashi et al. 2002
N-WASP	Cdc42 effector, interacts with Arp2/3	Inhibition: disrupted spine formation, increased filopodia number	Irie & Yamaguchi 2002
Oligophrenin-1	Rho-GAP	Inhibition: decreased spine length	Govek et al. 2004

(Continued)

TABLE 2 (Continued)

Protein	Functional type	Synaptic phenotype	Reference
PAK	Rac1 effector	Inhibition: decreased spine length, but increased head size, perforated spine number, and docked vesicle number. Enhanced synaptic transmission and LTP, reduced LTD, blockade of EphB mediated spine formation	Penzes et al. 2003, Hayashi et al. 2004
PIX	GEF for Cdc42 and Rac1	Inhibition: decreased spine number, dendritic protrusions, and synapses	Zhang et al. 2003
Profilin	ABP	Inhibition: elongated spines and loss of spine head	Ackermann & Matus 2003
Rac1	Rho GTPase	Activation: increased dendritic protrusions (filopodia and lamellipodia), decreased normal spines Inhibition: decreased spine numbers, and blockade of EphB-mediated spine formation	Luo et al. 1996, Nakayama et al. 2000, Penzes et al. 2003, Zhang et al. 2003
RhoA	Rho GTPase	Activation: dendritic simplification, reduced spine length and density	Nakayama et al. 2000, Govek et al. 2004
ROCK	RhoA effector	Activation: dendritic simplification	Nakayama et al. 2000
Rnd1	Rho GTPase	Activation: elongated spines Inhibition: decreased spine number and increased number of headless protrusions	Ishikawa et al. 2003

[1]RTK, receptor tyrosine kinase; GEF, guanine nucleotide exchange factor; GAP, GTPase activating protein.

et al. 2003). Furthermore, clustering of some postsynaptic proteins on dendritic spines is overtly affected by actin depolymerization. For example, the levels of the α-amino-3-hydroxy-5-methyl-4-isoxazolepropionic acid (AMPA) receptor subunit, GluR1, are significantly diminished from many dendritic spines after latrunculin treatment. In the absence of F-actin, α-actinin-2, drebrin, calcium/calmodulin-dependent protein kinase IIα (CaMKIIα), and NR1 [an N-methyl-D-aspartate (NMDA)-type glutamate receptor subunit] also disperse from spines, but PSD-95 does not (Allison et al. 1998, 2000).

Interestingly, one of these proteins, drebrin (an ABP that can cluster actin), has been shown to govern dendritic spine morphogenesis by allowing the recruitment of PSD-95 to the synapse (Takahashi et al. 2003). The observed loss of drebrin from spines upon F-actin depolymerization indicates that the mechanism by which actin participates in spine maintenance may share some properties with its role in spinogenesis. The differential sensitivity of drebrin and PSD-95 to latrunculin treatment, however, supports the developmental switch in some aspect of the organizational state of the postsynaptic scaffold from newly created spines to mature spines.

A ROLE FOR ACTIN IN THE SYNAPTIC VESICLE CYCLE

Although synapses become largely independent of actin for their structural preservation as development proceeds (Zhang & Benson 2001), actin remains enriched at synapses. What might be the physiological role for actin at mature synapses? In this section, we review the current state of our knowledge concerning the functions of actin during synaptic transmission. We limit our discussion to basal synaptic transmission and focus on the influence of actin at the presynaptic terminal. The contribution of actin in the postsynaptic terminal, which is closely associated with dendritic spine behavior, is covered in the following section on synaptic plasticity.

The presynaptic terminal is a highly evolved axonal specialization whose principal

Synaptic plasticity: the ability of a synapse to alter its strength or efficacy; synaptic strength can be increased or decreased in response to temporally coordinated patterns of activity

task is to coordinate the release of neurotransmitters with the arrival of action potentials. Synaptic vesicle clustering, mobilization, fusion, and subsequent recycling are therefore vital to neurotransmitter release, and much work has been carried out toward defining the molecular players involved in each step of the process (for recent reviews see Murthy & De Camilli 2003, Stevens 2003, Südhof 2004). The involvement of actin in regulating the synaptic vesicle cycle has been addressed over the years, and every physiological parameter of the cycle appears to be influenced in some way by the actin cytoskeleton (Doussau & Augustine 2000). Delineating the precise role(s) for actin at the presynaptic terminal, however, has been hampered by the inherent requirement for actin in cell survival and by the fact that its contribution varies according to synapse type, developmental stage, and history of synaptic activity.

Clustering of the Reserve Synaptic Vesicle Pool

At most nerve terminals, synaptic vesicles are organized into at least two functionally distinct pools: the reserve pool (RP) and the readily releasable pool (RRP; see **Figure 1** and Südhof & Scheller 2001, Südhof 2004 for reviews). Using morphological criteria, the RP is a cluster of vesicles present distally from the active zone. In contrast, the RRP constitutes a synaptic vesicle population docked at the active zone, which is primed to undergo membrane fusion (Stevens & Tsujimoto 1995, Rosenmund & Stevens 1996). Electrophysiologically, the RRP comprises a pool that is depleted upon strong nerve stimulation, and whose recovery is dependent on replenishment via recycling of exocytosed vesicles and/or by the recruitment of vesicles from the RP (Regehr & Stevens 2001, Zucker & Regehr 2002).

In the most popular model of presynaptic actin function, actin regulates the availability of the RP. As described above, actin surrounds the RP, but very little actin is found within the cluster itself (Landis et al. 1988, Hirokawa et al. 1989, Dunaevsky & Connor 2000, Shupliakov et al. 2002, Bloom et al. 2003, Sankaranarayanan et al. 2003, Richards et al. 2004). Actin may thus provide a corralling function by forming a physical barrier to impede vesicle dispersion (**Figure 1**). Synapsins, in turn, could stabilize the reserve vesicle cluster by their ability to link vesicles to the surrounding actin cytoskeleton (Greengard et al. 1994). In support of this notion, synapsin-1a disperses away from the presynaptic terminal upon neuronal stimulation at hippocampal boutons, when the synapse might require access to the RP of vesicles (Chi et al. 2001). Likewise, synapsin-1a disperses upon depolymerization of F-actin by latrunculin A treatment. The RP synaptic vesicle cluster, however, appears to remain intact following actin depolymerization, at least under resting conditions. Therefore, actin is more likely to act as a passive scaffold than a corral and exert a subtle control over modulating the availability of the RP at this central nervous system synapse (Sankaranarayanan et al. 2003). Similarly, at the resting frog NMJ, latrunculin A treatment has little effect on the organization of the RP examined at both the light and electron microscopy levels (Richards et al. 2004). A more pronounced requirement for actin in regulating the RP vesicle cluster has been reported for other synapses, however. At *Drosophila* NMJ, disrupting actin dynamics with cytochalasin D—a fungal metabolite that induces actin depolymerization by capping the fast-growing end of F-actin (see **Table 1**)—depletes an RP of vesicles that is accessed primarily during high-frequency stimulation (Kuromi & Kidokoro 1998). The diverse effects of actin depolymerization on the RP suggest that regulation of the reserve pool of vesicles becomes dependent on actin under conditions when the synapse is depleted of the RRP.

Delivering Synaptic Vesicles to the Active Zone

Prior to undergoing exocytosis, vesicles from the RP must be recruited to the active zone and enter the RRP (Südhof & Scheller 2001, Südhof 2004). Two distinct views for actin in promoting vesicle delivery to the RRP have been proposed. First, actin may provide cytoskeletal tracks that guide the movement of synaptic vesicles to the RRP via actin-based molecular motors (Prekeris & Terrian 1997, Evans et al. 1998). Second, actin could form a barrier between the RP and the RRP that is breached only by an activity-dependent signal that disassembles the barrier and allows access to the RP (**Figure 3**). Several pieces of evidence support the first model in which actin polymerization plays a facilitatory role in vesicle recruitment. For example, at the Calyx of Held, a giant synapse in the mammalian brainstem where the movement of vesicles between pools has been well characterized electrophysiologically, actin is instrumental in replenishing the RRP (Sakaba & Neher 2003). Here, latrunculin A causes a significant delay in vesicle recruitment that is prevented by stabilizing F-actin with phalloidin (see **Table 1**). Furthermore, studies from NMJs of *Xenopus* (Wang et al. 1996), snake (Cole et al. 2000), and *Drosophila* (Kuromi & Kidokoro 1998) all corroborate a model in which actin plays a facilitatory role in vesicle mobilization. In hippocampal boutons, however, latrunculin A does not affect the rate of refilling of the RRP, which indicates that vesicle mobilization does not require a dynamic actin network at this synapse (Morales et al. 2000). Similarly, in retinal bipolar cells, disruption of dynamic actin filaments has no discernible effects on vesicle cycling (Job & Lagnado 1998). In *Xenopus* NMJ, an opposite effect has been reported in that actin depolymerization leads to an increase in vesicle mobilization, indicating an inhibitory role for actin in vesicle recruitment (Wang et al. 1996). This study supports the second model in which actin forms a barrier between the RP and the RRP. In summary, how actin influences synaptic vesicle mobilization might be governed by the specific physiological requirement for sustained neurotransmitter release at each synapse type.

Synaptic Vesicle Exocytosis

The active zone is a highly specialized presynaptic junctional structure that caters to the exquisitely orchestrated exocytosis of synaptic vesicles. Actin filaments are present at the active zone (Hirokawa et al. 1989, Phillips et al. 2001), and accordingly, synaptic adhesion proteins that ultimately link to the actin cytoskeleton are enriched within or adjacent to the active zone (Südhof 2001). Does actin influence synaptic vesicle exocytosis at the active zone, and if so, how?

Several studies have reported changes in neurotransmitter release following pharmacological disruption of the actin cytoskeleton (Wang et al. 1996, Kuromi & Kidokoro 1998, Kim & Lisman 1999, Cole et al. 2000, Sankaranarayanan et al. 2003, Richards et al. 2004). Interpretation of these results, however, has been complicated by the involvement of actin in the steps preceding synaptic vesicle exocytosis and in endocytosis (see below). Nevertheless, a direct regulation of neurotransmitter release at the active zone by actin is supported by strong experimental evidence. In hippocampal synapses, latrunculin A treatment promotes neurotransmitter release as demonstrated by an increase in the frequency of miniature excitatory postsynaptic currents (mEPSCs) and the size of evoked EPSCs (Morales et al. 2000). Additionally, latrunculin A treatment increases the rate of exocytosis measured using FM1-43, a styryl dye that labels recycling synaptic vesicle membranes (Sankaranarayanan et al. 2003). Importantly, the RRP size and its rate of refilling are not altered by latrunculin A (Morales et al. 2000), which indicates that latrunculin A exerts its effect on vesicles that are already docked and primed at the active zone. Consistent with a facilitative

mEPSCs: excitatory synaptic responses elicited by neurotransmitter (glutamate) release from the spontaneous fusion of a single synaptic vesicle

EPSCs: excitatory postsynaptic responses elicited by an action potential

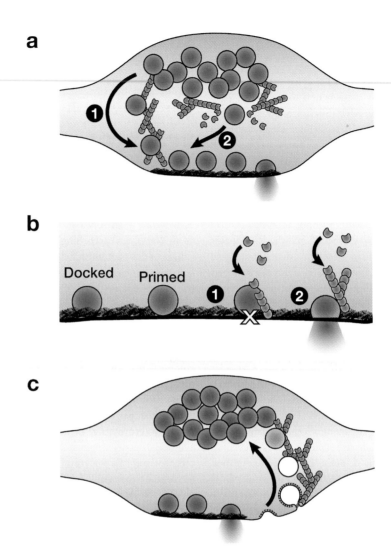

Figure 3

Putative functions for actin in the synaptic vesicle cycle. (*a*) Vesicle recruitment. Two divergent models for actin requirement in recruiting vesicles from the RP to the RRP are illustrated. (*1*) Actin filaments assemble between the two vesicles' pools and allow molecular motors, such as myosin, to propel and deliver synaptic vesicles to the RRP along actin tracks. (*2*) Actin acts as a barrier between the RP and the RRP. In this scenario, F-actin depolymerization dismantles the barrier and permits vesicles to replenish the RRP via diffusion or active transport. (*b*) Neurotransmitter release. Actin regulates a late step in synaptic vesicle exocytosis either negatively or positively. (*1*) In the first model, actin provides a clamp function, for example, by interfering with the facile molecular reorganization requisite for the assembly of the fusion machinery at the release site. (*2*) In the second model, actin facilitates neurotransmitter release by coupling to the fusion machinery in a manner that enhances its efficacy or by rendering the release site amenable to fusion. (*c*) Endocytosis. Actin plays a facilitatory role in synaptic vesicle endocytosis. Following exocytosis, synaptic vesicles are retrieved by clathrin-mediated endocytosis. Actin filaments assist in pinching off endocytosed vesicles from the plasma membrane, a process that involves the small GTPase dynamin. Intense neuronal stimulation causes extension of actin filaments from the endocytic zone, and these filaments act as tracks for newly endocytosed vesicles to return to the RP vesicle cluster. See **Figure 1** for legend.

effect of actin depolymerization on exocytosis, cytochalasin D increases the mEPSC frequency in mouse hippocampal slices (Meng et al. 2002). Nevertheless, in rat hippocampal slice preparations, actin depolymerizing agents either have no effect on basal synaptic transmission (Krucker et al. 2000) or reduce mEPSC size and frequency (Kim & Lisman 1999). The overall deleterious effects of actin-perturbing drugs, which reflect the universal requirement for actin in multiple aspects of cell physiology, may mask the enhancement of exocytosis in some experimental conditions or preparations.

The physiological relevance of actin turnover in modulating neurotransmitter release is further supported by the analysis of mice harboring a homozygous deletion of the LIM kinase 1 gene (*LIMK-1*). LIMK-1 is a downstream effector of the Rho family of small GTPases (see **Figure 2** and **Table 2**). This kinase modifies actin dynamics by phosphorylating and thereby inactivating ADF/cofilin, an ABP that promotes actin disassembly when nonphosphorylated. One would thus expect that in the absence of LIMK-1, actin turnover might be enhanced, a condition that could occlude the increase in neurotransmitter release triggered by actin-depolymerizing agents. Accordingly, hippocampal slices isolated from LIMK-1 knockout mice show a higher basal mEPSC rate compared with wild type mice, and cytochalasin D is ineffective in further increasing the mEPSC frequency (Meng et al. 2002). Taken together, these studies suggest that the actin cytoskeletal network negatively regulates neurotransmitter release from the RRP, perhaps by forming a structural barrier for exocytosis at or near the active zone (**Figure 3**). This barrier could be an integral part of the release machinery, could be regulated by LIMK-1 activity, and could be composed of actin itself or active zone protein(s) that link to the actin cytoskeleton.

The effects of signaling components that act on the actin cytoskeleton provide further evidence for a regulatory function for actin, or its related components, at a late stage in synaptic vesicle exocytosis. The use of bacterial toxins that target the Rho GTPases (see **Table 1**) have identified Rac1 as a key player in modulating acetylcholine release from synapses in the buccal ganglion of *Aplysia*. Specifically, Rac1 is involved in exocytosis by controlling the fusion competence of release sites (Doussau et al. 2000, Humeau et al. 2002). Among its many functions, Rac1 has been most notably associated with the assembly of actin filaments during lamellipodia formation in fibroblasts and neuronal growth cones (Ridley et al. 1992, Kozma et al. 1997; reviewed in Hall 1998, Etienne-Manneville & Hall 2002). Whether Rac1 regulates vesicle fusion in *Aplysia* by a similar modification of the actin cytoskeleton, or by some other means, such as modulation of the release site lipid composition, remains to be clarified (Humeau et al. 2001, Wenk & De Camilli 2004).

Curiously, the enhanced neurotransmitter release triggered by latrunculin A at hippocampal synapses is accompanied by a concomitant shape change of the synaptic junction. The active zone marker, bassoon, the presynaptic GFP-actin, and the postsynaptic glutamate receptor subunit signals all become elongated in parallel to the junctional interface (Morales et al. 2000). Perhaps the altered morphology represents a loosening of tensile forces normally exerted by the actin network at the intercellular junction. Moreover, the shape change of synaptic protein signals may reflect the requisite junctional reorganization for accommodating (or facilitating) enhanced exocytosis. In support of such a possibility, N-cadherin, an adhesion molecule that links the pre- and postsynaptic terminals, also spreads laterally upon depolarization, concurrent with the induction of exocytosis (Tanaka et al. 2000).

The contribution of actin in exocytosis also has been extensively studied in secretory cells, where the release of secretory granule contents is modulated by cortical actin—a dense network of filamentous actin that is present beneath the plasma membrane in most cells. Contrary to nerve terminals, neuroendocrine cells do not harbor a distinct subcellular structure that is dedicated to exocytosis. Instead, cortical actin likely attenuates exocytosis by presenting a passive diffusion barrier that restricts the access of secretory granules to the plasma membrane. Growing evidence, however, reveals the active participation of cortical actin filaments in exocytosis (Matter et al. 1989, Li et al. 1994, Pendleton & Koffer 2001, Trifaro et al. 2002). For instance, in PC12 cells, an immortalized neuroendocrine cell line, F-actin reorganization is a prerequisite for the regulated exocytosis of dense-core granules. Here, activation of Cdc42, a member of the Rho GTPase family, permits exocytosis by initiating actin filament assembly via one of its effector molecules—the neural Wiskott-Aldrich syndrome protein (N-WASP; Gasman et al. 2004). In chromaffin cells, stimulation induces the disassembly of cortical actin, which is driven by Ca^{2+}-dependent activation of scinderin (a protein that severs F-actin), and by protein kinase C, which in turn promotes actin remodeling through phosphorylation of MARCKS (myristoylated alanine-rich C kinase substrate) (Trifaro et al. 2002). To determine whether the same actin signaling pathways are used during neurotransmitter release at the active zone in neurons would be informative.

Synaptic Vesicle Endocytosis

Following exocytosis, synaptic vesicles are reclaimed for a further round of neurotransmitter release, and clathrin-mediated endocytosis is an important part of this process (reviewed in Slepnev & De Camilli 2000, Südhof 2004). Several studies in non-neuronal systems, including yeast (Kubler & Riezman 1993) and mammalian epithelial cells (Gottlieb et al. 1993), have established a close link between actin polymerization and endocytosis (reviewed in Qualmann et al. 2000, Engqvist-Goldstein & Drubin 2003). Given the high conservation of the fundamental features of clathrin-dependent endocytosis, it is not surprising that actin is also implicated in the retrieval of synaptic vesicles (see **Figure 3**). The best studied example involves endocytosis at the reticulospinal giant synapse of the lamprey (Shupliakov et al. 2002, Bloom et al. 2003). Here, in response to an exhaustive stimulation, a rich F-actin cytomatrix develops from the endocytic zone (the presynaptic membrane area surrounding the active zone) and reaches toward the internal synaptic vesicle cluster. The new actin filaments are associated with clathrin-coated vesicles (CCVs) at the plasma membrane and with uncoated synaptic vesicles along their length. Disrupting these filaments using the C2 toxin from *Clostridium botulinum* (see **Table 1**) triggers two striking changes in presynaptic organization: (*a*) an increase in the number of unconstricted clathrin-coated pits at the plasma membrane, and (*b*) a marked reduction in the size of the vesicle cluster in stimulated but not resting terminals. A similar activity-dependent depletion of the RP vesicles upon actin disruption has been reported at the frog and *Drosophila* NMJs (Kuromi & Kidokoro 1998, Richards et al. 2004).

These observations are consistent with the sequential participation of actin in two separate endocytic processes. First, actin might play a part in the fission of CCVs at the endocytic zone, a possibility supported by the link between actin and dynamin—a GTPase involved in the pinching off of CCVs from the plasma membrane (see for example Kuromi & Kidokoro 1998, reviewed in Engqvist-Goldstein & Drubin 2003). Note that the production of CCVs must be tightly coupled to exocytosis. The actin/dynamin binding protein, Abp1, may serve to coordinate the endocytic machinery with the presynaptic cytoskeletal matrix via its association with the active zone protein, Piccolo (Fenster et al. 2003). Alterations of the active zone resulting from membrane incorporation, which occur during exocytosis, could then be transduced to the endocytic machinery via Abp1. Second, once the CCVs have pinched off from the plasma membrane, actin might promote the transport of endocytosed vesicles to the internal RP cluster. This process could be driven by actin polymerization/treadmilling (Merrifield et al. 2002) using a mechanism similar to the propulsion of internalized pathogenic bacteria by actin-rich "comet-tails" (Frischknecht & Way 2001) or, alternatively, by actin-associated molecular motors such as myosin (Taunton 2001).

At small mammalian CNS synapses, the involvement of actin in endocytosis has been debated. The multiple modes of vesicle retrieval that operate at these synapses further complicate the problem. For example, in contrast to clathrin-mediated endocytosis, the kiss-and-run mode of neurotransmitter release (Ceccarelli et al. 1973) does not involve large membrane rearrangements and may be independent of actin. Perhaps different stimulation conditions recruit different modes of vesicle recycling, in accord with the state of depletion of different vesicle pools. In this case, actin does not uniformly participate in endocytic processes, and therefore, a requirement for actin would not always be apparent. Another variable is the synapse type. Larger synapses, for instance, might be more reliant on actin for replenishing the various vesicle pools because of the larger distances that vesicles are trafficked.

ACTIVITY-DEPENDENT SYNAPTIC PLASTICITY

Synapses have been generally conceived as stable structures, largely owing to extensive morphological characterizations of fixed tissue. A major conceptual change, however, has occurred in recent years with the observation that actin is an abundant component of the synapse (see above; Fifkova & Delay 1982, Matus et al. 1982, Landis & Reese 1983). Subsequent investigations, particularly over the past decade, have unequivocally demonstrated that the synapse is far from the rigid intercellular junction it was once thought to be. Rather, similar to other cellular regions enriched for actin (e.g., the leading edge of growth cones or motile fibroblasts), the synapse is a highly dynamic structure that can rapidly respond and adapt to different intrinsic or extrinsic cues. The development of fluorescent probes and novel advances in high resolution, real-time live cell imaging technologies have been instrumental in this conceptual advance, allowing investigators to map synapse morphology over long time periods with exquisite spatiotemporal precision.

In this section we examine how transient and stable morphological changes at the synapse are driven via regulation of the actin cytoskeleton, and we question how these structural changes are functionally coupled to alterations in synaptic strength and connectivity. To illustrate this idea we concentrate on the dendritic spines at excitatory synapses, which have been extensively studied in the context of experience-dependent plasticity (see Edwards 1995, Geinisman 2000, Matus 2000, Bonhoeffer & Yuste 2002, Nikonenko et al. 2002, Nimchinsky et al. 2002, Kasai et al. 2003 for reviews). The reciprocal relationship between changes in synaptic efficacy and morphological spine plasticity is providing insights into how synapse form and function cooperate. We highlight some of the signaling molecules involved in synaptic plasticity and address the less-appreciated phenomenon of actin-based plasticity at the presynaptic terminal.

Actin Dynamics and Short-Term Plasticity

That dendritic spines undergo spontaneous shape changes was first demonstrated by Matus and colleagues using dissociated hippocampal cultures transfected with GFP-actin (Fischer et al. 1998). Spines remodel by as much as 30% of their linear dimensions over a period of seconds. This surprisingly large amount of movement is driven by actin polymerization and occurs at spines innervated by a presynaptic partner (Fischer et al. 1998, Dunaevsky et al.

2001). Further investigations have confirmed the occurrence of amorphous spine motility in vivo (Lendvai et al. 2000) and in preparations that preserve the native circuitry (Dunaevsky et al. 1999). Other spontaneous actin-based changes of dendritic spine morphology have since been characterized and include (*a*) elongation, (*b*) splitting, (*c*) merging, (*d*) disappearance, (*e*) appearance, and (*f*) extensions of fine processes from the spine head (reviewed in Bonhoeffer & Yuste 2002).

The function of spontaneous spine movements has been addressed by examining the causal relationship between spine remodeling and synaptic activity. Spine motility is regulated by glutamate receptors (Fischer et al. 2000). Specifically, activation of AMPA-type glutamate receptors (AMPARs) transiently blocks spine movements and causes the spine head to round up. Similarly, NMDA receptor (NMDAR) activation prevents spine motility. These observations indicate that synaptic activation of AMPARs or NMDARs stabilizes spine motility, which might occur by a stimulus-induced decrease in actin turnover. Similarly, intense neuronal stimulation promotes an increase in the levels of spine F-actin in cultured hippocampal slices and in intact hippocampus (Fukazawa et al. 2003, Okamoto et al. 2004; also see Zhang & Benson 2001). Contrary to these findings, activation of AMPA and NMDA receptors in hippocampal cultures has also been reported to trigger the collapse of spines (Halpain et al. 1998, Hering & Sheng 2003). The degree of activation of individual synapses, the recruitment of extrasynaptic receptors, or the past history of synaptic activity presumably engages different signaling pathways, thereby biasing the spines to undergo either activity-dependent stabilization or destabilization (see below).

The molecular candidates that couple synaptic activity with changes in spine motility include synaptic proteins with links to the actin cytoskeleton and whose behavior is regulated by neuronal activity. For example, neuronal stimulation induces the translocation of the following actin-binding and/or structural proteins to or away from spines: profilin (Ackermann & Matus 2003), N-cadherin (Bozdagi et al. 2000), β-catenin (Murase et al. 2002), αN catenin (Abe et al. 2004), and cortactin (Hering & Sheng 2003). In the prevailing view, these molecular displacements contribute to the activity-induced localized change in synaptic actin dynamics, which then affects spine movements and/or shape. These molecules, in addition, could alter the activity of synaptic adhesion molecules that link to the actin cytoskeleton, thus retrogradely communicating the postsynaptic changes to the presynaptic terminal.

We have described how neuronal stimulation can drive the changes in dendritic spine motility, but how is spine motility and/or shape built into the mechanisms of synaptic transmission? Does activity that produces morphological plasticity in postsynaptic terminals also induce a parallel change in presynaptic terminals, for instance, by a reciprocal or retrograde transsynaptic mechanical communication? Analogous to spontaneous spine movements, the actin cytoskeleton in presynaptic terminals also undergoes submicron displacements over a time scale of seconds under resting conditions (Colicos et al. 2001). In contrast to these spontaneous movements that lack directional preference, physiological stimulation paradigms trigger a striking redistribution of actin within single boutons (Job & Lagnado 1998, Colicos et al. 2001, Huntley et al. 2002, Shupliakov et al. 2002, Sankaranarayanan et al. 2003). Two types of activity-induced change in actin morphology have been reported (**Figure 4**): (*a*) an overall increase in actin levels during stimulation (Sankaranarayanan et al. 2003), and (*b*) a directional condensation of actin within minutes of a single tetanic stimulus train that occurs independently of postsynaptic receptor activation (Colicos et al. 2001). In the latter case, actin condenses toward the presynaptic junction, as if to indent the dendritic spine, and then returns to the basal state after ~5 min poststimulation. How these two phenomena are related is not clear, although an increase in the overall level of presynaptic actin may be required for the directional condensation that ensues. These activity-induced presynaptic actin dynamics are mirrored postsynaptically by a seemingly coordinated and transient lateral expansion of actin in the spine head and movement of spine actin toward the dendrite shaft (**Figure 4**, Colicos et al. 2001). Okamoto and colleagues (2004) recently reported a similar reversible expansion of the spine head driven by F-actin polymerization, in cultured hippocampal slices, following a stimulus train of the sort that has been shown not to invoke significant changes in synaptic efficacy.

The intimate association of transient actin remodeling with synaptic activity implies a physiological role for the reorganization of the presynaptic actin cytoskeleton in neurotransmission. To meet the demands of maintaining neurotransmitter output during repetitive stimulation, remodeling of the presynaptic terminal could facilitate some aspect(s) of the synaptic vesicle cycle (**Figure 3**, see above), such as vesicle fusion at the active zone (Morales et al. 2000), vesicle recruitment to the RRP (Sakaba & Neher 2003), or clathrin-mediated endocytosis (Shupliakov et al. 2002).

Similarly, short-term activity-dependent modifications of spine actin dynamics and spine shape could also modify the postsynaptic receptor scaffold and the exo-endocytic traffic of glutamate receptors that have been implicated in regulating changes in synaptic strength (reviewed in Malinow & Malenka 2002). For instance, the stimulus-dependent widening of the spine head may represent the delivery of AMPA receptors and a transient increase in the postsynaptic responsiveness because available evidence supports a tight relationship between AMPAR density, synaptic strength, and spine size (Matsuzaki et al. 2001).

Stable Actin Remodeling and Long-Term Plasticity

In addition to short-term plasticity, temporally coordinated patterns of synaptic

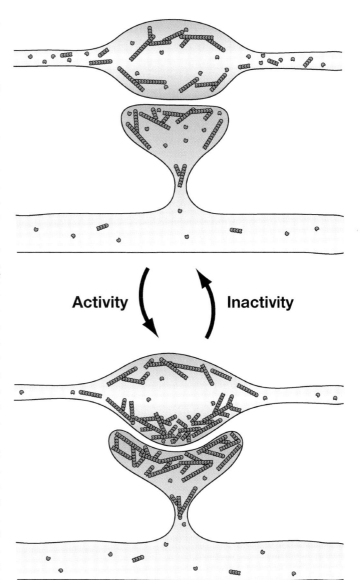

Figure 4

Schematic cartoon of activity-dependent changes in synaptic actin and spine shape. Synaptic stimulation increases F-actin in the pre- and postsynaptic terminals. Presynaptic F-actin is assembled and/or recruited from synaptic and axonal pools of actin and condenses toward the active zone. The spine head undergoes a lateral expansion and contracts toward the dendrite shaft in a process coordinated with the remodeling of presynaptic actin. Synaptic actin and spine morphology subsequently return to a basal state. See text and **Figure 1** legend for details.

activity can stably modify synaptic responses. The two representative forms of such activity-dependent long-term synaptic plasticity are long-term potentiation (LTP) and long-term depression (LTD). The durable form of LTP accompanies a stable reorganization of synapse structure, both pre- (Bozdagi et al. 2000) and postsynaptically (Engert & Bonhoeffer 1999, Maletic-Savatic et al. 1999), and the participation of actin in these processes has been rigorously examined during recent years. Here, we review the role of actin in LTP and LTD, extending the requirement for actin in driving transient structural remodeling and short-term plasticity discussed in the previous section.

Protocols that lead to the induction of LTP, such as repeated bursts of tetanic stimulation or brief application of glutamate, induce a number of characteristic changes in synaptic actin, which manifest themselves as (*a*) alterations at existing synapses or (*b*) formation of new synapses. At established dendritic spines, LTP induction shifts the F-actin/G-actin ratio in favor of F-actin within about 40 sec of a tetanic stimulus (Okamoto et al. 2004). Potential sources of the additional F-actin are the pool of synaptic or dendritic G-actin (Zhang & Benson 2002) and/or translation of β-actin mRNA in the spine (Tiruchinapalli et al. 2003). Unlike the transient remodeling of spine actin described above, the increase in F-actin is durable and is maintained for up to five weeks in the dentate gyrus in vivo (Fukazawa et al. 2003). Furthermore, F-actin enrichment in the spine parallels a robust and stable increase in the size of the spine head (**Figure 5**) (Matsuzaki et al. 2004, Okamoto et al. 2004). That actin is involved in sustaining the durable phase of LTP is supported by the specific inhibition of the late but not the early phase of LTP by actin-depolymerizing agents (Krucker et al. 2000, but see Kim & Lisman 1999). The LTP-dependent shift to F-actin in the spine is likely to involve a pathway linking LIMK-1 and ADF/cofilin to actin polymerization (**Figure 2**) because inhibiting the activity of ADF/cofilin impairs the durable expression of LTP (Fukazawa et al. 2003). Whether additional synaptic components also play a part in stabilizing actin and spine remodeling during LTP remains to be tested. One such candidate is CaMKIIβ, which shows actin-binding activity, and regulates several proteins involved in NMDAR-dependent LTP via phosphorylation (reviewed in Lisman et al. 2002).

In ultrastructural studies, the sizes of the pre- and postsynaptic compartments show a tight correlation (Harris & Stevens 1989). Thus the presynaptic terminal must undergo at least some degree of remodeling complementary to the changes in the postsynaptic terminal during LTP. For example, LTP induction could enlarge presynaptic terminal boutons and/or lengthen the active zone to match the increase in spine head size. Prominent alterations are more noticeable, however, and tetanic stimuli of the sort known to trigger LTP induce striking remodeling in the axons. These changes include the formation of new axonal varicosities (De Paola et al. 2003) and new axonal actin puncta, which arise either de novo or from the splitting of preexisting synaptic actin puncta (Colicos et al. 2001). Importantly, the new presynaptic actin puncta become associated with an actively recycling synaptic vesicle pool (**Figure 5**) (Colicos et al. 2001). Presynaptic actin remodeling has also been reported in *Aplysia* mechanosensory neurons during the growth of new synapses that accompanies long-term facilitation (Bailey et al. 1996, Hatada et al. 2000). That new synapse formation is associated with a stable increase in synaptic strength is further supported in studies involving chemical forms of LTP in dissociated hippocampal cultures (Antonova et al. 2001) and slices (Bozdagi et al. 2000). Here, chemical LTP induction triggers a rapid and pronounced increase in the number of pre- and postsynaptic clusters. Altogether, the activity-dependent increase in synaptic actin puncta and the number of pre- and postsynaptic assemblies share many mechanistic features with LTP, being

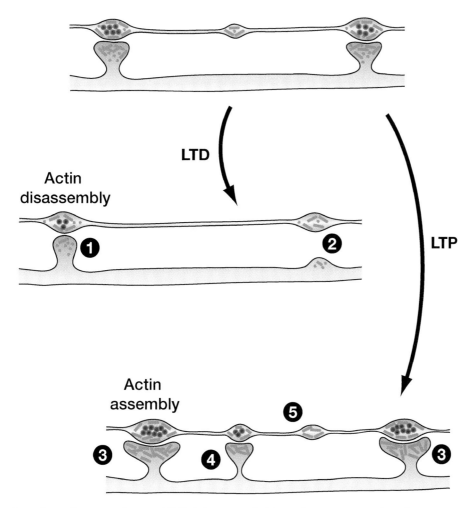

Figure 5
Model illustrating how actin remodeling mediates structural plasticity associated with LTD and LTP. LTD induces a shift in the F/G-actin ratio toward G-actin (*1*), and eventual regression of spines (*2*). LTP induction increases F-actin particularly in dendritic spines and causes them to widen, which is a hallmark feature of stimulus-induced reversible change (*3*; see also **Figure 4**). An enlarged synapse subsequently splits to give rise to a new synapse (*4*). LTP induction also promotes the formation of new presynaptic actin puncta (*5*), which eventually become associated with an actively recycling vesicle pool and spines.

dependent, for example, on NMDAR activation and calcium signaling (Antonova et al. 2001, Colicos et al. 2001). New synapse formation is likely to represent the requisite structural remodeling often associated with the consolidation of LTP to a more durable form.

On the assumption that synapse form and function are closely correlated, if a stable increase in synaptic strength promotes morphological changes in existing synapses, such as the widening of the spine head, then stable reduction of synaptic strength should also induce a change in synapse shape, for example, a reduction in spine head size. A recent report that examined the effect of LTP on spine actin dynamics also tested the effects of LTD induction. As predicted, LTD triggers a change in structural dynamics opposite to the changes induced by LTP: a shift in the spine F-actin/G-actin ratio to G-actin, and a concomitant decrease in spine head volume with the occasional disappearance of some spines (**Figure 5**) (Okamoto et al. 2004). Collectively, these findings establish actin as a molecular regulator of bidirectional morphological spine plasticity, where controlled cycles of actin polymerization and depolymerization coordinate the degree of synaptic activation by fine-tuning spine size (Okamoto et al. 2004). Estimates of the state of actin turnover in spines through monitoring of fluorescence recovery after photobleaching of GFP-actin suggests that a similar

ABP: actin binding protein

ADF: actin-depolymerizing factor

F-actin: filamentous actin

G-actin: globular actin

mEPSC: miniature excitatory postsynaptic current

PSD: postsynaptic density

LTD-inducing paradigm has an apparent stabilizing effect on actin (Star et al. 2002). This effect is impaired in neurons derived from mice containing a homozygous deletion of the gene encoding gelsolin, an ABP that severs and caps actin at its barbed end (**Table 2**). Although the findings of these two studies might not be immediately reconcilable, they nonetheless indicate that gelsolin is a potential player in the regulation of spine plasticity exhibited during LTD, perhaps by mediating the shift toward G-actin via severing of actin filaments, or alternatively, by capping and stabilizing a pool of actin in the spine that is important for maintaining the decrease in synaptic efficacy.

Learning and Memory

Learning and memory are associated with long-term structural change at synapses (Lendvai et al. 2000, Kleim et al. 2002, Trachtenberg et al. 2002; reviewed in Bailey et al. 1996, Lamprecht & LeDoux 2004). Given that changes in actin dynamics play a central role in synapse remodeling, synaptic actin likely contributes to learning and memory. A number of recent studies have tested a role for actin in behavioral learning paradigms. Intra-hippocampal injection of actin assembly inhibitors (latrunculin A or cytochalasin D) completely prevents both the acquisition and extinction of context-dependent fear responses, which demonstrates that actin rearrangements are required for these two distinct forms of learning process (Fischer et al. 2004). In addition, genetic disruption of Rho GTPase signaling pathways that remodel actin also impairs learning and memory in mice. For example, transgenic mice expressing a forebrain-specific dominant negative p21-activated kinase (PAK, a downstream effector of Rac1 and Cdc42 signaling to the cytoskeleton) (see **Table 2** and **Figure 2**) show deficits in hippocampal-dependent memory tasks (Hayashi et al. 2004). Likewise, mice lacking WAVE-1, an indirect effector of Rac1 signaling, or LIMK-1 exhibit reduced learning and memory when subjected to similar tasks (**Figure 2**) (Meng et al. 2002, Soderling et al. 2003). Furthermore, pharmacological inactivation of the RhoA pathway significantly impairs long-term memory in wild type rats (Dash et al. 2004).

The physiological relevance of actin regulation in neuronal circuit function is exemplified by the contribution of dysfunctional Rho GTPase signaling pathways to brain disorders involving mental retardation (recently reviewed in Chechlacz & Gleeson 2003). Mental retardation (MR) affects some 2%–3% of children and young adults, and patients suffer from a globally compromised cognitive functioning. A number of genes identified in patients carrying the heritable forms of MR encode modulators of Rho GTPase signaling pathways that modify actin dynamics (see **Figure 2**, **Table 2**) (Ramakers 2002, Chechlacz & Gleeson 2003). These genes include *ARHGEF6* (αPIX), *PAK3* (PAK3), *OPHN1* (Oligophrenin-1), *LIMK-1* (LIMK-1, which is associated with William's syndrome), and *FMR1* (FMR1, which is the most commonly mutated gene in fragile X syndrome and has been linked to Rac1 signaling; see Bardoni & Mandel 2002, Schenck et al. 2003). Interestingly, at the cellular level, neurons obtained from brains of MR patients or animal models of MR show impaired dendritic arborization and spine structure similar to the effects of molecular defects in actin regulatory pathways (listed in **Table 2**). These phenotypes include an overall reduction in spine densities and, among the spines present, a relative abundance of those with long tortuous necks that resemble immature spines. The close link between spine shape and synaptic efficacy suggests that pathogenesis of MR is related to these abnormal spine shapes and involves decreased neuronal connectivity. The identification of these MR genes supports the overall importance of deregulation of synaptic actin to dysfunction of neuronal circuits and stresses the urgency of delineating the molecular mechanisms by which actin dynamics modulate the activity of synapses.

CONCLUDING REMARKS

Prompted by the high degree of morphological dynamics exhibited by synapses, investigations into the functional basis of structural synaptic changes have begun to unravel the significant contribution of actin cytoskeletal regulation to basal synaptic transmission and to various forms of synaptic plasticity. An important underlying principle is the integration of synapse form and function. That structural remodeling of synapses, which is driven by changes in actin dynamics, is inseparable from the functional changes has been acknowledged best, thus far, in dendritic spines. The tight link between spine shape and synapse activity serves as a starting point for future studies aimed at delineating how molecular signaling machineries coordinate real-time changes in synapse activity and actin dynamics in pre- and postsynaptic terminals.

The seemingly matching activity-induced morphological behavior of presynaptic boutons and dendritic spines indicates that the coordination of synapse form and function is communicated across the two sides of the synapse. Of particular interest is the nature of such trans-synaptic antero- and retrograde signals, which couple patterns of synaptic activity with specific forms of cytoskeletal remodeling. How the coordination of functional and morphological plasticity at individual synapses influences neighboring synapses to affect the overall neuronal circuit behavior also warrants a detailed examination. The answers to these central questions will be derived in the coming years through the continued application of new quantitative live imaging technologies combined with direct pharmacogenetic targeting of individual proteins at the synapse.

ACKNOWLEDGMENTS

We gratefully acknowledge members of the Goda lab including Kevin Staras, Lorenzo Cingolani, Kevin Darcy, Takashi Okuda, and Tiago Branco for useful discussions and comments. We thank also Edward Koo and Anastasia Christakou for their critical comments. We apologize to authors whose primary research could not be cited directly owing to space constraints.

LITERATURE CITED

Abe K, Chisaka O, Van Roy F, Takeichi M. 2004. Stability of dendritic spines and synaptic contacts is controlled by alpha N-catenin. *Nat. Neurosci.* 7:357–63

Ackermann M, Matus A. 2003. Activity-induced targeting of profilin and stabilization of dendritic spine morphology. *Nat. Neurosci.* 6:1194–200

Allison DW, Chervin AS, Gelfand VI, Craig AM. 2000. Postsynaptic scaffolds of excitatory and inhibitory synapses in hippocampal neurons: maintenance of core components independent of actin filaments and microtubules. *J. Neurosci.* 20:4545–54

Allison DW, Gelfand VI, Spector I, Craig AM. 1998. Role of actin in anchoring postsynaptic receptors in cultured hippocampal neurons: differential attachment of NMDA versus AMPA receptors. *J. Neurosci.* 18:2423–36

Antonova I, Arancio O, Trillat AC, Wang HG, Zablow L, et al. 2001. Rapid increase in clusters of presynaptic proteins at onset of long-lasting potentiation. *Science* 294:1547–50

Bailey CH, Bartsch D, Kandel ER. 1996. Toward a molecular definition of long-term memory storage. *Proc. Natl. Acad. Sci. USA* 93:13445–52

Bamji SX, Shimazu K, Kimes N, Huelsken J, Birchmeier W, et al. 2003. Role of beta-catenin in synaptic vesicle localization and presynaptic assembly. *Neuron* 40:719–31

Barbieri JT, Riese MJ, Aktories K. 2002. Bacterial toxins that modify the actin cytoskeleton. *Annu. Rev. Cell. Dev. Biol.* 18:315–44

Bardoni B, Mandel JL. 2002. Advances in understanding of fragile X pathogenesis and FMRP function, and in identification of X linked mental retardation genes. *Curr. Opin. Genet. Dev.* 12:284–93

Bloom O, Evergren E, Tomilin N, Kjaerulff O, Low P, et al. 2003. Colocalization of synapsin and actin during synaptic vesicle recycling. *J. Cell. Biol.* 161:737–47

Bonhoeffer T, Yuste R. 2002. Spine motility. Phenomenology, mechanisms, and function. *Neuron* 35:1019–27

Boquet P, Lemichez E. 2003. Bacterial virulence factors targeting Rho GTPases: parasitism or symbiosis? *Trends Cell. Biol.* 13:238–46

Bozdagi O, Shan W, Tanaka H, Benson DL, Huntley GW. 2000. Increasing numbers of synaptic puncta during late-phase LTP: N-cadherin is synthesized, recruited to synaptic sites, and required for potentiation. *Neuron* 28:245–59

Busch C, Aktories K. 2000. Microbial toxins and the glycosylation of rho family GTPases. *Curr. Opin. Struct. Biol.* 10:528–35

Ceccarelli B, Hurlbut WP, Mauro A. 1973. Turnover of transmitter and synaptic vesicles at the frog neuromuscular junction. *J. Cell. Biol.* 57:499–524

Chechlacz M, Gleeson JG. 2003. Is mental retardation a defect of synapse structure and function? *Pediatr. Neurol.* 29:11–17

Chi P, Greengard P, Ryan TA. 2001. Synapsin dispersion and reclustering during synaptic activity. *Nat. Neurosci.* 4:1187–93

Cohen RS, Chung SK, Pfaff DW. 1985. Immunocytochemical localization of actin in dendritic spines of the cerebral cortex using colloidal gold as a probe. *Cell Mol. Neurobiol.* 5:271–84

Cole JC, Villa BR, Wilkinson RS. 2000. Disruption of actin impedes transmitter release in snake motor terminals. *J. Physiol.* 525(Pt. 3):579–86

Colicos MA, Collins BE, Sailor MJ, Goda Y. 2001. Remodeling of synaptic actin induced by photoconductive stimulation. *Cell* 107:605–16

Cooper JA. 1987. Effects of cytochalasin and phalloidin on actin. *J. Cell. Biol.* 105:1473–78

Cowan WM, Südhof TC, Stevens CF, eds. 2001. *Synapses*. Baltimore, MD: Johns Hopkins Univ. Press

Coyle IP, Koh YH, Lee WC, Slind J, Fergestad T, et al. 2004. Nervous wreck, an SH3 adaptor protein that interacts with Wsp, regulates synaptic growth in Drosophila. *Neuron* 41:521–34

da Silva JS, Dotti CG. 2002. Breaking the neuronal sphere: regulation of the actin cytoskeleton in neuritogenesis. *Nat. Rev. Neurosci.* 3:694–704

Dash PK, Orsi SA, Moody M, Moore AN. 2004. A role for hippocampal Rho-ROCK pathway in long-term spatial memory. *Biochem. Biophys. Res. Commun.* 322:893–98

De Paola V, Arber S, Caroni P. 2003. AMPA receptors regulate dynamic equilibrium of presynaptic terminals in mature hippocampal networks. *Nat. Neurosci.* 6:491–500

Dent EW, Gertler FB. 2003. Cytoskeletal dynamics and transport in growth cone motility and axon guidance. *Neuron* 40:209–27

dos Remedios CG, Chhabra D, Kekic M, Dedova IV, Tsubakihara M, et al. 2003. Actin binding proteins: regulation of cytoskeletal microfilaments. *Physiol. Rev.* 83:433–73

Doussau F, Augustine GJ. 2000. The actin cytoskeleton and neurotransmitter release: an overview. *Biochimie* 82:353–63

Doussau F, Gasman S, Humeau Y, Vitiello F, Popoff M, et al. 2000. A Rho-related GTPase is involved in Ca(2+)-dependent neurotransmitter exocytosis. *J. Biol. Chem.* 275:7764–70

Drenckhahn D, Frotscher M, Kaiser HW. 1984. Concentration of F-actin in synaptic formations of the hippocampus as visualized by staining with fluorescent phalloidin. *Brain Res.* 300:381–84

Drenckhahn D, Kaiser HW. 1983. Evidence for the concentration of F-actin and myosin in synapses and in the plasmalemmal zone of axons. *Eur. J. Cell. Biol.* 31:235–40

Dunaevsky A, Blazeski R, Yuste R, Mason C. 2001. Spine motility with synaptic contact. *Nat. Neurosci.* 4:685–86

Dunaevsky A, Connor EA. 2000. F-actin is concentrated in nonrelease domains at frog neuromuscular junctions. *J. Neurosci.* 20:6007–12

Dunaevsky A, Tashiro A, Majewska A, Mason C, Yuste R. 1999. Developmental regulation of spine motility in the mammalian central nervous system. *Proc. Natl. Acad. Sci. USA* 96:13438–43

Eaton BA, Fetter RD, Davis GW. 2002. Dynactin is necessary for synapse stabilization. *Neuron* 34:729–41

Edwards FA. 1995. Anatomy and electrophysiology of fast central synapses lead to a structural model for long-term potentiation. *Physiol. Rev.* 75:759–87

Engert F, Bonhoeffer T. 1999. Dendritic spine changes associated with hippocampal long-term synaptic plasticity. *Nature* 399:66–70

Engqvist-Goldstein AE, Drubin DG. 2003. Actin assembly and endocytosis: from yeast to mammals. *Annu. Rev. Cell. Dev. Biol.* 19:287–332

Ethell IM, Irie F, Kalo MS, Couchman JR, Pasquale EB, Yamaguchi Y. 2001. EphB/syndecan-2 signaling in dendritic spine morphogenesis. *Neuron* 31:1001–13

Etienne-Manneville S, Hall A. 2002. Rho GTPases in cell biology. *Nature* 420:629–35

Evans LL, Lee AJ, Bridgman PC, Mooseker MS. 1998. Vesicle-associated brain myosin-V can be activated to catalyze actin-based transport. *J. Cell. Sci.* 111 (Pt. 14):2055–66

Fenster SD, Kessels MM, Qualmann B, Chung WJ, Nash J, et al. 2003. Interactions between Piccolo and the actin/dynamin-binding protein Abp1 link vesicle endocytosis to presynaptic active zones. *J. Biol. Chem.* 278:20268–77

Fiala JC, Feinberg M, Popov V, Harris KM. 1998. Synaptogenesis via dendritic filopodia in developing hippocampal area CA1. *J. Neurosci.* 18:8900–11

Fifkova E, Delay RJ. 1982. Cytoplasmic actin in neuronal processes as a possible mediator of synaptic plasticity. *J. Cell. Biol.* 95:345–50

Fischer A, Sananbenesi F, Schrick C, Spiess J, Radulovic J. 2004. Distinct roles of hippocampal de novo protein synthesis and actin rearrangement in extinction of contextual fear. *J. Neurosci.* 24:1962–66

Fischer M, Kaech S, Knutti D, Matus A. 1998. Rapid actin-based plasticity in dendritic spines. *Neuron* 20:847–54

Fischer M, Kaech S, Wagner U, Brinkhaus H, Matus A. 2000. Glutamate receptors regulate actin-based plasticity in dendritic spines. *Nat. Neurosci.* 3:887–94

Frischknecht F, Way M. 2001. Surfing pathogens and the lessons learned for actin polymerization. *Trends Cell. Biol.* 11:30–38

Fukazawa Y, Saitoh Y, Ozawa F, Ohta Y, Mizuno K, Inokuchi K. 2003. Hippocampal LTP is accompanied by enhanced F-actin content within the dendritic spine that is essential for late LTP maintenance in vivo. *Neuron* 38:447–60

Furukawa K, Fu W, Li Y, Witke W, Kwiatkowski DJ, Mattson MP. 1997. The actin-severing protein gelsolin modulates calcium channel and NMDA receptor activities and vulnerability to excitotoxicity in hippocampal neurons. *J. Neurosci.* 17:8178–86

Gasman S, Chasserot-Golaz S, Malacombe M, Way M, Bader MF. 2004. Regulated exocytosis in neuroendocrine cells: a role for subplasmalemmal Cdc42/N-WASP-induced actin filaments. *Mol. Biol. Cell.* 15:520–31

Geinisman Y. 2000. Structural synaptic modifications associated with hippocampal LTP and behavioral learning. *Cereb. Cortex* 10:952–62

Goda Y, Davis GW. 2003. Mechanisms of synapse assembly and disassembly. *Neuron* 40:243–64

Gotow T, Miyaguchi K, Hashimoto PH. 1991. Cytoplasmic architecture of the axon terminal: filamentous strands specifically associated with synaptic vesicles. *Neuroscience* 40:587–98

Gottlieb TA, Ivanov IE, Adesnik M, Sabatini DD. 1993. Actin microfilaments play a critical role in endocytosis at the apical but not the basolateral surface of polarized epithelial cells. *J. Cell. Biol.* 120:695–710

Govek EE, Newey SE, Akerman CJ, Cross JR, Van der Veken L, Van Aelst L. 2004. The X-linked mental retardation protein oligophrenin-1 is required for dendritic spine morphogenesis. *Nat. Neurosci.* 7:364–72

Gray EG. 1983. Neurotransmitter release mechanisms and microtubules. *Proc. R. Soc. London B. Biol. Sci.* 218:253–58

Greengard P, Benfenati F, Valtorta F. 1994. Synapsin I, an actin-binding protein regulating synaptic vesicle traffic in the nerve terminal. *Adv. Second Messenger Phosphoprotein Res.* 29:31–45

Hall A. 1998. Rho GTPases and the actin cytoskeleton. *Science* 279:509–14

Halpain S, Hipolito A, Saffer L. 1998. Regulation of F-actin stability in dendritic spines by glutamate receptors and calcineurin. *J. Neurosci.* 18:9835–44

Harris KM, Stevens JK. 1989. Dendritic spines of CA 1 pyramidal cells in the rat hippocampus: serial electron microscopy with reference to their biophysical characteristics. *J. Neurosci.* 9:2982–97

Hatada Y, Wu F, Sun ZY, Schacher S, Goldberg DJ. 2000. Presynaptic morphological changes associated with long-term synaptic facilitation are triggered by actin polymerization at preexisting varicosities. *J. Neurosci.* 20:RC82, 1–5

Hayashi ML, Choi SY, Rao BS, Jung HY, Lee HK, et al. 2004. Altered cortical synaptic morphology and impaired memory consolidation in forebrain-specific dominant-negative PAK transgenic mice. *Neuron* 42:773–87

Henkemeyer M, Itkis OS, Ngo M, Hickmott PW, Ethell IM. 2003. Multiple EphB receptor tyrosine kinases shape dendritic spines in the hippocampus. *J. Cell. Biol.* 163:1313–26

Hering H, Sheng M. 2003. Activity-dependent redistribution and essential role of cortactin in dendritic spine morphogenesis. *J. Neurosci.* 23:11759–69

Hilfiker S, Pieribone VA, Czernik AJ, Kao HT, Augustine GJ, Greengard P. 1999. Synapsins as regulators of neurotransmitter release. *Philos. Trans R. Soc. London B. Biol. Sci.* 354:269–79

Hirokawa N, Sobue K, Kanda K, Harada A, Yorifuji H. 1989. The cytoskeletal architecture of the presynaptic terminal and molecular structure of synapsin 1. *J. Cell. Biol.* 108:111–26

Humeau Y, Popoff MR, Kojima H, Doussau F, Poulain B. 2002. Rac GTPase plays an essential role in exocytosis by controlling the fusion competence of release sites. *J. Neurosci.* 22:7968–81

Humeau Y, Vitale N, Chasserot-Golaz S, Dupont JL, Du G, et al. 2001. A role for phospholipase D1 in neurotransmitter release. *Proc. Natl. Acad. Sci. USA* 98:15300–5

Huntley GW, Benson DL, Colman DR. 2002. Structural remodeling of the synapse in response to physiological activity. *Cell* 108:1–4

Irie F, Yamaguchi Y. 2002. EphB receptors regulate dendritic spine development via intersectin, Cdc42 and N-WASP. *Nat. Neurosci.* 5:1117–18

Ishikawa Y, Katoh H, Negishi M. 2003. A role of Rnd1 GTPase in dendritic spine formation in hippocampal neurons. *J. Neurosci.* 23:11065–72

Job C, Lagnado L. 1998. Calcium and protein kinase C regulate the actin cytoskeleton in the synaptic terminal of retinal bipolar cells. *J. Cell. Biol.* 143:1661–72

Jontes JD, Smith SJ. 2000. Filopodia, spines, and the generation of synaptic diversity. *Neuron* 27:11–14

Kasai H, Matsuzaki M, Noguchi J, Yasumatsu N, Nakahara H. 2003. Structure-stability-function relationships of dendritic spines. *Trends Neurosci.* 26:360–68

Kim CH, Lisman JE. 1999. A role of actin filament in synaptic transmission and long-term potentiation. *J. Neurosci.* 19:4314–24

Kleim JA, Freeman JH Jr, Bruneau R, Nolan BC, Cooper NR, et al. 2002. Synapse formation is associated with memory storage in the cerebellum. *Proc. Natl. Acad. Sci. USA* 99:13228–31

Kozma R, Sarner S, Ahmed S, Lim L. 1997. Rho family GTPases and neuronal growth cone remodelling: relationship between increased complexity induced by Cdc42Hs, Rac1, and acetylcholine and collapse induced by RhoA and lysophosphatidic acid. *Mol. Cell. Biol.* 17:1201–11

Krucker T, Siggins GR, Halpain S. 2000. Dynamic actin filaments are required for stable long-term potentiation (LTP) in area CA1 of the hippocampus. *Proc. Natl. Acad. Sci. USA* 97:6856–61

Kubler E, Riezman H. 1993. Actin and fimbrin are required for the internalization step of endocytosis in yeast. *Embo J.* 12:2855–62

Kuromi H, Kidokoro Y. 1998. Two distinct pools of synaptic vesicles in single presynaptic boutons in a temperature-sensitive Drosophila mutant, shibire. *Neuron* 20:917–25

Lamprecht R, LeDoux J. 2004. Structural plasticity and memory. *Nat. Rev. Neurosci.* 5:45–54

Landis DM, Hall AK, Weinstein LA, Reese TS. 1988. The organization of cytoplasm at the presynaptic active zone of a central nervous system synapse. *Neuron* 1:201–9

Landis DM, Reese TS. 1983. Cytoplasmic organization in cerebellar dendritic spines. *J. Cell. Biol.* 97:1169–78

Lendvai B, Stern EA, Chen B, Svoboda K. 2000. Experience-dependent plasticity of dendritic spines in the developing rat barrel cortex in vivo. *Nature* 404:876–81

Lerm M, Schmidt G, Aktories K. 2000. Bacterial protein toxins targeting rho GTPases. *FEMS Microbiol. Lett.* 188:1–6

Li G, Rungger-Brandle E, Just I, Jonas JC, Aktories K, Wollheim CB. 1994. Effect of disruption of actin filaments by Clostridium botulinum C2 toxin on insulin secretion in HIT-T15 cells and pancreatic islets. *Mol. Biol. Cell.* 5:1199–213

Li L, Chin LS, Shupliakov O, Brodin L, Sihra TS, et al. 1995. Impairment of synaptic vesicle clustering and of synaptic transmission, and increased seizure propensity, in synapsin I-deficient mice. *Proc. Natl. Acad. Sci. USA* 92:9235–39

Li Z, Sheng M. 2003. Some assembly required: the development of neuronal synapses. *Nat. Rev. Mol. Cell. Biol.* 4:833–41

Lisman J, Schulman H, Cline H. 2002. The molecular basis of CaMKII function in synaptic and behavioural memory. *Nat. Rev. Neurosci.* 3:175–90

Luo L. 2002. Actin cytoskeleton regulation in neuronal morphogenesis and structural plasticity. *Annu. Rev. Cell. Dev. Biol.* 18:601–35

Luo L, Hensch TK, Ackerman L, Barbel S, Jan LY, Jan YN. 1996. Differential effects of the Rac GTPase on Purkinje cell axons and dendritic trunks and spines. *Nature* 379:837–40

Ma XM, Huang J, Wang Y, Eipper BA, Mains RE. 2003. Kalirin, a multifunctional Rho guanine nucleotide exchange factor, is necessary for maintenance of hippocampal pyramidal neuron dendrites and dendritic spines. *J. Neurosci.* 23:10593–603

Maletic-Savatic M, Malinow R, Svoboda K. 1999. Rapid dendritic morphogenesis in CA1 hippocampal dendrites induced by synaptic activity. *Science* 283:1923–27

Malinow R, Malenka RC. 2002. AMPA receptor trafficking and synaptic plasticity. *Annu. Rev. Neurosci.* 25:103–26

Matsuzaki M, Ellis-Davies GC, Nemoto T, Miyashita Y, Iino M, Kasai H. 2001. Dendritic spine geometry is critical for AMPA receptor expression in hippocampal CA1 pyramidal neurons. *Nat. Neurosci.* 4:1086–92

Matsuzaki M, Honkura N, Ellis-Davies GC, Kasai H. 2004. Structural basis of long-term potentiation in single dendritic spines. *Nature* 429:761–66

Matter K, Dreyer F, Aktories K. 1989. Actin involvement in exocytosis from PC12 cells: studies on the influence of botulinum C2 toxin on stimulated noradrenaline release. *J. Neurochem.* 52:370–76

Matus A. 2000. Actin-based plasticity in dendritic spines. *Science* 290:754–58

Matus A, Ackermann M, Pehling G, Byers HR, Fujiwara K. 1982. High actin concentrations in brain dendritic spines and postsynaptic densities. *Proc. Natl. Acad. Sci. USA* 79:7590–94

McGee AW, Bredt DS. 2003. Assembly and plasticity of the glutamatergic postsynaptic specialization. *Curr. Opin. Neurobiol.* 13:111–18

Meng Y, Zhang Y, Tregoubov V, Janus C, Cruz L, et al. 2002. Abnormal spine morphology and enhanced LTP in LIMK-1 knockout mice. *Neuron* 35:121–33

Meyer G, Feldman EL. 2002. Signaling mechanisms that regulate actin-based motility processes in the nervous system. *J. Neurochem.* 83:490–503

Morales M, Colicos MA, Goda Y. 2000. Actin-dependent regulation of neurotransmitter release at central synapses. *Neuron* 27:539–50

Muller D, Nikonenko I. 2003. Dynamic presynaptic varicosities: a role in activity-dependent synaptogenesis. *Trends Neurosci.* 26:573–75

Murai KK, Nguyen LN, Irie F, Yamaguchi Y, Pasquale EB. 2003. Control of hippocampal dendritic spine morphology through ephrin-A3/EphA4 signaling. *Nat. Neurosci.* 6:153–60

Murase S, Mosser E, Schuman EM. 2002. Depolarization drives beta-Catenin into neuronal spines promoting changes in synaptic structure and function. *Neuron* 35:91–105

Murthy VN, De Camilli P. 2003. Cell biology of the presynaptic terminal. *Annu. Rev. Neurosci.* 26:701–28

Nakayama AY, Harms MB, Luo L. 2000. Small GTPases Rac and Rho in the maintenance of dendritic spines and branches in hippocampal pyramidal neurons. *J. Neurosci.* 20:5329–38

Nikonenko I, Jourdain P, Alberi S, Toni N, Muller D. 2002. Activity-induced changes of spine morphology. *Hippocampus* 12:585–91

Nimchinsky EA, Sabatini BL, Svoboda K. 2002. Structure and function of dendritic spines. *Annu. Rev. Physiol.* 64:313–53

Okamoto KI, Nagai T, Miyawaki A, Hayashi Y. 2004. Rapid and persistent modulation of actin dynamics regulates postsynaptic reorganization underlying bidirectional plasticity. *Nat. Neurosci.* 7:1104–12

Pendleton A, Koffer A. 2001. Effects of latrunculin reveal requirements for the actin cytoskeleton during secretion from mast cells. *Cell Motil. Cytoskeleton* 48:37–51

Penzes P, Beeser A, Chernoff J, Schiller MR, Eipper BA, et al. 2003. Rapid induction of dendritic spine morphogenesis by trans-synaptic ephrinB-EphB receptor activation of the Rho-GEF kalirin. *Neuron* 37:263–74

Penzes P, Johnson RC, Sattler R, Zhang X, Huganir RL, et al. 2001. The neuronal Rho-GEF Kalirin-7 interacts with PDZ domain-containing proteins and regulates dendritic morphogenesis. *Neuron* 29:229–42

Phillips GR, Huang JK, Wang Y, Tanaka H, Shapiro L, et al. 2001. The presynaptic particle web: ultrastructure, composition, dissolution, and reconstitution. *Neuron* 32:63–77

Pieribone VA, Shupliakov O, Brodin L, Hilfiker-Rothenfluh S, Czernik AJ, Greengard P. 1995. Distinct pools of synaptic vesicles in neurotransmitter release. *Nature* 375:493–97

Pollard TD, Borisy GG. 2003. Cellular motility driven by assembly and disassembly of actin filaments. *Cell* 112:453–65

Prekeris R, Terrian DM. 1997. Brain myosin V is a synaptic vesicle-associated motor protein: evidence for a Ca^{2+}-dependent interaction with the synaptobrevin-synaptophysin complex. *J. Cell. Biol.* 137:1589–601

Qualmann B, Kessels MM, Kelly RB. 2000. Molecular links between endocytosis and the actin cytoskeleton. *J. Cell. Biol.* 150:F111–16

Ramakers GJ. 2002. Rho proteins, mental retardation and the cellular basis of cognition. *Trends Neurosci.* 25:191–99

Regehr WG, Stevens CF. 2001. Physiology of synaptic transmission and short-term plasticity. See Cowan et al. 2001, pp. 135–75

Revenu C, Athman R, Robine S, Louvard D. 2004. The co-workers of actin filaments: from cell structures to signals. *Nat. Rev. Mol. Cell. Biol.* 5:635–46

Richards DA, Rizzoli SO, Betz WJ. 2004. Effects of wortmannin and latrunculin A on slow endocytosis at the frog neuromuscular junction. *J. Physiol.* 557:77–91

Ridley AJ, Paterson HF, Johnston CL, Diekmann D, Hall A. 1992. The small GTP-binding protein rac regulates growth factor-induced membrane ruffling. *Cell* 70:401–10

Roos J, Hummel T, Ng N, Klambt C, Davis GW. 2000. Drosophila Futsch regulates synaptic microtubule organization and is necessary for synaptic growth. *Neuron* 26:371–82

Rosenmund C, Stevens CF. 1996. Definition of the readily releasable pool of vesicles at hippocampal synapses. *Neuron* 16:1197–207

Ruiz-Canada C, Ashley J, Moeckel-Cole S, Drier E, Yin J, Budnik V. 2004. New synaptic bouton formation is disrupted by misregulation of microtubule stability in aPKC mutants. *Neuron* 42:567–80

Saito SY, Watabe S, Ozaki H, Kobayashi M, Suzuki T, et al. 1998. Actin-depolymerizing effect of dimeric macrolides, bistheonellide A and swinholide A. *J. Biochem. (Tokyo)* 123:571–78

Sakaba T, Neher E. 2003. Involvement of actin polymerization in vesicle recruitment at the calyx of Held synapse. *J. Neurosci.* 23:837–46

Sankaranarayanan S, Atluri PP, Ryan TA. 2003. Actin has a molecular scaffolding, not propulsive, role in presynaptic function. *Nat. Neurosci.* 6:127–35

Schenck A, Bardoni B, Langmann C, Harden N, Mandel JL, Giangrande A. 2003. CYFIP/Sra-1 controls neuronal connectivity in Drosophila and links the Rac1 GTPase pathway to the fragile X protein. *Neuron* 38:887–98

Sheng M, Kim MJ. 2002. Postsynaptic signaling and plasticity mechanisms. *Science* 298:776–80

Shupliakov O, Bloom O, Gustafsson JS, Kjaerulff O, Low P, et al. 2002. Impaired recycling of synaptic vesicles after acute perturbation of the presynaptic actin cytoskeleton. *Proc. Natl. Acad. Sci. USA* 99:14476–81

Slepnev VI, De Camilli P. 2000. Accessory factors in clathrin-dependent synaptic vesicle endocytosis. *Nat. Rev. Neurosci.* 1:161–72

Soderling SH, Langeberg LK, Soderling JA, Davee SM, Simerly R, et al. 2003. Loss of WAVE-1 causes sensorimotor retardation and reduced learning and memory in mice. *Proc. Natl. Acad. Sci. USA* 100:1723–28

Spector I, Braet F, Shochet NR, Bubb MR. 1999. New anti-actin drugs in the study of the organization and function of the actin cytoskeleton. *Microsc. Res. Tech.* 47:18–37

Star EN, Kwiatkowski DJ, Murthy VN. 2002. Rapid turnover of actin in dendritic spines and its regulation by activity. *Nat. Neurosci.* 5:239–46

Stevens CF. 2003. Neurotransmitter release at central synapses. *Neuron* 40:381–88

Stevens CF, Tsujimoto T. 1995. Estimates for the pool size of releasable quanta at a single central synapse and for the time required to refill the pool. *Proc. Natl. Acad. Sci. USA* 92:846–49

Südhof TC. 2001. The synaptic cleft and synaptic cell adhesion. See Cowan et al. 2001, pp. 275–313

Südhof TC. 2004. The synaptic vesicle cycle. *Annu. Rev. Neurosci.* 27:509–47

Südhof TC, Scheller RH. 2001. Mechanism and regulation of neurotransmitter release. See Cowan et al. 2001, pp. 177–215

Takahashi H, Sekino Y, Tanaka S, Mizui T, Kishi S, Shirao T. 2003. Drebrin-dependent actin clustering in dendritic filopodia governs synaptic targeting of postsynaptic density-95 and dendritic spine morphogenesis. *J. Neurosci.* 23:6586–95

Takai Y, Shimizu K, Ohtsuka T. 2003. The roles of cadherins and nectins in interneuronal synapse formation. *Curr. Opin. Neurobiol.* 13:520–26

Tanaka H, Shan W, Phillips GR, Arndt K, Bozdagi O, et al. 2000. Molecular modification of N-cadherin in response to synaptic activity. *Neuron* 25:93–107

Tashiro A, Dunaevsky A, Blazeski R, Mason CA, Yuste R. 2003. Bidirectional regulation of hippocampal mossy fiber filopodial motility by kainate receptors: a two-step model of synaptogenesis. *Neuron* 38:773–84

Taunton J. 2001. Actin filament nucleation by endosomes, lysosomes and secretory vesicles. *Curr. Opin. Cell. Biol.* 13:85–91

Terry DR, Spector I, Higa T, Bubb MR. 1997. Misakinolide A is a marine macrolide that caps but does not sever filamentous actin. *J. Biol. Chem.* 272:7841–45

Tiruchinapalli DM, Oleynikov Y, Kelic S, Shenoy SM, Hartley A, et al. 2003. Activity-dependent trafficking and dynamic localization of zipcode binding protein 1 and beta-actin mRNA in dendrites and spines of hippocampal neurons. *J. Neurosci.* 23:3251–61

Togashi H, Abe K, Mizoguchi A, Takaoka K, Chisaka O, Takeichi M. 2002. Cadherin regulates dendritic spine morphogenesis. *Neuron* 35:77–89

Trachtenberg JT, Chen BE, Knott GW, Feng G, Sanes JR, et al. 2002. Long-term in vivo imaging of experience-dependent synaptic plasticity in adult cortex. *Nature* 420:788–94

Trifaro JM, Lejen T, Rose SD, Pene TD, Barkar ND, Seward EP. 2002. Pathways that control cortical F-actin dynamics during secretion. *Neurochem. Res.* 27:1371–85

Uehata M, Ishizaki T, Satoh H, Ono T, Kawahara T, et al. 1997. Calcium sensitization of smooth muscle mediated by a Rho-associated protein kinase in hypertension. *Nature* 389:990–94

Walker JH, Boustead CM, Witzemann V, Shaw G, Weber K, Osborn M. 1985. Cytoskeletal proteins at the cholinergic synapse: distribution of desmin, actin, fodrin, neurofilaments, and tubulin in Torpedo electric organ. *Eur. J. Cell. Biol.* 38:123–33

Wang XH, Zheng JQ, Poo MM. 1996. Effects of cytochalasin treatment on short-term synaptic plasticity at developing neuromuscular junctions in frogs. *J. Physiol.* 491(Pt. 1):187–95

Wenk MR, De Camilli P. 2004. Protein-lipid interactions and phosphoinositide metabolism in membrane traffic: insights from vesicle recycling in nerve terminals. *Proc. Natl. Acad. Sci. USA* 101:8262–69

Winder SJ. 2003. Structural insights into actin-binding, branching and bundling proteins. *Curr. Opin. Cell. Biol.* 15:14–22

Yarmola EG, Somasundaram T, Boring TA, Spector I, Bubb MR. 2000. Actin-latrunculin A structure and function. Differential modulation of actin-binding protein function by latrunculin A. *J. Biol. Chem.* 275:28120–27

Yuste R, Bonhoeffer T. 2004. Genesis of dendritic spines: insights from ultrastructural and imaging studies. *Nat. Rev. Neurosci.* 5:24–34

Zhang H, Webb DJ, Asmussen H, Horwitz AF. 2003. Synapse formation is regulated by the signaling adaptor GIT1. *J. Cell. Biol.* 161:131–42

Zhang W, Benson DL. 2001. Stages of synapse development defined by dependence on F-actin. *J. Neurosci.* 21:5169–81

Zhang W, Benson DL. 2002. Developmentally regulated changes in cellular compartmentation and synaptic distribution of actin in hippocampal neurons. *J. Neurosci. Res.* 69:427–36

Ziv NE, Garner CC. 2004. Cellular and molecular mechanisms of presynaptic assembly. *Nat. Rev. Neurosci.* 5:385–99

Zucker RS, Regehr WG. 2002. Short-term synaptic plasticity. *Annu. Rev. Physiol.* 64:355–405

Molecular Pathophysiology of Parkinson's Disease

Darren J. Moore,[1,2] Andrew B. West,[1,2] Valina L. Dawson,[1,2,3,4] and Ted M. Dawson[1,2,3]

Institute for Cell Engineering,[1] Departments of Neurology,[2] Neuroscience,[3] and Physiology,[4] Johns Hopkins University School of Medicine, Baltimore, Maryland 21205; email: dmoore20@jhmi.edu, abwest@jhmi.edu, vdawson@jhmi.edu, tdawson@jhmi.edu

Key Words

α-synuclein, parkin, mitochondrial complex-I, ubiquitin-proteasome system, oxidative stress

Abstract

Parkinson's disease (PD) is a progressive neurodegenerative movement disorder that results primarily from the death of dopaminergic neurons in the substantia nigra. Although the etiology of PD is incompletely understood, the recent discovery of genes associated with rare monogenic forms of the disease, together with earlier studies and new experimental animal models, has provided important and novel insight into the molecular pathways involved in disease pathogenesis. Increasing evidence indicates that deficits in mitochondrial function, oxidative and nitrosative stress, the accumulation of aberrant or misfolded proteins, and ubiquitin-proteasome system dysfunction may represent the principal molecular pathways or events that commonly underlie the pathogenesis of sporadic and familial forms of PD.

Contents

- INTRODUCTION ... 58
- THE HERITABILITY OF PD ... 58
- THE CONTRIBUTION OF GENES TO THE PATHOGENESIS OF PD ... 59
  - α-*Synuclein* (OMIM 163890; PARK1; PARK4) ... 59
  - *Parkin* (OMIM 602544; PARK2) ... 64
  - *UCH-L1* (OMIM 191342; PARK5; Neuron-Specific PGP9.5) ... 68
  - *PINK1* (OMIM 608309; PARK6; PTEN-Induced Putative Kinase 1) ... 69
  - *DJ-1* (OMIM 602533; PARK7) ... 69
- COMMON PATHWAYS UNDERLYING THE PATHOGENESIS OF PD ... 71
  - Mitochondrial Dysfunction and Oxidative Stress ... 71
  - Impairment of the Ubiquitin-Proteasome System ... 73
- CONCLUSIONS ... 74
- NOTE ADDED IN PROOF ... 76

INTRODUCTION

Parkinson's disease (PD) is a progressive neurodegenerative movement disorder that is estimated to affect approximately 1% of the population older than 65 years of age (Lang & Lozano 1998a,b). Clinically, most patients present with the cardinal symptoms of bradykinesia, resting tremor, rigidity, and postural instability. A number of patients also suffer from autonomic, cognitive, and psychiatric disturbances. The major symptoms of PD result from the profound and selective loss of dopaminergic (DA) neurons in the substantia nigra pars compacta (SNc), but there is widespread neuropathology with the SNc becoming involved later toward the middle stages of the disease (Braak et al. 2003). The pathological hallmarks of PD are round eosinophilic intracytoplasmic proteinaceous inclusions termed Lewy bodies (LBs) and dystrophic neurites (Lewy neurites) present in surviving neurons (Forno 1996). PD is primarily a sporadic disorder and its specific etiology is incompletely understood, but important new insights have recently been provided through studying the genetics, epidemiology, and neuropathology of PD, in addition to the development of new experimental models. Until recently, PD had been considered the prototypical nongenetic disorder. In the past seven years, the identification of distinct genetic loci responsible for rare Mendelian forms of PD has challenged this view and has provided us with vital clues to understanding the molecular pathogenesis of the more common sporadic forms of this disease. These genetic advances have revolutionized the way we think about PD and have opened up new and exciting areas of research. Despite such advances, much recent research has continued to focus on the contribution of nongenetic or environmental factors to the development of sporadic forms of PD.

THE HERITABILITY OF PD

For most of the twentieth century, genetic predisposition to Charcot's "la maladie de Parkinson" was thought to play a negligible role in development of the syndrome. The notion that environmental factors, not heredity, caused PD was further propagated following the post-encephalitic outbreak of a variant of the syndrome. Yet for more than one hundred years, clinicians noted that patients with PD often had an affected relative (Gowers 1900). Further studies corroborated these suspicions with the identification and characterization of families that inherited PD in a Mendelian fashion (Bell & Clark 1926). In contrast, a number of studies utilizing twin registries demonstrated a low rate of concordance in monozygotic and dizygotic twins, indicative of a lack of genetic susceptibility in PD (Marttila et al. 1988, Ward et al. 1983). These early twin studies seemed to shift the

Table 1 Loci and genes associated with familial PD or implicated in PD[1]

Locus	Chromosome location	Gene	Inheritance pattern	Typical phenotype	Reference
PARK1 & PARK4	4q21–q23	α-synuclein	AD	Earlier onset, features of DLB common	Polymeropolous et al. 1997, Singleton et al. 2003
PARK2	6q25.2–q27	parkin	usually AR	Earlier onset with slow progression	Kitada et al. 1998
PARK3	2p13	unknown	AD, IP	Classic PD, sometimes dementia	Gasser et al. 1998
PARK5	4p14	UCH-L1	unclear	Classic PD	Leroy et al. 1998
PARK6	1p35–p36	PINK1	AR	Earlier onset with slow progression	Valente et al. 2004a
PARK7	1p36	DJ-1	AR	Earlier onset with slow progression	Bonifati et al. 2003
PARK8	12p11.2–q13.1	LRRK2[2]	AD	Classic PD	Funayama et al. 2002
PARK10	1p32	unknown	unclear	Classic PD	Hicks et al. 2002
PARK11	2q36–q37	unknown	unclear	Classic PD	Pankratz et al. 2003
NA	5q23.1–q23.3	Synphilin-1	unclear	Classic PD	Marx et al. 2003
NA	2q22–q23	NR4A2	unclear	Classic PD	Le et al. 2003

[1] Abbreviations: NA, not assigned; AD, autosomal dominant; AR, autosomal recessive; IP, incomplete penetrance; DLB, dementia with Lewy bodies.
[2] See note added in proof.

direction of PD research away from genetics, and the concomitant identification of 1-methyl-4-phenyl-1,2,3,6-tetrahydropyridine (MPTP)-induced parkinsonism (Langston et al. 1983) further downplayed any potential genetic component to PD. Despite these past assumptions that genetics played little role in PD, it became clear that there is a significant genetic component to disease (Dawson & Dawson 2003, Hicks et al. 2002, Marder et al. 1996, Sveinbjornsdottir et al. 2000). Indeed, there are at least 10 distinct genetic loci associated with PD, and mutations have been identified in four genes that definitively and unambiguously cause familial forms of PD (**Table 1**).

THE CONTRIBUTION OF GENES TO THE PATHOGENESIS OF PD

Although monogenic and sporadic forms of PD are clinically and pathologically distinct from each other, they tend to share many overlapping features that include, most importantly, parkinsonism with nigrostriatal DA degeneration (Hardy et al. 2003), which perhaps implies that common pathogenic mechanisms may underlie disease. However, although the genes linked to different monogenic forms of PD do not necessarily fit into a common pathogenic pathway, they nevertheless promote our understanding of the specific molecular pathways that lead to DA neuronal degeneration in PD. This section discusses our current understanding of those gene products linked to monogenic forms of PD (PARK1, 2, 4, 5, 6, and 7), emphasizing, in particular, the normal function of each protein and how its dysfunction may contribute to disease pathogenesis.

α-Synuclein (OMIM 163890; PARK1; PARK4)

The first gene for familial PD was initially mapped to chromosome 4q21–q23 in an Italian American family (Contursi kindred) with more than 60 affected individuals spanning 5 generations with autosomal dominant disease inheritance (Polymeropoulos et al. 1996). An A53T missense mutation was isolated in affected individuals in the gene encoding the α-synuclein protein (Polymeropoulos et al.

1997). Subsequently, a second mutation in the α-*synuclein* gene (A30P) was found in a German family (Kruger et al. 1998), in addition to an E46K mutation in a Spanish family (Zarranz et al. 2004). Lastly, a genomic triplication of a region spanning the α-*synuclein* gene segregated with disease in the Iowa kindred (Singleton et al. 2003). Moreover, there is some suggestion that genetic variability in the α-*synuclein* promoter associates with sporadic PD, which implicates that variability of α-synuclein protein levels can predispose individuals to disease (Pals et al. 2004).

α-synuclein is a 140-amino-acid protein belonging to a family of related synucleins that include β- and γ-synuclein (Clayton & George 1998). Structurally, human α-synuclein consists of an N-terminal amphipathic region containing six imperfect repeats (with a KTKEGV consensus motif), a hydrophobic central region [containing the non-amyloid-β component (NAC) domain], and an acidic C-terminal region (**Figure 1**). α-synuclein is an intrinsically unstructured or natively unfolded protein but has significant conformational plasticity. For example,

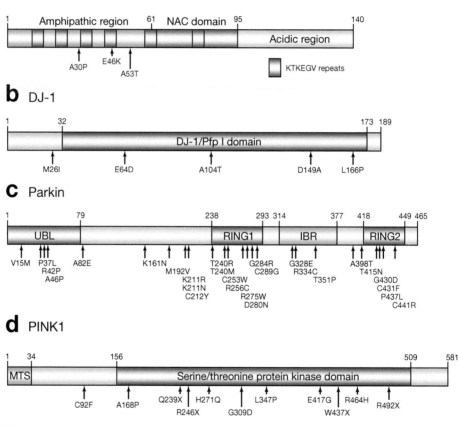

Figure 1

Domain architecture and familial mutations of proteins associated with PD. Protein domains of human (*A*) α-synuclein (140 aa), (*B*) DJ-1 (189 aa), (*C*) parkin (465 aa), and (*D*) PINK1 (581 aa). The position of amino acid missense mutations (*single letter code*) associated with familial forms of PD are indicated for each protein (*arrows*). Nonsense mutations are indicated by X, which denotes the introduction of a premature stop codon. Not all familial mutations are indicated for parkin. Refer to Mata et al. 2004 for detailed references to all reported parkin mutations. MTS, mitochondrial targeting sequence.

depending on the environment α-synuclein can remain unstructured, can form monomeric and oligomeric species, or can form amyloidogenic filaments (Uversky 2003). The physiological function of α-synuclein is unclear. α-synuclein is highly expressed throughout the mammalian brain and is enriched in presynaptic nerve terminals, where it can associate with membranes and vesicular structures (Irizarry et al. 1996, Kahle et al. 2000). Recent studies have shown that α-synuclein specifically associates with membrane microdomains known as lipid rafts, and this raft association may be required for its synaptic localization (Fortin et al. 2004). Analysis of mice with a targeted deletion of the α-*synuclein* gene suggest a role for α-synuclein in synaptic vesicle recycling and DA neurotransmission (Abeliovich et al. 2000). A role in synaptic vesicle recycling is further supported by observations in vitro that demonstrate that α-synuclein can bind to acidic phospholipid vesicles (Davidson et al. 1998) and can also bind to and inhibit the activity of mammalian phospholipase D (PLD) (Jenco et al. 1998). Furthermore, in cultured cells α-synuclein can regulate lipid metabolism by protecting lipid droplets from hydrolysis (Cole et al. 2002) and can regulate the size of presynaptic vesicular pools (Murphy et al. 2000). Studies in the model organism yeast further demonstrate that α-synuclein selectively associates with the plasma membrane, inhibits PLD activity, induces lipid droplet accumulation, and modulates vesicle trafficking (Outeiro & Lindquist 2003). Thus, α-synuclein may play an important role in regulating synaptic vesicle size and recycling with particular relevance to dopamine storage.

α-synuclein is considered to play a central role in the pathophysiology of PD, in part, on the basis of the identification of mutations or triplications of the α-*synuclein* gene associated with familial PD (Kruger et al. 1998, Polymeropoulos et al. 1997, Singleton et al. 2003). Moreover, the identification of fibrillar forms of the α-synuclein protein as a major structural component of LBs in PD and other synucleinopathies provides compelling evidence that α-synuclein plays a major role in the pathogenesis of PD (Spillantini et al. 1998). Mutations in α-*synuclein* cause PD through a toxic gain-of-function mechanism consistent with the dominant inheritance pattern of mutations. The genomic triplication of α-*synuclein* leads to an approximate doubling of expression, thereby demonstrating that overexpression of wild-type α-synuclein is sufficient to cause disease. The effects of α-synuclein missense mutations are not so obvious. Both the A30P and A53T mutant proteins display an increased propensity to self-aggregate to form oligomeric species and LB-like fibrils in vitro compared with wild-type α-synuclein (Conway et al. 1998). α-synuclein oligomers are the precursors for higher-order aggregates, such as amyloid-like fibrils, which precipitate as the filamentous structures observed in LBs and Lewy neurites (**Figure 2**). Some investigators proposed that oligomeric fibrillization intermediates of α-synuclein termed protofibrils, rather than the fibrils themselves, may be the pathogenic cytotoxic moiety. For example, the A53T and A30P mutations both share the capacity to promote the oligomerization, but not fibrillization, of α-synuclein (Conway et al. 2000). Furthermore, these mutants may form annular protofibrils that resemble a class of

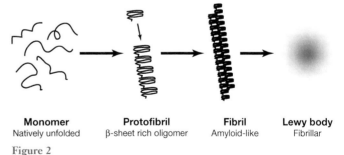

| Monomer | Protofibril | Fibril | Lewy body |
| Natively unfolded | β-sheet rich oligomer | Amyloid-like | Fibrillar |

Figure 2

Schematic of α-synuclein fibrillogenesis. Natively unfolded or disordered α-synuclein monomers form β-sheet rich oligomers that comprise a transient population of protofibrils of heterogenous structure that may include spheres, chains, or rings. The protofibrils may give rise to more stable amyloid-like fibrils. α-synuclein fibrils eventually aggregate and precipitate to form LBs in vivo. Figure adapted from Lansbury & Brice 2002.

pore-forming bacterial toxin (Lashuel et al. 2002), which suggests that protofibrils might cause inappropriate permeabilization of cellular membranes. Catecholamines, particularly dopamine, can react with α-synuclein to form covalent adducts that slow conversion of protofibrils to fibrils (Conway et al. 2001), thus promoting protofibril accumulation. Soluble oligomeric forms of α-synuclein have recently been observed in human brain tissue (Sharon et al. 2003), which suggests that oligomeric species are physiologically relevant.

The notion of cytotoxic protofibrils is further supported by studies in one line of α-synuclein transgenic mice, whereby motoric impairment and loss of DA nerve terminals are observed in the presence of non-fibrillar α-synuclein inclusions (Masliah et al. 2000). However, the failure of transgenic mice overexpressing the protofibrillogenic A30P mutant α-synuclein to exhibit neurodegeneration (Lee et al. 2002) suggests that protofibrils may not be the primary cytotoxic moiety. Indeed, only when the A30P mutant forms both fibrils and inclusions is neurodegeneration observed in transgenic mice and flies (Feany & Bender 2000, Kahle et al. 2001, Neumann et al. 2002). Consistent with the cytotoxicity of fibrillar α-synuclein is the observation that β-amyloid promotes the formation of α-synuclein fibrillar inclusions in bigenic mice overexpressing mutant amyloid precursor protein and α-synuclein, leading to a more severe α-synuclein-related pathological and behavioral phenotype (Masliah et al. 2001). At this stage the relative contributions of α-synuclein protofibrils and fibrils to PD pathogenesis are incompletely understood, but both likely contribute to disease pathogenesis.

The mechanism by which wild-type α-synuclein aggregates in sporadic forms of PD is poorly understood. A number of factors enhance α-synuclein aggregation or fibrillization in different systems. Mitochondrial complex-I inhibitors such as rotenone and paraquat clearly lead to aggregation and accumulation of α-synuclein in vitro and in animal models (Betarbet et al. 2000; Manning-Bog et al. 2002; Sherer et al. 2002b, 2003), and other forms of oxidative and nitrosative stress also promote α-synuclein aggregation (Ischiropoulos & Beckman 2003). Oxidative damage may play a pertinent role in the aggregation of α-synuclein in sporadic PD because there is selective tyrosine nitration of α-synuclein in lesions in PD and other synucleinopathies (Giasson et al. 2000). Indeed, recent studies suggest that tyrosine nitration of α-synuclein may potentiate fibril formation of unmodified α-synuclein and may decrease the rate of degradation by the 20S proteasome and the cysteine protease calpain I (Hodara et al. 2004). α-synuclein protein levels also increase with aging in human substantia nigra (Li et al. 2004). The stabilization of α-synuclein with aging may be a significant factor in the pathogenesis of α-synucleinopathies because it could lead to the accumulation of pathogenic protein modifications, such as oxidative damage (Li et al. 2004). Proteasomal inhibition is also associated with increases in α-synuclein fibrillization with formation of insoluble inclusions in primary neuronal cultures and in vivo (McNaught et al. 2004, Rideout et al. 2004). Endogenous cofactors may play a role in modulating α-synuclein fibrillogenesis. Recent evidence suggests that interactions between α-synuclein and tau synergistically promote the fibrillization of both proteins in vitro and in animal models (Giasson et al. 2003), which is supported by the co-occurrence of α-synuclein and tau pathology in some monogenic forms of PD, in other neurodegenerative disorders, and in α-synuclein transgenic mice. β-amyloid can also enhance α-synuclein fibrillization in vivo, and pathology related to both proteins can coexist in neurodegenerative diseases such as the LB variant of Alzheimer's disease (Masliah et al. 2001). The interaction of α-synuclein with amyloidogenic proteins could be one mechanism that drives the formation of pathological fibrillar inclusions in human neurodegenerative diseases. Conversely, non-amyloidogenic proteins such as β-synuclein can have the opposite effect to prevent α-synuclein fibrillization in vivo and may represent an endogenous negative regulator of fibrillization (Hashimoto et al. 2001). Mutations in the *β-synuclein* gene may predispose to dementia with LBs (DLB) (Ohtake et al. 2004). Posttranslational modifications may additionally contribute to the aggregation or fibrillization of α-synuclein because it is both selectively ubiquitinated and phosphorylated in lesions in sporadic PD and other synucleinopathies (Hasegawa et al. 2002).

Although the process of α-synuclein fibrillization may be the key pathogenic event in most forms of PD, the mechanism by which α-synuclein species exert their downstream neurotoxic effects is unclear. One proposed mechanism is through direct impairment of the ubiquitin-proteasome system (UPS). For example, pathogenic species of α-synuclein are more resistant to proteasomal degradation and can directly bind to 20/26S proteasomal subunits and impair proteolytic activity (Bennett et al. 1999, Lindersson et al. 2004, Snyder et al. 2003). Overexpression of mutant α-synuclein also sensitizes cultured cells to the toxicity associated with proteasome inhibitors, suggesting a prior level of proteasome impairment (Petrucelli et al. 2002, Tanaka et al. 2001). Thus, the reduced clearance of α-synuclein species and their direct inhibitory effect on proteasomal activity may contribute to disease pathogenesis. It is presently unclear whether α-synuclein turnover is physiologically regulated by the proteasome, although both ubiquitin-dependent and -independent mechanisms have been described (Tofaris et al. 2001, Webb et al. 2003). Furthermore, proteasomal inhibition produces LB-like α-synuclein inclusions in rodent models of PD and in cultured neurons (McNaught et al. 2004, Rideout et al. 2004). Alternative pathways of α-synuclein degradation have also been proposed including processing by the lysosome/autophagy pathway and by cytoplasmic proteases such as calpain I (Mishizen-Eberz et al. 2003, Webb et al. 2003). In addition to effects on the UPS, α-synuclein overexpression in cultured cells, particularly of mutant forms, has been linked to mitochondrial deficits (Hsu et al. 2000), defective cellular trafficking (Gosavi et al. 2002), apoptosis (Lee et al. 2001b), impaired chaperone-mediated autophagy (Cuervo et al. 2004), and increased sensitivity to oxidative stress (Ko et al. 2000) and dopamine-mediated toxicity (Tabrizi et al. 2000). Thus, the mechanism of α-synuclein-mediated toxicity likely affects numerous critical cellular pathways and highlights the complexity of disentangling the primary pathogenic events from the secondary events.

A role for dopamine in mediating α-synuclein toxicity is an attractive notion that would account for the relatively selective degeneration of DA neurons in PD. In cultured cells, the toxic effects of α-synuclein appear to be selective for DA neurons, which requires endogenous dopamine production and is mediated by reactive oxygen species (ROS) (Xu et al. 2002). Furthermore, α-synuclein can interact with and enhance the activity of the dopamine transporter, thereby accelerating cellular dopamine uptake and dopamine-induced apoptosis (Lee et al. 2001a). Overexpression of mutant α-synuclein can also downregulate the vesicular monoamine transporter 2, which perhaps promotes enhanced levels of cytoplasmic dopamine and increased oxidative stress (Lotharius et al. 2002). Thus, one effect of toxic α-synuclein species could be to increase cytosolic dopamine levels, which, owing to the high oxidative potential of dopamine metabolism, could promote oxidative stress and ensuing DA neuronal cell death. However, degeneration observed in *α-synuclein*-linked PD (PARK1 and PARK4) is not limited to DA neurons or even to catecholaminergic neurons but is widespread and may affect other neurotransmitter systems. This mechanism would account only for loss of DA neurons in PD, and α-synuclein toxicity may affect distinct neuronal populations through alternate mechanisms. Future research will help to delineate the precise mechanism of α-synuclein toxicity in specific neuronal populations that will hopefully give rise to an ordered pathway of events that lead to neuronal degeneration in PD.

Parkin (OMIM 602544; PARK2)

A large region spanning chromosome 6q25.2–q27 was initially linked to a rare form of autosomal recessive juvenile-onset parkinsonism (AR-JP), in consanguineous Japanese families (Matsumine et al. 1997). Subsequently, a homozygous deletion spanning a microsatellite marker was identified in an affected case, and the adjacent gene was cloned and named *parkin* (Kitada et al. 1998). Mutations in the *parkin* gene are relatively common in familial PD, with mutations found in 50% of familial early-onset cases compatible with recessive inheritance and 10% of all early-onset PD cases (Lucking et al. 2000). A genome-wide scan in families with early-onset PD revealed significant linkage only to *parkin*, thereby demonstrating a unique role for the *parkin* gene in the development of early-onset PD (Scott et al. 2001). However, mutations have also been described in apparent sporadic cases with a clinical presentation indistinguishable from late-onset idiopathic PD (Lucking et al. 1998). A wide variety of *parkin* mutations have been described in PD cases, ranging from deletions of a single nucleotide to hundreds of thousands of nucleotides, in addition to genomic multiplications and missense mutations (Mata et al. 2004, West & Maidment 2004). Most *parkin*-linked disease is inherited in an autosomal recessive fashion; however, there are several descriptions of families with *parkin* mutations and disease segregating in a mode incompatible with recessive inheritance, and some evidence indicates that haploinsufficiency in the *parkin* gene predisposes to PD (Farrer et al. 2001).

The *parkin* gene encodes a 465-amino-acid protein with a modular structure that contains an N-terminal ubiquitin-like (UBL) domain, a central linker region, and a C-terminal RING domain comprising two RING finger motifs separated by an in-between-RING (IBR) domain (**Figure 1**). Like many other RING finger–containing proteins, parkin can function as an E3 ubiquitin protein ligase (Shimura et al. 2000, Zhang et al. 2000). E3 ligases are an important part of the cellular machinery that covalently tags target proteins with ubiquitin (Glickman & Ciechanover 2002) (**Figure 3**). Ubiquitination of proteins results from the successive actions of ubiquitin-activating (E1), conjugating (E2), and ligase (E3) enzymes, respectively. Subsequent cycles add additional ubiquitin molecules to a previously ligated ubiquitin, resulting in the formation of a poly-ubiquitin chain containing four or more ubiquitin molecules. Such poly-ubiquitinated proteins are specifically recognized by the 26S proteasome and are subsequently targeted for degradation. The ubiquitination machinery can also tag proteins with single ubiquitin molecules or with alternatively linked poly-ubiquitin chains that do not signal proteasomal degradation but are implicated in numerous nondegradative cellular processes (**Figure 3**). E3 ligases typically confer substrate specificity to the ubiquitination process by simultaneously interacting with E2-ubiquitin and the substrate protein to catalyze the transfer of ubiquitin from E2 to a substrate lysine residue. Parkin interacts with the E2 enzymes, UbcH7 and UbcH8 (Shimura et al. 2000, Zhang et al. 2000), as well as with the endoplasmic reticulum–associated E2s, UBC6 and UBC7 (Imai et al. 2001). Parkin generally tends to interact with E2s and substrate proteins through its RING domain. The UBL domain of parkin can interact with subunits of the 19S cap of the 26S proteasomal complex, such as Rpn10, which most likely mediates transfer of poly-ubiquitinated substrates to the proteasome (Sakata et al. 2003). Parkin may function as part of a larger protein complex. The interaction of parkin with an SCF-like (Skp1-Cullin-F-box protein) complex, or with a complex containing Hsp70 and CHIP, can enhance its E3 ligase activity and its neuroprotective capacity (Imai et al. 2002, Staropoli et al. 2003). Most

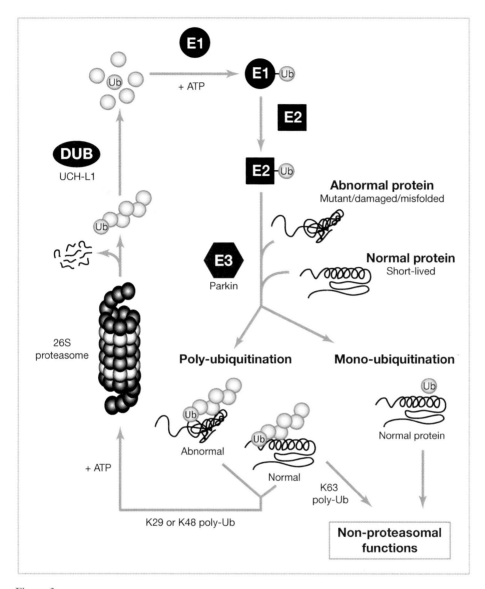

Figure 3

The ubiquitin-proteasome system. Ubiquitin (Ub) monomers are activated by the Ub-activating enzyme (E1) and are then transferred to a Ub-conjugating enzyme (E2). Normal or abnormal target proteins are recognized by a Ub protein ligase (E3), such as parkin, which mediates the transfer of Ub from the E2 enzyme to the target protein. The sequential covalent attachment of Ub monomers to a lysine (K) acceptor residue of the previous Ub results in the formation of a poly-Ub chain. Poly-Ub chains linked through K29 or K48 signal the target protein for degradation through the 26S proteasome in an ATP-dependent manner, resulting in the generation of small peptide fragments. The resulting poly-Ub chains are recycled to free Ub monomers by deubiquitinating (DUB) enzymes, such as UCH-L1, for subsequent rounds of ubiquitination. The addition of Ub also has other diverse roles. Normal proteins can be singly or multiply mono-ubiquitinated, or poly-ubiquitinated with K63-linked chains, which leads to nonproteasomal functions that include DNA repair, endocytosis, protein trafficking, and transcription (Glickman & Ciechanover 2002).

familial-associated mutations in parkin are considered to be loss-of-function and tend to impair the interaction of parkin with E2s or substrates and reduce or abolish parkin's E3 ligase activity or expression. Hence, parkin mutations are thought to result, in general, in the improper targeting of its substrates for proteasomal degradation leading to their potentially neurotoxic accumulation. For this reason, great importance has been placed on the identification of protein substrates of parkin and their possible role in DA neuron loss in PD.

A number of putative substrates have been reported for parkin on the basis of in vitro and cell culture experiments, including CDCrel-1 (Zhang et al. 2000), synphilin-1 (Chung et al. 2001b), a rare O-glycosylated form of α-synuclein (Shimura et al. 2001), the parkin-associated endothelin receptor-like receptor (Pael-R) (Imai et al. 2001), synaptotagmin XI (Huynh et al. 2003), cyclin E (Staropoli et al. 2003), the p38 subunit of the aminoacyl-tRNA synthetase complex (Corti et al. 2003), and α/β-tubulin (Ren et al. 2003). The large number of substrates identified is somewhat surprising because typically E3 ligases demonstrate a high level of specificity for one or a small number of substrates. Parkin substrates are diverse, widely distributed, and initially appear to have little in common, and it remains to be seen which, if any, of these putative substrates are relevant in vivo. Some of parkin's substrates have been implicated in enhancing neuronal cell death or toxicity, thus reinforcing a potential role in neuronal dysfunction in PD. CDCrel-1 is a synaptic vesicle–associated protein implicated in regulating neurotransmitter release and can specifically inhibit dopamine release (Dong et al. 2003). Overexpression of CDCrel-1 in SNc DA neurons of rats by virus-mediated gene transfer induces dopamine-dependent neurodegeneration (Dong et al. 2003). When overexpressed in cultured cells, Pael-R tends to become unfolded and insoluble, inducing the unfolded protein response that eventually leads to cell death (Imai et al. 2001). Further-more, panneuronal expression of Pael-R in *Drosophila* causes age-dependent selective degeneration of DA neurons (Yang et al. 2003). Similarly, overexpression of the p38 subunit results in aggresome-like inclusion formation and/or cell death, depending on the cell type (Corti et al. 2003). The deleterious accumulation of such toxic substrates in the absence of parkin in AR-JP patients may be one mechanism that eventually leads to dysfunction and death of susceptible neurons.

Some parkin substrates also have a tendency to become insoluble and form cytoplasmic inclusions when overexpressed in cells. For example, both Pael-R and p38 form insoluble inclusions in cells, and both have been detected in LBs in sporadic PD together with synphilin-1 and synaptotagmin XI (Corti et al. 2003, Huynh et al. 2003, Imai et al. 2001, Murakami et al. 2004, Wakabayashi et al. 2000). Synphilin-1 interacts with α-synuclein and coexpression of both proteins in cultured cells leads to the formation of insoluble LB-like cytoplasmic inclusions (Engelender et al. 1999). Although parkin can interact with and ubiquitinate synphilin-1, it can also localize to and ubiquitinate α-synuclein/synphilin-1 LB-like inclusions (Chung et al. 2001b). Furthermore, parkin can protect against the toxicity induced by α-synuclein/synphilin-1 overexpression following proteasome inhibition (Chung et al. 2004). A direct role for synphilin-1 in PD is suggested by the identification of a R621C mutation in the *synphilin-1* gene in two apparently sporadic PD patients of German origin (Marx et al. 2003). Overexpression of the R621C mutant reduced the number of inclusions formed following proteasome inhibition, compared with wild-type synphilin-1, and also sensitized cells to staurosporine-induced cell death (Marx et al. 2003). These findings support a causative role for R621C mutant synphilin-1 in PD and suggest that inclusion formation may be a protective event. Parkin, synphilin-1, and α-synuclein are therefore intimately linked in a common biochemical pathway that may contribute to the biogenesis of LBs and may play an important role in the pathogenesis of PD.

On the basis of strong evidence that loss of function of parkin is the underlying cause of AR-JP, mice with a targeted deletion of the *parkin* gene have been generated to model this form of PD. Surprisingly, parkin knockout mice do not develop a PD-like behavioral or neuropathological phenotype. Two of the knockouts show subtle behavioral deficits, mild alterations of dopaminergic and glutamatergic neurotransmission, and abnormalities in dopamine metabolism (Goldberg et al. 2003, Itier et al. 2003). A third line of knockout mice shows a reduced number of noradrenergic neurons in the locus coeruleus, accompanied by a marked reduction of the norepinephrine-dependent acoustic startle response (Von Coelln et al. 2004b). Furthermore, none of the putative parkin substrates have so far been reported to accumulate in the brains of parkin knockout mice, which has cast doubt on the authenticity of these substrates as well as the contribution of parkin to their turnover by the proteasome. Proteomic analysis has instead revealed reduced levels of several proteins involved in mitochondrial oxidative phosphorylation and protection from oxidative stress in the ventral midbrain of parkin knockout mice (Palacino et al. 2004). This was accompanied by decreases in mitochondrial respiratory capacity and age-dependent increases of oxidative damage. This suggests an unexpected role for parkin in the regulation of normal mitochondrial function. Recent studies in *Drosophila* models of PD support this notion. *Drosophila* parkin null mutants exhibit reduced lifespan, locomotor defects, and male sterility (Greene et al. 2003, Pesah et al. 2004). The locomotor defects derive from apoptotic cell death of muscle subsets, whereas male sterility results from a spermatid individualization defect. The earliest manifestation of muscle degeneration and defective spermatogenesis is mitochondrial pathology, and there are also accompanying signs of increased oxidative stress. The observation of mitochondrial defects and increased oxidative stress in parkin-deficient mice and flies in the absence of DA neuron loss suggests that mitochondrial impairment may be a primary pathogenic event in AR-JP that may eventually trigger selective neuronal degeneration. The absence of DA neuron loss in these models could suggest that neuronal degeneration proceeds over a prolonged time scale exceeding that obtainable in animal models or may require a further pathogenic "hit" to precipitate neuronal death.

The mechanism by which parkin confers neuroprotection, or specifically, promotes the survival of DA neurons, is a central unanswered question. Consistent with a role in maintaining mitochondrial integrity, overexpression of parkin in cultured cells confers resistance to stimuli that promote mitochondria-dependent apoptosis (Darios et al. 2003). Furthermore, a small proportion of parkin is localized to the outer membrane of mitochondria in cells (Darios et al. 2003). In cultured cells, parkin overexpression may confer protection against dopamine-mediated toxicity possibly by decreasing oxidative stress through an undetermined mechanism, thus potentially linking parkin to the survival of DA neurons (Jiang et al. 2004). Parkin can also confer protection against kainate-induced excitotoxicity in primary neuronal cultures, presumably by suppressing cyclin E accumulation (Staropoli et al. 2003). Similarly, parkin can also suppress the ensuing cell death induced by overexpression of Pael-R or the p38 subunit in cultured cells, possibly by ubiquitinating and promoting their degradation (Corti et al. 2003, Imai et al. 2001). In cell culture, overexpression of parkin protects against the toxicity induced by proteasomal inhibition and overexpression of mutant α-synuclein (Petrucelli et al. 2002). Furthermore, in *Drosophila*, parkin overexpression can protect against the selective loss of DA neurons induced by neuronal expression of α-synuclein or Pael-R (Yang et al. 2003). In the α-synuclein *Drosophila* model, parkin overexpression leads to a sharp reduction in the abundance of α-synuclein-positive LB-like inclusions, which suggests that parkin may act to clear specifically aberrant α-synuclein deposits. Because parkin does not directly interact with native α-synuclein, alterations in the structure of α-synuclein, either through post-translational modifications (including phosphorylation, O-glycosylation, ubiquitination, or nitration) or conformational changes (oligomers or fibrils), may promote an interaction of both proteins at some level, possibly by converging on a common molecular pathway. In this respect it is intriguing that parkin can suppress the toxic effects of α-synuclein species apparently through rescuing impaired proteasome function (Petrucelli et al. 2002). The surprisingly large number of putative parkin substrates, together with the capacity of parkin to confer cellular protection against a diversity of toxic insults, suggests that parkin may represent a multipurpose neuroprotectant (Feany & Pallanck 2003).

UCH-L1 (OMIM 191342; PARK5; Neuron-Specific PGP9.5)

Using a candidate gene-screening approach to identify new mutations in familial PD cases, a heterozygous I93M mutation in the gene encoding ubiquitin carboxyl-terminal hydrolase L1 (UCH-L1) was identified in an affected sibling pair in a German family (Leroy et al. 1998). However, the transmitting parent was asymptomatic, suggesting that either the I93M variant is nonpathogenic or that it causes disease with incomplete penetrance. An additional heterozygous M124L variant was described in an affected individual; however, this variant failed to segregate with disease, suggesting that rare polymorphisms in UCH-L1 may not be pathogenic (Farrer et al. 2000). Additional mutations have not been identified despite extensive genetic screening (Lincoln et al. 1999). Thus, the involvement of the UCH-L1 gene in familial PD is contentious. A common nonsynonymous polymorphism (S18Y) in the UCH-L1 gene was first reported as underrepresented in a European cohort of cases and controls (Maraganore et al. 1999), and a number of subsequent studies either confirmed or failed to replicate the original results. A meta-analysis of the literature provided some evidence to corroborate a potential protective effect of the S18Y variant in PD, which thereby suggests that genetic variability in the UCH-L1 gene plays a role in the development of late-onset idiopathic PD (Maraganore et al. 2004).

UCH-L1 is a highly abundant, neuron-specific protein that belongs to a family of deubiquitinating enzymes that are responsible for hydrolyzing polymeric ubiquitin chains to free ubiquitin monomers (Wilkinson et al. 1989). UCH-L1 might additionally function as a dimerization-dependent ubiquitin protein ligase (Liu et al. 2002) and can apparently maintain ubiquitin homeostasis by promoting the stability of ubiquitin monomers in vivo (Osaka et al. 2003). UCH-L1 has been localized to LBs in sporadic PD (Lowe et al. 1990). The mechanism by which UCH-L1 mutations cause PD is poorly understood. The I93M mutation decreased the in vitro hydrolytic activity of UCH-L1 (Leroy et al. 1998), which originally suggested that this form of PD results from a partial loss of function. However, this idea may be an oversimplified interpretation because the I93M mutation shows incomplete penetrance, and mutant mice lacking functional UCH-L1 (the gracile axonal dystrophy mouse) do not develop a parkinsonian phenotype (Leroy et al. 1998, Saigoh et al. 1999). Some investigators have proposed that UCH-L1 may recycle free ubiquitin by cleaving ubiquitinated peptides that are products of the proteasomal degradation of polyubiquitinated proteins (**Figure 3**). Thus, reduced UCH-L1 hydrolytic activity caused by the I93M mutation might impair the overall efficiency of the UPS by reducing the availability of free ubiquitin monomers, thereby leading to potentially deleterious protein accumulation (Leroy et al. 1998). Although a direct role in the UPS is an intriguing possibility, the function of UCH-L1 in vivo remains to be clarified. UCH-L1 can also promote the accumulation of α-synuclein in cultured cells presumably through the addition of Lys63-linked poly-ubiquitin chains that are not associated with proteasomal degradation (Liu et al. 2002). The potentially protective S18Y polymorphic variant of UCH-L1 has reduced ligase activity but exhibited comparable hydrolase activity to wild-type enzyme and does not promote α-synuclein accumulation (Liu et al. 2002). Thus, both the ubiquitin ligase and hydrolase activities of UCH-L1 may play a normal role in the UPS and may be relevant to the pathogenesis of PD.

PINK1 (OMIM 608309; PARK6; PTEN-Induced Putative Kinase 1)

A genome-wide homozygosity screen performed on a large Sicilian family with four early-onset PD patients revealed a shared 12.5 cM region on chromosome 1p35–p36 (Valente et al. 2001). Additional unrelated families with positive linkage to this region were described (Valente et al. 2002), and mutations in the *PINK1* gene were identified (Valente et al. 2004a). Initial screens for *PINK1* mutations in early-onset familial cases revealed a number of novel mutations; however, these mutations are less common than are alterations in the *parkin* gene in early-onset PD cases (Hatano et al. 2004, Valente et al. 2004b). Genetic variation in the *PINK1* gene did not influence the onset of idiopathic PD in a large European cohort of cases and controls (Healy et al. 2004).

PINK1 is a 581-amino-acid protein that contains a mitochondrial targeting sequence at its N-terminus and a highly conserved protein kinase domain similar to serine/threonine kinases of the Ca^{2+}-calmodulin family (Valente et al. 2004a) (**Figure 1**). Accordingly, overexpressed PINK1 is localized to mitochondria in cultured cells (Valente et al. 2004a). Studies of PINK1 are at an early stage, and currently, little is known about the physiological function of PINK1. Although PINK1 is considered to be a mitochondrial protein kinase, the kinase activity of PINK1 has not yet been demonstrated, and as such, no putative mitochondrial substrates or interacting proteins have been identified. Mutations in *PINK1* are thought to cause PD through loss of function of PINK1 activity with most mutations clustering in or around the putative kinase domain (Hatano et al. 2004; Valente et al. 2004a,b), which perhaps suggests that the loss of PINK1 kinase activity directly causes PD. Initial studies in cultured cells suggest that PINK1 may afford some protection against mitochondrial dysfunction and apoptosis induced by proteasomal inhibition, although the mechanism for this action is not understood (Valente et al. 2004a). The G309D mutation, identified in a Spanish family with PD, impairs this protective effect of PINK1, and homology modeling reveals that this mutation is located in the ADP binding site of PINK1 and may therefore interfere with ADP binding and kinase activity (Bossy-Wetzel et al. 2004, Valente et al. 2004a). Thus, the loss of the putative kinase activity of PINK1 likely affects mitochondrial function. It has been suggested that PINK1 phosphorylates mitochondrial proteins, in response to cellular stress, to prevent mitochondrial dysfunction (Valente et al. 2004a), although alternatively an inability to normally phosphorylate mitochondrial proteins could actually lead to mitochondrial dysfunction. Although such a mechanism remains to be formally demonstrated, what is clear instead is that a probable role for PINK1 in mitochondrial function links for the first time a primary defect of mitochondria to the molecular pathogenesis of PD. This may have important implications for sporadic forms of PD in which deficits in mitochondrial function have long been proposed.

DJ-1 (OMIM 602533; PARK7)

Homozygosity mapping in a family with early-onset parkinsonism and multiple consanguinity loops demonstrated significant evidence for linkage on chromosome 1p36 (van Duijn et al. 2001). Mutations within the gene encoding DJ-1 were identified in both an Italian and Dutch family (Bonifati et al. 2003). Additional mutations, including splice site alterations, missense mutations, and small deletions were identified in a number of other familial PD cases (Bonifati et al. 2004). However mutations are extremely rare in early-onset PD cases, present in perhaps less than 1% of PD cohorts (Lockhart et al. 2004). To date, evidence suggests that genetic variability within the *DJ-1* gene does not contribute to the onset of idiopathic PD (Morris et al. 2003).

The *DJ-1* gene encodes a highly conserved protein of 189 amino acids that belongs to the DJ-1/ThiJ/PfpI superfamily (**Figure 1**). DJ-1 is ubiquitously and abundantly expressed in most mammalian tissues, including the brain, where it is localized to both neurons and glia (Bandopadhyay et al. 2004, Olzmann et al. 2004). DJ-1 does not appear to be localized to LBs in sporadic PD and other synucleinopathies but does colocalize with tau-positive inclusions in a number of neurodegenerative tauopathies and with α-synuclein-positive glial inclusions in multiple system atrophy (Neumann et al. 2004, Rizzu et al. 2004), which suggests that DJ-1 may play a diverse role in seemingly distinct neurodegenerative diseases. Furthermore, insoluble forms of DJ-1 are dramatically increased in the brains of sporadic PD patients (Moore et al. 2005) perhaps also implicating DJ-1 in sporadic forms of this disease. The crystal structure of human DJ-1 has been resolved and reveals a flavodoxin-like fold similar to the bacterial protease PH1704 and the stress-inducible molecular chaperone Hsp31, from *Escherichia coli* and yeast (Tao & Tong 2003, Wilson et al. 2003). The crystal structure also shows that DJ-1 exists as a dimer in solution, which has been confirmed in cultured cells (Miller et al. 2003, Moore et al. 2003b). A putative active site has been identified near the dimer interface with similarities to the active site catalytic triad (Cys-His-Asp/Glu) of cysteine proteases, involving residues Cys106, His126, and perhaps Glu18, although these residues do not show an orientation favorable for proton transfer that is typical of cysteine protease catalysis (Tao & Tong 2003). Consistent with these observations DJ-1 may possess chaperone-like activity as well as weak proteolytic activity in vitro against synthetic model substrates (Lee et al. 2003, Olzmann et al. 2004). At present no substrates have been definitively identified for DJ-1.

The physiological function of DJ-1 is unclear although many lines of evidence suggest that DJ-1 may function as an anti-oxidant protein or as a sensor of oxidative stress. For example, DJ-1 demonstrates an acidic shift in isoelectric point in cultured cells following oxidative stress owing mainly to oxidation of cysteine residues, particularly Cys106, which can be converted to a cysteine sulfinic acid (Cys-SO_2H) (Canet-Aviles et al. 2004, Mitsumoto et al. 2001). DJ-1 can also eliminate hydrogen peroxide in vitro by oxidizing itself suggesting that it may function, in part, as a direct scavenger of ROS (Taira et al. 2004). In cultured cells, overexpression of DJ-1 protects against oxidative injury whereas knockdown of DJ-1 by short interfering RNA enhances the susceptibility to oxidative stress (Taira et al. 2004). A recent study has shown that oxidative stress promotes cysteine sulfinic acid-driven mitochondrial localization of DJ-1 and subsequent protection against mitochondria-dependent cell death (Canet-Aviles et al. 2004). Oxidative stress also promotes an interaction of DJ-1 with parkin in cultured cells (Moore et al. 2005), perhaps linking both proteins in a common neuroprotective pathway. Thus, DJ-1 may play a critical role in both sensing and conferring protection against a range of oxidative stressors. DJ-1 may also confer protection against endoplasmic reticulum stress, proteasomal inhibition, and the toxicity induced by overexpression of Pael-R (Yokota et al. 2003). These diverse cellular insults all share the capacity to induce protein misfolding and aggregation perhaps suggesting that DJ-1 is a component of the UPS and may confer protection by functioning as a molecular chaperone or protease to refold or promote the degradation of misfolded or aggregated proteins. The possibility that DJ-1 possesses dual enzymatic function i.e., chaperone and protease activities, should not be discounted at this early stage. However, the precise mechanism by which DJ-1 confers neuroprotection awaits further clarification. Familial-associated mutations in *DJ-1* are considered to cause PD through a loss-of-function mechanism, consistent with the recessive inheritance pattern in two *DJ-1*-linked families with PD (Bonifati et al. 2003). The manner in which missense mutations cause loss of DJ-1 function are beginning to be clarified. The L166P mutation, identified in an Italian kindred with PD, drastically destabilizes the DJ-1 protein by promoting the unfolding of its C-terminal region, leading to a loss of dimerization, and subsequently enhancing its degradation by the proteasome (Miller et al. 2003, Moore et al. 2003b, Olzmann et al. 2004). Parkin, CHIP and Hsp70 may play a role in these events since they interact robustly with DJ-1 harboring the L166P mutation (Moore et al. 2005). The L166P mutation also impairs the neuroprotective function of DJ-1 against a range of oxidative stimuli in cultured cells, probably as a direct consequence of its instability (Taira et al. 2004). Other missense mutations may similarly reduce the ability of DJ-1 to protect against selective forms of oxidative stress (Takahashi-Niki et al. 2004), and such mutations also share the capacity to reduce DJ-1 dimerization (Moore et al. 2005). Further clarifying the properties of such mutations may provide novel insight into the neuroprotective function of DJ-1.

COMMON PATHWAYS UNDERLYING THE PATHOGENESIS OF PD

Prior to the identification of genes underlying monogenic forms of PD, both mitochondrial dysfunction and oxidative stress were considered to play a prominent role in the pathogenesis of sporadic PD. Both genetic and non-genetic studies have further implicated these pathways, but they have also highlighted protein mishandling due to UPS dysfunction as a major pathway leading to neuronal degeneration in PD. This section reviews accumulating evidence that mitochondrial dysfunction, oxidative stress, and impairment of the UPS may represent the principal molecular pathways that commonly underlie the pathogenesis of both sporadic and familial forms of PD.

Mitochondrial Dysfunction and Oxidative Stress

Post-mortem studies have consistently implicated oxidative damage in the pathogenesis of PD, and in particular, oxidative damage to lipids, proteins, and DNA has been observed in the SNc of sporadic PD brains (Jenner 2003). Oxidative stress is considered to compromise the integrity of vulnerable neurons and thus to contribute to neuronal degeneration. The source of this increased oxidative stress is unclear but may include mitochondrial dysfunction, increased dopamine metabolism that can yield excess hydrogen peroxide and other ROS, an increase in reactive iron, and impaired antioxidant defense pathways (Jenner 2003). Mitochondria are exposed to a highly oxidative environment, and the process of oxidative phosphorylation is associated with the production of ROS. Much evidence suggests a major role for mitochondrial dysfunction in the pathogenesis of PD, and in particular, defects in mitochondrial complex-I (complex-I) of the respiratory chain. A complex-I defect could most obviously contribute to neuronal degeneration in PD through decreased ATP synthesis as well as damage caused by excess production of ROS. There are consistent findings of decrements in complex-I activity in the SNc of sporadic PD patients (Schapira et al. 1990), although the cause of this is unknown. Complex-I activity is also reduced in cytoplasmic hybrid (cybrids) cell lines that contain mitochondrial DNA (mtDNA) from sporadic PD patients, which indicates that deficits in complex-I can be stably transmitted, although it is unclear whether such defects in mtDNA arise somatically or are due to inherited mutations (Swerdlow et al. 1996). Rare maternal patterns of inheritance of PD in some families are consistent with the notion of mitochondrial inheritance (Wooten et al. 1997). Indeed, cybrid cell lines derived from maternal descendents of these families similarly exhibit reduced complex-I activity, increased ROS production, and increased radical scavenging enzyme activities (Swerdlow et al. 1998). Genetic evidence that alterations in complex-I activity play a role in the pathogenesis of sporadic PD is provided in part by the observation that a single nucleotide polymorphism in the gene encoding the NADH dehydrogenase 3 enzyme of complex-I, causing an amino acid change from threonine to alanine, leads to a significantly reduced risk of developing PD in Caucasian populations (van der Walt et al. 2003). However, as yet no specific disease-related mutations have been detected in sporadic PD in mitochondrial or nuclear genes that encode complex-I proteins, and pathogenic mutations in mtDNA also have not been identified.

Several epidemiological studies suggest that pesticides and other environmental toxins that inhibit complex-I are involved in the pathogenesis of sporadic PD (Sherer et al. 2002a). MPTP inhibits complex-I and replicates most features of PD in humans and in animal models (Dauer & Przedborski 2003). MPTP was identified as a contaminant of the manufacture of a synthetic opiate, and drug users who accidentally injected MPTP developed a syndrome resembling PD (Langston et al. 1983). The selectivity of MPTP for DA neurons is due to its conversion in astrocytes by monoamine oxidase B to the active metabolite, 1-methyl-4-phenyl pyridinium (MPP$^+$), which is taken up by DA neurons via the dopamine transporter, where it inhibits complex-I and ultimately leads to cell death (Dauer & Przedborski 2003). In aged nonhuman primates, MPTP treatment produces intracellular proteinaceous inclusions resembling immature LBs that are filamentous and contain α-synuclein (Forno et al. 1988). Coadministration of the widely used herbicide paraquat (1,1'-dimethyl-4,4'-5 bipyridinium) and the fungicide maneb (manganese ethylenepistithiocarbamate) leads to the pronounced and selective loss of nigrostriatal DA neurons in mice (Thiruchelvam et al. 2000). Paraquat is a complex-I inhibitor with structural similarity to MPP$^+$ and, when administered alone, can also induce selective degeneration of DA neurons together with upregulation and aggregation of α-synuclein in the SNc of mice (Manning-Bog et al. 2002, McCormack et al. 2002). Chronic systemic complex-I inhibition caused by exposure to rotenone, a common insecticide and fish poison, induces parkinsonism in rats, including selective nigrostriatal DA degeneration and the formation of LB-like intraneuronal filamentous inclusions containing α-synuclein and ubiquitin (Betarbet et al. 2000, Sherer et al. 2003). In contrast to MPTP and paraquat, rotenone is not selectively taken up by DA neurons but can still induce selective DA neuronal degeneration, which implies that DA neurons are especially vulnerable to deficits in complex-I. These findings suggest that deficits in complex-I may be central to the pathogenesis of sporadic PD and imply that environmental factors may contribute to PD pathogenesis.

How might the gene products linked to monogenic forms of PD be associated with mitochondrial dysfunction observed in sporadic forms of the disease? Current evidence suggests an intriguing connection between α-synuclein and mitochondria. Complex-I inhibition both in vitro and in vivo consistently leads to the accumulation of LB-like α-synuclein-positive inclusions, which suggests that α-synuclein aggregation is a downstream consequence of mitochondrial dysfunction and might be an effector of neuronal cell death (Betarbet et al. 2000, Forno et al. 1988, Manning-Bog et al. 2002). This idea is supported in part by the observation that cybrid lines derived from members of the Contursi kindred with PD fail to manifest complex-I deficiency (Swerdlow et al. 2001). Furthermore, α-synuclein knockout mice are resistant to the neurotoxic effects of MPTP, whereas α-synuclein transgenic mice show enhanced toxicity (Dauer et al. 2002, Song et al. 2004), which thus implies that α-synuclein is required for mediating the deleterious downstream effects of complex-I inhibition. α-synuclein itself may further contribute to mitochondrial dysfunction induced by complex-I inhibition because α-synuclein transgenic mice develop enhanced SNc mitochondrial pathology following exposure to MPTP, compared with wild-type mice (Song et al. 2004). Overexpression of mutant α-synuclein can sensitize cultured cells to mitochondrial-dependent apoptosis (Tanaka et al. 2001), which suggests that mutant forms of α-synuclein can also impair mitochondrial function.

Other gene products linked to monogenic forms of PD also appear to be implicated in mitochondrial function. On the basis of studies in mice and *Drosophila*, parkin may have an unexpected role in the regulation of normal mitochondrial function, perhaps linking mitochondria with UPS function (Greene et al. 2003, Palacino et al. 2004). Mitochondrial dysfunction is probably the leading source of increased oxidative and nitrosative stress observed in the brains of sporadic PD patients (Ischiropoulos & Beckman 2003, Jenner 2003). Such stress can promote the *S*-nitrosylation of parkin through its RING domain both in vitro and in cultured cells, and this *S*-nitrosylation can impair parkin's ubiquitin ligase activity and its neuroprotective function (Chung et al. 2004). Furthermore, parkin is selectively *S*-nitrosylated in brains from MPTP-treated mice in a nitric oxide–dependent manner and in patients with sporadic PD and DLB (Chung et al. 2004). Thus, inhibition of parkin's E3 ligase activity by *S*-nitrosylation could contribute to the pathogenesis of PD and related disorders by impairing the ubiquitination of parkin substrates. Because the level of *S*-nitrosylated proteins is increased in PD and DLB brains in general (Chung et al. 2004), *S*-nitrosylation of proteins due to oxidative and nitrosative stress may play a prominent role in disease pathogenesis. DJ-1 may play a role in mitochondrial function because a proportion of DJ-1 is normally localized to mitochondria, whereas oxidative stress induced by complex-I inhibition can enhance DJ-1 mitochondrial localization (Canet-Aviles et al. 2004). Finally, PINK1, a putative mitochondrial kinase, is the first gene to directly link mitochondria to PD, and accordingly, PINK1 can partially protect against mitochondrial dysfunction induced by proteasome inhibition (Valente et al. 2004a). Future analysis of post-mortem brain tissue from monogenic forms of PD and genetic animal models, including DJ-1 or PINK1 knockout mice, may help to clarify further the molecular pathway linking PD with mitochondrial dysfunction and oxidative stress.

Impairment of the Ubiquitin-Proteasome System

Emerging evidence suggests that impairment of the UPS and protein mishandling may also underlie the molecular pathogenesis of familial and sporadic forms of PD (Giasson & Lee 2003, Moore et al. 2003a). Consistent with this notion there are both structural and functional deficits in the 20/26S proteasome in the SNc of sporadic PD patients (McNaught et al. 2002, 2003). Systemic exposure of rats to naturally occurring and synthetic proteasome inhibitors closely recapitulates many key features of PD, including progressive parkinsonism, selective neurodegeneration of the nigrostriatal pathway and specific brainstem nuclei, and the formation of LB-like intracytoplasmic inclusions containing α-synuclein and ubiquitin (McNaught et al. 2004). If these studies are independently replicated and confirmed, it suggests that perhaps proteasomal dysfunction may be a common end point that precipitates DA neuronal degeneration in PD. The accumulation and aggregation of potentially cytotoxic proteins, including α-synuclein, in LBs in DA neurons in sporadic PD strongly suggest generalized protein mishandling and subsequent proteolytic stress, which perhaps implies impaired UPS function (Chung et al. 2001a). In *Drosophila*, transgenic or pharmacologically induced overexpression of molecular chaperones rescues the motoric and pathological features induced by transgenic overexpression of normal or mutant α-synuclein (Auluck & Bonini 2002, Auluck et al. 2002), which further suggests a role for protein mishandling in PD. Parkin overexpression has similar effects in *Drosophila* models of PD (Yang et al. 2003). A role for chaperones and other components of the UPS, including UCH-L1, proteasomal subunits, and ubiquitin, in sporadic PD is also supported by their presence in LBs in postmortem brain tissue (Auluck et al. 2002, Forno 1996, Ii et al. 1997, Lowe et al. 1990).

Perhaps the most compelling evidence supporting a role for the UPS in the pathogenesis of PD is the association of *parkin* with familial forms of this disease. Disease-linked mutations in parkin are thought to cause defects in normal UPS function with subsequent proteolytic stress due to aberrant protein accumulation, perhaps leading to the eventual demise of DA neurons (Chung et al. 2001a, Von Coelln et al. 2004a). On the basis of structural comparison with Hsp31, DJ-1 may also participate in the UPS as a redox-sensitive molecular chaperone to alleviate protein misfolding by interacting with early unfolding intermediates (Quigley et al. 2003). As discussed previously, α-synuclein species may be associated with UPS dysfunction through binding and inhibiting the 20/26S proteasome (Snyder et al. 2003), and mutated or aggregated forms of α-synuclein may also overwhelm the degradative capacity of the proteasome, leading to further impairment (Bence et al. 2001, Petrucelli et al. 2002). The consistent presence of fibrillar α-synuclein as a major component of LBs in PD (Spillantini et al. 1998), and the formation of LB-like inclusions containing α-synuclein following proteasome inhibition in vivo (McNaught et al. 2004), tends to support this notion. LBs are a pathological hallmark of sporadic and some familial forms of PD and indicate the involvement of protein mishandling in disease pathogenesis, although we do not know whether LB formation is a primary or secondary event. The role of LB formation in PD is the subject of some controversy with both pathogenic and protective mechanisms being proposed (Bence et al. 2001, Chung et al. 2001a). Recent studies suggest that the biogenesis of LBs may be akin to the formation of aggresomes (Olanow et al. 2004, Tanaka et al. 2004). Aggresomes are cytoprotective proteinaceous inclusions formed at the centrosome to sequester and aid in the degradation of excess, possibly deleterious proteins (Kopito 2000). One might hypothesize that the formation of LBs is in direct response to proteolytic stress due to proteasomal impairment (Chung et al. 2001a). Parkin has been implicated in promoting LB formation because the majority of patients with *parkin* mutations lack LB pathology (Mori et al. 1998). Alternatively, parkin-mediated neurodegeneration may proceed through mechanisms distinct from those that cause classic PD with LBs, or parkin may be downstream of α-synuclein aggregation, thus bypassing the formation of LBs. The idea that LBs are protective is supported in part by the observation that *parkin*-linked AR-JP patients, which lack LBs, exhibit an earlier onset and more aggressive disease compared with patients with classic PD with LBs (Hardy et al. 2003). The occurrence of LBs in other monogenic forms of PD, including PARK 5, 6, and 7, is not yet known, but future neuropathological analysis may shed light on the significance of LB formation in PD.

CONCLUSIONS

Mitochondrial dysfunction, oxidative stress, and impairment of the UPS may underlie the molecular pathogenesis of familial and sporadic PD, and these pathways may be linked together at multiple levels (**Figure 4**). This

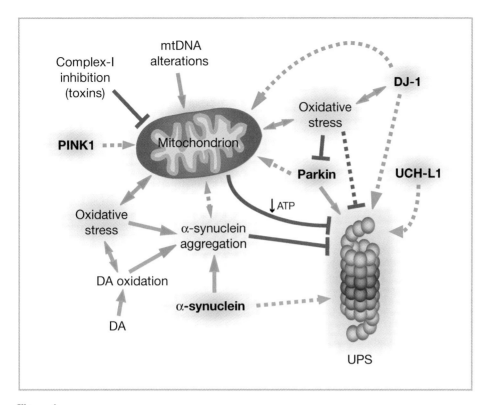

Figure 4

Common pathways underlying PD pathogenesis. Mutations in five genes encoding α-*synuclein*, *parkin*, *UCH-L1*, *PINK1*, and *DJ-1* are associated with familial forms of PD through pathogenic pathways that may commonly lead to deficits in mitochondrial and UPS function. PINK1, parkin, and DJ-1 may play a role in normal mitochondrial function, whereas parkin, UCH-L1, and DJ-1 may be involved in normal UPS function. α-synuclein fibrillization and aggregation is promoted by pathogenic mutations, oxidative stress, and oxidation of cytosolic dopamine (DA), leading to impaired UPS function and possibly mitochondrial damage. α-synuclein may normally be degraded by the UPS. Some environmental toxins and pesticides can inhibit complex-I and lead to mitochondrial dysfunction, whereas alterations in mitochondrial DNA (mtDNA) may influence mitochondrial function. Impaired mitochondrial function leads to oxidative stress, deficits in ATP synthesis, and α-synuclein aggregation, which may contribute to UPS dysfunction. Oxidative and nitrosative stress may also influence the antioxidant function of DJ-1, can impair parkin function through *S*-nitrosylation, and may promote dopamine oxidation. Excess dopamine metabolism may further promote oxidative stress. Mitochondrial and UPS dysfunction, oxidative stress, and α-synuclein aggregation ultimately contribute to the demise of DA neurons in PD. Red lines indicate inhibitory effects, green arrows depict defined relationships between components or systems, and blue dashed arrows indicate proposed or putative relationships.

idea is supported by the fact that α-synuclein, parkin, and DJ-1 share the capacity to influence both mitochondrial and UPS function, perhaps providing the beginnings of a common pathway involved in neuronal degeneration in PD. However, it is currently unclear whether dysfunction of mitochondria and the UPS tends to converge on a common pathway or factor, whether they are part of the same pathway, or whether they cause PD through distinct mechanisms. Because the aggregation of α-synuclein is clearly downstream from complex-I inhibition and such aggregation can also inhibit or overwhelm proteasomal function, a putative model for the pathogenesis of sporadic PD emerges from these

AR-JP: autosomal recessive juvenile-onset parkinsonism

DA: dopaminergic

DLB: dementia with Lewy bodies

LB: Lewy body

MPTP: 1-methyl-4-phenyl-1,2,3,6-tetrahydropyridine

PD: Parkinson's disease

Pael-R: parkin-associated endothelin receptor-like receptor

PINK1: PTEN-induced putative kinase 1

ROS: reactive oxygen species

SNc: substantia nigra pars compacta

UCH-L1: ubiquitin carboxyl-terminal hydrolase L1

UPS: ubiquitin-proteasome system

observations. If complex-I inhibition is central to PD pathogenesis, it would set in motion a series of events that lead to α-synuclein aggregation, increased oxidative stress, and deficits in ATP synthesis, all of which can impair normal UPS function. Inhibition of the UPS would lead to the accumulation of proteins otherwise targeted for degradation, some of which may be cytotoxic; that in combination with oxidative damage would ultimately lead to the demise of DA neurons. Parkin, UCH-L1, and DJ-1 may be involved in maintaining normal UPS function, whereas PINK1, together with parkin and DJ-1, may regulate normal mitochondrial function; disease-linked mutations in these genes would lead to a similar set of events precipitating in the demise of DA neurons. However, this pathway of events is by no means straightforward because proteasomal inhibition alone may potentially cause a PD-like phenotype including α-synuclein aggregation (McNaught et al. 2004), and furthermore, proteasomal inhibition can also reciprocally impair mitochondrial function (Sullivan et al. 2004). These observations suggest a large degree of cross-talk between mitochondria and the UPS, and dysfunction in either or both systems may lead to the common end point of DA neuronal degeneration. Future studies of the monogenic forms of PD and their identified gene products, together with experimental animal models of complex-I or UPS dysfunction, will help to determine whether common or distinct molecular pathways contribute to the pathogenesis of familial and sporadic PD.

NOTE ADDED IN PROOF

Recently, two studies have identified the gene associated with PARK8-linked PD. The PARK8 locus was originally identified by parametric two-point analysis of a large Japanese family with autosomal dominant parkinsonism, which yielded significant linkage to chromosome 12p11.2–q13.1, and haplotype analysis further reduced the linked region to a 13.6-cM interval (PARK8, OMIM 607060) (Funayama et al. 2002). Several Caucasian families consistent with autosomal dominant parkinsonism also demonstrated linkage to the PARK8 region, suggesting that genetic variation in PARK8 may be a significant cause of autosomal dominant PD (Zimprich et al. 2004a). Upon screening candidate genes within the linked region, heterozygous mutations within the leucine-rich repeat kinase 2 (*LRRK2*) gene were identified in a number of families (Paisán-Ruíz et al. 2004, Zimprich et al. 2004a). In particular, Paisán-Ruíz et al. (2004) identified two different mutations that segregated with disease in four Basque families (R1396G) and in one large family from the United Kingdom (Y1654C). Subsequent analysis of a population of apparently unrelated Spanish PD patients revealed that approximately 8% of cases carried the R1396G mutation (Paisán-Ruíz et al. 2004). Zimprich et al. (2004b) identified mutations in the same two amino acid residues of the *LRRK2* gene in two large families of German-Canadian (Y1699C) and probable English (Western Nebraska kindred; R1441C) origin; the difference in numbering resulted from the inclusion of a 45 amino acid sequence that may correspond to exon 6. Three distinct mutations (I1122V and I2020T, and the putative splice site variant L1114L [3342A > G]) as well as the R1441C variant were identified in four families by analysis of an additional 32 families with typical late-onset PD, compatible with dominant transmission. These initial findings suggest that mutations in the *LRRK2* gene are, to date, the most common identified genetic cause of late-onset PD.

The *LRRK2* gene contains 51 exons that are predicted to encode a 2527 amino acid protein that has been named dardarin (Paisán-Ruíz et al. 2004). The predicted LRRK2/dardarin protein contains numerous highly conserved domains, including multiple leucine-rich repeats, a Ras-like/small GTPase superfamily domain, a tyrosine kinase-like domain, and a WD40 domain. Mutations in LRRK2/dardarin tend to be distributed throughout these functional domains, and thus it is currently unclear which domains are relevant for neurodegeneration. Although most PARK8-linked families described to date exhibit a clinical phenotype of classic PD, neuropathology in those affected individuals examined so far ranges from pure nigral degeneration without LBs to nigral degeneration associated with brainstem LBs typical of PD, widespread LBs consistent with DLB, or neurofibrillary tau-positive tangles suggestive of tauopathy (Funayama et al. 2002, Wszolek et al. 2004, Zimprich et al. 2004b). At present, it is unclear how mutations in LRRK2 cause PD, but it will be of particular interest to determine whether LRRK2/dardarin is related to other gene products or molecular pathways associated with PD. It is tempting to speculate that the putative kinase activity of LRRK2/dardarin (and perhaps that of PINK1) may be involved in the phosphorylation of proteins implicated in PD pathogenesis, such as the phosphoproteins α-synuclein and tau, possibly leading to alterations in their expression, turnover, processing, conformation, cellular localization, or protein-protein interactions.

ACKNOWLEDGMENTS

This work was supported by grants from the USPHS NS38377, NS457565, NS48206, and NS43691. D.J.M. and A.B.W. are supported by the Herbert Freidberg Fellowship. T.M.D. is the Leonard and Madlyn Abramson Professor in Neurodegenerative Diseases at the Johns Hopkins University School of Medicine.

LITERATURE CITED

Abeliovich A, Schmitz Y, Farinas I, Choi-Lundberg D, Ho WH, et al. 2000. Mice lacking α-synuclein display functional deficits in the nigrostriatal dopamine system. *Neuron* 25:239–52

Auluck PK, Bonini NM. 2002. Pharmacological prevention of Parkinson disease in *Drosophila*. *Nat. Med.* 8:1185–86

Auluck PK, Chan HY, Trojanowski JQ, Lee VM, Bonini NM. 2002. Chaperone suppression of α-synuclein toxicity in a *Drosophila* model for Parkinson's disease. *Science* 295:865–68

Bandopadhyay R, Kingsbury AE, Cookson MR, Reid AR, Evans IM, et al. 2004. The expression of DJ-1 (PARK7) in normal human CNS and idiopathic Parkinson's disease. *Brain* 127:420–30

Bell J, Clark AJ. 1926. A pedigree of paralysis agitans. *Ann. Eugen.* 1:455–62

Bence NF, Sampat RM, Kopito RR. 2001. Impairment of the ubiquitin-proteasome system by protein aggregation. *Science* 292:1552–55

Bennett MC, Bishop JF, Leng Y, Chock PB, Chase TN, Mouradian MM. 1999. Degradation of α-synuclein by proteasome. *J. Biol. Chem.* 274:33855–58

Betarbet R, Sherer TB, MacKenzie G, Garcia-Osuna M, Panov AV, Greenamyre JT. 2000. Chronic systemic pesticide exposure reproduces features of Parkinson's disease. *Nat. Neurosci.* 3:1301–6

Bonifati V, Oostra BA, Heutink P. 2004. Linking DJ-1 to neurodegeneration offers novel insights for understanding the pathogenesis of Parkinson's disease. *J. Mol. Med.* 82:163–74

Bonifati V, Rizzu P, van Baren MJ, Schaap O, Breedveld GJ, et al. 2003. Mutations in the DJ-1 gene associated with autosomal recessive early-onset parkinsonism. *Science* 299:256–59

Bossy-Wetzel E, Schwarzenbacher R, Lipton SA. 2004. Molecular pathways to neurodegeneration. *Nat. Med.* 10:S2–9

Braak H, Del Tredici K, Rub U, de Vos RA, Jansen Steur EN, Braak E. 2003. Staging of brain pathology related to sporadic Parkinson's disease. *Neurobiol. Aging* 24:197–211

Canet-Aviles RM, Wilson MA, Miller DW, Ahmad R, McLendon C, et al. 2004. The Parkinson's disease protein DJ-1 is neuroprotective due to cysteine-sulfinic acid-driven mitochondrial localization. *Proc. Natl. Acad. Sci. USA* 101:9103–8

Chung KK, Dawson VL, Dawson TM. 2001a. The role of the ubiquitin-proteasomal pathway in Parkinson's disease and other neurodegenerative disorders. *Trends Neurosci.* 24:S7–14

Chung KK, Thomas B, Li X, Pletnikova O, Troncoso JC, et al. 2004. S-nitrosylation of parkin regulates ubiquitination and compromises parkin's protective function. *Science* 304:1328–31

Chung KK, Zhang Y, Lim KL, Tanaka Y, Huang H, et al. 2001b. Parkin ubiquitinates the α-synuclein-interacting protein, synphilin-1: implications for Lewy-body formation in Parkinson disease. *Nat. Med.* 7:1144–50

Clayton DF, George JM. 1998. The synucleins: a family of proteins involved in synaptic function, plasticity, neurodegeneration and disease. *Trends Neurosci.* 21:249–54

Cole NB, Murphy DD, Grider T, Rueter S, Brasaemle D, Nussbaum RL. 2002. Lipid droplet binding and oligomerization properties of the Parkinson's disease protein α-synuclein. *J. Biol. Chem.* 277:6344–52

Conway KA, Harper JD, Lansbury PT. 1998. Accelerated in vitro fibril formation by a mutant α-synuclein linked to early-onset Parkinson disease. *Nat. Med.* 4:1318–20

Conway KA, Lee SJ, Rochet JC, Ding TT, Williamson RE, Lansbury PT Jr. 2000. Acceleration of oligomerization, not fibrillization, is a shared property of both α-synuclein mutations linked to early-onset Parkinson's disease: implications for pathogenesis and therapy. *Proc. Natl. Acad. Sci. USA* 97:571–76

Conway KA, Rochet JC, Bieganski RM, Lansbury PT Jr. 2001. Kinetic stabilization of the α-synuclein protofibril by a dopamine-α-synuclein adduct. *Science* 294:1346–49

Corti O, Hampe C, Koutnikova H, Darios F, Jacquier S, et al. 2003. The p38 subunit of the aminoacyl-tRNA synthetase complex is a Parkin substrate: linking protein biosynthesis and neurodegeneration. *Hum. Mol. Genet.* 12:1427–37

Cuervo AM, Stefanis L, Fredenburg R, Lansbury PT, Sulzer D. 2004. Impaired degradation of mutant α-synuclein by chaperone-mediated autophagy. *Science* 305:1292–95

Darios F, Corti O, Lucking CB, Hampe C, Muriel MP, et al. 2003. Parkin prevents mitochondrial swelling and cytochrome c release in mitochondria-dependent cell death. *Hum. Mol. Genet.* 12:517–26

Dauer W, Kholodilov N, Vila M, Trillat AC, Goodchild R, et al. 2002. Resistance of α-synuclein null mice to the parkinsonian neurotoxin MPTP. *Proc. Natl. Acad. Sci. USA* 99:14524–29

Dauer W, Przedborski S. 2003. Parkinson's disease: mechanisms and models. *Neuron* 39:889–909

Davidson WS, Jonas A, Clayton DF, George JM. 1998. Stabilization of α-synuclein secondary structure upon binding to synthetic membranes. *J. Biol. Chem.* 273:9443–49

Dawson TM, Dawson VL. 2003. Rare genetic mutations shed light on the pathogenesis of Parkinson disease. *J. Clin. Invest.* 111:145–51

Dong Z, Ferger B, Paterna JC, Vogel D, Furler S, et al. 2003. Dopamine-dependent neurodegeneration in rats induced by viral vector-mediated overexpression of the parkin target protein, CDCrel-1. *Proc. Natl. Acad. Sci. USA* 100:12438–43

Engelender S, Kaminsky Z, Guo X, Sharp AH, Amaravi RK, et al. 1999. Synphilin-1 associates with α-synuclein and promotes the formation of cytosolic inclusions. *Nat. Genet.* 22:110–14

Farrer M, Chan P, Chen R, Tan L, Lincoln S, et al. 2001. Lewy bodies and parkinsonism in families with parkin mutations. *Ann. Neurol.* 50:293–300

Farrer M, Destee T, Becquet E, Wavrant-De Vrieze F, Mouroux V, et al. 2000. Linkage exclusion in French families with probable Parkinson's disease. *Mov. Disord.* 15:1075–83

Feany MB, Bender WW. 2000. A *Drosophila* model of Parkinson's disease. *Nature* 404:394–98

Feany MB, Pallanck LJ. 2003. Parkin: a multipurpose neuroprotective agent? *Neuron* 38:13–16

Forno LS. 1996. Neuropathology of Parkinson's disease. *J. Neuropathol. Exp. Neurol.* 55:259–72

Forno LS, Langston JW, DeLanney LE, Irwin I. 1988. An electron microscopic study of MPTP-induced inclusion bodies in an old monkey. *Brain Res.* 448:150–57

Fortin DL, Troyer MD, Nakamura K, Kubo S, Anthony MD, Edwards RH. 2004. Lipid rafts mediate the synaptic localization of α-synuclein. *J. Neurosci.* 24:6715–23

Funayama M, Hasegawa K, Kowa H, Saito M, Tsuji S, Obata F. 2002. A new locus for Parkinson's disease (PARK8) maps to chromosome 12p11.2–q13.1. *Ann. Neurol.* 51:296–301

Gasser T, Muller-Myhsok B, Wszolek ZK, Oehlmann R, Calne DB, et al. 1998. A susceptibility locus for Parkinson's disease maps to chromosome 2p13. *Nat. Genet.* 18:262–65

Giasson BI, Duda JE, Murray IV, Chen Q, Souza JM, et al. 2000. Oxidative damage linked to neurodegeneration by selective α-synuclein nitration in synucleinopathy lesions. *Science* 290:985–89

Giasson BI, Forman MS, Higuchi M, Golbe LI, Graves CL, et al. 2003. Initiation and synergistic fibrillization of tau and α-synuclein. *Science* 300:636–40

Giasson BI, Lee VM. 2003. Are ubiquitination pathways central to Parkinson's disease? *Cell* 114:1–8

Glickman MH, Ciechanover A. 2002. The ubiquitin-proteasome proteolytic pathway: destruction for the sake of construction. *Physiol. Rev.* 82:373–428

Goldberg MS, Fleming SM, Palacino JJ, Cepeda C, Lam HA, et al. 2003. Parkin-deficient mice exhibit nigrostriatal deficits but not loss of dopaminergic neurons. *J. Biol. Chem.* 278:43628–35

Gosavi N, Lee HJ, Lee JS, Patel S, Lee SJ. 2002. Golgi fragmentation occurs in the cells with prefibrillar α-synuclein aggregates and precedes the formation of fibrillar inclusion. *J. Biol. Chem.* 277:48984–92

Gowers WR. 1900. *A Manual of Diseases of the Nervous System. Vol. I. Diseases of the Nerves and Spinal Cord.* Philadelphia: Blakiston's Son

Greene JC, Whitworth AJ, Kuo I, Andrews LA, Feany MB, Pallanck LJ. 2003. Mitochondrial pathology and apoptotic muscle degeneration in *Drosophila* parkin mutants. *Proc. Natl. Acad. Sci. USA* 100:4078–83

Hardy J, Cookson MR, Singleton A. 2003. Genes and parkinsonism. *Lancet Neurol.* 2:221–28

Hasegawa M, Fujiwara H, Nonaka T, Wakabayashi K, Takahashi H, et al. 2002. Phosphorylated α-synuclein is ubiquitinated in alpha-synucleinopathy lesions. *J. Biol. Chem.* 277:49071–76

Hashimoto M, Rockenstein E, Mante M, Mallory M, Masliah E. 2001. β-synuclein inhibits α-synuclein aggregation: a possible role as an anti-parkinsonian factor. *Neuron* 32:213–23

Hatano Y, Li Y, Sato K, Asakawa S, Yamamura Y, et al. 2004. Novel PINK1 mutations in early-onset parkinsonism. *Ann. Neurol.* 56:424–27

Healy DG, Abou-Sleiman PM, Ahmadi KR, Muqit MM, Bhatia KP, et al. 2004. The gene responsible for PARK6 Parkinson's disease, PINK1, does not influence common forms of parkinsonism. *Ann. Neurol.* 56:329–35

Hicks AA, Petursson H, Jonsson T, Stefansson H, Johannsdottir HS, et al. 2002. A susceptibility gene for late-onset idiopathic Parkinson's disease. *Ann. Neurol.* 52:549–55

Hodara R, Norris EH, Giasson BI, Mishizen-Eberz AJ, Lynch DR, et al. 2004. Functional consequences of α-synuclein tyrosine nitration: diminished binding to lipid vesicles and increased fibril formation. *J. Biol. Chem.* 279:47746–53

Hsu LJ, Sagara Y, Arroyo A, Rockenstein E, Sisk A, et al. 2000. α-synuclein promotes mitochondrial deficit and oxidative stress. *Am. J. Pathol.* 157:401–10

Huynh DP, Scoles DR, Nguyen D, Pulst SM. 2003. The autosomal recessive juvenile Parkinson disease gene product, parkin, interacts with and ubiquitinates synaptotagmin XI. *Hum. Mol. Genet.* 12:2587–97

Ii K, Ito H, Tanaka K, Hirano A. 1997. Immunocytochemical co-localization of the proteasome in ubiquitinated structures in neurodegenerative diseases and the elderly. *J. Neuropathol. Exp. Neurol.* 56:125–31

Imai Y, Soda M, Hatakeyama S, Akagi T, Hashikawa T, et al. 2002. CHIP is associated with Parkin, a gene responsible for familial Parkinson's disease, and enhances its ubiquitin ligase activity. *Mol. Cell.* 10:55–67

Imai Y, Soda M, Inoue H, Hattori N, Mizuno Y, Takahashi R. 2001. An unfolded putative transmembrane polypeptide, which can lead to endoplasmic reticulum stress, is a substrate of Parkin. *Cell* 105:891–902

Irizarry MC, Kim TW, McNamara M, Tanzi RE, George JM, et al. 1996. Characterization of the precursor protein of the non-A β component of senile plaques (NACP) in the human central nervous system. *J. Neuropathol. Exp. Neurol.* 55:889–95

Ischiropoulos H, Beckman JS. 2003. Oxidative stress and nitration in neurodegeneration: cause, effect, or association? *J. Clin. Invest.* 111:163–69

Itier JM, Ibanez P, Mena MA, Abbas N, Cohen-Salmon C, et al. 2003. Parkin gene inactivation alters behaviour and dopamine neurotransmission in the mouse. *Hum. Mol. Genet.* 12:2277–91

Jenco JM, Rawlingson A, Daniels B, Morris AJ. 1998. Regulation of phospholipase D2: selective inhibition of mammalian phospholipase D isoenzymes by α- and β-synucleins. *Biochemistry* 37:4901–9

Jenner P. 2003. Oxidative stress in Parkinson's disease. *Ann. Neurol.* 53:S26–36

Jiang H, Ren Y, Zhao J, Feng J. 2004. Parkin protects human dopaminergic neuroblastoma cells against dopamine-induced apoptosis. *Hum. Mol. Genet.* 13:1745–54

Kahle PJ, Neumann M, Ozmen L, Muller V, Jacobsen H, et al. 2000. Subcellular localization of wild-type and Parkinson's disease-associated mutant α-synuclein in human and transgenic mouse brain. *J. Neurosci.* 20:6365–73

Kahle PJ, Neumann M, Ozmen L, Muller V, Odoy S, et al. 2001. Selective insolubility of α-synuclein in human Lewy body diseases is recapitulated in a transgenic mouse model. *Am. J. Pathol.* 159:2215–25

Kitada T, Asakawa S, Hattori N, Matsumine H, Yamamura Y, et al. 1998. Mutations in the parkin gene cause autosomal recessive juvenile parkinsonism. *Nature* 392:605–8

Ko L, Mehta ND, Farrer M, Easson C, Hussey J, et al. 2000. Sensitization of neuronal cells to oxidative stress with mutated human α-synuclein. *J. Neurochem.* 75:2546–54

Kopito RR. 2000. Aggresomes, inclusion bodies and protein aggregation. *Trends Cell Biol.* 10:524–30

Kruger R, Kuhn W, Muller T, Woitalla D, Graeber M, et al. 1998. Ala30Pro mutation in the gene encoding α-synuclein in Parkinson's disease. *Nat. Genet.* 18:106–8

Lang AE, Lozano AM. 1998a. Parkinson's disease. First of two parts. *N. Engl. J. Med.* 339:1044–53

Lang AE, Lozano AM. 1998b. Parkinson's disease. Second of two parts. *N. Engl. J. Med.* 339:1130–43

Langston JW, Ballard P, Tetrud JW, Irwin I. 1983. Chronic Parkinsonism in humans due to a product of meperidine-analog synthesis. *Science* 219:979–80

Lansbury PT Jr, Brice A. 2002. Genetics of Parkinson's disease and biochemical studies of implicated gene products. *Curr. Opin. Cell. Biol.* 14:653–60

Lashuel HA, Hartley D, Petre BM, Walz T, Lansbury PT Jr. 2002. Neurodegenerative disease: amyloid pores from pathogenic mutations. *Nature* 418:291

Lee FJ, Liu F, Pristupa ZB, Niznik HB. 2001a. Direct binding and functional coupling of α-synuclein to the dopamine transporters accelerate dopamine-induced apoptosis. *Faseb J.* 15:916–26

Lee M, Hyun D, Halliwell B, Jenner P. 2001b. Effect of the overexpression of wild-type or mutant α-synuclein on cell susceptibility to insult. *J. Neurochem.* 76:998–1009

Lee MK, Stirling W, Xu Y, Xu X, Qui D, et al. 2002. Human α-synuclein-harboring familial Parkinson's disease-linked Ala-53 → Thr mutation causes neurodegenerative disease with α-synuclein aggregation in transgenic mice. *Proc. Natl. Acad. Sci. USA* 99:8968–73

Lee SJ, Kim SJ, Kim IK, Ko J, Jeong CS, et al. 2003. Crystal structures of human DJ-1 and *Escherichia coli* Hsp31, which share an evolutionarily conserved domain. *J. Biol. Chem.* 278:44552–59

Le WD, Xu P, Jankovic J, Jiang H, Appel SH, et al. 2003. Mutations in NR4A2 associated with familial Parkinson disease. *Nat. Genet.* 33:85–89

Leroy E, Boyer R, Auburger G, Leube B, Ulm G, et al. 1998. The ubiquitin pathway in Parkinson's disease. *Nature* 395:451–52

Li W, Lesuisse C, Xu Y, Troncoso JC, Price DL, Lee MK. 2004. Stabilization of α-synuclein protein with aging and familial Parkinson's disease-linked A53T mutation. *J. Neurosci.* 24:9400–9

Lincoln S, Vaughan J, Wood N, Baker M, Adamson J, et al. 1999. Low frequency of pathogenic mutations in the ubiquitin carboxy-terminal hydrolase gene in familial Parkinson's disease. *Neuroreport* 10:427–29

Lindersson E, Beedholm R, Hojrup P, Moos T, Gai W, et al. 2004. Proteasomal inhibition by α-synuclein filaments and oligomers. *J. Biol. Chem.* 279:12924–34

Liu Y, Fallon L, Lashuel HA, Liu Z, Lansbury PT Jr. 2002. The UCH-L1 gene encodes two opposing enzymatic activities that affect α-synuclein degradation and Parkinson's disease susceptibility. *Cell* 111:209–18

Lockhart PJ, Lincoln S, Hulihan M, Kachergus J, Wilkes K, et al. 2004. DJ-1 mutations are a rare cause of recessively inherited early onset parkinsonism mediated by loss of protein function. *J. Med. Genet.* 41:e22

Lotharius J, Barg S, Wiekop P, Lundberg C, Raymon HK, Brundin P. 2002. Effect of mutant α-synuclein on dopamine homeostasis in a new human mesencephalic cell line. *J. Biol. Chem.* 277:38884–94

Lowe J, McDermott H, Landon M, Mayer RJ, Wilkinson KD. 1990. Ubiquitin carboxyl-terminal hydrolase (PGP 9.5) is selectively present in ubiquitinated inclusion bodies characteristic of human neurodegenerative diseases. *J. Pathol.* 161:153–60

Lucking CB, Abbas N, Durr A, Bonifati V, Bonnet AM, et al. 1998. Homozygous deletions in parkin gene in European and North African families with autosomal recessive juvenile parkinsonism. The European Consortium on Genetic Susceptibility in Parkinson's Disease and the French Parkinson's Disease Genetics Study Group. *Lancet* 352:1355–56

Lucking CB, Durr A, Bonifati V, Vaughan J, De Michele G, et al. 2000. Association between early-onset Parkinson's disease and mutations in the parkin gene. French Parkinson's Disease Genetics Study Group. *N. Engl. J. Med.* 342:1560–67

Manning-Bog AB, McCormack AL, Li J, Uversky VN, Fink AL, Di Monte DA. 2002. The herbicide paraquat causes up-regulation and aggregation of α-synuclein in mice: paraquat and α-synuclein. *J. Biol. Chem.* 277:1641–44

Maraganore DM, Farrer MJ, Hardy JA, Lincoln SJ, McDonnell SK, Rocca WA. 1999. Case-control study of the ubiquitin carboxy-terminal hydrolase L1 gene in Parkinson's disease. *Neurology* 53:1858–60

Maraganore DM, Lesnick TG, Elbaz A, Chartier-Harlin MC, Gasser T, et al. 2004. UCHL1 is a Parkinson's disease susceptibility gene. *Ann. Neurol.* 55:512–21

Marder K, Tang MX, Mejia H, Alfaro B, Cote L, et al. 1996. Risk of Parkinson's disease among first-degree relatives: a community-based study. *Neurology* 47:155–60

Marttila RJ, Kaprio J, Koskenvuo M, Rinne UK. 1988. Parkinson's disease in a nationwide twin cohort. *Neurology* 38:1217–19

Marx FP, Holzmann C, Strauss KM, Li L, Eberhardt O, et al. 2003. Identification and functional characterization of a novel R621C mutation in the synphilin-1 gene in Parkinson's disease. *Hum. Mol. Genet.* 12:1223–31

Masliah E, Rockenstein E, Veinbergs I, Mallory M, Hashimoto M, et al. 2000. Dopaminergic loss and inclusion body formation in α-synuclein mice: implications for neurodegenerative disorders. *Science* 287:1265–69

Masliah E, Rockenstein E, Veinbergs I, Sagara Y, Mallory M, et al. 2001. β-amyloid peptides enhance α-synuclein accumulation and neuronal deficits in a transgenic mouse model linking Alzheimer's disease and Parkinson's disease. *Proc. Natl. Acad. Sci. USA* 98:12245–50

Mata IF, Lockhart PJ, Farrer MJ. 2004. Parkin genetics: one model for Parkinson's disease. *Hum. Mol. Genet.* 13:R127–33

Matsumine H, Saito M, Shimoda-Matsubayashi S, Tanaka H, Ishikawa A, et al. 1997. Localization of a gene for an autosomal recessive form of juvenile Parkinsonism to chromosome 6q25.2-27. *Am. J. Hum. Genet.* 60:588–96

McCormack AL, Thiruchelvam M, Manning-Bog AB, Thiffault C, Langston JW, et al. 2002. Environmental risk factors and Parkinson's disease: selective degeneration of nigral dopaminergic neurons caused by the herbicide paraquat. *Neurobiol. Dis.* 10:119–27

McNaught KS, Belizaire R, Isacson O, Jenner P, Olanow CW. 2003. Altered proteasomal function in sporadic Parkinson's disease. *Exp. Neurol.* 179:38–46

McNaught KS, Belizaire R, Jenner P, Olanow CW, Isacson O. 2002. Selective loss of 20S proteasome α-subunits in the substantia nigra pars compacta in Parkinson's disease. *Neurosci. Lett.* 326:155–58

McNaught KS, Perl DP, Brownell AL, Olanow CW. 2004. Systemic exposure to proteasome inhibitors causes a progressive model of Parkinson's disease. *Ann. Neurol.* 56:149–62

Miller DW, Ahmad R, Hague S, Baptista MJ, Canet-Aviles R, et al. 2003. L166P mutant DJ-1, causative for recessive Parkinson's disease, is degraded through the ubiquitin-proteasome system. *J. Biol. Chem.* 278:36588–95

Mishizen-Eberz AJ, Guttmann RP, Giasson BI, Day GA 3rd, Hodara R, et al. 2003. Distinct cleavage patterns of normal and pathologic forms of α-synuclein by calpain I in vitro. *J. Neurochem.* 86:836–47

Mitsumoto A, Nakagawa Y, Takeuchi A, Okawa K, Iwamatsu A, Takanezawa Y. 2001. Oxidized forms of peroxiredoxins and DJ-1 on two-dimensional gels increased in response to sublethal levels of paraquat. *Free Radic. Res.* 35:301–10

Moore DJ, Dawson VL, Dawson TM. 2003a. Role for the ubiquitin-proteasome system in Parkinson's disease and other neurodegenerative brain amyloidoses. *Neuromolecular Med.* 4:95–108

Moore DJ, Zhang L, Dawson TM, Dawson VL. 2003b. A missense mutation (L166P) in DJ-1, linked to familial Parkinson's disease, confers reduced protein stability and impairs homo-oligomerization. *J. Neurochem.* 87:1558–67

Moore DJ, Zhang L, Troncoso J, Lee MK, Hattori N, et al. 2005. Association of DJ-1 and parkin mediated by pathogenic DJ-1 mutations and oxidative stress. *Hum. Mol. Genet.* 14:71–84

Mori H, Kondo T, Yokochi M, Matsumine H, Nakagawa-Hattori Y, et al. 1998. Pathologic and biochemical studies of juvenile parkinsonism linked to chromosome 6q. *Neurology* 51:890–92

Morris CM, O'Brien KK, Gibson AM, Hardy JA, Singleton AB. 2003. Polymorphism in the human DJ-1 gene is not associated with sporadic dementia with Lewy bodies or Parkinson's disease. *Neurosci. Lett.* 352:151–53

Murakami T, Shoji M, Imai Y, Inoue H, Kawarabayashi T, et al. 2004. Pael-R is accumulated in Lewy bodies of Parkinson's disease. *Ann. Neurol.* 55:439–42

Murphy DD, Rueter SM, Trojanowski JQ, Lee VM. 2000. Synucleins are developmentally expressed, and α-synuclein regulates the size of the presynaptic vesicular pool in primary hippocampal neurons. *J. Neurosci.* 20:3214–20

Neumann M, Kahle PJ, Giasson BI, Ozmen L, Borroni E, et al. 2002. Misfolded proteinase K-resistant hyperphosphorylated α-synuclein in aged transgenic mice with locomotor deterioration and in human α-synucleinopathies. *J. Clin. Invest.* 110:1429–39

Neumann M, Muller V, Gorner K, Kretzschmar HA, Haass C, Kahle PJ. 2004. Pathological properties of the Parkinson's disease-associated protein DJ-1 in α-synucleinopathies and tauopathies: relevance for multiple system atrophy and Pick's disease. *Acta Neuropathol. (Berlin)* 107:489–96

Ohtake H, Limprasert P, Fan Y, Onodera O, Kakita A, et al. 2004. β-synuclein gene alterations in dementia with Lewy bodies. *Neurology* 63:805–11

Olanow CW, Perl DP, DeMartino GN, McNaught KS. 2004. Lewy-body formation is an aggresome-related process: a hypothesis. *Lancet Neurol.* 3:496–503

Olzmann JA, Brown K, Wilkinson KD, Rees HD, Huai Q, et al. 2004. Familial Parkinson's disease-associated L166P mutation disrupts DJ-1 protein folding and function. *J. Biol. Chem.* 279:8506–15

Osaka H, Wang YL, Takada K, Takizawa S, Setsuie R, et al. 2003. Ubiquitin carboxy-terminal hydrolase L1 binds to and stabilizes monoubiquitin in neuron. *Hum. Mol. Genet.* 12:1945–58

Outeiro TF, Lindquist S. 2003. Yeast cells provide insight into alpha-synuclein biology and pathobiology. *Science* 302:1772–75

Paisán-Ruíz C, Jain S, Evans EW, Gilks WP, Simon J, et al. 2004. Cloning of the gene containing mutations that cause PARK8-linked Parkinson's disease. *Neuron* 44:595–600

Palacino JJ, Sagi D, Goldberg MS, Krauss S, Motz C, et al. 2004. Mitochondrial dysfunction and oxidative damage in parkin-deficient mice. *J. Biol. Chem.* 279:18614–22

Pals P, Lincoln S, Manning J, Heckman M, Skipper L, et al. 2004. α-synuclein promoter confers susceptibility to Parkinson's disease. *Ann. Neurol.* 56:591–95

Pankratz N, Nichols WC, Uniacke SK, Halter C, Rudolph A, et al. 2003. Significant linkage of Parkinson disease to chromosome 2q36–37. *Am. J. Hum. Genet.* 72:1053–57

Pesah Y, Pham T, Burgess H, Middlebrooks B, Verstreken P, et al. 2004. *Drosophila* parkin mutants have decreased mass and cell size and increased sensitivity to oxygen radical stress. *Development* 131:2183–94

Petrucelli L, O'Farrell C, Lockhart PJ, Baptista M, Kehoe K, et al. 2002. Parkin protects against the toxicity associated with mutant α-synuclein: Proteasome dysfunction selectively affects catecholaminergic neurons. *Neuron* 36:1007–19

Polymeropoulos MH, Higgins JJ, Golbe LI, Johnson WG, Ide SE, et al. 1996. Mapping of a gene for Parkinson's disease to chromosome 4q21–q23. *Science* 274:1197–99

Polymeropoulos MH, Lavedan C, Leroy E, Ide SE, Dehejia A, et al. 1997. Mutation in the α-synuclein gene identified in families with Parkinson's disease. *Science* 276:2045–47

Quigley PM, Korotkov K, Baneyx F, Hol WG. 2003. The 1.6-A crystal structure of the class of chaperones represented by *Escherichia coli* Hsp31 reveals a putative catalytic triad. *Proc. Natl. Acad. Sci. USA* 100:3137–42

Ren Y, Zhao J, Feng J. 2003. Parkin binds to α/β-tubulin and increases their ubiquitination and degradation. *J. Neurosci.* 23:3316–24

Rideout HJ, Dietrich P, Wang Q, Dauer WT, Stefanis L. 2004. α-synuclein is required for the fibrillar nature of ubiquitinated inclusions induced by proteasomal inhibition in primary neurons. *J. Biol. Chem.* 279:46915–20

Rizzu P, Hinkle DA, Zhukareva V, Bonifati V, Severijnen LA, et al. 2004. DJ-1 colocalizes with tau inclusions: a link between parkinsonism and dementia. *Ann. Neurol.* 55:113–18

Saigoh K, Wang YL, Suh JG, Yamanishi T, Sakai Y, et al. 1999. Intragenic deletion in the gene encoding ubiquitin carboxy-terminal hydrolase in gad mice. *Nat. Genet.* 23:47–51

Sakata E, Yamaguchi Y, Kurimoto E, Kikuchi J, Yokoyama S, et al. 2003. Parkin binds the Rpn10 subunit of 26S proteasomes through its ubiquitin-like domain. *EMBO Rep.* 4:301–6

Schapira AH, Cooper JM, Dexter D, Clark JB, Jenner P, Marsden CD. 1990. Mitochondrial complex I deficiency in Parkinson's disease. *J. Neurochem.* 54:823–27

Scott WK, Nance MA, Watts RL, Hubble JP, Koller WC, et al. 2001. Complete genomic screen in Parkinson disease: evidence for multiple genes. *JAMA* 286:2239–44

Sharon R, Bar-Joseph I, Frosch MP, Walsh DM, Hamilton JA, Selkoe DJ. 2003. The formation of highly soluble oligomers of α-synuclein is regulated by fatty acids and enhanced in Parkinson's disease. *Neuron* 37:583–95

Sherer TB, Betarbet R, Greenamyre JT. 2002a. Environment, mitochondria, and Parkinson's disease. *Neuroscientist* 8:192–97

Sherer TB, Betarbet R, Stout AK, Lund S, Baptista M, et al. 2002b. An in vitro model of Parkinson's disease: linking mitochondrial impairment to altered α-synuclein metabolism and oxidative damage. *J. Neurosci.* 22:7006–15

Sherer TB, Kim JH, Betarbet R, Greenamyre JT. 2003. Subcutaneous rotenone exposure causes highly selective dopaminergic degeneration and α-synuclein aggregation. *Exp. Neurol.* 179:9–16

Shimura H, Hattori N, Kubo S, Mizuno Y, Asakawa S, et al. 2000. Familial Parkinson disease gene product, parkin, is a ubiquitin-protein ligase. *Nat. Genet.* 25:302–5

Shimura H, Schlossmacher MG, Hattori N, Frosch MP, Trockenbacher A, et al. 2001. Ubiquitination of a new form of α-synuclein by parkin from human brain: implications for Parkinson's disease. *Science* 293:263–69

Singleton AB, Farrer M, Johnson J, Singleton A, Hague S, et al. 2003. α-synuclein locus triplication causes Parkinson's disease. *Science* 302:841

Snyder H, Mensah K, Theisler C, Lee J, Matouschek A, Wolozin B. 2003. Aggregated and monomeric α-synuclein bind to the S6′ proteasomal protein and inhibit proteasomal function. *J. Biol. Chem.* 278:11753–59

Song DD, Shults CW, Sisk A, Rockenstein E, Masliah E. 2004. Enhanced substantia nigra mitochondrial pathology in human α-synuclein transgenic mice after treatment with MPTP. *Exp. Neurol.* 186:158–72

Spillantini MG, Crowther RA, Jakes R, Hasegawa M, Goedert M. 1998. α-synuclein in filamentous inclusions of Lewy bodies from Parkinson's disease and dementia with lewy bodies. *Proc. Natl. Acad. Sci. USA* 95:6469–73

Staropoli JF, McDermott C, Martinat C, Schulman B, Demireva E, Abeliovich A. 2003. Parkin is a component of an SCF-like ubiquitin ligase complex and protects postmitotic neurons from kainate excitotoxicity. *Neuron* 37:735–49

Sullivan PG, Dragicevic NB, Deng JH, Bai Y, Dimayuga E, et al. 2004. Proteasome inhibition alters neural mitochondrial homeostasis and mitochondria turnover. *J. Biol. Chem.* 279:20699–707

Sveinbjornsdottir S, Hicks AA, Jonsson T, Petursson H, Gugmundsson G, et al. 2000. Familial aggregation of Parkinson's disease in Iceland. *N. Engl. J. Med.* 343:1765–70

Swerdlow RH, Parks JK, Cassarino DS, Binder DR, Bennett JP Jr, et al. 2001. Biochemical analysis of cybrids expressing mitochondrial DNA from Contursi kindred Parkinson's subjects. *Exp. Neurol.* 169:479–85

Swerdlow RH, Parks JK, Davis JN 2nd, Cassarino DS, Trimmer PA, et al. 1998. Matrilineal inheritance of complex I dysfunction in a multigenerational Parkinson's disease family. *Ann. Neurol.* 44:873–81

Swerdlow RH, Parks JK, Miller SW, Tuttle JB, Trimmer PA, et al. 1996. Origin and functional consequences of the complex I defect in Parkinson's disease. *Ann. Neurol.* 40:663–71

Tabrizi SJ, Orth M, Wilkinson JM, Taanman JW, Warner TT, et al. 2000. Expression of mutant α-synuclein causes increased susceptibility to dopamine toxicity. *Hum. Mol. Genet.* 9:2683–89

Taira T, Saito Y, Niki T, Iguchi-Ariga SM, Takahashi K, Ariga H. 2004. DJ-1 has a role in antioxidative stress to prevent cell death. *EMBO Rep.* 5:213–18

Takahashi-Niki K, Niki T, Taira T, Iguchi-Ariga SM, Ariga H. 2004. Reduced anti-oxidative stress activities of DJ-1 mutants found in Parkinson's disease patients. *Biochem. Biophys. Res. Commun.* 320:389–97

Tanaka M, Kim YM, Lee G, Junn E, Iwatsubo T, Mouradian MM. 2004. Aggresomes formed by α-synuclein and synphilin-1 are cytoprotective. *J. Biol. Chem.* 279:4625–31

Tanaka Y, Engelender S, Igarashi S, Rao RK, Wanner T, et al. 2001. Inducible expression of mutant α-synuclein decreases proteasome activity and increases sensitivity to mitochondria-dependent apoptosis. *Hum. Mol. Genet.* 10:919–26

Tao X, Tong L. 2003. Crystal structure of human DJ-1, a protein associated with early onset Parkinson's disease. *J. Biol. Chem.* 278:31372–79

Thiruchelvam M, Richfield EK, Baggs RB, Tank AW, Cory-Slechta DA. 2000. The nigrostriatal dopaminergic system as a preferential target of repeated exposures to combined paraquat and maneb: implications for Parkinson's disease. *J. Neurosci.* 20:9207–14

Tofaris GK, Layfield R, Spillantini MG. 2001. α-synuclein metabolism and aggregation is linked to ubiquitin-independent degradation by the proteasome. *FEBS Lett.* 509:22–26

Uversky VN. 2003. A protein-chameleon: conformational plasticity of α-synuclein, a disordered protein involved in neurodegenerative disorders. *J. Biomol. Struct. Dyn.* 21:211–34

Valente EM, Abou-Sleiman PM, Caputo V, Muqit MM, Harvey K, et al. 2004a. Hereditary early-onset Parkinson's disease caused by mutations in PINK1. *Science* 304:1158–60

Valente EM, Bentivoglio AR, Dixon PH, Ferraris A, Ialongo T, et al. 2001. Localization of a novel locus for autosomal recessive early-onset parkinsonism, PARK6, on human chromosome 1p35–p36. *Am. J. Hum. Genet.* 68:895–900

Valente EM, Brancati F, Ferraris A, Graham EA, Davis MB, et al. 2002. PARK6-linked parkinsonism occurs in several European families. *Ann. Neurol.* 51:14–18

Valente EM, Salvi S, Ialongo T, Marongiu R, Elia AE, et al. 2004b. PINK1 mutations are associated with sporadic early-onset parkinsonism. *Ann. Neurol.* 56:336–41

van der Walt JM, Nicodemus KK, Martin ER, Scott WK, Nance MA, et al. 2003. Mitochondrial polymorphisms significantly reduce the risk of Parkinson disease. *Am. J. Hum. Genet.* 72:804–11

van Duijn CM, Dekker MC, Bonifati V, Galjaard RJ, Houwing-Duistermaat JJ, et al. 2001. Park7, a novel locus for autosomal recessive early-onset parkinsonism, on chromosome 1p36. *Am. J. Hum. Genet.* 69:629–34

Von Coelln R, Dawson VL, Dawson TM. 2004a. Parkin-associated Parkinson's disease. *Cell Tissue Res.* 318:175–84

Von Coelln R, Thomas B, Savitt JM, Lim KL, Sasaki M, et al. 2004b. Loss of locus coeruleus neurons and reduced startle in parkin null mice. *Proc. Natl. Acad. Sci. USA* 101:10744–49

Wakabayashi K, Engelender S, Yoshimoto M, Tsuji S, Ross CA, Takahashi H. 2000. Synphilin-1 is present in Lewy bodies in Parkinson's disease. *Ann. Neurol.* 47:521–23

Ward CD, Duvoisin RC, Ince SE, Nutt JD, Eldridge R, Calne DB. 1983. Parkinson's disease in 65 pairs of twins and in a set of quadruplets. *Neurology* 33:815–24

Webb JL, Ravikumar B, Atkins J, Skepper JN, Rubinsztein DC. 2003. α-synuclein is degraded by both autophagy and the proteasome. *J. Biol. Chem.* 278:25009–13

West AB, Maidment NT. 2004. Genetics of parkin-linked disease. *Hum. Genet.* 114:327–36

Wilkinson KD, Lee KM, Deshpande S, Duerksen-Hughes P, Boss JM, Pohl J. 1989. The neuron-specific protein PGP 9.5 is a ubiquitin carboxyl-terminal hydrolase. *Science* 246:670–73

Wilson MA, Collins JL, Hod Y, Ringe D, Petsko GA. 2003. The 1.1-A resolution crystal structure of DJ-1, the protein mutated in autosomal recessive early onset Parkinson's disease. *Proc. Natl. Acad. Sci. USA* 100:9256–61

Wooten GF, Currie LJ, Bennett JP, Harrison MB, Trugman JM, Parker WD Jr. 1997. Maternal inheritance in Parkinson's disease. *Ann. Neurol.* 41:265–68

Wszolek ZK, Pfeiffer RF, Tsuboi Y, Uitti RJ, McComb RD, et al. 2004. Autosomal dominant parkinsonism associated with variable synuclein and tau pathology. *Neurology* 62:1619–22

Xu J, Kao SY, Lee FJ, Song W, Jin LW, Yankner BA. 2002. Dopamine-dependent neurotoxicity of α-synuclein: a mechanism for selective neurodegeneration in Parkinson disease. *Nat. Med.* 8:600–6

Yang Y, Nishimura I, Imai Y, Takahashi R, Lu B. 2003. Parkin suppresses dopaminergic neuron-selective neurotoxicity induced by Pael-R in *Drosophila*. *Neuron* 37:911–24

Yokota T, Sugawara K, Ito K, Takahashi R, Ariga H, Mizusawa H. 2003. Down regulation of DJ-1 enhances cell death by oxidative stress, ER stress, and proteasome inhibition. *Biochem. Biophys. Res. Commun.* 312:1342–48

Zarranz JJ, Alegre J, Gomez-Esteban JC, Lezcano E, Ros R, et al. 2004. The new mutation, E46K, of α-synuclein causes Parkinson and Lewy body dementia. *Ann. Neurol.* 55:164–73

Zhang Y, Gao J, Chung KK, Huang H, Dawson VL, Dawson TM. 2000. Parkin functions as an E2-dependent ubiquitin-protein ligase and promotes the degradation of the synaptic vesicle-associated protein, CDCrel-1. *Proc. Natl. Acad. Sci. USA* 97:13354–59

Zimprich A, Biskup S, Leitner P, Lichtner P, Farrer M, et al. 2004b. Mutations in *LRRK2* cause autosomal-dominant parkinsonism with pleomorphic pathology. *Neuron* 44:601–7

Zimprich A, Muller-Myhsok B, Farrer M, Leitner P, Sharma M, et al. 2004a. The PARK8 locus in autosomal dominant parkinsonism: confirmation of linkage and further delineation of the disease-containing interval. *Am. J. Hum. Genet.* 74:11–19

Large-Scale Genomic Approaches to Brain Development and Circuitry

Mary E. Hatten[1] and Nathaniel Heintz[2]

Laboratory of Developmental Neurobiology,[1] Laboratory of Molecular Biology, Howard Hughes Medical Institute,[2] The Rockefeller University, New York, NY 10021; email: hatten@rockefeller.edu, heintz@rockefeller.edu

Key Words

spontaneous mutant mice, ENU mutagenesis, gene trap vectors, BAC transgenic expression vectors, screens for novel CNS phenotypes and genotypes

Abstract

Over the past two decades, molecular genetic studies have enabled a common conceptual framework for the development and basic function of the nervous system. These studies, and the pioneering efforts of mouse geneticists and neuroscientists to identify and clone genes for spontaneous mouse mutants, have provided a paradigm for understanding complex processes of the vertebrate brain. Gene cloning for human brain malformations and degenerative disorders identified other important central nervous system (CNS) genes. However, because many debilitating human disorders are genetically complex, phenotypic screens are difficult to design. This difficulty has led to large-scale, genomic approaches to discover genes that are uniquely expressed in brain circuits and regions that control complex behaviors. In this review, we summarize current phenotype- and genotype-driven approaches to discover novel CNS-expressed genes, as well as current approaches to carry out large-scale, gene-expression screens in the CNS.

Contents

SYSTEMATIC SCREENS FOR
GENE FUNCTION IN THE
CENTRAL NERVOUS SYSTEM 90
  Spontaneous Neurological Mutant
    Mice 90
  ENU Mutagenesis 92
  Dominant and Semi-Dominant
    Genome-Wide ENU
    Mutagenesis Screens 93
  Recessive Genome-Wide ENU
    Mutagenesis Screens 94
  Genotype-Driven Screens 95
  Gene-Trap Screening for CNS
    Expressed Genes 95
  Large-Scale Screens Using Gene
    Targeting 96
SYSTEMATIC SCREENS OF CNS
GENE EXPRESSION 97
  Large-Scale Expression Studies of
    Single Genes 98
  In Situ Hybridization 98
  Reporter Gene Assays 99
  Transcriptional Profiling Using
    Microarray 101
  CNS Proteomics 102
CONCLUSIONS 102

SYSTEMATIC SCREENS FOR GENE FUNCTION IN THE CENTRAL NERVOUS SYSTEM

Spontaneous Neurological Mutant Mice

The characterization of spontaneous neurological mutant mice, primarily at the Jackson Labs, has provided innumerable mouse models of inherited disruptions of brain development and function. In the postgenomic era, an increasing number of the ~300 characterized spontaneous mouse mutants have been mapped and cloned, providing one of the most important resources for understanding the molecular pathways of primary brain functions. Beginning with the work of Richard Sidman and colleagues in the 1960s (Sidman 1983), characterization of scores of mouse lines harboring spontaneous neurological mutations has led to basic insights on the genetic and epigenetic regulation of mammalian brain development. It is noteworthy that many of these mutations are gain-of-function alleles that would not normally be constructed by simple gene targeting approaches.

Insight into early steps of brain development, such as neural crest migration and development of the enteric nervous system, has been obtained from studies of spontaneous mutations in the *splotch* (Goulding et al. 1993) and *piebald* mice (Hosoda et al. 1994), identifying the *Pax3* and *Ednrb* genes as critical for these processes. Upon closure of the neural tube, genes such as *dreher* (*Lmx1a*) (Millonig et al. 2000) are critical to the formation of the roof plate and establishment of dorso-ventral patterning. As cortical regions of the brain emerge, a host of spontaneous mutants affecting the cerebellar cortex have revealed basic principles concerning the establishment and patterning of the cerebellar territory. For example, genetic influences on cerebellar patterning and foliation have emerged from studies of the *meander tail* (*Mea*) (Ross et al. 1990) and *rostral cerebellar malformation* (*Rcm*; *Unc5h3*) genes (Ackerman et al. 1997). A role of specific ion channels in granule cell development was revealed in studies on the *weaver* (*wv*) mouse, which results from a gain-of-function mutation in the *Girk2* gene (Slesinger et al. 1996). This mutation causes a constitutive inward Ca^{2+} current leading to cell death (Kofuji et al. 1996) just as the granule cells commence differentiation and inward migration along the Bergmann glial fiber system to establish the granule cell layer of the cerebellar cortex. The importance of spontaneous mouse mutants to our understanding of the development of cortical lamination provides one of the most important illustrations of the epigenic control of cortical histogenesis. This is best illustrated in the neocortex, where the mutants *reeler* (*Reln*), *scrambler* (*Dab1*), and *Yotari* (*Dab1*) define a principal genetic pathway for neuronal positioning in the cortex. In these mutants, an early structure in cortical development, the preplate, falls to split into the cortical plate and subplate, an event required for the subsequent development of the neuronal laminae. The details of the reeler pathway have been reviewed elsewhere (e.g., see Rice & Curran 2001).

Two principal classes of neurons exist in the cerebellar cortex: the granule neuron and the Purkinje neuron. The Purkinje cell is the synaptic target of the granule neuron, which receives mossy fiber inputs from a variety of brain regions and acts to modulate the activity of the Purkinje cell, the only output neuron of the cerebellar circuit. Purkinje cell axons project to the cerebellar deep nuclei, which in turn form connections with the vestibular system, the thalamus, and the corticospinal tract axons, among others. Thus the firing of the Purkinje cell plays a critical role in maintaining balance, motor coordination, and eye movements, as well as some aspects of learning motor skills. As a result of the critical role of cerebellar Purkinje cells in motor coordination, spontaneous neurological mutations that affect Purkinje cell function are easy to detect, even for nonspecialists. Spontaneous neurological mutations that act in the Purkinje cell include classical mutations that have been cloned and studied in detail, such as *tottering* [*Tg* (*la*)], *staggerer* (*sg*) (Hamilton et al. 1996), *Lurcher* (*Lc*) (Zuo et al. 1997), *Purkinje cell degeneration* (*Pcd*), and *ataxia* [*ax* (*J*)], and as yet unpublished spontaneous neurological mutations such as *pogostick* (*pogo*) (Jeong & Hyun 2000, Jeong et al. 2000), *nervous* (*nr*) (Landis 1973), and *stumbler* (*stu*) (Caddy & Sidman 1981). The phenotypic evaluation of each of these mouse lines and the identification of the causative mutations provide a rich resource for identifying pathways critical for the development and function of cerebellar Purkinje neurons. The Purkinje cell thus provides a model for genetic approaches to the development and function of central nervous system (CNS) synaptic circuitry.

The first insights into the contributions of the RORα nuclear hormone receptor and thyroid hormone to Purkinje cell development came from studies of the *staggerer* (*sg*) mutation (Hamilton et al. 1996). Homozygous *staggerer* (*sg*) mice show severe ataxia due to defects in the development of Purkinje cells, including immature dendritic arborization, synapse formation, and gene expression. Cloning of the *Sg* revealed a defect in the gene encoding RORα, a member of the nuclear hormone-receptor superfamily. The *staggerer* mutation prevents translation of the ligand-binding homology domain of RORα (Hamilton et al. 1996). Subsequent experiments by Mason and colleagues (Heuer & Mason 2003) proved this model correct because the addition of thyroid hormone to isolated Purkinje cells in vitro promotes their maturation into cells that resemble those seen in vivo. A second classical cerebellar mutation that affects Purkinje cells is the *Lurcher* gene. Although Purkinje cell development and axon target interactions are normal in homozygous *Lurcher* mice, soon after the cells begin to mature, they enter a degeneration pathway. Cloning of the gene responsible for *Lurcher* revealed a defect in a potassium channel, which became active only after neurotransmitter-mediated synaptic activity in young adult mice (De Jager & Heintz 1998). Further studies on *Lurcher* by Heintz and colleagues provided the first genetic evidence linking neurotransmitter receptor activation and autophagy with neurodegeneration (Yue et al. 2002, Zuo et al. 1997). *Purkinje cell degeneration* (*Pcd*), another classical spontaneous neurological mutation, exhibits adult-onset degeneration of Purkinje neurons. Cloning of the gene mutated in *Pcd* reveals a mutation in the axotomy-induced gene *Nna1*, a nuclear receptor thought to be induced in spinal cord motor neurons following axotomy (Fernandez-Gonzalez et al. 2002). Thus, studies on *Pcd* provide the first genetic connection between neurodegeneration and regeneration. Finally, spinocerebellar synaptic defects in mice homozygous for a mutation in the *ataxia* (ax^J) gene, which causes adult-onset tremors and hind limb paralysis, result from mutations in a ubiquitin-specific protease 14 (Usp14). In ataxia mice, the mutant Usp14 fails to process polyubiquitin, providing the first indication that ubiquitin proteases play a role in CNS synaptic activity regulation (Wilson et al. 2002).

Analyses of spontaneous mouse mutants have also been important in understanding mechanisms of human disease. Defects in neuronal migration, together with defects in ion channel function in the neocortex, define one of the most common set of primary diseases of the developing human brain, the epilepsies. Mutations in more than 70 genes now define the pathways leading to the episodic abnormalities in the synchronization of cortical circuitry in epilepsy. As reviewed recently by Noebels (Noebels 1999, 2003), some of the inherited errors lead to basic changes in membrane excitability and synaptic transmission, whereas others act indirectly by altering the balance of assemblies of inhibitory and excitatory circuits. Although the present collection of spontaneous mouse mutants has been invaluable for deciphering many aspects of brain function, this approach is limited by the low frequency at which spontaneous mouse mutants occur (5×10^{-6} per locus), and by the fact that the only mutations noticed easily by animal handlers in large colonies are those that influence motor and balance function systems. For complex traits, sophisticated behavioral assessments are being developed to screen for mutant animals in large colonies of inbred strains of mice. N-ethyl-N-nitrosourea (ENU) mutagenesis is the method of choice to obtain multi-allelic series of single genes at complex loci and to further define genetic pathways and gain insight on gene function in the development of complex neurobiological pathways.

ENU Mutagenesis

ENU has become the principal mutagen for large-scale mouse mutagenesis projects because it produces mutation rates at specific loci at a rate of 1×10^{-3} per gamete. The unique strength of ENU mutagenesis is its remarkable efficiency. At optimal ENU dose regimes, a screen of only 100 ENU-treated lines induces 10,000 new mutations to be evaluated. The potency of ENU thus opens up a variety of approaches with the mouse, previously possible in model organisms such as *D. melanogaster* or *D. rerior* (zebrafish). At this mutation rate, genome-wide screens for both recessive and dominant mutations can be carried out. A study by Peters et al. (1986) indicates that recessive, loss-of-function mutations are recovered four times as often as are dominant gain-of-function mutations. The ability to carry out screens for dominant mutations, however, is necessary to reveal phenotypes not evident in recessive screens, including dominant, gain-of-function mutations, dominant-negative mutations, and mutations arising from haplo-insufficiency. Because the vast majority of human disease genes are autosomal dominant mutations (McKusick 1997), dominant ENU mutagenesis screens offer one of the surest routes to the discovery of mouse mutants that model human disease.

Three general categories of screens are being carried out with ENU mutagenesis. First, ENU can be used to isolate an allelic series for a known gene (locus). Second, ENU mutagenesis can be used to generate mouse mutations with phenotypes similar to phenotypes of cloned genes for which there is functional information. Finally, ENU mutagenesis can be used to generate a series of mutations in a cloned gene for which there is no prior functional annotation. Some of the mutant mice in such a new series of mutations could potentially have phenotypes that provide insight into the function of the gene under study (reviewed in Brown & Nolan 1998).

Dominant and Semi-Dominant Genome-Wide ENU Mutagenesis Screens

Dominant and semidominant genome-wide screens rely on a simple mating protocol, where ENU-treated males are bred with wild-type females and the offspring of these matings are scored for dominant and semidominant mutations. From Bode's original calculations (Bode 1984), ENU mutagenesis generates a mutation frequency of 1 in 1500. Thus, if one in five mutagenized males has sperm mutant at any given locus, each animal can produce 1500/5 or approximately 300 different mutants. This fact makes it feasible to carry out large-scale genomic screens with a manageable number of mice. Mice with phenotypes of interest can be bred and intercrossed to characterize semi-dominance.

The identification of the mouse *clock* mutation by Takahashi and colleagues is considered the paradigm ENU mutagenesis experiment. Although circadian period mutations had been identified in *Drosophila*, none of the spontaneous neurological mutant mice provided any clues as to the identity of mammalian genes for circadian rhythm. Takahashi and colleagues used a dominant screen (discussed below) for circadian locomotor activity to identify mutant mice among the progeny of ENU-mutagenized mice (Vitaterna et al. 1994). This screen identified the mouse mutant clock, which had a prolonged circadian period of activity in heterozygote animals with an even longer period of activity in homozygous animals. Positional cloning methods were then used to identify a point mutation in a candidate novel gene (King et al. 1997). The defective circadian rhythm in clock mutant mice was rescued by crossing the animals with *clock* bacterial artificial chromosomes (BAC)-transgenic animals (Antoch et al. 1997). This elegant series of experiments illustrates the importance of ENU mutagenesis to the discovery of novel phenotypes critical to understanding CNS function (Shimomura et al. 2001, Vitaterna et al. 2001).

The first large scale dominant mutation screens were carried out by Brown and colleagues in the UK (Nolan et al. 2002) and by Balling, Hrabe de Angelis and colleagues in Germany (Hrabe de Angelis et al. 2000). Jointly these projects identified more than 300 new confirmed mouse mutations and several thousand new inherited phenotypes. The Hartwell group in the UK has focused more attention on assessing mutants with neurological and behavioral defects (see below), while the German group focused on hematological, immunological and allergy defects. Both of the screens, however, included phenotypic testing for new deafness mutants.

Many of the new deafness genes identified in these genome-wide dominant screens mapped to regions of the mouse chromosome where spontaneous and targeted mouse mutants had not been reported. A number of the mutants had defects in the organization of hair cells in the inner ear, including *shalom* and *headturner* (Kiernan et al. 2001, Tsai et al. 2001), and in the vestibular apparatus. Surprisingly, both of these mutants encode mutations in *Jag1*, a ligand for Notch signaling. In humans, JAG1 mutants have been described as having the Alagille syndrome, a condition that includes defects in the semicircular canal in the vestibular apparatus (Oda et al. 1997). In addition, other mutants, *Jeff* and *Jumbo*, have hearing impairment resulting from chronic middle ear inflammatory disease. These models provide the first hope of new insight into *otitis media*, one of the most common causes of hearing loss in children (Hardisty et al. 2003). The remarkable success of this screen in identifying novel deafness genes underscores the importance of phenotype-driven screens for neurological mutations.

In addition to these exciting deafness genes, the ENU screen at Hartwell identified several new mutants that are allelic to the *trembler* locus (Isaacs et al. 2000). The *trembler* locus carries mutations in the peripheral myelin protein *Pmp22*. Two of the new mutants have alterations in amino acids that mutated in the most severe human peripheral neuropathy, Dejerine Sottas syndrome. This set of new ENU mutants illustrates the importance of having an allelic series for a given locus (Brown & Hardisty 2003).

Recessive Genome-Wide ENU Mutagenesis Screens

To identify new recessive neurological mutants, the protocol described above for dominant genome screens has to be extended to three generations. Thus, ENU-treated males are bred to wild-type females, and the F1 progeny (G1) are then mated to wild-type mice again to establish litters of siblings sharing a common mutation (G2). These animals have to be back-crossed to the F1s to generate G3 progeny to screen for recessive phenotypes. Thus, whereas recessive mutations occur four times more frequently than dominant mutations, the husbandry requirements to carry out large-scale recessive genome-wide screens are far more demanding than those needed for dominant genome screens.

A more practical approach is to design recessive screens for specific regions of particular chromosomes, using stocks of animals carrying deletions in chromosomes of interest (Justice et al. 1997). This approach is obviously suited for directed screens aimed at generating recessive mutants for chromosomal regions with candidate loci for given traits. Because mouse deletion stocks are presently available for approximately 15% of the mouse genome, this approach is practical for a large number of genes. A primary advantage of this approach is the reduction in mouse breeding required because an ENU-treated male is mated to wild-type females, and F1 offspring are bred to animals carrying a given deletion. As the deletion mutations carry a marker that can be used to identify offspring with mutations in the targeted region of the genome, recessive lethal mutations can be detected easily by the loss of animals with a mutant phenotype in the second generation. One of the primary advantages of this approach is the information about the location of the mutation, which allows one both to assess candidate genes in the deletion region and to use BAC technologies to clone novel genes more easily.

A number of new molecular genetic approaches, including *cre-lox* strategies and strategies involving selectable markers in embryonic stem (ES) cells, to engineer deletions have been developed by Bradley and colleagues, among others (Brown & Balling 2001, Brown & Hardisty 2003, Justice et al. 1999, van der Weyden et al. 2002). Advances in targeting strategies for large deletions in specific chromosomes will increase the catalog of viable mouse stocks carrying region-specific deletions. Targeted recessive screens can then be carried out to identify novel phenotypes with interesting CNS phenotypes or to characterize existing spontaneous mouse mutants with compelling phenotypes.

The efficiency of ENU mutagenesis opens up novel approaches to identify genes that are critical for the histogenesis of the nervous system and for the development of synaptic circuitry. A new screen to use ENU-mutagenesis to identify genes critical for the earliest stages of development has been carried out by Anderson and colleagues, who used several key strategies to identify genes important for early embryonic development steps. First, they used region-based screens to treat animals heterozygous for a recessive point mutation allele of albino and hence have albino coat color. F1 and then G2 animals with mutations near albino are easy to detect because there are no viable albino offspring. Breeding the heterozygote therefore allows identification of recessive, embryonic lethal phenotypes. Alternatively, ENU can be used in genome-wide recessive screens to identify recessive mutations identified by a morphological abnormality rather than by embryonic lethality (Anderson 2000). Some of these mutations caused neural tube defects, one of the most important classes of human birth defects.

One of the earliest studies using ENU mutagenesis provides an elegant example of the use of recessive ENU screens to generate an allelic series for loci with interesting neurological phenotypes. In 1984, Bode evaluated the potential use of ENU mutagenesis in mice to isolate new alleles of the *t* region of chromosome 17 (Bode 1984). His initial paper provided the doses for ENU mutagenesis of male mice and the strategy for designing a screen to isolate new alleles for three loci in the *t* region: *tailless*, *quaking*, and *tufted*. The neurological mutant mouse *quaking* (qk^v), first characterized in 1964, has demyelination in the CNS and peripheral nervous system (PNS), as well as severe clonic/tonic seizures in viable adult mice (Sidman et al. 1964). In a series of ENU screens (Bode 1984, Justice & Bode 1986), Bode and colleagues identified an allelic series for the *quaking* locus that included embryonic lethal mutants. Although the homozygous ENU-induced mutants died as embryos, a series of phenotypes was made by generating compound heterozygote, with the original *qk* allele. Of the five new alleles generated by ENU mutagenesis, two harbored mis-sense mutations that led to the structure function studies on an identified candidate gene that encoded *Qki*, an RNA binding protein with at least three protein isoforms (Cox et al. 1999). Although the *Qki* gene is intact in the $qk^{vmutant}$ mouse, the defect is caused by a low-level expression of an alternatively spiced form of the gene that lacks the RNA-binding domain.

As discussed above, the development of sophisticated protocols for screening the behavior of mutant animals is a prerequisite for the discovery of new classes of genes that control CNS functions. Toward that end, Paigen & Eppig have developed a systematic and hierarchical protocol for phenotype assessment (SHIRPA) (Paigen & Eppig 2000). This protocol consists of a series of primary, secondary, and tertiary screens specific for deficits in muscle, spinal motor neuron, and spino-cerebellar, sensory, autonomic, and neuropsychological functions. Crawley & Paylor (Crawley et al. 1997) have designed even more sophisticated screens for neurological and neuropsychological deficits in mutant mice. A number of other new strategies, based on computerized behavior monitoring, are being developed to refine further the neurobehavioral assessments of mutant animals (Tecott & Nestler 2004). The advent of new, more sophisticated phenotypic screens for mouse mutants with novel neurological defects are a prerequisite to developing models for the manifold pathways underlying the development of brain structure and circuitry.

Genotype-Driven Screens

A complementary approach to genome-wide and region-specific screens based on phenotypes is the use of ENU mutagenesis in ES cells to generate allelic series of mutations. This approach, pioneered by Magnuson and colleagues (Chen et al. 2000), emphasizes genotype-driven genetics using sequence information to generate mutations. Their experiments demonstrated that ENU-induced loss-of-function mutants in the X-linked *Hypoxanthine phosphoribosyl transferase* (*Hprt*) locus occur at a frequency of 1 in 1000. Allelic series are especially powerful tools for dissecting gene function as alleles of different strengths enable a fine-tuned dissection of structure/function and often provide models for mutations seen in human disease. This methodology thus provides a powerful addition to the ENU mutagenesis protocols described above and will undoubtedly expand the repertoire of neurological mutants in the mouse. This approach has already been used to generate a very large allelic series of mutations in *Smad2* and *Smad4*, two important genes in the transforming growth factor beta superfamily that cannot be selected for easily in phenotype-based screens. (Vivian et al. 2002). Several of the mutants in the series had defects in neuroectoderm formation, an early critical step in CNS histogenesis.

Gene-Trap Screening for CNS Expressed Genes

The gene-trap approach enables a prescreen of mouse embryonic stem cells for insertional mutations in genes encoding secreted and membrane-spanning proteins. The "trap" depends on the selective activation of the N-terminal signal sequence of an endogenous gene to generate an active β-galactosidase fusion protein that captures secreted proteins via their signal sequence. The gene-trap method is based on the transfection of murine ES cells with gene-trap vectors that contain a splice acceptor site upstream from a promoter-less reporter gene such as *lacZ* (Skarnes 1990, Skarnes et al. 1992). With this method the reporter gene integrates randomly into the host genome with expression of the reporter gene limited to integration into a transcriptionally active gene. This approach is mutagenic as the endogenous gene is disrupted, but cloning of the gene of interest is facilitated by generation of the cDNA from the reporter gene transcript by combined 5′ rapid amplification of complementary DNA ends (5′RACE) and polymerase chain reaction (PCR) (Skarnes et al. 1995). One drawback of this general approach is the need to generate large numbers of mice from the pool of ES cell clones to identify mice with interesting phenotypes.

The gene-trap method has evolved as different types of trap vectors—enhancer, promoter, and gene-trap—which can be introduced into ES cells by either electroporation or retroviral transfer, have emerged. Among these, the gene-trap vectors have proved to be the most efficient. Although a number of creative selection methods have been devised to reduce the burden of large numbers of animals per screen, one limitation of this approach is the general finding that there are "hot and cold" genomic spots for gene-trap vector insertions. Thus, considerable scholarly concern has been raised about whether this approach will report insertions into all classes of genes at all stages of development. To circumvent these problems, Skarnes and Beddington (Skarnes 2000) and others developed gene-trap strategies to detect specific classes of proteins, including secreted proteins, membrane proteins, signaling pathway proteins, and axonal proteins (Skarnes et al. 2004). The latter screen was the first to use the gene-trap approach to identify genes that mark axonal pathways, i.e., brain wiring patterns. To achieve the latter goal, Leighton et al. (2001) added an axonal marker, human placental alkaline phosphatase (PLAP). As with ENU mutagenesis, several groups have carried out large-scale, phenotype-driven, gene-trap screens (Wurst et al. 1995). Interestingly, in these large screens, two thirds of the generated mice had reporter gene expression in the CNS; only 15% lacked any CNS expression. The apparently high number of trappable genes in the CNS makes this an attractive approach for screens for novel genes and novel phenotypes in brain development and function.

Two approaches have been used to generate large-scale genotype-based screens. The first, developed by Ruley and colleagues (Hicks et al. 1997) uses a retroviral promoter trap retroviral shuttle to disrupt genes in ES cells. A second group has used a gene trap vector with two functional units, resulting in a bank of mutated ES genes called the Omnibank, containing 2000 genes (Zambrowicz et al. 1998). Soriano and colleagues have also developed a novel reporter gene resulting from the fusion of enhanced green fluorescent protein (EGFP) to bacterial nitroreductase (GFNR) (Medico et al. 2001). The emergence of new technologies, together with the combined use of ENU mutagenized ES cells and gene-trap vectors, promises to generate a number of interesting new tools for gene discovery in the nervous system.

Large-Scale Screens Using Gene Targeting

The ENU mutagenesis and gene-trap strategies described above rely on efficient identification of mutations in the mouse genome that result from (nearly) random events. With the advent of methodology for targeted mutagenesis of the mouse genome using homologous recombination in ES cells (Bradley et al. 1992, Capecchi 1989, Koller & Smithies 1992) came the potential to produce mutations in all mouse genes and to assess resulting phenotypes. Although the construction of vectors for gene targeting was time consuming and rate limiting during the initial phase of this technology, studies of mice carrying targeted mutations (or knockout mice) revealed the enormous scientific impact of this technology. The completion of the human and mouse genome sequences allowed definition of the mammalian gene set and the capability for efficient targeted mutagenesis. This has resulted in several large-scale projects that employ gene targeting to complete the mutagenesis of the mouse genome.

The first large-scale effort to use gene targeting to characterize mouse genes (Valenzuela et al. 2003) employed homologous recombination in *Escherichia coli* (Yang et al. 1997) to manipulate BACs resulting in targeting constructs that both disrupt the gene of interest and insert a lacZ reporter gene for expression analysis. Using this approach, mutations were generated in hundreds of genes, representing ~0.5%–1.0% of the mouse genome. Furthermore, analysis of lacZ expression in heterozygous animals derived from these targeting events revealed many interesting and novel expression patterns. Although this study covered a relatively small fraction of the mouse genome, it was the first to demonstrate that gene targeting could be applied efficiently to create a library of mutations across the mouse genome.

A second effort to update traditional gene-targeting approaches involves the indexing of nearly 100,000 insertional targeting vectors (Adams et al. 2004). In this case, the site of insertion of each targeting vector is mapped onto the mouse genome, allowing one to choose a targeting construct from the library that has a very high probability of inactivating the targeted gene. Combination of the 5′ and 3′ targeting vectors described in this study further offers the potential to insert loxP sites into the genome for large-scale deletions and chromosome engineering (Adams & Bradley 2002). Although this study reported generation of relatively few targeted alleles from the insertional vector libraries, the preparation and mapping of these vectors represents an important resource for the mouse genetics community. Two large-scale projects to create and distribute mutant alleles for all mouse genes are now being initiated (Austin et al. 2004, Auwerx et al. 2004).

SYSTEMATIC SCREENS OF CNS GENE EXPRESSION

Although analysis of phenotypic mutation effects is a primary approach toward understanding gene function, it has become clear over the past decade that for many genes, perhaps most, visible or behavioral mutant phenotypes are difficult to discern. For example, in fruit flies lethal and/or easily observable phenotypes are evident for only about one third of the genes in the genome (Celniker & Rubin 2003). As a consequence, large-scale screens to reveal the expression patterns of genes in the mammalian nervous system have been initiated. These screens are critical both to understand observed phenotypes and to focus attention on appropriate structures for discovery of more subtle phenotypic consequences of mutation. Systematic screens of this type can also be used to direct attention to genes likely to play important roles in CNS function or dysfunction.

The mammalian CNS contains, minimally, several thousand functionally distinct cell types. Each of these cell types arises at a specific time and place in the developing nervous system and incorporates into developing CNS circuitry on a distinct schedule. Given this complexity, and the assembly of neurons into circuits that control function and behavior, interpretation of gene function is difficult in the absence of precise knowledge of the regions and cell types in which it is expressed. Early efforts to screen for genes expressed in specific classes of CNS cells at particular stages of development used a variety of subtractive hybridization (Miller et al. 1987, Nordquist et al. 1988, Oberdick et al. 1989, Porteus et al. 1992) and immunologic approaches (Kuhar et al. 1993). In most cases, these approaches yielded small numbers of interesting novel genes. Although these studies cannot be considered comprehensive, subsequent studies of genes isolated solely on the basis of their expression patterns in the CNS have repeatedly illustrated the value of this approach toward identifying important CNS functions.

Large-Scale Expression Studies of Single Genes

Analysis of gene expression in the CNS is presently conducted by two different strategies: analysis of single-gene expression patterns at cellular resolution using in situ hybridization or reporter gene assays, and transcriptional profiling of dissected CNS tissue samples or single cells using microarrays. The data collected in these studies provide different insights into gene function as a consequence of the nature of the assays. Analysis of the expression patterns of single genes over the course of development can often provide an initial hypothesis concerning gene function. Perhaps the most famous example of this strategy is the discovery of mammalian odorant receptors, whose function was inferred from the nature and complexity of the gene family, and the particular patterns of expression of putative odorant receptor mRNAs in the olfactory epithelium and bulb (Buck & Axel 1991). In contrast, studies of gene expression using microarray data generate comprehensive profiles of genes expressed in the cell type or tissue of interest, often providing insights into suites of genes that can contribute to development and disease (Mirnics & Pevsner 2004). The importance of both single-gene expression analysis and microarray transcriptional profiling has led to large-scale efforts to determine the anatomic and temporal specificity of CNS gene expression.

In Situ Hybridization

A number of high-throughput efforts are under way to systematically analyze the expression patterns of CNS-expressed genes in the nervous system. All these projects aim to provide fundamental information concerning temporal and spatial expression of specific genes throughout CNS development, although they each have a different focus that reflects the technical limitations of the methodology used or the interests of the research group performing the experiments. Given the scope of these large projects, the published papers serve mostly as an introduction to the project and its technical basis. To evaluate the data produced in each project and its presentation to the public, it is most useful to log in to the various Web sites and browse the databases for genes of interest.

At present, there are at least four ongoing large-scale in situ hybridization (ISH) projects to analyze CNS-expressed genes that have released their data to the public. These include Genepaint (Visel et al. 2004; **http://www.genepaint.org**), Emage (Baldock et al. 2003; **http://genex.hgu.mrc.ac.uk/Emage**), Allen Brain Atlas (**http://www.brainatlas.com**); and BGEM (Brain Gene Expression Map; **http://www.stjudebgem.org**), a component of GENSAT (Gene Expression Nervous System Atlas; **http://www.ncbi.nlm.nih.gov/projects/gensat**). Genepaint, Emage, and the Allen Brain Atlas are presently using colorimetric ISH protocols in either whole-mount tissue or brain sections to generate maps of CNS gene expression. Although in situ hybridization methodology offers the potential to visualize specific cell types that express the transcript, the localization of most mRNAs to the cell soma prevents visualization of the arbors of the cell and their definitive identification. Colorimetric ISH methods normally incorporate immunologic amplification steps into the experimental protocol, complicating efforts to obtain quantitative information concerning expression levels. The BGEM project uses radioactive probes and photographic emulsions to generate darkfield images of gene expression in brain sections. These studies provide quantitative gene-expression levels because they detect isotopic decay of the probe, which is a linear process. In aggregate, these projects represent an impressive initiative to use high throughput ISH to provide detailed information of mRNA abundance in the developing and adult brain.

Reporter Gene Assays

The advantages of reporter gene strategies for ultrahigh resolution mapping of gene expression, and for identification of vectors and animals that offer experimental access to specific cell populations, was first appreciated by invertebrate geneticists (Celniker & Rubin 2003). Since then, reporter genes have been widely used in mammalian studies to both disrupt and mark genes using gene targeting and gene trapping (see above). More recently, the ability to generate BAC transgenic mice that accurately express reporter genes has also become routine (Gong et al. 2002). These experiments make use of a variety of reporter genes, e.g., β-galactosidase (*lacZ*), firefly luciferase (*Luc*), human placental alkaline phosphatase (PLAP), green fluorescence protein (GFP), etc., to localize expression; each gene has particular advantages and limitations. The most versatile of these reporters are members of the GFP family of fluorescent proteins, first isolated from jellyfish (Prasher et al. 1992) and first employed as reporter genes in *C. elegans* (Chalfie et al. 1994), which now include a large variety of spectral and functional variants (Zhang et al. 2002). Although reporter gene assays offer a number of important advantages over ISH techniques (high sensitivity, visualization of cellular morphology, vital imaging, etc.), they are less efficient and more expensive than ISH assays.

Reporter genes have been incorporated into all the gene trapping and high-throughput targeted mutagenesis projects referred to above, and a large number of tagged genes have been generated in these studies. However, CNS expression data is publicly available for only a small minority of the targeted genes. In most cases, the data are obtained from the lacZ reporter, which has the advantage of extremely high sensitivity but cannot reveal detailed cellular morphology unless expressed at very high levels and cannot be detected in living tissue. In most cases, expression studies using *lacZ* as the reporter gene, do not reveal detailed cellular morphology unless the lacZ is expressed at very high levels. In addition, the lacZ reporter cannot be detected in living tissue, precluding its use as a vital marker.

The Gene Expression Nervous System Atlas (GENSAT; http://www.ncbi.nlm.nih.gov/projects/gensat) project is specifically designed to take advantage of EGFP reporter genes to both map gene expression in the developing and adult brain and to provide experimental access to CNS cell populations (Gong et al. 2003). The GENSAT BAC Transgenic Project is the first large-scale effort to utilize both the mammalian genome sequence and the BAC clones, which provide the basis of the genome physical map to generate anatomic data and experimental resources for use by the neuroscience community. The two-stage approach employed by GENSAT incorporates a high throughput ISH screen for CNS gene expression (BGEM, see above), followed by the use of BAC transgenic reporter gene analysis (Gong et al. 2003; http://www.gensat.org). In this way, large numbers of genes can be assayed for regional expression in the CNS, and those of particularly high interest can be studied at ultrahigh resolution using the EGFP reporter gene. To date, the GENSAT Project has assayed more than 1000 genes by ISH and nearly 400 using BAC transgenic reporter genes. The systematic data-collection strategy and the high-definition cellular morphology that can be achieved using the EGFP reporter gene (**Figure 1**) have allowed GENSAT to identify specific neuronal and glial cell types according to the classical anatomic literature. In addition, the ability of EGFP to

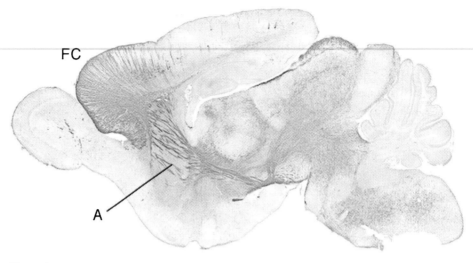

Figure 1

Dopamine receptor 4 (*Dr4*) expression in prefrontal cortex. *Dr4* BAC transgenic mice were generated in the GENSAT project, as described in the text. Sagittal cryostat section of postnatal day 7 brain, immunostained with antibodies against the reporter EGFP (DAB method), reveal Drd4+ Pyramidal neurons in the frontal cortex and their axonal projections (A, *arrow*) in the striatum.

be transported into the axonal projections of the labeled cells has allowed visualization of many subpopulations of these cell types on the basis of their axonal projection patterns. For example, cortical pyramidal cells projecting to distinct subcortical sites (e.g., *Drd4*, *Etv1*; http://www.gensat.org) can be distinguished. These anatomic data, and the ability of BAC vectors to reproducibly target expression in transgenic animals, provide genetic access to an increasing variety of specific CNS cell types. Furthermore, the most of the GENSAT BAC transgenic lines express EGFP at sufficiently high levels for vital imaging, and they have been made freely available to the public for further genetic, anatomic, physiologic, and cell biologic analyses. GENSAT is a CNS-biased gene-expression project that seeks to incorporate information from the literature, other large-scale projects, and a panel of advisors with a broad range of expertise (**http://www.ncbi.nlm.nih.gov/projects/gensat**). In addition to generating a gene-expression atlas, the GENSAT Project produced several important discoveries regarding brain development and circuitry. The key discoveries described in Gong et al. (2003) are summarized below.

Genes Involved in Axon-Target Interactions—Sema3B. A key step CNS development is the outgrowth of specific sets of neurons and their interactions with particular targets. In the GENSAT screen, the Semaphorin 3B (Sema3B) BAC transgenic mice provide an example of the ability of BAC constructs to reveal the morphology and topographic projections of neurons that express a particular guidance molecule. Thus, retinal ganglion cells expressing Sema3B reveal an interesting pattern of terminations in superior colliculus and establish that this molecule is expressed in mossy fiber afferents to the cerebellum.

Genes that Reveal Tangential Migratory Patterns—Lhx6 and Pde1c. The migration of immature neurons from germinal zones to positions where they form synaptic connections is one of the hallmarks of cortical development. The BAC transgenic mice provide a means to visualize these migrations in real time and to follow the same cells over the developmental epoch of neural layer formation. GABAergic interneurons, which constitute 20% of the neurons in cortex, originate in the basal forebrain and migrate into the neocortex during cortical histogenesis. In the BAC screen, Lhx6 provided a marker for this population of migrating interneurons. In addition to generating lines of mice that confirm existing models of cell migrations in the cortex, lines were identified that revealed new pathways of migration. One of these was the Phosphodiesterase 1C (*Pde1c*) gene. In the embryonic cortex, analysis of Pde1c BAC transgenic mice revealed labeled cells spreading across the cortical surface in a tangential migratory pattern, which matched the pattern proposed for subpial granule (SGL) neurons. This population includes the Cajal Retzius cells, previously thought to arise exclusively in the ventricular zone of the cortex.

The hundreds of BAC transgenic mice now available on **http://www.gensat.org** to the neuroscience community include animals with novel developmental programs and circuitry that provide access to cell migration, axon-target projections, and functions of specific brain regions (e.g., frontal cortex). The reagents of this project, including BAC transgenic mice and modified BACs, are freely available to researchers.

Transcriptional Profiling Using Microarray

Oligonucleotide microarray (Gene Chip) technology provides a powerful means to identify genes expressed in distinct anatomical regions of the brain. To date, four prominent CNS regions have been profiled using this general strategy: the cerebellar cortex, hippocampal formation, amygdala, and hypothalamus. These areas are of particular interest, owing to their functions in spatial and emotional memories. A broad study by Anderson and colleagues, using commercial microarray, revealed that among 34,000 genes detected, only ∼200 genes were differentially expressed in only 1 of the 5 regions queried (Zirlinger et al. 2001). In the amygdala, the majority of 13 genes found to mark specific subnuclei with known functions in emotional memory were unknown genes [expressed sequence tags (ESTs)], making it difficult to assess potential functions. To refine the starting tissue more precisely, and to finely dissect the sub regions of the amygdala prior to microarray analysis, they used laser capture methods (Zirlinger & Anderson 2003). This screen generated 8 new amygdala enriched genes, bringing the current total to 21. Similar studies have identified genes enriched in hippocampal formation and olfactory bulb (Lein et al. 2004). A second gene-expression screening approach is serial expression analysis (SAGE) (Velculescu et al. 1995). The SAGE method is based on the isolation of unique sequence tags from individual transcripts and the concatenation of tags serially into long DNA molecules sequenced to quantify tag frequency in the population. SAGE tag frequency is directly proportional to the originating mRNA copy number and is therefore a reliable measure of the abundance of a given mRNA. The sequence TAG also allows identification of enriched genes for further analysis. The microSAGE strategy (Blackshaw et al. 2003) employs oligo-dT-coated tubes or magnetic beads to capture polyadenylated RNA directly from cell lysates and to allow reverse transcription, cDNA synthesis, and the subsequent SAGE steps to occur sequentially in the same vessel. This approach allows gene-expression analyses to be performed with the small amounts of material that might be harvested from a particular brain area. Cepko and colleagues have used this approach (Blackshaw et al. 2004) extensively in genes specific to different classes of retinal neurons (Blackshaw et al. 2001), stages of development (Blackshaw et al. 2004), or disease processes (Sharon et al. 2002).

Gauge and colleagues have used this approach to profile gene expression in discrete

ENU: N-ethyl-N-nitrosourea

BAC: bacterial artificial chromosome

EGFP: enhanced green fluorecent protein

ISH: in situ hybridization

GENSAT: Gene Expression Nervous System Atlas

domains of the hippocampal formation (Zhao et al. 2001). More recent, comparative SAGE analyses of 78,784 tags from the hypothalamus with 125,296 tags from the neocortex reveal that unique sets of RNA are expressed in the two brain regions. Chromosomal mapping of the unique genes in the two regions suggested that genes that are differentially expressed for these two brain regions are in very close proximity, suggesting possible coregulation.

A direct comparison of the microarray and SAGE screening methods by Akil and collaborators (Evans et al. 2002) indicates that Affymetrix Gene Chips, like SAGE, reliably detect only medium-to-high-abundance transcripts and that detection of low-abundance transcripts, many of which have great relevance to biological function in brain, is inconsistent. They estimated that no more than 30% of the total transcriptosome of the hippocampus is detected in these methods. Further, they estimated that only genes expressed at a general level of 1–5 per cell are detected by these methods. This is a serious limitation. Akil points out that some critical genes such as gonadotropin-releasing hormone (GnRH) (Yellon et al. 1990) are expressed only in a few thousand cells dispersed throughout the hypothalamus. On average, current microarray methods detect GnRH less than 50% of the time (Evans et al. 2002).

CNS Proteomics

Proteomics is the large-scale, comparative analysis of proteins expressed in specific cell types or tissues under varying experimental conditions. The methodology required for proteomic analysis has been the subject of many excellent, recent reviews (de Hoog & Mann 2004, Fountoulakis 2004) and is not discussed here. Although proteomic studies of the mammalian brain have not yet reached maturity, these studies offer important opportunities for advancing our understanding of molecular mechanisms of CNS function that are not available using nucleic acid–based methodologies.

At least three important advantages of large-scale proteomics analysis are likely to have an impact on CNS studies. First, for most genes expressed in the CNS, alternative splicing can result in the generation of multiple mRNAs and, consequently, proteins. Although these splice variants can be detected using RNA-based methods, the functional properties of each of the protein species must be determined in protein-based biochemical assays. Thus, the use of functional proteomics to address the biochemical properties of novel proteins expressed in the CNS will be an important complement to the phenotypic analyses described above. Second, many proteins are regulated by posttranslational modification. The sensitive mass spectrometry assays (McLachlin & Chait 2001) now employed in proteomics studies can allow discovery and analysis of these modifications, which can lead to molecular and genetic tests of their in vivo significance. Third, proteomic analysis of protein assemblies and organelles can lead to a deep appreciation of the functional properties of specialized structures critical to the nervous system. For example, analysis of synaptic proteins (Cho et al. 1992, Kennedy 2000) or proteins at synapses that express NMDA receptors (Choudhary & Grant 2004, Husi et al. 2000) have led to the discovery of a large number of novel synaptic proteins and have elicited an appreciation for the biochemical and functional complexity of this crucial structure. These are just three of the many important reasons why we can expect an increase in the number and variety of proteomics studies focusing on the brain.

CONCLUSIONS

The importance of phenotypic and anatomic analyses for deciphering gene function within the CNS was first demonstrated by the characterization of spontaneous neurological mutant mice with interesting CNS phenotypes. More recently, development of efficient procedures for ENU mutagenesis, gene trapping, and gene targeting by homologous recombination demonstrate that in the next decade mutant alleles will be identified for virtually every mouse gene. Concurrently, large-scale efforts are underway to map the expression of all CNS-expressed genes and to provide genetic access to the wide variety of CNS cell types. The impact of these large-scale projects on neuroscience research is just beginning to become evident. These efforts will lead to an increased efficiency in collection of valuable data and to intellectual and experimental opportunities that would not otherwise present themselves. The foresight required to invest in these large-scale efforts will be richly rewarded by the use of the information and materials provided to advance the research efforts of individual laboratories in the field of neuroscience.

ACKNOWLEDGMENTS

The authors were supported by NIH NINDS grant NS R01 15429-25 (M.E.H.), NIH NINDS Contract N01 NS02331 (N.H., M.E.H.), and The Howard Hughes Medical Institute (N.H.). We are grateful to Dr. Kathryn Zimmerman for critical comments on the manuscript and to Ms. Judy Walsh for help preparing the manuscript.

LITERATURE CITED

Ackerman SL, Kozak LP, Przyborski SA, Rund LA, Boyer BB, Knowles BB. 1997. The mouse rostral cerebellar malformation gene encodes an UNC-5-like protein. *Nature* 386:838–42

Adams DJ, Biggs PJ, Cox T, Davies R, dvan er Weyden L, et al. 2004. Mutagenic insertion and chromosome engineering resource (MICER). *Nat. Genet.* 36:867–71

Adams DJ, Bradley A. 2002. Induced mitotic recombination: a switch in time. *Nat. Genet.* 30:6–7

Anderson KV. 2000. Finding the genes that direct mammalian development: ENU mutagenesis in the mouse. *Trends Genet.* 16:99–102

Antoch MP, Song EJ, Chang AM, Vitaterna MH, Zhao Y, et al. 1997. Functional identification of the mouse circadian Clock gene by transgenic BAC rescue. *Cell* 89:655–67

Austin CP, Battey JF, Bradley A, Bucan M, Capecchi M, et al. 2004. The knockout mouse project. *Nat. Genet.* 36:921–24

Auwerx J, Avner P, Baldock R, Ballabio A, Balling R, et al. 2004. The European dimension for the mouse genome mutagenesis program. *Nat. Genet.* 36:925–27

Baldock RA, Bard JB, Burger A, Burton N, Christiansen J, et al. 2003. EMAP and EMAGE: a framework for understanding spatially organized data. *Neuroinformatics* 1:309–25

Blackshaw S, Fraioli RE, Furukawa T, Cepko CL. 2001. Comprehensive analysis of photoreceptor gene expression and the identification of candidate retinal disease genes. *Cell* 107:579–89

Blackshaw S, Harpavat S, Trimarchi J, Cai L, Huang H, et al. 2004. Genomic analysis of mouse retinal development. *PLoS Biol.* 2:E247

Blackshaw S, Kuo WP, Park PJ, Tsujikawa M, Gunnersen JM, et al. 2003. MicroSAGE is highly representative and reproducible but reveals major differences in gene expression among samples obtained from similar tissues. *Genome Biol.* 4:R17

Bode VC. 1984. Ethylnitrosourea mutagenesis and the isolation of mutant alleles for specific genes located in the T region of mouse chromosome 17. *Genetics* 108:457–70

Bradley A, Ramirez-Solis R, Zheng H, Hasty P, Davis A. 1992. Genetic manipulation of the mouse via gene targeting in embryonic stem cells. *Ciba Found. Symp.* 165:256–69; discussion pp. 69–76

Brown SD, Balling R. 2001. Systematic approaches to mouse mutagenesis. *Curr. Opin. Genet. Dev.* 11:268–73

Brown SD, Hardisty RE. 2003. Mutagenesis strategies for identifying novel loci associated with disease phenotypes. *Semin. Cell. Dev. Biol.* 14:19–24

Brown SD, Nolan PM. 1998. Mouse mutagenesis-systematic studies of mammalian gene function. *Hum. Mol. Genet.* 7:1627–33

Buck L, Axel R. 1991. A novel multigene family may encode odorant receptors: a molecular basis for odor recognition. *Cell* 65:175–87

Caddy KW, Sidman RL. 1981. Purkinje cells and granule cells in the cerebellum of the Stumbler mutant mouse. *Brain Res.* 227:221–36

Capecchi MR. 1989. Altering the genome by homologous recombination. *Science* 244:1288–92

Celniker SE, Rubin GM. 2003. The Drosophila melanogaster genome. *Annu. Rev. Genomics Hum. Genet.* 4:89–117

Chalfie M, Tu Y, Euskirchen G, Ward WW, Prasher DC. 1994. Green fluorescent protein as a marker for gene expression. *Science* 263:802–5

Chen Y, Yee D, Dains K, Chatterjee A, Cavalcoli J, et al. 2000. Genotype-based screen for ENU-induced mutations in mouse embryonic stem cells. *Nat. Genet.* 24:314–17

Cho KO, Hunt CA, Kennedy MB. 1992. The rat brain postsynaptic density fraction contains a homolog of the Drosophila discs-large tumor suppressor protein. *Neuron* 9:929–42

Choudhary J, Grant SG. 2004. Proteomics in postgenomic neuroscience: the end of the beginning. *Nat. Neurosci.* 7:440–45

Cox RD, Hugill A, Shedlovsky A, Noveroske JK, Best S, et al. 1999. Contrasting effects of ENU induced embryonic lethal mutations of the quaking gene. *Genomics* 57:333–41

Crawley JN, Belknap JK, Collins A, Crabbe JC, Frankel W, et al. 1997. Behavioral phenotypes of inbred mouse strains: implications and recommendations for molecular studies. *Psychopharmacology (Berl.)* 132:107–24

de Hoog C, Mann M. 2004. Proteomics. *Annu. Rev. Genomics Hum. Genet.* 5:267–93

De Jager PL, Heintz N. 1998. The lurcher mutation and ionotropic glutamate receptors: contributions to programmed neuronal death in vivo. *Brain Pathol.* 8:795–807

Evans SJ, Datson NA, Kabbaj M, Thompson RC, Vreugdenhil E, et al. 2002. Evaluation of Affymetrix Gene Chip sensitivity in rat hippocampal tissue using SAGE analysis. Serial Analysis of Gene Expression. *Eur. J. Neurosci.* 16:409–13

Fernandez-Gonzalez A, La Spada AR, Treadaway J, Higdon JC, Harris BS, et al. 2002. Purkinje cell degeneration (pcd) phenotypes caused by mutations in the axotomy-induced gene, Nna1. *Science* 295:1904–6

Fountoulakis M. 2004. Application of proteomics technologies in the investigation of the brain. *Mass Spectrom. Rev.* 23:231–58

Gong S, Yang XW, Li C, Heintz N. 2002. Highly efficient modification of bacterial artificial chromosomes (BACs) using novel shuttle vectors containing the R6Kgamma origin of replication. *Genome Res.* 12:1992–98

Gong S, Zheng C, Doughty ML, Losos K, Didkovsky N, et al. 2003. A gene expression atlas of the central nervous system based on bacterial artificial chromosomes. *Nature* 425:917–25

Goulding M, Sterrer S, Fleming J, Balling R, Nadeau J, et al. 1993. Analysis of the Pax-3 gene in the mouse mutant splotch. *Genomics* 17:355–63

Hamilton BA, Frankel WN, Kerrebrock AW, Hawkins TL, FitzHugh W, et al. 1996. Disruption of the nuclear hormone receptor RORalpha in staggerer mice. *Nature* 379:736–39

Hardisty RE, Erven A, Logan K, Morse S, Guionaud S, et al. 2003. The deaf mouse mutant Jeff (Jf) is a single gene model of otitis media. *J. Assoc. Res. Otolaryngol.* 4:130–38

Heuer H, Mason CA. 2003. Thyroid hormone induces cerebellar Purkinje cell dendritic development via the thyroid hormone receptor alpha1. *J. Neurosci.* 23:10604–12

Hicks GG, Shi EG, Li XM, Li CH, Pawlak M, Ruley HE. 1997. Functional genomics in mice by tagged sequence mutagenesis. *Nat. Genet.* 16:338–44

Hosoda K, Hammer RE, Richardson JA, Baynash AG, Cheung JC, et al. 1994. Targeted and natural (piebald-lethal) mutations of endothelin-B receptor gene produce megacolon associated with spotted coat color in mice. *Cell* 79:1267–76

Hrabe de Angelis MH, Flaswinkel H, Fuchs H, Rathkolb B, Soewarto D, et al. 2000. Genome-wide, large-scale production of mutant mice by ENU mutagenesis. *Nat. Genet.* 25:444–47

Husi H, Ward MA, Choudhary JS, Blackstock WP, Grant SG. 2000. Proteomic analysis of NMDA receptor-adhesion protein signaling complexes. *Nat. Neurosci.* 3:661–69

Isaacs AM, Davies KE, Hunter AJ, Nolan PM, Vizor L, et al. 2000. Identification of two new Pmp22 mouse mutants using large-scale mutagenesis and a novel rapid mapping strategy. *Hum. Mol. Genet.* 9:1865–71

Jeong YG, Hyun BH. 2000. Abnormal synaptic organization between granule cells and Purkinje cells in the new ataxic mutant mouse, pogo. *Neurosci. Lett.* 294:77–80

Jeong YG, Hyun BH, Hawkes R. 2000. Abnormalities in cerebellar Purkinje cells in the novel ataxic mutant mouse, pogo. *Brain Res. Dev. Brain Res.* 125:61–67

Justice MJ, Bode VC. 1986. Induction of new mutations in a mouse t-haplotype using ethyl-nitrosourea mutagenesis. *Genet. Res.* 47:187–92

Justice MJ, Noveroske JK, Weber JS, Zheng B, Bradley A. 1999. Mouse ENU mutagenesis. *Hum. Mol. Genet.* 8:1955–63

Justice MJ, Zheng B, Woychik RP, Bradley A. 1997. Using targeted large deletions and high-efficiency N-ethyl-N-nitrosourea mutagenesis for functional analyses of the mammalian genome. *Methods* 13:423–36

Kennedy MB. 2000. Signal-processing machines at the postsynaptic density. *Science* 290:750–54

Kiernan AE, Ahituv N, Fuchs H, Balling R, Avraham KB, et al. 2001. The Notch ligand Jagged1 is required for inner ear sensory development. *Proc. Natl. Acad. Sci. USA* 98:3873–78

King DP, Zhao Y, Sangoram AM, Wilsbacher LD, Tanaka M, et al. 1997. Positional cloning of the mouse circadian clock gene. *Cell* 89:641–53

Kofuji P, Hofer M, Millen KJ, Millonig JH, Davidson N, et al. 1996. Functional analysis of the weaver mutant GIRK2 K+ channel and rescue of weaver granule cells. *Neuron* 16:941–52

Koller BH, Smithies O. 1992. Altering genes in animals by gene targeting. *Annu. Rev. Immunol.* 10:705–30

Kuhar SG, Feng L, Vidan S, Ross ME, Hatten ME, Heintz N. 1993. Changing patterns of gene expression define four stages of cerebellar granule neuron differentiation. *Development* 117:97–104

Landis SC. 1973. Ultrastructural changes in the mitochondria of cerebellar Purkinje cells of nervous mutant mice. *J. Cell Biol.* 57:782–97

Leighton PA, Mitchell KJ, Goodrich LV, Lu X, Pinson K, et al. 2001. Defining brain wiring patterns and mechanisms through gene trapping in mice. *Nature* 410:174–79

Lein ES, Zhao X, Gage FH. 2004. Defining a molecular atlas of the hippocampus using DNA microarrays and high-throughput in situ hybridization. *J. Neurosci.* 24:3879–89

McKusick V. 1997. *Mendelian Inheritance in Man*. Baltimore, MD: Johns Hopkins Univ. Press

McLachlin DT, Chait BT. 2001. Analysis of phosphorylated proteins and peptides by mass spectrometry. *Curr. Opin. Chem. Biol.* 5:591–602

Medico E, Gambarotta G, Gentile A, Comoglio PM, Soriano P. 2001. A gene trap vector system for identifying transcriptionally responsive genes. *Nat. Biotechnol.* 19:579–82

Miller FD, Naus CC, Higgins GA, Bloom FE, Milner RJ. 1987. Developmentally regulated rat brain mRNAs: molecular and anatomical characterization. *J. Neurosci.* 7:2433–44

Millonig JH, Millen KJ, Hatten ME. 2000. The mouse Dreher gene Lmx1a controls formation of the roof plate in the vertebrate CNS. *Nature* 403:764–69

Mirnics K, Pevsner J. 2004. Progress in the use of microarray technology to study the neurobiology of disease. *Nat. Neurosci.* 7:434–39

Noebels JL. 1999. Single-gene models of epilepsy. *Adv. Neurol.* 79:227–38

Noebels JL. 2003. The biology of epilepsy genes. *Annu. Rev. Neurosci.* 26:599–625

Nolan PM, Hugill A, Cox RD. 2002. ENU mutagenesis in the mouse: application to human genetic disease. *Brief Funct. Genomic Proteomic* 1:278–89

Nordquist DT, Kozak CA, Orr HT. 1988. cDNA cloning and characterization of three genes uniquely expressed in cerebellum by Purkinje neurons. *J. Neurosci.* 8:4780–89

Oberdick J, Levinthal F, Levinthal C. 1989. A purkinje cell differentiation marker shows a partial dna sequence homology to the cellular sis/pdgf2 gene. *Neuron* 3:386

Oda T, Elkahloun AG, Pike BL, Okajima K, Krantz ID, et al. 1997. Mutations in the human Jagged1 gene are responsible for Alagille syndrome. *Nat. Genet.* 16:235–42

Paigen K, Eppig JT. 2000. A mouse phenome project. *Mamm. Genome* 11:715–17

Peters J, Ball ST, Andrews SJ. 1986. The detection of gene mutations by electrophoresis, and their analysis. *Prog. Clin. Biol. Res.* 209B:367–74

Porteus MH, Brice AE, Bulfone A, Usdin TB, Ciaranello RD, Rubenstein JL. 1992. Isolation and characterization of a library of cDNA clones that are preferentially expressed in the embryonic telencephalon. *Brain Res. Mol. Brain Res.* 12:7–22

Prasher DC, Eckenrode VK, Ward WW, Prendergast FG, Cormier MJ. 1992. Primary structure of the Aequorea victoria green-fluorescent protein. *Gene* 111:229–33

Rice DS, Curran T. 2001. Role of the reelin signaling pathway in central nervous system development. *Annu. Rev. Neurosci.* 24:1005–39

Ross ME, Fletcher C, Mason CA, Hatten ME, Heintz N. 1990. Meander tail reveals a discrete developmental unit in the mouse cerebellum. *Proc. Natl. Acad. Sci. USA* 87:4189–92

Sharon D, Blackshaw S, Cepko CL, Dryja TP. 2002. Profile of the genes expressed in the human peripheral retina, macula, and retinal pigment epithelium determined through serial analysis of gene expression (SAGE). *Proc. Natl. Acad. Sci. USA* 99:315–20

Shimomura K, Low-Zeddies SS, King DP, Steeves TD, Whiteley A, et al. 2001. Genome-wide epistatic interaction analysis reveals complex genetic determinants of circadian behavior in mice. *Genome Res.* 11:959–80

Sidman RL. 1983. Experimental neurogenetics. *Res. Publ. Assoc. Res. Nerv. Ment. Dis.* 60:19–46

Sidman RL, Dickie MM, Appel SH. 1964. Mutant mice (Quaking and Jimpy) with deficient myelination in the central nervous system. *Science* 144:309–11

Skarnes WC. 1990. Entrapment vectors: a new tool for mammalian genetics. *Biotechnology (NY)* 8:827–31

Skarnes WC. 2000. Gene trapping methods for the identification and functional analysis of cell surface proteins in mice. *Methods Enzymol.* 328:592–615

Skarnes WC, Auerbach BA, Joyner AL. 1992. A gene trap approach in mouse embryonic stem cells: the lacZ reported is activated by splicing, reflects endogenous gene expression, and is mutagenic in mice. *Genes Dev.* 6:903–18

Skarnes WC, Moss JE, Hurtley SM, Beddington RS. 1995. Capturing genes encoding membrane and secreted proteins important for mouse development. *Proc. Natl. Acad. Sci. USA* 92:6592–96

Skarnes WC, von Melchner H, Wurst W, Hicks G, Nord AS, et al. 2004. A public gene trap resource for mouse functional genomics. *Nat. Genet.* 36:543–44

Slesinger PA, Patil N, Liao YJ, Jan YN, Jan LY, Cox DR. 1996. Functional effects of the mouse weaver mutation on G protein-gated inwardly rectifying K+ channels. *Neuron* 16:321–31

Tecott LH, Nestler EJ. 2004. Neurobehavioral assessment in the information age. *Nat. Neurosci.* 7:462–66

Tsai H, Hardisty RE, Rhodes C, Kiernan AE, Roby P, et al. 2001. The mouse slalom mutant demonstrates a role for Jagged1 in neuroepithelial patterning in the organ of Corti. *Hum. Mol. Genet.* 10:507–12

Valenzuela DM, Murphy AJ, Frendewey D, Gale NW, Economides AN, et al. 2003. High-throughput engineering of the mouse genome coupled with high-resolution expression analysis. *Nat. Biotechnol.* 21:652–59

van der Weyden L, Adams DJ, Bradley A. 2002. Tools for targeted manipulation of the mouse genome. *Physiol. Genomics* 11:133–64

Velculescu VE, Zhang L, Vogelstein B, Kinzler KW. 1995. Serial analysis of gene expression. *Science* 270:484–87

Visel A, Thaller C, Eichele G. 2004. GenePaint.org: an atlas of gene expression patterns in the mouse embryo. *Nucleic Acids Res.* 32:D552–56 (Database issue)

Vitaterna MH, King DP, Chang AM, Kornhauser JM, Lowrey PL, et al. 1994. Mutagenesis and mapping of a mouse gene, Clock, essential for circadian behavior. *Science* 264:719–25

Vitaterna MH, Takahashi JS, Turek FW. 2001. Overview of circadian rhythms. *Alcohol Res. Health* 25:85–93

Vivian JL, Chen Y, Yee D, Schneider E, Magnuson T. 2002. An allelic series of mutations in Smad2 and Smad4 identified in a genotype-based screen of N-ethyl-N-nitrosourea-mutagenized mouse embryonic stem cells. *Proc. Natl. Acad. Sci. USA* 99:15542–47

Wilson SM, Bhattacharyya B, Rachel RA, Coppola V, Tessarollo L, et al. 2002. Synaptic defects in ataxia mice result from a mutation in Usp14, encoding a ubiquitin-specific protease. *Nat. Genet.* 32:420–25

Wurst W, Rossant J, Prideaux V, Kownacka M, Joyner A, et al. 1995. A large-scale gene-trap screen for insertional mutations in developmentally regulated genes in mice. *Genetics* 139:889–99

Yang XW, Model P, Heintz N. 1997. Homologous recombination based modification in Escherichia coli and germline transmission in transgenic mice of a bacterial artificial chromosome. *Nat. Biotechnol.* 15:859–65

Yellon SM, Lehman MN, Newman SW. 1990. The gonadotropin-releasing hormone neuronal system of the male Djungarian hamster: distribution from the olfactory tubercle to the medial basal hypothalamus. *Neuroendocrinology* 51:219–25

Yue Z, Horton A, Bravin M, DeJager PL, Selimi F, Heintz N. 2002. A novel protein complex linking the delta 2 glutamate receptor and autophagy: implications for neurodegeneration in lurcher mice. *Neuron* 35:921–33

Zambrowicz BP, Friedrich GA, Buxton EC, Lilleberg SL, Person C, Sands AT. 1998. Disruption and sequence identification of 2,000 genes in mouse embryonic stem cells. *Nature* 392:608–11

Zhang J, Campbell RE, Ting AY, Tsien RY. 2002. Creating new fluorescent probes for cell biology. *Nat. Rev. Mol. Cell Biol.* 3:906–18

Zhao X, Lein ES, He A, Smith SC, Aston C, Gage FH. 2001. Transcriptional profiling reveals strict boundaries between hippocampal subregions. *J. Comp. Neurol.* 441:187–96

Zirlinger M, Anderson D. 2003. Molecular dissection of the amygdala and its relevance to autism. *Genes Brain Behav.* 2:282–94

Zirlinger M, Kreiman G, Anderson DJ. 2001. Amygdala-enriched genes identified by microarray technology are restricted to specific amygdaloid subnuclei. *Proc. Natl. Acad. Sci. USA* 98:5270–75

Zuo J, De Jager PL, Takahashi KA, Jiang W, Linden DJ, Heintz N. 1997. Neurodegeneration in Lurcher mice caused by mutation in delta2 glutamate receptor gene. *Nature* 388:769–73

Autism: A Window Onto the Development of the Social and the Analytic Brain

Simon Baron-Cohen and Matthew K. Belmonte

Autism Research Centre, Department of Psychiatry, University of Cambridge, Cambridge, CB2 2AH, United Kingdom; email: sb205@cam.ac.uk, belmonte@mit.edu

Key Words

social cognition, attention, frontal lobe, network connectivity

Abstract

Although the neurobiological understanding of autism has been increasing exponentially, the diagnosis of autism spectrum conditions still rests entirely on behavioral criteria. Autism is therefore most productively approached using a combination of biological and psychological theory. The triad of behavioral abnormalities in social function, communication, and restricted and repetitive behaviors and interests can be explained psychologically by an impaired capacity for empathizing, or modeling the mental states governing the behavior of people, along with a superior capacity for systemizing, or inferring the rules governing the behavior of objects. This empathizing-systemizing theory explains other psychological models such as impairments of executive function or central coherence, and may have a neurobiological basis in abnormally low activity of brain regions subserving social cognition, along with abnormally high activity of regions subserving lower-level, perceptual processing—a pattern that may result from a skewed balance of local versus long-range functional connectivity.

Contents

INTRODUCTION 110
EMPATHIZING AND
    SYSTEMIZING 110
EXECUTIVE FUNCTION 111
CENTRAL COHERENCE 111
THE SOCIAL BRAIN 112
AUTISM AS A
    DEVELOPMENTAL END
    POINT 113
THE ANALYTICAL BRAIN 114
A BASIS IN NEURAL
    CONNECTIVITY 116
ABNORMAL EARLY BRAIN
    DEVELOPMENT 117
GENETICS AND THE BROADER
    AUTISM PHENOTYPE 118

INTRODUCTION

Autism is diagnosed when a child or adult has abnormalities in a triad of behavioral domains: social development, communication, and repetitive behaviors and obsessive interests (Am. Psychol. Assoc. 1994, World Health Organ. 1994). Autism can occur at any point on the intelligence quotient (IQ) continuum, and IQ (Rutter 1978) and level of language function by age six (Szatmari et al. 2003) are strong predictors of clinical outcome. Whereas autism *per se* is invariably accompanied by language delay, people with Asperger syndrome (AS) share with autism the social impairments and restricted behaviors, but without serious language impairment, and with verbal and nonverbal IQ in the average range or above (Wing 1981). This review introduces the main cognitive theories of autism and relates these theories to key neurobiological findings.

EMPATHIZING AND SYSTEMIZING

The empathizing-systemizing theory of autism (Baron-Cohen 2002) proposes that autism spectrum conditions involve deficits in the normal process of empathy, relative to mental age. These deficits can occur by degrees. The term empathizing encompasses a range of other terms: theory of mind, mind-reading, empathy, and taking the intentional stance (Dennett 1987). Empathy comprises two major elements: (*a*) attribution of mental states to oneself and others, as a natural way to make sense of the actions of agents (Baron-Cohen 1994, Leslie 1995, Premack 1990); and (*b*) emotional reactions that are appropriate to others' mental states.

Since the first test of mind-blindness was administered to children with autism (Baron-Cohen et al. 1985), more than thirty experimental tests have been developed. The vast majority of these tests have revealed profound impairments in the development of empathizing (Baron-Cohen 1995, Baron-Cohen et al. 1993). The residual effects of such impairments can be subtle but nevertheless significant; some children and adults with AS, for example, perform at ceiling on simple tests of inference of others' beliefs but are impaired at inferring others' complex emotions (Baron-Cohen et al. 2001). This deficit in empathizing is thought to underlie the difficulties that such children experience in social and communicative development (Baron-Cohen 1988, Tager-Flusberg 1993), and in the imagination of others' minds (Baron-Cohen 1987, Leslie 1987). We can think of these symptoms as the triad of deficits.

Although autism is most often conceptualized as a syndrome of deficits, its altered developmental emphases can also lead to remarkable cognitive strengths. The pattern of cognitive superiorities found in autism can be explained by the concept of systemizing: the drive to analyze objects and events to understand their structure and to predict their future behavior. Systems are ubiquitous in the environment: technical systems (such as machines and tools), natural systems (such as biological and geographical phenomena), and abstract systems (such as mathematics or computer programs). We make sense of such systems by observing the regularities in their behavior and inferring the rules that govern the system via an analysis of input-operation-output relationships (Baron-Cohen 2002). The empathizing-systemizing (E-S) theory holds that alongside deficits in empathizing, in autism systemizing is either intact or superior (Baron-Cohen et al. 2002). Several studies indicate that systemizing in autism is at least in line with mental age, or superior (Baron-Cohen et al. 1986, 2003; Lawson et al. 2004). Systemizing may underlie a different set of behavioral features in autism that we refer to as the triad of strengths (**Figure 1**).

EXECUTIVE FUNCTION

People with autism spectrum conditions show unusually strong repetitive behaviors, a desire for routines, and a need for sameness. One cognitive account of this aspect of the syndrome is the executive dysfunction theory (Ozonoff et al. 1991), which posits that autism involves a form of frontal lobe pathology leading to perseveration or inability to shift focus. Although evidence for such executive deficits does exist (Pennington et al. 1997, Russell 1997), the high variance in measures of executive function in autism spectrum conditions (Liss et al. 2001), along with the lack of correlation between measures of executive function and measures of reciprocal social interaction and repetitive behaviors (Joseph & Tager-Flusberg 2004), suggests that executive dysfunction is unlikely to be a core feature of autism spectrum conditions.

The executive account has also traditionally ignored the content of repetitive behaviors. The E-S theory in contrast draws attention to the fact that much repetitive behavior involves the child's obsessional or strong interests, the foci of which cluster in the domain of strongly regular systems (Baron-Cohen & Wheelwright 1999). Rather than primary executive dysfunction, these behaviors may reflect an unusually strong interest in systems.

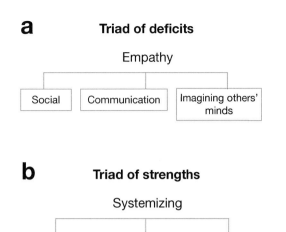

Figure 1
Complementary triads of deficits and strengths in autism.

CENTRAL COHERENCE

The concept of weak central coherence (CC) (Frith 1989, Happé 1996) refers to an abnormally weak tendency to bind local details into global percepts. Weak central coherence in autism has been demonstrated in the context of superior performance on visuomotor tasks such as the Embedded Figures Test (EFT) (Jolliffe & Baron-Cohen 1997, Shah & Frith 1983), the Wechsler Block Design subtest (Shah & Frith 1993), tasks of visual discrimination (Plaisted et al. 1998a) and visual search (Plaisted et al. 1998b, O'Riordan et al. 2001), as well as impaired performance on more abstract tasks such as arranging sentences to form a coherent context (Jolliffe & Baron-Cohen 2000). The general pattern is one of stronger-than-normal segmentation of stimuli and attention to detail within stimuli.

The CC theory thus predicts that people with autism spectrum conditions will perform best on (and, by implication, will be most driven by) tasks and occupations in which piecemeal analysis of individual details is an asset, whereas the E-S theory predicts that people with autism spectrum conditions will be most driven by tasks and occupations that involve analysis of rule-based systems. To a great extent, these predictions overlap:

Systemizing demands excellent attention to detail to isolate parameters that may then be tested individually for their effects on the system's output.

However, differences in theoretical predictions arise when one considers the case of complex systems in which manipulations of inputs produce widespread effects on outputs, or in which differences in outputs are evoked by complex interactions among widely separated inputs. The CC theory, taken by itself, predicts that people with autism will fail to learn such systems because mastering them requires a global view of the interrelations between large sets of inputs and outputs, as opposed to simple systems that can be understood in terms of relations between one or a few inputs and outputs. Conversely the E-S theory, taken by itself, predicts that (relative to their mental age) people with autism will be able to learn how any sort of regular system works, regardless of its complexity, so long as it is highly lawful.

A third prediction arises, though, if one considers E-S and CC not as mutually exclusive explanations of autistic behavior, but as complementary ones that can be developmentally unified. Specifically, the attention to detail described by weak central coherence may be one of the earliest manifestations of a strong drive toward systemizing, or vice versa, interest in systemizing may arise as a consequence of attention to detail. As cognitive capacities mature, strong "systemizers" may begin to apply a sort of engineering methodology, in which even complex systems are understood by successive local observations in which one input at a time is manipulated while all others are held constant, and effects on the outputs are observed in a similarly sequential manner. Thus the ultimate effect of the cognitive style described as weak central coherence may be, at least in high-functioning cases of autism, not a lack of ability to understand global relationships but rather a difference in the process by which global relationships are established. Experimental comparison of the ability to make inferences about complex systems, between people with and without autism spectrum conditions and across different stages of development or levels of functioning, may lead to the recognition of E-S as an elaboration of the CC model, one that may make more precise and more accurate predictions about the behavior of people with autism when confronted with complex systems.

Although both central coherence and systemizing are useful psychological models to explain many aspects of autistic behavior, a complete explanation of autism will require that these psychological models be joined with neurobiological substrates—a process complicated by the fact that neither capacity is likely to be atomic in neurobiological terms.

THE SOCIAL BRAIN

A neural basis of empathy has built on a model first proposed by Brothers (1990). She suggested, on the basis of animal lesion studies (Kling & Brothers 1992), single-cell recording studies (Brothers et al. 1990), and neurological studies, that social intelligence was a function of three regions: the amygdala, orbitofrontal and medial frontal cortices, and superior temporal sulcus and gyrus. Together, she termed these interacting regions the social brain (**Figure 2**). It is perhaps no coincidence that abnormalities in autism have been found in the amygdala, in orbito-frontal cortex, and in medial frontal cortex.

Four lines of evidence converge on the hypothesis of an amygdala deficit in autism (Baron-Cohen et al. 2000). Histopathologically, cell packing density in the amygdala is increased in autism (Bauman & Kemper 1985, 1994). Behaviorally, people with autism show a similar pattern of deficits to those seen in patients with amygdala lesions (Adolphs et al. 2001, Howard et al. 2000). In terms of gross anatomy, magnetic resonance imaging (MRI) morphometry suggests abnormal volume of the amygdala, although there is disagreement as to whether amygdala volume in autism may be reduced (Abell et al. 1999, Aylward et al.

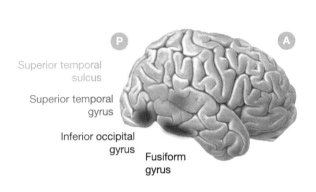
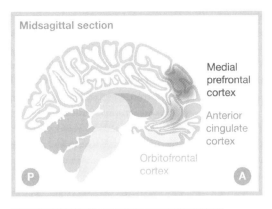
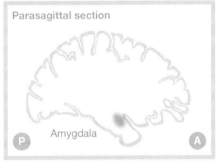

Figure 2
The social brain. Medial and inferior frontal and superior temporal cortices, along with the amygdala, form a network of brain regions that implement computations relevant to social processes. Perceptual inputs to these social computations may arise in part from regions in the fusiform gyrus and from the adjacent inferior occipital gyrus that activate in response to faces. This social computational network has been implicated in autism. (Adapted from Talairach & Tournoux 1998 and from an illustration by C. Ashwin.)

1999, Pierce et al. 2001) or enlarged (Howard et al. 2000, Sparks et al. 2002). Physiologically, adults with autism spectrum conditions manifest abnormally low activation of the amygdala during tasks of inferring emotion from pictures of the eyes (Baron-Cohen et al. 1999) or of the whole face (Wang et al. 2004), and during passive processing of facial expressions of unfamiliar faces (Critchley et al. 2000, Pierce et al. 2001) but not familiar faces (Pierce et al. 2004). In other areas of the social brain, reduced activity has been found in left medial frontal cortex during an empathizing (theory of mind) task (Happé et al. 1996), in orbitofrontal cortex during recognition of mental state words (Baron-Cohen et al. 1994), and in superior temporal sulcus during passive listening to speech sounds as compared with nonspeech sounds (Gervais et al. 2004).

AUTISM AS A DEVELOPMENTAL END POINT

These studies of the social brain have shown that by the time the autistic brain is fully developed, key processes and structures associated with social cognition are abnormal. Though this correspondence between abnormal brain function and behavioral impairments may seem a compelling basis for a neurobiological theory of autism as a deficit of empathizing, in the analysis of developmental conditions it is crucial to realize that the symptoms that are most diagnostically significant and clinically debilitating are not necessarily the most etiologically primary (Belmonte 2004a). Primary dysfunctions can be masked by the evolution of compensatory processing strategies that normalize behavior (Rubia 2002), and also by the induction of activity-dependent secondary dysfunctions (Akshoomoff et al. 2002, Courchesne et al. 1994a), which disrupt behavior in new ways.

A tacit assumption in almost all functional imaging studies, historically, has been that the behavioral or cognitive capacity of interest can be localized to one or a few discrete brain structures, just as clinical impairments in cases of acquired brain disorders can be associated with anatomically circumscribed lesions. Although this lesion model is experimentally convenient, it is likely to be inappropriate to the study of developmental conditions in which dysfunctions are likely to comprise not single regions but complex networks of anatomically and functionally distant areas (Johnson et al. 2002). Localized abnormalities therefore may provide some insight into developmental end points but not directly into developmental processes. The eventual understanding of autism will hinge on our accepting the classification "developmental disorder" as one of not merely taxonomic but also etiologic significance.

THE ANALYTICAL BRAIN

A more complete picture of brain function and dysfunction in autism can be constructed if one examines not only the deficits associated with impaired empathizing but also the more subtle cognitive differences and even superiorities associated with strong systemizing. For instance, functional imaging during performance of the Embedded Figures Test—on which behavioral performance of people with autism is superior to normal—reveals unusually high activation in ventral occipital areas and abnormally low activation in prefrontal and parietal areas (Ring et al. 1999).

This theme of abnormally high activation in unimodal or low-level processing regions alongside low activation in frontal and other integrative regions (**Figure 3**) recurs in findings of heightened activity during face processing in peristriate cortex (Critchley et al. 2000), inferior temporal gyrus (Schultz et al. 2000), medial occipital gyrus and superior parietal lobule (Hubl et al. 2003), precuneus (Wang et al. 2004), and other areas outside the fusiform face area (Pierce et al. 2001); by comparison, fusiform activity is abnormally low. Similarly, a visual attention task evokes heightened activity in ventral occipital cortex and abnormally low activations in parietal and prefrontal cortices (Belmonte & Yurgelun-Todd 2003). In a sentence processing task, activation is greater than normal in Wernicke's area and less than normal in Broca's area, which suggests that processing is enhanced at the level of single words and impoverished at the level of sentential context (Just et al. 2004). In addition, activity in superior temporal gyrus during inference of mental state from pictures of eyes is heightened (Baron-Cohen et al. 1999), and connectivity between extrastriate and prefrontal and temporal cortices during attribution of mental states from movements of animated shapes is weakened (Castelli et al. 2002), while prefrontal and medial temporal activations are abnormally low. Cardiovascular, neuroendocrine, and neurochemical indices of arousal in novel and stressful situations are consistent with this idea of abnormal excitability or arousal (Hirstein et al. 2001, Tordjman et al. 1997).

Perhaps as a consequence of such high excitability in response to both relevant and irrelevant stimuli, neural response as a function of selective attention is abnormally modulated. In adults with autism, the P1 evoked potential is either abnormally augmented in response to stimuli at the attended location, or abnormally generalized to stimuli distant from the attended location (Townsend & Courchesne 1994). During shifts of attention between hemifields, the normal, spatiotopically selective augmentation of the visual steady-state evoked potential is absent, and instead both hemispheres activate indiscriminately during shifts of attention into either hemifield (Belmonte 2000, Belmonte & Yurgelun-Todd 2003). In children with autism, the visual N2 to novel stimuli is augmented during task performance, even when these stimuli are not relevant to the task (Kemner et al. 1994). When a response is required to an auditory stimulus, the P3 in these same children with autism is reported to be abnormally generalized to occipital sites overlying visual processing areas (Kemner et al. 1995). In general, both in children and in adults, perceptual filtering in autism seems to occur in an all-or-none manner; there is little specificity in selecting for the location of the stimulus, for the behavioral relevance of the stimulus, or even for the sensory modality in which the stimulus appears. Although compensatory cognitive mechanisms may substitute for this neural inefficiency during periods when attention is statically focused (Belmonte & Yurgelun-Todd 2003), such mechanisms may be unable to update quickly,

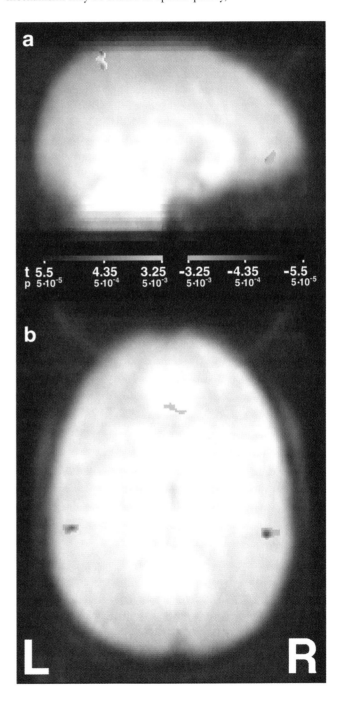

Figure 3

High activation at low levels of processing and low activation at high levels of processing in the autistic brain. (*a*) This functional magnetic resonance image in a parasagittal plane through the right hemisphere ($x = 13$) illustrates hyperactivation in intraparietal sulcus and hypoactivation in middle frontal gyrus in children with autism spectrum disorders, as compared with normal children performing a difficult visual discrimination task (M.K. Belmonte, M. Gomot & S. Baron-Cohen, unpublished data). Simultaneously attending to the centers of two 3×3 arrays of oriented, colored sine-wave gratings, one in the left hemifield and one in the right hemifield, subjects indicated via a forced-choice response whether a target orientation was present in the central stimulus in one hemifield along with a target color in the central stimulus in the other hemifield, while ignoring the surrounding distractors. Stimuli were presented for 167 ms, with a 3 s response period. The functional correlation compares event-related activity when distractors are incongruent to the attended stimuli with event-related activity when distractors are congruent. (*b*) This axial slice ($z = +18$) illustrates abnormally low functional correlations between anterior cingulate cortex [approximate Talairach coordinates (0, 40, 0), not shown on this slice] and bilateral anterior superior temporal gyri in the autism group.

leading to the behavioral symptom of impairment in rapid shifts of attention between modalities (Courchesne et al. 1994b), between spatial locations (Belmonte 2000; Harris et al. 1999; Townsend et al. 1996a,b, 1999; Wainwright-Sharp & Bryson 1993, 1996), and between object features (Courchesne et al. 1994c, Rinehart et al. 2001).

One should consider the possible developmental relationship between abnormal attention and perception in autism and the diagnostic feature of restricted and repetitive behaviors. Normal perception can be thought of as viewing a film in a sort of Cartesian cinema: As the actors and events in the film convey information in multiple modalities and at multiple locations on the screen, the viewer integrates these sources rapidly and effortlessly. If, however, neural properties preclude shifting attention among these many foci and integrating information analyzed by various perceptual and cognitive subsystems, the mind's ability to construct an internal narration of the film's events will be impaired. Consider trying to make sense of a film from only the audio track or only a small corner of the picture.

To make matters worse, films in the Cartesian cinema normally have very short runs because the events and experiences of every day and every hour are unique. A natural response when faced with this difficulty in integrating external events into coherent internal representations would be to replay the same film over and over again, in hopes that over many screenings one will be able to piece together all the details. Considered in the context of abnormal attention and perception, reliance on restricted and repetitive behaviors may thus be an adaptive mechanism (Belmonte et al. 2004b), a way of decreasing environmental variance so that the social world can be reduced more effectively to a regular, predictable, and systemizable set of scripts. Excessive repetition or "need for sameness" may be a sign of hypersystemizing (Baron-Cohen 2004). In this regard, some of the complex behavioral abnormalities in autism can be viewed as the developmental reaction to an atypical perceptual and cognitive style.

A BASIS IN NEURAL CONNECTIVITY

The combination of sensory hyperarousal and abnormal attentional selectivity suggests that autism may involve overconnected neural networks, in which signal is insufficiently differentiated from noise or irrelevant information and in which information capacity is therefore reduced (Belmonte et al. 2004a,b; Rubenstein & Merzenich 2003). This idea is consistent with genetic and neurochemical results, such as linkage to the 15q11–13 region, which contains a cluster of γ-amino-butyric acid (GABA) receptor genes (Buxbaum et al. 2002), low GABA receptor binding in hippocampus (Blatt et al. 2001), and low GABA levels in blood platelets (Rolf et al. 1993), and with the substantial comorbidity of epilepsy with autism (Ballaban-Gil & Tuchman 2000). In addition, high levels of noise in neural networks could explain autistic psychophysical anomalies such as high visual motion coherence thresholds (Milne et al. 2002) and broad tuning of auditory filters (Plaisted et al. 2003). Photomicrographic examination of neurons in several cortical regions suggests a reduction in the size of cortical minicolumns in the autistic brain and an increase in cell dispersion within minicolumns (Casanova et al. 2002a,b)—characteristics that could alter local network properties along with long-range projections (Courchesne & Pierce 2005). Magnetic resonance spectrographic findings of reduced N-acetyl aspartate and other neuronal metabolites (Chugani et al. 1999, Otsuka et al. 1999, Hisaoka et al. 2001, Friedman et al. 2003) are consistent with this finding of more widely dispersed neurons.

This notion of abnormally high neural connectivity in autism seems inconsistent with functional imaging results showing low correlations between levels of activation in widely separated brain regions (Castelli et al. 2002, Just et al. 2004). Indeed, some investigators have proposed that autism involves not a surfeit of connectivity but rather a deficit (Brock et al. 2002, Just et al. 2004)—a dysfunction that would explain and unify psychological theories of deficits in executive function (Ozonoff et al. 1991), complex information processing (Minshew et al. 1997), central coherence (Frith 1989), and empathizing (Baron-Cohen et al. 2002), all of which depend on the rapid and integrated operation of many separate neural systems.

The apparent contradiction between theories of overconnectivity and underconnectivity in autism may arise because of the multiple connotations of the term connectivity. Conceptually we can differentiate local connectivity within neural assemblies from long-range connectivity between functional brain regions. On another axis, we can also separate physical connectivity (associated with synapses and tracts) from computational connectivity (associated with information transfer). Physically, in the autistic brain, high local connectivity may develop in tandem with low long-range connectivity (Just et al. 2004, Belmonte et al. 2004a, Courchesne & Pierce 2005)—perhaps as a consequence of widespread alterations in synapse elimination and/or formation that skew the computationally optimal balance between local and long-range connections (Sporns et al. 2000). A decrease in network entropy due to indiscriminately high connectivity within local networks could yield abnormally low information capacity and may develop in tandem with abnormally low computational connectivity with other regions.

ABNORMAL EARLY BRAIN DEVELOPMENT

How can this hypothesis of abnormally high local connectivity and abnormally low long-range connectivity be tested? Certainly, a large amount of information is likely to be found eventually in large-scale histopathological studies of the autistic brain across many brain regions and developmental periods. Until such detailed microscopic data are available, investigation can be guided by in vivo imaging studies. Both MRI volumetric analysis and simple measures of head circumference indicate that autism involves transient postnatal macrencephaly (Courchesne 2002, Courchesne et al. 2003). Neonates later diagnosed with autism or PDD-NOS (pervasive developmental disorder—not otherwise specified) have normal head circumference; but by 2–4 years of age, 90% of these have larger-than-average brain volumes (Aylward et al. 2002, Carper & Courchesne 2000, Courchesne et al. 2001, Piven et al. 1995, Sparks et al. 2002).

This enlargement occurs specifically within cerebellar and cerebral white matter and cerebral gray matter (Courchesne et al. 2001, Herbert et al. 2003). Anatomical parcellation of white matter reveals that the enlargement occurs in short-range, radiate fibers but not in deeper, long-range fibers (Herbert et al. 2004). Indeed, the posterior corpus callosum in autism is actually smaller than normal (Egaas et al. 1995), and the degree of this callosal hypoplasia correlates with the degree of frontal hyperplasia (Lewis et al. 2004). This compartmental specificity of white matter hyperplasia is consistent with the idea of differential effects on local and long-range connections. Regionally, frontal lobes show the greatest degree of enlargement, and occipital lobes show the least (Carper et al. 2002, Piven 2004); within the frontal lobe, the dorsolateral convexity and medial frontal gyrus—areas that figure prominently in the social brain—show significant overgrowth, whereas precentral gyrus and orbital cortex are not robustly affected (Carper & Courchesne 2005). Thus the cortical areas most affected are precisely those broadly projecting, phylogenetically and ontogenetically late-developing regions that are essential to complex cognitive functions such as social behavior and language. This frontal emphasis in the pattern of overgrowth seems to be the likely cause of anterior and superior shifting of major sulci in the frontal

AS: Asperger syndrome

CC: central coherence

E-S: empathizing-systemizing

EEG: electroencephalogram

GABA: γ-amino-butyric acid

PDD-NOS: Pervasive Developmental Disorder—Not Otherwise Specified

and temporal lobes in autism (Levitt et al. 2003). Cerebellar and cerebral white matter volumes, combined with cerebellar vermis size, can discriminate 95% of toddlers with autism from normal controls and can predict whether a child with autism will be high or low functioning (Courchesne et al. 2001).

Anatomical findings outside the cerebrum also may bear on the issue of connectivity. The cerebellum, in particular, is one of the most consistently abnormal structures in the autistic brain (Courchesne 1997, Hashimoto et al. 1995). The number of Purkinje cells in cerebellar cortex is abnormally low (Bauman & Kemper 1985, 1994; Ritvo et al. 1986; Williams et al. 1980). Purkinje cells are the sole inhibitory influence on the deep cerebellar nuclei, which project via the thalamus to most of the cerebral cortex. Absence of Purkinje cell output could therefore induce abnormal excitability in cerebro-ponto-cerebellar-thalamic control loops (Courchesne 1997), increasing the level of noise in cortical networks. Interestingly, people with autism show abnormally low activation of cerebellar cortex in a visual vigilance task (Allen & Courchesne 2003), but abnormally high activation during a purely motor task of self-paced button-pressing (Allen et al. 2004, Allen & Courchesne 2003), and the degree of abnormally high cerebellar motor activation correlates with the anatomical deficit in cerebellar volume (Allen et al. 2004).

GENETICS AND THE BROADER AUTISM PHENOTYPE

Ultimately the cognitive and neural abnormalities in autism spectrum conditions are likely to be due to genetic factors. The sibling recurrence risk for autism is approximately 4.5%, or a tenfold increase over general population rates (Jorde et al. 1991). An epidemiologic study of same-sex autistic twins found that 60% of monozygotic pairs were concordant for autism versus 0% of dizygotic pairs (Bailey et al. 1995). When a broader phenotype (of related cognitive or social abnormalities) was considered, 92% of monozygotic pairs were concordant versus 10% of dizygotic pairs. The high concordance in monozygotic twins indicates a high degree of genetic influence, and the risk to a monozygotic co-twin can be estimated at more than 200 times the general population rate. The cognitive profile in siblings of people with autism includes superior spatial and verbal span, poor set-shifting, poor planning, and poor verbal fluency (Hughes et al. 1999). Parents of children with autism spectrum conditions perform above normal on the Embedded Figures Test (Baron-Cohen & Hammer 1997) and other tasks that demand strong systemizing skills (Happé et al. 2001), but they perform below normal when inferring mental state from facial expression (Baron-Cohen & Hammer 1997), a task that requires empathizing. In addition, parents show impairments in pragmatic language (Folstein et al. 1999) and executive function (Hughes et al. 1997) and display autistic personality characteristics such as rigidity, aloofness, and anxiety (Piven et al. 1997). In many first-degree relatives, performance IQ is lower than verbal IQ (Folstein et al. 1999, Fombonne et al. 1997, Piven & Palmer 1997) owing to impairment on Picture Arrangement and Picture Completion, both of which demand attention to global, contextual information.

These subtle characteristics of the broader autism phenotype are visible in the systemizing skills that relatives tend to develop: Occupations in engineering are overrepresented in the fathers and grandfathers of people with autism (Baron-Cohen et al. 1997), and conversely, the incidence of autism is increased in the families of engineers, mathematicians, and physicists (Baron-Cohen et al. 1998). These subtle abnormalities in nonautistic family members may be particularly informative as to the etiology of autism because they may represent primary differences that have not been masked by the full syndrome of autism (Belmonte et al. 2004b).

Although investigators have not yet agreed on the genetic loci involved in autism, two regions stand out in several (but not all) studies. These are 15q11–13, near the $GABA_A\beta_3$ receptor subunit gene (*GABRB3*), and 17q11.2, near the serotonin transporter gene (*SLC6A4*). The latter is of interest because of reports of elevated serotonin (5HT) levels of platelets in autism (Anderson 1990). Serotonin innervates the limbic system and so plausibly plays a role in emotion recognition and empathy. In mice, mothers homozygous for *GABRB3* knockout fail to engage in normal nurturing behavior and have epileptiform electroencephalograms (EEG) (Homanics et al. 1997, DeLorey et al. 1998), suggesting abnormal cortical excitability. A recent association of autism with the X-linked neuroligin genes *NLGN3* and *NLGN4* (Jamain et al. 2003) is consistent with the notion of abnormal neural connectivity, as well as potentially helping to explain the 4:1 ratio of males to females affected by autism. Immunogenetic and immunological abnormalities and infectious etiologies in autism (van Gent et al. 1997) are another domain likely to have an impact on neural connectivity because normal axonal development and synaptic pruning depend on expression of major histocompatibility complex proteins (Huh et al. 2000). The overlap in behavioral symptoms between autism and Fragile X syndrome, a disorder in which synaptic structure is affected, is also consistent with an abnormality of connectivity at the neural level (Belmonte et al. 2004ab). Although several reviews of the genetics of autism are available (Cook 2001, Folstein & Rosen-Sheidley 2001, Lauritsen & Ewald 2001), this is a rapidly developing field.

Future progress in autism research will depend on identifying the ways in which large numbers of genetic biases and environmental factors converge to affect neural structure and dynamics, and the ways in which such abnormalities at the neural level diverge through activity-dependent development to produce an end state in which empathizing abilities are impaired and systemizing abilities are augmented. Such a unified understanding of the psychobiology of autism will offer targets for intervention on many levels.

ACKNOWLEDGMENTS

During the period of this work, the authors were supported by the Medical Research Council (United Kingdom), the National Alliance for Autism Research, the Nancy Lurie Marks Family Foundation, and Cure Autism Now. Portions of this paper are based on Baron-Cohen (2004).

LITERATURE CITED

Abell F, Krams M, Ashburner J, Passingham R, Friston K, et al. 1999. The neuroanatomy of autism: a voxel-based whole brain analysis of structural scans. *J. Cogn. Neurosci.* 10:1647–51

Adolphs R, Sears L, Piven J. 2001. Abnormal processing of social information from faces in autism. *J. Cogn. Neurosci.* 13:232–40

Akshoomoff NA, Pierce K, Courchesne E. 2002. The neurobiological basis of autism from a developmental perspective. *Dev. Psychopathol.* 14:613–34

Allen G, Courchesne E. 2003. Differential effects of developmental cerebellar abnormality on cognitive and motor functions in the cerebellum: an fMRI study of autism. *Am. J. Psychiatry* 160:262–73

Allen G, Müller R-A, Courchesne E. 2004. Cerebellar function in autism: functional magnetic resonance image activation during a simple motor task. *Biol. Psychiatry* 56:269–78

Am. Psychiatr. Assoc. 1994. *Diagnostic and Statistical Manual of Mental Disorders*. Washington, DC: Am. Psychiatr. Assoc. 4th ed.

Anderson GM, Horne WC, Chatterjee D, Cohen DJ. 1990. The hyperserotonemia of autism. *Ann. NY Acad. Sci.* 600:331–40

Aylward EH, Minshew NJ, Field K, Sparks BF, Singh N. 2002. Effects of age on brain volume and head circumference in autism. *Neurology* 59:175–83

Aylward EH, Minshew NJ, Goldstein G, Honeycutt NA, Augustine AM, et al. 1999. MRI volumes of amygdala and hippocampus in non-mentally retarded autistic adolescents and adults. *Neurology* 53:2145–50

Bailey A, Le Couteur A, Gottesman I, Bolton P, Simmonoff E, et al. 1995. Autism as a strongly genetic disorder: evidence from a British twin study. *Psychol. Med.* 25:63–77

Ballaban-Gil K, Tuchman R. 2000. Epilepsy and epileptiform EEG: association with autism and language disorders. *Ment. Retard. Dev. Disabil. Res. Rev.* 6:300–8

Baron-Cohen S. 1987. Autism and symbolic play. *Br. J. Dev. Psychol.* 5:139–48

Baron-Cohen S. 1988. Social and pragmatic deficits in autism: cognitive or affective? *J. Autism Dev. Disord.* 18:379–402

Baron-Cohen S. 1994. How to build a baby that can read minds: cognitive mechanisms in mindreading. *Cah. Psychol. Cogn./Curr. Psychol. Cogn.* 13:513–52

Baron-Cohen S. 1995. *Mindblindness: An Essay on Autism and Theory of Mind*. Cambridge, MA: MIT Press

Baron-Cohen S. 2002. The extreme male brain theory of autism. *Trends Cogn. Sci.* 6:248–54

Baron-Cohen S. 2004. The cognitive neuroscience of autism. *J. Neurol. Neurosurg. Psychiatry* 75:945–48

Baron-Cohen S, Bolton P, Wheelwright S, Scahill V, Short L, et al. 1998. Autism occurs more often in families of physicists, engineers, and mathematicians. *Autism* 2:296–301

Baron-Cohen S, Hammer J. 1997. Parents of children with Asperger syndrome: What is the cognitive phenotype? *J. Cogn. Neurosci.* 9:548–54

Baron-Cohen S, Leslie AM, Frith U. 1985. Does the autistic child have a 'theory of mind'? *Cognition* 21:37–46

Baron-Cohen S, Leslie AM, Frith U. 1986. Mechanical, behavioural and intentional understanding of picture stories in autistic children. *Br. J. Dev. Psychol.* 4:113–25

Baron-Cohen S, Richler J, Bisarya D, Gurunathan N, Wheelwright S. 2003. The Systemising Quotient (SQ): an investigation of adults with Asperger syndrome or high functioning autism and normal sex differences. *Phil. Trans. R. Soc. B* 358:361–74

Baron-Cohen S, Ring H, Bullmore E, Wheelwright S, Ashwin C, Williams S. 2000. The amygdala theory of autism. *Neurosci. Biobehav. Rev.* 24:355–64

Baron-Cohen S, Ring H, Moriarty J, Shmitz P, Costa D, Ell P. 1994. Recognition of mental state terms. Clinical findings in children with autism and a functional neuroimaging study of normal adults. *Br. J. Psychiatry* 165:640–49

Baron-Cohen S, Ring H, Wheelwright S, Bullmore ET, Brammer MJ, et al. 1999. Social intelligence in the normal and autistic brain: an fMRI study. *Eur. J. Neurosci.* 11:1891–98

Baron-Cohen S, Tager-Flusberg H, Cohen D, eds. 1993. *Understanding Other Minds: Perspectives from Autism*. Oxford: Oxford Univ. Press

Baron-Cohen S, Wheelwright S. 1999. Obsessions in children with autism or Asperger Syndrome: a content analysis in terms of core domains of cognition. *Br. J. Psychiatry* 175:484–90

Baron-Cohen S, Wheelwright S, Griffin R, Lawson J, Hill J. 2002. The exact mind: empathising and systemising in autism spectrum conditions. In *Handbook of Cognitive Development*, ed. U Goswami, pp. 491–508. Oxford: Blackwell

Baron-Cohen S, Wheelwright S, Hill J, Raste Y, Plumb I. 2001. The 'Reading the Mind in the Eyes' test revised version: a study with normal adults, and adults with Asperger Syndrome or High-Functioning autism. *J. Child Psychol. Psychiatry* 42:241–52

Baron-Cohen S, Wheelwright S, Stott C, Bolton P, Goodyer I. 1997. Is there a link between engineering and autism? *Autism* 1:153–63

Bauman ML, Kemper TL. 1985. Histoanatomic observation of the brain in early infantile autism. *Neurology* 35:866–74

Bauman ML, Kemper TL. 1994. Neuroanatomic observations of the brain in autism. In *The Neurobiology of Autism*, ed. ML Bauman, TL Kemper, pp. 119–45. Baltimore, MD: John Hopkins Univ. Press

Belmonte MK. 2000. Abnormal attention in autism shown by steady-state visual evoked potentials. *Autism* 4:269–85

Belmonte MK, Allen G, Beckel-Mitchener A, Boulanger LM, Carper RA, Webb SJ. 2004a. Autism and abnormal development of brain connectivity. *J. Neurosci.* 24:9228–31

Belmonte MK, Cook Jr EH, Anderson GM, Rubenstein JLR, Greenough WT, et al. 2004b. Autism as a disorder of neural information processing: directions for research and targets for therapy. *Mol. Psychiatry* 9:646–63; Unabridged ed. http://www.cureautismnow.org/media/3915.pdf

Belmonte MK, Yurgelun-Todd DA. 2003. Functional anatomy of impaired selective attention and compensatory processing in autism. *Cogn. Brain Res.* 17:651–64

Blatt GJ, Fitzgerald CM, Guptill JT, Booker AB, Kemper TL, Bauman ML. 2001. Density and distribution of hippocampal neurotransmitter receptors in autism: an autoradiographic study. *J. Autism Dev. Disord.* 31:537–43

Brock J, Brown CC, Boucher J, Rippon G. 2002. The temporal binding deficit hypothesis of autism. *Dev. Psychopathol.* 14:209–24

Brothers L. 1990. The social brain: a project for integrating primate behaviour and neurophysiology in a new domain. *Concepts Neurosci.* 1:27–51

Brothers L, Ring B, Kling A. 1990. Responses of neurons in the macaque amygdala to complex social stimuli. *Behav. Brain Res.* 41:199–213

Buxbaum JD, Silverman JM, Smith CJ, Greenberg DA, Kilifarski M, et al. 2002. Association between a GABRB3 polymorphism and autism. *Mol. Psychiatry* 7:311–16

Carper RA, Courchesne E. 2000. Inverse correlation between frontal lobe and cerebellum sizes in children with autism. *Brain* 123:836–44

Carper RA, Courchesne E. 2005. Localized enlargement of the frontal cortex in early autism. *Biol. Psychiatry*. 57:126–33

Carper RA, Moses P, Tigue ZD, Courchesne E. 2002. Cerebral lobes in autism: early hyperplasia and abnormal age effects. *NeuroImage* 16:1038–51

Casanova MF, Buxhoeveden DP, Switala AE, Roy E. 2002a. Asperger's syndrome and cortical neuropathology. *J. Child Neurol.* 17:142–45

Casanova MF, Buxhoeveden DP, Switala AE, Roy E. 2002b. Minicolumnar pathology in autism. *Neurology* 58:428–32

Castelli F, Frith C, Happé F, Frith U. 2002. Autism, Asperger syndrome and brain mechanisms for the attribution of mental states to animated shapes. *Brain* 125:1839–49

Chugani DC, Sundram BS, Behen M, Lee ML, Moore GJ. 1999. Evidence of altered energy metabolism in autistic children. *Prog. Neuropsychopharmacol. Biol. Psychiatry* 23:635–41

Cook Jr EH. 2001. Genetics of autism. *Child Adolesc. Psychiatr. Clin. N. Am.* 10:333–50

Courchesne E. 1997. Brainstem, cerebellar and limbic neuroanatomical abnormalities in autism. *Curr. Opin. Neurobiol.* 7:269–78

Courchesne E. 2002. Abnormal early brain development in autism. *Mol. Psychiatry* 7:21–23

Courchesne E, Carper RA, Akshoomoff NA. 2003. Evidence of brain overgrowth in the first year of life in autism. *JAMA* 290:337–44

Courchesne E, Chisum H, Townsend J. 1994a. Neural activity-dependent brain changes in development: implications for psychopathology. *Dev. Psychopathol.* 6:697–722

Courchesne E, Karns CM, Davis HR, Ziccardi R, Carper RA, et al. 2001. Unusual brain growth patterns in early life of patients with autistic disorder. *Neurology* 57:245–54

Courchesne E, Pierce K. 2005. Brain overgrowth in autism during a critical time in development: implications for frontal pyramidal neuron and interneuron development and connectivity. *Int. J. Dev. Neurosci.* 23:153–70

Courchesne E, Townsend J, Akshoomof NA, Saitoh O, Yeung-Courchesne R, et al. 1994b. Impairment in shifting attention in autistic and cerebellar patients. *Behav. Neurosci.* 108:848–65

Courchesne E, Townsend J, Akshoomoff NA, Yeung-Courchesne R, Lincoln AJ, et al. 1994c. A new finding: impairment in shifting attention in autistic and cerebellar patients. In *Atypical Cognitive Deficits in Developmental Disorders: Implications for Brain Function*, ed. SH Broman, J Grafman, pp. 101–37. Hillsdale, NJ: Erlbaum

Critchley HD, Daly EM, Bullmore ET, Williams SCR, Van Amelsvoort T, et al. 2000. The functional neuroanatomy of social behaviour: changes in cerebral blood flow when people with autistic disorder process facial expressions. *Brain* 123:2203–12

DeLorey TM, Handforth A, Anagnostaras SG, Homanics GE, Minassian BA, et al. 1998. Mice lacking the beta-3 subunit of the $GABA_A$ receptor have the epilepsy phenotype and many of the behavioral characteristics of Angelman syndrome. *J. Neurosci.* 18:8505–14

Dennett D. 1987. *The Intentional Stance*. Cambridge, MA: MIT Press

Egaas B, Courchesne E, Saitoh O. 1995. Reduced size of corpus callosum in autism. *Arch. Neurol.* 52:794–801

Folstein SE, Gilman SE, Landa R, Hein J, Santangelo SL, et al. 1999. Predictors of cognitive test patterns in autism families. *J. Child Psychol. Psychiatry* 40:1117–28

Folstein SE, Rosen-Sheidley B. 2001. Genetics of autism: complex aetiology for a heterogeneous disorder. *Nat. Rev. Genet.* 2:943–55

Fombonne E, Bolton P, Prior J, Jordan H, Rutter M. 1997. A family study of autism: cognitive patterns and levels in parents and siblings. *J. Child Psychol. Psychiatry* 38:667–83

Friedman SD, Shaw DW, Artru AA, Richards TL, Gardner J, et al. 2003. Regional brain chemical alterations in young children with autism spectrum disorder. *Neurology* 60:100–7

Frith U. 1989. *Autism: Explaining the Enigma*. Oxford: Blackwell

Gervais H, Belin P, Boddaert N, Leboyer M, Coez A, et al. 2004. Abnormal cortical voice processing in autism. *Nat. Neurosci.* 7:801–2

Happé F. 1996. Studying weak central coherence at low levels: children with autism do not succumb to visual illusions. A research note. *J. Child Psychol. Psychiatry* 37:873–77

Happé F, Briskman J, Frith U. 2001. Exploring the cognitive phenotype of autism: weak "central coherence" in parents and siblings of children with autism: I. Experimental tests. *J. Child Psychol. Psychiatry* 42:299–307

Happé F, Ehlers S, Fletcher P, Frith U, Johansson M, et al. 1996. Theory of mind in the brain. Evidence from a PET scan study of Asperger syndrome. *NeuroReport* 8:197–201

Harris NS, Courchesne E, Townsend J, Carper RA, Lord C. 1999. Neuroanatomic contributions to slowed orienting of attention in children with autism. *Cogn. Brain Res.* 8:61–71

Hashimoto T, Tayama M, Murakawa K, Yoshimoto T, Miyazaki M, et al. 1995. Development of the brainstem and cerebellum in autistic patients. *J. Autism Dev. Disord.* 25:1–17

Herbert MR, Zeigler DA, Deutsch CK, O'Brien LM, Lange N, et al. 2003. Dissociations of cerebral cortex, subcortical and cerebral white matter volumes in autistic boys. *Brain* 126:1182–92

Herbert MR, Zeigler DA, Makris N, Filipek PA, Kemper TL, et al. 2004. Localization of white matter volume increase in autism and developmental language disorder. *Ann. Neurol.* 55:530–40

Hirstein W, Iversen P, Ramachandran VS. 2001. Autonomic responses of autistic children to people and objects. *Proc. R. Soc. London B* 268:1883–88

Hisaoka S, Harada M, Nishitani H, Mori K. 2001. Regional magnetic resonance spectroscopy of the brain in autistic individuals. *Neuroradiology* 43:496–98

Homanics GE, DeLorey TM, Firestone LL, Quinlan JJ, Handforth A, et al. 1997. Mice devoid of gamma-aminobutyrate type A receptor beta-3 subunit have epilepsy, cleft palate and hypersensitive behavior. *Proc. Natl. Acad. Sci. USA* 94:4143–48

Howard MA, Cowell PE, Boucher J, Broks P, Mayes A, et al. 2000. Convergent neuroanatomical and behavioural evidence of an amygdala hypothesis of autism. *Neuroreport* 11:2931–35

Hubl D, Bölte S, Feineis-Matthews S, Lanfermann H, Federspiel A, et al. 2003. Functional imbalance of visual pathways indicates alternative face processing strategies in autism. *Neurology* 61:1232–37

Hughes C, Leboyer M, Bouvard M. 1997. Executive function in parents of children with autism. *Psychol. Med.* 27:209–20

Hughes C, Plumet MH, Leboyer M. 1999. Towards a cognitive phenotype for autism: increased prevalence of executive dysfunction and superior spatial span amongst siblings of children with autism. *J. Child Psychol. Psychiatry* 40:705–18

Huh GS, Boulanger LM, Du H, Riquelme PA, Brotz TM, Shatz CJ. 2000. Functional requirement for class I MHC in CNS development and plasticity. *Science* 290:2155–59

Jamain S, Quach H, Betancur C, Råstam M, Colineaux C, et al. 2003. Mutations of the X-linked genes encoding neuroligins NLGN3 and NLGN4 are associated with autism. *Nat. Genet.* 34:27–29

Johnson MH, Halit H, Grice SJ, Karmiloff-Smith A. 2002. Neuroimaging of typical and atypical development: a perspective from multiple levels of analysis. *Dev. Psychopathol.* 14:521–36

Jolliffe T, Baron-Cohen S. 1997. Are people with autism or Asperger's syndrome faster than normal on the Embedded Figures Task? *J. Child Psychol. Psychiatry* 38:527–34

Jolliffe T, Baron-Cohen S. 2000. Linguistic processing in high-functioning adults with autism or Asperger syndrome: Can global coherence be achieved? A further test of central coherence theory. *Psychol. Med.* 30:1169–87

Jorde L, Hasstedt S, Ritvo E, Mason-Brothers A, Freeman B, et al. 1991. Complex segregation analysis of autism. *Am. J. Hum. Genet.* 49:932–38

Joseph RM, Tager-Flusberg H. 2004. The relationship of theory of mind and executive functions to symptom type and severity in children with autism. *Dev. Psychopathol.* 16:137–55

Just MA, Cherkassky VL, Keller TA, Minshew NJ. 2004. Cortical activation and synchronization during sentence comprehension in high-functioning autism: evidence of underconnectivity. *Brain* 127:1811–21

Kemner C, Verbaten MN, Cuperus JM, Camfferman G, van Engeland H. 1994. Visual and somatosensory event-related brain potentials in autistic children and three different control groups. EEG Clin. *Neurophysiol.* 92:225–37

Kemner C, Verbaten MN, Cuperus JM, Camfferman G, van Engeland H. 1995. Auditory event-related brain potentials in autistic children and three different control groups. *Biol. Psychiatry* 38:150–65

Kling A, Brothers L. 1992. The amygdala and social behavior. In *Neurobiological Aspects of Emotion, Memory, and Mental Dysfunction*, ed. J Aggleton, pp. 353–77. New York: Wiley

Lauritsen M, Ewald H. 2001. The genetics of autism. *Acta Psychiatr. Scand.* 103:411–27

Lawson J, Baron-Cohen S, Wheelwright S. 2004. Empathising and systemising in adults with and without Asperger syndrome. *J. Autism Dev. Disord.* 34:301–10

Leslie A. 1995. ToMM, ToBy, and Agency: core architecture and domain specificity. In *Domain Specificity in Cognition and Culture*, ed. L Hirschfeld, S Gelman, pp. 119–48. New York: Cambridge Univ. Press

Leslie AM. 1987. Pretence and representation: the origins of "theory of mind." *Psychol. Rev.* 94:412–26

Levitt JG, Blanton RE, Smalley S, Thompson PM, Guthrie D, et al. 2003. Cortical sulcal maps in autism. *Cereb. Cortex* 13:728–35

Lewis JD, Elman JL, Courchesne E. 2004. Patterns of brain growth predict patterns of long-distance connectivity. Program No. 1029.9. *2004 Abstract Viewer/Itinerary Planner*. Washington, DC: Soc. Neurosci.

Liss M, Fein D, Allen D, Dunn M, Feinstein C, et al. 2001. Executive functioning in high-functioning children with autism. *J. Child Psychol. Psychiatry* 42:261–70

Milne E, Swettenham J, Hansen P, Campbell R, Jeffries H, Plaisted K. 2002. High motion coherence thresholds in children with autism. *J. Child Psychol. Psychiatry* 43:255–63

Minshew NJ, Goldstein G, Siegel DJ. 1997. Neuropsychologic functioning in autism: profile of a complex information processing disorder. *J. Int. Neuropsychol. Soc.* 3:303–16

O'Riordan MA, Plaisted KC, Driver J, Baron-Cohen S. 2001. Superior visual search in autism. *J. Exp. Psychol. Hum. Percept. Perform.* 27:719–30

Otsuka H, Harada M, Mori K, Hisaoka S, Nishitani H. 1999. Brain metabolites in the hippocampus-amygdala region and cerebellum in autism: an 1H-MR spectroscopy study. *Neuroradiology* 41:517–19

Ozonoff S, Pennington B, Rogers SJ. 1991. Executive function deficits in high-functioning autistic individuals: relationship to theory of mind. *J. Child Psychol. Psychiatry* 32:1081–85

Pennington B, Rogers S, Bennetto L, Griffith E, Reed D, Shyu V. 1997. Validity test of the executive dysfunction hypothesis of autism. In *Autism as an Executive Disorder*, ed. J Russell, pp. 143–78. Oxford: Oxford Univ. Press

Pierce K, Haist F, Sedaghat F, Courchesne E. 2004. The brain response to personally familiar faces in autism: findings of fusiform activity and beyond. *Brain* 127:2703–16

Pierce K, Müller R-A, Ambrose J, Allen G, Courchesne E. 2001. Face processing occurs outside the fusiform 'face area' in autism; evidence from functional MRI. *Brain* 124:2059–73

Piven J. 2004. *Longitudinal MRI study of 18–35 month olds with autism*. Presented at the Int. Meet. for Autism Res., Sacramento, CA

Piven J, Arndt S, Bailey J, Havercamp S, Andreasen NC, Palmer P. 1995. An MRI study of brain size in autism. *Am. J. Psychiatry* 152:1145–49

Piven J, Palmer P. 1997. Cognitive deficits in parents from multiple-incidence autism families. *J. Child Psychol. Psychiatry* 38:1011–21

Piven J, Palmer P, Landa R, Santangelo S, Jacobi D, Childress D. 1997. Personality and language characteristics in parents from multiple-incidence autism families. *Am. J. Med. Genet.* 74:398–411

Plaisted K, O'Riordan M, Baron-Cohen S. 1998a. Enhanced discrimination of novel, highly similar stimuli by adults with autism during a perceptual learning task. *J. Child Psychol. Psychiatry* 39:765–75

Plaisted K, O'Riordan M, Baron-Cohen S. 1998b. Enhanced visual search for a conjunctive target in autism: a research note. *J. Child Psychol. Psychiatry* 39:777–83

Plaisted K, Saksida L, Alcantara JI, Weisblatt EJL. 2003. Towards an understanding of the mechanisms of weak central coherence effects: experiments in visual configural learning and auditory perception. *Phil. Trans. R. Soc. London B* 358:375–86

Premack D. 1990. The infant's theory of self-propelled objects. *Cognition* 36:1–16

Rinehart NJ, Bradshaw JL, Moss SA, Brereton AV, Tonge BJ. 2001. A deficit in shifting attention present in high-functioning autism but not Asperger's disorder. *Autism* 5:67–80

Ring H, Baron-Cohen S, Williams S, Wheelwright S, Bullmore E, et al. 1999. Cerebral correlates of preserved cognitive skills in autism. A functional MRI study of Embedded Figures task performance. *Brain* 122:1305–15

Ritvo ER, Freeman BJ, Scheibel AB, Doung T, Robinson H, et al. 1986. Lower Purkinje cell counts in the cerebella of four autistic subjects: initial findings of the UCLA-NSAC autopsy research report. *Am. J. Psychiatry* 143:862–66

Rolf LH, Haarmann FY, Grotemeyer KH, Kehrer H. 1993. Serotonin and amino acid content in platelets of autistic children. *Acta Psychiatr. Scand.* 87:312–16

Rubenstein JLR, Merzenich MM. 2003. Model of autism: increased ratio of excitation/inhibition in key neural systems. *Genes Brain Behav.* 2:255–67

Rubia K. 2002. The dynamic approach to neurodevelopmental psychiatric disorders: use of fMRI combined with neuropsychology to elucidate the dynamics of psychiatric disorders, exemplified in ADHD and schizophrenia. *Behav. Brain Res.* 130:47–56

Russell J. 1997. How executive disorders can bring about an inadequate theory of mind. In *Autism as an Executive Disorder*, ed. J Russell, pp. 256–304. Oxford: Oxford Univ. Press

Rutter M. 1978. Language disorder and infantile autism. In *Autism: A Reappraisal of Concepts and Treatment*, ed. M Rutter M, E Schopler, pp. 85–104. New York: Plenum

Schultz R, Gauthier I, Klin A, Fulbright R, Anderson A, et al. 2000. Abnormal ventral temporal cortical activity among individuals with autism and Asperger syndrome during face discrimination. *Arch. Gen. Psychiatry* 57:331–40

Shah A, Frith U. 1983. An islet of ability in autism: a research note. *J. Child Psychol. Psychiatry* 24:613–20

Shah A, Frith U. 1993. Why do autistic individuals show superior performance on the block design test? *J. Child Psychol. Psychiatry* 34:1351–64

Sparks BF, Friedman SD, Shaw DW, Aylward EH, Echelard D, et al. 2002. Brain structural abnormalities in young children with autism spectrum disorder. *Neurology* 59:184–92

Sporns O, Tononi G, Edelman GM. 2000. Theoretical neuroanatomy: relating anatomical and functional connectivity in graphs and cortical connection matrices. *Cereb. Cortex* 10:127–41

Szatmari P, Bryson SE, Boyle MH, Streiner DL, Duku E. 2003. Predictors of outcome among high functioning children with autism and Asperger syndrome. *J. Child Psychol. Psychiatry* 44:520–28

Tager-Flusberg H. 1993. What language reveals about the understanding of minds in children with autism. In *Understanding Other minds: Perspectives from Autism*, ed. S Baron-Cohen, H Tager-Flusberg, DJ Cohen, pp. 138–57. Oxford: Oxford Univ. Press

Talairach J, Tournoux P. 1988. *A Coplanar Stereotactic Atlas of the Human Brain*. Stuttgart, Germany: Thieme Verlag

Tordjman S, Anderson GM, McBride PA, Hertzig ME, Snow ME, et al. 1997. Plasma beta-endorphin, adrenocorticotropin hormone and cortisol in autism. *J. Child Psychol. Psychiatry* 38:705–15

Townsend J, Courchesne E. 1994. Parietal damage and narrow "spotlight" spatial attention. *J. Cogn. Neurosci.* 6:220–32

Townsend J, Courchesne E, Covington J, Westerfield M, Harris NS, et al. 1999. Spatial attention deficits in patients with acquired or developmental cerebellar abnormality. *J. Neurosci.* 19:5632–43

Townsend J, Courchesne E, Egaas B. 1996a. Slowed orienting of covert visual-spatial attention in autism: specific deficits associated with cerebellar and parietal abnormality. *Dev. Psychopathol.* 8:563–84

Townsend J, Singer-Harris N, Courchesne E. 1996b. Visual attention abnormalities in autism: delayed orienting to location. *J. Int. Neurospsychol. Soc.* 2:541–50

van Gent T, Heijnen CJ, Treffers PD. 1997. Autism and the immune system. *J. Child Psychol. Psychiatry* 38:337–49

Wainwright-Sharp JA, Bryson SE. 1993. Visual orienting deficits in high-functioning people with autism. *J. Autism Dev. Disord.* 23:1–13

Wainwright-Sharp JA, Bryson SE. 1996. Visual-spatial orienting in autism. *J. Autism Dev. Disord.* 26:423–38

Wang AT, Dapretto M, Hariri AR, Sigman M, Bookheimer SY. 2004. Neural correlates of facial affect processing in children and adolescents with autism spectrum disorder. *J. Am. Acad. Child Adolesc. Psychiatry* 43:481–90

Williams RS, Hauser SL, Purpura DP, Delong GR, Swisher CN. 1980. Autism and mental retardation: neuropathologic studies performed in four retarded persons with autistic behaviour. *Arch. Neurol.* 37:749–53

Wing L. 1981. Asperger's syndrome: a clinical account. *Psychol. Med.* 11:115–29

World Health Organ. 1994. *International Classification of Diseases*. Geneva, Switz.: World Health Organ. 10th ed.

Axon Retraction and Degeneration in Development and Disease

Liqun Luo[1] and Dennis D.M. O'Leary[2]

[1]Department of Biological Sciences, Neurosciences Program, Stanford University, Stanford, California 94305; email: lluo@stanford.edu

[2]Molecular Neurobiology Laboratory, The Salk Institute, La Jolla, California 92037; email: doleary@salk.edu

Key Words

axon elimination, visual system development, cytoskeletal signaling, Rho GTPases, Wallerian degeneration, neurodegenerative diseases

Abstract

The selective elimination of axons, dendrites, axon and dendrite branches, and synapses, without loss of the parent neurons, occurs during normal development of the nervous system as well as in response to injury or disease in the adult. The widespread developmental phenomena of exuberant axonal projections and synaptic connections require both small-scale and large-scale axon pruning to generate precise adult connectivity, and they provide a mechanism for neural plasticity in the developing and adult nervous system, as well as a mechanism to evolve differences between species in a projection system. Such pruning is also required to remove axonal connections damaged in the adult, to stabilize the affected neural circuits, and to initiate their repair. Pruning occurs through either retraction or degeneration. Here we review examples of these phenomena and consider potential cellular and molecular mechanisms that underlie axon retraction and degeneration and how they might relate to each other in development and disease.

Contents

- INTRODUCTION 128
- AXON ELIMINATION DURING DEVELOPMENT 129
- SMALL-SCALE DEVELOPMENTAL PRUNING OF AXONAL CONNECTIONS 129
  - Synapse Elimination at the Neuromuscular Junction 130
  - Axon Pruning in Development of Eye-Specific Retinal Connections 131
- LARGE-SCALE DEVELOPMENTAL PRUNING OF AXONAL CONNECTIONS 132
  - Target Selection by Interstitial Axon Branching Followed by Axon Elimination 133
  - Axon Elimination to Develop Topographic Maps in a Target 136
  - Phylogenic Differences in Developmental Axon Elimination May Generate Species-Unique Axon Projection Patterns 138
  - Neuronal Remodeling During Insect Metamorphosis 139
- AXON DEGENERATION IN NEUROLOGICAL DISEASES 139
  - Wallerian Degeneration 139
  - Dying Back Degeneration 139
  - Neurodegenerative Diseases 140
- MECHANISMS OF AXON RETRACTION 140
  - Cytoskeleton 140
  - Intracellular Signaling Pathways 141
  - Extracellular Cues and Their Receptors 142
- MECHANISMS OF AXON DEGENERATION 142
  - Degeneration of *Drosophila* Mushroom Body γ Neuron Axons During Development 142
  - Mechanisms of Wallerian Degeneration 143
  - Similarities Between MB γ Axon Pruning and Wallerian Degeneration 146
  - Many Degeneration Diseases May Share Similarities with Wallerian Degeneration 147
  - Potential Connection with Axonal Transport 148
- SUMMARY AND PERSPECTIVE 148
- NOTE ADDED IN PROOF 149

INTRODUCTION

The selective elimination of axons, dendrites, and synaptic connections without death of the parent neurons is a crucial common theme in many biological processes central to normal neural function. Development of precise axonal connectivity between appropriate sets of neurons or to peripheral targets such as muscles and organs requires the elimination of exuberant neuronal processes. Maintenance of functional circuits involves plasticity in their connectivity to allow for growth, learning, and memory. Finally, responses of axons to injury or disease that remove or stabilize affected circuits can promote their repair.

In both developmental and adult axon elimination, the phenomena can be divided into two general classes on the basis of the relative scale of the events: small-scale events such as elimination of synaptic connections and the local pruning of an axonal or dendritic arbor, and large-scale events such as the elimination of a significant length of the primary axon and major axon collaterals. The first section of this article describes examples of these phenomena during development and

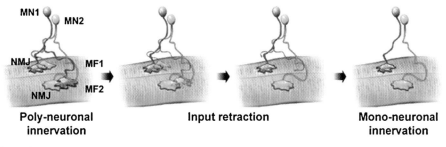

Figure 1

Small-scale axon elimination during development of the vertebrate nervous system. Small-scale axon elimination exemplified by synapse elimination at a "generic" vertebrate neuromuscular junction (NMJ). Input from a motor neuron (MN) to a muscle fiber (MF) forms a neuromuscular junction. In an adult, a typical neuromuscular junction is innervated by an axon from a single motor neuron. However, during development, a junction is multiply innervated by axons originating from two or more motor neurons. Competition between the inputs results in progressive changes in synaptic structure and strength, and the "loser" eventually retracts its terminal ending and the branch that gives rise to it.

mechanisms that selectively pattern developmental axon elimination. The second section focuses on axon elimination in the adult nervous system, which occurs during disease, aging, and injury. Learning and memory are not considered here, although structural changes underlying learning and memory may share mechanisms similar to those described here.

In general, the mode of axon elimination occurs through one of two distinct phenomena that relate to the scale of the event: Small-scale elimination typically occurs by retraction, whereas large-scale elimination appears to occur primarily by degeneration. The final section of the article considers potential cellular and molecular mechanisms that underlie axon retraction and degeneration and how these mechanisms might relate in development and disease.

AXON ELIMINATION DURING DEVELOPMENT

A common feature of the development of neural connections throughout the nervous system, particularly in vertebrates, is that compared with the adult, axons connect to more targets and targets are contacted by more axons during development. This early phase of exuberance in the formation of connections requires that the initial axonal connections be pruned through retraction or degeneration to develop the precise connectivity found in the adult. These strategies ensure the establishment of proper functional connections and provide a substrate for neural plasticity. Following, we present examples of small-scale (**Figure 1**) and large-scale axon elimination (**Figure 2**).

SMALL-SCALE DEVELOPMENTAL PRUNING OF AXONAL CONNECTIONS

Small-scale events include synapse elimination, reduction of poly-neuronal innervation of a target cell, and local pruning of an axonal arbor. Most if not all neural systems exhibit each of these overlapping phenomena (Katz & Shatz 1996, Lohof et al. 1996, Wong & Lichtman 2003). Classic examples of synapse elimination per se include the neuromuscular junction and climbing fiber inputs to cerebellar Purkinje cells, each of which reduces innervation of the target cell from poly-neuronal to mono-neuronal (**Figure 1**). Classic examples of local arbor pruning include segregation of visual input into eye-specific patterns in the retinogeniculate and geniculocortical projections, each of which reduces input to a target cell from binocular to monocular (**Figure 3**). In each of these instances,

Extension and overshoot of primary axons

Interstitial branching along primary axons

Arborization of interstitial branches

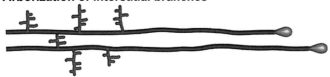

Degeneration of overshooting primary axons and ectopic branches and arbors

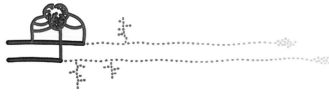

Figure 2

Large-scale axon elimination during development of the vertebrate nervous system. Large-scale axon elimination exemplified by a generic example. In the developing vertebrate brain, the primary axon typically extends well past its targets, which are later innervated by collateral branches that form de novo interstitially along the primary axon. Later, excess or inappropriate collaterals as well as a considerable length of the primary axon that overshoots the most distal target are eliminated by a process that appears to rely primarily on degeneration (indicated by broken axons).

synapses are eliminated early in the developmental maturation of the projection system, and the total number of synapses, or the size of the neuromuscular junction (NMJ), exhibits a progressive and substantial increase after the reduction in poly-neuronal input.

Studies that have directly visualized these events, or analyzed systems that mimic them, have provided considerable information about the process and the sequence of changes that underlie them. Time-lapse video imaging of marked axons in numerous systems has shown that developing axons, even after they have established synaptic contacts, are extremely dynamic and typically probe their locale through ongoing extensions and retractions of thin processes (O'Rourke et al. 1994, Alsina et al. 2001, Ruthazer et al. 2003, Walsh & Lichtman 2003, Kasthuri & Lichtman 2004). Further, these studies have directly revealed that small-scale elimination events appear to be due to retraction of selective axon segments and terminal endings or selective branches that form an axon arbor (Bernstein & Lichtman 1999). Time-lapse analysis of the developing neuromuscular junction revealed that the vacated postsynaptic site is quickly reinnervated by a persistant axon enhancing that axon's synaptic strength (Walsh & Lichtman 2003); similar events likely occur at CNS synapses (Mariani & Changeux 1981, Chen & Regehr 2000). Both the relative level of neural activity and correlation in activity patterns between axons that innervate the same target cells have pivotal roles in determining the competitive interactions between axons that drive these elimination events (Wiesel 1982, Personius & Balice-Gordon 2001, Feller 2002, Wong & Lichtman 2003). What axons compete for remains ambiguous, but it is widely assumed to be trophic support provided by the target cells.

Synapse Elimination at the Neuromuscular Junction

In the adult, a typical NMJ is innervated by a single motor axon. However, during development, a junction is usually multiply innervated by axons originating from two or more motor neurons (**Figure 1**). At developing neuromuscular synapses in vertebrates, different motor axons connecting to a muscle fiber compete for maintenance of their synapses. Competition results in progressive changes in synaptic structure and strength that lead to the physiologic weakening and retraction of some inputs, a process termed synapse elimination. At the same time, the remaining, "victorious" input is strengthened as it takes over vacated postsynaptic sites and is subsequently maintained throughout adult life (Colman et al.

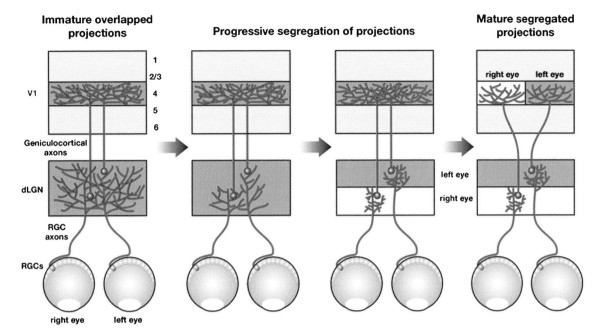

Figure 3

Small-scale axon elimination to develop eye-specific patterned connections. In adult mammals, inputs from the two eyes projecting to the same target are segregated into eye-specific domains. Axons of retinal ganglion cells (RGCs) from each eye are segregated into distinct layers or patches in their major diencephalic target, the dorsal lateral geniculate nucleus (dLGN). In the primary visual area (V1) of the cortex, geniculocortical axons from the eye-specific layers of the dLGN are segregated into eye-specific stripes or columns in layer 4. In each projection system, the two monocular inputs are initially overlapped and gradually segregate from one another by the selective local pruning of overlapping parts of axonal arbors to form the adult pattern of connections through a competitive process driven by correlated neural activity.

1997, Walsh & Lichtman 2003). Postsynaptic sites not reinnervated quickly lose receptors for neurotransmitter signaling, i.e., acetylcholine receptors concentrated at the postsynaptic site on the muscle. Competition between axons at neuromuscular synapses is based on the degree of correlation in their spiking patterns; uncorrelated activity enhances competition, whereas correlated activity diminishes competition (Personius & Balice-Gordon 2001, Kasthuri & Lichtman 2003). In addition, the relative level of activity between competing axons appears to be a critical parameter, where decreasing synaptic efficacy diminishes the competitive edge of the affected synapses (Balice-Gordon & Lichtman 1994, Buffelli et al. 2003). Activity-dependent modulation of axonal competition is considered in more detail at CNS synapses in the following section.

Axon Pruning in Development of Eye-Specific Retinal Connections

The interdigitation of eye-specific axon connections is a fundamental organization of the mammalian visual system (Reid 2003). Eye-specific connections are formed initially through the direct projections of retinal ganglion cells (RGCs) to diencephalon and midbrain, and they are subsequently reiterated through higher-order projections in neocortex. Eye-specific connections are patterned to segregate inputs from the two eyes, which allows for binocular inputs to selective sets of higher-order neurons to generate stereopsis.

In the adult brain, inputs from the two eyes projecting to the same target are segregated into eye-specific domains. In the major diencephalic target of RGC axons, the dorsal lateral geniculate nucleus (dLGN), RGC axons from each eye are segregated into distinct layers or patches, whereas in the primary visual area of the cortex, dLGN axons from the eye-specific layers are segregated into stripes or columns within layer 4. In each projection system, the two monocular inputs are initially overlapped and gradually segregate from one another by the selective local pruning of overlapping parts of axonal arbors to form the adult pattern of connections (**Figure 3**). The preferential stabilization of some axonal connections and the elimination of others that result in the segregation of eye-specific inputs were postulated decades ago to be controlled by a Hebbian mechanism (Hebb 1949) on the basis of a postsynaptic detection of correlated (i.e., appropriate) versus noncorrelated (inappropriate) inputs (Stent 1973). Neural activity has since been shown to drive eye-specific segregation; critical features of this activity-dependent process have been identified (Wiesel 1982, Feller 2002, Wong & Lichtman 2003) and the NMDA receptor has been implicated as a postsynaptic detector of correlated activity in some systems (Debski & Cline 2002, Zhou et al. 2003). In the absence of retinal activity, RGC inputs to the dLGN do not segregate into eye-specific layers (Penn et al. 1998, Rossi et al. 2001). Altering spontaneous activity in one retina relative to the other alters the normal pattern of eye-specific connections, which indicates that local arbor pruning is a competitive process between RGC axons from the two eyes (Penn et al. 1998, Stellwagen & Shatz 2002).

Although patterned vision would correlate activity among neighboring RGCs relative to distant ones, segregation is largely complete prior to the onset of visually evoked activity (Wong 1999, Tian & Copenhagen 2003). However, this form of arbor pruning is coincident with retinal waves that spontaneously propagate across the retina in a stochastic manner and are generated by a network of cholinergic amacrine cells that interconnect RGCs expressing nicotinic acetylcholine receptors (nAChRs). Retinal waves strongly correlate patterns of spontaneous activity among neighboring RGCs even before photoreceptors are generated and become synaptically connected to RGCs through bipolar cells (Meister et al. 1991, Wong et al. 1993). Local pruning of RGC axon arbors, which leads to eye-specific layers, is delayed in mice deficient for the $\beta 2$ subunit of the neuronal nAChR that lack cholinergic-mediated retinal waves, and in contrast to wild type mice, $\beta 2$ mutant mice projections from the two eyes to the dLGN remain overlapped at the end of the first postnatal week (Rossi et al. 2001, Muir-Robinson et al. 2002). Subsequently, though, correlated RGC activity due to ionotropic glutamate receptor–mediated waves and patterned vision do drive the pruning of RGC inputs from the two eyes into eye-specific patches, albeit in an abnormal distribution (Muir-Robinson et al. 2002). These findings indicate that correlated RGC activity is required and sufficient to drive the competitive interactions between RGC axons from the two eyes, which leads to small-scale axon elimination and eye-specific segregation of retinogeniculate arborizations.

LARGE-SCALE DEVELOPMENTAL PRUNING OF AXONAL CONNECTIONS

In the vertebrate brain, most neurons innervate multiple, widely separated targets by long axon collaterals extended from a primary axon, and therefore they face a unique problem of target selection during development. This process of target selection is accomplished by a distinct mechanism referred to as delayed interstitial axon branching. As the primary axon elongates, its growth cone makes navigational decisions in response to guidance molecules to direct the axon through its pathway along which its targets are positioned at varying distances, some adjacent to the path and others at a considerable distance. Later, collateral branches form along the primary axon and extend toward their target, where upon reaching the target the collaterals themselves extend branches that go on to arborize. Initially, the number of collaterals formed and the number of targets to which they project is larger than that found in the adult, but both are later reduced through a process of large-scale axon elimination (**Figure 2**). This mechanism is used to generate the adult patterns of all vertebrate projection systems analyzed, including all major projections of the mammalian neocortex—callosal, intracortical, and subcortical projections (**Figure 4**) (O'Leary 1992, O'Leary & Koester 1993). In addition, large-scale axon elimination is used to remodel projections within a large target, as in, for example, the elimination of the distal component of the primary axon and entire ectopically positioned arbors, as during the development of topographic mapping of RGC axons in the midbrain (**Figure 5**; McLaughlin et al. 2003a).

In a sense, large-scale axon elimination is an effective means of post hoc axonal pathfinding and target selection. Given that during vertebrate development large-scale axon elimination typically involves the rapid pruning of many millimeters of primary axon along with its distal interstitial branches and the arbors that they may have formed, degeneration seems to be a more likely mechanism over retraction. Although limited, morphological evidence obtained using anterograde axon fate labeling that spans the time course of axon pruning indicates that large-scale axon elimination is indeed accomplished by axon degeneration, at least in the development of layer 5 subcortical projections (**Figure 4**) (Reinoso & O'Leary 1989) and retinotopic mapping in the midbrain (**Figure 5**) (Nakamura & O'Leary 1989). In these systems, the portions of the fate-labeled axons undergoing elimination have a blebbed, fragmented morphology coincidentally along their length that closely resembles the appearance of γ neuron axons in the *Drosophila* mushroom body (MB) pruned via a mechanism of local degeneration, as described in a later section.

Target Selection by Interstitial Axon Branching Followed by Axon Elimination

A classic example of neurons that employ delayed interstitial branching and subsequent axon elimination to develop their functionally appropriate connections are layer 5 neurons in the mammalian neocortex (O'Leary et al. 1990, O'Leary & Koester 1993) (**Figure 4**). Layer 5 neurons form the major output projection of the cortex and establish connections with numerous nuclei in the midbrain and hindbrain, as well as with a specific level of the spinal cord. During development, layer 5 axons extend out of the cortex along a spinally directed pathway; their growth cones ignore all eventual targets as they grow past them and continue to extend caudally through the corticospinal tract (O'Leary & Terashima 1988). Axon collaterals extended by layer 5 axons later innervate the brainstem and spinal targets by a dynamic process that involves the repeated extensions and retractions of short branches before a branch is stabilized and extends as a long collateral into the target where it arborizes (Bastmeyer & O'Leary 1996). Initially, layer 5 neurons project more broadly and form collateral projections to a larger set of layer 5 targets than they will retain in the adult (**Figure 4**). This widespread, or exuberant, projection pattern of layer 5 axons seems to be elaborated according to a specific axonal growth program characteristic of a general class of subcortically projecting, layer 5 neurons. However, the organization of the adult neocortex into specialized areas characterized by unique functions and axonal connections requires that each area establish projections to specific subsets of targets in the brainstem and spinal cord.

The patterns of layer 5 projections characteristic of the adult are pruned from the initial widespread pattern by selective axon elimination, which in rats occurs during the second

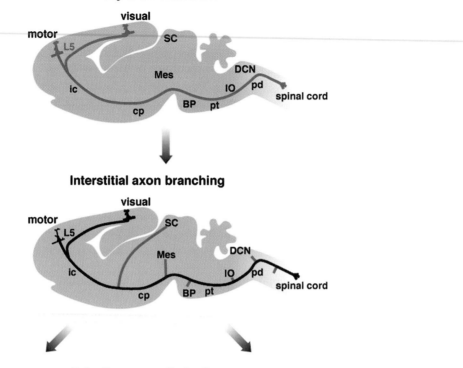

Figure 4

Large-scale axon elimination to develop area-specific subcortical projections of layer 5 neurons of the neocortex. The three major sequential steps in the development of subcortical projections of mammalian layer 5 neurons are illustrated in schematics of a sagittal view of the developing rodent brain. Primary axon extension: Layer 5 neurons (L5) extend a primary axon out of the cortex along a pathway that directs them toward the spinal cord passing by their subcortical targets. Delayed collateral branch formation: Subcortical targets are later contacted exclusively by axon collaterals that develop by a delayed extension of collateral branches interstitially along the spinally directed primary axon. As a population, layer 5 neurons in essentially all areas of rodent neocortex develop branches to a common set of targets. Selective axon elimination: As illustrated for visual and motor areas, specific collateral branches (some up to 2 mm in length) or segments of the primary axon (often more than 5 mm in length) are selectively eliminated, apparently by degeneration (*dashed lines*), to generate the mature projections functionally appropriate for the area of neocortex in which the layer 5 neuron is located. Abbreviations: BP, basilar pons; cp, cerebral peduncle; DCN, dorsal column nuclei; ic, internal capsule; IO, inferior olive; Mes, mesencephalon; pd, pyramidal decussation; pt, pyramidal tract; SC, superior colliculus (adapted from O'Leary & Koester 1993).

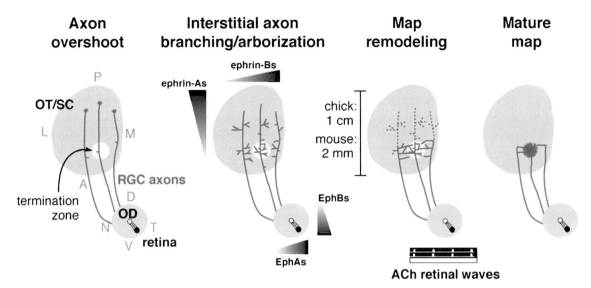

Figure 5

Large-scale axon elimination to develop retinotopic maps. Illustrated are the progressive stages in the development of a retinotopic map in the optic tectum (OT) of chicks or its mammalian homologue, the superior colliculus (SC) of mice. Initially, RGC axons enter the OT/SC and overshoot the location of their future termination zone (*white circle*), often by several millimeters. In addition, RGC axons are distributed over a broad lateral (L) to medial (M) extent of the target. The low to high anterior (A) to posterior (P) gradient of the axon repellents, ephrin-As, stops the posterior extension of RGC axons and generates topographic specificity in the subsequent interstitial axon branching and branch arborization. RGCs express EphAs in a high temporal (T) to low nasal (N) gradient. Interstitial branches extend laterally or medially toward their future TZ, guided in part by attractant and repellent actions of graded ephrin-Bs in the target mediated by a high ventral (V) to low dorsal (D) expression of EphB receptors by RGCs. Inappropriate arbors and branches, and long segments of the primary axon posterior to a sustained branch and its arbor, are rapidly eliminated over a few days largely by degeneration through a process dependent on spontaneous waves of activity mediated by a network of cholinergic (ACh) amacrine cells that propagate across the retina and strongly correlate to the spiking patterns of neighboring RGCs (*white spike trains inside black bar*). Ephs and ephrins also likely contribute to aspects of map remodeling independent of correlated activity, including the elimination of overshooting primary axons. The lengths of the A-P axis of the chick OT and mouse SC during map development are indicated. OD, optic disk. Adapted from Hindges et al. (2002).

and third postnatal weeks, after functional identities of cortical areas are established (Stanfield et al. 1982, O'Leary & Koester 1993). The selection of collaterals or distal parts of the primary axon to be eliminated versus retained is dictated by the functional specialization of the cortical area in which the layer 5 neuron is located. For example, layer 5 neurons in motor areas, which control muscle movements, lose their collateral branch to the superior colliculus but retain branches to other targets involved in motor control including the basilar pons, inferior olive, dorsal column nuclei, and spinal gray matter. In contrast, layer 5 neurons in visual areas, which are specialized to process vision, lose all collateral projections caudal to the basilar pons, and even the entire segment of their primary axon directed to the spinal cord, while retaining branches to the basilar pons and superior colliculus. The elimination process is very effective; for example, for a visual layer 5 neuron, elimination results in the loss of the distal part of the primary axon in its entirety up to the branch point in the axon tract where a "sustaining" collateral extends from it and arborizes in the basilar pons (O'Leary & Terashima 1988).

Progress has been slow in defining the molecular and cellular mechanisms that control large-scale axon elimination, but some intriguing insights are available, including a role for the homeodomain transcription factor Otx1 in the pruning of layer 5 axons. Otx1 protein, which localizes to layer 5 neurons, undergoes a translocation from the cytoplasm to the nucleus coincident with the normal pruning of initially exuberant layer 5 axons (Weimann et al. 1999, Zhang et al. 2002). In adult mice with a targeted deletion of Otx1, layer 5 neurons in occipital cortex (e.g., visual areas) retain their normally transient primary axon to the spinal cord (Weimann et al. 1999). In addition, the collateral projection of occipital layer 5 axons to the superior colliculus aberrantly extends into the inferior colliculus, and the collateral projections to the basilar pons resemble those formed by motor areas. These findings suggest that Otx1 provides a late onset of nuclear transcriptional regulation to turn on genes essential for axon pruning and/or to repress genes that are negative regulators of the pruning machinery.

Another example of plasticity in layer 5 axon pruning, possibly related to Otx1 function, is changes in the selectivity of area-specific axon elimination by layer 5 neurons following heterotopic transplantation of pieces of developing cortex (Stanfield & O'Leary 1985, O'Leary & Stanfield 1989). Layer 5 neurons transplanted from visual areas to motor areas permanently retain their normally transient spinal axon, and layer 5 neurons transplanted from motor areas to visual areas lose their normally permanent spinal axon and retain their transient axon collateral to the superior colliculus. Thus, the projections retained by the transplanted layer 5 neurons are appropriate for the cortical area to which they are transplanted rather than where they were born. These findings indicate that features associated with the local environment in which a layer 5 neuron matures contribute to selectivity of axon elimination. It is interesting to speculate that changes in Otx1 expression or defects in translocation of Otx1 protein by transplanted layer 5 neurons may contribute to their plasticity in axon pruning.

The mechanisms that control large-scale axon elimination may well differ between axonal projections, but studies of other systems suggest a potential contribution of thalamocortical input or the modality of sensory input received by the transplanted layer 5 neurons to changes in pruning selectivity. As with layer 5 projection neurons, the limited adult distribution of callosal cortical neurons is also pruned from an initally widepread distribution by selective axon elimination during development (Innocenti 1981, O'Leary et al. 1981). The selectivity in the process of callosal axon elimination is perturbed by numerous peripheral manipulations of either visual or somatosensory input that have a common feature of altering patterns of neural activity (e.g., strabismus; Shatz 1977, Lund et al. 1978, Koralek & Killackey 1990) or levels of activity (e.g., diminished activity by dark-rearing or eyelid suture, or silencing of retinal activity by enucleation or pharmacological manipulations; Innocenti 1986, Dehay et al. 1989, Frost et al. 1990, Zufferey et al. 1999). These treatments lead to retention of callosal connections between parts of cortex that would normally lose them. Altering sensory input or neural activity also perturbs selective axon elimination required to develop appropriate patterns of horizontal, intracortical connections within the cortex (Katz & Callaway 1992, Katz & Shatz 1996, Zufferey et al. 1999). Thus sensory input has a crucial role in influencing the selectivity of axon elimination required to develop the adult patterns of callosal and intracortical connections, and possibly layer 5 subcortical projections.

Axon Elimination to Develop Topographic Maps in a Target

Most axonal projections within the CNS establish in their target an orderly arrangement of connections termed a topographic map, which reflects the spatial order of their parent neurons. In the visual system, this map is retinotopic and projects the visual field from the retina onto targets in the brain, initially through direct RGC projections to dorsal thalamus and midbrain. In birds and mammals, the formation of a retinotopic map in the major midbrain target, the optic tectum or superior colliculus, respectively, involves the establishment of an initial, very coarse map, which subsequently undergoes large-scale pruning to generate a refined map (McLaughlin et al. 2003a) (**Figure 5**). All arbors are established by primary branches that form in a topographically biased manner interstitially along RGC axons that overshoot their correct termination zone along the anterior-posterior axis of the target. The interstitial branches often extend long distances along the medial-lateral axis of the target before forming complex arborizations at the topographically correct location.

Receptors of the Eph family of tyrosine kinases and their ephrin ligands have crucial roles in retinotopic mapping, with EphA-ephrin-A signaling graded along the anterior-posterior axis of the target, and EphB-ephrin-B signaling graded along its medial-lateral axis (McLaughlin et al. 2003a). EphA-ephrin-A signaling controls in part RGC axon mapping along the anterior-posterior axis by inhibiting the formation of primary branches and their arborization posterior to the correct termination zone (Yates et al. 2001). Complementing this action, ephrin-B1 acts bifunctionally as an axon attractant and repellent through EphB forward signaling to direct branches along the L-M axis of the target to their topographically correct site and to delimit arbors through branch inhibition (Hindges et al. 2002, McLaughlin et al. 2003b). Computational modeling predicts that the increase in total ephrin-A repellent, due to the progressive increase of ephrin-As on RGC branches and arbors added to that of ephrin-As expressed by target cells, acts later to prune overshooting axons and aberrant arbors (Yates et al. 2004). In support of this modeling is a report that in vitro, ephrin-As induce the degeneration of EphA-expressing hippocampal axons, implicating ephrin-As in the pruning of hippocampal axons required to generate topographic order in the hippocamposeptal projection (Gao et al. 1999).

Although computer simulations show that gradients of guidance molecules alone can develop a considerable degree of topographic order and refinement, including axon elimination, they are not sufficient to develop a completely refined projection (Yates et al. 2004). In addition, the generation of a properly refined map requires a parameter that resembles an assumed role for near-neighbor-correlated neural activity. Similar to the local arbor pruning that generates eye-specific connections in the dLGN, the large-scale axon elimination required to develop topographic order in the mouse retinocollicular projection occurs during the first postnatal week before the onset of visually evoked activity (Hindges et al. 2002) but is coincident with cholinergic retinal waves (Wong et al. 1993). Mice deficient for the $\beta 2$ subunit of the nAChR, which selectively lack cholinergic retinal waves (Bansal et al. 2000), fail to develop a refined topographic map and retain an aberrantly broad distribution of ectopic arbors and axons (McLaughlin et al. 2003c). During the second postnatal week, glutamatergic retinal waves correlate RGC firing, and later patterned visual activity does also (Bansal et al. 2000), but the $\beta 2$-deficient mice fail to develop a refined retinotopic map and retain an aberrant pattern of broadly distributed ectopic arbors and axons into adulthood (McLaughlin et al. 2003c). Thus, the competitive interactions that lead to large-scale axon elimination required to remodel the mammalian retinocollicular projection into a refined map require correlated RGC activity during a critical period spanning the first postnatal week.

Developmental critical periods for changes in axonal connectivity including axon pruning have been defined in many neural connections; in general, maturational changes that affect the parent neurons and their targets are believed to define the end of a critical period and the diminished capacity of axons for large-scale morphological plasticity at later ages (Hensch 2004). Interestingly, although later-correlated RGC activity patterns are insufficient to generate a refined retinotopic map (McLaughlin et al. 2003c), they are sufficient to drive the local pruning of RGC axon arbors, which leads to eye-specific segregation (Muir-Robinson et al. 2002, Ruthazer et al. 2003). Thus, RGC axon projections to the dLGN retain the capacity for the small-scale axon elimination required for their development of eye-specific patterns in the dLGN after the time that the same axons lose their ability to undergo large-scale axon elimination required to generate a retinotopic map in the midbrain superior colliculus. Although in mice, RGC input to the dLGN is formed by collateral branches of RGC axons that project to the SC (essentially all RGCs project to the SC, and about a third project to the dLGN), the difference in the critical periods between the retinogeniculate and retinocollicular projections may reflect differences in plasticity between classes of RGCs, with the subset of RGCs that forms the retinogeniculate projection retaining a greater capacity for activity-dependent remodeling.

Phylogenic Differences in Developmental Axon Elimination May Generate Species-Unique Axon Projection Patterns

The developmental phenomenon of transiently exuberant axonal projections may provide a substrate for functional sparing or recovery following neural insults during development (Innocenti et al. 1999). In addition, this mechanism may provide a substrate for evolutionary change and contribute to differences between species in axonal connections (O'Leary 1992, Innocenti 1995, Deacon 2000). A compelling example of this possibility is provided by studies of a substantial projection that arises in the subiculum of the hippocampal formation, passes through the fornix, a major forebrain axon tract, and terminates in the mammillary bodies of the hypothalamus. In adult rats, subicular axons do not extend caudally beyond the mammillary bodies. However, in developing rats, subicular axons extend through the fornix and continue caudally past the mammillary bodies in dense fascicles for considerable distances into the midbrain and hindbrain tegmentum, and some fascicles cross to the opposite side. Later, these axons form collateral projections to the mammillary bodies and lose the long segment of the primary axon distal to it (Stanfield et al. 1987).

Although the postmammillary component of the fornix is not maintained in adult rats, it is present to varying degrees in adult brains of other mammalian species (O'Leary 1992). In some species, such as the squirrel monkey, the postmammillary component of the fornix is minor, whereas in others— for example, the cat, some strains of rabbits, and the African elephant—it is a major axonal projection that extends prominently into the midbrain tegmentum with only a proportion of the axons appearing to connect to the mammillary bodies. These projection patterns in adults of other species closely resemble the distribution of transient postmammillary axons in developing rats. This similarity suggests that mammals initially develop a common pattern of postmammillary trajectories, but that the postmammillary components of the projection are later differentially elaborated and maintained across species, including complete elimination in some. Thus, major phylogenetic differences in axonal projections, and their functional manifestations, could emerge through the evolution of species-specific pruning (or stabilization) of a developmentally similar projection pattern. This phenomenon is not likely limited to the development of the fornix; differences among adults of various species in the distribution of other homologous axonal pathways may well be found as the result in part of distinctions in axon elimination.

Neuronal Remodeling During Insect Metamorphosis

Axon and dendrite pruning also occur extensively in the development of invertebrate nervous systems. The most extensively studied example is neuronal remodeling during insect metamorphosis. Holometabolous insects possess two distinct nervous systems at the larval and adult stages. Many neurons are born during larval and pupal stages to function specifically in the larger, more complex adult nervous system. Other neurons are used only in the simpler larval nervous system and die during metamorphosis. But a third class of neurons is morphologically differentiated and likely to function in both the larval and the adult systems; these neurons do so by reorganizing their dendrites and axons during metamorphosis (Truman 1990, Tissot & Stocker 2000). With regard to the scale of pruning, although the absolute length of processes to be pruned is relatively small compared with larger brains of vertebrates, the relative length compared with the size of their neurons or their brains would clearly put these prunings into the large-scale category. Interestingly, at least in one most extensively studied example, the elimination of larval-specific axon branches of MB γ neurons, pruning occurs by a degenerative mechanism (see the section on Mechanisms of Axon Degeneration for more details).

AXON DEGENERATION IN NEUROLOGICAL DISEASES

Axon elimination occurs in many neurological diseases or as a consequence of injury. In many cases it is unclear whether retraction or degeneration is the primary cause of axon elimination, so we follow the terminology in medical literature in which degeneration is more commonly used.

Wallerian Degeneration

Wallerian degeneration (Waller 1850) refers to a series of events that occur in distal axons when they are severed from the cell body. The following reasons highlight the importance of studying Wallerian degeneration: (*a*) It is a model for spinal cord or other nerve injury; (*b*) many other forms of axon degeneration in various neurological diseases share similarities with Wallerian degeneration at the final stages; (*c*) Wallerian degeneration can be studied both in the intact animal and in cultured explant in vitro; (*d*) a mouse mutant that significantly delays Wallerian degeneration has provided an entry point into the mechanisms of axon degeneration (see the section on Mechanisms of Axon Degeneration for more details).

Griffin et al. (1995) provided a detailed description of the events following axon transection. After the axons are transected, there is a variable period of time during which the only visible events are an apparent accumulation of materials at the proximal end of the distal axon stump, presumably owing to a block of retrograde axonal transport. What follows next is a rapid breakdown of the axonal cytoskeleton in an all-or-none fashion with few intermediate stages. Breakdown of neurofilaments via ion-sensitive proteases such as calpain causes cytoskeletal breakdown. Axonal degeneration then triggers a series of responses from surrounding cells, including changes in glial cells surrounding the axons and recruitment of microglia and macrophages to the severed axons, presumably to phagocytose the debris of degenerated axons and myelin (Griffin et al. 1995). This stereotyped sequence of axon degeneration may efficiently remove the damaged tissues, which would facilitate regeneration and repair.

Dying Back Degeneration

A variety of neurological disorders are characterized by initial degeneration of the distal regions of long axons, followed by distal-to-proximal progression, and therefore this process is termed dying back degeneration (Cavanagh 1964). These disorders include amyotrophic lateral sclerosis (Lou Gehrig's disease), spinal muscular atrophy, spinocerebellar disorders, peripheral neuropathies, nutritional neurological disorders, various intoxications, and AIDS (Cavanagh 1964, Berger & Schaumburg 1995, Raff et al. 2002). The cause and onset of these dying back diseases vary, but the final stage of each disease exhibits axon fragmentation resembling that of Wallerian degeneration. Other diseases described below that exhibit axon atrophies also may share this dying back mechanism, but they have not been studied to sufficient detail.

Wallerian-like degeneration can also be mimicked, at least morphologically, by neurotrophin deprivation. In a classic set of in vitro experiments, Campenot (1982) used a culture dish insert with isolated chambers to show that local bath application of nerve growth factor (NGF) to distal portions of axons of sympathetic ganglia was required to maintain the axons. When NGF was withdrawn from the distal chamber, the axons degenerated to their proximal portions that, along with their parent neurons, were maintained by NGF in an environmentally separate chamber.

Neurodegenerative Diseases

Many neurodegenerative diseases including Alzheimer's, Parkinson's, Huntington's, and prion diseases exhibit axonal and dendritic atrophies. The important question is whether these neuritic atrophies precede neuronal degeneration, or whether they are a secondary consequence of neuronal death. As long as axon or dendrite degeneration occurs prior to neuronal death, these neuritic atrophies should contribute significantly to the clinical symptoms whether they are the primary causes of neuronal death. Because it is difficult to determine the sequence of events from postmortem brain, investigators must establish animal models for these diseases. These animal models could then lead to systematic time-course studies of disease progression with high temporal and anatomical resolution. In the case of Huntington's disease, for instance, Li et al. (2001) showed that, in an animal model, axonal accumulation of the Huntingtin protein, defects of axonal transport, and axonal degeneration all precede neuronal death (Li et al. 2001). In vivo imaging also indicated that amyloid deposits in a mouse model of Alzheimer's disease induced breakage of axonal and dendritic branches (Tsai et al. 2004). Delaying or preventing axon degeneration could alleviate the clinical symptoms of some of these diseases (see the section on Mechanisms of Axon Degeneration for more details).

MECHANISMS OF AXON RETRACTION

In principle, axon retraction (as well as axon degeneration, considered in the following section) could be due to (*a*) a direct induction of an intrinsic program for retraction (or degeneration); (*b*) activation of a default retraction (or degeneration) program, owing to its diminished repression; or (*c*) a default, owing to an inadequate maintenance of the axonal cytoskeleton, for example because of diminished trophic support. Compared with axon growth and guidance, relatively little is known about the mechanisms of axon retraction. Because axon retraction is a change of cell shape just as is axon elongation or turning, it is useful to adopt a similar paradigm to consider the process of axon retraction, namely that these changes of cell shapes are based on modifications of the cytoskeleton. These modifications are regulated by signaling pathways that receive input from the extracellular environment or from cell-intrinsic programs that reflect the maturational status of the neurons.

Cytoskeleton

Numerous studies have implicated the integrity of the microtubule cytoskeleton in maintaining the axon stability (e.g., Yamada et al. 1970). Interestingly, whereas inhibiting microtubule polymerization alone results in axon retraction in cultured neurons, simultaneous inhibition of actin polymerization or deprivation of ATP blocks axon retraction caused by inhibition of microtubule assembly, which suggests that axon retraction is an active process that requires intimate interaction between actin and the microtubule cytoskeleton (Solomon & Magendantz 1981). Recently Ahmad et al. (2000) showed that motor proteins may play important roles in axon retraction. For example, inhibition of microtubule motor dynein caused axon retraction in the presence of intact microtubules; this effect was reversed by depletion of the actin cytoskeleton or inhibition of myosin motors. Thus, myosin-actin and dynein-microtubule cytoskeletons were proposed to provide counter-balance of forces in regulating axon stability.

Eaton et al. (2002) showed that disruption of the dynein/dynactin complex leads to retraction of synaptic terminals in *Drosophila* NMJ in vivo, which indicates that microtubule motors are also essential for stability of presynaptic terminals.

Intracellular Signaling Pathways

Given the importance of cytoskeleton in maintaining axon integrity, signaling pathways that lead to destabilization of microtubules, actin, or interactions between the two cytoskeletal systems could be involved in axon retraction. However, few signaling mechanisms have been investigated thoroughly with the exception of a pathway involving the small GTP-binding protein RhoA.

Activation of RhoA leads to neurite retraction in cultured neuroblastoma cell lines (Jalink et al. 1994, Kozma et al. 1997) or prevention of axon initiation in primary neurons (Bito et al. 2000). Activation of RhoA in maturing neurons in brain slices or in vivo results in retraction of dendritic processes (Li et al. 2000, Nakayama et al. 2000, Wong et al. 2000). Further, activation of RhoA is also implicated in mediating the activity of myelin-associated inhibitors in blocking axon regeneration (He & Koprivica 2004). RhoA has many downstream effectors. One serine-threonine kinase, Rho kinase (Rok or ROCK), is most critical in RhoA-mediated axon and dendrite retraction because inhibiting this kinase activity completely reverted RhoA-mediated retraction in a number of cases tested in vitro, for example, in cultured hippocampal pyramidal neurons (Nakayama et al. 2000). Many substrates have been identified for Rok, but the most important in the context of axon retraction is the regulation of myosin regulatory light chain phosphorylation (Luo 2002), consistent with the importance of myosin activity in maintaining axon integrity, discussed above.

Does such an axon retraction pathway exist in mature neurons? A genetic study in *Drosophila* MB neurons provided some insight. Inhibition of p190RhoGAP, a negative regulator for RhoA, results in axon retraction of MB axon branches. Genetic analysis indicated that this effect is mediated by RhoA, through activation of Rok and phosphorylation of myosin regulatory light chain (Billuart et al. 2001). These findings suggested that in mature neurons the RhoA-mediated axon retraction pathway is intact but largely repressed by negative regulators such as p190RhoGAP. A recent genome-wide screen of RhoGEFs in *Drosophila* identified two specific RhoGEFs, positive regulators of Rho GTPases, which act to counterbalance p190RhoGAP's effect in regulating axon stability (A. Goldstein, S. Hakeda-Suzuki, A. Maresh, B. Dickson, L. Luo, submitted). Thus, extracellular signals that regulate the activity of these RhoGEFs and RhoGAPs could influence structural plasticity of neurons.

Why should mature neurons maintain an intact axon retraction pathway, only to repress it with negative regulators? One possibility is to use such a pathway for structural remodeling of neurons in response to experience, learning, or injury. For instance, local inactivation of a negative regulator such as p190RhoGAP could result in removal of specific axonal or dendritic branches. Supporting this structural plasticity hypothesis, p190RhoGAP could be regulated by molecules known to regulate neural plasticity such as integrin or src family tyrosine kinases (Billuart et al. 2001), and regulation of p190RhoGAP and Rok activity has been associated with fear conditioning in mammalian neurons (Lamprecht et al. 2002).

RhoA signaling pathway is most likely one of several signaling pathways whose final readout is destabilization of microtubules or actin or disruption of their interaction, which could lead to axon retraction in mature neurons. Analyses of factors known to regulate microtubule and actin stability could be fruitful entry points for future investigations of signaling pathways that mediate axon retraction. Other potential studies would include testing whether signaling pathways implicated in growth inhibition or repulsion (as in the case of RhoA) play a role in axon retraction in mature neurons.

Extracellular Cues and Their Receptors

Most of the extracellular factors that affect axon behavior are studied in the context of axon guidance, branching, and arborization. Given the potential similarities between repulsive axon guidance and axon retraction, repulsive guidance molecules such as ephrins, Slits, or Semaphorins, all of which signal through RhoA in the context of axon guidance or inhibition of axon growth (Luo 2002, Guan & Rao 2003, He & Koprivica 2004), could also signal axon retraction. Ephrins are strong candidates to promote axon pruning in the context of map refinement, as discussed in a previous section. A recent study also implicates a Semaphorin in regulating axon pruning.

Bagri et al. (2003) found that the stereotypic pruning of a specific component of hippocampal mossy fibers, the infrapyramidal bundle (IPB), is defective in mice mutant for the Semaphorin receptors Plexin-A3 or Neuropilin-2. Semaphorin 3F is expressed in cells along the path of the IPB when it undergoes pruning, and genetic analysis indicates that Plexin-A3 acts cell autonomously to regulate IPB pruning. IPB pruning appears to occur via axon retraction because no evidence for degeneration was found in vivo, and Semaphorins can induce axon shortening and retraction in explant cultures or dissociated neurons in a Plexin A3-dependent manner. Thus, it seems likely that Semaphorins serve as axon retraction signals, and the patterns of Semaphorin expression dictate the extent of axon pruning (Bagri et al. 2003).

MECHANISMS OF AXON DEGENERATION

Axon degeneration, or fragmentation of intact axons into pieces, may involve cell biological mechanisms quite different from axon retraction or any other biological processes. Very little is known about these mechanisms. We focus on two systems for axon degeneration that have given the most insights: developmental axon pruning in insect metamorphosis and Wallerian degeneration of injured axons in mammals. Although these two systems differ in species (invertebrate versus mammal), stage of neuronal maturation, and causes, they exhibit interesting similarities, suggesting the exciting possibility of an evolutionarily conserved common mechanism of axon degeneration.

Degeneration of *Drosophila* Mushroom Body γ Neuron Axons During Development

The γ neurons of the *Drosophila* mushroom body (MB), a structure essential for olfactory learning and memory, form functional circuits in both larvae and adult. During early hours of metamorphosis, γ neurons prune their larval-specific dorsal and medial branches (Lee et al. 1999; see also Technau & Heisenberg 1982). A detailed time course analysis of γ axon pruning at the single-cell level using fluorescence microscopy revealed that γ axon pruning is achieved via a local degeneration mechanism: Axons appear to break into pieces without obvious distal-to-proximal retraction (Watts et al. 2003) (**Figure 6a**). This degeneration was confirmed by an independent study using a different fluorescent axon marker (Awasaki & Ito 2004) and a genetically encoded EM marker for electron microscopy (Watts et al. 2004). Because of the genetic accessibility, MB γ axon pruning offers a good system to dissect mechanisms of axon degeneration (**Figure 6b**).

MB γ axon pruning requires an intrinsic transcriptional program controlled by the steroid hormone ecdysone. Two complementary experiments concluded that MB γ axon pruning requires an intrinsic transcriptional program controlled by ecdysone. First, single-cell γ neuron clones homozygous mutant for a coreceptor for ecdysone do not prune their axons, despite the fact that all neighboring axons undergo normal pruning (Lee et al. 2000). Second, when the whole brain is mutant for ecdysone receptor (isoform B1), both metamorphosis and γ neuron pruning are arrested. However, if ecdysone receptor is supplied back in γ neurons only, their pruning is restored despite the mutant brain environment (Lee et al. 2000). These experiments demonstrated that γ axon pruning requires a cell-autonomous action of the ecdysone receptor. Recent experiments indicated that ecdysone receptor B1 expression itself is transcriptionally regulated by a TGF-β signaling pathway (Zheng et al. 2003).

MB γ axon pruning requires cell-autonomous action of the ubiquitin-proteasome system (UPS). Watts et al. (2003), who used a gene knock-out in small populations of MB neurons in a wild-type background, concluded that MB γ axon pruning requires cell-autonomous action of the UPS. Remarkably, MB neurons homozygous mutant for E1, the first enzyme in the cascade of the UPS, or either of the two tested proteasome subunits essential for UPS function, exhibit normal axon growth and guidance in larvae. However, they do not prune their larval-specific axon branches, despite the fact that all neighboring MB axons exhibit normal pruning (**Figure 7b-c**, compared with **Figure 7a**). Dendritic pruning is also blocked by inhibition of UPS (Watts et al. 2003) or ecdysone reception (Lee et al. 2000), suggesting a mechanistic link to axon pruning.

MB γ axon pruning requires glia to engulf degenerating axon fragments. Using a genetically encoded EM marker that allows for the distinction of cellular profiles in electron micrographs, Watts et al. (2004) showed that during early metamorphosis of the fragments of MB γ axons are engulfed by glia for endosome-lysosome-mediated degradation. An interesting question arises as to whether glia simply react to axon fragmentation and serve as scavengers or whether glia actively assist axon fragmentation, perhaps by providing spatial cues that determine which parts of an axon are to degenerate. This question has not been resolved, but some evidence suggests that glial action is not purely passive and that there is extensive interaction between neurons and glia. First, glia cell numbers exhibit a selective increase near degenerating MB axons independent of axon fragmentation (Watts et al. 2004); second, glial infiltration of axon lobes is dependent on an ecdysone-induced signal from γ neurons (Awasaki & Ito 2004); third, transient inhibition of glial function by transiently blocking endocytosis in glial cells arrests axon pruning (Awasaki & Ito 2004). Therefore, glia may actively participate in directing axon pruning.

Mechanisms of Wallerian Degeneration

Lunn et al. (1989) reported a remarkable strain of mice that exhibit much slower Wallerian degeneration when axons are severed. Instead of rapid degeneration in a few days,

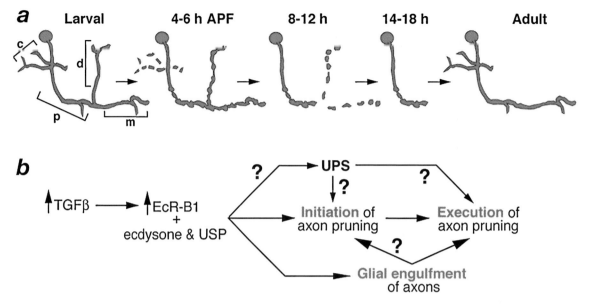

Figure 6

Developmental axon pruning for *Drosophila* mushroom body (MB) γ neurons. (*a*) Schematic illustration of developmental axon pruning of MB γ neurons. Larval γ neurons consist of dendrites near the cell body in the calyx (c) and axons that extend down the peduncle (p) and branch to form the dorsal (d) and medial (m) axon lobes. Four to six hours after puparium formation (APF), dendrites begin to fragment and axons begin to swell. From 8 to 12 h APF, axons in dorsal and medial lobes undergo fragmentation, whereas dendrite fragments disappear. The remaining axon fragments disappear from 14 to 18 h APF. After 18 h APF, axons re-extend into only the medial lobe. Modified from Watts et al. (2004). (*b*) A model for molecular pathways involved in MB γ axon pruning. TGFβ/Activin signaling is required for the γ neuron–specific expression of ecdysone steroid hormone receptor (EcRB1), which acts as a heterodimer with its coreceptor Ultraspiracle (USP) to initiate a gene expression program for axon pruning in response to ecdysone signaling. Axon pruning requires cell-autonomous action of the ubiquitin proteasome system (UPS) to initiate axon pruning by degrading key regulators of axon pruning and/or to execute the pruning process by degrading essential structural components. In addition to this intrinsic program of axon pruning, glial cells are specifically enriched at axon lobes where they engulf degenerating axon fragments. Question marks denote the links that remain hypothetical.

axons from Wld^s (Wallerian degeneration slow) mice survived for 1–2 weeks. This remarkable finding opened doors for mechanistic studies of Wallerian degeneration.

Wallerian degeneration is an active process of axon self-destruction. The notion that distal portions of the transected axons die passively for lack of nutritional support from the soma was shattered by the discovery of Wld^s mice, which show that distal axons can survive for a long time after transection. Distal axons degenerate soon after severing most likely because a degeneration program is triggered and actively executed. The availability of Wld^s mice also allowed transplantation experiments showing that severed axons of Wld^s genotype surrounded by wild type cells live as long as those in Wld^s mice. These findings demonstrate that the neuro-protective effect of Wld^s is autonomous to neurons (Glass et al. 1993).

Wld^s acts by regulating NAD metabolism to protect axons. Wld^s acts as a dominant mutation and was found to be caused by a

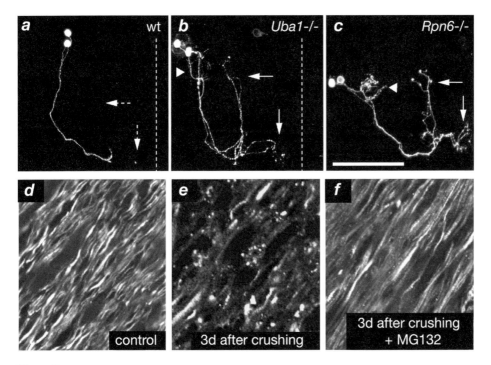

Figure 7

Comparing *Drosophila* MB γ axon pruning and Wallerian degeneration. (*a–c*) At 18 h APF, wild-type γ neurons have pruned their larval specific branches (*a*), but MB γ neurons homozygous mutant for a ubiquitin-activating enzyme Uba1 (*b*), or a proteasome subunit Rpn6 (*c*), fail to prune their larval-specific axons, which demonstrates that the ubiquitin proteasome system is cell-autonomously required for MB γ axon pruning. Labeled wild-type or homozygous mutant Uba1 and Rpn6 clones were generated by the MARCM (mosaic analysis with a repressible cell marker) technique (Lee & Luo 1999). (*d–f*) Three days after crushing optic nerve in rats, distal axons of retinal ganglion cells undergo characteristic Wallerian degeneration as indicated by broken axon fragments (*e*), compared with uncrushed control (*d*). In the presence of proteasome inhibitor MG132, Wallerian degeneration is significantly delayed (*f*). Axons are stained with anti-tubulin antibody. Panels *a–c* are from Watts et al. (2003), and panels *d–f* are from Zhai et al. (2003).

triplication resulting in the generation of a fusion protein consisting of the first 70 amino acids of UFD2/E4, an evolutionarily conserved protein used in protein polyubiquitination, and the full-length nicotinamide mononucleotide adenylyltransferase (Nmnat), an enzyme involved in NAD metabolism (Conforti et al. 2000). Transgenic expression of Wld^s protects distal axon degeneration after severing in a dose-dependent fashion (Mack et al. 2001). Because the first 70 amino acids of UFD2 do not contain the enzymatic domain for polyubiquitination, investigators proposed that Wld^s protects axons by either increasing the activity of Nmnat or by serving as a dominant negative protein that interferes with UFD2 and thereby the ubiquitin-proteosome system (Coleman & Perry 2002, Mack et al. 2001).

A recent report determined that the Nmnat portion of the Wld^s fusion protein has axon-protective activity. Using dorsal root ganglia explants as an in vitro model for Wallerian degeneration, Araki et al. (2004) found that expression of Wld^s and full-length Nmnat conferred axon-protective activity; expression of the first 70 amino acids or a dominant negative form of UFD2, or Nmnat with a point mutation in the catalytic domain, did not have an axon protection effect. These findings indicate the importance of a NAD-dependent process in axon protection. Curiously, although overexpression of Nmnat did not increase NAD level in the cells, incubating neurons with NAD 24 h prior to severing their axons (but not after the severing) also has the axon-protection effect. These findings, coupled with the predominant nuclear localization of Wld^s, suggested that Nmnat exerts its neuro-protective function indirectly in the nucleus by regulating gene expression, likely mediated by a Sirt1-dependent chromatin deacytelation, a known effector for NAD metabolism in regulating cellular aging (Araki et al. 2004).

Wallerian degeneration involves regulation of the ubiquitin-proteasome system. The fact that Wld^s fusion protein does not protect axons through interfering with UPS does not exclude involvement of UPS in Wallerian degeneration. In fact inhibition of UPS through proteasome inhibitors, or expression of a deubiquitination enzyme, significantly delayed axon degeneration in an in vitro model of Wallerian degeneration. Application of proteasome inhibitors also delayed the in vivo Wallerian degeneration of the optic nerve after severing (**Figure 7d–f**). Interestingly, to have the protective effector, proteasome inhibitors must be applied to neural explants 1–3 h prior to severing the axons, which suggests that UPS regulates an early step of Wallerian degeneration (Zhai et al. 2003).

Similarities Between MB γ Axon Pruning and Wallerian Degeneration

Cell-autonomous programs for axon self-destruction. As summarized above, both MB γ axon pruning and Wallerian degeneration appear to invoke cell-autonomous action of a genetic program. Available evidence from other systems also supports this notion. For instance, during developmental pruning of layer 5 subcortical projections (see Large-Scale Developmental Pruning of Axonal Connections), Otx1 appears to induce transcription of an essential component of the axon pruning machinery or downregulate an inhibitor of axon pruning. In addition, extensive work from *Manduca* and *Drosophila* has shown that the insect metamorphosis hormone ecdysone plays a key role in regulating neuronal remodeling, both pruning of larval-specific axonal and dendritic branches and likely re-extension of adult-specific branches (Levine et al. 1995). Dendritic pruning of peptidergic interneurons also requires cell autonomous action of ecdysone receptors (Schubiger et al. 1998, 2003). Even synaptic partners of a circuit independently require cell-autonomous action of ecdysone receptors to prune their axonal and dendritic branches (Marin et al. 2005).

Microtubule breakdown in early stages of axon degeneration. Systematic search for the sequence of cellular events identified degradation of the microtubule cytoskeleton as the earliest sign of MB axon pruning (Watts et al. 2003). In the in vitro explant model of Wallerian degeneration, breaks in the microtubule cytoskeleton were also identified as the earliest sign (Zhai et al. 2003), preceding the previous earliest sign of neurofilament breakdown. In neither case has a causal relationship between microtubule breakdown and axon degeneration been established. Given the importance of the microtubule cytoskeleton in maintaining axon integrity, this will be an important avenue for future research.

Caspase-independent processes. The self-destructive nature of axon degeneration raises an intriguing possibility that axon degeneration may share mechanisms similar to apoptosis. Caspase activation is a hallmark of apoptosis. However, inhibition of caspase activity using a variety of inhibitors did not prevent or delay distal axon degeneration after transection, even though proximal cell body apoptosis was prevented in the same experiment (Finn et al. 2000). Likewise, deletion of apoptotic activators or expression of a caspase inhibitor in *Drosophila* MB neurons had no effect on axon pruning (Watts et al. 2003).

Regulation by the ubiquitin-proteasome system. Genetic analysis demonstrated that UPS is cell-autonomously required for MB γ axon pruning and suggested that it may be involved in the initiation of the axon degeneration program by removing one or more negative regulators (Watts et al. 2003). This finding is consistent with an early involvement of UPS in Wallerian degeneration (Zhai et al. 2003). UPS is also critical in structural modification of synapses (DiAntonio & Hicke 2004, Steward & Schuman 2003). If indeed there is a conserved neuro-protective protein that needs to be degraded for the initiation of axon degeneration program, identifying the specific E3 ligase(s) and the neuro-protective substrate(s) will be a major step forward in our understanding of the axon degeneration program and will perhaps offer insights into possible therapies for neurodegenerative diseases.

Although we have discussed the remarkable similarities between certain forms of developmental axon pruning and Wallerian degeneration in both phenomenology and potential mechanisms, a word of caution should be added that they should not be regarded as identical. After all, developmental axon pruning and Wallerian degeneration are triggered by very different events.

Many Degeneration Diseases May Share Similarities with Wallerian Degeneration

The clues from studies of Wallerian degeneration allow researchers to test whether the mechanisms involved in Wallerian degeneration apply to neurodegeneration of other causes. For instance, just like delay of Wallerian degeneration, UPS inhibition could also delay degeneration in vitro owing to neurotrophin deprivation (Zhai et al. 2003). Among the most remarkable findings in recent years are those that show the power of Wld^s in protecting axon degeneration arising from different causes.

Mice homozygous for a naturally occurring recessive mutation *progressive motor neuropathy* (*pmn*) exhibit muscle weakness, motor axon degeneration in a "dying back" mode, and eventual organismal death by six weeks after birth (Schmalbruch et al. 1991). The cause of the disease is a point mutation in the *tubulin specific chaperone E* (*Tbce*) locus that encodes a protein essential for folding of α and β tubulin dimers (Bommel et al. 2002, Martin et al. 2002). The symptoms and pathology of *pmn* mice mimic severe forms of human spinal muscular atrophy (SMA), so *pmn* mice have been used as an animal model for studying SMA and other motor neuron diseases. Early work showed that Bcl-2 overexpression or glia-derived neurotrophic factor (GDNF) treatment of *pmn* mice could prevent or delay motor neuron death but could not prevent axon degeneration and did not increase life span (Sagot et al. 1995, Sagot et al. 1996). These studies suggest that axon degeneration, rather than motor neuron death, is the primary cause of the disease. Remarkably, crossing in the *Wld^s* mutation into *pmn/pmn* mice not only delayed axon degeneration and rescued motor neuron death, but also attenuated the symptoms and significantly increased the lifespan of *pmn/pmn* mice (Ferri et al. 2003). This finding suggested the existence of mechanistic similarities between Wallerian degeneration and dying back diseases and demonstrated the utility of inhibition of axon degeneration in treating motor neuron diseases.

The protective effect of Wld^s has been demonstrated also in two other cases. Axon atrophy contributes to clinical symptoms of demyelination diseases such as multiple sclerosis (Bjartmar et al. 2003). Mice homozygous mutant for myelin basic protein P0 exhibit myelin-related axon loss. Introduction of *Wld^s* into homozygous P0 mutants reduced axon loss and increased motor neuron survival and function at least at early ages (Samsam et al. 2003). The neuroprotective effect of

dLGN: dorsal lateral geniculate nucleus

IPB: infrapyramidal bundle

MB: mushroom body

nAChRs: nicotinic acetylcholine receptors

NGF: nerve growth facto

pmn: progressive motor neuropathy

RGC: retinal ganglion cell

UPS: ubiquitin-proteasome system

Wld^s: Wallerian degeneration slow

Wld^s was examined also in an animal model of Parkinson's disease (Sajadi et al. 2004). Stereotaxic injection of 6-hydroxydopamine (6-OHDA), a catecholaminergic neurotoxin, into the nigrostriatal pathway of wild-type mice results in degeneration of dopaminergic axons and death of dopaminergic neurons. A strong protective effect for dopaminergic axons and their function was seen when similar injections were performed in Wld^s mice. However, this protective effect was limited to injections of distal but not proximal nigrostriatal pathways, and dopaminergic neuron death was not protected. These results suggested that axon segments distal to the toxin injection site may undergo degeneration similar to Wallerian degeneration, which can be protected by Wld^s (Sajadi et al. 2004).

Taken together, these recent studies on animal models suggested that protection of axon degeneration could be an important therapeutic strategy for neurological diseases (Coleman & Perry 2002, Raff et al. 2002). The fact that a delay of axon degeneration can at least alleviate symptoms of motor neuron disease, Parkinson's disease, or myelin disease suggests that axon degeneration is a critical part of their pathogenesis. Finally, these studies support the notion that a common axon degeneration pathway, akin to Wallerian degeneration, is shared in diseases of many different origins.

Potential Connection with Axonal Transport

Defects in axonal transport could be causes of many neurological disorders including Alzheimer's, Huntington's, and motor neuron diseases (Gunawardena & Goldstein 2004). Long-term transport blockade of nutrients, survival factors, or any other proteins essential for neuronal function should lead to neuronal dysfunction and degeneration. Axon transection can be viewed as an extreme case of axonal transport block, and therefore we speculate that the program for axon self-destruction in Wallerian degeneration may contribute also to degeneration owing to block of axonal transport. Indeed, disruption of the microtubule cytoskeleton appears to be the earliest sign of degeneration, as discussed before, which could further exacerbate the defects in axonal transport. On the other hand, a transport defect could lead to a lack of receiving survival factors that could trigger both developmental axon pruning and dying back diseases.

SUMMARY AND PERSPECTIVE

Pruning of exuberant neuronal connections is a fundamental and widespread mechanism to develop the diversity and specificity evident in axonal connections of the adult vertebrate brain. The phenomenon of developing exuberant axonal projections followed by selective pruning could provide a substrate for plasticity during both development and evolution of the nervous system. Alterations in pruning are a likely source of functional sparing or recovery following insults during development and may contribute to speciation. Large-scale reorganizations of neural circuits through pruning also occur in invertebrates, the best example of which is during metamorphosis of holometabolous insects. Although we have focused on pruning of axons, dendritic pruning is equally prevalent, which also contributes indirectly to synapse or axon elimination.

Developmental axon pruning occurs through retraction, degeneration, or a combination of both processes. Both retraction and degeneration appear to involve cell-intrinsic programs that execute these events. These axon destruction programs likely persist in adult neurons. The existence of these axon-destruction programs in developing and mature neurons requires tight regulation of these programs to maintain the stability of neuronal connections. We speculate that the function of these axon-destruction programs in the adult is to enable neurons to rapidly change their connections in response to learning, experience, or injury.

Axon degeneration and neuritic atrophies also occur widely in neurological and neurodegenerative diseases, and are the primary causes for some diseases. Protecting axon degeneration could be useful to ameliorate clinical symptoms. Axon degeneration in diseases may utilize mechanisms similar to those used during developmental axon pruning. We speculate that mis-regulation of axon-destruction programs could contribute significantly to the pathogenesis of certain neurological diseases, and interference of the axon self-destruction program may be a fruitful therapeutic approach.

We are only beginning to explore the mechanisms of axon degeneration and retraction. Given their relevance to wiring, plasticity, repair, and dysfunction of the nervous system, we feel that relatively little is known about the molecular and cellular mechanisms that govern these fundamental events, particularly during development. We hope that our review will inspire talented young scientists to work on these fascinating problems.

NOTE ADDED IN PROOF

A recent technically elegant study of axon elimination at the developing NMJ has revealed that as motor axons retract they shed membrane-enclosed axosomes, which are engulfed by surrounding Schwann cells, leading to a mixing of axonal and glial cytoplasm (Bishop et al. 2004). This mechanism is different from the previous view that axon retraction from a NMJ occurs by a progressive distal to proximal "resorption" of the retracting axon (Bernstein & Lichtman 1999, Walsh & Lichtman 2003) and is also distinct from a classical Wallerian degeneration, where no proximal to distal retraction is evident (see text) but shares some features of both. Whether this provocative finding proves to be unique for synaptic elimination at NMJ, and whether branch retraction occurring in the CNS is also accompanied by such resorption, remains to be determined.

ACKNOWLEDGMENTS

We thank Ben Barres, Zhigang He, Todd McLaughlin, Ryan Watts, Eric Hoopfer, and Oren Schuldiner for comments on this review. Research in our laboratories is supported by grants from the NIH.

LITERATURE CITED

Ahmad FJ, Hughey J, Wittmann T, Hyman A, Greaser M, Baas PW. 2000. Motor proteins regulate force interactions between microtubules and microfilaments in the axon. *Nat. Cell. Biol.* 2:276–80

Alsina B, Vu T, Cohen-Cory S. 2001. Visualizing synapse formation in arborizing optic axons in vivo: dynamics and modulation by BDNF. *Nat. Neurosci.* 4:1093–101

Araki T, Sasaki Y, Milbrandt J. 2004. Increased nuclear NAD biosynthesis and SIRT1 activation prevent axonal degeneration. *Science* 305:1010–13

Awasaki T, Ito K. 2004. Engulfing action of glial cells is required for programmed axon pruning during *Drosophila* metamorphosis. *Curr. Biol.* 14:668–77

Bagri A, Cheng HJ, Yaron A, Pleasure SJ, Tessier-Lavigne M. 2003. Stereotyped pruning of long hippocampal axon branches triggered by retraction inducers of the semaphorin family. *Cell* 113:285–99

Balice-Gordon RJ, Lichtman JW. 1994. Long-term synapse loss induced by focal blockade of postsynaptic receptors. *Nature* 372:519–524

Bansal A, Singer JH, Hwang B, Feller MB. 2000. Mice lacking specific nAChR subunits exhibit dramatically altered spontaneous activity patterns and reveal a limited role for retinal waves in forming ON/OFF circuits in the inner retina. *J. Neurosci.* 20:7672–81

Bastmeyer M, O'Leary DDM. 1996. Dynamics of target recognition by interstitial axon branching along developing cortical axons. *J. Neurosci.* 16:1450–59

Berger AR, Schaumburg HH. 1995. Human peripheral nerve disease (peripheral neuropathies). In *The Axon: Structure, Function and Pathophysiology*, ed. SG Waxman, JD Kocsis, PK Sytys, pp. 648–60. New York: Oxford Univ. Press

Bernstein M, Lichtman JW. 1999. Axonal atrophy: the retraction reaction. *Curr. Opin. Neurobio.* 9:364–70

Billuart P, Winter CG, Maresh A, Zhao X, Luo L. 2001. Regulating axon branch stability: the role of p190 RhoGAP in repressing a retraction signaling pathway. *Cell* 107:195–207

Bishop DL, Misgeld T, Walsh MK, Gan W-B, Lichtman JW. 2004. Axon branch removal at developing synapses by axosome shedding. *Neuron* 44:651–61

Bito H, Furuyashiki T, Ishihara H, Shibasaki Y, Ohashi K, et al. 2000. A critical role for a Rho-associated kinase, p160ROCK, in determining axon outgrowth in mammalian CNS neurons. *Neuron* 26:431–41

Bjartmar C, Wujek JR, Trapp BD. 2003. Axonal loss in the pathology of MS: consequences for understanding the progressive phase of the disease. *J. Neurol. Sci.* 206:165–71

Bommel H, Xie G, Rossoll W, Wiese S, Jablonka S, et al. 2002. Missense mutation in the tubulin-specific chaperone E (Tbce) gene in the mouse mutant progressive motor neuronopathy, a model of human motoneuron disease. *J. Cell. Biol.* 159:563–69

Buffelli M, Burgess RW, Feng G, Lobe CG, Lichtman JW, Sanes JR. 2003. Genetic evidence that relative synaptic efficacy biases the outcome of synaptic competition. *Nature* 424(6947):430–34

Campenot RB. 1982. Development of sympathetic neurons in compartmentalized cultures. II. Local control of neurite survival by nerve growth factor. *Dev. Biol.* 93:13–21

Cavanagh JB. 1964. The significance of "dying back" processes in human and experimental neurological diseases. *Int. Rev. Exp. Pathol.* 3:21967

Chen C, Regehr WG. 2000. Developmental remodeling of the retinogeniculate synapse. *Neuron* 28:955–66

Cohen-Cory S, Fraser SE. 1995. Effects of brain-derived neurotrophic factor on optic axon branching and remodelling in vivo. *Nature* 378:192–96

Cohen-Cory S. 1999. BDNF modulates, but does not mediate, activity-dependent branching and remodeling of optic axon arbors in vivo. *J. Neurosci.* 19:9996–10003

Coleman M, Perry V. 2002. Axon pathology in neurological disease: a neglected therapeutic target. *Trends Neurosci.* 25:532–37

Colman H, Nabekura J, Lichtman JW. 1997. Alterations in synaptic strength preceding axon withdrawal. *Science* 275:356–61

Conforti L, Tarlton A, Mack TG, Mi W, Buckmaster EA, et al. 2000. A Ufd2/D4Cole1e chimeric protein and overexpression of Rbp7 in the slow Wallerian degeneration (WldS) mouse. *Proc. Natl. Acad. Sci. USA* 97:11377–82

Deacon TW. 2000. Evolutionary perspectives on language and brain plasticity. *J. Comm. Disorders* 33:273–91

Debski EA, Cline HT. 2002. Activity-dependent mapping in the retinotectal projection. *Curr. Opin. Neurobio.* 12:93–99

Dehay C, Horsburgh G, Berland M, Killackey H, Kennedy H. 1989. Maturation and connectivity of the visual cortex in monkey is altered by prenatal removal of retinal input. *Nature* 337:265–67

DiAntonio A, Hicke L. 2004. Ubiquitin-dependent regulation of the synapse. *Annu. Rev. Neurosci.* 27:223–46

Eaton BA, Fetter RD, Davis GW. 2002. Dynactin is necessary for synapse stabilization. *Neuron* 34:729–41

Feller MB. 2002. The role of nAChR-mediated spontaneous retinal activity in visual system development. *J. Neurobiol.* 53:556–67

Ferri A, Sanes JR, Coleman MP, Cunningham JM, Kato AC. 2003. Inhibiting axon degeneration and synapse loss attenuates apoptosis and disease progression in a mouse model of motoneuron disease. *Curr. Biol.* 13:669–73

Finn JT, Weil M, Archer F, Siman R, Srinivasan A, Raff MC. 2000. Evidence that Wallerian degeneration and localized axon degeneration induced by local neurotrophin deprivation do not involve caspases. *J. Neurosci.* 20:1333–41

Frost DO, Moy YP, Smith DC. 1990. Effects of alternating monocular occlusion on the development of visual callosal connections. *Exp. Brain Res.* 83:200–9

Gao PP, Yue Y, Cerretti DP, Dreyfus C, Zhou R. 1999. Ephrin-dependent growth and pruning of hippocampal axons. *Proc. Natl. Acad. Sci. USA* 96(7):4073–77

Glass JD, Brushart TM, George EB, Griffin JW. 1993. Prolonged survival of transected nerve fibres in C57BL/Ola mice is an intrinsic characteristic of the axon. *J. Neurocytol.* 22:311–21

Griffin JW, George EB, Hsieh S-T, Glass JD. 1995. Axonal degeneration and disorders of the axonal cytoskeleton. In *The Axon: Structure, Function and Pathophysiology*, ed. SG Waxman, JD Kocsis, PK Sytys, pp. 375–90. New York: Oxford Univ. Press

Guan KL, Rao Y. 2003. Signalling mechanisms mediating neuronal responses to guidance cues. *Nat. Rev. Neurosci.* 4:941–56

Gunawardena S, Goldstein LS. 2004. Cargo-carrying motor vehicles on the neuronal highway: transport pathways and neurodegenerative disease. *J. Neurobiol.* 58:258–71

Hebb DO. 1949. *The Organization of Behavior: A Neuropsychological Theory*. New York: Wiley

He Z, Koprivica V. 2004. The Nogo signaling pathway for regeneration block. *Annu. Rev. Neurosci.* 27:341–68

Hensch TK. 2004. Critical period regulation. *Annu. Rev. Neurosci.* 27:549–79

Hindges R, McLaughlin T, Genoud N, Henkemeyer M, O'Leary DDM. 2002. EphB forward signaling controls directional branch extension and arborization required for dorsal ventral retinotopic mapping. *Neuron* 35:475–87

Huberman AD, Stellwagen D, Chapman B. 2002. Decoupling eye-specific segregation from lamination in the lateral geniculate nucleus. *J. Neurosci.* 22:9419–29

Innocenti GM. 1981. Growth and reshaping of axons in the establishment of visual callosal connections. *Science* 212:824–27

Innocenti GM. 1986. General organization of callosal connections in the cerebral cortex. In *Cerebral Cortex*, ed. EG Jones, A Peters, pp. 291–353. New York: Plenum. Vol. 5

Innocenti GM. 1995. Exuberant development of connections, and its possible permissive role in cortical evolution. *Trends Neurosci.* 18:397–402

Innocenti GM, Kiper DC, Knyazeva MG, Deonna TW. 1999. On nature and limits of cortical developmental plasticity after an early lesion in a child. *Restor. Neurol. Neurosci.* 15:219–27

Innocenti GM, Ansermet F, Parnas J. 2003. Schizophrenia, neurodevelopment and corpus callosum. *Mol. Psychiatry* 8:261–74

Jalink K, van Corven EJ, Hengeveld T, Morii N, Narumiya S, Moolenaar WH. 1994. Inhibition of lysophosphatidate- and thrombin-induced neurite retraction and neuronal cell rounding by ADP ribosylation of the small GTP-binding protein Rho. *J. Cell. Biol.* 126:801–10

Kasthuri N, Lichtman JW. 2003. The role of neuronal identity in synaptic competition. *Nature* 424(6947):426–30

Kasthuri N, Lichtman JW. 2004. Structural dynamics of synapses in living animals. *Curr. Opin. Neurobiol.* 14(1): 105–11

Katz LC, Callaway EM. 1992. Development of local circuits in mammalian visual cortex. *Ann. Rev. Neurosci.* 15:31–56

Katz LC, Shatz CJ. 1996. Synaptic activity and the construction of cortical circuits. *Science* 274:1133–38

Koralek KA, Killackey HP. 1990. Callosal projections in rat somatosensory cortex are altered by early removal of afferent input. *Proc. Natl. Acad. Sci. USA* 87:1396–1400

Kozma R, Sarner S, Ahmed S, Lim L. 1997. Rho family GTPases and neuronal growth cone remodelling: relationships between increased complexity induced by Cdc42Hs, Rac1, and acetylcholine and collapse induced by RhoA and lysophosphatidic acid. *Mol. Cell. Biol.* 17:1201–11

Lamprecht R, Farb CR, LeDoux JE. 2002. Fear memory formation involves p190 RhoGAP and ROCK proteins through a GRB2-mediated complex. *Neuron* 36:727–38

Lee T, Lee A, Luo L. 1999. Development of *Drosophila* mushroom bodies: sequential generation of three distinct types of neurons from a single neuroblast. *Development* 126:4065–76

Lee T, Luo L. 1999. Mosaic analysis with a repressible cell marker for studies of gene function in neuronal morphogenesis. *Neuron* 22:451–61

Lee T, Marticke S, Sung C, Robinow S, Luo L. 2000. Cell-autonomous requirement of the USP/EcR-B ecdysone receptor for mushroom body neuronal remodeling in *Drosophila*. *Neuron* 28:807–18

Levine RB, Morton DB, Restifo LL. 1995. Remodeling of the insect nervous system. *Curr. Opin. Neurobiol.* 5:28–35

Li H, Li SH, Yu ZX, Shelbourne P, Li XJ. 2001. Huntingtin aggregate-associated axonal degeneration is an early pathological event in Huntington's disease mice. *J. Neurosci.* 21:8473–81

Li Z, Van Aelst L, Cline HT. 2000. Rho GTPases regulate distinct aspects of dendritic arbor growth in Xenopus central neurons in vivo. *Nat. Neurosci.* 3:217–25

Lohof AM, Delhaye-Bouchaud N, Mariani J. 1996. Synapse elimination in the central nervous system: functional significance and cellular mechanisms. *Rev. Neurosci.* 7:85–101

Luo L. 2002. Actin cytoskeleton regulation in neuronal morphogenesis and structural plasticity. *Annu. Rev. Cell Dev. Biol.* 18:601–35

Lund RD, Mitchell DE, Henry GH. 1978. Squint-induced modification of callosal connections in cats. *Brain Res.* 144:169–72

Lunn ER, Perry VH, Brown MC, Rosen H, Gordon S. 1989. Absence of Wallerian degeneration does not hinder regeneration in peripheral nerve. *Eur. J. Neurosci.* 1:27–33

Mack TG, Reiner M, Beirowski B, Mi W, Emanuelli M, et al. 2001. Wallerian degeneration of injured axons and synapses is delayed by a Ube4b/Nmnat chimeric gene. *Nat. Neurosci.* 4:1199–206

Mariani J, Changeux JP. 1981. Ontogenesis of olivocerebellar relationships. I. Studies by intracellular recordings of the multiple innervation of Purkinje cells by climbing fibers in the developing rat cerebellum. *J. Neurosci.* 1:696–702

Marin EC, Watts RJ, Takana NK, Ito K, Luo L. 2005. Developmentally programmed remodeling of the *Drosophila* olfactory circuit. *Development* 132:725–37

Martin N, Jaubert J, Gounon P, Salido E, Haase G, et al. 2002. A missense mutation in Tbce causes progressive motor neuronopathy in mice. *Nat. Genet.* 32:443–47

Mason CA, Sretavan DW. 1997. Glia, neurons, and axon pathfinding during optic chiasm development. *Curr. Opin. Neurobiol.* 7:647–53

McLaughlin T, Hindges R, O'Leary DDM. 2003a. Regulation of axial patterning of the retina and its topographic mapping in the brain. *Curr. Opin. Neurobiol.* 13:57–69

McLaughlin T, Hindges R, Yates PA, O'Leary DDM. 2003b. Bifunctional action of ephrin-B1 as a repellent and attractant to control bidirectional branch extension in dorsal-ventral retinotopic mapping. *Development* 130:2407–18

McLaughlin T, Torborg CL, Feller MB, O'Leary DDM. 2003c. Retinotopic map refinement requires spontaneous retinal waves during a brief critical period of development. *Neuron* 40:1147–60

Meister M, Pine J, Baylor DA. 1994. Multi-neuronal signals from the retina: acquisition and analysis. *J. Neurosci. Methods* 51:95–106

Meister M, Wong RO, Baylor DA, Shatz CJ. 1991. Synchronous bursts of action potentials in ganglion cells of the developing mammalian retina. *Science* 252:939–43

Muir-Robinson G, Hwang BJ, Feller MB. 2002. Retinogeniculate axons undergo eye-specific segregation in the absence of eye-specific layers. *J. Neurosci.* 22:5259–64

Nakamura H, O'Leary DDM. 1989. Inaccuracies in initial growth and arborization of chick retinotectal axons followed by course corrections and axon remodeling to develop topographic order. *J. Neurosci.* 9:3776–95

Nakayama AY, Harms MB, Luo L. 2000. Small GTPases Rac and Rho in the maintenance of dendritic spines and branches in hippocampal pyramidal neurons. *J. Neurosci.* 20:5329–38

O'Leary DDM. 1992. Development of connectional diversity and specificity in the mammalian brain by the pruning of collateral projections. *Curr. Opin. Neurobiol.* 2:70–77

O'Leary DDM, Bicknese AR, De Carlos JA, Heffner CD, Koester SE, et al. 1990. Target selection by cortical axons: alternative mechanisms to establish axonal connections in the developing brain. *Cold Spring Hbr. Symp. Quant. Biol.* 55:453–68

O'Leary DDM, Koester SE. 1993. Development of projection neuron types, axonal pathways and patterned connections of the mammalian cortex. *Neuron* 10:991–1006

O'Leary DDM, Stanfield BB. 1989. Selective elimination of axons extended by developing cortical neurons is dependent on regional locale. Experiments utilizing fetal cortical transplants. *J. Neurosci.* 9:2230–46

O'Leary DDM, Terashima T. 1988. Cortical axons branch to multiple subcortical targets by interstitial axon budding: implications for target recognition and "waiting periods." *Neuron* 1:901–10

O'Leary DDM, Stanfield BB, Cowan WM. 1981. Evidence that early postnatal restriction of the cells of origin of the callosal projection is due to the elimination of axonal collaterals rather than to the death of neurons. *Dev. Brain Res.* 1:607–17

O'Rourke NA, Cline HT, Fraser SE. 1994. Rapid remodeling of retinal arbors in the tectum with and without blockade of synaptic transmission. *Neuron* 12:921–34

Penn AA, Riquelme PA, Feller MB, Shatz CJ. 1998. Competition in retinogeniculate patterning driven by spontaneous activity. *Science* 279:2108–12

Personius KE, Balice-Gordon RJ. 2000. Activity-dependent editing of neuromuscular synaptic connections. *Brain Res. Bull.* 53(5):513–22

Personius KE, Balice-Gordon RJ. 2001. Loss of correlated motor neuron activity during synaptic competition at developing neuromuscular synapses. *Neuron* 31:395–408

Raff MC, Whitmore AV, Finn JT. 2002. Axonal self-destruction and neurodegeneration. *Science* 296:868–71

Reid RC. 2003. Vision. See Squire et al. 2003, pp. 469–98

Reinoso BS, O'Leary DDM. 1989. Extension, branching and regression of visual corticospinal axons in rat. *Soc. Neurosci. Abstr.* 15:1337

Rossi FM, Pizzorusso T, Porciatti V, Marubio LM, Maffei L, Changeux JP. 2001. Requirement of the nicotinic acetylcholine receptor beta 2 subunit for the anatomical and functional development of the visual system. *Proc. Natl. Acad. Sci. USA* 98:6453–58

Ruthazer ES, Akerman CJ, Cline HT. 2003. Control of axon branch dynamics by correlated activity in vivo. *Science* 301:66–70

Sagot Y, Dubois-Dauphin M, Tan SA, de Bilbao F, Aebischer P, et al. 1995. Bcl-2 overexpression prevents motoneuron cell body loss but not axonal degeneration in a mouse model of a neurodegenerative disease. *J. Neurosci.* 15:7727–33

Sagot Y, Tan SA, Hammang JP, Aebischer P, Kato AC. 1996. GDNF slows loss of motoneurons but not axonal degeneration or premature death of pmn/pmn mice. *J. Neurosci.* 16:2335–41

Sajadi A, Schneider BL, Aebischer P. 2004. Wlds-mediated protection of dopaminergic fibers in an animal model of Parkinson disease. *Curr. Biol.* 14:326–30

Samsam M, Mi W, Wessig C, Zielasek J, Toyka KV, et al. 2003. The Wlds mutation delays robust loss of motor and sensory axons in a genetic model for myelin-related axonopathy. *J. Neurosci.* 23:2833–39

Schmalbruch H, Jensen HJ, Bjaerg M, Kamieniecka Z, Kurland L. 1991. A new mouse mutant with progressive motor neuronopathy. *J. Neuropathol. Exp. Neurol.* 50:192–204

Schubiger M, Tomita S, Sung C, Robinow S, Truman JW. 2003. Isoform specific control of gene activity in vivo by the *Drosophila* ecdysone receptor. *Mech. Dev.* 120:909–18

Schubiger M, Wade AA, Carney GE, Truman JW, Bender M. 1998. *Drosophila* EcR-B ecdysone receptor isoforms are required for larval molting and for neuron remodeling during metamorphosis. *Development* 125:2053–62

Shatz CJ. 1977. Anatomy of interhemispheric connections in the visual system of Boston Siamese and ordinary cats. *J. Comp. Neurol.* 173:497–518

Solomon F, Magendantz M. 1981. Cytochalasin separates microtubule disassembly from loss of asymmetric morphology. *J. Cell. Biol.* 89:157–61

Squire LR, Bloom FE, McConnell SK, Roberts JL, Spitzer NC, Zigmond MJ, eds. 2003. *Fundamentals of Neuroscience*. San Diego, CA: Academic

Stanfield BB, Nahin B, O'Leary DDM. 1987. A transient post-mammillary component of the rat fornix during development: implications for interspecific differences in mature axonal projections. *J. Neurosci.* 7:3350–61

Stanfield BB, O'Leary DDM. 1985. Fetal occipital cortical neurons transplanted to rostral cortex develop and maintain a pyramidal tract axon. *Nature* 313:135–37

Stanfield BB, O'Leary DDM. 1988. Neurons in the subiculum with transient postmammillary collaterals during development maintain projections to the mammillary complex. *Exp. Brain Res.* 72:185–90

Stanfield BB, O'Leary DDM, Fricks C. 1982. Selective collateral elimination in early postnatal development restricts cortical distribution of rat pyramidal tract neurones. *Nature* 298:371–73

Stellwagen D, Shatz CJ. 2002. An instructive role for retinal waves in the development of retinogeniculate connectivity. *Neuron* 33:357–67

Stent GS. 1973. A physiological mechanism for Hebb's postulate of learning. *Proc. Natl. Acad. Sci. USA* 70:997–1001

Steward O, Schuman EM. 2003. Compartmentalized synthesis and degradation of proteins in neurons. *Neuron* 40:347–59

Technau G, Heisenberg M. 1982. Neural reorganization during metamorphosis of the corpora pedunculata in *Drosophila melanogaster*. *Nature* 295:405–7

Tian N, Copenhagen DR. 2003. Visual stimulation is required for refinement of ON and OFF pathways in postnatal retina. *Neuron* 39:85–96

Tissot M, Stocker RF. 2000. Metamorphosis in *Drosophila* and other insects: the fate of neurons throughout the stages. *Prog. Neurobiol.* 62:89–111

Truman JW. 1990. Metamorphosis of the central nervous system of *Drosophila*. *J. Neurobiol.* 21:1072–84

Tsai J, Grutzendler J, Duff K, Gan WB. 2004. Fibrillar amyloid deposition leads to local synaptic abnormalities and breakage of neuronal branches. *Nat. Neurosci.* 7:1181–83

Waller A. 1850. Experiments on the section of glossopharyngeal and hypoglossal nerves of the frog and observations of the alternatives produced thereby in the structure of their primitive fibers. *Phil. Trans. R. Soc. London* 140:423–29

Walsh MK, Lichtman JW. 2003. In vivo time-lapse imaging of synaptic takeover associated with naturally occurring synapse elimination. *Neuron* 37:67–73

Watts RJ, Hoopfer ED, Luo L. 2003. Axon pruning during *Drosophila* metamorphosis: evidence for local degeneration and requirement of the ubiquitin-proteasome system. *Neuron* 38:871–85

Watts RJ, Schuldiner O, Perrino J, Larsen C, Luo L. 2004. Glia engulf degenerating axons during developmental axon pruning. *Curr. Biol.* 14:678–84

Weeks JC, Truman JW. 1985. Independent steroid control of the fates of metamorphosis and their muscles during insect metamorphosis. *J. Neurosci.* 5:2290–300

Wiesel TN. 1982. Postnatal development of the visual cortex and the influence of environment. *Nature* 29:583–91

Weimann JM, Zhang YA, Levin ME, Devine WP, Brulet P, McConnell SK. 1999. Cortical neurons require Otx1 for the refinement of exuberant axonal projections to subcortical targets. *Neuron* 24:819–31

Wong RO. 1999. Retinal waves and visual system development. *Annu. Rev. Neurosci.* 22:29–47

Wong RO, Meister M, Shatz CJ. 1993. Transient period of correlated bursting activity during development of the mammalian retina. *Neuron* 11:923–38

Wong ROL, Lichtman JW. 2003. Synapse elimination. See Squire et al. 2003, pp. 533–54

Wong WT, Faulkner-Jones B, Sanes JR, Wong ROL. 2000. Rapid dendritic remodeling in the developing retina: dependence on neurotransmission and reciprocal regulation by Rac and Rho. *J. Neurosci.* 20:5024–36

Yamada KM, Spooner BS, Wessells NK. 1970. Axon growth: roles of microfilaments and microtubules. *Proc. Natl. Acad. Sci. USA* 66:1206–12

Yates PA, Holub AD, McLaughlin T, Sejnowski TJ, O'Leary DDM. 2004. Computational modeling of retinotopic map development to define contributions of EphA-ephrin-A gradients, axon-axon interactions and patterned activity. *J. Neurobiol.* 59:95–113

Yates PA, Roskies AR, McLaughlin T, O'Leary DDM. 2001. Topographic specific axon branching controlled by ephrin-As is the critical event in retinotectal map development. *J. Neurosci.* 21:8548–63

Zhai Q, Wang J, Kim A, Liu Q, Watts R, et al. 2003. Involvement of the ubiquitin-proteasome system in the early stages of Wallerian degeneration. *Neuron* 39:217–25

Zhang YA, Okada A, Lew CH, McConnell SK. 2002. Regulated nuclear trafficking of the homeodomain protein Otx1 in cortical neurons. *Mol. Cell. Neurosci.* 19:430–46

Zheng X, Wang J, Haerry TE, Wu AY, Martin J, et al. 2003. TGF-beta signaling activates steroid hormone receptor expression during neuronal remodeling in the *Drosophila* brain. *Cell* 112:303–15

Zhou Q, Tao HW, Poo MM. 2003. Reversal and stabilization of synaptic modifications in a developing visual system. *Science* 300(5627):1953–57

Zufferey PD, Jin F, Nakamura H, Tettoni L, Innocenti GM. 1999. The role of pattern vision in the development of cortico-cortical connections. *Eur. J. Neurosci.* 11:2669–88

Structure and Function of Visual Area MT

Richard T. Born[1] and David C. Bradley[2]

[1]Department of Neurobiology, Harvard Medical School, Boston, Massachusetts 02115-5701; email: rborn@hms.harvard.edu

[2]Department of Psychology, University of Chicago, Chicago, Illinois 60637; email: bradley@uchicago.edu

Key Words

extrastriate, motion perception, center-surround antagonism, magnocellular, structure-from-motion, aperture problem

Abstract

The small visual area known as MT or V5 has played a major role in our understanding of the primate cerebral cortex. This area has been historically important in the concept of cortical processing streams and the idea that different visual areas constitute highly specialized representations of visual information. MT has also proven to be a fertile culture dish—full of direction- and disparity-selective neurons—exploited by many labs to study the neural circuits underlying computations of motion and depth and to examine the relationship between neural activity and perception. Here we attempt a synthetic overview of the rich literature on MT with the goal of answering the question, What does MT do?

Contents

INTRODUCTION 158
MT WAS A KEY PART OF THE
    EARLY EXPLORATION OF
    EXTRASTRIATE CORTEX 158
CONNECTIONS 159
FUNCTIONAL ORGANIZATION 162
BASIC TUNING PROPERTIES ... 164
SURROUND MECHANISMS 165
THE COMPUTATION OF
    VELOCITY 168
NOISE REDUCTION 173
SEGMENTATION 173
THE COMPUTATION OF
    STRUCTURE 174
EXTRARETINAL EFFECTS 176
PERCEPTUAL CORRELATES
    AND POPULATION CODES ... 177
    Single-Neuron Sensitivity 177
    Vector Summation versus
        Winner-Take-All 178
    Distributed Speed and Acceleration
        Codes 179
CONCLUSIONS 179

INTRODUCTION

The middle temporal visual area (MT or V5) of the macaque monkey possesses a number of attributes that have made it particularly attractive to systems neuroscientists. This region is typical of extrastriate cortex but is still readily identifiable both anatomically and functionally. Though extrastriate, it is still quite close to the retina—its principle inputs as few as five synapses from the photoreceptors—a feature which means, among other things, that the mechanisms by which its receptive field properties arise can be profitably studied. And, although MT neurons are near enough to the inputs to be mechanistically tractable, they are also close enough to some outputs—in particular, those involved in eye movements—to provide an easily measurable, continuous readout of computations performed in this pathway. Finally, MT neurons are concerned with visual motion, which is of obvious ethological importance, which has been extensively characterized psychophysically, and for which there are well-defined mathematical descriptions. Much of the work on MT has focused on its role in visual motion processing, though, as we hope to make clear in what follows, MT plays a richer and more varied role in vision.

MT WAS A KEY PART OF THE EARLY EXPLORATION OF EXTRASTRIATE CORTEX

Part of MT's significance is historical; it played an important role in the discovery of new extrastriate visual areas (Felleman & Van Essen 1991) and in the idea that they constitute specialized representations of the visual world (Zeki 1978, Barlow 1986).

At the beginning of the twentieth century, primate visual cortex was thought to consist of only three architectonically distinct fields (Brodmann 1909). Beginning in the late 1940s, however, it became clear that considerably more of the cortex was involved in vision. The first demonstration came from temporal lobe lesions that produced visual impairment (Mishkin 1954, Mishkin & Pribram 1954) unaccompanied by deficits in other sensory modalities (Weiskrantz & Mishkin 1958, Brown 1963). Mapping studies using surface electrodes also revealed visually responsive regions well anterior to those traditionally associated with vision (Talbot & Marshall 1941, Clare & Bishop 1954, Woolsey et al. 1955). In addition, new anatomical techniques (Nauta & Gygax 1954) permitted the labeling of connections after lesions of striate cortex (Kuypers et al. 1965, Cragg & Ainsworth 1969, Zeki 1969), which revealed a direct striate (V1) projection zone situated on the posterior bank of the superior temporal sulcus (STS).

MT was discovered at roughly the same time by two different groups. In England, Dubner & Zeki (1971) were able to record

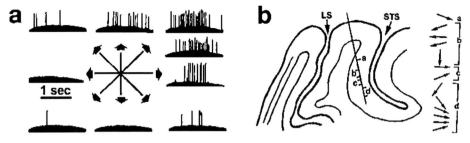

Figure 1

First demonstration of direction selectivity in macaque MT/V5 by Dubner & Zeki (1971). (*a*) Neuronal responses to a bar of light swept across the receptive field in different directions (modified from figure 1 of Dubner & Zeki 1971). Each trace shows the spiking activity of the neuron as the bar was swept in the direction indicated by the arrow. The neuron's preferred direction was up and to the right. (*b*) Oblique penetration through MT (modified from figure 3 of Dubner & Zeki 1971) showing the shifts in preferred direction indicative of the direction columns subsequently demonstrated by Albright et al. (1984). See also **Figure 4**.

visual responses from the V1-projection zone in anesthetized macaques, in so doing establishing a number of physiological hallmarks, particularly their direction-selective responses (**Figure 1***a*). Quite presciently, they also suggested a columnar organization for direction-selective neurons (**Figure 1***b*) and a role for MT signals in guiding pursuit eye movements, both subsequently confirmed (Albright et al. 1984, Lisberger et al. 1987). Around the same time, Allman & Kaas (1971) were recording from owl monkeys and using a different approach. They made systematic rows of microelectrode penetrations across the entire cortex, mapping receptive fields as they went, thus discovering a large number of retinotopically organized maps. One of these, which they named MT for middle temporal, mapped onto a well-defined region of dense myelination in the lower layers and contained neurons that responded better to drifting bars than to flashed spots. The myelination was also later shown to be characteristic of the macaque motion area (Van Essen et al. 1981), which Zeki subsequently named V5. This histochemical feature has been an underappreciated factor in contributing to the detail with which MT has been studied because it has permitted reliable comparisons across different studies.

Following the first studies, a series of papers confirmed that MT contained a high concentration of direction-selective neurons in several species of both New and Old World monkeys (Zeki 1974, 1980; Baker et al. 1981; Van Essen et al. 1981; Maunsell & Van Essen 1983a,b; Felleman & Kaas 1984). These studies indicated that MT was both unique as a cortical area highly specialized for visual motion and, at the same time, common to a number of different primate species.

CONNECTIONS

Like every other cortical area, MT has a rich set of interconnections with other regions of the cortex as well as with numerous subcortical structures. These connections have been discussed in previous publications (Felleman & Van Essen 1991, Orban 1997, Lewis & Van Essen 2000), so we do not recapitulate them here. From a broad perspective, MT's corticocortical connections identify it as one of the main inputs into the dorsal or posterior parietal processing stream (Ungerleider & Mishkin 1982, Maunsell & Newsome 1987), and its key outputs target structures that are implicated in the analysis of optic flow (e.g., MST, VIP) and the generation of eye movements (e.g., LIP, FEF, SC, dorsolateral

Preferred direction: the direction of motion eliciting the greatest response from a given neuron

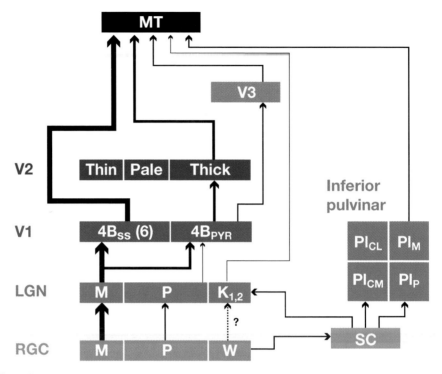

Figure 2

Gestalt map of major routes into MT in the manner of Felleman & Van Essen (1991). Line thickness is roughly proportional to the magnitude of the inputs, on the basis of a combination of projection neuron numbers and, where data are available, the characteristics of their axon terminals (see **Figure 3**). The thickest lines represent the direct cortical pathway emphasized in the text. Following are important caveats: The pathways shown are those discussed in the text and omit a number of known feedforward cortical inputs that appear lesser in magnitude (V3A, VP, PIP) as well as many subcortical inputs. The sources of the direct and indirect projections from V1 are probably not defined purely by cell morphology (i.e., spiny stellate versus pyramidal; see Elston & Rosa 1997), though they are largely distinct; the largest 4B cells contribute to the direct pathway (Sincich & Horton 2003). The precise nature of the retinal inputs to K1,2 is not known, though their response properties are W-like in the galago (Irvin et al. 1986). Also, the proposed input to MT from the SC via the pulvinar is rendered problematic by the finding that, in owl monkey pulvinar, the principle target of SC terminals (PI_{CM}) is different from the main source of MT projections (PI_M) (see Stepniewska et al. 1999). Abbreviations: $4B_{SS}$, spiny stellate neurons in layer 4B; $4B_{PYR}$, pyramidal neurons in layer 4B; LGN, lateral geniculate nucleus; M, magnocellular stream; P, parvocellular stream; K, koniocellular layers of LGN; PI_{CL}, central lateral nucleus of the inferior pulvinar; PI_{CM}, central medial nucleus of the inferior pulvinar; PI_M, medial nucleus of the inferior pulvinar; PI_P, posterior nucleus of the inferior pulvinar; RGC, retinal ganglion cells; SC, superior colliculus; VP, ventral posterior area.

pons). Because we desire to address how MT neurons acquire their unique visual response properties and discuss the role they play in motion computations, we focus the present discussion on MT's major inputs (**Figure 2**) and their functional implications. In particular, we argue that the most important input to MT is from a magnocellular-dominated projection originating from layer 4B of V1.

To a first approximation, MT is dominated by cortical rather than subcortical inputs. Nevertheless, unlike some other extrastriate areas such as V2 (Schiller & Malpeli

1977, Girard & Bullier 1989) and V4 (Girard et al. 1991), which are completely dependent on input from striate cortex, at least some MT neurons remain both visually responsive and even direction-selective following removal or inactivation of V1 (Rodman et al. 1989, Girard et al. 1992). In some cases, residual MT function may have been conferred by callosal connections from the intact hemisphere (Girard et al. 1992) and, in other cases, via the superior colliculus (SC) (Rodman et al. 1990). However, although SC lesions completely eliminated residual MT responses after V1 lesions, alone they produced no significant changes in MT responses (Rodman et al. 1990).

Also of interest in this regard is a small, direct LGN input to MT, mainly from koniocellular neurons (Stepniewska et al. 1999, Sincich et al. 2004). Although such a projection could, in theory, explain remaining function in MT after V1 lesions, it is unclear how these LGN cells would convey direction selectivity to MT or why SC lesions would abolish it. Rodman et al. (1990) raised the possibility that extrastriate-projecting LGN neurons did not receive a direct retinal input—supported by the anatomical study by Benevento & Yoshida (1981) in which intraocular injections of tritiated amino acids failed to produce labeling in extrastriate cortex—but instead were dependent on SC inputs for their visual responsiveness. This idea remains an interesting possibility, which awaits a direct test.

Cortical feedforward inputs to MT come from several areas, including V1, V2, V3, V3A, VP, and PIP (Maunsell & Van Essen 1983c, Felleman & Van Essen 1991); those from V2, V1, and V3 are the largest inputs, judging from the numbers of labeled neurons in each area after MT injections (Maunsell & Van Essen 1983c). However, single axon data suggest that the most potent input is probably from V1. These studies show that some V1 inputs to MT are highly specialized (**Figure 3**): They have larger axons (up to 3 μm in diameter, versus 1 μm for other corticocortical axons; Rockland 1989, 1995) and terminal boutons that are both larger and more complex than those from V2, often forming multiple synapses on a single MT neuron (Rockland 1989, 1995, Anderson et al. 1998, Anderson & Martin 2002). These specializations, which appear unique to the V1-to-MT projection, should combine to provide fast and secure synaptic transmission, though this idea has not received a direct test.

Of the inputs directly from V1, those from layer 4B predominate, at least numerically. After injections of retrograde tracers into macaque MT, more than 90% of the labeled V1 neurons are found in layer 4B; the remaining are found in the large cells of Meynert near the boundary of layers 5 and 6 (Tigges et al. 1981, Maunsell & Van Essen 1983c, Shipp & Zeki 1989a) (**Figure 3a**). These MT-projecting 4B neurons are predominantly spiny stellate in morphology (Shipp & Zeki 1989a) (though, see also Elston & Rosa 1997), are the largest cells in this layer (Sincich & Horton 2003), and appear to receive exclusively M-inputs via layer 4Cα (Yabuta et al. 2001). In addition to this direct V1-MT connection, there are important indirect cortical inputs via V3 (Maunsell & Van Essen 1983c) and the thick cytochrome oxidase stripes of V2 (DeYoe & Van Essen 1985; Shipp & Zeki 1985, 1989b) (**Figure 3a**). These indirect inputs also originate in V1 but from a mostly separate population of neurons within layer 4B (Sincich & Horton 2003) that receives a mixed M and P input (though still predominantly M by a margin of about 2.5:1) (Yabuta et al. 2001) and is preferentially distributed beneath interblob regions (Sincich & Horton 2002). Overall, this anatomical picture is consistent with functional studies, showing that reversible inactivation of the M-layers of the LGN nearly completely abolishes the visual responsiveness of MT neurons, whereas P-layer inactivation has a much smaller, though measurable, effect (Maunsell et al. 1990), the latter presumably mediated by the indirect pathway.

FUNCTIONAL ORGANIZATION

MT is retinotopically organized, each hemisphere containing a more or less complete map of the contralateral visual hemi-field, with a marked emphasis on the fovea [the central 15° of the visual field occupies over half of MT's surface area (Van Essen et al. 1981)] and a bias toward the lower quadrant of the visual field (Maunsell & Van Essen 1987). Within this relatively crude retinotopic map,

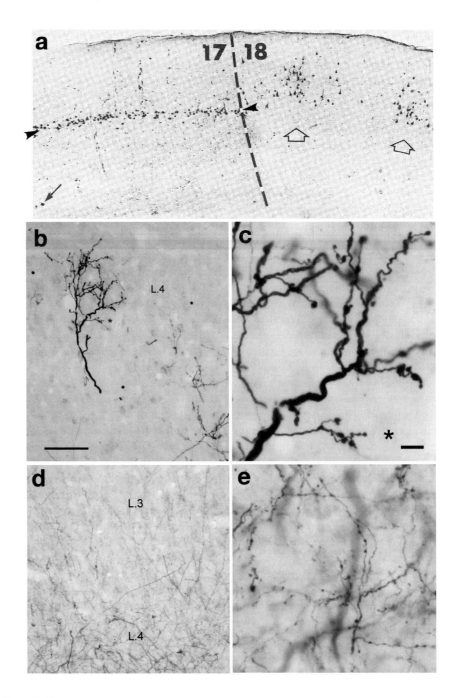

there exist, at finer spatial scales, several other organizations concerning neural tuning for different stimulus parameters.

MT was the first extrastriate visual area for which clear-cut evidence of a columnar organization was discovered. The possibility of direction columns was raised in the initial publication of Dubner & Zeki (1971) but was not shown definitively until more than 10 years later (Albright et al. 1984). Evidence for a columnar organization consisted of relatively smooth changes in the preferred directions of neurons during oblique penetrations through MT, with direction sequences interrupted occasionally by sudden jumps and, in particular, by jumps of 180° more often than would be predicted by chance. To account for their observations, Albright and coworkers proposed a model in which columns of smoothly varying preferred directions ran side by side with a corresponding set of columns preferring the locally opposite direction. This was subsequently supported by functional labeling studies using 2-deoxyglucose (Geesaman et al. 1997). DeAngelis & Newsome later showed a strong columnar organization in terms of tuning for binocular disparity (DeAngelis & Newsome 1999) coexisting with the direction columns (**Figure 4**). There is also a clustering of neurons by speed preference, but the organization is not strictly columnar (Liu & Newsome 2003b).

In the macaque, neurons whose receptive fields possess antagonistic surrounds are more common in supragranular layers, whereas those lacking such surrounds are found predominantly in the input layers (Lagae et al. 1989, Raiguel et al. 1995). In the owl monkey, where direction columns are not well defined (Malonek et al. 1994) and tuning for binocular disparity is rare (Felleman & Kaas 1984, Born 2000), a robust columnar organization exists with respect to center-surround interactions: groups of neurons having antagonistic surrounds interdigitated with neurons that lack such surrounds and therefore respond optimally to wide-field motion (Born & Tootell 1992, Berezovskii & Born 2000, Born 2000). The evidence for such an organization in the macaque is more equivocal: Some investigators have found no consistent organization (DeAngelis & Newsome 1999), and others report a tangential clustering (i.e., parallel to the cortical surface) (Raiguel et al. 1995) in addition to the laminar segregation noted

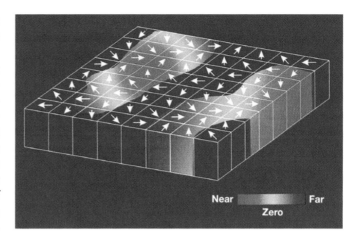

Figure 4

Functional organization of macaque MT (from DeAngelis & Newsome 1999). Superimposed on the model of direction columns originally proposed by Albright et al. (1984) are the columnar zones of strong (*rainbows*) and weak (*blue*) binocular disparity tuning. Within the zones of strong disparity tuning, the preferred disparities vary in a smooth manner, similar to the direction columns and to orientation columns in V1.

Figure 3

Comparison of two of the major cortical inputs to MT. (*a*) Labeled neurons in V1 (17) and V2 (18) after a large injection of HRP into MT of a squirrel monkey (from figure 8 of Tigges et al. 1981; 40 × magnification). The far greater number of labeled V1 cells in layer 4B (*arrowheads*) as compared with layer 6 (*arrow*) has also been found in the macaque monkey (Maunsell & Van Essen 1983c, Shipp & Zeki 1989a). Note the two well-defined clusters in the upper layers of V2 (*open arrows*), subsequently shown to be confined to the thick cytochrome oxidase stripes (DeYoe & Van Essen 1985, Shipp & Zeki 1985). (*b–e*) High resolution tracers (BDA) demonstrate differences in size and shape of terminal arbors and boutons from V1 (*b, c*) and V2 (*d, e*) within macaque MT (K. Rockland, unpublished data). (*b, c*) Example of a large-caliber V1 axon with large boutons. Note, however, the mix of large and small boutons. (*d, e*) Field of terminations from V2. Note more uniformly small boutons. The scale bar is 100 μm in panels *b* and *d*, and 10 μm in panels *c* and *e*. See also Rockland (2002).

Binocular disparity: a difference in the relative position of a stimulus on each of the two retinas. This disparity is the basis of stereoscopic depth perception.

above. This clustering may occur predominantly in the upper (output) layers of macaque MT because a segregation of center-surround properties appears to exist in subdivisions of one of MT's main projection zones, MSTd versus MSTl (Tanaka et al. 1986, Komatsu & Wurtz 1988, Eifuku & Wurtz 1998). The finding that microstimulation in macaque MT has qualitatively different effects on smooth pursuit eye movements depending on the nature of center-surround interactions at the stimulation site (Born et al. 2000) is consistent with the idea of segregation and, to date, is the only direct experimental evidence to support a role for MT center-surround interactions in the type of figure-ground comparisons originally suggested by Allman and his colleagues (Allman et al. 1985b).

BASIC TUNING PROPERTIES

The visual responses of MT neurons are determined principally by five properties of the stimulus: (*a*) retinal position, (*b*) direction of motion, (*c*) speed of motion, (*d*) binocular disparity, and (*e*) stimulus size (due to surround suppression). In the following paragraphs we expand on selected aspects of the above description. The basic visual-response properties of MT neurons have been addressed extensively in excellent reviews by Orban (1997) and Britten (2003), so we focus on more recent studies and on those most relevant to the subsequent discussion. In particular, we attempt, where possible, to compare MT properties with those of its principle inputs, with an eye toward understanding what MT contributes to cortical motion processing.

The principal V1 inputs to MT described above are themselves strongly direction selective (Movshon & Newsome 1996) and also are tuned for speed (Orban et al. 1986). Moreover, given that the most direction-selective V1 cells also tend to be highly selective for binocular disparity (Prince et al. 2000), many of MT's V1 inputs are also likely disparity tuned. Conceivably then, MT neurons inherit three of their most important properties from V1.

So what, then, does MT add? It is chastening to note up front that more than 30 years of physiology have not yielded a clear-cut answer to this question. In a number of cases, early evidence pointed to differences that were theoretically attractive but that, upon closer and more quantitative analysis, failed to materialize. Because these examples are informative, we first consider several of them, before moving on to other differences that have been borne out.

MT receptive fields are much larger than those in V1—a ballpark figure is tenfold greater in linear dimensions—so one might suppose that MT neurons can compute motion and disparity over larger spatial ranges than can V1 cells. This idea was particularly attractive for directional interactions because studies of human perception have revealed at least two different motion-sensing mechanisms that operate over different spatial scales and also differ with respect to other properties, such as contrast invariance (Nakayama 1985). As a result, investigators proposed that MT neurons might inherit short-range motion sensitivity from their V1 inputs but would, in addition, compute motion over longer spatial displacements, commensurate with the size of their receptive fields. Indeed, initial experiments by Mikami and colleagues (1986), using sequences of flashed bars, suggested that MT neurons did produce directional signals to larger spatial separations, on average, than did V1 cells. However, a recent reexamination of this issue, using more directly comparable stimuli, revealed very similar upper limits for V1 and MT (Churchland et al. 2004). In fact, the directional interactions of neurons from both areas take place over extremely small spatial ranges—fractions of a degree in receptive fields that are, in MT, many degrees wide (**Figure 5***d–f*)—and they reverse direction for contrast-inverting sequences (Livingstone et al. 2001, Livingstone & Conway 2003), both of which are characteristics of the short-range perceptual process (Braddick 1974, Anstis & Rogers 1975). Finally, when short-range ($\Delta x \approx 0.13°$, $\Delta t \approx 17$ ms) and long-range ($\Delta x \approx 1°$, $\Delta t \approx 125$ ms) apparent motion cues were pitted against each other in opposing directions, the direction of the short-range motion dominated the responses of MT neurons (Shadlen et al. 1993), whereas the long-range direction clearly dominates the visual percept.

A similar comparison has been made with respect to the spatial scale of interactions for binocular disparity (**Figure 5a–c**). The vast majority of MT neurons are sensitive to the relative position of visual stimuli on the two retinas (Maunsell & Van Essen 1983b, Bradley et al. 1995, Bradley & Andersen 1998, DeAngelis & Newsome 1999, DeAngelis & Uka 2003)—a comparison essential for stereoscopic depth perception (see Cumming & DeAngelis 2001 for review)—and, indeed, MT activity has been linked to this perceptual function as firmly as it has been linked to motion perception (see below). In this case, the spatial scale of the binocular interactions in MT does appear significantly coarser than that of V1 (DeAngelis & Uka 2003) but may not be that different from the scale in V2 (Thomas et al. 2002) (**Figure 5**). Given the clustering of disparity-tuned neurons in V2 thick cytochrome oxidase stripes (Hubel & Livingstone 1987, Peterhans & von der Heydt 1993), which are known to project to MT (Shipp & Zeki 1985, DeYoe & Van Essen 1985), it is thus plausible that MT inherits its disparity tuning as well.

Overall, a number of recent studies concluded that although differences exist in some of the population properties of V1 and MT neurons with respect to direction, speed, and disparity tuning, the more carefully and quantitatively these parameters have been studied and compared, the subtler the differences have become. But this is not to say that MT simply mirrors its V1 inputs. MT is now known to be involved in a number of visual functions that are complex and, in at least some cases, are linked to perception and behavior. We discuss these functions in the following paragraphs, starting with center-surround interactions, followed by the computation of velocity, and then operations related to segmentation and structure in three-dimensional space. We conclude with extraretinal effects and discuss perceptual correlates and the mechanisms by which the MT population might be decoded by other brain centers.

SURROUND MECHANISMS

About half of the neurons in MT have receptive fields with antagonistic surrounds (Allman et al. 1985a, Tanaka et al. 1986, Raiguel et al. 1995, Bradley & Andersen 1998, Born 2000, DeAngelis & Uka 2003). These neurons respond well to a centrally placed visual stimulus, such as a small patch of moving dots; however, if the stimulus is made larger so that it invades the surrounding region, the response decreases (**Figure 6a**).

In general, the surround effects are such that maximal suppression occurs when the surround stimulus moves in the same direction and at the same disparity as that in the center (Allman et al. 1985a, Bradley & Andersen 1998). As such, the center-surround apparatus would act as a differentiator over at least two dimensions, direction and depth, bestowing on MT firing rates the quality of salience. The more a stimulus sticks out in terms of direction and depth, the larger the neuron's response will be; in fact, the effects combine roughly linearly (Bradley & Andersen 1998). Surround suppression also depends on speed, but surprisingly, suppression is not consistently maximal when surround speed matches center speed; in fact, results have been quite mixed (Allman et al. 1985a, Tanaka et al. 1986, Orban 1997). So there may not be a simple differencing mechanism based on center-surround speed comparisons.

Although we have discussed MT surround effects in terms of direction, speed, and disparity relative to the center stimulus, effects are not relative, at least for direction and speed (Born 2000). Surround stimuli modulate the

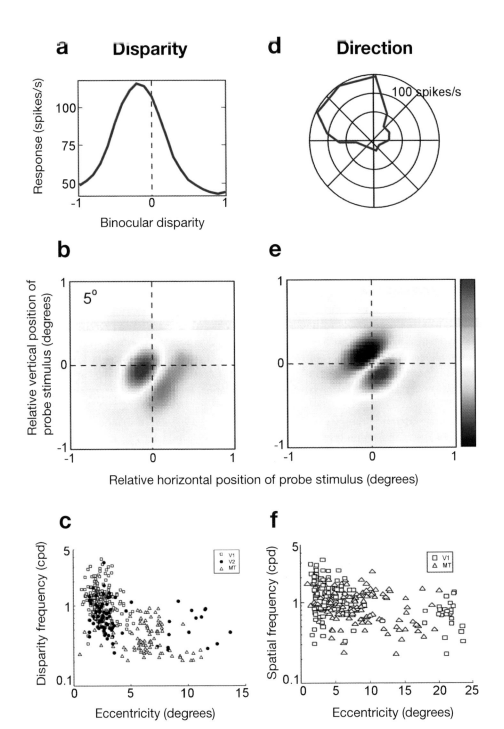

magnitude of responses to central stimuli but do not laterally shift tuning peaks for direction and speed. In the same way, we would expect disparity to have an absolute rather than relative effect, but we are unaware of any experiments that have tested this.

Recent data suggest that MT surrounds are actually quite complex. For example, center-surround interactions behave quite differently for low- and high-contrast stimuli: Area summation prevailed in the former case and suppression prevailed in the latter (**Figure 6a,b**; Pack et al. 2005). These results are consistent with psychophysical results demonstrating improved motion integration at low contrasts (Tadin et al. 2003). The change in center-surround interactions with contrast is interesting because it points to a strategy of the visual system, first suggested by Marr (1982), to integrate for increased sensitivity when the signal is weak, but to exploit the high information content of image discontinuities by differentiating when the signal is strong. A similar effect of contrast on surround effects has been observed in both V1 (Levitt & Lund 1997, Polat et al. 1998, Kapadia et al. 1999, Sceniak et al. 1999) and the LGN (Solomon et al. 2002), so this strategy may be general.

Another level of complexity concerning MT surrounds is their spatial organization (Raiguel et al. 1995, Xiao et al. 1997a,b). The Orban lab has used small patches of moving dots to probe various positions within the surround while the center was stimulated optimally with a separate dot patch. Although about 20% of the antagonistic surrounds were circularly symmetric, 50% were asymmetric with most of the suppression being confined to a single location on one side of the preferred-null direction axis, and another 25% showed bilaterally symmetric zones of "end" suppression that tended to lie along the same axis (**Figure 6c**). The neurons having asymmetric and bilaterally symmetric surrounds appear well suited to calculate directional derivatives of the first and second order, respectively, and computational studies have indicated that such computations are potentially useful for encoding important surface features, such as slant and tilt (first

---

Figure 5

Comparison of the spatial scales at which interactions for binocular disparity (*a–c*) and direction selectivity (*d–f*) are computed. Panels *a* and *d* show conventional tuning curves for a single MT cell whose receptive field was located 4.4° to the right of and 2.7° above the fovea (∼5° eccentricity). This cell preferred crossed disparities (*a*) and responded optimally to motion up and to the left (*d*). Panels *b* and *e* show two-spot interaction maps for the same cell. Orange indicates facilitation, and blue indicates suppression. Panel *b* plots the probability of spiking as a function of the relative position of a probe stimulus presented to the left eye relative to the position of a simultaneously presented reference spot in the right eye (defined as 0,0 on the map). Consistent with the standard disparity tuning curve, the cell was facilitated (*orange*) by spots in the right eye appearing to the left of those in the left eye (crossed disparity) and suppressed (*blue*) by uncrossed disparities. Panel *e* plots the relative positions of the two spots on successive frames ($\Delta t = 13$ ms), revealing the facilitation for probe (*preceding*) spots down and to the right of the reference spot. For both two-spot maps, note that the relevant interactions take place over a very small spatial range. To quantify this, a gabor function was fit to the two-spot map, and the spatial frequency of the sinusoid was used as a measure of the coarseness of the interactions. For this cell, the disparity spatial frequency was 1.7 c/deg and the directional spatial frequency was 1.67 c/deg. Panels *c* and *f* show population data on the coarseness of disparity (*c*) and directional (*f*) interactions for a population of neurons in V1 and MT (and V2, for disparity). Because we did not have a large sample of disparity maps, we have used the data from figure 8 of Cumming & DeAngelis (2001). Their measure of coarseness was different from that described above; however, we believe both measures reflect the same underlying substructure. Note that, at a given eccentricity, the range of spatial scales for V1 and MT largely overlap and the interactions for MT cells are at a much finer spatial scale than the size of their receptive fields. Panels *a*, *b*, *d*, and *e* are from Pack et al. (2003); data for panel *f* are from C.C. Pack, M.S. Livingstone, B.R. Conway, & R.T. Born (unpublished observations).

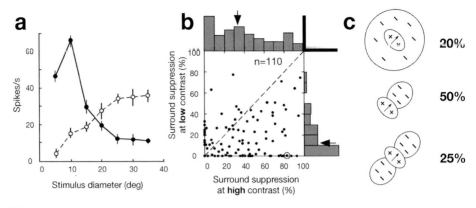

Figure 6

Center-surround interactions in MT. (*A*) Effect of contrast on center-surround interactions for one MT neuron. When tested with high-contrast random dots (RMS contrast 9.8 cd/m^2) the neuron responded optimally to a circular dot patch 10° in diameter and was strongly suppressed by larger patterns. The same test using a low-contrast dot pattern (0.7 cd/m^2) revealed strong area summation with increasing size. (*B*) Population of 110 MT neurons showing the strength of surround suppression measured at both high and low contrast. Surround suppression was quantified as the percent reduction in response between the largest dot patch (35° diameter) and the stimulus eliciting the maximal response. Each dot represents data from one neuron; the dashed diagonal is the locus of points for which the surround suppression was unchanged by contrast. The circled dot is the cell from panel *A*. (*C*) Asymmetries in the spatial organization of the suppressive surround (after Xiao et al. 1997). Different kinds of surround geometry are potentially useful for calculating spatial changes in flow fields that may be involved in the computation of structure from motion. Neurons whose receptive fields have circularly symmetric surrounds (*top*) are postulated to underlie figure-ground segregation. The first- (*middle*) and second-order (*bottom*) directional derivatives can be used to determine surface tilt (or slant) and surface curvature, respectively (Buracas & Albright 1996). Panels *A* and *B* are from Pack et al. 2005.

> **Null direction:** the direction of motion opposite that eliciting the greatest response from a given neuron; e.g., if a neuron responds optimally to rightward motion, its preferred direction is right and its null direction is left.

order) or curvature (second order) (Droulez & Cornilleau-Peres 1990, Koenderink & van Doorn 1992, Buracas & Albright 1996). This potential role of the surround in structural computations is discussed further in the next section.

The source of MT surrounds remains unclear. One possibility is that the surrounds are already present in the inputs to MT. Though center-surround interactions for motion have been reported in V1 (Gulyas et al. 1987, Levitt & Lund 1997), the relative paucity of such interactions in the input layers of MT (Lagae et al. 1989, Raiguel et al. 1995, Born 2000) and the very large size of MT surrounds—at least several-fold larger than their centers (Allman et al. 1985a, Tanaka et al. 1986, Raiguel et al. 1995)—make this an unlikely explanation. It may be that surrounds reflect feedback from higher areas such as MST or are created by horizontal connections within MT (Malach et al. 1997).

THE COMPUTATION OF VELOCITY

By "velocity" we mean the vector representation of the direction and speed of retinal motion. As discussed above, MT adds little to the raw direction and speed tuning already found in V1, but researchers still think it plays a role in computing the motion of whole objects or patterns. The nature of that role is the subject of this section. We first discuss some theoretical considerations and outline the roles MT might play.

For a rigid object, it would seem trivial to compute pattern motion because one would expect every part of the object to have the same velocity. But the measurements obtained depend critically on how the local motion is sampled. Because of the "aperture problem," moving edges seen through small apertures appear to move orthogonally to their length; therefore, for an object sampled at high resolution (through small apertures), different parts of the object appear to move in different directions, depending on the orientation of the sampled edges (Fennema & Thompson 1979) (**Figure 7***a*). In the primate, visual motion is first computed in V1, and these neurons have small receptive fields; therefore, investigators generally think that V1 neurons see normal (orthogonal) velocities (Hubel & Wiesel 1968, Movshon et al. 1985, Heeger 1987). The problem then is to compute two-dimensional (2D) pattern motion on the basis of these local, one-dimensional (1D) samples. In theory, two local samples are sufficient to compute the pattern motion; the geometric solution to the problem is called the intersection of constraints (IOC), which is illustrated in **Figure 7***b*.

As we address below, some MT neurons report 2D motion, but the computation of 2D motion itself does not need to occur in MT; indeed, it could take place anywhere along a continuum between two extremes. At one extreme is the idea that the V1 stage is linear and the informative (nonlinear) computation occurs in MT (Adelson & Movshon 1982, Albright 1984, Movshon et al. 1985, Heeger 1987, Simoncelli & Heeger 1998). At the other extreme, 2D motion is extracted in V1 through nonlinearities, such as endstopping, and all that is required of MT is to pool the V1 inputs (Wilson et al. 1992, Barth & Watson 2000). Below, we discuss physiological evidence supporting each of these ideas. At present, there is no definitive evidence for either model, and indeed, we conclude by suggesting that both may be operative, depending on the stimulus conditions.

According to models of the first type, the aperture problem is built into the system by virtue of the linearity of the motion

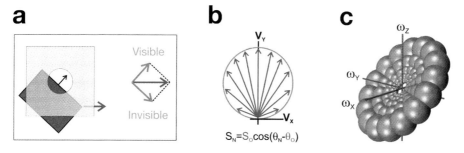

Figure 7

The problem of two-dimensional motion detection. (*A*) The aperture problem. A moving edge seen through an aperture appears to move perpendicularly to itself because the object's motion, in this case to the right, can be decomposed into vector components, one parallel to the edge and one perpendicular. The parallel component is invisible because there is no contrast parallel to an edge, so only the perpendicular component remains. (*B*) The relationship between the component (apertured) vector samples and the global motion of an object is cosinusoidal; that is, the speed of the samples is the object speed times the cosine of the difference between the object direction and the direction of the vector sample, as shown in the equation. In the equation, S_N and θ_N are the normal (sampled) speed and direction, and θ_O and S_O are the speed and direction of the object. Because there are two unknowns, the object direction and speed (*shown in red in equation*), two samples are needed to solve for the object velocity. This requirement is usually referred to as the intersection of constraints (IOC). It is often visualized differently (Movshon et al. 1985). (*C*) A possible neural algorithm for solving the IOC, as suggested by Simoncelli & Heeger (1998). For a rigid, translating object, all local spatiotemporal frequencies must lie on a plane in frequency space. To detect this, one could create a linear filter, represented by blobs in the diagram, for each location on the plane, then sum the energy passed through the filters.

detection stage of V1 simple cells. Consider the frequency representation of moving objects (Watson & Ahumada 1983). For any rigid moving object, the spatiotemporal frequency of all local samples must lie on a plane in frequency space (**Figure 7c**). V1 simple cells, to a first approximation, act like linear space-time filters (McLean & Palmer 1989, Reid et al. 1987), and complex cells are thought to differ mainly in their phase-insensitivity but not in their frequency selectivity (Movshon et al. 1978). So V1 neurons effectively measure the amount of motion energy in their passband (Adelson & Bergen 1985). In other words, such neurons do not really know what the stimulus direction is; they see only the motion component within their frequency band. This is the expression of the aperture problem in frequency space (see **Figure 8**).

If V1 neurons see component motion, they are ignorant, in a sense; MT neurons would need to be relatively intelligent in combining V1 inputs to recover the true, 2D direction of motion. As described above, the IOC construction is the basic rule needed to compute pattern velocity from component (local) velocities. But what physiological mechanism could do this? Heeger (1987) and Simoncelli & Heeger (1998), building on important theoretical (Adelson & Movshon 1982, Watson & Ahumada 1983, Movshon et al. 1985) and physiological (Hubel & Wiesel 1962, Movshon et al. 1985, Reid et al. 1987) groundwork, described what we refer to as the F-plane model. It assumes a front end made of linear V1 cells whose outputs are summed over a plane in frequency space by an MT pattern cell. This planar summation is an instantiation of the IOC rule (**Figure 7c**). The model is able to explain a number of physiological results, including MT responses to variable coherence stimuli (Newsome et al. 1989) and to plaid patterns (Adelson & Movshon 1982). There are other important models of MT computation, however, which we regrettably do not have space to discuss here (Wilson et al. 1992, Nowlan & Sejnowski 1995, Lisberger & Movshon 1999).

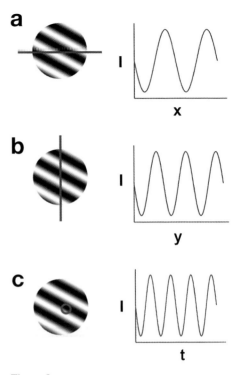

Figure 8
Cutting a horizontal slice through an oblique sine wave grating reveals a sinusoidal modulation of intensity vs. x position (*A*). Cutting a vertical slice through the same grating reveals an identical intensity modulation versus y position (*B*). Looking at a fixed location as the grating moves reveals a sinusoidal modulation of intensity versus time (*C*). Changing the angle (direction) of the grating changes the relative x and y sinusoidal frequencies, and changing the grating speed changes the temporal frequency. Thus, the velocity (direction and speed) of the grating is completely characterized by a single, three-dimensional frequency, $(\omega_x, \omega_y, \omega_t)$. Just as any sound is the sum of its harmonics, any moving object can be represented in terms of its component frequencies, each equivalent to a single, moving sinusoidal grating. If the object is rigid and not rotating, all of its component frequencies will lie on a plane in $\omega_x, \omega_y, \omega_t$ space. The aperture "problem," stated in these terms, arises insofar as V1 neurons are linear $\omega_x, \omega_y, \omega_t$ filters, each detecting a certain component frequency. The relative blindness, or ignorance, implied by the aperture problem is in the linearity of these cells; they are presumably unaffected by frequencies outside their passband. Thus, the "component" cells of Movshon et al. (1985; see text) see only one grating—one frequency.

Note that even if V1 neurons act like linear filters, MT neurons might not necessarily combine their outputs according to an IOC rule; a vector average or sum, for example, would also be a plausible combination (Mingolla et al. 1992, Wilson et al. 1992, Rubin & Hochstein 1993).

An alternative to the F-plane model is that 2D-motion information is computed in V1 through specific nonlinear operations. For example, image squaring can be used to extract periodic elements, which can be low-pass filtered from component frequencies (Wilson et al. 1992), and luminance minima and maxima are reliable features to track, as well (Bowns 1996). We see below that endstopping in V1 neurons amounts to feature detection, which has been formally related to the more general notion of curvature (Barth & Watson 2000). Because features always move in the object direction and at the object speed, MT would not really need to elaborate on V1's output, other than perhaps spatial pooling and noise reduction. At this stage, we cannot rule out that such pooling is the only role played by MT in pattern motion detection.

Having outlined the different possible computational mechanisms, we now discuss evidence for and against them. The first breakthrough came from experiments by Movshon and colleagues (1985), who tested MT and V1 neurons with stimuli known as plaids. These stimuli were formed by superimposing two sine gratings moving in different directions, in this case 135° apart. Treating the summed gratings as a single, plaid stimulus, they measured direction tuning in MT neurons and found a continuum of different tuning curves. At one extreme, there were two peaks in the direction tuning curve, corresponding to the two-pattern directions that resulted in one of the gratings moving in the neuron's preferred direction. At the other extreme, the tuning curve was unimodal, peaking where the pattern as a whole moved in the neuron's preferred direction. Movshon and colleagues (1985) coined the terms pattern and component cell to represent these two types of response. They found that 25% of the MT neurons yielded tuning curves significantly more like the pattern response and 40% more like the component response; the remaining 35% of the neurons were unclassified. In V1, nearly all of the cells displayed component behavior, and none matched the pattern prediction. These authors proposed a two-stage mechanism where pattern cells combine the outputs from component cells to compute pattern direction and suggested the possibility of an IOC construction. Because the plaids were symmetric, the direction of the IOC and the direction of a vector average (for example) would come out the same, so these experiments did not prove the operation of an IOC mechanism.

However, Albright showed that for some MT neurons, which he called Type II cells, the preferred orientation of a static bar was parallel to the preferred direction of motion (Albright 1984). This behavior is consistent with the IOC rule because the vector component 90° off the object direction has zero speed (see **Figure 7b**). These Type II cells were later shown to correspond to Movshon et al.'s pattern cells (Rodman & Albright 1989). Other studies have found evidence that MT pattern cells are bimodally tuned for bars (Okamoto et al. 1999) and gratings (Simoncelli et al. 1996). The IOC model predicts this also, assuming the stimulus is moving beneath the neuron's preferred speed. In that case, the neuron should have two preferred directions, one for each time the stimulus crosses the cosine-shaped function relating direction to speed (see **Figure 7b**). Unfortunately, both studies were based on small samples.

Other evidence suggests that a two-stage model may not be required because the aperture problem is overcome in V1. Layer 4B neurons, which dominate the V1 input to MT, are heavily end-stopped, typically exhibiting around 75% suppression by extended contours as compared with their maximal response (Sceniak et al. 2001). This effect is largely independent of the orientation of the stimulus in the surround; therefore, these neurons respond well to line terminators moving in their preferred direction, regardless of the orientation of the contour (Pack et al. 2003). Although not tested directly, these cells likely show a similar preference for corners, dots, and other features compared with extended contours. Because features always move in the object direction, and because end-stopping suppresses neurons not responding to features, end-stopping tends to have the important effect of defeating the aperture problem. Whether end-stopping actually produces a pattern-motion computation in the naturally functioning visual system remains to be seen.

But some evidence suggests that end-stopping does play a critical role in solving the aperture problem. Pack & Born (2001) tested MT neurons with a pattern of line segments drifting coherently such that the motion components—the velocities normal to the segments—differed by 45° from the direction of the pattern as a whole. The neuronal responses were initially strongly biased by the component directions and then evolved to the pattern direction over the course of about 80 ms. In a separate study of V1 neurons, the end-stopping effect was shown to also develop over a similar time course (Pack et al. 2003). Thus, MT neurons may have computed the motion of the line pattern by integrating the output of a feature-based mechanism afforded by end-stopping in V1.

But the slow time course observed in MT by Pack & Born (2001) does not need to reflect end-stopping in V1. Smith et al. (2005) recently measured the time course over which MT neurons develop pattern selectivity and also found it to be gradual. Because they used sine wave plaids, the observed time course may have had nothing to do with end-stopping, reflecting instead mechanisms within MT or even top-down effects. Of course, one could argue that the overlap regions of the plaids constitute features that could be tracked by end-stopped (or other feature-selective) neurons; in that case, though, one would expect to find pattern cells in the V1 inputs to MT. Movshon & Newsome studied this in 12 MT-projecting V1 cells and did not find any (Movshon & Newsome 1996), but this sample may be too small to tell. Only 9 of these cells were tested with plaids, and only 6 of the 12 cells were in layer 4B. Because 90% of MT's V1 input comes from 4B, it is premature to conclude that the V1 cells feeding MT do not have pattern behavior. In fact, Tinsley et al. (2003) and Guo et al. (2004) both found a small number of pattern-selective cells in V1. Clearly, the issue of whether substantial pattern selectivity occurs in primate V1 remains unresolved. Still, under conditions where Movshon et al. did not find V1 pattern cells, they did find MT pattern cells (Movshon et al. 1985); therefore, the pattern mechanism under those conditions cannot be explained solely in terms of feature tracking.

In summary, there is substantial evidence for both feature tracking in V1 and a two-stage mechanism involving linear filtering in V1 followed by a nonlinear process, something like IOC or vector averaging, in MT. We note that IOC and feature tracking mechanisms are not necessarily incompatible; in fact, there is perceptual evidence that different sorts of computation are at play under different circumstances (Weiss et al. 2002). In the absence of end-stopping, V1 neurons are assumed to extract motion energy at the various spatiotemporal frequencies in the stimulus. For the F-plane model to be robust, it is best to have energy broadly distributed over the frequency plane defined by the object's velocity; otherwise, it is difficult for the MT population to determine the orientation of this plane. This may be why sinusoidal plaid stimuli, whose energy occupies only localized portions of the frequency plane, are not perceived as moving in the IOC direction unless the grating directions are roughly symmetric about the pattern direction (Yo & Wilson 1992). When end-stopping is operational, neurons with extended contours in their receptive fields are suppressed; so the remaining input consists mainly of features that have energy that is well dispersed over the frequency plane. Therefore, the overall effect would be to provide MT with a well-balanced, broad-spectrum input. At low contrast, where end-stopping becomes weak or even nonexistent (Sceniak et al. 1999), the system would revert to a more linear mode of operation, using signals derived from moving contours as well as features. The idea of representing moving objects in terms of sparse features at high contrast and diffuse frequency components at low contrast is consistent with theoretical studies demonstrating the advantage of minimizing redundancy when signal-to-noise is large, and at the same time it emphasizes sensitivity and noise reduction using broad pooling mechanisms at low contrast (Field 1987, van Hateren 1992). Of course, other nonlinear mechanisms besides end-stopping could also be used to track features (Wilson et al. 1992, Bowns 1996).

NOISE REDUCTION

Regardless of how pattern velocity is computed, the visual motion system is compelled at some point to filter noise from the processing stream. MT neurons appear to serve this function in a way that is akin to common mode rejection by a differential amplifier (Snowden et al. 1991, Qian & Andersen 1994, Qian et al. 1994, Bradley et al. 1995). Snowden et al. showed that MT neurons responding to dot patterns moving in their preferred direction were suppressed when dot patterns moving in nonpreferred directions were simultaneously introduced (Snowden et al. 1991). This suppression, termed motion opponency, is not a guaranteed result; V1 neurons, for instance, are not substantially affected by nonpreferred motion directions, consistent with their approximately linear behavior (Qian & Andersen 1995). Snowden et al. demonstrated that nonpreferred motions exhibited a suppression in MT that was roughly divisive; that is, the nonpreferred pattern tended to reduce the response gain for the preferred-direction pattern. Later studies by Qian & Andersen showed that suppression of MT by nonpreferred directions is particularly strong when opposing motions occur within $\sim0.5°$ of each other, roughly the scale of V1 receptive fields (Qian & Andersen 1994). The fact that V1 cells do not themselves show appreciable motion opponency suggests that there is strong mutual inhibition at the stage of V1 inputs onto the dendrites of MT neurons. Because a flash emits motion energy simultaneously in all directions, a mechanism that cancels opposite-direction signals on a local scale could be a particularly effective way of reducing responses to flicker.

There is no reason to assume that the only function of motion opponency in MT is to reduce noise. In fact, opponency is likely at least partly a manifestation of gain normalization, the process by which neural responses are scaled according to the total amount of neural activity in their immediate vicinity (Simoncelli & Heeger 1998, Heuer & Britten 2002). But there is little doubt that opponency would tend to reduce the responses to motion noise. Some behavioral evidence also exists: Perceptual studies in a patient with bilateral damage to a region corresponding to MT/V5 found that performance was normal for various motion tasks except when noise was added to the stimulus, in which case performance fell to chance (Zihl et al. 1983, Marcar et al. 1997). This, together with the suppressive effects of nonpreferred motion discussed above, suggests that MT has a basic role in noise reduction within the motion processing stream.

SEGMENTATION

Although pooling is an important stage in motion processing, it introduces its own problems. It is not uncommon for more than one moving object to appear in the same part of visual space, and pooling these movements would obviously be inappropriate because there is nothing meaningful about the joint (say, average) velocity of two independent objects. Therefore, pooling mechanisms need to be accompanied by parsing mechanisms that distinguish groups of signals to be integrated separately (Nowlan & Sejnowski 1995, Hildreth et al. 1995).

Several lines of evidence suggest that MT neurons are equipped with response properties that allow them to carry out, or at least contribute to, this parsing. For one thing, the suppression of MT responses normally exerted by nonpreferred motion is relaxed when the nonpreferred and preferred motions occur in separate depth planes (Bradley et al. 1995). Opposing movements in the same part of 2D visual space usually are separated in depth (otherwise they would collide), which would tend to prevent the pooling and cancellation of motion signals emanating from different objects. Also, most MT neurons have an antagonistic surround that is least suppressive under conditions where center and surround stimuli move in different directions and speeds and at different stereoscopic depths (see Surround Mechanisms). This could be a mechanism for emphasizing the relative motion of an object against its background. Other studies suggest that more complex mechanisms may be in place. For example, Albright and colleagues superimposed square wave gratings and manipulated the luminance at the overlaps, according to physical transmittance rules, to produce stimuli that looked either transparent (separate gratings) or coherent (single plaid) in their static state. They then demonstrated that single MT neurons could exhibit either pattern- or component-like direction tuning, depending on whether the stimulus was in its coherent or transparent configuration, respectively (Stoner & Albright 1992, Stoner & Albright 1996). These results could reflect external parsing mechanisms that influence MT, but they could also reflect low-level mechanisms where the overlap regions of the plaids are detected by nonlinear mechanisms in V1, leading to a change in the distribution of inputs to MT. Finally, MT neurons are better able to extract signal (motion) from noise when the two have different isoluminant colors (Croner & Albright 1997, 1999). Altogether the evidence rather strongly suggests that MT neurons are critically involved in segmenting an image into separately moving parts.

THE COMPUTATION OF STRUCTURE

One of the most important discoveries since MT was first located was the binocular disparity tuning of its neurons. Maunsell & Van Essen (1983a) showed, using stereoscopic moving bars, that a majority of MT cells were disparity selective, although it is now known that moving dots, which have a broader frequency content, reveal selectivity in almost all MT cells. The tuning resembles sigmoids, Gaussians, and shapes in between. Curves sometimes peak near zero disparity, but usually they peak or flatten out well to the left or right. Therefore, with some exceptions, these neurons are tuned "near" or "far," with about a twofold preponderance of the near-tuned (Bradley & Andersen 1998). Importantly, Maunsell & Van Essen showed that MT neurons are not tuned for motion through 3D space. They are simply tuned for a certain 2D direction in a plane a certain distance from the fixation point. This remained something of an anomaly until the mid-1990s when experimenters began searching for possible functions of disparity in MT. First, they showed that null-direction motion, which normally suppresses MT activity, becomes less suppressive if it occurs outside the preferred depth plane of the neuron. The suppression is itself thought to have a role in noise reduction, so this depth constraint provides a way of confining the filtering mechanism to a particular surface. Later it was shown that surround inhibition is also modulated by disparity, predictably in such a way as to minimize inhibition when the surround stimulus is outside the depth plane containing the center stimulus (Bradley & Andersen 1998). This could be a mechanism for segmenting an object from its background. So there is some evidence that disparity is wired into MT in a way that facilitates the processing of visual motion, presumably for purposes of noise reduction, segmentation, and probably other functions.

However, subsequent studies suggested that MT is involved in the perception of depth itself (Bradley et al. 1998, DeAngelis et al. 1998, Dodd et al. 2001). Two groups of investigators trained monkeys to view revolving cylinders whose direction of revolution was unspecified in the stimulus, but which nevertheless appeared to rotate (**Figure 9**). The perceived rotation was bistable, and monkeys reported their percept on each trial. Both groups found a clear correlation between MT firing rates and the monkeys' judgments. In terms of motion, both percepts meant seeing two, opposite directions; the only difference was the order of these directions in depth. So these were the first neural correlates of the perception of depth. In fact, Dodd et al. (2001) computed choice probability, a measure of the correlation between perception and neural activity (Britten et al. 1996), and found the strongest neuro-perceptual correlation of any MT study, to date.

This role for MT in depth perception has only been strengthened by additional experiments. In fact, of the criteria linking neurons to perception proposed by Parker & Newsome (1998), the only important one that has yet to be met is that of MT lesions affecting disparity judgments. Thus far, this research has rigorously demonstrated (*a*) neuronal selectivity for disparity (Maunsell & Van Essen 1983b, DeAngelis & Uka 2003), (*b*) neuronal disparity sensitivity that is sufficient to account for the abilities of monkeys to perform coarse disparity discrimination tasks (Uka & DeAngelis 2003), (*c*) a predictive relationship between MT neuronal activity and monkeys' perceptual decisions concerning depth (Uka & DeAngelis 2004), and (*d*) the ability to predictably bias monkey's disparity judgments by microstimulation of disparity columns favoring a given depth (DeAngelis et al. 1998).

Another hint that MT neurons may be involved in the extraction of surface properties is the presence of inhomogeneities within the receptive field with respect to tuning for both speed (Treue & Andersen 1996) and binocular disparity (Nguyenkim & DeAngelis 2003). As a moving textured surface in the fronto-parallel plane is either tilted forward or backward or slanted to the right or left, the retinal projection of local motion and depth vectors will form a gradient in both speed and disparity. If MT neuronal receptive fields possessed systematic variation in these tuning properties as a function of position within their receptive fields, the resulting structure might serve as a template for a particular 3D surface orientation. Indeed, both of the above groups have demonstrated tuning for surface orientation on the basis of either cue alone. The group testing speed gradients confined their stimuli to the receptive field center, but it appears that, if asymmetries of the surround described above are included, the tuning to speed gradients may become even more marked (Xiao et al. 1997a). Furthermore, it now appears that at least some MT neurons are selective for surface orientation defined by both cues. In such cases, the selectivity is generally consistent and reenforcing, in that tuning is sharper to gradients defined by the combination than by gradients defined by either cue alone (Nguyenkim & DeAngelis 2004).

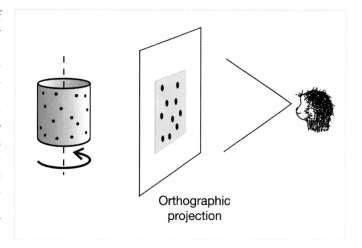

Figure 9

Schematic illustrating cylinder experiments first used to link depth perception to MT. A monkey views the orthographic projection, or shadow, of dots revolving on a transparent, cylindric surface. The planar image contains only dots moving in opposite directions, but the observer perceives a three-dimensional revolving cylinder. Monkeys are trained to report the perception of structure (Siegel & Andersen 1988) or the direction of revolution (Bradley et al. 1998, Dodd et al. 2001).

EXTRARETINAL EFFECTS

For most behavior, visual information must be integrated with other, nonvisual information. An example of this is the use of copies of the signals used to generate eye movements, so-called efference copy or corollary discharge, to aid in disambiguating retinal motion caused by the eye movements themselves from that due to motion of an object. In ascending the hierarchy of visual areas, at least in the dorsal stream, such extraretinal influences tend to become increasingly powerful. Thus, whereas investigators have reported small-to-modest effects of eye position (Bremmer et al. 1997), saccadic eye movements (Thiele et al. 2002), and attention (Treue & Maunsell 1996) for some MT neurons, much larger and more prevalent effects of these signals have been found in areas above MT, particularly MST and LIP. So, at least to a first approximation, MT is a relatively faithful representation of events occurring on the retina.

The issue of extraretinal influences in the dorsal stream was first examined by Wurtz's group at the National Institutes of Health (NIH) in the context of smooth-pursuit eye movements (Newsome et al. 1988). MT neuronal signals are important for the initiation of pursuit (see Lisberger et al. 1987 for review), and many MT neurons with foveal receptive fields are also active during ongoing pursuit. However, when Newsome et al. (1988) eliminated retinal motion, either by briefly extinguishing the target or by using the recorded eye movements to stabilize the target's image on the retina, the MT responses disappeared. This result stood in marked contrast to many neurons in MST that showed continued directional responses under the same conditions.

Subsequent groups have found some tendency for the position of the eye in the orbit to affect the overall level of responsiveness of some MT neurons, without affecting their tuning for direction (Bremmer et al. 1997). But, in line with the idea of hierarchy, the effects in MT were both rarer and smaller in magnitude than those found in MST, one of the next higher-tier motion-processing areas to which MT projects. And it is telling that this type of effect was initially discovered in an area still further up the hierarchy—7a of posterior parietal cortex (Andersen et al. 1985), where the effects of eye position are even more profound and where vestibular, auditory, and somatosensory information are also integrated (Andersen et al. 1997). None of these latter signals are known to influence neurons in MT, and anatomical data would suggest they are unlikely to do so.

Perhaps the most prevalent nonretinal influence on MT neurons is that of attention. This is manifest as an enhanced neuronal response to visual stimuli when an animal is attending to either the spatial location of the neuron's receptive field (Seidemann & Newsome 1999, Treue & Maunsell 1999) or to some preferred neuronal feature, such as a particular direction of motion (Treue & Martinez-Trujillo 1999). The gain increase appears to be a straightforward multiplication of the responses, without any changes in the shape of the direction tuning curve (Treue & Maunsell 1999) nor in the nature of the underlying motion computations (Cook & Maunsell 2004). However, the magnitude of the gain may change both as a function of stimulus contrast (Martinez-Trujillo & Treue 2002) and, for feature-based attention, as a function of the similarity between the attended feature and that preferred by the neuron, actually becoming negative as the attended feature approaches the antipreferred feature of the neuron (Martinez-Trujillo & Treue 2004). As for other extraretinal effects, however, the general magnitude of attentional modulation in MT appears to follow the cortical hierarchy: Attentional modulation strengths range from less than 10% in V1 (McAdams & Maunsell 1999), to a ballpark figure of 20%–30% in MT (Treue & Maunsell 1999), to values well over 50% in higher-tier areas, such as MST and 7a (Maunsell & Cook 2002).

PERCEPTUAL CORRELATES AND POPULATION CODES

Single-Neuron Sensitivity

In classic experiments at Stanford University, Newsome and colleagues trained macaques to watch moving dot patterns and then indicate which of two opposite directions they had seen (see Parker & Newsome 1998 for review). The strength of the motion signal was controlled by varying the fraction of dots moving coherently, and the remaining dots moved in random directions. The task was executed simultaneously with the recording of single MT neurons, in each case aligning the coherent motion axis with the preferred direction of the recorded neuron. Using methods from signal detection theory, Newsome and colleagues were able to compare neuronal sensitivity directly with that of the monkey itself. Remarkably, they discovered that most MT neurons were at least as sensitive as the monkey itself. Later, the same group defined the choice probability (CP), an extension of detection theory, which reflects the correlation between a subject's judgments and random fluctuations in a single neuron's firing rate. For their task, the mean CP in macaques was only 0.55; chance was 0.50 and perfect correlation was 1. In a computational analysis, Shadlen et al. (1996) concluded that roughly 70–100 neurons would have been involved in the decision pool for the task, far more than would seem necessary considering the exquisite sensitivity of single cells.

Shadlen et al. (1996) suggested several possible reasons for the discrepancy, including correlated noise in MT's inputs, which would limit the benefits obtained by pooling, noise in downstage decision processes, and the likelihood that, owing to limitations in the precision of cortical connectivity, signals from relatively insensitive neurons would also be included, thus degrading the calculation. But this coarse pooling may not occur under all conditions. For example, a recent study at the University of Chicago showed that when monkeys performed fine direction discrimination, neuronal sensitivity was at best 2–3 times worse than the observer as a whole, and no amount of pooling could account for the psychophysical data unless the most sensitive neurons were selectively pooled (Purushothaman & Bradley 2005). The different results obtained by the two groups probably reflects important differences in the tasks they used. The Stanford task used a large direction difference embedded in noise, whereas the Chicago stimuli were noise-free but the monkeys had to discriminate very small direction differences. One would expect the former task to reflect the sensitivity and filtering capability of the neurons, whereas the latter task should depend more on the slopes of the direction tuning curves. Of course, both tasks probe critical aspects of MT processing. The point is that a neuron's sensitivity relative to the observer is likely to depend on the task, and in particular the computational role the neuron plays in the decision process.

Other examples reinforce this idea. Uka & DeAngelis (2004) trained monkeys in a near/far-depth-judgment task and compared their sensitivity with that of MT neurons. In each case the near and far depths were set at the worst and best values for the MT cell being studied, much like testing a neuron with its preferred and antipreferred directions. Though depth differences were large, the task was made difficult by adding noise to the stimulus. They found that single neurons were on average as sensitive as the observers. In contrast, Liu & Newsome (2003a, 2005) trained monkeys to discriminate small speed differences and found that MT neurons were much less sensitive than the observers. Overall, it would seem that single MT neurons are exquisitely sensitive when the task is to detect a large direction or disparity difference that is heavily corrupted by noise, but they are less sensitive (relative to the observer) when the task requires fine discrimination of the relevant cue (e.g., direction or speed).

Yet another important factor in interpreting such studies is the role of temporal integration. The Stanford experiments used stimulus presentation intervals of two seconds, after which time monkeys indicated their decision. If the monkeys actually made their decisions much earlier, however, this could affect the ratio of neuron-to-observer sensitivity because the signal detection analysis used the entire two seconds to compute neuronal sensitivity (Mazurek & Shadlen 2002). Cook & Maunsell (2002) found some support for this idea in monkeys trained in a reaction time task with variable-coherence stimuli. When integration times were commensurate for neurons and observers, the neuronal sensitivity was found to be substantially less than observers. Some caution is warranted, however, because the task was not identical to that used in the Stanford experiments. In contrast, Britten et al. (1992) showed that psychophysical and neural thresholds declined similarly with shorter stimulus presentations, which would produce similar neuron/observer-sensitivity ratios regardless of the particular time window chosen. Uka & DeAngelis (2004) also found a similarly small effect of integration interval on neuronal sensitivity in their disparity task, probably because of serial correlation (Osborne et al. 2004). Overall, the bulk of the evidence suggests that observers probably do make decisions in less than two seconds, and that information continues to accumulate for two seconds in spike trains, assuming it is extracted with the type of signal detection analysis used in the detection and discrimination experiments discussed above. That said, the effects probably are not dramatic, and the basic claim of the early Stanford experiments, that monkeys could perform the task in question with a relatively small number of neurons—compared with the hundreds or thousands of neurons one might suspect—is probably correct.

Vector Summation versus Winner-Take-All

For any simple moving stimulus, the response distribution in MT assumes a roughly Gaussian shape. By "simple" we mean that the stimulus is translating, rigid, and not rotating; by "distribution" we refer to the mean firing plotted on the ordinate versus the preferred direction of the neurons on the abscissa. Under the same visual circumstances that create such a distribution in MT, subjects accurately perceive the direction of the moving stimulus. Given the mountain of evidence linking MT to direction perception, we seek to determine how the response distribution is linked to the direction perception. More formally, we recognize that the response distribution is characterized by many numbers—the firing rates of many neurons—but the perceived direction is a single number. So there must be a code, a rule, for the conversion of many to one. But what is it?

Two simple possibilities come to mind. First, the direction percept may derive from the peak of the distribution; that is, the preferred direction of the most active neurons is taken by decision networks as the direction signal. Second, the overall distribution mean could constitute the direction signal. The Newsome lab has also attempted to probe these two potential mechanisms using microstimulation (Groh et al. 1997, Nichols & Newsome 2002). The basic experiments involved a moving visual target paired with the electrical stimulation of neurons tuned to a different direction, and animals were required either to pursue the visual target or to report simply the direction of motion they perceived. The authors' rationale was that if direction percepts derive from activity peaks, then the animals' answers should center on either the direction of the visual target or the direction encoded by the stimulated neurons. In comparison, if direction percepts depend on the activity distribution as a whole, then the perceived direction should lie somewhere in between. The general finding of these experiments was that direction percepts are usually somewhere between the two extremes, suggesting that direction percepts derived from pooled activity rather than from activity peaks. But a potential confound in these experiments is that animals might have perceived two directions at one time. Forced to choose a single direction, they may have settled on a strategy of splitting the difference. There is, however, no evidence for this.

Besides the winner-take-all and vector-average hypotheses, Weiss et al. (2002) have proposed a Bayesian model of motion perception, which could be implemented with appropriate weighting of MT responses. Unlike vector-average and winner-take-all schemes, the model would address the MT response distribution probabilistically; as such, it could take into account the effects of noise on uncertainty and allow the introduction of priors (biases). The model can explain a remarkable array of psychophysical observations, so it will be interesting to test the model with MT recordings as well.

Distributed Speed and Acceleration Codes

Maunsell & Van Essen (1983b) found that MT cells were generally broadly speed tuned but, by and large, tended to have distinct preferred speeds. Lagae et al. (1993) later distinguished MT neurons in terms of their speed tuning as being low-pass, high-pass, or bandpass. Because high-pass neurons generally increased firing as stimulus speed increased, they pointed out that the overall mean activity of these cells could be used to gauge the stimulus speed. Recently, both Churchland et al. (2001) and Priebe & Lisberger (2004) found that speed percepts in macaques were consistent with the firing rate-weighted average preferred speed of MT neurons. Lisberger & Movshon (1999) successfully used a similar population average to account for changes in stimulus speed in anesthetized monkeys. Because the population average can shift only if single neurons change their activity, one would expect MT firing rates to correlate with speed judgments. Indeed, preliminary evidence suggests that they do (Liu & Newsome 2005).

Lisberger & Movshon (1999) studied MT responses to accelerating random dot patterns. Judging from the neurons' sustained firing rates, there was no evidence for single-neuron encoding of acceleration. However, using a measure of the neurons' transient responses and taking their weighted average, the authors could accurately predict the target's acceleration. This result was important not only because it was the first evidence for acceleration coding in MT, but also because it revealed a signal quite hidden in the population response. One wonders how many other kinds of information are encoded in such distributed forms.

CONCLUSIONS

Cortical physiologists often claim—whether or not it is true—that in studying a particular area they hope to discover general principles of cortical function. One could argue that more information has been learned from MT than from any other visual area. We make this bold claim because MT has, in many ways, disappointed its explorers, who initially expected a number of obvious contributions to motion processing but did not find them. Although MT may slightly extend motion detection to longer ranges and higher speeds, these are clearly not its main functions; nor does it make the obvious step from speed tuning (as in V1) to acceleration tuning, at least in single neurons. So it is a testament to the persistence and ingenuity of researchers in the field that visual motion research has reexamined itself and reapproached the study of MT in a variety of novel and fruitful ways. These new approaches and their results have been the focus of this review.

Overall, MT does not appear to detect or measure visual motion; this computation occurs in V1. It also does not elaborate substantially on this basic signal; for example,

BDA: biotinylated dextran amine

FEF: frontal eye field

LGN: lateral geniculate nucleus

LIP: lateral intraparietal area

MT: middle temporal area

MST: medial superior temporal area

MSTd: dorsal subdivision of MST

MSTl: lateral subdivision of MST

PIP: posterior intraparietal area

SC: superior colliculus

V1: primary visual cortex, striate cortex or area 17

V2: second visual area

V3: third visual area

V3A: V3 accessory area

V4: fourth visual area

V5: MT

VIP: ventral intraparietal area

VP: ventral posterior area

direction tuning is not much sharper in MT, and speed tuning is not much broader. One of MT's main functions—that is, above and beyond what is done in V1—concerns integration and segmentation. Obviously, its large receptive fields combine information over space, and it integrates V1 inputs and combines them, at least under some conditions, to compute pattern motion. Its opponent mechanisms probably have a noise-reducing effect. But with integration, new problems arise, in particular, the inappropriate merging of independent moving objects. MT appears to have built-in mechanisms to deal with this. For example, opponency has a disparity constraint, and possibly other constraints, to limit integration to a particular depth. And the center-surround apparatus is direction- and disparity-constrained in such a way that neurons tend to segment motion from its background.

The idea that MT deals only with segmentation and integration might have sufficed up until the mid 1990s, but since then, a series of remarkable studies has made it clear that MT is involved in the computation of structure (although the first evidence for this fact dates back to Siegel & Andersen 1988). The sensitivity of MT neurons to speed gradients; the correlation between MT responses and the perception of 3D cylinders; and the remarkable integration of direction, speed, and disparity gradients all make a compelling case that MT is processing motion but doing more than computing the direction and speed of motion.

One of the important, largely unexplored questions about MT is how its population response is read out for different tasks. Several studies have addressed this question; however, there are many ways to decode a population (Seung & Sompolinsky 1993, Weiss et al. 2002, Ben Hamed et al. 2003), and many experiments will be required before we understand the mathematics between MT activities and downstream behavior. Because its response properties are well understood and because of its tight links to perception, MT is an ideal place to evaluate models of population decoding.

ACKNOWLEDGMENTS

We thank numerous colleagues who aided in the preparation of this review: M. Churchland, G. DeAngelis, K.P. Hoffmann, J. Liu, J.A. Movshon, W. Newsome, K. Rockland, A. Thiele, and S. Treue for providing preliminary data and figure material; V. Berezovskii for help with anatomical tables; E. Adelson, J. Allman, T. Albright, D. Heeger, J. Kaas, N. Priebe, E. Simoncelli, and S. Zeki for helpful discussions and comments on earlier drafts; and C. Pack and G. Purushothaman for invaluable insight. Supported by EY11379 (R.T.B.) and NS40690 and EY13138 (D.C.B.).

LITERATURE CITED

Adelson EH, Bergen JR. 1985. Spatiotemporal energy models for the perception of motion. *J. Opt. Soc. Am. A* 2(2):284–99

Adelson EH, Movshon JA. 1982. Phenomenal coherence of moving visual patterns. *Nature* 300(5892):523–25

Albright TD. 1984. Direction and orientation selectivity of neurons in visual area MT of the macaque. *J. Neurophysiol.* 52(6):1106–30

Albright TD, Desimone R, Gross CG. 1984. Columnar organization of directionally selective cells in visual area MT of the macaque. *J. Neurophysiol.* 51(1):16–31

Allman JM, Kaas JH. 1971. A representation of the visual field in the caudal third of the middle temporal gyrus of the owl monkey (*Aotus trivirgatus*). *Brain Res.* 31:85–105

Allman JM, Miezin F, McGuinness E. 1985a. Direction- and velocity-specific responses from beyond the classical receptive field in the middle temporal visual area (MT). *Perception* 14(2):105–26

Allman JM, Miezin F, McGuinness E. 1985b. Stimulus specific responses from beyond the classical receptive field: neurophysiological mechanisms for local-global comparisons in visual neurons. *Annu. Rev. Neurosci.* 8:407–30

Andersen RA, Essick GK, Siegel RM. 1985. Encoding of spatial location by posterior parietal neurons. *Science* 230(4724):456–58

Andersen RA, Snyder LH, Bradley DC, Xing J. 1997. Multimodal representation of space in the posterior parietal cortex and its use in planning movements. *Annu. Rev. Neurosci.* 20:303–30

Anderson JC, Binzegger T, Martin KA, Rockland KS. 1998. The connection from cortical area V1 to V5: a light and electron microscopic study. *J. Neurosci.* 18(24):10525–40

Anderson JC, Martin KA. 2002. Connection from cortical area V2 to MT in macaque monkey. *J. Comp. Neurol.* 443(1):56–70

Anstis SM, Rogers BJ. 1975. Illusory reversal of visual depth and movement during changes of contrast. *Vision Res.* 15:957–61

Baker JF, Petersen SE, Newsome WT, Allman JM. 1981. Visual response properties of neurons in four extrastriate visual areas of the owl monkey (Aotus trivirgatus): a quantitative comparison of medial, dorsomedial, dorsolateral, and middle temporal areas. *J. Neurophysiol.* 45(3):397–416

Barlow HB. 1986. Why have multiple cortical areas? *Vision Res.* 26(1):81–90

Barth E, Watson AB. 2000. A geometric framework for nonlinear visual coding. *J. Opt. Soc. Am. A* 7(4):155–65

Ben Hamed S, Page W, Duffy C, Pouget A. 2003. MSTd neuronal basis functions for the population encoding of heading direction. *J. Neurophysiol.* 90(2):549–58

Benevento LA, Yoshida K. 1981. The afferent and efferent organization of the lateral geniculo-prestriate pathways in the macaque monkey. *J. Comp. Neurol.* 203(3):455–74

Berezovskii VK, Born RT. 2000. Specificity of projections from wide-field and local motion-processing regions within the middle temporal visual area of the owl monkey. *J. Neurosci.* 20(3):1157–69

Born RT. 2000. Center-surround interactions in the middle temporal visual area of the owl monkey. *J. Neurophysiol.* 84(5):2658–69

Born RT, Groh JM, Zhao R, Lukasewycz SJ. 2000. Segregation of object and background motion in visual area MT: effects of microstimulation on eye movements. *Neuron* 26(3):725–34

Born RT, Tootell RB. 1992. Segregation of global and local motion processing in primate middle temporal visual area. *Nature* 357(6378):497–99

Bowns L. 1996. Evidence for a feature tracking explanation of why type II plaids move in the vector sum direction at short durations. *Vision Res.* 36(22):3685–94

Braddick O. 1974. A short-range process in apparent motion. *Vision Res.* 14(7):519–27

Bradley DC, Andersen RA. 1998. Center-surround antagonism based on disparity in primate area MT. *J. Neurosci.* 18(18):7552–65

Bradley DC, Chang GC, Andersen RA. 1998. Encoding of three-dimensional structure-from-motion by primate area MT neurons. *Nature* 392(6677):714–17

Bradley DC, Qian N, Andersen RA. 1995. Integration of motion and stereopsis in middle temporal cortical area of macaques. *Nature* 373(6515):609–11

Bremmer F, Ilg UJ, Thiele A, Distler C, Hoffmann KP. 1997. Eye position effects in monkey cortex. I. Visual and pursuit-related activity in extrastriate areas MT and MST. *J. Neurophysiol.* 77(2):944–61

Britten KH. 2003. The middle temporal area: motion processing and the link to perception. In *The Visual Neurosciences*, ed. LM Chalupa, JF Werner, pp. 1203–16. Cambridge, MA: MIT Press

Britten KH, Newsome WT, Shadlen MN, Celebrini S, Movshon JA. 1996. A relationship between behavioral choice and the visual responses of neurons in macaque MT. *Vis. Neurosci.* 13(1):87–100

Britten KH, Shadlen MN, Newsome WT, Movshon JA. 1992. The analysis of visual motion: a comparison of neuronal and psychophysical performance. *J. Neurosci.* 12(12):4745–65

Brodmann K. 1909. *Vergleichende Lokalisationlehre der Grosshirnrinde in ihren Prinzipien dargestellt auf Grund des Zellenbaues*. Leipzig: Barth

Brown TS. 1963. Olfactory and visual discrimination in the monkey after selective lesions of the temporal lobe. *J. Comp. Physiol. Psychol.* 56:764–68

Buracas GT, Albright TD. 1996. Contribution of area MT to perception of three-dimensional shape: a computational study. *Vision Res.* 36(6):869–87

Churchland MM, Lisberger SG. 2001. Shifts in the population response in the middle temporal visual area parallel perceptual and motor illusions produced by apparent motion. *J. Neurosci.* 21(23):9387–402

Churchland MM, Priebe NJ, Lisberger SG. 2005. Comparison of the spatial limits on direction selectivity in visual areas MT and V1. *J. Neurophysiol.* 93(3):1235–45

Clare MH, Bishop HG. 1954. Responses from an association area secondarily activated from optic cortex. *J. Neurophysiol.* 17:271–77

Cook EP, Maunsell JH. 2002. Dynamics of neuronal responses in macaque MT and VIP during motion detection. *Nat. Neurosci.* 5(10):985–94

Cook EP, Maunsell JH. 2004. Attentional modulation of motion integration of individual neurons in the middle temporal visual area. *J. Neurosci.* 24(36):7964–77

Cragg BG, Ainsworth A. 1969. The topography of the afferent projections in the circumstriate visual cortex of the monkey studied by the Nauta method. *Vision Res.* 9:733–47

Croner LJ, Albright TD. 1997. Image segmentation enhances discrimination of motion in visual noise. *Vision Res.* 37(11):1415–27

Croner LJ, Albright TD. 1999. Segmentation by color influences responses of motion-sensitive neurons in the cortical middle temporal visual area. *J. Neurosci.* 19(10):3935–51

Cumming BG, DeAngelis GC. 2001. The physiology of stereopsis. *Annu. Rev. Neurosci.* 24:203–38

DeAngelis GC, Cumming BG, Newsome WT. 1998. Cortical area MT and the perception of stereoscopic depth. *Nature* 394(6694):677–80

DeAngelis GC, Newsome WT. 1999. Organization of disparity-selective neurons in macaque area MT. *J. Neurosci.* 19(4):1398–415

DeAngelis GC, Uka T. 2003. Coding of horizontal disparity and velocity by MT neurons in the alert macaque. *J. Neurophysiol.* 89(2):1094–111

DeYoe EA, Van Essen DC. 1985. Segregation of efferent connections and receptive field properties in visual area V2 of the macaque. *Nature* 317(6032):58–61

Dodd JV, Krug K, Cumming BG, Parker AJ. 2001. Perceptually bistable three-dimensional figures evoke high choice probabilities in cortical area MT. *J. Neurosci.* 21(13):4809–21

Droulez J, Cornilleau-Peres V. 1990. Visual perception of surface curvature. The spin variation and its physiological implications. *Biol. Cybern.* 62(3):211–24

Dubner R, Zeki SM. 1971. Response properties and receptive fields of cells in an anatomically defined region of the superior temporal sulcus in the monkey. *Brain Res.* 35(2):528–32

Eifuku S, Wurtz RH. 1998. Response to motion in extrastriate area MSTl: center-surround interactions. *J. Neurophysiol.* 80(1):282–96

Elston GN, Rosa MG. 1997. The occipitoparietal pathway of the macaque monkey: comparison of pyramidal cell morphology in layer III of functionally related cortical visual areas. *Cereb. Cortex* 7(5):432–52

Felleman DJ, Kaas JH. 1984. Receptive-field properties of neurons in middle temporal visual area (MT) of owl monkeys. *J. Neurophysiol.* 52(3):488–513

Felleman DJ, Van Essen DC. 1991. Distributed hierarchical processing in the primate cerebral cortex. *Cereb. Cortex* 1(1):1–47

Fennema CL, Thompson WB. 1979. Velocity determination in scenes containing several moving images. *Comput. Graphics Image Process.* 9:301–15

Field DJ. 1987. Relations between the statistics of natural images and the response properties of cortical cells. *J. Opt. Soc. Am. A* 4(12):2379–94

Gattass R, Gross CG. 1981. Visual topography of striate projection zone (MT) in posterior superior temporal sulcus of the macaque. *J. Neurophysiol.* 46(3):621–38

Geesaman BJ, Born RT, Andersen RA, Tootell RB. 1997. Maps of complex motion selectivity in the superior temporal cortex of the alert macaque monkey: a double-label 2-deoxyglucose study. *Cereb. Cortex* 7(8):749–57

Girard P, Bullier J. 1989. Visual activity in area V2 during reversible inactivation of area 17 in the macaque monkey. *J. Neurophysiol.* 62(6):1287–302

Girard P, Salin PA, Bullier J. 1991. Visual activity in macaque area V4 depends on area 17 input. *Neuroreport* 2(2):81–84

Girard P, Salin PA, Bullier J. 1992. Response selectivity of neurons in area MT of the macaque monkey during reversible inactivation of area V1. *J. Neurophysiol.* 67(6):1437–46

Groh JM, Born RT, Newsome WT. 1997. How is a sensory map read out? Effects of microstimulation in visual area MT on saccades and smooth pursuit eye movements. *J. Neurosci.* 17(11):4312–30

Gulyas B, Orban GA, Duysens J, Maes H. 1987. The suppressive influence of moving textured backgrounds on responses of cat striate neurons to moving bars. *J. Neurophysiol.* 57(6):1767–91

Guo K, Benson PJ, Blakemore C. 2004. Pattern motion is present in V1 of awake but not anaesthetized monkeys. *Eur. J. Neurosci.* 19(4):1055–66

Heeger DJ. 1987. Model for the extraction of image flow. *J. Opt. Soc. Am. A* 4(8):1455–71

Heuer HW, Britten KH. 2002. Contrast dependence of response normalization in area MT of the rhesus macaque. *J. Neurophysiol.* 88(6):3398–408

Hildreth EC, Ando H, Andersen RA, Treue S. 1995. Recovering three-dimensional structure from motion with surface reconstruction. *Vision Res.* 35(1):117–37

Hubel DH, Livingstone MS. 1987. Segregation of form, color, and stereopsis in primate area 18. *J. Neurosci.* 7(11):3378–415

Hubel DH, Wiesel TN. 1962. Receptive fields, binocular interaction and functional architecture in the cat's visual cortex. *J. Physiol. (London)* 160:106–54

Hubel DH, Wiesel TN. 1968. Receptive fields and functional architecture of monkey striate cortex. *J. Physiol.* 195(1):215–43

Irvin GE, Norton TT, Sesma MA, Casagrande VA. 1986. W-like response properties of interlaminar zone cells in the lateral geniculate nucleus of a primate (Galago crassicaudatus). *Brain Res.* 362(2):254–70

Kapadia MK, Westheimer G, Gilbert CD. 1999. Dynamics of spatial summation in primary visual cortex of alert monkeys. *Proc. Natl. Acad. Sci. USA* 96(21):12073–78

Koenderink JJ, van Doorn AJ. 1992. Second-order optic flow. *J. Opt. Soc. Am. A* 9:530–38

Komatsu H, Wurtz RH. 1988. Relation of cortical areas MT and MST to pursuit eye movements. I. Localization and visual properties of neurons. *J. Neurophysiol.* 60(2):580–603

Kuypers HGJM, Szwarcbart MK, Mishkin M, Rosvold HE. 1965. Occipito-temporal cortico-cortical connections in the rhesus monkey. *Exp. Neurol.* 11:245–62

Lagae L, Gulyas B, Raiguel S, Orban GA. 1989. Laminar analysis of motion information processing in macaque V5. *Brain Res.* 496(1–2):361–67

Lagae L, Raiguel S, Orban GA. 1993. Speed and direction selectivity of macaque middle temporal neurons. *J. Neurophysiol.* 69(1):19–39

Levitt JB, Lund JS. 1997. Contrast dependence of contextual effects in primate visual cortex. *Nature* 387(6628):73–76

Lewis JW, Van Essen DC. 2000. Corticocortical connections of visual, sensorimotor, and multimodal processing areas in the parietal lobe of the macaque monkey. *J. Comp. Neurol.* 428(1):112–37

Lisberger SG, Morris EJ, Tychsen L. 1987. Visual motion processing and sensory-motor integration for smooth pursuit eye movements. *Annu. Rev. Neurosci.* 10:97–129

Lisberger SG, Movshon JA. 1999. Visual motion analysis for pursuit eye movements in area MT of macaque monkeys. *J. Neurosci.* 19(6):2224–46

Liu J, Newsome WT. 2003a. Correlation between MT activity and perceptual judgments of speed. *Soc. Neurosci. Abstr.* 29:438

Liu J, Newsome WT. 2003b. Functional organization of speed tuned neurons in visual area MT. *J. Neurophysiol.* 89(1):246–56

Liu J, Newsome WT. 2005. Correlation between speed perception and neural activity. *J. Neurosci.* 25:711–22

Livingstone MS, Conway BR. 2003. Substructure of direction-selective receptive fields in macaque V1. *J. Neurophysiol.* 89(5):2743–59

Livingstone MS, Pack CC, Born RT. 2001. Two-dimensional substructure of MT receptive fields. *Neuron* 30(3):781–93

Malach R, Schirman TD, Harel M, Tootell RBH, Malonek D. 1997. Organization of intrinsic connections in owl monkey area MT. *Cereb. Cortex* 7(4):386–93

Malonek D, Tootell RB, Grinvald A. 1994. Optical imaging reveals the functional architecture of neurons processing shape and motion in owl monkey area MT. *Proc. R. Soc. London B Biol. Sci.* 258(1352):109–19

Marcar VL, Zihl J, Cowey A. 1997. Comparing the visual deficits of a motion blind patient with the visual deficits of monkeys with area MT removed. *Neuropsychologia* 35(11):1459–65

Marr D. 1982. *Vision: A Computational Investigation into the Human Representation and Processing of Visual Information.* New York: Freeman

Martinez-Trujillo J, Treue S. 2002. Attentional modulation strength in cortical area MT depends on stimulus contrast. *Neuron* 35(2):365–70

Martinez-Trujillo JC, Treue S. 2004. Feature-based attention increases the selectivity of population responses in primate visual cortex. *Curr. Biol.* 14(9):744–51

Maunsell JH, Cook EP. 2002. The role of attention in visual processing. *Philos. Trans. R. Soc. London B Biol. Sci.* 357(1424):1063–72

Maunsell JH, Nealey TA, DePriest DD. 1990. Magnocellular and parvocellular contributions to responses in the middle temporal visual area (MT) of the macaque monkey. *J. Neurosci.* 10(10):3323–34

Maunsell JH, Newsome WT. 1987. Visual processing in monkey extrastriate cortex. *Annu. Rev. Neurosci.* 10:363–401

Maunsell JH, Van Essen DC. 1983a. Functional properties of neurons in middle temporal visual area of the macaque monkey. I. Selectivity for stimulus direction, speed, and orientation. *J. Neurophysiol.* 49(5):1127–47

Maunsell JH, Van Essen DC. 1983b. Functional properties of neurons in middle temporal visual area of the macaque monkey. II. Binocular interactions and sensitivity to binocular disparity. *J. Neurophysiol.* 49(5):1148–67

Maunsell JH, Van Essen DC. 1983c. The connections of the middle temporal visual area (MT) and their relationship to a cortical hierarchy in the macaque monkey. *J. Neurosci.* 3(12):2563–86

Maunsell JH, Van Essen DC. 1987. Topographic organization of the middle temporal visual area in the macaque monkey: representational biases and the relationship to callosal connections and myeloarchitectonic boundaries. *J. Comp. Neurol.* 266(4):535–55

Mazurek ME, Shadlen MN. 2002. Limits to the temporal fidelity of cortical spike rate signals. *Nat. Neurosci.* 5(5):463–71

McAdams CJ, Maunsell JH. 1999. Effects of attention on orientation-tuning functions of single neurons in macaque cortical area V4. *J. Neurosci.* 19(1):431–41

McLean J, Palmer LA. 1989. Contribution of linear spatiotemporal receptive field structure to velocity selectivity of simple cells in area 17 of cat. *Vision Res.* 29(6):675–79

Mikami A, Newsome WT, Wurtz RH. 1986. Motion selectivity in macaque visual cortex. II. Spatiotemporal range of directional interactions in MT and V1. *J. Neurophysiol.* 55(6):1328–39

Mingolla E, Todd JT, Norman JF. 1992. The perception of globally coherent motion. *Vision Res.* 32(6):1015–31

Mishkin M. 1954. Visual discrimination performance following partial ablations of the temporal lobe. II. Ventral surface vs. hippocampus. *J. Comp. Physiol. Psychol.* 47(3):187–93

Mishkin M, Pribram KH. 1954. Visual discrimination performance following partial ablations of the temporal lobe. I. Ventral vs. lateral. *J. Comp. Physiol. Psychol.* 47(1):14–20

Movshon JA, Adelson EH, Gizzi MS, Newsome WT. 1985. The analysis of moving visual patterns. In *Pattern Recognition Mechanisms*, ed. C Chagas, R Gattass, C Gross, pp. 117–51. Rome: Vatican

Movshon JA, Newsome WT. 1996. Visual response properties of striate cortical neurons projecting to area MT in macaque monkeys. *J. Neurosci.* 16(23):7733–41

Movshon JA, Thompson ID, Tolhurst DJ. 1978. Receptive field organization of complex cells in the cat's striate cortex. *J. Physiol.* 283:79–99

Nakayama K. 1985. Biological image motion processing: a review. *Vision Res.* 25(5):625–60

Nauta WJH, Gygax PA. 1954. Silver impregnation of degenerating axons in the central nervous system. A modified technique. *Stain Technol.* 29:91–94

Newsome WT, Britten KH, Movshon JA. 1989. Neuronal correlates of a perceptual decision. *Nature* 341(6237):52–54

Newsome WT, Wurtz RH, Komatsu H. 1988. Relation of cortical areas MT and MST to pursuit eye movements. II. Differentiation of retinal from extraretinal inputs. *J. Neurophysiol.* 60(2):604–20

Nguyenkim JD, DeAngelis GC. 2003. Disparity-based coding of three-dimensional surface orientation by macaque middle temporal neurons. *J. Neurosci.* 23(18):7117–28

Nguyenkim JD, DeAngelis GC. 2004. Macaque MT neurons are selective for 3D surface orientation defined by multiple cues. *Soc. Neurosci. Abstr.* 30:368–412

Nichols MJ, Newsome WT. 2002. Middle temporal visual area microstimulation influences veridical judgments of motion direction. *J. Neurosci.* 22(21):9530–40

Nowlan SJ, Sejnowski TJ. 1995. A selection model for motion processing in area MT of primates. *J. Neurosci.* 15(2):1195–214

Okamoto H, Kawakami S, Saito H, Hida E, Odajima K, et al. 1999. MT neurons in the macaque exhibited two types of bimodal direction tuning as predicted by a model for visual motion detection. *Vision Res.* 39(20):3465–79

Orban GA. 1997. Visual processing in macaque area MT/V5 and its satellites (MSTd and MSTv). In *Extrastriate Cortex in Primates*, ed. KS Rockland, JF Kaas, A Peters, pp. 359–434. New York: Plenum

Orban GA, Kennedy H, Bullier J. 1986. Velocity sensitivity and direction selectivity of neurons in areas V1 and V2 of the monkey: influence of eccentricity. *J. Neurophysiol.* 56(2):462–80

Osborne LC, Bialek W, Lisberger SG. 2004. Time course of information about motion direction in visual area MT of macaque monkeys. *J. Neurosci.* 24(13):3210–22

Pack CC, Born RT. 2001. Temporal dynamics of a neural solution to the aperture problem in visual area MT of macaque brain. *Nature* 409(6823):1040–42

Pack CC, Born RT, Livingstone MS. 2003. Two-dimensional substructure of stereo and motion interactions in macaque visual cortex. *Neuron* 37(3):525–35

Pack CC, Gartland AJ, Born RT. 2004. Integration of contour and terminator signals in visual area MT of alert macaque. *J. Neurosci.* 24(13):3268–80

Pack CC, Livingstone MS, Duffy KR, Born RT. 2003. End-stopping and the aperture problem: two-dimensional motion signals in macaque V1. *Neuron* 39(4):671–80

Pack CC, Hunter JN, Born RT. 2005. Contrast dependence of suppressive influences in cortical area MT of alert macaque. *J. Neurophysiol.* 93(3):1809–15

Parker AJ, Newsome WT. 1998. Sense and the single neuron: probing the physiology of perception. *Annu. Rev. Neurosci.* 21:227–77

Peterhans E, von der Heydt R. 1993. Functional organization of area V2 in the alert macaque. *Eur. J. Neurosci.* 5(5):509–24

Polat U, Mizobe K, Pettet MW, Kasamatsu T, Norcia AM. 1998. Collinear stimuli regulate visual responses depending on cell's contrast threshold. *Nature* 391(6667):580–84

Priebe NJ, Lisberger SG. 2004. Estimating target speed from the population response in visual area MT. *J. Neurosci.* 24(8):1907–16

Prince SJ, Pointon AD, Cumming BG, Parker AJ. 2000. The precision of single neuron responses in cortical area V1 during stereoscopic depth judgments. *J. Neurosci.* 20(9):3387–400

Purushothaman G, Bradley DC. 2005. Neural population code for fine perceptual decisions in area MT. *Nat. Neurosci.* 8(1):99–106

Qian N, Andersen RA. 1994. Transparent motion perception as detection of unbalanced motion signals. II. Physiology. *J. Neurosci.* 14(12):7367–80

Qian N, Andersen RA. 1995. V1 responses to transparent and nontransparent motions. *Exp. Brain Res.* 103(1):41–50

Qian N, Andersen RA, Adelson EH. 1994. Transparent motion perception as detection of unbalanced motion signals. III. Modeling. *J. Neurosci.* 14(12):7381–92

Raiguel S, Van Hulle MM, Xiao DK, Marcar VL, Orban GA. 1995. Shape and spatial distribution of receptive fields and antagonistic motion surrounds in the middle temporal area (V5) of the macaque. *Eur. J. Neurosci.* 7(10):2064–82

Recanzone GH, Wurtz RH. 1999. Shift in smooth pursuit initiation and MT and MST neuronal activity under different stimulus conditions. *J. Neurophysiol.* 82(4):1710–27

Recanzone GH, Wurtz RH. 2000. Effects of attention on MT and MST neuronal activity during pursuit initiation. *J. Neurophysiol.* 83(2):777–90

Reid RC, Soodak RE, Shapley RM. 1987. Linear mechanisms of directional selectivity in simple cells of cat striate cortex. *Proc. Natl. Acad. Sci. USA* 84(23):8740–44

Rockland KS. 1989. Bistratified distribution of terminal arbors of individual axons projecting from area V1 to middle temporal area (MT) in the macaque monkey. *Vis. Neurosci.* 3(2):155–70

Rockland KS. 1995. Morphology of individual axons projecting from area V2 to MT in the macaque. *J. Comp. Neurol.* 355(1):15–26

Rockland KS. 2002. Visual cortical organization at the single axon level: a beginning. *Neurosci Res.* 42(3):155–66

Rodman HR, Albright TD. 1989. Single-unit analysis of pattern-motion selective properties in the middle temporal visual area (MT). *Exp. Brain Res.* 75(1):53–64

Rodman HR, Gross CG, Albright TD. 1989. Afferent basis of visual response properties in area MT of the macaque. I. Effects of striate cortex removal. *J. Neurosci.* 9(6):2033–50

Rodman HR, Gross CG, Albright TD. 1990. Afferent basis of visual response properties in area MT of the macaque. II. Effects of superior colliculus removal. *J. Neurosci.* 10(4):1154–64

Rubin N, Hochstein S. 1993. Isolating the effect of one-dimensional motion signals on the perceived direction of moving two-dimensional objects. *Vision Res.* 33(10):1385–96

Sceniak MP, Hawken MJ, Shapley R. 2001. Visual spatial characterization of macaque V1 neurons. *J. Neurophysiol.* 85(5):1873–87

Sceniak MP, Ringach DL, Hawken MJ, Shapley R. 1999. Contrast's effect on spatial summation by macaque V1 neurons. *Nat. Neurosci.* 2(8):733–39

Schiller PH, Malpeli JG. 1977. The effect of striate cortex cooling on area 18 cells in the monkey. *Brain Res.* 126(2):366–69

Seidemann E, Newsome WT. 1999. Effect of spatial attention on the responses of area MT neurons. *J. Neurophysiol.* 81(4):1783–94

Seung HS, Sompolinsky H. 1993. Simple models for reading neuronal population codes. *Proc. Natl. Acad. Sci. USA* 90(22):10749–53

Shadlen MN, Britten KH, Newsome WT, Movshon JA. 1996. A computational analysis of the relationship between neuronal and behavioral responses to visual motion. *J. Neurosci.* 16(4):1486–510

Shadlen MN, Zohary E, Britten KH, Newsome WT. 1993. Directional properties of MT neurons examined with motion energy filtered apparent motion stimuli. *Invest. Ophthalmol. Vis. Sci. Suppl.* 34(4):908

Shipp S, Zeki S. 1985. Segregation of pathways leading from area V2 to areas V4 and V5 of macaque monkey visual cortex. *Nature* 315(6017):322–25

Shipp S, Zeki S. 1989a. The organization of connections between areas V5 and V1 in macaque monkey visual cortex. *Eur. J. Neurosci.* 1(4):309–32

Shipp S, Zeki S. 1989b. The organization of connections between areas V5 and V2 in macaque monkey visual cortex. *Eur. J. Neurosci.* 1(4):333–54

Siegel RM, Andersen RA. 1988. Perception of three-dimensional structure from motion in monkey and man. *Nature* 331(6153):259–61

Simoncelli E, Bair W, Cavanaugh J, Movshon T. 1996. Testing and refining a computational model of neural responses in area MT. *Invest. Opthalmol. Vis. Sci. Suppl.* 37:916

Simoncelli EP, Heeger DJ. 1998. A model of neuronal responses in visual area MT. *Vision Res.* 38(5):743–61

Sincich LC, Horton JC. 2002. Divided by cytochrome oxidase: a map of the projections from V1 to V2 in macaques. *Science* 295(5560):1734–37

Sincich LC, Horton JC. 2003. Independent projection streams from macaque striate cortex to the second visual area and middle temporal area. *J. Neurosci.* 23(13):5684–92

Sincich LC, Park KF, Wohlgemuth MJ, Horton JC. 2004. Bypassing V1: a direct geniculate input to area MT. *Nat. Neurosci.* 7(10):1123–28

Smith MA, Majaj N, Movshon JA. 2005. Dynamics of motion signaling by neurons in macaque area MT. *Nat. Neurosci.* 8(2):220–28

Snowden RJ, Treue S, Erickson RG, Andersen RA. 1991. The response of area MT and V1 neurons to transparent motion. *J. Neurosci.* 11(9):2768–85

Solomon SG, White AJ, Martin PR. 2002. Extraclassical receptive field properties of parvocellular, magnocellular, and koniocellular cells in the primate lateral geniculate nucleus. *J. Neurosci.* 22(1):338–49

Stepniewska I, Qi HX, Kaas JH. 1999. Do superior colliculus projection zones in the inferior pulvinar project to MT in primates? *Eur. J. Neurosci.* 11(2):469–80

Stoner GR, Albright TD. 1992. Neural correlates of perceptual motion coherence. *Nature* 358(6385):412–14

Stoner GR, Albright TD. 1996. The interpretation of visual motion: evidence for surface segmentation mechanisms. *Vision Res.* 36(9):1291–310

Tadin D, Lappin JS, Gilroy LA, Blake R. 2003. Perceptual consequences of centre-surround antagonism in visual motion processing. *Nature* 424(6946):312–15

Talbot SA, Marshall WH. 1941. Physiological studies on neural mechanisms of visual localization and discrimination. *Am. J. Ophthal.* 24:1255–64

Tanaka K, Hikosaka K, Saito H, Yukie M, Fukada Y, Iwai E. 1986. Analysis of local and wide-field movements in the superior temporal visual areas of the macaque monkey. *J. Neurosci.* 6(1):134–44

Thiele A, Henning P, Kubischik M, Hoffmann KP. 2002. Neural mechanisms of saccadic suppression. *Science* 295(5564):2460–62

Thomas OM, Cumming BG, Parker AJ. 2002. A specialization for relative disparity in V2. *Nat. Neurosci.* 5(5):472–78

Tigges J, Tigges M, Anschel S, Cross NA, Letbetter WD, McBride RL. 1981. Areal and laminar distribution of neurons interconnecting the central visual cortical areas 17, 18, 19, and MT in squirrel monkey (Saimiri). *J. Comp. Neurol.* 202(4):539–60

Tinsley CJ, Webb BS, Barraclough NE, Vincent CJ, Parker A, Derrington AM. 2003. The nature of V1 neural responses to 2D moving patterns depends on receptive-field structure in the marmoset monkey. *J. Neurophysiol.* 90(2):930–37

Treue S, Andersen RA. 1996. Neural responses to velocity gradients in macaque cortical area MT. *Vis. Neurosci.* 13(4):797–804

Treue S, Martinez-Trujillo JC. 1999. Feature-based attention influences motion processing gain in macaque visual cortex. *Nature* 399(6736):575–79

Treue S, Maunsell JH. 1996. Attentional modulation of visual motion processing in cortical areas MT and MST. *Nature* 382(6591):539–41

Treue S, Maunsell JH. 1999. Effects of attention on the processing of motion in macaque middle temporal and medial superior temporal visual cortical areas. *J. Neurosci.* 19(17):7591–602

Uka T, DeAngelis GC. 2003. Contribution of middle temporal area to coarse depth discrimination: comparison of neuronal and psychophysical sensitivity. *J. Neurosci.* 23(8):3515–30

Uka T, DeAngelis GC. 2004. Contribution of area MT to stereoscopic depth perception: choice-related response modulations reflect task strategy. *Neuron* 42(2):297–310

Ungerleider LG, Mishkin M. 1982. Two cortical visual systems. In *The Analysis of Visual Behavior*, ed. DJ Ingle, RJW Mansfield, MS Goodale, pp. 549–86. Cambridge, MA: MIT Press.

Van Essen DC, Maunsell JH, Bixby JL. 1981. The middle temporal visual area in the macaque: myeloarchitecture, connections, functional properties and topographic organization. *J. Comp. Neurol.* 199(3):293–326

van Hateren JH. 1992. A theory of maximizing sensory information. *Biol. Cybern.* 68(1):23–29

Watson AB, Ahumada AJ. 1983. A look at motion in the frequency domain. *NASA Tech. Memorandum* 84352

Weiskrantz L, Mishkin M. 1958. Effects of temporal and frontal cortical lesions on auditory discrimination in monkeys. *Brain* 81(3):406–14

Weiss Y, Simoncelli EP, Adelson EH. 2002. Motion illusions as optimal percepts. *Nat. Neurosci.* 5(6):598–604

Wilson HR, Ferrera VP, Yo C. 1992. A psychophysically motivated model for two-dimensional motion perception. *Vis. Neurosci.* 9(1):79–97

Woolsey CN, Akert K, Benjamin RM, Leibowitz H, Welker WI. 1955. Visual cortex of the marmoset. *Fed. Proc.* 14:166

Xiao DK, Marcar VL, Raiguel SE, Orban GA. 1997a. Selectivity of macaque MT/V5 neurons for surface orientation in depth specified by motion. *Eur. J. Neurosci.* 9(5):956–64

Xiao DK, Raiguel S, Marcar V, Orban GA. 1997b. The spatial distribution of the antagonistic surround of MT/V5 neurons. *Cereb. Cortex* 7(7):662–77

Yabuta NH, Sawatari A, Callaway EM. 2001. Two functional channels from primary visual cortex to dorsal visual cortical areas. *Science* 292(5515):297–300

Yo C, Wilson HR. 1992. Perceived direction of moving two-dimensional patterns depends on duration, contrast and eccentricity. *Vision Res.* 32(1):135–47

Zeki SM. 1969. Representation of central visual fields in prestriate cortex of monkey. *Brain Res.* 14(2):271–91

Zeki SM. 1974. Functional organization of a visual area in the posterior bank of the superior temporal sulcus of the rhesus monkey. *J. Physiol.* 236(3):549–73

Zeki SM. 1978. Functional specialisation in the visual cortex of the rhesus monkey. *Nature* 274(5670):423–28

Zeki SM. 1980. The response properties of cells in the middle temporal area (area MT) of owl monkey visual cortex. *Proc. R. Soc. London B Biol. Sci.* 207(1167):239–48

Zihl J, von Cramon D, Mai N. 1983. Selective disturbance of movement vision after bilateral brain damage. *Brain* 106:313–40

Growth and Survival Signals Controlling Sympathetic Nervous System Development

Natalia O. Glebova and David D. Ginty

Department of Neuroscience, Howard Hughes Medical Institute, The Johns Hopkins University School of Medicine, Baltimore, Maryland 21205, USA; email: nglebov@jhmi.edu; dginty@jhmi.edu

Key Words

axon, dendrite, cholinergic switch, neurotrophin, p75, TrkA

Abstract

The precise coordination of the many events in nervous system development is absolutely critical for the correct establishment of functional circuits. The postganglionic sympathetic neuron has been an amenable model for studying peripheral nervous system formation. Factors that control several developmental events, including multiple stages of axon extension, neuron survival and death, dendritogenesis, synaptogenesis, and establishment of functional diversity, have been identified in this neuron type. This knowledge allows us to integrate the various intricate processes involved in the formation of a functional sympathetic nervous system and thereby create a paradigm for understanding neuronal development in general.

Contents

INTRODUCTION 192
AXON GROWTH 192
   Initiation 193
   Proximal Axon Extension 193
   Target Innervation 197
   Growth Inhibition 198
   Growth Mechanisms 199
SURVIVAL 200
DEATH 202
DENDRITE FORMATION 204
SYNAPTOGENESIS 207
DIVERSIFICATION 208
CONCLUSION 210

INTRODUCTION

The complexity of nervous system development is astounding: Not only must the organism generate neuron numbers appropriate for the needs of targets being innervated, but also it must instruct these neurons to extend axons, elaborate dendrites, and form synapses to ensure the exquisite specificity of connectivity necessary for normal physiological function. Understanding this elaborate progression of intricate events may seem an insurmountable feat. In the case of the sympathetic nervous system (**Figure 1**), however, recent progress has significantly advanced our understanding of the development of this part of the peripheral nervous system since the first use of the term sympathetic by Winslow in 1732 (Kuntz 1934). We thus have the unique opportunity to integrate the distinct temporal and spatial elements of sympathetic neuron development into a coordinated process that leads from neuron specification to axon growth to maturation, and ultimately to create a unified paradigm for neuron development.

Postganglionic sympathetic neurons arise from neural crest cells, which migrate ventrally from the margins of the dorsal neural tube to form a column of sympathetic ganglion primordia near the dorsal aorta, undergo specification, and commence acquisition of noradrenergic properties. These neuroblasts then coalesce to form the definitive sympathetic ganglia, migrating rostrally to establish the superior cervical ganglion (SCG) and ventrally to form prevertebral ganglia, with the remainder of the column becoming the sympathetic chain. Sympathetic neuroblasts in the ganglia complete proliferation and differentiation and begin to elaborate axons and dendrites. Several of the factors that play a role in these early events are known, including neuregulin-1 in ventral migration of neural crest cells (Britsch et al. 1998); hepatocyte growth factor (HGF) in neuroblast survival and differentiation (Maina et al. 1998); Semaphorin 3A in neural crest cell migration and sympathetic ganglion formation (Kawasaki et al. 2002); bone morphogenetic protein (BMP) family members BMP-2 and BMP-7 derived from the dorsal aorta in induction of noradrenergic differentiation; and the transcription factors Mash1, Phox2a and b, Cash1, dHand, and GATA-3 in specification of the noradrenergic phenotype (discussed recently in excellent reviews by Anderson et al. 1997, Francis & Landis 1999, Anderson 2000, Ernsberger 2001, Goridis & Rohrer 2002, Rohrer 2003, Howard 2005). This review thus does not address sympathetic neuron migration and specification but focuses on recent work that illuminates the developmental events experienced by postmitotic postganglionic sympathetic neurons.

AXON GROWTH

Axon growth presents several enigmatic questions regarding the transformation of a round cell into a morphologically complex, process-bearing neuron. How is axon extension initiated? Do extrinsic factors emerging in the developing ganglion prompt the neuron to commence axogenesis? Or do neuron-intrinsic cues that appear at a certain time during neuron specification trigger the onset of axon growth? Perhaps both of these mechanisms play important roles, with instructive environmental signals acting on neurons expressing factors permissive for initiation of axon extension. Once axons begin to grow, they encounter the challenge of finding their way through the maze of dynamically changing tissues of a developing embryo. What factors control proximal and distal sympathetic axon extension, and how does signaling from these cues eventually converge to mediate the cytoskeletal changes that result in axon growth? Many of these complex questions have been explored in sympathetic neurons, and the discoveries that have emerged thus far reveal an intricate integration of successive steps of axon growth initiation followed by proximal and then distal axon extension.

Initiation

Sympathetic neurons begin to extend solitary axons early in development: The first axons appear during formation of sympathetic ganglia (**Figure 2**) (Rubin 1985a). The molecular mechanisms underlying axon initiation have not been elucidated, but one possibility is a local autocrine growth factor loop in which HGF produced by neurons triggers axon extension (Yang et al. 1998b, Maina & Klein 1999). Further investigation is required to determine whether autocrine HGF induces sympathetic axon growth in vivo, and whether HGF acts selectively on the neuron that produced it or signals to neighboring axons as well. It is clear, however, that the regulation of axon initiation is distinct from that of subsequent stages of axon outgrowth since molecules that control axon extension at later times during development have been found not to be responsible for prompting axon initiation (discussed below).

Proximal Axon Extension

The mystery of axon initiation notwithstanding, it has long been known that proximal axon projections from several sympathetic ganglia en route to final target organs occur along arterial vasculature (Kuntz 1934, Rubin

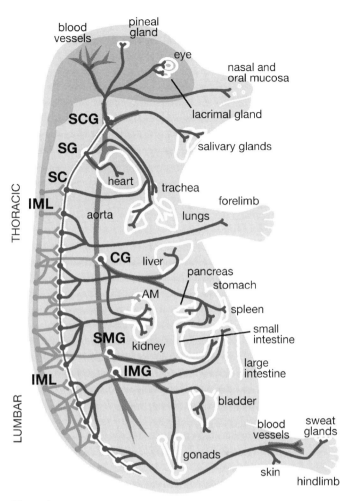

Figure 1

Schematic representation of sympathetic nervous system anatomy in the mouse. Preganglionic sympathetic neurons (*green*) with cell bodies located in the intermediolateral column (IML) at thoracic and lumbar levels of the spinal cord synapse on postganglionic neurons in sympathetic ganglia (*yellow*). Paravertebral ganglia include the superior cervical ganglion (SCG), the stellate ganglion (SG), and the sympathetic chain (SC). Prevertebral ganglia include the celiac ganglion (CG), the superior mesenteric ganglion (SMG), and the inferior mesenteric ganglion (IMG). Selected target organs innervated by postganglionic sympathetic neurons (*blue*) are shown. The adrenal medulla (AM) receives preganglionic sympathetic innervation. Portions of arterial vasculature along which postganglionic sympathetic axons from several ganglia project are shown. The number of sympathetic chain ganglia, the distribution of preganglionic inputs into the ganglia, the convergence of preganglionic and postganglionic projections, and the thoracolumbar levels of ganglia innervating specific organs are all approximations.

SYMPATHETIC NERVOUS SYSTEM FUNCTION

The sympathetic nervous system is part of the autonomic nervous system, which is responsible for maintaining homeostasis. Postganglionic sympathetic neurons innervate endocrine and exocrine glands, cardiac muscle, and smooth muscle throughout the body (**Figure 1**), carrying out vasomotor, pilomotor, secretomotor, sudomotor, and pupilodilator functions. The sympathetic nervous system regulates many physiological processes, including body temperature, blood pressure, respiration, cardiac output, blood glucose levels, gastrointestinal peristalsis, micturition, and ejaculation. In response to an external or internal environmental change such as stress or a change in body posture, activation of the sympathetic nervous system affects multiple organ systems. For example, blood pressure is elevated owing to increases in heart rate, contractility, and peripheral vascular resistance; glucose is mobilized from the liver; pupils and bronchi are dilated; gastrointestinal motility is decreased, and sphincter tone is increased. Although an animal can survive without a sympathetic nervous system, it can do so only in conditions of constant ambient temperature and in the absence of stress. The importance of normal sympathetic function is underscored by the pathological states associated with sympathetic dysfunction, such as hypertension, congestive heart failure, dysautonomia, various neuropathies, and other disorders (De Quattro & Feng 2002, Goldstein et al. 2002, Low 2002).

1985a). However, only recently have factors involved in mediating proximal axon outgrowth along blood vessels been identified. Artemin is one such molecule, as this member of the GDNF family is expressed in smooth muscle cells of blood vessels in a spatiotemporal pattern developmentally appropriate for a role in mediating sympathetic axon growth along the vasculature at a time of initial axon extension (Baloh et al. 1998, Nishino et al. 1999, Enomoto et al. 2001, Honma et al. 2002). Furthermore, artemin induces sympathetic axon growth in vitro (Enomoto et al. 2001, Yan et al. 2003) and attracts sympathetic axons in vitro and in vivo (Enomoto et al. 2001, Honma et al. 2002, Yan et al. 2003). SCG and sympathetic chain ganglia of mice deficient for *artemin* exhibit abnormally short and misdirected proximal axonal projections (Honma et al. 2002). Mice with targeted null mutations in the artemin coreceptors *Ret* or *GFRα3* (Baloh et al. 1998, Airaksinen & Saarma 2002) exhibit phenotypes essentially identical to *artemin*$^{-/-}$ animals (Schuchardt et al. 1994, Nishino et al. 1999, Enomoto et al. 2001, Honma et al. 2002). Thus, artemin signaling has a critical function in proximal sympathetic axon extension (**Figure 3a**).

However, the role of artemin in proximal axon extension is complicated by the requirement of this factor for sympathetic neuroblast migration: In *artemin*$^{-/-}$ mice, SCG neuron precursors cannot complete the last rostral migration step to the site of the definitive SCG (Honma et al. 2002). Therefore, although the defects in proximal axonal projections are clear in *artemin*$^{-/-}$ mice, the requirement of artemin for proper neuroblast migration in vivo is a confounding phenomenon, since it is not known whether the final migration step occurs before or after proximal axon extension. Some SCG neurons may indeed extend axons before final migration is complete (Rubin 1985a). Soma translocation may thus be a passive event secondary to axon growth, with axons leading the way for migrating cell bodies, in contrast to active soma-initiated movement. The migration deficits seen in the absence of artemin may therefore be reflective of aberrant axon growth, or conversely, deficient axon extension may be secondary to improper migration. The precise role of artemin in sympathetic axon growth thus awaits in vivo analysis of its function independent of its effects on migration.

Although artemin has a function in proximal sympathetic axon growth, be it direct or indirect, it cannot be the only factor mediating axon extension along blood vessels, since despite the abnormalities seen in embryonic *artemin*$^{-/-}$, *GFRα3*$^{-/-}$, or *Ret*$^{-/-}$ mice, the majority of these knockout mice do achieve at least partial sympathetic innervation of targets in adulthood (Enomoto et al. 2001, Honma et al. 2002). This suggests that another

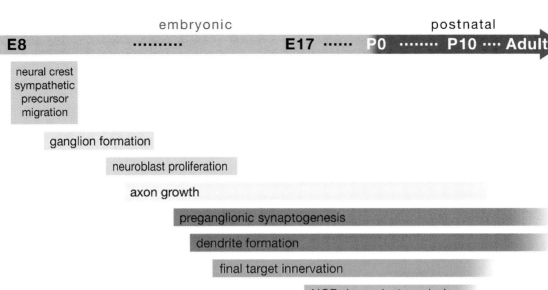

Figure 2
Order of events during sympathetic nervous system development in mouse. Approximate embryonic (E) and postnatal (P) time points are indicated (based on Chun & Patterson 1977; Rubin 1985a,b,c; Easton et al. 1997; Ernsberger & Rohrer 1999; Farinas 1999; Goridis & Rohrer 2002; Honma et al. 2002; Kawasaki et al. 2002; Glebova & Ginty 2004; Kuruvilla et al. 2004).

unidentified factor(s) in addition to artemin controls proximal axon extension. NT-3 is a good candidate for such a function, since it is expressed in blood vessels and can induce sympathetic axon growth in vitro (Maisonpierre et al. 1990b, Scarisbrick et al. 1993, Francis et al. 1999, Kuruvilla et al. 2004). This member of the neurotrophin family indeed plays a role in the development of proximal sympathetic axon projections, as axon extension along blood vessels and eventually into peripheral targets is markedly reduced for both prevertebral and paravertebral ganglia in $NT\text{-}3^{-/-}$ mice (Kuruvilla et al. 2004). However, although many sympathetic axons do not project as far along the vasculature in the absence of NT-3 as they do in control animals, sympathetic neurons do extend short proximal axons, and most targets receive at least some sympathetic innervation in $NT\text{-}3^{-/-}$ mice (ElShamy et al. 1996, Francis et al. 1999, Kuruvilla et al. 2004). Thus, although NT-3 is required for proximal sympathetic axon growth (**Figure 3a**), other factors control axon initiation.

The role of NT-3 as a mediator of proximal sympathetic axon extension differs from this neurotrophin's previously suggested function in sympathetic nervous system development. NT-3 has been implicated in sympathetic neuroblast survival in vitro (Birren et al. 1993, DiCicco-Bloom et al. 1993, Verdi et al. 1996) and in vivo (ElShamy et al. 1996). However, sympathetic ganglion formation (Farinas et al. 1996, Wilkinson et al. 1996) and the number of sympathetic neurons generated in $NT\text{-}3^{-/-}$ mice are not different from wild type, with neuron loss occurring late in development concurrently with the neuron loss seen in $NGF^{-/-}$ mice (Wyatt et al. 1997, Francis

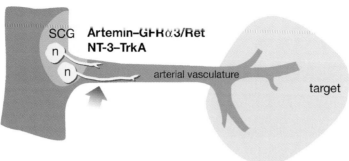

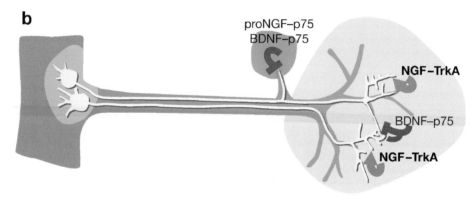

Figure 3

A model for control of sympathetic axon growth during development illustrated for two sympathetic neurons (n) in the SCG. (*a*) Artemin–GFRα3/Ret and NT-3–TrkA signaling induce proximal axon (*arrow*) extension along arterial vasculature toward the final target. (*b*) NGF–TrkA signaling, on the other hand, promotes distal axon extension and perhaps branching later in development. Final target innervation may also be under inhibitory control by p75. In the event that an axon collateral reaches an incorrect target (*blue*) not expressing nerve growth factor (NGF) but expressing a p75 ligand such as BDNF or proNGF, p75 signaling may inhibit the growth of that collateral, thereby preventing inappropriate target innervation. If a collateral arrives late at the correct target and is thus unable to obtain enough NGF to sustain TrkA activation, BDNF or proNGF from the target or from the neuron may inhibit the growth of this collateral, thus limiting axon density in the target.

et al. 1999, Kuruvilla et al. 2004). NT-3 thus does not mediate sympathetic neuroblast survival in vivo but is instead responsible for proximal sympathetic axon extension. The 50% loss of SCG neurons observed in $NT\text{-}3^{-/-}$ mice (Ernfors et al. 1994b, Farinas et al. 1994) may therefore be a consequence of deficits in early axon growth and the subsequent impaired ability of these neurons to obtain the target-derived NGF crucial for their survival (Kuruvilla et al. 2004, discussed below).

The relative contributions of NT-3, artemin, and possibly other unidentified molecules to sympathetic axon growth along blood vessels during development warrant further inquiry. For instance, because substantial amounts of both proximal axon growth and target innervation remain in the absence of either artemin or NT-3, an assessment of axonal projections in mice lacking both factors will be necessary to establish whether other blood vessel–derived growth factors contribute to proximal axon extension. Furthermore, the mechanisms of artemin and NT-3 action need to be ascertained, including whether these factors are permissive or instructive for sympathetic axon guidance in vivo. Some evidence suggests that artemin and NT-3 may serve as attractants: Artemin expression shifts from central to peripheral blood vessels in parallel with axon extension along these vessels, and artemin attracts sympathetic axons in mouse embryo cultures (Honma et al. 2002), whereas NT-3 is an attractive cue for Xenopus spinal neurons in vitro (Ming et al. 1997). The molecular mechanisms by which artemin–GFRα3/Ret and NT-3–TrkA signaling in sympathetic neurons translate into axon growth are also important topics for investigation. Do these two factors share signaling mechanisms to the cytoskeleton? Or do they use independent pathways? And if so, how are neuronal responses to these extracellular factors integrated? Another intriguing issue is that blood vessels, in addition to being intermediate routes for sympathetic axon extension, are also final sympathetic neuron targets and require sympathetic innervation for normal function (Kuntz 1934). How is it that some sympathetic axons functionally innervate blood vessels, while others continue to extend, ignoring these potential targets and using them instead as intermediate conduits en route to their final target organs? Although this question remains enigmatic, some of the factors that mediate proximal sympathetic axon growth are now known. The control of distal sympathetic axon extension into final targets has also been investigated, and one of the factors crucial for this process has been found.

Target Innervation

Sympathetic neurons extend axons along the vasculature and other intermediate targets with the goal of reaching and innervating their peripheral target organs. Whereas artemin and NT-3 mediate proximal sympathetic axon extension, final target innervation is controlled by another neurotrophin, nerve growth factor (NGF). NGF has been a leading candidate for mediating final target innervation since its discovery as the first neurotrophin more than 50 years ago (Levi-Montalcini 1987). Extensive evidence suggests that NGF controls sympathetic axon extension toward peripheral targets; NGF is produced by sympathetic target organs (Korsching 1993), and sympathetic innervation density corresponds to the amount of NGF produced by the organ (Korsching & Thoenen 1983, Shelton & Reichardt 1984). Furthermore, NGF promotes sympathetic axon growth in vitro (Cohen et al. 1954, Mains & Patterson 1973), and exogenous NGF induces sympathetic axon extension in vivo (Levi-Montalcini & Booker 1960b, Albers et al. 1994). However, because NGF is also required for sympathetic neuron survival (see below), it has only recently become possible to ascertain whether it is necessary for sympathetic axon growth into peripheral targets. The strategy of eliminating the proapoptotic factor Bax concurrently with a neurotrophin (Patel et al. 2000, 2003; Glebova & Ginty 2004; Kuruvilla et al. 2004) has allowed an assessment of the role of NGF in sympathetic axon extension in vivo by circumventing the excess apoptosis that otherwise occurs in $NGF^{-/-}$ or $NT\text{-}3^{-/-}$ mice. Analysis of $NGF^{-/-}$; $Bax^{-/-}$ mice indicates that NGF is required for final sympathetic target innervation in vivo but not for proximal axon projection from sympathetic ganglia (Glebova & Ginty 2004). Interestingly, whereas some targets exhibit a complete loss of sympathetic innervation in $NGF^{-/-}$; $Bax^{-/-}$ mice, others show partial defects, with certain organs demonstrating an NGF requirement for parenchymal innervation but not for axon growth along vasculature into the organ (Glebova & Ginty 2004). Thus, although NGF is required for peripheral innervation of some targets (**Figure 3b**), other factors are also involved in distal sympathetic axon extension.

The molecules responsible for the residual innervation of several target tissues in the absence of NGF remain to be identified, but several factors known to promote sympathetic axon growth in vitro are good candidates for future in vivo analysis. HGF, for example, in addition to mediating neuroblast survival and differentiation in vivo, enhances NGF-mediated sympathetic axon extension (Maina et al. 1998, Thompson et al. 2004) and promotes axon growth directly in vitro (Thompson et al. 2004). GDNF, expressed in sympathetic target tissues (Henderson et al. 1994; Trupp et al. 1995, 1997; Hellmich et al. 1996; Golden et al. 1999), induces sympathetic neurite outgrowth in vitro (Yan et al. 2003). Another excellent candidate is NT-3, which is expressed in some targets (Maisonpierre et al. 1990a) and required in vivo for proximal sympathetic axon growth along certain intermediate targets (Kuruvilla et al. 2004). Some extracellular matrix molecules also stimulate sympathetic axon extension in vitro (Lein & Higgins 1989, Reichardt et al. 1990, Lein et al. 1991). In fact, injection of antibodies to vascular cell-adhesion molecule-1 (VCAM-1), which is expressed in the heart, or to its receptor $\alpha 4\beta 1$ integrin on sympathetic neurons, results in decreased cardiac sympathetic innervation, although this may be due to a defect in axon maintenance and not growth (Wingerd et al. 2002). Identification of the factors that mediate sympathetic innervation of target organs innervated independently of NGF remains a future challenge.

The mechanisms by which these extrinsic factors control distal sympathetic axon growth need to be addressed as well. In particular, how do instructive and permissive cues coordinate axon extension into target organs? Could some factors, for example, NGF, be permissive, mediating process outgrowth and allowing axons to enter target fields, with organ-specific instructive cues subsequently superseding to regulate innervation of specific target cells? NGF may also function as a guidance cue: It is a chemoattractant for chick sympathetic neurons in vitro (Levi-Montalcini et al. 1954) and in vivo (Levi-Montalcini & Hamburger 1951, Menesini-Chen et al. 1978), and localized NGF induces turning of sensory neuron axons in culture (Gallo et al. 1997, Gundersen & Barrett 1979). Considering the complexity of axon guidance control in some, if not all, neuronal populations (Huber et al. 2003), the paucity of our knowledge regarding sympathetic axon guidance merits attention. The role of target-derived factors in sympathetic axon guidance is uncertain, however: In *tabby* mutant mice, which lack sweat glands, sympathetic innervation appears transiently where sweat glands would be located in wild-type mice and then eventually disappears (Guidry & Landis 1995). This suggests a role for target-produced factors in maintenance of innervation but not in ingrowth of axons. Perhaps this observation is an example of a distinct kind of axon growth regulation, consistent with the fact that during development, as the embryo grows, the axons that were the first to make contact with their target must grow accordingly to maintain the innervation they had established earlier. Still, sympathetic axons must be guided through the maze of the developing organism to find their correct targets. Future efforts will establish whether the factors that control sympathetic axon growth also act as guidance cues, and whether additional molecules that direct sympathetic axon guidance exist.

Growth Inhibition

Specific attractive guidance cues are essential for correct nervous system development, but accurate target innervation also depends on repulsive factors (Huber et al. 2003). Sympathetic neurons express receptors for repulsive cues of the Semaphorin family, neuropilins-1 and -2 (Kawakami et al. 1996; Kolodkin et al. 1997; Chen et al. 1997, 1998; Giger et al. 1998). Furthermore, Semaphorin 3F, for example, repels sympathetic axons in vitro (Chen et al. 1997; 2000; Giger et al. 1998, 2000) and induces sympathetic growth cone collapse in the presence of NGF, at least in part by antagonizing NGF-TrkA signaling (Atwal et al. 2003). Another mechanism of sympathetic axon growth inhibition involves signaling through the neurotrophin receptor p75, which binds all neurotrophins with similar affinity and can modulate Trk binding of and signaling in response to specific neurotrophins by forming a heteromeric receptor complex with Trks (Bibel & Barde 2000). In the absence of Trk receptors, p75 signals to inhibit axon growth: It directly interacts with Rho-GDI, thus releasing RhoA and activating this small GTPase to inhibit axon elongation (Yamashita & Tohyama 2003). Although this intracellular pathway for axon growth–inhibitory p75 signaling has been characterized, whether any p75 ligand functions to inhibit sympathetic axon extension during development is unknown.

One candidate for such an inhibitory p75 ligand in sympathetic neurons is BDNF, which signals exclusively through p75 in these neurons because they do not express the BDNF Trk receptor, TrkB (Fagan et al. 1996). In fact, exogenous BDNF inhibits sympathetic axon outgrowth, and endogenous BDNF signals in an autocrine manner through p75 to inhibit TrkA-mediated neurite extension in vitro (Kohn et al. 1999). In vivo, BDNF is present in at least one sympathetic target, the pineal gland, which is hyperinnervated in $BDNF^{-/-}$ mice (Kohn et al. 1999). These findings suggest that target-derived (Maisonpierre et al. 1990a, Kohn et al. 1999) and/or autocrine (Schecterson & Bothwell 1992, Causing et al. 1997) BDNF–p75 signaling could act to inhibit sympathetic axon growth during development. For this to occur, TrkA would have to be activated weakly, if at all, in the neuron because TrkA activation would suppress p75-mediated signaling (Huang & Reichardt 2003). Such a situation could arise if a sympathetic axon or a collateral, after obtaining NGF from a correct target, reaches an incorrect target not expressing NGF, or arrives late at an NGF-producing target, at a time when this neurotrophin's availability is diminished. BDNF from such a target or from other sympathetic neurons in the ganglion could then activate p75 to inhibit growth of inappropriate axons or collaterals or to limit axon density in the target (**Figure 3b**). These are all provocative scenarios prompting further investigation that will reveal the in vivo function of BDNF–p75 signaling, as well as the identities of other factors that may direct repulsive guidance and the termination of axon growth in correct target organs.

Growth Mechanisms

The variety of molecules known to control sympathetic axon growth in vivo and the potential diversity of unidentified factors involved in this process raise the question of how such different modes of signaling translate into the cytoskeletal changes that lead to axon extension (Dent & Gertler 2003, Huber et al. 2003). Do signals from GDNF and neurotrophin family members converge on the same intracellular pathways to regulate a common set of effectors that in turn mediate the assembly of mictrotubules, actin, and other structural components of the cytoskeleton? If so, how is regulation of distinct phases of axon extension accomplished? Interestingly, NT-3 and NGF signal through the same receptor TrkA in sympathetic neurons to mediate proximal and distal axonal projection, respectively (Belliveau et al. 1997, Glebova & Ginty 2004, Kuruvilla et al. 2004). This suggests that these two neurotrophins use the same signaling mechanisms to promote axon extension at different stages in development. But is that the case, and what are the intracellular TrkA effectors that act in response to NT-3 and NGF?

Although NGF-induced axon outgrowth has for many years been known to require local NGF signaling within axons (Campenot 1977), the molecular mechanisms involved in this process are just now becoming understood. Local growth signals from NGF are transduced within sympathetic axons through the mitogen-associated protein kinase (MAPK) kinase (MEK) and PI3 kinase pathways (Atwal et al. 2000), which may directly regulate the stability of microtubules and actin (Toker & Cantley 1997, Veeranna et al. 1998). In sensory neurons, NGF signals locally at the growth cone to induce axon extension through a cascade that includes PI3K, GSK-3β, and APC, a microtubule plus end-binding protein that stabilizes microtubules (Zhou et al. 2004). Sympathetic neurons may employ a similar mechanism of NGF-dependent axon extension. In addition to local TrkA activation, NGF signals retrogradely to the cell body to generate changes in gene transcription (Riccio et al. 1997, Lonze et al. 2002). This retrograde signal transduction may also contribute to sympathetic axon growth, as activation of specific transcriptional programs plays a role in neurotrophin-dependent sensory axon growth (Lonze et al. 2002, Graef et al. 2003).

Other evidence suggests that at least some forms of retrograde neurotrophin signaling may not be necessary for sympathetic axon extension: Although both NT-3 and NGF signal through TrkA to mediate sympathetic axon growth (Davies et al. 1995, Fagan et al. 1996, Belliveau et al. 1997, Tessarollo et al. 1997, Glebova & Ginty 2004, Kuruvilla et al. 2004), only NGF can support retrograde TrkA signaling in sympathetic neurons (Kuruvilla et al. 2004). The mechanism by which these two ligands differentially regulate trafficking of the same receptor is unclear but may involve distinct kinetics of TrkA activation by NGF versus NT-3 or different TrkA effectors utilized by the two neurotrophins (Belliveau et al. 1997). What is clear, however, is that both NGF and NT-3, despite the dissimilarity in their abilities to signal retrogradely, support sympathetic axon extension. Interestingly, once sympathetic axons reach their targets, final target-derived NGF induces p75 expression in developing sympathetic neurons, which in turn reduces the sensitivity of these neurons to NT-3 by modulating the ability of NT-3, but not NGF, to activate TrkA (Kuruvilla et al. 2004). This hierarchical signaling cascade thus ensures the appropriate spatiotemporal control of sympathetic neuron development as axons pass intermediate NT-3-expressing targets and enter distal NGF-producing fields.

The difference in TrkA trafficking raises the possibility that NT-3 and NGF use distinct molecular mechanisms to stimulate two types of axon growth, proximal and distal, respectively. The possibility that the fundamental nature of axon growth is distinct between these two developmental stages is intriguing. In fact, axon extension within final targets is different from proximal axon growth: Sympathetic axons do not branch until they reach their targets (Rubin 1985a). Furthermore, NGF is more effective than NT-3 at inducing branching complexity in adult sympathetic neurons in culture (Orike et al. 2001b), and localized sources of NGF promote sensory axon collateral sprouting in vitro (Gallo & Letourneau 1998). Therefore, NGF in the target field may be able to stimulate sympathetic axon branching in addition to extension, whereas NT-3, an intermediate target-derived factor, may mediate sympathetic axon extension only. Whether this is true in sympathetic neuron development in vivo, whether local axonal protein synthesis (Brittis et al. 2002) plays a role in NT-3- and/or NGF-mediated axon extension, and how NT-3, NGF, artemin, and other unidentified molecules that control sympathetic axon extension signal to the cytoskeleton are questions that will be answered in the future.

SURVIVAL

While much has been learned about the regulation of sympathetic axon growth, the molecular pathways responsible for sympathetic neuron survival during development have been even better characterized. Astonishingly, more than 50% of the neurons generated in the embryo are killed by apoptosis during normal development (Oppenheim 1991).

Why would such a seemingly inefficient phenomenon be a part of normal development? What determines whether neurons die or survive, and how is regulation of survival accomplished at the molecular level? These questions have been extensively investigated in the sympathetic nervous system, beginning with the pioneering studies by Victor Hamburger and Rita Levi-Montalcini who, more than 50 years ago, found that neuronal survival is critically dependent on targets (Levi-Montalcini 1987, Cowan 2001). These investigations laid the foundation for the neurotrophic hypothesis, which stipulates that to ensure the optimal amount of target innervation, neurons are initially overproduced during development and, once they innervate their targets, compete for limiting amounts of target-derived neurotrophic factors. Those neurons that do not obtain sufficient amounts of neurotrophins consequently die by apoptosis (Barde 1989, Oppenheim 1991, Davies 1996).

In accordance with the neurotrophic hypothesis (Oppenheim 1991), NGF, the classic target-derived neurotrophic factor (Levi-Montalcini 1987, Cowan 2001), is synthesized by target organs and binds to TrkA on axon terminals to initiate local as well as retrograde signaling to cell bodies. NGF is retrogradely transported from target fields to neuronal cell bodies within sympathetic ganglia through postganglionic axons in vivo (Hendry et al. 1974, Johnson et al. 1978) and in vitro (Ure & Campenot 1997). In vitro, NGF–TrkA signaling leads to endocytosis of a ligand-receptor complex and the formation of signaling endosomes. These vesicles are then retrogradely transported to the cell body where phosphorylated TrkA accumulates in the soma, resulting in activation of signaling events that support neuronal survival and gene expression (**Figure 4**) (Riccio et al. 1997, 1999; Watson et al. 1999; Kuruvilla et al. 2000; Lonze et al. 2002; Delcroix et al. 2003; Ye et al. 2003; Heerssen et al. 2004). Other modes of NGF retrograde signaling have been suggested, but direct evidence supporting such mechanisms is lacking (Senger & Campenot 1997, Miller & Kaplan 2001b, Ginty & Segal 2002, MacInnis & Campenot 2002, Campenot & MacInnis 2004). Thus, through retrograde axonal signaling, NGF activates several well-characterized prosurvival intracellular signaling pathways, which have been reviewed recently (Huang & Reichardt 2001, Sofroniew et al. 2001) and are not discussed here.

The importance of NGF signaling in sympathetic neuron development is revealed in the drastic reduction of sympathetic neuron numbers upon treatment of animals with anti-NGF antibodies (Levi-Montalcini & Booker 1960a, Angeletti & Levi-Montalcini 1971) and in the virtually complete loss of sympathetic neurons in $NGF^{-/-}$ mice (Crowley et al. 1994). However, this phenomenon of excess cell death is not limited to $NGF^{-/-}$ animals: 50% of sympathetic neurons that normally survive die in NT-$3^{-/-}$ mice (Ernfors et al. 1994b, Farinas et al. 1994). While this could imply that NT-3 mediates retrograde survival similarly to NGF, NT-3 in fact cannot signal retrogradely in and support retrograde survival of sympathetic neurons in vitro, even though it uses the same receptor, TrkA, as does NGF in these neurons (Kuruvilla et al. 2004). How NGF and NT-3 both signal through TrkA to mediate sympathetic axon growth while concurrently differing in their abilities to support survival is not yet clear (Cordon-Cardo et al. 1991, Belliveau et al. 1997, Kuruvilla et al. 2004). Why this difference in retrograde TrkA signaling exists at all is another intriguing question. The neurotrophic hypothesis suggests one answer: The inability of the neurotrophin that controls proximal axon growth (NT-3) to support retrograde survival would ensure that the final target-derived neurotrophin (NGF) exclusively mediates neuronal survival, and that the number of neurons generated during development thus precisely matches the innervation requirements of the final target (Kuruvilla et al. 2004). The excess cell death in NT-$3^{-/-}$ mice may therefore be due to

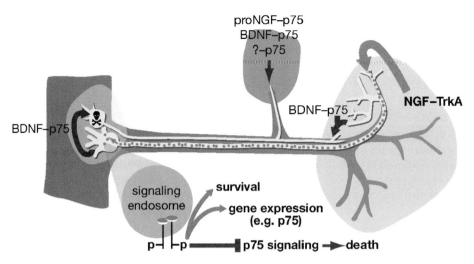

Figure 4
A model for control of sympathetic neuron survival and death during development. NGF expressed by the target binds TrkA on axon terminals, where it is endocytosed as a ligand-receptor complex and transported to the cell body in a signaling endosome (*green*) to support survival, in part by regulating gene expression. One of the genes upregulated by NGF is p75, which may have a proapoptotic function. TrkA activation suppresses prodeath p75 signaling, thus ensuring cell survival. However, if a neuron does not obtain adequate NGF to sustain prosurvival TrkA signaling after p75 induction, such suppression will not occur. p75 may then promote cell death in response to ligands such as BDNF or proNGF from the target or from the neurons themselves. An axon collateral innervating an incorrect target not expressing NGF may also obtain a p75 ligand that will promote cell death in the absence of adequate TrkA activation.

deficits in proximal axon projections resulting in the inability of these neurons to obtain the target-derived NGF necessary for their survival (Kuruvilla et al. 2004). Interestingly, postnatal sympathetic neurons gradually lose their dependency on NGF for survival (Angeletti et al. 1971; Lazarus et al. 1976; Chun & Patterson 1977; Goedert et al. 1978; Orike et al. 2001a,b; Tsui-Pierchala et al. 2002; Thompson et al. 2004), but how this phenomenon occurs is not well understood.

DEATH

The support of sympathetic neuron survival by target-derived NGF during development is a well-established phenomenon. However, it does not exclude the possibility that target-derived prodeath factors in addition to prosurvival factors are also involved in neuronal development (Oppenheim 1991). In fact, there is evidence that sympathetic neurons are capable of receiving active proapoptotic signals. The neurotrophin receptor p75, which binds all neurotrophins with similar affinity (Rodriguez-Tebar et al. 1992), can promote cell death: In postnatal sympathetic neurons, which express high levels of TrkA and p75, addition of BDNF, which signals through p75 but not TrkA in these neurons, induces sympathetic neuron apoptosis (Bamji et al. 1998). In support of this in vitro finding, the number of sympathetic neurons is increased in neonatal $p75^{-/-}$ mice (Bamji et al. 1998, Brennan et al. 1999). However, sympathetic neuron number is not significantly different in adult $p75^{-/-}$ mice as compared with controls (Lee et al. 1992, Bamji et al. 1998, Brennan et al. 1999). Thus, although sympathetic neurons still die in the absence of p75, this neurotrophin receptor is necessary for the induction of rapid cell death during development as apoptosis is delayed in the absence of p75 both in vitro and in vivo (Bamji et al. 1998).

What are the ligand(s) that may bind to p75 to actively promote cell death? BDNF is one excellent candidate because, as mentioned above, it can induce sympathetic neuron apoptosis in vitro, and the number of SCG neurons has been shown to be elevated in $BDNF^{-/-}$ mice (Bamji et al. 1998). Other studies performed at similar postnatal ages, however, have reported that sympathetic neuron numbers are not significantly different in $BDNF^{-/-}$ mice as compared with controls (Ernfors et al. 1994a, 1995; Liu et al. 1995). The in vivo evidence in support of a role for BDNF in the induction of rapid sympathetic neuron death during development is thus somewhat complicated, but several models have been proposed for the function of BDNF–p75 signaling during sympathetic neuron development (**Figure 4**). Rapid cell death induced by BDNF may eliminate neurons that innervate inappropriate targets expressing BDNF but not NGF. However, BDNF is also expressed in a subset of sympathetic targets (Maisonpierre et al. 1990a, Kohn et al. 1999), so it may function to remove neurons that innervate correct targets but fail to obtain sufficient amounts of NGF to activate TrkA and thereby suppress the proapoptotic activation of p75 by BDNF. Another possibility based on BDNF expression in sympathetic neurons themselves (Schecterson & Bothwell 1992, Causing et al. 1997) is that autocrine BDNF–p75 signaling may eliminate neurons in such situations (Majdan & Miller 1999). An argument against this autocrine BDNF–p75 signaling model, however, is that SCG neuron numbers are not decreased in transgenic mice overexpressing BDNF in sympathetic neurons (Causing et al. 1997). Furthermore, BDNF activation of p75 may not be necessary for this process, as more neurons survive in $p75^{-/-}$ mice than in $BDNF^{-/-}$ mice, which suggests the existence of additional ligand(s) for activating apoptosis through p75 (Majdan & Miller 1999).

ProNGF, a precursor form of NGF, is another candidate ligand for p75 in the induction of rapid death of sympathetic neurons (Chao & Bothwell 2002). ProNGF binds p75 with greater affinity than does mature NGF and is less effective than NGF in promoting TrkA signaling in sympathetic neurons. Furthermore, treatment of SCG neurons with proNGF in culture results in cell death (Lee et al. 2001, Nykjaer et al. 2004). Interestingly, proNGF is secreted by SCG neurons themselves, which suggests an autocrine or paracrine loop involving induction of cell death that would be effective in neurons with inadequate TrkA activation and the resulting inability to override the prodeath p75 signal (Hasan et al. 2003). The in vivo relevance of proNGF signaling during development, its expression, and regulation of proteolytic cleavage, as well as the potential functions of other proneurotrophins, remain to be determined. It is possible, however, that proapoptotic p75 signaling may not require a ligand at all, as unliganded p75 itself may mediate apoptosis (Rabizadeh et al. 1993), and overexpression of the intracellular domain of p75 in transgenic mice results in increased sympathetic neuron loss both during development and in the adult (Majdan et al. 1997). In vivo genetic evidence, however, has suggested that p75 does not actively stimulate cell death in sympathetic neurons, but rather that the delayed apoptosis in $p75^{-/-}$ mice is the result of more efficient NT-3 signaling (Brennan et al. 1999). Other findings contradict this model: Sympathetic neuron apoptosis is delayed in $p75^{-/-}$ mice that also lack *TrkA* (Majdan et al. 2001), suggesting that NT-3 does not signal to increase survival in the absence of p75, but rather that p75 mediates prodeath signaling that is suppressed by TrkA activation.

How would such putative proapoptotic signals be relevant during sympathetic nervous system development? Any situation involving a p75-mediated prodeath signal requires that the neuron reaches its target (so that it obtains the NGF necessary for upregulating p75 expression) but fails to obtain adequate NGF to sustain prosurvival TrkA activation and thus silence the proapoptotic p75 signal (Majdan et al. 2001). This could occur if an axon extends beyond the correct target and thus reaches an inappropriate one, or if it arrives late at the correct target when NGF has become more limiting (**Figure 4**). Still, the absence of prodeath signaling through p75 only delays sympathetic neuron apoptosis but does not prevent it, as adult $p75^{-/-}$ mice have normal numbers of sympathetic neurons (Lee et al. 1992, Bamji et al. 1998, Brennan et al. 1999). So what is the effect of this delayed apoptosis on sympathetic nervous system development? Although most sympathetic targets are innervated in adult $p75^{-/-}$ mice, some show reduced innervation (Lee et al. 1994, Brennan et al. 1999, Kuruvilla et al. 2004). But if p75 is a prodeath molecule, how is delayed sympathetic neuron apoptosis associated with reduced target innervation? Perhaps increased competition between the greater number of neurons surviving during development in the absence of p75 severely limits the availability of axon extension–promoting factors, thus leading to deficiencies in target innervation. Alternatively, p75 may play multiple separate roles during sympathetic neuron development, influencing both survival and growth of these neurons. Current evidence supports the latter model, in which p75 functions to (*a*) modulate NT-3's actions as an axonal growth factor, which in turn facilitates target innervation and the subsequent acquisition of the target-derived survival factor NGF; and (*b*) promote apoptosis in response to one or more ligands under conditions of limiting NGF–TrkA signaling. More investigation is needed to determine the ligands that activate p75 to induce rapid apoptosis, the signaling pathways involved (Aloyz et al. 1998, Miller & Kaplan 2001a, Barker 2004), and the role of other molecules such as TNFα in mediating active cell death (Barker et al. 2001), as well as the significance of this phenomenon in sympathetic nervous system development. The mechanisms by which sympathetic neuron survival and death during development are controlled thus still present surprising discoveries more than half a century after the proposal of the neurotrophic hypothesis, and the sympathetic nervous system endures as a paradigm for studying neuronal survival during development.

DENDRITE FORMATION

Another critical event in nervous system development is elaboration of dendrites that will come to receive inputs from presynaptic partners and thus play an essential role in the propagation of signals that will ultimately control target cell function. What factors regulate the generation of the remarkably complex and diverse dendritic arbors? Target-derived signals play a critical role in dendrite development, as target size influences the size and complexity of sympathetic dendritic arbors (Voyvodic 1989, Andrews et al. 1996). An increase in the size of the target field leads to an expansion of the dendritic field (Voyvodic 1989), whereas postganglionic axotomy results in dendrite retraction (Purves 1975, Yawo 1987). NGF, one target-derived factor, promotes dendrite growth in neonatal (Snider 1988) and adult (Ruit et al. 1990) mice. Additional target-produced molecules crucial for sympathetic dendritogenesis are likely to exist and may include members of the BMP family (discussed below).

Although sympathetic targets play a role in dendrite elaboration, they may not be necessary for dendrite initiation, since dendrites begin to grow in vivo after sympathetic neurons extend axons but before they reach their targets (**Figure 2**) (Rubin 1985a). Furthermore, NGF is not sufficient to trigger dendrite growth in vitro (Bruckenstein & Higgins 1988, Lein et al. 1995, Vaillant et al. 2002). One factor that may control initiation of dendrite growth is preganglionic activity, since the first preganglionic axons enter sympathetic ganglia prior to dendrite appearance (**Figure 2**) (Rubin 1985a,b) and neuronal activity promotes dendrite initiation in vitro (Vaillant et al. 2002). Preganglionic neuronal activity, however, may not be required for extension of dendrites once

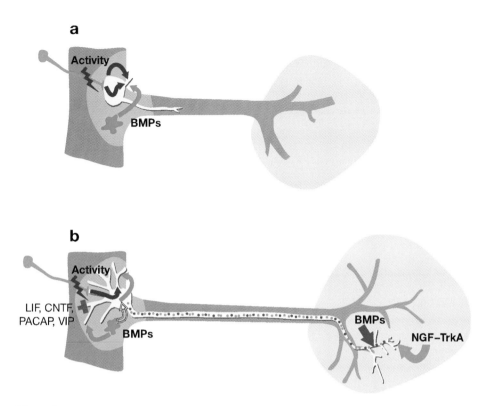

Figure 5

A model for control of dendrite formation for a sympathetic neuron in the SCG. (*a*) Activity (*purple*) in the form of preganglionic input or spontaneous activity in the postganglionic neuron may induce dendrite initiation. BMPs from glia may also promote dendritogenesis (*blue*). (*b*) Target-derived factors such as NGF or BMPs may stimulate dendrite elaboration, whereas mediators of growth inside the ganglion may include preganglionic neuronal activity (*purple*); glia through production of BMPs and/or induction of BMP synthesis in neurons (*blue*); and inhibitory factors such as leukemia inhibitory factor (LIF), ciliary neurotrophic factor (CNTF), pituitary adenylate cyclase-activating polypeptide (PACAP), and vasoactive intestinal peptide (VIP) (*red*).

growth has been initiated, as preganglionic denervation does not affect dendrite growth in neonatal animals at a time when dendrite initiation has already occurred (Voyvodic 1987). Activity may thus trigger dendrite initiation, with target-derived factors such as NGF acting later in development to stimulate dendrite growth. Preganglionic activity may not be the only mediator controlling dendrite initiation, however. Sponataneous activity in postganglionic neurons, for example, may also support this process (**Figure 5a**).

Furthermore, dendrite initiation may be under molecular control of members of the BMP family. BMP-5, -6, and -7, for example, stimulate dendritogenesis in sympathetic neurons in culture in the presence of NGF (Lein et al. 1995, Beck et al. 2001, Lein et al. 2002). In vivo, these BMPs are expressed in the SCG in a spatiotemporal pattern consistent with a role in dendritogenesis (Beck et al. 2001, Lein et al. 2002). Both glia and sympathetic neurons produce BMP-5, -6 and -7 (Beck et al. 2001, Lein et al. 2002), and glia also promote sympathetic dendrite development in vitro (Tropea et al. 1988), at least partially through BMPs (Lein et al. 2002). BMP signaling in sympathetic neuron dendritogenesis involves the translocation of Smad1, a transcription factor, from neuron cytoplasm to the nucleus (Guo et al. 2001, Lein et al. 2002) and increased expression of MAP2, with subsequent microtubule stabilization (Guo et al. 1998). Glia may also induce sympathetic dendrite growth by promoting BMP synthesis in neurons or in glial cells themselves (Lein et al. 2002) (**Figure 5**). Interestingly, BMP-mediated induction of gene expression changes in sympathetic neurons is modulated by neuronal activity in vitro (Fann & Patterson 1994), which suggests that activity and BMPs may collaborate in dendrite induction. The details of BMP signaling in sympathetic dendritogenesis, including receptor expression in neurons, how BMPs regulate dendritogenesis at the level of transcription (Lein & Higgins 1991, Lein et al. 1995), and the identities of genes whose expression may be controlled by Smad1 in a manner critical for dendritogenesis (Nohe et al. 2004), are not yet clear.

One interesting potential mechanism by which BMPs may promote dendrite growth is integrative nuclear FGFR1 signaling (IFNS), which is thought to integrate extracellular and intracellular signaling (Stachowiak et al. 2003). In this pathway, the typically plasma membrane–associated FGFR1 is located in the cytosol and, in response to a variety of signals (including BMP-7), translocates to the nucleus together with its ligand, FGF-2, to regulate transcription (Stachowiak et al. 2003). In cultured sympathetic neurons, BMP-7 treatment induces FGF2 expression and an increase in the amount of FGFR1 in the nucleus. Moreover, interference with IFNS specifically inhibits dendrite, but not axon, extension (Horbinski et al. 2002), which suggests that BMP-7 may use the IFNS pathway to control dendritogenesis. However, BMP-7-dependent dendrite growth is not completely inhibited by IFNS antagonism; thus, BMP-7 may signal through additional pathways to regulate sympathetic dendrite extension (Horbinski et al. 2002). Interestingly, the MEK-ERK (extracellular signal–regulated kinase) pathway may impinge on BMP signaling, as inhibition of MEK-ERK signaling potentiates dendrite growth and increases nuclear accumulation of Smad1 in BMP-7-treated sympathetic neurons in vitro (Kim et al. 2004). This is in direct opposition to the mechanism of activity-induced dendrite extension, where induction of sympathetic dendrite formation by activity involves activation of the MEK-ERK pathway, resulting in increased stabilization of microtubules by MAP2 (Vaillant et al. 2002). MEK-ERK inhibition in that paradigm thus blocks activity-induced dendrite extension (Vaillant et al. 2002). Furthermore, NGF itself also activates the MEK-ERK pathway (Huang & Reichardt 2001) and stimulates dendrite growth in vivo (Purves et al. 1988, Snider 1988). The conundrum of how MEK-ERK signaling is antagonistic to BMP-7-induced dendritogenesis, but activated during, and necessary for, dendrite growth stimulated by activity, underscores the complexity of numerous signaling pathways converging in the developing neuron and modulating each other's functions.

This intricate interplay between growth factors, neuronal activity, and signaling pathways that regulate dendritogenesis must also incorporate the multitude of molecules that actively mediate dendrite retraction. Several factors inhibit BMP-7-induced dendrite growth in vitro, including LIF and CNTF (Guo et al. 1997; 1999), PACAP and VIP (Drahushuk et al. 2002), retinoic acid (Chandrasekaran et al. 2000), and interferon γ (Kim et al. 2002). The in vivo relevance of these dendrite growth inhibitors to sympathetic dendrite formation remains to be established, as do the sources of these molecules and the signaling pathways that result in inhibition of dendrite growth but not axon extension.

Several fundamental questions regarding in vivo control and mechanisms of initiation, growth, stabilization, retraction, and maintenance of sympathetic dendritic arbor morphology thus remain. First and foremost is the question of the relative contributions of preganglionic input, neuronal activity, ganglion-derived factors, and target-derived factors to the in vivo initiation and elaboration of dendritic arbors (**Figure 5**). The establishment of distinct dendritic morphologies of sympathetic neuron subtypes projecting to specific targets (Andrews et al. 1996) is another interesting topic for investigation. Is dendritic arborization pattern development controlled extrinsically by inputs from specific preganglionic neuron types that then specify the postganglionic neuron identity and thus the target innervated? Or do target-derived factors define the sympathetic neuron subtype, which in turn regulates dendritic morphology via factors intrinsic to the neuron? Furthermore, the specific transcriptional programs that control dendritic tree architecture critical for proper neuron function remain to be identified in sympathetic neurons (Jan & Jan 2003, Miller & Kaplan 2003). The molecular mechanisms of dendrite extension are also enigmatic, although one molecule that may establish the mixed polarity orientation of microtubules characteristic of dendrites but not axons (Baas 1999) has been found in sympathetic neurons (Sharp et al. 1997). As complex a process as it may be during development, dendritogenesis continues into adulthood as well (Voyvodic 1987), and the control of this continual remodeling that includes extension, retraction, and de novo formation (Purves et al. 1986) is understood even less than embryonic dendrite development. Thus, although some factors regulating dendritogenesis in vitro have been identified, much future investigation is required for a full understanding of the in vivo control of this fascinating developmental process.

SYNAPTOGENESIS

The ultimate goal of sympathetic neuron development is creation of circuits that will mediate autonomic function. The postganglionic sympathetic neuron must therefore connect with the preganglionic neuron to form an interneuronal synapse as well as establish functional contact with its target. Although some progress has been made in the investigation of these two types of synaptogenesis, our understanding of sympathetic synapse formation mechanisms is far from complete. Cholinergic preganglionic synapses start developing early in embryogenesis, even before dendrite formation, and initially form on postganglionic somata, eventually occurring on dendrites (**Figure 2**) (Forehand 1985, Rubin 1985c). Target tissues retrogradely affect preganglionic synapse development (Devay et al. 1999, Majewski 1999, Rosenberg et al. 2002), while presynaptic input and activity are also crucial for sympathetic interneuronal synaptogenesis (Smolen 1981, Role 1988, Moss and Role 1993, Rosenberg et al. 2002). Some specific factors that may play a role in this process are BDNF (Causing et al. 1997), neuregulin (Yang et al. 1998a), NT-4 (Roosen et al. 2001), and agrin (Gingras et al. 2002).

While preganglionic synaptogenesis remains a poorly understood process, the formation of contacts between postganglionic sympathetic neurons and their targets is even more of a mystery. Postganglionic sympathetic axonal connections with their diverse target cells are not classical synapses; rather, they are large varicosities that lack archetypal postsynaptic specializations and possess presynaptic release sites located as far as micrometers away from target cells, thus necessitating significant neurotransmitter diffusion (Hillarp 1946, Paton & Vizi 1969, Smolen 1988, Elenkov et al. 2000). This is in contrast to classical chemical synapses, where the gap between pre- and postsynaptic cells is on the order of tens of nanometers (Elenkov et al. 2000). The mechanisms controlling development of these large sympathetic varicosities are largely unknown, although target-derived factors such as NGF may promote their formation (Lockhart et al. 2000). How postganglionic sympathetic neurons recognize their target cells and establish varicosities, and how preganglionic and postganglionic innervation of specific targets is coordinated (Ernsberger 2001, Schober & Unsicker 2001), are also interesting questions. Preganglionic neurons may recognize specific postganglionic neuron types innervating appropriate targets and then undergo synaptogenesis, or preganglionic synapse formation may play a role in directing target-specific postganglionic contact. Analysis of these developmental events in the experimentally amenable sympathetic nervous system will expand our understanding of the establishment of nervous system circuitry.

DIVERSIFICATION

The vast variety of dendritic arbor morphologies underscores the diversity of postganglionic sympathetic neurons. These cells exhibit considerable interganglionic and intraganglionic heterogeneity, with differences in dendrite and soma morphology, neuropeptide expression, and electrophysiological properties related to their various functions in target organs (Gibbins 1991, Andrews et al. 1996, Jobling & Gibbins 1999). How is this diversity established? Is it stochastic? Or are certain sympathetic neuroblasts predetermined early in embryogenesis to become particular subtypes of mature neurons and then innervate their specific targets? Perhaps the diversification occurs after target innervation, with target tissue sending a retrograde signal that will determine neuronal characteristics. Alternatively, the sequential temporal progression of neuron production and differentiation, combined with dynamically changing environmental cues, may be responsible for specifying the neuronal subtype (White et al. 2001).

The answers to these questions are not yet clear, but progress has been made in understanding the development of the subset of sympathetic neurons that use acetylcholine as the neurotransmitter (**Figure 6**) (Ernsberger & Rohrer 1999, Francis & Landis 1999). Although the majority of postganglionic sympathetic neurons are noradrenergic, a small subset are cholinergic and innervate eccrine sweat glands (Stevens & Landis 1987), periosteum (Asmus et al. 2000), and skeletal muscle vasculature (Ernsberger & Rohrer 1999) [although not in rodents (Guidry & Landis 2000)]. Interestingly, these cholinergic sympathetic neurons first develop an adrenergic phenotype, then undergo a transient stage with both adrenergic and cholinergic properties, and finally become fully cholinergic (Landis & Keefe 1983, Guidry & Landis 1998). Transplantation and coculture experiments with sweat glands have shown that this switch is target-dependent (Schotzinger & Landis 1988, 1990; Habecker & Landis 1994), and studies in *tabby* mutant mice lacking sweat glands have revealed the requirement of the target for the switch (Guidry & Landis 1995). The production of the target-secreted factor(s) (Rao & Landis 1990, Habecker & Landis 1994) in turn has been shown in vitro to require sympathetic innervation for sweat glands (Habecker et al. 1995b) but not for periosteum (Asmus et al. 2001). However, a study of $TH^{-/-}$ mice that are either pigmented or albino [also lacking tyrosinase, an alternative enzyme in catecholamine biosynthesis (Rios et al. 1999)] has demonstrated that catecholamines are required for functional maturation of sweat glands but not for the cholinergic switch in innervating neurons (Tian et al. 2000).

The identity of the target-derived molecule(s) that induce this cholinergic switch has remained elusive, although several neuropoietic cytokines, including LIF, CNTF, and CT-1 can promote cholinergic differentiation in vitro and in vivo (Saadat et al. 1989, Yamamori et al. 1989, Bamber et al. 1994, Pennica et al. 1995). However, in mice lacking *LIF* (Rao et al. 1993), *CNTF* (Masu et al. 1993, Francis et al. 1997), or both (Francis et al. 1997), the cholinergic switch is intact. And CT-1 is not required for the sweat gland extract-induced cholinergic differentiation in culture (Habecker et al. 1995a). Still, signaling through the heterodimer of LIFRβ and gp130, receptors for these members of the neurokine family, is required for the cholinergic switch in culture (Habecker et al. 1997). The target-derived

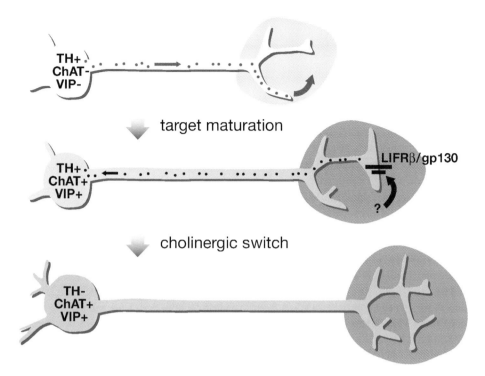

Figure 6

Schematic representation of sympathetic cholinergic phenotype acquisition. Sympathetic neurons that innervate sweat glands are initially noradrenergic and express TH but not cholinergic markers such as choline acetyltransferase (ChAT) or VIP. These neurons stimulate functional maturation of their target, which then induces cholinergic differentiation of the innervating neurons by secreting an unidentified neurokine that signals through LIFRβ/gp130. The cholinergic switch includes a transient stage during which the neurons express both noradrenergic [tyrosine hydroxylase (TH)] and cholinergic (ChAT and VIP) characteristics, but they eventually become fully cholinergic (TH−, ChAT+, VIP+).

cholinergic differentiation factor may thus be an unidentified member of the neurokine family. NT-3 is another candidate, as it induces cholinergic properties in chick sympathetic neurons in vitro (Brodski et al. 2000). Other potential cholinergic differentiation factors are GDNF and neurturin, which promote cholinergic differentiation of chick sympathetic neurons in vitro (Brodski et al. 2002). Analysis of $GFR\alpha 2^{-/-}$ mice suggests that neurturin–GFRα2 signaling is trophic for cholinergic sympathetic neurons and necessary for postnatal cholinergic sweat gland innervation, although it is not clear whether this is due to a role in axon growth or maintenance (Hiltunen & Airaksinen 2004). The definitive identification of the cholinergic switch–inducing factor(s) thus remains a future challenge.

The story of cholinergic sympathetic differentiation is not so simple, however: In addition to the cholinergic switch that occurs late in development, some sympathetic neurons undergo early target-independent cholinergic differentiation. Even before sympathetic ganglia fully form, cholinergic sympathetic neurons are present in the primordial chain, persist after the ganglia form, and are present throughout development (Schafer et al. 1997) into adulthood (Morales et al. 1995). While BMPs are involved in noradrenergic specification (Rohrer 2003), the factor(s)

APC: adenomatous polyposis coli

BDNF: brain-derived neurotrophic factor

BMP: bone morphogenetic protein

ChAT: choline acetyltransferase

CNTF: ciliary neurotrophic factor

CT-1: cardiotropin-1

ERK: extracellular signal-regulated kinase

FGFR1: fibroblast growth factor receptor-1

GDNF: glial cell line—derived neurotrophic factor

GSK-3β: glycogen synthase kinase 3β

HGF: hepatocyte growth factor

IFNS: integrative nuclear FGFR1 signaling

LIF: leukemia inhibitory factor

MAP2: microtubule-associated protein-2

MEK: MAP kinase kinase

NGF: nerve growth factor

NT-3: neurotrophin-3

NT-4: neurotrophin-4

PACAP: pituitary adenylate cyclase–activating polypeptide

PI3K: phosphatidylinositol 3-Kinase

Rho-GDI: Rho GDP dissociation inhibitor

SCG: superior cervical ganglion

Smad1: SMA- and MAD-related protein 1

TH: tyrosine hydroxylase

Trk: tropomyosin-related kinase

VCAM-1: vascular cell-adhesion molecule-1

VIP: vasoactive intestinal peptide

regulating this early acquisition of the cholinergic phenotype are unknown. Since the cholinergic neurons that innervate sweat glands are originally adrenergic and switch after they innervate the target (Guidry & Landis 1998), these early cholinergic neurons may innervate a different target, such as the periosteum (Asmus et al. 2000). However, this target exhibits cholinergic switching via a signaling pathway similar to that seen in sweat glands (Asmus et al. 2000, 2001). Another cholinergic target, skeletal muscle vasculature, has been recently shown not to receive cholinergic innervation in rodents (Guidry & Landis 2000). Moreover, cholinergic axon fibers are not detected in their known targets until postnatally (Guidry & Landis 1998), which implies that the early cholinergic sympathetic neurons have a function distinct from the neurons that undergo the late cholinergic switch. What is this function? Do these neurons innervate still unidentified targets? Does this early cholinergic property acquisition occur via the same signaling mechanisms as does the late, target-dependent one? Analysis of $LIFR\beta^{-/-}$ mice indicates that, in contrast to the late cholinergic switch, acquisition of early embryonic cholinergic properties in sympathetic neurons does not require LIFRβ signaling (Stanke et al. 2000). Thus, whereas sympathetic neurons that become cholinergic late in development clearly require target-derived signals, the early cholinergic neurons may not. The function of these cholinergic sympathetic neurons and the mechanisms of their differentiation are unknown.

The cholinergic sympathetic neuron is just one example of the vast phenotypic diversity of sympathetic neurons, which also includes differences among postganglionic neurons innervating different targets and along the rostrocaudal axis, as well as variations in neurotransmitter phenotype and neuropeptide content (Ernsberger 2001). In the case of the cholinergic switch, the target dependency is clear, but how early cholinergic neurons are specified is still a mystery. The regulation of acquisition of other phenotypic differences is even more enigmatic, and whether sympathetic neuron heterogeneity is predetermined, stochastic, or dependent on target remains a fascinating and as yet unanswered question.

CONCLUSION

The postganglionic sympathetic neuron has been a prototypical experimental system for studying neuronal development for more than 50 years, and much has been learned about the molecular control of its development. Many factors involved in sympathetic developmental events have been identified, including those that play roles in migration and specification of sympathetic neuron fate; proximal and peripheral axon extension; target-controlled neuron survival and death; dendrite formation and growth; synaptogenesis; and acquisition of phenotypic heterogeneity. What has emerged from these studies is an appreciation that multiple members of multiple families of growth factors from different sources with diverse and sometimes coincident signaling mechanisms collaborate to orchestrate the development of one remarkably complex cell, the sympathetic neuron. Several neurotrophins, including NT-3, NGF, and BDNF, mediate a number of distinct developmental processes such as proximal axon extension, final target innervation, growth inhibition, survival, death, and dendrite elongation. Multiple BMP family members act in sympathetic neuron differentiation and dendrite formation, and neuronal activity has a function in dendritogenesis and synaptogenesis. A GDNF family member, artemin, plays roles in neuroblast migration and proximal axon growth, while cytokine signaling through LIFRβ and gp130 is critical for mature phenotype acquisition in a subset of sympathetic neurons. However, many developmental events critical for the establishment of a functioning sympathetic nervous system remain unclear. Unanswered questions in specification, ganglion formation, neuronal death and survival, axon growth and branching, dendrite extension, axon guidance, target recognition, synaptogenesis, acquisition of diverse but target-appropriate phenotypes, interactions between neuronal and glial development, coordination of preganglionic and postganglionic neuron development, and maintenance of established circuits in the adult animal all await future investigation. The sympathetic neuron will continue to serve as an amenable paradigm to answer these fundamental questions, thus establishing the principles that form the framework for our thinking about how all populations of neurons and their complex circuits are formed and maintained.

ACKNOWLEDGMENTS

The authors' work is supported by the Medical Scientist Training Program (N.G.), NIH grant R01NS34814 (D.G.), and the Howard Hughes Medical Institute (D.G.). We thank Christopher Deppmann, Alex Kolodkin, Rejji Kuruvilla, Pamela Lein, Michael Rutlin, Shanthini Sockanathan, and Larry Zweifel for helpful comments on the manuscript.

LITERATURE CITED

Airaksinen MS, Saarma M. 2002. The GDNF family: signalling, biological functions and therapeutic value. *Nat. Rev. Neurosci.* 3:383–94

Albers KM, Wright DE, Davis BM. 1994. Overexpression of nerve growth factor in epidermis of transgenic mice causes hypertrophy of the peripheral nervous system. *J. Neurosci.* 14:1422–32

Aloyz RS, Bamji SX, Pozniak CD, Toma JG, Atwal J, et al. 1998. p53 is essential for developmental neuron death as regulated by the TrkA and p75 neurotrophin receptors. *J. Cell. Biol.* 143:1691–703

Anderson DJ. 2000. Genes, lineages and the neural crest: a speculative review. *Philos. Trans. R. Soc. London B Biol. Sci.* 355:953–64

Anderson DJ, Groves A, Lo L, Ma Q, Rao M, et al. 1997. Cell lineage determination and the control of neuronal identity in the neural crest. *Cold Spring Harb. Symp. Quant. Biol.* 62:493–504

Andrews TJ, Thrasivoulou C, Nesbit W, Cowen T. 1996. Target-specific differences in the dendritic morphology and neuropeptide content of neurons in the rat SCG during development and aging. *J. Comp. Neurol.* 368:33–44

Angeletti PU, Levi-Montalcini R. 1971. Growth regulation of the sympathetic nervous system: immunosympathectomy and chemical sympathectomy. *Rev. Eur. Etud. Clin. Biol.* 16:866–74

Angeletti PU, Levi-Montalcini R, Caramia F. 1971. Analysis of the effects of the antiserum to the nerve growth factor in adult mice. *Brain Res.* 27:343–55

Asmus SE, Parsons S, Landis SC. 2000. Developmental changes in the transmitter properties of sympathetic neurons that innervate the periosteum. *J. Neurosci.* 20:1495–504

Asmus SE, Tian H, Landis SC. 2001. Induction of cholinergic function in cultured sympathetic neurons by periosteal cells: cellular mechanisms. *Dev. Biol.* 235:1–11

Atwal JK, Massie B, Miller FD, Kaplan DR. 2000. The TrkB-Shc site signals neuronal survival and local axon growth via MEK and P13-kinase. *Neuron* 27:265–77

Atwal JK, Singh KK, Tessier-Lavigne M, Miller FD, Kaplan DR. 2003. Semaphorin 3F antagonizes neurotrophin-induced phosphatidylinositol 3-kinase and mitogen-activated protein kinase kinase signaling: a mechanism for growth cone collapse. *J. Neurosci.* 23:7602–9

Baas PW. 1999. Microtubules and neuronal polarity: lessons from mitosis. *Neuron* 22:23–31

Baloh RH, Tansey MG, Lampe PA, Fahrner TJ, Enomoto H, et al. 1998. Artemin, a novel member of the GDNF ligand family, supports peripheral and central neurons and signals through the GFRalpha3-RET receptor complex. *Neuron* 21:1291–302

Bamber BA, Masters BA, Hoyle GW, Brinster RL, Palmiter RD. 1994. Leukemia inhibitory factor induces neurotransmitter switching in transgenic mice. *Proc. Natl. Acad. Sci. USA* 91:7839–43

Bamji SX, Majdan M, Pozniak CD, Belliveau DJ, Aloyz R, et al. 1998. The p75 neurotrophin receptor mediates neuronal apoptosis and is essential for naturally occurring sympathetic neuron death. *J. Cell. Biol.* 140:911–23

Barde YA. 1989. Trophic factors and neuronal survival. *Neuron* 2:1525–34

Barker PA. 2004. p75NTR is positively promiscuous: novel partners and new insights. *Neuron* 42:529–33

Barker V, Middleton G, Davey F, Davies AM. 2001. TNFalpha contributes to the death of NGF-dependent neurons during development. *Nat. Neurosci.* 4:1194–98

Beck HN, Drahushuk K, Jacoby DB, Higgins D, Lein PJ. 2001. Bone morphogenetic protein-5 (BMP-5) promotes dendritic growth in cultured sympathetic neurons. *BMC Neurosci.* 2:12–23

Belliveau DJ, Krivko I, Kohn J, Lachance C, Pozniak C, et al. 1997. NGF and neurotrophin-3 both activate TrkA on sympathetic neurons but differentially regulate survival and neuritogenesis. *J. Cell. Biol.* 136:375–88

Bibel M, Barde YA. 2000. Neurotrophins: key regulators of cell fate and cell shape in the vertebrate nervous system. *Genes Dev.* 14:2919–37

Birren SJ, Lo L, Anderson DJ. 1993. Sympathetic neuroblasts undergo a developmental switch in trophic dependence. *Development* 119:597–610

Brennan C, Rivas-Plata K, Landis SC. 1999. The p75 neurotrophin receptor influences NT-3 responsiveness of sympathetic neurons in vivo. *Nat. Neurosci.* 2:699–705

Britsch S, Li L, Kirchhoff S, Theuring F, Brinkmann V, et al. 1998. The ErbB2 and ErbB3 receptors and their ligand, neuregulin-1, are essential for development of the sympathetic nervous system. *Genes Dev.* 12:1825–36

Brittis PA, Lu Q, Flanagan JG. 2002. Axonal protein synthesis provides a mechanism for localized regulation at an intermediate target. *Cell* 110:223–35

Brodski C, Schnurch H, Dechant G. 2000. Neurotrophin-3 promotes the cholinergic differentiation of sympathetic neurons. *Proc. Natl. Acad. Sci. USA* 97:9683–88

Brodski C, Schaubmar A, Dechant G. 2002. Opposing functions of GDNF and NGF in the development of cholinergic and noradrenergic sympathetic neurons. *Mol. Cell. Neurosci.* 19:528–38

Bruckenstein DA, Higgins D. 1988. Morphological differentiation of embryonic rat sympathetic neurons in tissue culture. I. Conditions under which neurons form axons but not dendrites. *Dev. Biol.* 128:324–36

Campenot RB. 1977. Local control of neurite development by nerve growth factor. *Proc. Natl. Acad. Sci. USA* 74:4516–19

Campenot RB, MacInnis BL. 2004. Retrograde transport of neurotrophins: fact and function. *J. Neurobiol.* 58:217–29

Causing CG, Gloster A, Aloyz R, Bamji SX, Chang E, et al. 1997. Synaptic innervation density is regulated by neuron-derived BDNF. *Neuron* 18:257–67

Chandrasekaran V, Zhai Y, Wagner M, Kaplan PL, Napoli JL, Higgins D. 2000. Retinoic acid regulates the morphological development of sympathetic neurons. *J. Neurobiol.* 42:383–93

Chao MV, Bothwell M. 2002. Neurotrophins: to cleave or not to cleave. *Neuron* 33:9–12

Chen H, Bagri A, Zupicich JA, Zou Y, Stoeckli E, et al. 2000. Neuropilin-2 regulates the development of selective cranial and sensory nerves and hippocampal mossy fiber projections. *Neuron* 25:43–56

Chen H, Chedotal A, He Z, Goodman CS, Tessier-Lavigne M. 1997. Neuropilin-2, a novel member of the neuropilin family, is a high affinity receptor for the semaphorins Sema E and Sema IV but not Sema III. *Neuron* 19:547–59

Chen H, He Z, Bagri A, Tessier-Lavigne M. 1998. Semaphorin-neuropilin interactions underlying sympathetic axon responses to class III semaphorins. *Neuron* 21:1283–90

Chun LL, Patterson PH. 1977. Role of nerve growth factor in the development of rat sympathetic neurons in vitro. II. Developmental studies. *J. Cell. Biol.* 75:705–11

Cohen S, Levi-Montalcini R, Hamburger V. 1954. A nerve growth-stimulating factor isolated from Sarcomas 37 and 180. *Proc. Natl. Acad. Sci. USA* 40:1014–18

Cordon-Cardo C, Tapley P, Jing SQ, Nanduri V, O'Rourke E, et al. 1991. The trk tyrosine protein kinase mediates the mitogenic properties of nerve growth factor and neurotrophin-3. *Cell* 66:173–83

Cowan WM. 2001. Viktor Hamburger and Rita Levi-Montalcini: the path to the discovery of nerve growth factor. *Annu. Rev. Neurosci.* 24:551–600

Crowley C, Spencer SD, Nishimura MC, Chen KS, Pitts-Meek S, et al. 1994. Mice lacking nerve growth factor display perinatal loss of sensory and sympathetic neurons yet develop basal forebrain cholinergic neurons. *Cell* 76:1001–11

Davies AM. 1996. The neurotrophic hypothesis: Where does it stand? *Philos. Trans. R. Soc. London B Biol. Sci.* 351:389–94

Davies AM, Minichiello L, Klein R. 1995. Developmental changes in NT3 signalling via TrkA and TrkB in embryonic neurons. *Embo. J.* 14:4482–89

Delcroix JD, Valletta JS, Wu C, Hunt SJ, Kowal AS, Mobley WC. 2003. NGF signaling in sensory neurons: evidence that early endosomes carry NGF retrograde signals. *Neuron* 39:69–84

Dent EW, Gertler FB. 2003. Cytoskeletal dynamics and transport in growth cone motility and axon guidance. *Neuron* 40:209–27

DeQuattro V, Feng M. 2002. The sympathetic nervous system: the muse of primary hypertension. *J. Hum. Hypertens.* 16(Suppl. 1): S64–69

Devay P, McGehee DS, Yu CR, Role LW. 1999. Target-specific control of nicotinic receptor expression at developing interneuronal synapses in chick. *Nat. Neurosci.* 2:528–34

DiCicco-Bloom E, Friedman WJ, Black IB. 1993. NT-3 stimulates sympathetic neuroblast proliferation by promoting precursor survival. *Neuron* 11:1101–11

Drahushuk K, Connell TD, Higgins D. 2002. Pituitary adenylate cyclase-activating polypeptide and vasoactive intestinal peptide inhibit dendritic growth in cultured sympathetic neurons. *J. Neurosci.* 22:6560–69

Easton RM, Deckwerth TL, Parsadanian AS, Johnson EM Jr. 1997. Analysis of the mechanism of loss of trophic factor dependence associated with neuronal maturation: a phenotype indistinguishable from Bax deletion. *J. Neurosci.* 17:9656–66

Elenkov IJ, Wilder RL, Chrousos GP, Vizi ES. 2000. The sympathetic nerve—an integrative interface between two supersystems: the brain and the immune system. *Pharmacol. Rev.* 52:595–638

ElShamy WM, Linnarsson S, Lee KF, Jaenisch R, Ernfors P. 1996. Prenatal and postnatal requirements of NT-3 for sympathetic neuroblast survival and innervation of specific targets. *Development* 122:491–500

Enomoto H, Crawford PA, Gorodinsky A, Heuckeroth RO, Johnson EM Jr, Milbrandt J. 2001. RET signaling is essential for migration, axonal growth and axon guidance of developing sympathetic neurons. *Development* 128:3963–74

Ernfors P, Lee KF, Jaenisch R. 1994a. Mice lacking brain-derived neurotrophic factor develop with sensory deficits. *Nature* 368:147–50

Ernfors P, Lee KF, Kucera J, Jaenisch R. 1994b. Lack of neurotrophin-3 leads to deficiencies in the peripheral nervous system and loss of limb proprioceptive afferents. *Cell* 77:503–12

Ernfors P, Kucera J, Lee KF, Loring J, Jaenisch R. 1995. Studies on the physiological role of brain-derived neurotrophic factor and neurotrophin-3 in knockout mice. *Int. J. Dev. Biol.* 39:799–807

Ernsberger U. 2001. The development of postganglionic sympathetic neurons: coordinating neuronal differentiation and diversification. *Auton. Neurosci.* 94:1–13

Ernsberger U, Rohrer H. 1999. Development of the cholinergic neurotransmitter phenotype in postganglionic sympathetic neurons. *Cell Tissue Res.* 297:339–61

Fagan AM, Zhang H, Landis S, Smeyne RJ, Silos-Santiago I, Barbacid M. 1996. TrkA, but not TrkC, receptors are essential for survival of sympathetic neurons in vivo. *J. Neurosci.* 16:6208–18

Fann MJ, Patterson PH. 1994. Neuropoietic cytokines and activin A differentially regulate the phenotype of cultured sympathetic neurons. *Proc. Natl. Acad. Sci. USA* 91:43–47

Farinas I. 1999. Neurotrophin actions during the development of the peripheral nervous system. *Microsc. Res. Tech.* 45:233–42

Farinas I, Jones KR, Backus C, Wang XY, Reichardt LF. 1994. Severe sensory and sympathetic deficits in mice lacking neurotrophin-3. *Nature* 369:658–61

Farinas I, Yoshida CK, Backus C, Reichardt LF. 1996. Lack of neurotrophin-3 results in death of spinal sensory neurons and premature differentiation of their precursors. *Neuron* 17:1065–78

Forehand CJ. 1985. Density of somatic innervation on mammalian autonomic ganglion cells is inversely related to dendritic complexity and preganglionic convergence. *J. Neurosci.* 5:3403–8

Francis N, Farinas I, Brennan C, Rivas-Plata K, Backus C, et al. 1999. NT-3, like NGF, is required for survival of sympathetic neurons, but not their precursors. *Dev. Biol.* 210:411–27

Francis NJ, Asmus SE, Landis SC. 1997. CNTF and LIF are not required for the target-directed acquisition of cholinergic and peptidergic properties by sympathetic neurons in vivo. *Dev. Biol.* 182:76–87

Francis NJ, Landis SC. 1999. Cellular and molecular determinants of sympathetic neuron development. *Annu. Rev. Neurosci.* 22:541–66

Gallo G, Lefcort FB, Letourneau PC. 1997. The trkA receptor mediates growth cone turning toward a localized source of nerve growth factor. *J. Neurosci.* 17:5445–54

Gallo G, Letourneau PC. 1998. Localized sources of neurotrophins initiate axon collateral sprouting. *J. Neurosci.* 18:5403–14

Gibbins IL. 1991. Vasomotor, pilomotor and secretomotor neurons distinguished by size and neuropeptide content in superior cervical ganglia of mice. *J. Auton. Nerv. Syst.* 34:171–83

Giger RJ, Cloutier JF, Sahay A, Prinjha RK, Levengood DV, et al. 2000. Neuropilin-2 is required in vivo for selective axon guidance responses to secreted semaphorins. *Neuron* 25:29–41

Giger RJ, Urquhart ER, Gillespie SK, Levengood DV, Ginty DD, Kolodkin AL. 1998. Neuropilin-2 is a receptor for semaphorin IV: insight into the structural basis of receptor function and specificity. *Neuron* 21:1079–92

Gingras J, Rassadi S, Cooper E, Ferns M. 2002. Agrin plays an organizing role in the formation of sympathetic synapses. *J. Cell. Biol.* 158:1109–18

Ginty DD, Segal RA. 2002. Retrograde neurotrophin signaling: Trk-ing along the axon. *Curr. Opin. Neurobiol.* 12:268–74

Glebova NO, Ginty DD. 2004. Heterogeneous requirement of NGF for sympathetic target innervation in vivo. *J. Neurosci.* 24:743–51

Goedert M, Otten U, Thoenen H. 1978. Biochemical effects of antibodies against nerve growth factor on developing and differentiated sympathetic ganglia. *Brain Res.* 148:264–68

Golden JP, DeMaro JA, Osborne PA, Milbrandt J, Johnson EM Jr. 1999. Expression of neurturin, GDNF, and GDNF family-receptor mRNA in the developing and mature mouse. *Exp. Neurol.* 158:504–28

Goldstein DS, Robertson D, Esler M, Straus SE, Eisenhofer G. 2002. Dysautonomias: clinical disorders of the autonomic nervous system. *Ann. Intern. Med.* 137:753–63

Goridis C, Rohrer H. 2002. Specification of catecholaminergic and serotonergic neurons. *Nat. Rev. Neurosci.* 3:531–41

Graef IA, Wang F, Charron F, Chen L, Neilson J, et al. 2003. Neurotrophins and netrins require calcineurin/NFAT signaling to stimulate outgrowth of embryonic axons. *Cell* 113:657–70

Guidry G, Landis SC. 1995. Sympathetic axons pathfind successfully in the absence of target. *J. Neurosci.* 15:7565–74

Guidry GL, Landis SC. 1998. Developmental regulation of neurotransmitters in sympathetic neurons. *Adv. Pharmacol.* 42:895–98

Guidry G, Landis SC. 2000. Absence of cholinergic sympathetic innervation from limb muscle vasculature in rats and mice. *Auton. Neurosci.* 82:97–108

Gundersen RW, Barrett JN. 1979. Neuronal chemotaxis: chick dorsal-root axons turn toward high concentrations of nerve growth factor. *Science* 206:1079–80

Guo X, Chandrasekaran V, Lein P, Kaplan PL, Higgins D. 1999. Leukemia inhibitory factor and ciliary neurotrophic factor cause dendritic retraction in cultured rat sympathetic neurons. *J. Neurosci.* 19:2113–21

Guo X, Lin Y, Horbinski C, Drahushuk KM, Kim IJ, et al. 2001. Dendritic growth induced by BMP-7 requires Smad1 and proteasome activity. *J. Neurobiol.* 48:120–30

Guo X, Metzler-Northrup J, Lein P, Rueger D, Higgins D. 1997. Leukemia inhibitory factor and ciliary neurotrophic factor regulate dendritic growth in cultures of rat sympathetic neurons. *Brain Res. Dev. Brain Res.* 104:101–10

Guo X, Rueger D, Higgins D. 1998. Osteogenic protein-1 and related bone morphogenetic proteins regulate dendritic growth and the expression of microtubule-associated protein-2 in rat sympathetic neurons. *Neurosci. Lett.* 245:131–34

Habecker BA, Landis SC. 1994. Noradrenergic regulation of cholinergic differentiation. *Science* 264:1602–4

Habecker BA, Pennica D, Landis SC. 1995a. Cardiotrophin-1 is not the sweat gland-derived differentiation factor. *Neuroreport* 7:41–44

Habecker BA, Symes AJ, Stahl N, Francis NJ, Economides A, et al. 1997. A sweat gland-derived differentiation activity acts through known cytokine signaling pathways. *J. Biol. Chem.* 272:30421–28

Habecker BA, Tresser SJ, Rao MS, Landis SC. 1995b. Production of sweat gland cholinergic differentiation factor depends on innervation. *Dev. Biol.* 167:307–16

Hasan W, Pedchenko T, Krizsan-Agbas D, Baum L, Smith PG. 2003. Sympathetic neurons synthesize and secrete pro-nerve growth factor protein. *J. Neurobiol.* 57:38–53

Heerssen HM, Pazyra MF, Segal RA. 2004. Dynein motors transport activated Trks to promote survival of target-dependent neurons. *Nat. Neurosci.* 7:596–604

Hellmich HL, Kos L, Cho ES, Mahon KA, Zimmer A. 1996. Embryonic expression of glial cell-line derived neurotrophic factor (GDNF) suggests multiple developmental roles in neural differentiation and epithelial-mesenchymal interactions. *Mech. Dev.* 54:95–105

Henderson CE, Phillips HS, Pollock RA, Davies AM, Lemeulle C, et al. 1994. GDNF: a potent survival factor for motoneurons present in peripheral nerve and muscle. *Science* 266:1062–64

Hendry IA, Stockel K, Thoenen H, Iversen LL. 1974. The retrograde axonal transport of nerve growth factor. *Brain Res.* 68:103–21

Hillarp N. 1946. Structure of the synapse and the peripheral innervation apparatus of the autonomic nervous system. *Acta. Anat.* Suppl. IV:1–153

Hiltunen PH, Airaksinen MS. 2004. Sympathetic cholinergic target innervation requires GDNF family receptor GFR alpha 2. *Mol. Cell. Neurosci.* 26:450–57

Honma Y, Araki T, Gianino S, Bruce A, Heuckeroth R, et al. 2002. Artemin is a vascular-derived neurotropic factor for developing sympathetic neurons. *Neuron* 35:267–82

Horbinski C, Stachowiak EK, Chandrasekaran V, Miuzukoshi E, Higgins D, Stachowiak MK. 2002. Bone morphogenetic protein-7 stimulates initial dendritic growth in sympathetic neurons through an intracellular fibroblast growth factor signaling pathway. *J. Neurochem.* 80:54–63

Howard MJ. 2005. Mechanisms and perspectives on differentiation of autonomic neurons. *Dev. Biol.* 277:271–86

Huang EJ, Reichardt LF. 2001. Neurotrophins: roles in neuronal development and function. *Annu. Rev. Neurosci.* 24:677–736

Huang EJ, Reichardt LF. 2003. Trk receptors: roles in neuronal signal transduction. *Annu. Rev. Biochem.* 72:609–42

Huber AB, Kolodkin AL, Ginty DD, Cloutier JF. 2003. Signaling at the growth cone: ligand-receptor complexes and the control of axon growth and guidance. *Annu. Rev. Neurosci.* 26:509–63

Jan YN, Jan LY. 2003. The control of dendrite development. *Neuron* 40:229–42

Jobling P, Gibbins IL. 1999. Electrophysiological and morphological diversity of mouse sympathetic neurons. *J. Neurophysiol.* 82:2747–64

Johnson EM Jr, Andres RY, Bradshaw RA. 1978. Characterization of the retrograde transport of nerve growth factor (NGF) using high specific activity [125I] NGF. *Brain Res.* 150:319–31

Kawakami A, Kitsukawa T, Takagi S, Fujisawa H. 1996. Developmentally regulated expression of a cell surface protein, neuropilin, in the mouse nervous system. *J. Neurobiol.* 29:1–17

Kawasaki T, Bekku Y, Suto F, Kitsukawa T, Taniguchi M, et al. 2002. Requirement of neuropilin 1-mediated Sema3A signals in patterning of the sympathetic nervous system. *Development* 129:671–80

Kim IJ, Beck HN, Lein PJ, Higgins D. 2002. Interferon gamma induces retrograde dendritic retraction and inhibits synapse formation. *J. Neurosci.* 22:4530–39

Kim IJ, Drahushuk KM, Kim WY, Gonsiorek EA, Lein P, et al. 2004. Extracellular signal-regulated kinases regulate dendritic growth in rat sympathetic neurons. *J. Neurosci.* 24:3304–12

Kohn J, Aloyz RS, Toma JG, Haak-Frendscho M, Miller FD. 1999. Functionally antagonistic interactions between the TrkA and p75 neurotrophin receptors regulate sympathetic neuron growth and target innervation. *J. Neurosci.* 19:5393–408

Kolodkin AL, Levengood DV, Rowe EG, Tai YT, Giger RJ, Ginty DD. 1997. Neuropilin is a semaphorin III receptor. *Cell* 90:753–62

Korsching S. 1993. The neurotrophic factor concept: a reexamination. *J. Neurosci.* 13:2739–48

Korsching S, Thoenen H. 1983. Nerve growth factor in sympathetic ganglia and corresponding target organs of the rat: correlation with density of sympathetic innervation. *Proc. Natl. Acad. Sci. USA* 80:3513–16

Kuntz A. 1934. *The Autonomic Nervous System*. Philadelphia: Lea and Febiger. 697 pp.

Kuruvilla R, Ye H, Ginty DD. 2000. Spatially and functionally distinct roles of the PI3-K effector pathway during NGF signaling in sympathetic neurons. *Neuron* 27:499–512

Kuruvilla R, Zweifel LS, Glebova NO, Lonze BE, Valdez G, et al. 2004. A neurotrophin signaling cascade coordinates sympathetic neuron development through differential control of TrkA trafficking and retrograde signaling. *Cell* 118:243–55

Landis SC, Keefe D. 1983. Evidence for neurotransmitter plasticity in vivo: developmental changes in properties of cholinergic sympathetic neurons. *Dev. Biol.* 98:349–72

Lazarus KJ, Bradshaw RA, West NR, Bunge P. 1976. Adaptive survival or rat sympathetic neurons cultured without supporting cells or exogenous nerve growth factor. *Brain Res.* 113:159–64

Lee KF, Bachman K, Landis S, Jaenisch R. 1994. Dependence on p75 for innervation of some sympathetic targets. *Science* 263:1447–49

Lee KF, Li E, Huber LJ, Landis SC, Sharpe AH, et al. 1992. Targeted mutation of the gene encoding the low affinity NGF receptor p75 leads to deficits in the peripheral sensory nervous system. *Cell* 69:737–49

Lee R, Kermani P, Teng KK, Hempstead BL. 2001. Regulation of cell survival by secreted proneurotrophins. *Science* 294:1945–48

Lein P, Johnson M, Guo X, Rueger D, Higgins D. 1995. Osteogenic protein-1 induces dendritic growth in rat sympathetic neurons. *Neuron* 15:597–605

Lein PJ, Beck HN, Chandrasekaran V, Gallagher PJ, Chen HL, et al. 2002. Glia induce dendritic growth in cultured sympathetic neurons by modulating the balance between bone morphogenetic proteins (BMPs) and BMP antagonists. *J. Neurosci.* 22:10377–87

Lein PJ, Higgins D. 1989. Laminin and a basement membrane extract have different effects on axonal and dendritic outgrowth from embryonic rat sympathetic neurons in vitro. *Dev. Biol.* 136:330–45

Lein PJ, Higgins D. 1991. Protein synthesis is required for the initiation of dendritic growth in embryonic rat sympathetic neurons in vitro. *Brain Res. Dev. Brain Res.* 60:187–96

Lein PJ, Higgins D, Turner DC, Flier LA, Terranova VP. 1991. The NC1 domain of type IV collagen promotes axonal growth in sympathetic neurons through interaction with the alpha 1 beta 1 integrin. *J. Cell. Biol.* 113:417–28

Levi-Montalcini R. 1987. The nerve growth factor 35 years later. *Science* 237:1154–62

Levi-Montalcini R, Booker B. 1960a. Destruction of the sympathetic ganglia in mammals by an antiserum to a nerve-growth protein. *Proc. Natl. Acad. Sci. USA* 46:384–91

Levi-Montalcini R, Booker B. 1960b. Excessive growth of the sympathetic ganglia evoked by a protein isolated from mouse salivary glands. *Proc. Natl. Acad. Sci. USA* 46:373–84

Levi-Montalcini R, Hamburger V. 1951. Selective growth stimulating effects of mouse sarcoma on the sensory and sympathetic nervous system of the chick embryo. *J. Exp. Zool.* 116:321–61

Levi-Montalcini R, Meyer H, Hamburger V. 1954. In vitro experiments on the effects of mouse sarcomas 180 and 37 on the spinal and sympathetic ganglia of the chick embryo. *Cancer Res.* 14:49–57

Liu X, Ernfors P, Wu H, Jaenisch R. 1995. Sensory but not motor neuron deficits in mice lacking NT4 and BDNF. *Nature* 375:238–41

Lockhart ST, Mead JN, Pisano JM, Slonimsky JD, Birren SJ. 2000. Nerve growth factor collaborates with myocyte-derived factors to promote development of presynaptic sites in cultured sympathetic neurons. *J. Neurobiol.* 42:460–76

Lonze BE, Riccio A, Cohen S, Ginty DD. 2002. Apoptosis, axonal growth defects, and degeneration of peripheral neurons in mice lacking CREB. *Neuron* 34:371–85

Low PA. 2002. Autonomic neuropathies. *Curr. Opin. Neurol.* 15:605–9

MacInnis BL, Campenot RB. 2002. Retrograde support of neuronal survival without retrograde transport of nerve growth factor. *Science* 295:1536–39

Maina F, Hilton MC, Andres R, Wyatt S, Klein R, Davies AM. 1998. Multiple roles for hepatocyte growth factor in sympathetic neuron development. *Neuron* 20:835–46

Maina F, Klein R. 1999. Hepatocyte growth factor, a versatile signal for developing neurons. *Nat. Neurosci.* 2:213–17

Mains RE, Patterson PH. 1973. Primary cultures of dissociated sympathetic neurons. I. Establishment of long-term growth in culture and studies of differentiated properties. *J. Cell. Biol.* 59:329–45

Maisonpierre PC, Belluscio L, Friedman B, Alderson RF, Wiegand SJ, et al. 1990a. NT-3, BDNF, and NGF in the developing rat nervous system: parallel as well as reciprocal patterns of expression. *Neuron* 5:501–9

Maisonpierre PC, Belluscio L, Squinto S, Ip NY, Furth ME, et al. 1990b. Neurotrophin-3: a neurotrophic factor related to NGF and BDNF. *Science* 247:1446–51

Majdan M, Lachance C, Gloster A, Aloyz R, Zeindler C, et al. 1997. Transgenic mice expressing the intracellular domain of the p75 neurotrophin receptor undergo neuronal apoptosis. *J. Neurosci.* 17:6988–98

Majdan M, Miller FD. 1999. Neuronal life and death decisions functional antagonism between the Trk and p75 neurotrophin receptors. *Int. J. Dev. Neurosci.* 17:153–61

Majdan M, Walsh GS, Aloyz R, Miller FD. 2001. TrkA mediates developmental sympathetic neuron survival in vivo by silencing an ongoing p75NTR-mediated death signal. *J. Cell. Biol.* 155:1275–85

Majewski M. 1999. Synaptogenesis and structure of the autonomic ganglia. *Folia Morphol. (Warsz.)* 58:65–99

Masu Y, Wolf E, Holtmann B, Sendtner M, Brem G, Thoenen H. 1993. Disruption of the CNTF gene results in motor neuron degeneration. *Nature* 365:27–32

Menesini Chen MG, Chen JS, Levi-Montalcini R. 1978. Sympathetic nerve fibers ingrowth in the central nervous system of neonatal rodent upon intracerebral NGF injections. *Arch. Ital. Biol.* 116:53–84

Miller FD, Kaplan DR. 2001a. Neurotrophin signalling pathways regulating neuronal apoptosis. *Cell. Mol. Life Sci.* 58:1045–53

Miller FD, Kaplan DR. 2001b. On Trk for retrograde signaling. *Neuron* 32:767–70

Miller FD, Kaplan DR. 2003. Signaling mechanisms underlying dendrite formation. *Curr. Opin. Neurobiol.* 13:391–98

Ming G, Lohof AM, Zheng JQ. 1997. Acute morphogenic and chemotropic effects of neurotrophins on cultured embryonic Xenopus spinal neurons. *J. Neurosci.* 17:7860–71

Morales MA, Holmberg K, Xu ZQ, Cozzari C, Hartman BK, et al. 1995. Localization of choline acetyltransferase in rat peripheral sympathetic neurons and its coexistence with nitric oxide synthase and neuropeptides. *Proc. Natl. Acad. Sci. USA* 92:11819–23

Moss BL, Role LW. 1993. Enhanced ACh sensitivity is accompanied by changes in ACh receptor channel properties and segregation of ACh receptor subtypes on sympathetic neurons during innervation in vivo. *J. Neurosci.* 13:13–28

Nishino J, Mochida K, Ohfuji Y, Shimazaki T, Meno C, et al. 1999. GFR alpha3, a component of the artemin receptor, is required for migration and survival of the superior cervical ganglion. *Neuron* 23:725–36

Nohe A, Keating E, Knaus P, Petersen NO. 2004. Signal transduction of bone morphogenetic protein receptors. *Cell Signal.* 16:291–99

Nykjaer A, Lee R, Teng KK, Jansen P, Madsen P, et al. 2004. Sortilin is essential for proNGF-induced neuronal cell death. *Nature* 427:843–48

Oppenheim RW. 1991. Cell death during development of the nervous system. *Annu. Rev. Neurosci.* 14:453–501

Orike N, Middleton G, Borthwick E, Buchman V, Cowen T, Davies AM. 2001a. Role of PI 3-kinase, Akt and Bcl-2-related proteins in sustaining the survival of neurotrophic factor-independent adult sympathetic neurons. *J. Cell. Biol.* 154:995–1005

Orike N, Thrasivoulou C, Wrigley A, Cowen T. 2001b. Differential regulation of survival and growth in adult sympathetic neurons: an in vitro study of neurotrophin responsiveness. *J. Neurobiol.* 47:295–305

Patel TD, Jackman A, Rice FL, Kucera J, Snider WD. 2000. Development of sensory neurons in the absence of NGF/TrkA signaling in vivo. *Neuron* 25:345–57

Patel TD, Kramer I, Kucera J, Niederkofler V, Jessell TM, et al. 2003. Peripheral NT3 signaling is required for ETS protein expression and central patterning of proprioceptive sensory afferents. *Neuron* 38:403–16

Paton WD, Vizi ES. 1969. The inhibitory action of noradrenaline and adrenaline on acetylcholine output by guinea-pig ileum longitudinal muscle strip. *Br. J. Pharmacol.* 35:10–28

Pennica D, Shaw KJ, Swanson TA, Moore MW, Shelton DL, et al. 1995. Cardiotrophin-1. Biological activities and binding to the leukemia inhibitory factor receptor/gp130 signaling complex. *J. Biol. Chem.* 270:10915–22

Purves D. 1975. Functional and structural changes in mammalian sympathetic neurones following interruption of their axons. *J. Physiol.* 252:429–63

Purves D, Hadley RD, Voyvodic JT. 1986. Dynamic changes in the dendritic geometry of individual neurons visualized over periods of up to three months in the superior cervical ganglion of living mice. *J. Neurosci.* 6:1051–60

Purves D, Snider WD, Voyvodic JT. 1988. Trophic regulation of nerve cell morphology and innervation in the autonomic nervous system. *Nature* 336:123–28

Rabizadeh S, Oh J, Zhong LT, Yang J, Bitler CM, et al. 1993. Induction of apoptosis by the low-affinity NGF receptor. *Science* 261:345–48

Rao MS, Landis SC. 1990. Characterization of a target-derived neuronal cholinergic differentiation factor. *Neuron* 5:899–910

Rao MS, Sun Y, Escary JL, Perreau J, Tresser S, et al. 1993. Leukemia inhibitory factor mediates an injury response but not a target-directed developmental transmitter switch in sympathetic neurons. *Neuron* 11:1175–85

Reichardt LF, Bossy B, Carbonetto S, de Curtis I, Emmett C, et al. 1990. Neuronal receptors that regulate axon growth. *Cold Spring Harb. Symp. Quant. Biol.* 55:341–50

Riccio A, Ahn S, Davenport CM, Blendy JA, Ginty DD. 1999. Mediation by a CREB family transcription factor of NGF-dependent survival of sympathetic neurons. *Science* 286:2358–61

Riccio A, Pierchala BA, Ciarallo CL, Ginty DD. 1997. An NGF-TrkA-mediated retrograde signal to transcription factor CREB in sympathetic neurons. *Science* 277:1097–100

Rios M, Habecker B, Sasaoka T, Eisenhofer G, Tian H, et al. 1999. Catecholamine synthesis is mediated by tyrosinase in the absence of tyrosine hydroxylase. *J. Neurosci.* 19:3519–26

Rodriguez-Tebar A, Dechant G, Gotz R, Barde YA. 1992. Binding of neurotrophin-3 to its neuronal receptors and interactions with nerve growth factor and brain-derived neurotrophic factor. *EMBO. J.* 11:917–22

Rohrer H. 2003. The role of bone morphogenetic proteins in sympathetic neuron development. *Drug News Perspect.* 16:589–96

Role LW. 1988. Neural regulation of acetylcholine sensitivity in embryonic sympathetic neurons. *Proc. Natl. Acad. Sci. USA* 85:2825–29

Roosen A, Schober A, Strelau J, Bottner M, Faulhaber J, et al. 2001. Lack of neurotrophin-4 causes selective structural and chemical deficits in sympathetic ganglia and their preganglionic innervation. *J. Neurosci.* 21:3073–84

Rosenberg MM, Blitzblau RC, Olsen DP, Jacob MH. 2002. Regulatory mechanisms that govern nicotinic synapse formation in neurons. *J. Neurobiol.* 53:542–55

Rubin E. 1985a. Development of the rat superior cervical ganglion: ganglion cell maturation. *J. Neurosci.* 5:673–84

Rubin E. 1985b. Development of the rat superior cervical ganglion: ingrowth of preganglionic axons. *J. Neurosci.* 5:685–96

Rubin E. 1985c. Development of the rat superior cervical ganglion: initial stages of synapse formation. *J. Neurosci.* 5:697–704

Ruit KG, Osborne PA, Schmidt RE, Johnson EM Jr, Snider WD. 1990. Nerve growth factor regulates sympathetic ganglion cell morphology and survival in the adult mouse. *J. Neurosci.* 10:2412–19

Saadat S, Sendtner M, Rohrer H. 1989. Ciliary neurotrophic factor induces cholinergic differentiation of rat sympathetic neurons in culture. *J. Cell. Biol.* 108:1807–16

Scarisbrick IA, Jones EG, Isackson PJ. 1993. Coexpression of mRNAs for NGF, BDNF, and NT-3 in the cardiovascular system of the pre- and postnatal rat. *J. Neurosci.* 13:875–93

Schafer MK, Schutz B, Weihe E, Eiden LE. 1997. Target-independent cholinergic differentiation in the rat sympathetic nervous system. *Proc. Natl. Acad. Sci. USA* 94:4149–54

Schecterson LC, Bothwell M. 1992. Novel roles for neurotrophins are suggested by BDNF and NT-3 mRNA expression in developing neurons. *Neuron* 9:449–63

Schober A, Unsicker K. 2001. Growth and neurotrophic factors regulating development and maintenance of sympathetic preganglionic neurons. *Int. Rev. Cytol.* 205:37–76

Schotzinger RJ, Landis SC. 1988. Cholinergic phenotype developed by noradrenergic sympathetic neurons after innervation of a novel cholinergic target in vivo. *Nature* 335:637–39

Schotzinger RJ, Landis SC. 1990. Acquisition of cholinergic and peptidergic properties by sympathetic innervation of rat sweat glands requires interaction with normal target. *Neuron* 5:91–100

Schuchardt A, D'Agati V, Larsson-Blomberg L, Costantini F, Pachnis V. 1994. Defects in the kidney and enteric nervous system of mice lacking the tyrosine kinase receptor Ret. *Nature* 367:380–83

Senger DL, Campenot RB. 1997. Rapid retrograde tyrosine phosphorylation of trkA and other proteins in rat sympathetic neurons in compartmented cultures. *J. Cell. Biol.* 138:411–21

Sharp DJ, Yu W, Ferhat L, Kuriyama R, Rueger DC, Baas PW. 1997. Identification of a microtubule-associated motor protein essential for dendritic differentiation. *J. Cell. Biol.* 138:833–43

Shelton DL, Reichardt LF. 1984. Expression of the beta-nerve growth factor gene correlates with the density of sympathetic innervation in effector organs. *Proc. Natl. Acad. Sci. USA* 81:7951–55

Smolen AJ. 1981. Postnatal development of ganglionic neurons in the absence of preganglionic input: morphological observations on synapse formation. *Brain Res.* 227:49–58

Smolen AJ. 1988. Morphology of synapses in the autonomic nervous system. *J. Electron Microsc. Tech.* 10:187–204

Snider WD. 1988. Nerve growth factor enhances dendritic arborization of sympathetic ganglion cells in developing mammals. *J. Neurosci.* 8:2628–34

Sofroniew MV, Howe CL, Mobley WC. 2001. Nerve growth factor signaling, neuroprotection, and neural repair. *Annu. Rev. Neurosci.* 24:1217–81

Stachowiak MK, Fang X, Myers JM, Dunham SM, Berezney R, et al. 2003. Integrative nuclear FGFR1 signaling (INFS) as a part of a universal "feed-forward-and-gate" signaling module that controls cell growth and differentiation. *J. Cell Biochem.* 90:662–91

Stanke M, Geissen M, Gotz R, Ernsberger U, Rohrer H. 2000. The early expression of VAChT and VIP in mouse sympathetic ganglia is not induced by cytokines acting through LIFRbeta or CNTFRalpha. *Mech. Dev.* 91:91–96

Stevens LM, Landis SC. 1987. Development and properties of the secretory response in rat sweat glands: relationship to the induction of cholinergic function in sweat gland innervation. *Dev. Biol.* 123:179–90

Tessarollo L, Tsoulfas P, Donovan MJ, Palko ME, Blair-Flynn J, et al. 1997. Targeted deletion of all isoforms of the trkC gene suggests the use of alternate receptors by its ligand neurotrophin-3 in neuronal development and implicates trkC in normal cardiogenesis. *Proc. Natl. Acad. Sci. USA* 94:14776–81

Thompson J, Dolcet X, Hilton M, Tolcos M, Davies AM. 2004. HGF promotes survival and growth of maturing sympathetic neurons by PI-3 kinase- and MAP kinase-dependent mechanisms. *Mol. Cell. Neurosci.* 27: 441–52

Tian H, Habecker B, Guidry G, Gurtan A, Rios M, et al. 2000. Catecholamines are required for the acquisition of secretory responsiveness by sweat glands. *J. Neurosci.* 20:7362–69

Toker A, Cantley LC. 1997. Signalling through the lipid products of phosphoinositide-3-OH kinase. *Nature* 387:673–76

Tropea M, Johnson MI, Higgins D. 1988. Glial cells promote dendritic development in rat sympathetic neurons in vitro. *Glia* 1:380–92

Trupp M, Belluardo N, Funakoshi H, Ibanez CF. 1997. Complementary and overlapping expression of glial cell line-derived neurotrophic factor (GDNF), c-ret proto-oncogene, and GDNF receptor-alpha indicates multiple mechanisms of trophic actions in the adult rat CNS. *J. Neurosci.* 17:3554–67

Trupp M, Ryden M, Jornvall H, Funakoshi H, Timmusk T, et al. 1995. Peripheral expression and biological activities of GDNF, a new neurotrophic factor for avian and mammalian peripheral neurons. *J. Cell. Biol.* 130:137–48

Tsui-Pierchala BA, Milbrandt J, Johnson EM Jr. 2002. NGF utilizes c-Ret via a novel GFL-independent, inter-RTK signaling mechanism to maintain the trophic status of mature sympathetic neurons. *Neuron* 33:261–73

Ure DR, Campenot RB. 1997. Retrograde transport and steady-state distribution of 125I-nerve growth factor in rat sympathetic neurons in compartmented cultures. *J. Neurosci.* 17:1282–90

Vaillant AR, Zanassi P, Walsh GS, Aumont A, Alonso A, Miller FD. 2002. Signaling mechanisms underlying reversible, activity-dependent dendrite formation. *Neuron* 34:985–98

Veeranna, Amin ND, Ahn NG, Jaffe H, Winters CA, et al. 1998. Mitogen-activated protein kinases (Erk1,2) phosphorylate Lys-Ser-Pro (KSP) repeats in neurofilament proteins NF-H and NF-M. *J. Neurosci.* 18:4008–21

Verdi JM, Groves AK, Farinas I, Jones K, Marchionni MA, et al. 1996. A reciprocal cell-cell interaction mediated by NT-3 and neuregulins controls the early survival and development of sympathetic neuroblasts. *Neuron* 16:515–27

Voyvodic JT. 1987. Development and regulation of dendrites in the rat superior cervical ganglion. *J. Neurosci.* 7:904–12

Voyvodic JT. 1989. Peripheral target regulation of dendritic geometry in the rat superior cervical ganglion. *J. Neurosci.* 9:1997–2010

Watson FL, Heerssen HM, Moheban DB, Lin MZ, Sauvageot CM, et al. 1999. Rapid nuclear responses to target-derived neurotrophins require retrograde transport of ligand-receptor complex. *J. Neurosci.* 19:7889–900

White PM, Morrison SJ, Orimoto K, Kubu CJ, Verdi JM, Anderson DJ. 2001. Neural crest stem cells undergo cell-intrinsic developmental changes in sensitivity to instructive differentiation signals. *Neuron* 29:57–71

Wilkinson GA, Farinas I, Backus C, Yoshida CK, Reichardt LF. 1996. Neurotrophin-3 is a survival factor in vivo for early mouse trigeminal neurons. *J. Neurosci.* 16:7661–69

Wingerd KL, Goodman NL, Tresser JW, Smail MM, Leu ST, et al. 2002. Alpha 4 integrins and vascular cell adhesion molecule-1 play a role in sympathetic innervation of the heart. *J. Neurosci.* 22:10772–80

Wyatt S, Pinon LG, Ernfors P, Davies AM. 1997. Sympathetic neuron survival and TrkA expression in NT3-deficient mouse embryos. *Embo. J.* 16:3115–23

Yamamori T, Fukada K, Aebersold R, Korsching S, Fann MJ, Patterson PH. 1989. The cholinergic neuronal differentiation factor from heart cells is identical to leukemia inhibitory factor. *Science* 246:1412–16

Yamashita T, Tohyama M. 2003. The p75 receptor acts as a displacement factor that releases Rho from Rho-GDI. *Nat. Neurosci.* 6:461–67

Yan H, Newgreen DF, Young HM. 2003. Developmental changes in neurite outgrowth responses of dorsal root and sympathetic ganglia to GDNF, neurturin, and artemin. *Dev. Dyn.* 227:395–401

Yang X, Kuo Y, Devay P, Yu C, Role L. 1998a. A cysteine-rich isoform of neuregulin controls the level of expression of neuronal nicotinic receptor channels during synaptogenesis. *Neuron* 20:255–70

Yang XM, Toma JG, Bamji SX, Belliveau DJ, Kohn J, et al. 1998b. Autocrine hepatocyte growth factor provides a local mechanism for promoting axonal growth. *J. Neurosci.* 18:8369–81

Yawo H. 1987. Changes in the dendritic geometry of mouse superior cervical ganglion cells following postganglionic axotomy. *J. Neurosci.* 7:3703–11

Ye H, Kuruvilla R, Zweifel LS, Ginty DD. 2003. Evidence in support of signaling endosome-based retrograde survival of sympathetic neurons. *Neuron* 39:57–68

Zhou FQ, Zhou J, Dedhar S, Wu YH, Snider WD. 2004. NGF-induced axon growth is mediated by localized inactivation of GSK-3beta and functions of the microtubule plus end binding protein APC. *Neuron* 42:897–912

Adult Neurogenesis in the Mammalian Central Nervous System

Guo-li Ming and Hongjun Song

Institute for Cell Engineering, Departments of Neurology and Neuroscience, Johns Hopkins University School of Medicine, Baltimore, Maryland 21205; email: gming1@bs.jhmi.edu, shongju1@bs.jhmi.edu

Key Words

neural stem cell, progenitor, development, regeneration, plasticity

Abstract

Forty years since the initial discovery of neurogenesis in the postnatal rat hippocampus, investigators have now firmly established that active neurogenesis from neural progenitors continues throughout life in discrete regions of the central nervous systems (CNS) of all mammals, including humans. Significant progress has been made over the past few years in understanding the developmental process and regulation of adult neurogenesis, including proliferation, fate specification, neuronal maturation, targeting, and synaptic integration of the newborn neurons. The function of this evolutionarily conserved phenomenon, however, remains elusive in mammals. Adult neurogenesis represents a striking example of structural plasticity in the mature CNS environment. Advances in our understanding of adult neurogenesis will not only shed light on the basic principles of adult plasticity, but also may lead to strategies for cell replacement therapy after injury or degenerative neurological diseases.

Contents

INTRODUCTION 224
HISTORIC OVERVIEW 225
METHODOLOGIES FOR
   INVESTIGATION OF ADULT
   NEUROGENESIS 226
   Analysis of Endogenous Adult
      Neurogenesis In Vivo 226
   Analysis of Adult Neurogenesis
      In Vitro and Ex Vivo 229
NEUROGENESIS IN THE
   INTACT ADULT
   MAMMALIAN CNS 231
   The Adult Neural Stem Cell, Its
      Proliferation, and Neuronal
      Fate Specification 231
   Neuronal Migration and Nerve
      Guidance 234
   Neuronal Maturation, Synapse
      Formation, and Plasticity 235
MODULATION OF
   ENDOGENOUS ADULT
   NEUROGENESIS 236
ADULT NEUROGENESIS
   UNDER PATHOLOGICAL
   STIMULATIONS 239
   Ischemic Brain Injury 239
   Seizures 239
   Radiation Injury 240
   Degenerative Neurological
      Diseases 240
POTENTIAL FUNCTIONS OF
   ADULT NEUROGENESIS 241
PERSPECTIVE 242

Adult neural stem cells: a type of unspecified precursor cells that has the capacity to proliferate in generating more of itself and to make new neurons, astrocytes, and oligodendrocytes

INTRODUCTION

Neurogenesis, a process of generating functionally integrated neurons from progenitor cells, was traditionally believed to occur only during embryonic stages in the mammalian CNS (Ramon y Cajal 1913). Only recently has it become generally accepted that new neurons are indeed added in discrete regions of the adult mammalian CNS (Gross 2000, Kempermann & Gage 1999, Lie et al. 2004). In most mammals, active neurogenesis occurs throughout life in the subventricular zone (SVZ) of the lateral ventricle and in the subgranular zone (SGZ) of the dentate gyrus in the hippocampus. Neurogenesis outside these two regions appears to be extremely limited, or nonexistent, in the intact adult mammalian CNS. After pathological stimulation, such as brain insults, adult neurogenesis appears to occur in regions otherwise considered to be nonneurogenic. Our understanding of adult neurogenesis in mammals has progressed significantly over the past decade, and we now know a great deal more about the biology of this biological phenomenon, from the identity and location of adult neural stem cells, and proliferation and fate specification of neural progenitors, to migration, nerve guidance, neuronal maturation, and synaptic integration of newborn neurons in the adult CNS environment (Alvarez-Buylla & Lim 2004, Gage 2000, Goh et al. 2003, Kempermann & Gage 1999, Lie et al. 2004). We have also gained significant knowledge of how adult neurogenesis is regulated in the normal and abnormal CNS (Duman et al. 2001, Fuchs & Gould 2000, Kempermann 2002). Advances in our understanding of adult neurogenesis have been facilitated by the isolation and in vitro analysis of multipotent neural progenitors derived from the adult CNS (Gottlieb 2002). The demonstration of active adult neurogenesis also opens possibilities to repair the adult CNS after injury or degenerative neurological diseases using cell replacement therapy in the near future (Lindvall et al. 2004, Rossi & Cattaneo 2002). In this review on adult neurogenesis in the mammalian CNS systems, we start with milestone discoveries and methodologies in the field of adult neurogenesis. We then summarize the current understanding of adult neurogenesis and finish with a discussion of potential functions of mammalian adult neurogenesis. Interested readers may consult several recent comprehensive reviews on this topic (Alvarez-Buylla & Lim 2004, Goh et al. 2003, Kempermann

& Gage 1999, Lie et al. 2004, Lindvall et al. 2004, Parent 2003, Rossi & Cattaneo 2002).

HISTORIC OVERVIEW

In addition to proving the neuron doctrine using the Golgi technique, Santiago Ramon y Cajal (1913) also concluded that neurons are generated exclusively during the prenatal phase of development. Although suggestions regarding the existence of dividing cells in the postnatal CNS were raised (Allen 1912, Hamilton 1901), it was impossible, using methods of the time, to trace the fate of those rare dividing cells and to prove that the newborn cells were in fact neurons rather than glia (Ramon y Cajal 1913). Since then, "no new neurons after birth" became a central dogma in neuroscience for almost a century (Gross 2000). In the late 1950s, a new method was developed to label dividing cells with [H^3]-thymidine, which incorporates into the replicating DNA during the S-phase of the cell cycle and can be detected with autoradiography (Sidman et al. 1959). The generation of new neurons was first reported using this technique in three-day old mouse brains (Smart 1961). Soon after, Altman and colleagues published a series of papers reporting [H^3]-thymidine evidence for new neurons in various regions of adult rats, including the dentate gyrus of the hippocampus (Altman & Das 1965), neocortex (Altman 1966) and olfactory bulb (Altman 1969). However, little attention was given to these studies, perhaps because they were considered to lack functional relevance. The issue of adult neurogenesis was revisited in the late 1970s when Kaplan & Hinds (1977) demonstrated that newborn neurons in the hippocampus survived for a long period of time. These new neurons also appeared to receive synaptic inputs (Kaplan & Bell 1983) and extend axon projections to their target area (Stanfield & Trice 1988). Meanwhile, a series of studies of adult neurogenesis in songbirds started to provide evidence for functional roles of postnatal neurogenesis in seasonal song learning

The predominant repair mechanisms in the adult CNS were traditionally thought to be very limited and restricted to a postmitotic state, such as sprouting of axon terminals and synaptic reorganization. The recent demonstration of functional neurogenesis and isolation of multipotent neural stem cells from the adult mammalian CNS, including that of humans, has important implications for the development of new strategies for the treatment of injury and neurodegenerative diseases in the adult CNS. First, stem cells with the potential to give rise to new neurons appear to reside in many different regions of the adult CNS. These findings raise the possibility that endogenous neural stem cells can be mobilized for the replacement of dying neurons in neurodegenerative diseases. Indeed, recent reports have provided evidence that in some injury models limited neuronal replacement occurs in the CNS. Second, the complete process of neuronal development can be recapitulated in certain areas of the adult CNS, resulting in integration of functional new neurons into the existing circuits. Exploration of the underlying mechanisms of adult neurogenesis may lead to strategies to support functional neuronal replacements in other areas with either endogenous progenitors or neuronal progeny of stem cells from other sources, such as embryonic stem cells.

(Nottebohm 2004). Adult neural stem cells, the sources of new neurons, were first isolated from the adult CNS of rodents (Reynolds & Weiss 1992) and later from humans (Kukekov et al. 1999). The field was revolutionized by the introduction of bromodeoxyuridine (BrdU), a synthetic thymidine analogue, as another S-phase marker of the cell cycle (Gratzner 1982). Because BrdU can be detected by immunocytochemistry for phenotypic analysis and stereological quantification, this approach remains the most commonly used technique in the field. Before the end of the twentieth century, adult neurogenesis was observed with BrdU incorporation in all mammals examined, including samples from human patients (Eriksson et al. 1998). Combined retroviral-based lineage tracing (Price et al. 1987, Sanes et al. 1986) and electrophysiological studies provided the most convincing evidence so far that newborn neurons in the adult mammalian CNS are

Neurogenesis: a process of generating functional neurons from progenitor cells, including proliferation and neuronal fate specification of neural progenitors, and maturation and functional integration of neuronal progeny into neuronal circuits

indeed functional and synaptically integrated (Belluzzi et al. 2003, Carleton et al. 2003, van Praag et al. 2002). A central question in the field of adult neurogenesis remains to be answered for years to come: What is the functional significance of this biological phenomenon in mammals?

METHODOLOGIES FOR INVESTIGATION OF ADULT NEUROGENESIS

Analysis of Endogenous Adult Neurogenesis In Vivo

The field of adult neurogenesis has been propelled by technical advances to facilitate identification of newborn neurons among the billions of existing neurons in the adult CNS. Three approaches have been explored so far (**Figure 1**).

Analysis based on the incorporation of nucleotide analogs during cell division. During DNA replication in the S-phase of the cell cycle, exogenous nucleotides such as [H^3]-thymidine or BrdU are incorporated into newly synthesized DNA and then passed on to cell progeny (**Figure 1***A*). Two different analogs can be used sequentially to measure the cell-cycle length (Cameron & McKay 2001, Hayes & Nowakowski 2002). By varying the pulsing paradigm and the examination time points after pulsing, this simple technique allows quantitative analysis of proliferation, differentiation, and survival of newborn cells (Kempermann et al. 1997, Miller & Nowakowski 1988). [H^3]-thymidine requires autoradiographic detection and has good stoichiometry if consistent exposure times and development procedures are used (Rogers 1973). BrdU, on the other hand, can be detected with immunohistochemistry (not stoichiometric) and allows both phenotypic analysis and stereological quantification of new cells. There are several limitations to this approach. First, it requires tissue fixation and DNA denaturing and therefore is not suitable for analyzing live cells. Second, labeling is restricted to the nucleus and requires careful confocal microscopy to confirm colocalization with cell-type-specific markers (Rakic 2002). Third, the amount of analogs incorporated from a single injection is diluted to undetectable levels after several rounds of cell division (Hayes & Nowakowski 2002). In addition to these technical limitations, we must also pay attention to additional caveats of this approach. BrdU or [H^3]-thymidine incorporation is an indication of DNA synthesis only, not cell division. Nucleotide analogs are also incorporated into nicked, damaged DNA undergoing repair, albeit on a smaller scale than during DNA replication (Selden et al. 1993). Thus, the dose and duration of BrdU pulsing, as well as the detection of BrdU, need to be appropriately controlled to avoid misidentification of repairing/dying cells as newborn cells. Furthermore, the possibility of cell cycle reentry by postmitotic neurons as a prelude to apoptosis after brain injury (Kuan et al. 2004) needs to excluded. Demonstrating neurogenesis after brain injury requires not only BrdU uptake and mature neuronal markers but also evidence showing the absence of apoptotic markers. In addition, we need to be cautious when interpreting results from experimental manipulations that can potentially change the accessibility, stability, or diffusion of the analogs, which might affect their incorporation, instead of directly affecting cell proliferation.

Analysis based on genetic marking with retroviruses. The expression of transgenes from retroviruses requires viral integration into the host genome (Lewis & Emerman 1994). For retroviruses that lack nuclear import mechanisms, such as the Muloney murine leukemia virus, viral integration occurs only when the nuclear membrane breaks down during mitosis (Lewis & Emerman 1994), thus making it a good indicator of cell division (**Figure 1***B*). Expression of a live reporter, such as green fluorescent protein (GFP), allows direct visualization and analysis

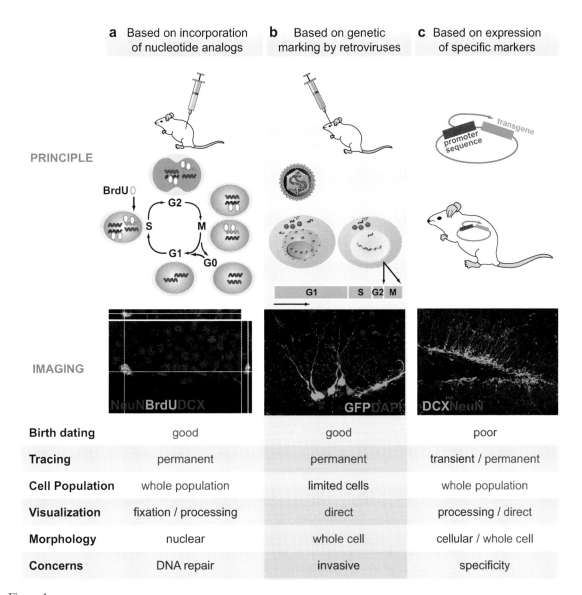

Figure 1

Methodologies for analysis of adult neurogenesis in vivo. Three different approaches to identify newborn neurons of the adult CNS in vivo are illustrated. (*A*) Analysis based on the incorporation of nucleotide analogs during DNA replication in the S-phase of the cell cycle. A sample picture shows confocal analysis to confirm the colocalization of bromodeoxyuridine (BrdU) and cell-type-specific markers. (*B*) Analysis based on retroviral genetic marking. The expression of transgenes from some retroviruses requires integration of the retroviral genome into the host genome, which occurs only during mitosis (M) when the nuclear membrane breaks down. A sample picture shows the expression of green fluorescent protein (GFP) in newborn cells two weeks after stereotaxic injection of the retrovirus into the hilus region of the adult mouse hippocampus. This approach allows direct visualization of the morphology of newborn cells. (*C*) Analysis based on the expression of specific markers for immature neurons. Transgenic mice can also be made to express reporters under specific promoters. A sample picture shows the expression of markers for immature neurons (DCX) and mature neurons (NeuN) in the adult mouse dentate gyrus of the hippocampus. Some of the advantages and disadvantages of these three approaches are listed.

of living newborn cells. This approach, however, requires invasive stereotaxic injection into specific brain regions. Future combination with site-specific recombinase and siRNA systems will make this "single-cell genetic" approach very powerful to investigate mechanisms underlying adult neurogenesis.

Analysis based on expression of specific markers. Developing neurons express distinct markers during their maturation process (Kempermann et al. 2004a). Common markers used for immature neurons include PSA-NCAM (poly-sialylated-neural cell-adhesion molecule), Tuj1 (β-tubulin isoform III), CRMP (collaspin response-mediated protein 4, also known as TOAD4), and DCX (doublecortin). Among the markers used for mature neurons are MAP-2ab (microtuble-associated protein-2 a and b isoforms) and NeuN (neuronal nuclei). Newborn neurons can be identified by the presence of immature markers and absence of mature markers of neurons. This approach can be used when birth dating with nucleotide analogs or retroviruses is impractical, such as for human tissue studies. The caveat for this approach relates to the specificity of the markers used for immature neurons. Some antibodies to these immature markers (e.g., Tuj1) also stain nonneuronal cells (Katsetos et al. 2001), and some markers (e.g., PSA-NCAM) are reexpressed in preexisting neurons (Charles et al. 2002) under certain conditions.

One exciting development in the field of adult neurogenesis is the generation of animal models that allow visualization and specific manipulation of newborn neurons in the adult CNS (**Figure 1C**). Several transgenic mice have been generated to express specific genes of interest under a desired promoter (Overstreet et al. 2004, Yamaguchi et al. 2000). For example, adult mice expressing GFP under the control of the regulatory regions of the nestin gene reveal both neural progenitors and some immature neurons (Yamaguchi et al. 2000). In the transgenic mice expressing GFP under the transcriptional control of the proopiomelanocortin (POMC) genomic sequences, a population of newly born granule cells of the dentate gyrus is selectively labeled (Overstreet et al. 2004). Because the expression of the reporter in these mice is transient (Overstreet et al. 2004, Yamaguchi et al. 2000), it is not possible to track the same population of cells along their maturation process. The next generation of animal models should take advantage of the available inducible systems in mice, such as inducible Cre recombinase (e.g., Cre-ER) and tetracycline-regulated systems (Branda & Dymecki 2004). These approaches may allow manipulation of a specific population of adult-generated neurons in a temporally and spatially precise manner for mechanistic and functional analysis of adult neurogenesis in vivo.

The phenotypic analysis of newborn cells requires examination of the colocalization of the cell-type specific markers and the lineage tracer. Because cells are tightly associated with each other in the adult CNS, the current standard is to perform three-dimensional reconstruction with confocal microscopy (**Figure 1A**). Electron microscopy has also been used to reveal the ultrastructure of newborn cells. In addition, unbiased stereological techniques are used for quantification of the newborn cells and their phenotypes (Kempermann et al. 1997). Generally, the development of newborn cells is followed by expression of different cell-type-specific markers. Using these immature neuronal markers in conjunction with lineage tracers helps to avoid mis-identifying dying/repaired neurons as newborn neurons (Magavi et al. 2000).

The functional analysis of adult neurogenesis has been carried out at three different levels: individual synapses and cells, neuronal circuits, and whole animals. For example, individual newborn neurons have been analyzed using electrophysiology in acute slice preparations (Overstreet et al. 2004, van Praag et al. 2002). Functional integration of new neurons also has been examined using virus-based transsynaptic neuronal tracing and analysis of c-Fos expression induced by neuronal activity (Carlen et al. 2002). At the neuronal-circuit and whole-animal levels, consequences of manipulating the extent (increase or decrease) of adult neurogenesis have been examined with electrophysiology and behavior analysis. Unfortunately, all current manipulations also affect other physiological processes in addition to neurogenesis. Therefore, functions of adult neurogenesis remain elusive.

In the near future, we will see sophisticated animal models that allow specific genetic marking or silencing (through expression of toxic proteins) of newborn neurons in a temporally and spatially controlled manner. Rapid technical advances in multiphoton confocal microscopy for in vivo imaging of adult CNS neurons over periods of milliseconds to months will allow direct visualization and analysis of adult neurogenesis in greater detail (Mizrahi et al. 2004, Mizrahi & Katz 2003). Our efforts in understanding adult neurogenesis will also be facilitated by neural modeling approaches (Cecchi et al. 2001, Chambers et al. 2004, Deisseroth et al. 2004).

Analysis of Adult Neurogenesis In Vitro and Ex Vivo

Multipotent neural progenitors have been isolated from various regions of the adult mammalian CNS, including human CNS regions (Gottlieb 2002). These cells can be expanded and genetically modified and still maintain their multipotentiality over many passages. Because of easy access and defined culture conditions, manipulation of adult neural progenitors in monolayer (Deisseroth et al. 2004, Song et al. 2002a) or slice cultures (Bolteus & Bordey 2004, Raineteau et al. 2004) allows precise analysis of the intrinsic and extrinsic mechanisms that control the various steps of neurogenesis, including proliferation, survival, fate specification, neuronal migration, maturation, and synapse formation.

Cultures of neural progenitors from the adult CNS are largely established on the basis of their preferential growth over other cell types when grown in defined media with specific growth factors (Gottlieb 2002) (**Figure 2**). Cells dissociated from specific tissues are plated either directly (Reynolds & Weiss 1992) or after partial purification steps to remove major contaminants (Palmer et al. 1999). Prospective isolation methods using fluorescence-activated cell sorting (FACS) have recently been developed on the basis of cell properties and/or cell-surface antigens (Rietze et al. 2001, Temple 2001, Uchida et al. 2000) or the expression of reporters under specific promoters (Roy et al. 2000). Two types of progenitor cultures are commonly used (Gottlieb 2002). In the neurosphere culture, individual neural progenitors proliferate on a nonadhesive substrate and generate suspended clusters of cells (Reynolds & Weiss 1992). In the adhesive culture, neural progenitors grow as a monolayer on coated substrates, such as laminin (Ray et al. 1993). The most commonly used growth factors to maintain self-renewal of cultured adult neural progenitors are epidermal growth factor (EGF) and basic fibroblast growth factor (FGF-2). Other growth factors also appear to be effective, such as Sonic hedgehog (Shh) (Lai et al. 2003) and amphiregulin (Falk & Frisen 2002). It remains to be examined whether neurogenesis from neural progenitors expanded under different conditions will have the same properties.

Neurogenesis from these neural progenitors can be examined in culture or after transplantation using immunocytochemistry, calcium and FM-imaging, electron microscopy, and electrophysiological techniques (**Figure 2**). Whereas factors that promote neuronal differentiation of multipotent neural progenitors are largely unknown, coculture of adult neural stem cells with hippocampal or SVZ astrocytes has been shown to promote neuronal differentiation (Lim & Alvarez-Buylla 1999, Song et al. 2002a). Neural progenitors also have been transplanted into early embryos and the fetal and adult CNS to examine their development (**Figure 2**).

> **Self-renewal:** a process of cell division to generate more cells of the same type. It is a hallmark property of stem cells.

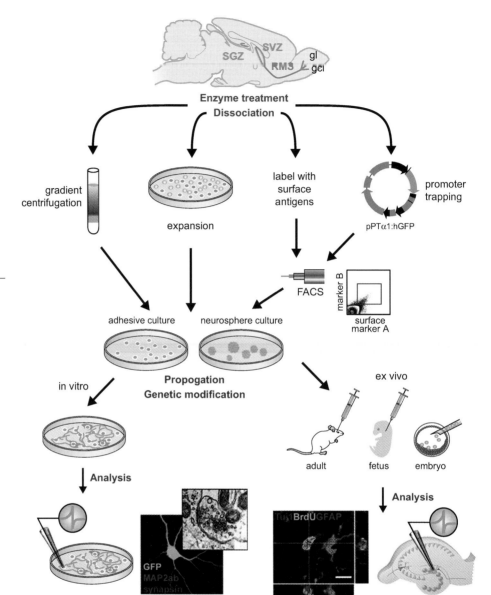

Figure 2
Methodology for analysis of adult neurogenesis in vitro and ex vivo. Multipotent neural progenitors have been cultured from various regions of the adult CNS. Here we illustrate several approaches that have been used to derive neural progenitors from the adult CNS and to analyze neurogenesis from these multipotent progenitors both in vitro and ex vivo. FACS, fluorescence-activated cell sorting.

During the past few years, many reports of isolation of adult cells with neurogenic potentials in vivo and in vitro have appeared, surprisingly, with some from nonneural tissues (Raff 2003). In many cases, single or multiple antibody markers were used as the only criteria to determine if the differentiated cells were neurons (e.g., Tuj1 or NeuN). Numerous studies have shown that morphological and immunochemical appearances are not necessarily predictive of physiological properties, thus functional analysis using electrophysiology to determine the neuronal properties is indispensable (Song et al. 2002b). In addition, cell fusion between progenitors and mature cell types occurs both in vitro and in vivo (Raff 2003). Therefore, it is also essential to distinguish between neuronal differentiation and cell-fusion events.

NEUROGENESIS IN THE INTACT ADULT MAMMALIAN CNS

Active neurogenesis occurs only in discrete regions of the intact adult CNS (Alvarez-Buylla & Lim 2004, Lie et al. 2004). From rodents to primates, neurons are generated continuously in the SVZ and migrate anteriorly through the rostral migratory stream (RMS) into the olfactory bulb to become interneurons (**Figure 3**). In the dentate gyrus of the hippocampus, new granule neurons are continuously born locally in all mammals examined, including humans (**Figure 4**).

The Adult Neural Stem Cell, Its Proliferation, and Neuronal Fate Specification

The identity of neural stem cells that give rise to new neurons in the adult CNS has been a subject of hot debate for the past few years. Several cell types have been proposed as the resident adult neural stem cell, including astrocytes (Doetsch et al. 1999), multiciliated ependymal cells (Johansson et al. 1999) and subependymal cells (Morshead et al. 1994). The astrocyte hypothesis is currently a prevalent view (Alvarez-Buylla & Lim 2004). In the SVZ of rodents (**Figure 3**), a subset of astrocytes ($GFAP^+$) gives rise to rapidly proliferating transient amplifying cells ($GFAP^-Dlx2^+$), which in turn generate migrating neuroblasts ($GFAP^-Dlx2^+PSA-NCAM^+$). In adult humans, some astrocytes lining the lateral ventrical divide in vivo and behave like multipotent neural progenitors in vitro (Sanai et al. 2004). It remains to be determined, however, whether these astrocytes indeed give rise to neurons in adult humans. In the SGZ (**Figure 4**), a subset of astrocytes has also been proposed as the stem cells that give rise to new granule neurons (Seri et al. 2001). These astrocytes ($GFAP^+$), with their cell bodies in the SGZ, have radial processes going through the granule cell layer and short tangential processes extending along the border of the granule cell layer and hilus (**Figure 4**). These cells divide and give rise to immature neurons ($DCX^+PSA-NCAM^+$). Investigators do not know whether these astrocytes also give rise to new neurons found in the adult human dentate gryus (Eriksson et al. 1998).

The factors that regulate in vivo proliferation of adult neural stem cells in the SVZ and the SGZ have not been fully characterized. Members of the FGF and EGF growth factor families are primary mitogens used to propagate adult neural progenitors in vitro and are likely to perform similar functions in vivo (Lie et al. 2004). The transient amplifying cells of the SVZ appear to express the EGF receptor (Doetsch et al. 2002, Morshead et al. 1994). In the SVZ, infusion of EGF or FGF-2 increases cell proliferation (Kuhn et al. 1997), whereas knockout of TGFα, a ligand for the EGF receptor, leads to a significant decrease in cell proliferation (Tropepe et al. 1997). Shh also promotes proliferation of adult neural stem cells both in vitro and in vivo (Lai et al. 2003, Machold et al. 2003). Interestingly neural progenitors and brain tumors share many common features (Oliver & Wechsler-Reya 2004). They express common sets of markers (e.g., nestin, CD133, bim-1, and sox-2) and share pathways regulating their proliferation (e.g., Shh, PTEN). Future comparative studies of neural progenitors and cancer cells will facilitate the understanding of the self-renewal of adult neural stem cells and of the origins of brain tumors.

What makes the SVZ and SGZ special in supporting the proliferation and neuronal differentiation of multipotent neural progenitors is an area of intensive investigation (Doetsch 2003, Lie et al. 2004). Investigators have postulated that endothelial cells and some special astrocytes provide a unique neurogenic niche (Doetsch 2003, Lie et al. 2004, Palmer et al. 2000). Astrocytes from the SVZ and hippocampus promote proliferation and neuronal fate specification of co-cultured adult neural progenitors (Lim & Alvarez-Buylla 1999, Song et al. 2002a). In contrast, astrocytes from the adult spinal cord,

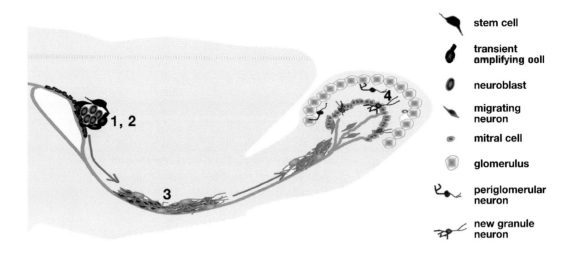

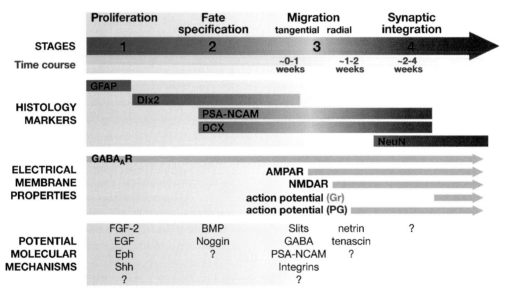

Figure 3

Generation of new interneurons in the olfactory bulb from neural stem cells in the subventricular zone (SVZ). Adult neurogenesis in the SVZ/olfactory systems undergoes four developmental stages. Stage 1. Proliferation: stem cells (*blue*) in the SVZ of the lateral ventricles give rise to transient amplifying cells (*light blue*). Stage 2. Fate specification: transient amplifying cells differentiate into immature neurons (*green*). Adjacent ependymal cells (*gray*) of the lateral ventricle are essential for neuronal fate specification by providing inhibitors of gliogenesis. Stage 3. Migration: Immature neurons (*green*) migrate with each other in chains through the rostral migratory stream (RMS) to the olfactory bulb. The migrating neurons are ensheathed by astrocytes. Once reaching the bulb, new neurons then migrate radially to the outer cell layers. Stage 4. Synaptic integration: Immature neurons differentiate into either granule neurons (Gr, *orange*) or periglomerular neurons (PG, *red*). These unusual interneurons lack an axon and instead release their neurotransmitter from the dendritic spines at specialized reciprocal synapses to dendrites of mitral or tufted cells. The specific properties of each stage are summarized below, mainly on the basis of studies in adult mice.

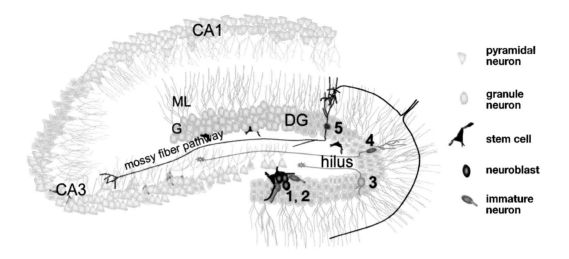

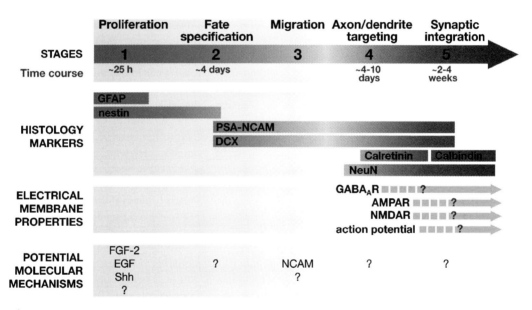

Figure 4

Generation of new granular neurons in the dentate gyrus of the hippocampus from neural stem cells in the subgranular zone (SGZ). Adult neurogenesis in the dentate gyrus of the hippocampus undergoes five developmental stages. Stage 1. Proliferation: Stem cells (*blue*) with their cell bodies located within the subgranular zone in the dentate gyrus have radial processes that project through the granular cell layer and short tangential processes that extend along the border of the granule cell layer and hilus. These stem cells give rise to transient amplifying cells (*light blue*). Stage 2. Differentiation: transient amplifying cells differentiate into immature neurons (*green*). Proliferating progenitors in the SGZ are tightly associated with astrocytes and vascular structures. Stage 3. Migration: Immature neurons (*light green*) migrate a short distance into the granule cell layer. Stage 4. Axon/dendrite targeting: Immature neurons (*orange*) extend their axonal projections along mossy fiber pathways to the CA3 pyramidal cell layer. They send their dendrites in the opposite direction toward the molecular layer. Stage 5. Synaptic integration: New granule neurons (*red*) receive inputs from the entorhinal cortex and send outputs to the CA3 and hilus regions. The specific properties of each stage are summarized below, mainly on the basis of studies in adult mice. DG, dentate gyrus region; ML, molecular cell layer; GL, granular cell layer.

a nonneurogenic region, do not promote neuronal differentiation (Song et al. 2002a). In vivo hot spots of cell proliferation in the SGZ are found to be in close proximity to capillaries (Palmer et al. 2000) and astrocytes (Seri et al. 2004, Song et al. 2002a). During fetal development, astrocytes, however, are born after most neurons and are thus unable to provide neurogenic signals for multipotent stem cells (Temple 2001). It becomes clear that astrocytes in the adult CNS are not merely supporting cells as traditionally believed. Just like neurons, astrocytes have a broad diversity of subtypes and functions; some behave like stem cells (Doetsch et al. 1999, Seri et al. 2001), some provide neurogenic signals (Lim & Alvarez-Buylla 1999, Song et al. 2002a), and some provide synaptogenic factors (Song et al. 2002b, Ullian et al. 2004).

The molecular mechanisms underlying neuronal fate specification during adult neurogenesis are just beginning to be elucidated. The fate choice is influenced by a cohort of proliferating, gliogenic, and neurogenic signals within the niche. Bone morphogenic protein (BMP) signaling has been shown to instruct adult neural progenitors to adopt a glial fate (Lim et al. 2000). Neuronal differentiation from adult neural progenitors in the neurogenic niche proceeds partially because of the local presence of BMP antagonists. Ependymal cells in the SVZ secrete Noggin (Lim et al. 2000), and astrocytes in the SGZ secrete neurogenesin-1 (Ueki et al. 2003), to serve as the BMP antagonist, respectively. We have recently identified Wnt-signaling as one of the candidate pathways that regulate neurogenic differentiation of adult neural stem cells both in vitro and in vivo (D.C. Lie, S.A. Colamarino, H. Song, L. Desire & F.H. Gage, unpublished observations). These extracellular signaling mechanisms act in part by interacting with cellular epigenetic mechanisms (Hsieh & Gage 2004), including interaction of chromatin remodeling enzymes with neurogenic factors, maintaining genomic stability (Zhao et al. 2003b), and regulating the fate choice of adult neural progenitors by noncoding RNA (Kuwabara et al. 2004).

Neuronal Migration and Nerve Guidance

In the olfactory system, newborn neurons go through extensive migration, first migrating tangentially along the wall of the lateral ventricle, then traveling anterior along the RMS to the olfactory bulb in close association with each other, and finally dispersing radially as individual cells into the outer cell layers in the bulb (**Figure 3**). This extensive migration occurs in species from rodents to primates but not in humans (Sanai et al. 2004). The migration along the RMS is a very unique process called chain migration in which neuroblasts migrate closely associated with each other in a tube-like structure formed by glial cells (Lois et al. 1996). Six days after injection of a retrovirus-expressing GFP into the SVZ of adult mice, GFP$^+$ neuroblasts were observed in the core of the olfactory bulb (Carleton et al. 2003). By 14 days after viral injection GFP$^+$ neurons with well-developed dendritic arbors and spines were observed. Studies over the past few years have started to unravel the mechanisms that support and direct the migration of newborn neurons from the SVZ to the olfactory bulb. The motility of chain migration is regulated by a cohort of factors, including PSA-NCAM (Hu et al. 1996), EphB2/ephrin-B2 (Conover et al. 2000), netrin/DCC (Murase & Horwitz 2002), GABA$_A$ receptor activation (Bolteus & Bordey 2004), and some integrins (Murase & Horwitz 2002). The directionality of chain migration is influenced by netrin/DCC (Murase & Horwitz 2002) and Slits/Robos signaling (Nguyen-Ba-Charvet et al. 2004, Wu et al. 1999). Once migrating neuroblasts reach the olfactory bulb, reelin acts as a detachment signal (Hack et al. 2002) and tenascin-R then initiates the detachment of the neuroblasts from the chains and directs radial migration to their target areas (Saghatelyan et al. 2004). Two types of interneurons are continuously generated in the olfactory bulb: granule neurons and periglomerular neurons. These unusual interneurons lack an axon and instead have reciprocal dendro-dendritic synapses with mitral or tufted cells (Shepherd et al. 2004).

In the dentate gyrus, newly generated neurons migrate only a short distance to the inner granule cell layer to become granule neurons (**Figure 4**). These newborn neurons rapidly extend long axonal projects along the mossy fiber pathway and reach their target CA3 pyramidal cell layer within 4 to 10 days after division (Hastings & Gould 1999). The dendrites of these neurons grow in the opposite direction of the axons, reaching the molecular layer within two weeks and maintaining growth to increase in complexity over months (van Praag et al. 2002). The molecular mechanisms underlying nerve growth and axon/dendrite guidance of adult-generated granular neurons are largely unknown. These newborn neurons may not express receptors for factors that normally inhibit axon regeneration (He & Koprivica 2004). Alternatively, they may have different internal states, such as high cytoplasmic levels of cAMP/cGMP, thus rendering them inert to inhibitory cues (Song & Poo 1999, 2001). Many developmental guidance cues, such as semaphorins, retain their expression in adulthood (Huber et al. 2003). It would be interesting to investigate whether these cues also guide newborn neurons in the adult CNS.

The existence of extensive neuronal migration, nerve growth, and axon/dendritic targeting in the adult CNS environment that otherwise is inhibitory for mature neurons (He & Koprivica 2004) provides a unique model system to investigate basic principles and mechanisms of neuronal navigation in adulthood. Advances in this field may also lead to novel strategies for repairing the adult CNS after injury or degenerative neurological diseases.

Neuronal Maturation, Synapse Formation, and Plasticity

Many aspects of the maturation process of adult-generated neurons are surprisingly different from what occurs during fetal development. In the olfactory system, tangentially migrating neuroblasts express extrasynaptic $GABA_A$ receptors first and then AMPA receptors; NMDA receptors are expressed last in the radially migrating neurons (Bolteus & Bordey 2004, Carleton et al. 2003). In contrast, NMDA receptors are often detected before AMPA receptors in developing neonate brains (Durand et al. 1996). Shortly after the completion of radial migration, maturing new neurons start to receive synaptic inputs with GABAergic inputs ahead of glutamatergic ones (Belluzzi et al. 2003, Carleton et al. 2003). Surprisingly, new granule neurons, but not new periglomerular neurons (Belluzzi et al. 2003), appear to acquire the ability to fire action potentials after synaptic inputs are made (Carleton et al. 2003), which is also different from what occurs in developing brains. This unique developmental sequence may allow them to integrate readily into a mature brain without altering existing cognitive processes.

In the dentate gyrus, the sequential events that occur during neuronal maturation and synapse formation have not yet been fully characterized using electrophysiology. Newborn granule neurons in the dentate gyrus appear to first receive GABAergic synaptic inputs around one week after birth and then glutamatergic inputs by two weeks (S. Ge, G. Ming & H. Song, unpublished observations). The synaptogenesis process appears to be quite prolonged. During development, the dendritic spines, the major sites of excitatory synaptic transmission, reach a plateau in their density around one month at $2-4/\mu m$ dendritic length (Nimchinsky et al. 2002). Adult-generated granule neurons have a mean spine density of $0.8/\mu m$ dendritic length at one month after birth and continue to grow, reaching $1.2/\mu m$ at six months (van Praag

Plasticity: the ability of the brain to reorganize neural pathways on the basis of new experiences. It can be reflected by changes in the efficacy of synaptic communication as well as by morphological changes

Long-term potentiation: long-lasting, activity-dependent changes in the efficacy of synaptic communication. Investigators believe that LTP provides an important key to understanding the cellular and molecular mechanisms by which memories are formed and stored.

et al. 2002). Young granule neurons also differ substantially from neighboring mature granule cells in both their active and passive membrane properties (Schmidt-Hieber et al. 2004, Wang et al. 2000). Neurons appear to have a high input resistance and a subthreshold Ca^{2+}-conductance, which finally enable action potential firing with very small excitatory currents (Schmidt-Hieber et al. 2004). The enhanced excitability might be important for the young neurons when only a few excitatory contacts have been formed. Furthermore, newborn neurons exhibit special properties in synaptic plasticity, such as having a lower threshold for the induction of long-term potentiation (LTP) (Schmidt-Hieber et al. 2004, Wang et al. 2000) and long-term-depression (LTD) (J. Bischofberger, personal communication) than do mature neurons. This enhanced synaptic plasticity appears to be present at least during the time period (Schmidt-Hieber et al. 2004) when the newly generated neurons express PSA-NCAM after mitosis (**Figure 4**). However, it is unclear whether these neurons exhibit such properties beyond this time window and whether other parameters change during their developmental course. Answers to these questions will define when and how new neurons contribute to the plasticity of the local circuitry and ultimately how this affects the animal behaviors.

The outputs of newborn neurons have rarely been examined by electrophysiological studies owing to the technical difficulties in finding pairs of connected neurons in vivo. New granule neurons in the dentate gyrus are likely to release glutamate as do mature granule neurons. Early studies have suggested that a small population of granule neurons in the dentate gyrus can also release GABA, and this GABAergic signaling is upregulated after seizures (Walker et al. 2001). It would be interesting to examine how much adult-generated neurons contribute to this GABAergic signaling during their development. In one report, some GABAergic basket cells in the dentate gyrus were found to incorporate BrdU and form inhibitory synapses with the granule cells (Liu et al. 2003). Other studies, however, did not observe incorporation of either BrdU or [^{3}H]-thymidine into basket cells (Seri et al. 2004). In the olfactory bulb, granule neurons release GABA, whereas periglomerular cells release GABA and sometimes also dopamine (Shepherd et al. 2004). Whether newborn interneurons in the olfactory bulb release the same types of neurotransmitters remains to be examined with electrophysiological studies.

The cellular and molecular mechanisms that regulate the integration of newly born neurons into existing neuronal circuits in the adult CNS are unknown. Will many identified developmental mechanisms also operate for these newborn neurons? Understanding these extreme examples of structural plasticity not only will shed light on the basic mechanisms of adult plasticity in the CNS, but also may provide strategies to integrate transplanted neuronal cell types for cell replacement therapy after injury or degenerative neurological diseases.

MODULATION OF ENDOGENOUS ADULT NEUROGENESIS

Adult neurogenesis in the hippocampus and the olfactory bulb is an extremely dynamic process (Duman et al. 2001, Fuchs & Gould 2000, Kempermann 2002). Extensive studies have shown that both intrinsic and extrinsic factors regulate adult neurogenesis at different stages, including proliferation, fate specification, migration, integration, and survival (**Table 1**).

The intrinsic genetic background influences SGZ neurogenesis in adult mice, and significant differences were found in the proliferation, survival, and differentiation of neural progenitors between several wild-derived and inbred laboratory mice (Kempermann & Gage 2002). Adult neurogenesis in both the SGZ (Kuhn et al. 1996) and SVZ (Enwere et al. 2004, Jin et al. 2003) is also reduced during aging (**Table 1**). Adrenal steroids may contribute to the aging-associated decline of

TABLE 1 Regulation of adult neurogenesis in the subgranular zone (SGZ) and subventricular zone (SVZ)*

Regulatory factors	Proliferation		Survival		Neuronal differentiation		Potential mechanisms
	SVZ	SGZ	SVZ	SGZ	SVZ	SGZ	
Mice strain		+/−		+/−		+/−	
Gender	n.c.	+/−	n.c.	n.c.	n.c.	n.c.	
Aging	−	−		n.c.		−	EGFR signaling − Corticosteroids levels +
Hormones							
Corticosterone		−					
Estrogen	n.c.	+		n.c.		n.c.	
Oestrogen		+					Serotonin?
Pregnancy	+	n.c.					Prolactin
Afferents, neurotransmitters							
Dopamine	−	−					D2L receptors
Serotonin	+	+					
Acetylcholine		−		−		−	
Glutamate		−		n.c.			mGluR, NMDAR
Norepinephrine	n.c.	+		n.c.		n.c.	PAC1 and PKC
PACAP	+	+					
Nitric oxide	−	n.c./−	n.c.		+/n.c.	n.c.	
Growth factors							
FGF-2	+	n.c.					
EGF	+	n.c.				−	
IGF-1		+		+		+	
Behavior							
Enriched environment	n.c.	+/n.c.	n.c.	+/n.c.	n.c.	n.c./+	VEGF
Enriched odor exposure	n.c.	n.c.	+	n.c.			
Physical activity	n.c.	+	n.c.	+			VEGF
Learning							
Water maze		n.c.		+/n.c.		n.c.	
Blink reflex		n.c.		+			
Dietary restriction		n.c.		+			BDNF, NT-3
Stress		−		n.c.			Glucocorticoids +
Drugs							
Antidepressants		+					Serotonin
Opiates		−					
Methamphetamine		−					
Lithium		+				n.c.	
Pathological stimulations							
Ischemia	+	+/−		+	+	+	NMDAR, CREB
Seizures	+	+/−		+/n.c.		+/n.c.	
Inflammation	+/−	−		−		−	IL-6, TNFα

(Continued)

TABLE 1 (*Continued*)

Regulatory factors	Proliferation		Survival		Neuronal differentiation		Potential mechanisms
	SVZ	SGZ	SVZ	SGZ	SVZ	SGZ	
Degenerative diseases							
AD/HD/PD	+	+					
Diabetes		−					

*The table is based on results from peer-reviewed publications on adult neurogenesis. It is intended to give a general overview of the diverse regulation of adult neurogenesis. We provide partial lists of examples under each category. Owing to space limitations, please see the supplementary material (follow the Supplemental Material link from the Annual Reviews home page at **http://www.annualreviews.org**) for references related to this table. It should be noted that the listed effects may not be direct. "+": increase; "−" decrease; "n.c.": no change. Unmarked indicates "not examined." Interested readers can consult several recent comprehensive reviews on this topic (Duman et al. 2001, Fuchs & Gould 2000, Kempermann 2002). AD, Alzheimer's disease; HD, Huntington's disease; PD, Parkinson's disease.

neurogenesis in the SGZ (Cameron & Gould 1994). Reducing corticosteroid levels in aged rats can restore the rate of cell proliferation, which suggests that aged neural progenitors retain their proliferation capacity as in younger adult animals (Cameron & McKay 1999). Other hormones, including estrogen and prolactin, also regulate adult neurogenesis (**Table 1**).

Electrical activity can serve as a common link between the internal and external stimuli by regulating, either directly or indirectly, different aspects of adult neurogenesis. For example, excitatory stimuli can be sensed by proliferating SVZ neural progenitors via L-type Ca^{2+} channels and NMDA receptors to inhibit glial fate specification and promote neuronal differentiation (Deisseroth et al. 2004). Activity also regulates the expression of tenascin-R, which guides the radial migration of newborn olfactory neurons (Saghatelyan et al. 2004). Many afferent inputs and various neurotransmitters, including classic (e.g., dopamine, serotonin, acetylcholine, and glutamate), peptide (e.g., PACAP), and gaseous (e.g., nitric oxide) neurotransmitters, have been implicated in regulating adult neurogenesis (Kempermann 2002) (**Table 1**).

Environmental stimuli can greatly affect the proliferation and survival of newborn neurons in the adult CNS (**Table 1**). Exposure of rodents to an enriched environment increases the survival of newborn neurons in the SGZ without affecting SVZ neurogenesis (Brown et al. 2003, Kempermann et al. 1997, Nilsson et al. 1999). Physical exercise, such as running, promotes SGZ neurogenesis by increasing cell proliferation and survival of the new granule neurons (van Praag et al. 1999a, van Praag et al. 1999b). Vascular endothelial growth factor (VEGF) signaling may be responsible for the increased neurogenesis by both enriched environment and running (Cao et al. 2004, Fabel et al. 2003). Hippocampus-dependent learning, such as blink reflex or water maze learning, appears to increase the survival of new granule neurons that have been generated only at a particular time window before the training (Gould et al. 1999a, Leuner et al. 2004, van Praag et al. 1999a). Similarly, enriched odor exposure increases the SVZ cell proliferation but not the SGZ neurogenesis (Rochefort et al. 2002).

Both physical and psychosocial stress paradigms, as well as some animal models of depression, lead to a decrease in cell proliferation in the SGZ (Duman et al. 2001, Fuchs & Gould 2000). This decrease results from the activation of the hypothalamic-pituitary-adrenal axis, which is known to inhibit adult neurogenesis (Cameron & Gould 1994). Interestingly, adult neurogenesis is also regulated by psychotropic drugs (Duman et al. 2001, Fuchs & Gould 2000). Long-term administration of different classes of antidepressants, including serotonin and norepinephrine-selective reuptake inhibitors, increases cell proliferation and adult SGZ neurogenesis. In contrast, several drugs of abuse decrease cell proliferation and neurogenesis in the SGZ (Eisch et al. 2000). Alcohol intoxication also decreases SGZ neurogenesis by inhibiting both proliferation and newborn cell survival (Crews et al. 2003).

The dynamic and selective regulation of neurogenesis in the hippocampus and olfactory bulb by a variety of stimulations points to the functional significance of this biological phenomenon. Understanding the mechanisms underlying these regulations will not only significantly enrich our general knowledge of adult neurogenesis, but also may shed light on the etiology and pathophysiology of some mental illness, such as depression (Kempermann 2002).

ADULT NEUROGENESIS UNDER PATHOLOGICAL STIMULATIONS

Injury and pathological stimulations not only affect different aspects of adult neurogenesis in neurogenic regions, but also have an impact in otherwise non-neurogenic regions (Arlotta et al. 2003, Parent 2003) (**Table 1**). Most brain injuries lead to increased proliferation of progenitors in the SGZ and the SVZ after a latent period and sometimes cause migration of newborn neurons to injury sites. Specific types of injury also appear to lead to neurogenesis from endogenous neural progenitors in regions where adult neurogenesis is extremely limited or nonexisting (Magavi et al. 2000). Whether these new neurons become functionally integrated remains to be determined.

Ischemic Brain Injury

Ischemic brain insults potently stimulate progenitor proliferation in both the SGZ and SVZ of adult rodents as shown by BrdU incorporation (Kokaia & Lindvall 2003, Parent 2003). In an experimental stroke model immature neurons also migrate from the SVZ to the damaged striatal area where they start to express markers for striatal medium-sized spiny neurons, the phenotype of most of the dead neurons (Arvidsson et al. 2002, Parent et al. 2002b). Most of these new neurons died between two and five weeks after the stroke (Arvidsson et al. 2002). These studies suggest that the local environment, although providing cues for attracting immature neurons and inducing neuronal subtype differentiation, is not adequate for long-term survival of the new neurons. In another study, intraventricular infusion of EGF and FGF-2 after global ischemia led to increased proliferation and neuronal differentiation of progenitors located in the caudal extension of the SVZ adjacent to the hippocampus (Nakatomi et al. 2002). These new neurons then migrated and integrated into the CA1 region of the hippocampus, apparently receiving synaptic input and sending outputs. Importantly, these animals also exhibited partial recovery of their synaptic responses and better performance in water maze tests.

Seizures

Studies of adult rodent models of limbic epileptogenesis or acute seizures showed that seizure or seizure-induced injury stimulates neurogenesis in both the SGZ and SVZ (Parent 2003). In the dentate gyrus, epilepsy increases the proliferation of progenitors five- to tenfold after a latent period (Parent et al. 1997). Most of these newborn cells differentiate into granule neurons, some of which mislocate in the hilus region. These ectopic granule-like cells maintain the basic electrophysiological characteristics of dentate granule neurons but fire abnormal bursts in synchrony with the CA3 pyramidal cells (Scharfman et al. 2000). Newborn neurons also participate in aberrant network reorganization in the epileptic hippocampus, with aberrant mossy fiber recurrent connections and persistent basal dendrites (Parent et al. 1997). In the SVZ, proliferation is significantly increased in a pilocarpine model of limbic epileptogenesis (Parent et al. 2002a). These neuroblasts also showed more rapid migration to the olfactory bulb, and some appeared to exit the RMS prematurely and migrated into injured forebrain regions.

Radiation Injury

Studies in animal models have shown that exposure to therapeutic doses of radiation leads to ablation of adult SGZ neurogenesis but not gliogenesis (Kempermann & Neumann 2003). This dramatic reduction of adult neurogenesis results from the combined effects of acute cell death, decreased proliferation, and neuronal differentiation of the neural progenitors. A striking feature of radiation exposure is a massive migroglial inflammatory response in the dentate gyrus, which by itself inhibits neurogenesis (Ekdahl et al. 2003, Monje et al. 2003). Recent studies showed that pharmacological blocking of inflammation elicited by irradiation, injection of bacterial lipopolysaccharide (Monje et al. 2003), or experimentally induced seizures (Ekdahl et al. 2003) can restore hippocampal neurogenesis. Proinflammatory mediators released by microglia, such as interleukin-6 (IL-6), seem to be important contributors to the inhibition of SGZ neurogenesis (Monje et al. 2003, Vallieres et al. 2002). Microglia also release trophic factors (Batchelor et al. 1999), such as brain-derived neurotrophic factor (BDNF), that promote neurogenesis (Benraiss et al. 2001). Thus, microglia may have both positive and negative effects on adult neurogenesis depending on the context.

Degenerative Neurological Diseases

Adult neurogenesis is significantly altered in chronic degenerative neurological diseases. Brains of Huntington's disease patients showed a significant increase in cell proliferation in the subependymal layer, revealed by the cell-cycle marker proliferating cell nuclear antigen (PCNA) (Curtis et al. 2003). Some of these PCNA$^+$ cells were also Tuj1$^+$, suggesting the existence of dividing new neurons in the diseased brain. In a lesion rat model of HD, BrdU incorporation in the SVZ significantly increased, and some BrdU$^+$ neurons migrated to the lesioned stratum (Tattersfield et al. 2004). Brains of Alzheimer's disease patients also showed increased expression of immature neuronal markers, such as DCX and PSA-NCAM, in the SGZ and the CA1 region of Ammon's horn (Jin et al. 2004b). In a transgenic mouse model of Alzheimer's disease, there was an approximately twofold increase in BrdU incorporation and expression of immature neuronal markers in the SGZ and SVZ even before the neuronal loss and deposition of amyloid (Jin et al. 2004a). In the case of Parkinson's disease patients, the proliferation of progenitors in the SGZ and SVZ is impaired, presumably as a consequence of dopaminergic denervation (Hoglinger et al. 2004). Experimental depletion of dopamine decreases the proliferation of progenitors in both SVZ and SGZ in rodents (Baker et al. 2004, Hoglinger et al. 2004). In the 6-hydroxydopamine mice model of Parkinson's disease, proliferation in the SVZ was reduced by ∼40% (Baker et al. 2004). The same lesion model also leads to the generation of a few new neurons in the substantia nigra where neurogenesis is extremely limited (Zhao et al. 2003a) or nonexistent under normal conditions (Lie et al. 2002).

Emerging evidence suggests that adult neurogenesis may be an intrinsic compensatory response to self-repair the adult CNS. Adult neurogenesis is also a very dynamic process under the regulation of both positive and negative influences that change rapidly over time after injury or during the progress of chronic diseases. The cellular and molecular mechanisms underlying injury-induced cell proliferation, differentiation, and migration are largely unknown. Several factors—including FGF-2, BDNF, and erythropoietin (Kokaia & Lindvall 2003)—have been implicated in modulating neurogenesis after insults. Many pressing questions remain before we can take advantage of this limited neurogenesis after injury to functionally repair the nervous system.

POTENTIAL FUNCTIONS OF ADULT NEUROGENESIS

Despite decades of intensive research, we still search for definitive evidence for the functions of adult neurogenesis in mammals (Fuchs & Gould 2000, Kempermann et al. 2004b). After the initial discovery of neurogenesis in the postnatal rat hippocampus, Altman (1967) suggested that new neurons are crucial in learning and memory. Since then much of what we have learned about the functional importance of neurogenesis in the adult brain comes first from studies of songbirds (Nottebohm 2004). For example, the first evidence that neurons generated in the adult brain can be recruited into functional circuits came from electrophysiological studies of new neurons in adult songbird in response to sound stimulation (Paton & Nottebohm 1984). During the past decade, rapidly accumulating correlative evidence supports the notion that adult neurogenesis and specific behaviors are affected in a reciprocal fashion both in songbirds and in mammals (Gould et al. 1999b). In the dentate gyrus, for example, running-induced increase of SGZ neurogenesis in rodents is associated with enhanced spatial learning in the Morris water maze task and with enhanced LTP in the dentate gyrus either in acute slices or in vivo (Farmer et al. 2004, van Praag et al. 1999a). By comparison, decreased SGZ neurogenesis by pharmacological manipulations or radiation led to defects in specific behavior tests and a reduction of LTP in the dentate gyrus (Shors et al. 2001, 2002; Snyder et al. 2001). Genetic and radiological approaches to disrupt antidepressant-induced neurogenesis also blocked behavioral responses to antidepressants (Santarelli et al. 2003). In the olfactory bulb, an odor-enriched environment enhances neurogenesis and improves olfactory memory without upregulating hippocampal neurogenesis (Rochefort et al. 2002). Conversely, reduced olfactory neurogenesis in genetically modified mice showed an impairment of discrimination between odors (Enwere et al. 2004, Gheusi et al. 2000). Unfortunately, all of the current approaches also affect other physiological processes besides neurogenesis.

An attractive and testable hypothesis is that newly generated neurons in the adult CNS exhibit unique physiological properties at specific stages during their maturation process that allow them to serve as major mediators for structural plasticity. In turn, this active structural plasticity is important for associative learning and memory, and possibly mood (depression). Recent studies already showed that new neurons exhibit different passive and active properties as compared with mature neurons. New neurons also exhibit the striking ability to migrate and extend axons and dendrites in a hostile environment largely inhibitory for mature neurons. Of particular interest is the finding that new neurons exhibit a lower threshold for the induction of LTP/LTD and have larger amplitudes of LTP (Schmidt-Hieber et al. 2004, Wang et al. 2000). The keys to elucidating the function(s) of adult neurogenesis may then rely on our future comparative studies between newborn neurons of known ages and existing mature neurons at the cellular physiology level.

Many tasks remain before we can better understand the fundamental biological significance of adult neurogenesis. First, we have to know more about the physiological properties of the newborn neurons, especially those involved in synaptic plasticity. This will require more rigorous electrophysiological analysis of the development profiles of these new neurons. When will these neurons contribute to the plasticity of the circuits? Do adult-generated neurons exhibit special properties during a limited time window or permanently? Second, we need to know how extensively these neurons are involved in the

BMP: bone morphogenic protein

BrdU: bromodeoxyuridine

DCX: doublecortin

EGF: epidermal growth factor

FGF-2: basic fibroblast growth factor

GFAP: glial fibrillary acidic protein

GFP: green fluorescent protein

LTP: long-term potentiation

LTD: long-term depression

MAP2ab: microtubule-associated protein 2 isoform a and b

PACAP: pituitary adenylate cyclase-activating polypeptide

PSA-NCAM: polysialylated-neural cell-adhesion molecule

RMS: rostral migratory stream

SGZ: subgranular zone

SVZ: subventricular zone

Tuj1: β-tublin isoform III

VEGF: vascular endothelial growth factor

existing neuronal circuits. This will require systematic anatomical analysis. What are the sources of the inputs these new neurons receive? Do adult-generated neurons form circuits different from those produced during development? Third, to get a clean analysis of behavioral relevance of adult neurogenesis we will need to develop animal models that can silence identified populations of adult-generated neurons in a temporally and spatially precise manner. Fourth, we need to explore which aspects of behaviors are affected by, or are affecting, adult neurogenesis. We need diverse behavior analysis. Because adult neurogenesis is evolutionarily conserved (Zupanc 2001), it should be beneficial to compare the functions of adult neurogenesis in different species. Our studies can also be guided by neural modeling. For example, theoretical modeling predicts significant advantages of new neurons over mature neurons for both temporary storage and clearance of memories (Cecchi et al. 2001).

PERSPECTIVE

The past decade has witnessed the falling of a century-old dogma and the introduction of the new field of adult neurogenesis. Now the focus of the newly formed field has shifted from documentation of the existence of adult neurogenesis to understanding its regulatory mechanisms and functions. One of the most exciting and unique features of adult neurogenesis is that the complete process of neuronal development is recapitulated in the mature CNS, an environment quite different from the embryonic CNS, where neural development has been traditionally investigated (Temple 2001). Many basic mechanisms regulating adult neurogenesis are still unknown. Much less is known about those mechanisms under pathological conditions. Intensive studies in the past decade have revealed an array of factors and signaling mechanisms regulating neuronal fate specification, migration, nerve growth, guidance, and synaptogenesis during fetal CNS development (Anderson 2001, Huber et al. 2003, Temple 2001). These studies have also identified inhibitory mechanisms underlying the extremely limited regeneration of adult CNS neurons (David & Lacroix 2003, Filbin 2003, Grandpre & Strittmatter 2001, He & Koprivica 2004, Schwab 2004). The task now is to investigate whether similar mechanisms will also govern neurogenesis in a mature CNS environment. Because in nonmammalian vertebrates new neurons are generated continuously in many regions of the adult CNS (Zupanc 2001), comparisons of adult neurogenesis between a broad range of species is likely to yield new insights into the evolution and functions of this phenomenon.

The demonstration of active neurogenesis in adult humans not only shows the unforeseen regenerative capacity of the mature CNS, but also raises hopes for repairing the damaged adult CNS after injury or degenerative neurological diseases. Understanding the basic mechanisms regulating adult neurogenesis under normal and abnormal conditions will provide the foundation for cell replacement therapy, using either endogenous adult neural stem cells or transplanted cells from different sources. As history shows, the field of adult neurogenesis is propelled by technical advances. Aided by novel technologies in live imaging, single-cell genetics with retroviruses, new animal models, and neural modeling, the best is yet to come.

ACKNOWLEDGMENTS

We apologize to all whose original work could not be cited owing to space limitations. We thank Fred Gage, Josef Bischofberger, Gerd Kempermann, Kurt Sailor, and Janet Sailor for their comments and suggestions. The work was supported by funding of Charles E. Culpeper Scholarship in Medical Science from Rockefeller Brothers Fund and Goldman Philanthropic Partnerships, Basal O'Connor Starter Scholar Research Award from March of Dimes; the

National Institutes of Health (NIH, NS048271) and Whitehall Foundation to G.-L.M.; and NIH NS047344 and AG024984, The Robert Packard Center for ALS Research at Johns Hopkins University, and Klingenstein Fellowship Awards in the Neurosciences to H.S.

LITERATURE CITED

Allen E. 1912. The cessation of mitosis in the central nervous system of albino rat. *J. Comp. Neurol.* 19:547–68

Altman J. 1966. Autoradiographic and histological studies of postnatal neurogenesis. II. A longitudinal investigation of the kinetics, migration and transformation of cells incorporating tritiated thymidine in infant rats, with special reference to postnatal neurogenesis in some brain regions. *J. Comp. Neurol.* 1966:431–74

Altman J. 1967. In *The Neurosciences, Second Study Program*, ed. GC Quarton, T Melnechuck, FO Schmitt, pp. 723–43. New York: Rockefeller Univ. Press

Altman J. 1969. Autoradiographic and histological studies of postnatal neurogenesis. IV. Cell proliferation and migration in the anterior forebrain, with special reference to persisting neurogenesis in the olfactory bulb. *J. Comp. Neurol.* 137:433–57

Altman J, Das GD. 1965. Autoradiographic and histological evidence of postnatal hippocampal neurogenesis in rats. *J. Comp. Neurol.* 124:319–35

Alvarez-Buylla A, Lim DA. 2004. For the long run: maintaining germinal niches in the adult brain. *Neuron* 41:683–86

Anderson DJ. 2001. Stem cells and pattern formation in the nervous system: the possible versus the actual. *Neuron* 30:19–35

Arlotta P, Magavi SS, Macklis JD. 2003. Induction of adult neurogenesis: molecular manipulation of neural precursors in situ. *Ann. N.Y. Acad. Sci.* 991:229–36

Arvidsson A, Collin T, Kirik D, Kokaia Z, Lindvall O. 2002. Neuronal replacement from endogenous precursors in the adult brain after stroke. *Nat. Med.* 8:963–70

Baker SA, Baker KA, Hagg T. 2004. Dopaminergic nigrostriatal projections regulate neural precursor proliferation in the adult mouse subventricular zone. *Eur. J. Neurosci.* 20:575–79

Batchelor PE, Liberatore GT, Wong JY, Porritt MJ, Frerichs F, et al. 1999. Activated macrophages and microglia induce dopaminergic sprouting in the injured striatum and express brain-derived neurotrophic factor and glial cell line-derived neurotrophic factor. *J. Neurosci.* 19:1708–16

Belluzzi O, Benedusi M, Ackman J, LoTurco JJ. 2003. Electrophysiological differentiation of new neurons in the olfactory bulb. *J. Neurosci.* 23:10411–18

Benraiss A, Chmielnicki E, Lerner K, Roh D, Goldman SA. 2001. Adenoviral brain-derived neurotrophic factor induces both neostriatal and olfactory neuronal recruitment from endogenous progenitor cells in the adult forebrain. *J. Neurosci.* 21:6718–31

Bolteus AJ, Bordey A. 2004. GABA release and uptake regulate neuronal precursor migration in the postnatal subventricular zone. *J. Neurosci.* 24:7623–31

Branda CS, Dymecki SM. 2004. Talking about a revolution: the impact of site-specific recombinases on genetic analyses in mice. *Dev. Cell.* 6:7–28

Brown J, Cooper-Kuhn CM, Kempermann G, Van Praag H, Winkler J, et al. 2003. Enriched environment and physical activity stimulate hippocampal but not olfactory bulb neurogenesis. *Eur. J. Neurosci.* 17:2042–46

Cameron HA, Gould E. 1994. Adult neurogenesis is regulated by adrenal steroids in the dentate gyrus. *Neuroscience* 61:203–9

Cameron HA, McKay RD. 1999. Restoring production of hippocampal neurons in old age. *Nat. Neurosci.* 2:894–97

Cameron HA, McKay RD. 2001. Adult neurogenesis produces a large pool of new granule cells in the dentate gyrus. *J. Comp. Neurol.* 435:406–17

Cao L, Jiao X, Zuzga DS, Liu Y, Fong DM, et al. 2004. VEGF links hippocampal activity with neurogenesis, learning and memory. *Nat. Genet.* 36:827–35

Carlen M, Cassidy RM, Brismar H, Smith GA, Enquist LW, Frisen J. 2002. Functional integration of adult-born neurons. *Curr. Biol.* 12:606–8

Carleton A, Petreanu LT, Lansford R, Alvarez-Buylla A, Lledo PM. 2003. Becoming a new neuron in the adult olfactory bulb. *Nat. Neurosci.* 6:507–18

Cecchi GA, Petreanu LT, Alvarez-Buylla A, Magnasco MO. 2001. Unsupervised learning and adaptation in a model of adult neurogenesis. *J. Comput. Neurosci.* 11:175–82

Chambers RA, Potenza MN, Hoffman RE, Miranker W. 2004. Simulated apoptosis/neurogenesis regulates learning and memory capabilities of adaptive neural networks. *Neuropsychopharmacology* 29:747–58

Charles P, Reynolds R, Seilhean D, Rougon G, Aigrot MS, et al. 2002. Re-expression of PSA-NCAM by demyelinated axons: an inhibitor of remyelination in multiple sclerosis? *Brain* 125:1972–79

Conover JC, Doetsch F, Garcia-Verdugo JM, Gale NW, Yancopoulos GD, Alvarez-Buylla A. 2000. Disruption of Eph/ephrin signaling affects migration and proliferation in the adult subventricular zone. *Nat. Neurosci.* 3:1091–97

Crews FT, Miller MW, Ma W, Nixon K, Zawada WM, Zakhari S. 2003. Neural stem cells and alcohol. *Alcohol. Clin. Exp. Res.* 27:324–35

Curtis MA, Penney EB, Pearson AG, van Roon-Mom WM, Butterworth NJ, et al. 2003. Increased cell proliferation and neurogenesis in the adult human Huntington's disease brain. *Proc. Natl. Acad. Sci. USA* 100:9023–27

David S, Lacroix S. 2003. Molecular approaches to spinal cord repair. *Annu. Rev. Neurosci.* 26:411–40

Deisseroth K, Singla S, Toda H, Monje M, Palmer TD, Malenka RC. 2004. Excitation-neurogenesis coupling in adult neural stem/progenitor cells. *Neuron* 42:535–52

Doetsch F. 2003. A niche for adult neural stem cells. *Curr. Opin. Genet. Dev.* 13:543–50

Doetsch F, Caille I, Lim DA, Garcia-Verdugo JM, Alvarez-Buylla A. 1999. Subventricular zone astrocytes are neural stem cells in the adult mammalian brain. *Cell* 97:703–16

Doetsch F, Petreanu L, Caille I, Garcia-Verdugo JM, Alvarez-Buylla A. 2002. EGF converts transit-amplifying neurogenic precursors in the adult brain into multipotent stem cells. *Neuron* 36:1021–34

Duman RS, Malberg J, Nakagawa S. 2001. Regulation of adult neurogenesis by psychotropic drugs and stress. *J. Pharmacol. Exp. Ther.* 299:401–7

Durand GM, Kovalchuk Y, Konnerth A. 1996. Long-term potentiation and functional synapse induction in developing hippocampus. *Nature* 381:71–75

Eisch AJ, Barrot M, Schad CA, Self DW, Nestler EJ. 2000. Opiates inhibit neurogenesis in the adult rat hippocampus. *Proc. Natl. Acad. Sci. USA* 97:7579–84

Ekdahl CT, Claasen JH, Bonde S, Kokaia Z, Lindvall O. 2003. Inflammation is detrimental for neurogenesis in adult brain. *Proc. Natl. Acad. Sci. USA* 100:13632–37

Enwere E, Shingo T, Gregg C, Fujikawa H, Ohta S, Weiss S. 2004. Aging results in reduced epidermal growth factor receptor signaling, diminished olfactory neurogenesis, and deficits in fine olfactory discrimination. *J. Neurosci.* 24:8354–65

Eriksson PS, Perfilieva E, Bjork-Eriksson T, Alborn AM, Nordborg C, et al. 1998. Neurogenesis in the adult human hippocampus. *Nat. Med.* 4:1313–17

Fabel K, Tam B, Kaufer D, Baiker A, Simmons N, et al. 2003. VEGF is necessary for exercise-induced adult hippocampal neurogenesis. *Eur. J. Neurosci.* 18:2803–12

Falk A, Frisen J. 2002. Amphiregulin is a mitogen for adult neural stem cells. *J. Neurosci. Res.* 69:757–62

Farmer J, Zhao X, van Praag H, Wodtke K, Gage FH, Christie BR. 2004. Effects of voluntary exercise on synaptic plasticity and gene expression in the dentate gyrus of adult male Sprague-Dawley rats in vivo. *Neuroscience* 124:71–79

Filbin MT. 2003. Myelin-associated inhibitors of axonal regeneration in the adult mammalian CNS. *Nat. Rev. Neurosci.* 4:703–13

Fuchs E, Gould E. 2000. Mini-review: In vivo neurogenesis in the adult brain: regulation and functional implications. *Eur. J. Neurosci.* 12:2211–14

Gheusi G, Cremer H, McLean H, Chazal G, Vincent JD, Lledo PM. 2000. Importance of newly generated neurons in the adult olfactory bulb for odor discrimination. *Proc. Natl. Acad. Sci. USA* 97:1823–28

Goh EL, Ma D, Ming GL, Song H. 2003. Adult neural stem cells and repair of the adult central nervous system. *J. Hematother. Stem. Cell Res.* 12:671–79

Gottlieb DI. 2002. Large-scale sources of neural stem cells. *Annu. Rev. Neurosci.* 25:381–407

Gould E, Beylin A, Tanapat P, Reeves A, Shors TJ. 1999a. Learning enhances adult neurogenesis in the hippocampal formation. *Nat. Neurosci.* 2:260–65

Gould E, Tanapat P, Hastings NB, Shors TJ. 1999b. Neurogenesis in adulthood: a possible role in learning. *Trends Cogn. Sci.* 3:186–92

Grandpre T, Strittmatter SM. 2001. Nogo: a molecular determinant of axonal growth and regeneration. *Neuroscientist* 7:377–86

Gratzner HG. 1982. Monoclonal antibody to 5-bromo- and 5-iododeoxyuridine: a new reagent for detection of DNA replication. *Science* 218:474–75

Gross CG. 2000. Neurogenesis in the adult brain: death of a dogma. *Nat. Rev. Neurosci.* 1:67–73

Hack I, Bancila M, Loulier K, Carroll P, Cremer H. 2002. Reelin is a detachment signal in tangential chain-migration during postnatal neurogenesis. *Nat. Neurosci.* 5:939–45

Hamilton A. 1901. The division of differentiated cells in the central nervous system of the white rat. *J. Comp. Neurol.* 11:297–320

Hastings NB, Gould E. 1999. Rapid extension of axons into the CA3 region by adult-generated granule cells. *J. Comp. Neurol.* 413:146–54

Hayes NL, Nowakowski RS. 2002. Dynamics of cell proliferation in the adult dentate gyrus of two inbred strains of mice. *Brain Res. Dev. Brain Res.* 134:77–85

He Z, Koprivica V. 2004. The Nogo signaling pathway for regeneration block. *Annu. Rev. Neurosci.* 27:341–68

Hoglinger GU, Rizk P, Muriel MP, Duyckaerts C, Oertel WH, et al. 2004. Dopamine depletion impairs precursor cell proliferation in Parkinson disease. *Nat. Neurosci.* 7:726–35

Hsieh J, Gage FH. 2004. Epigenetic control of neural stem cell fate. *Curr. Opin. Genet. Dev.* 14:461–69

Hu H, Tomasiewicz H, Magnuson T, Rutishauser U. 1996. The role of polysialic acid in migration of olfactory bulb interneuron precursors in the subventricular zone. *Neuron* 16:735–43

Huber AB, Kolodkin AL, Ginty DD, Cloutier JF. 2003. Signaling at the growth cone: ligand-receptor complexes and the control of axon growth and guidance. *Annu. Rev. Neurosci.* 26:509–63

Jin K, Galvan V, Xie L, Mao XO, Gorostiza OF, et al. 2004a. Enhanced neurogenesis in Alzheimer's disease transgenic (PDGF-APPSw,Ind) mice. *Proc. Natl. Acad. Sci. USA* 101:13363–67

Jin K, Peel AL, Mao XO, Xie L, Cottrell BA, et al. 2004b. Increased hippocampal neurogenesis in Alzheimer's disease. *Proc. Natl. Acad. Sci. USA* 101:343–47

Jin K, Sun Y, Xie L, Batteur S, Mao XO, et al. 2003. Neurogenesis and aging: FGF-2 and HB-EGF restore neurogenesis in hippocampus and subventricular zone of aged mice. *Aging Cell* 2:175–83

Johansson CB, Svensson M, Wallstedt L, Janson AM, Frisen J. 1999. Neural stem cells in the adult human brain. *Exp. Cell Res.* 253:733–36

Kaplan MS, Bell DH. 1983. Neuronal proliferation in the 9-month-old rodent-radioautographic study of granule cells in the hippocampus. *Exp. Brain Res.* 52:1–5

Kaplan MS, Hinds JW. 1977. Neurogenesis in the adult rat: electron microscopic analysis of light radioautographs. *Science* 197:1092–94

Katsetos CD, Del Valle L, Geddes JF, Assimakopoulou M, Legido A, et al. 2001. Aberrant localization of the neuronal class III beta-tubulin in astrocytomas. *Arch. Pathol. Lab Med.* 125:613–24

Kempermann G. 2002. Regulation of adult hippocampal neurogenesis—implications for novel theories of major depression. *Bipolar Disord.* 4:17–33

Kempermann G, Gage FH. 1999. New nerve cells for the adult brain. *Sci. Am.* 280:48–53

Kempermann G, Gage FH. 2002. Genetic influence on phenotypic differentiation in adult hippocampal neurogenesis. *Brain Res. Dev. Brain Res.* 134:1–12

Kempermann G, Jessberger S, Steiner B, Kronenberg G. 2004a. Milestones of neuronal development in the adult hippocampus. *Trends Neurosci.* 27:447–52

Kempermann G, Kuhn HG, Gage FH. 1997. More hippocampal neurons in adult mice living in an enriched environment. *Nature* 386:493–95

Kempermann G, Neumann H. 2003. Neuroscience. Microglia: the enemy within? *Science* 302:1689–90

Kempermann G, Wiskott L, Gage FH. 2004b. Functional significance of adult neurogenesis. *Curr. Opin. Neurobiol.* 14:186–91

Kokaia Z, Lindvall O. 2003. Neurogenesis after ischaemic brain insults. *Curr. Opin. Neurobiol.* 13:127–32

Kuan CY, Schloemer AJ, Lu A, Burns KA, Weng WL, et al. 2004. Hypoxia-ischemia induces DNA synthesis without cell proliferation in dying neurons in adult rodent brain. *J. Neurosci.* 24:10763–72

Kuhn HG, Dickinson-Anson H, Gage FH. 1996. Neurogenesis in the dentate gyrus of the adult rat: age-related decrease of neuronal progenitor proliferation. *J. Neurosci.* 16:2027–33

Kuhn HG, Winkler J, Kempermann G, Thal LJ, Gage FH. 1997. Epidermal growth factor and fibroblast growth factor-2 have different effects on neural progenitors in the adult rat brain. *J. Neurosci.* 17:5820–29

Kukekov VG, Laywell ED, Suslov O, Davies K, Scheffler B, et al. 1999. Multipotent stem/progenitor cells with similar properties arise from two neurogenic regions of adult human brain. *Exp. Neurol.* 156:333–44

Kuwabara T, Hsieh J, Nakashima K, Taira K, Gage FH. 2004. A small modulatory dsRNA specifies the fate of adult neural stem cells. *Cell* 116:779–93

Lai K, Kaspar BK, Gage FH, Schaffer DV. 2003. Sonic hedgehog regulates adult neural progenitor proliferation in vitro and in vivo. *Nat. Neurosci.* 6:21–27

Leuner B, Mendolia-Loffredo S, Kozorovitskiy Y, Samburg D, Gould E, Shors TJ. 2004. Learning enhances the survival of new neurons beyond the time when the hippocampus is required for memory. *J. Neurosci.* 24:7477–81

Lewis PF, Emerman M. 1994. Passage through mitosis is required for oncoretroviruses but not for the human immunodeficiency virus. *J. Virol.* 68:510–16

Lie DC, Dziewczapolski G, Willhoite AR, Kaspar BK, Shults CW, Gage FH. 2002. The adult substantia nigra contains progenitor cells with neurogenic potential. *J. Neurosci.* 22:6639–49

Lie DC, Song H, Colamarino SA, Ming GL, Gage FH. 2004. Neurogenesis in the adult brain: new strategies for central nervous system diseases. *Annu. Rev. Pharmacol. Toxicol.* 44:399–421

Lim DA, Alvarez-Buylla A. 1999. Interaction between astrocytes and adult subventricular zone precursors stimulates neurogenesis. *Proc. Natl. Acad. Sci. USA* 96:7526–31

Lim DA, Tramontin AD, Trevejo JM, Herrera DG, Garcia-Verdugo JM, Alvarez-Buylla A. 2000. Noggin antagonizes BMP signaling to create a niche for adult neurogenesis. *Neuron* 28:713–26

Lindvall O, Kokaia Z, Martinez-Serrano A. 2004. Stem cell therapy for human neurodegenerative disorders-how to make it work. *Nat. Med.* 10(Suppl.):S42–50

Liu S, Wang J, Zhu D, Fu Y, Lukowiak K, Lu YM. 2003. Generation of functional inhibitory neurons in the adult rat hippocampus. *J. Neurosci.* 23:732–36

Lois C, Garcia-Verdugo JM, Alvarez-Buylla A. 1996. Chain migration of neuronal precursors. *Science* 271:978–81

Machold R, Hayashi S, Rutlin M, Muzumdar MD, Nery S, et al. 2003. Sonic hedgehog is required for progenitor cell maintenance in telencephalic stem cell niches. *Neuron* 39:937–50

Magavi SS, Leavitt BR, Macklis JD. 2000. Induction of neurogenesis in the neocortex of adult mice. *Nature* 405:951–55

Miller MW, Nowakowski RS. 1988. Use of bromodeoxyuridine-immunohistochemistry to examine the proliferation, migration and time of origin of cells in the central nervous system. *Brain Res.* 457:44–52

Mizrahi A, Crowley JC, Shtoyerman E, Katz LC. 2004. High-resolution in vivo imaging of hippocampal dendrites and spines. *J. Neurosci.* 24:3147–51

Mizrahi A, Katz LC. 2003. Dendritic stability in the adult olfactory bulb. *Nat. Neurosci.* 6:1201–7

Monje ML, Toda H, Palmer TD. 2003. Inflammatory blockade restores adult hippocampal neurogenesis. *Science* 302:1760–65

Morshead CM, Reynolds BA, Craig CG, McBurney MW, Staines WA, et al. 1994. Neural stem cells in the adult mammalian forebrain: a relatively quiescent subpopulation of subependymal cells. *Neuron* 13:1071–82

Murase S, Horwitz AF. 2002. Deleted in colorectal carcinoma and differentially expressed integrins mediate the directional migration of neural precursors in the rostral migratory stream. *J. Neurosci.* 22:3568–79

Nakatomi H, Kuriu T, Okabe S, Yamamoto S, Hatano O, et al. 2002. Regeneration of hippocampal pyramidal neurons after ischemic brain injury by recruitment of endogenous neural progenitors. *Cell* 110:429–41

Nguyen-Ba-Charvet KT, Picard-Riera N, Tessier-Lavigne M, Baron-Van Evercooren A, Sotelo C, Chedotal A. 2004. Multiple roles for slits in the control of cell migration in the rostral migratory stream. *J. Neurosci.* 24:1497–506

Nilsson M, Perfilieva E, Johansson U, Orwar O, Eriksson PS. 1999. Enriched environment increases neurogenesis in the adult rat dentate gyrus and improves spatial memory. *J. Neurobiol.* 39:569–78

Nimchinsky EA, Sabatini BL, Svoboda K. 2002. Structure and function of dendritic spines. *Annu. Rev. Physiol.* 64:313–53

Nottebohm F. 2004. The road we travelled: discovery, choreography, and significance of brain replaceable neurons. *Ann. N.Y. Acad. Sci.* 1016:628–58

Oliver TG, Wechsler-Reya RJ. 2004. Getting at the root and stem of brain tumors. *Neuron* 42:885–88

Overstreet LS, Hentges ST, Bumaschny VF, de Souza FS, Smart JL, et al. 2004. A transgenic marker for newly born granule cells in dentate gyrus. *J. Neurosci.* 24:3251–59

Palmer TD, Markakis EA, Willhoite AR, Safar F, Gage FH. 1999. Fibroblast growth factor-2 activates a latent neurogenic program in neural stem cells from diverse regions of the adult CNS. *J. Neurosci.* 19:8487–97

Palmer TD, Willhoite AR, Gage FH. 2000. Vascular niche for adult hippocampal neurogenesis. *J. Comp. Neurol.* 425:479–94

Parent JM. 2003. Injury-induced neurogenesis in the adult mammalian brain. *Neuroscientist* 9:261–72

Parent JM, Valentin VV, Lowenstein DH. 2002a. Prolonged seizures increase proliferating neuroblasts in the adult rat subventricular zone-olfactory bulb pathway. *J. Neurosci.* 22:3174–88

Parent JM, Vexler ZS, Gong C, Derugin N, Ferriero DM. 2002b. Rat forebrain neurogenesis and striatal neuron replacement after focal stroke. *Ann. Neurol.* 52:802–13

Parent JM, Yu TW, Leibowitz RT, Geschwind DH, Sloviter RS, Lowenstein DH. 1997. Dentate granule cell neurogenesis is increased by seizures and contributes to aberrant network reorganization in the adult rat hippocampus. *J. Neurosci.* 17:3727–38

Paton JA, Nottebohm FN. 1984. Neurons generated in the adult brain are recruited into functional circuits. *Science* 225:1046–48

Price J, Turner D, Cepko C. 1987. Lineage analysis in the vertebrate nervous system by retrovirus-mediated gene transfer. *Proc. Natl. Acad. Sci. USA* 84:156–60

Raff M. 2003. Adult stem cell plasticity: fact or artifact? *Annu. Rev. Cell Dev. Biol.* 19:1–22

Raineteau O, Rietschin L, Gradwohl G, Guillemot F, Gahwiler BH. 2004. Neurogenesis in hippocampal slice cultures. *Mol. Cell Neurosci.* 26:241–50

Rakic P. 2002. Neurogenesis in adult primate neocortex: an evaluation of the evidence. *Nat. Rev. Neurosci.* 3:65–71

Ramon y Cajal S. 1913. *Degeneration and Regeneration of the Nervous System*. London: Oxford Univ. Press

Ray J, Peterson DA, Schinstine M, Gage FH. 1993. Proliferation, differentiation, and long-term culture of primary hippocampal neurons. *Proc. Natl. Acad. Sci. USA* 90:3602–6

Reynolds BA, Weiss S. 1992. Generation of neurons and astrocytes from isolated cells of the adult mammalian central nervous system. *Science* 255:1707–10

Rietze RL, Valcanis H, Brooker GF, Thomas T, Voss AK, Bartlett PF. 2001. Purification of a pluripotent neural stem cell from the adult mouse brain. *Nature* 412:736–39

Rochefort C, Gheusi G, Vincent JD, Lledo PM. 2002. Enriched odor exposure increases the number of newborn neurons in the adult olfactory bulb and improves odor memory. *J. Neurosci.* 22:2679–89

Rogers AW. 1973. *Techniques of Autoradiography*. New York: Elsevier Sci.

Rossi F, Cattaneo E. 2002. Opinion: Neural stem cell therapy for neurological diseases: dreams and reality. *Nat. Rev. Neurosci.* 3:401–9

Roy NS, Benraiss A, Wang S, Fraser RA, Goodman R, et al. 2000. Promoter-targeted selection and isolation of neural progenitor cells from the adult human ventricular zone. *J. Neurosci. Res.* 59:321–31

Saghatelyan A, de Chevigny A, Schachner M, Lledo PM. 2004. Tenascin-R mediates activity-dependent recruitment of neuroblasts in the adult mouse forebrain. *Nat. Neurosci.* 7:347–56

Sanai N, Tramontin AD, Quinones-Hinojosa A, Barbaro NM, Gupta N, et al. 2004. Unique astrocyte ribbon in adult human brain contains neural stem cells but lacks chain migration. *Nature* 427:740–44

Sanes JR, Rubenstein JL, Nicolas JF. 1986. Use of a recombinant retrovirus to study post-implantation cell lineage in mouse embryos. *EMBO J.* 5:3133–42

Santarelli L, Saxe M, Gross C, Surget A, Battaglia F, et al. 2003. Requirement of hippocampal neurogenesis for the behavioral effects of antidepressants. *Science* 301:805–9

Scharfman HE, Goodman JH, Sollas AL. 2000. Granule-like neurons at the hilar/CA3 border after status epilepticus and their synchrony with area CA3 pyramidal cells: functional implications of seizure-induced neurogenesis. *J. Neurosci.* 20:6144–58

Schmidt-Hieber C, Jonas P, Bischofberger J. 2004. Enhanced synaptic plasticity in newly generated granule cells of the adult hippocampus. *Nature* 429:184–87

Schwab ME. 2004. Nogo and axon regeneration. *Curr. Opin. Neurobiol.* 14:118–24

Selden JR, Dolbeare F, Clair JH, Nichols WW, Miller JE, et al. 1993. Statistical confirmation that immunofluorescent detection of DNA repair in human fibroblasts by measurement of bromodeoxyuridine incorporation is stoichiometric and sensitive. *Cytometry* 14:154–67

Seri B, Garcia-Verdugo JM, Collado-Morente L, McEwen BS, Alvarez-Buylla A. 2004. Cell types, lineage, and architecture of the germinal zone in the adult dentate gyrus. *J. Comp. Neurol.* 478:359–78

Seri B, Garcia-Verdugo JM, McEwen BS, Alvarez-Buylla A. 2001. Astrocytes give rise to new neurons in the adult mammalian hippocampus. *J. Neurosci.* 21:7153–60

Shepherd GM, Chen WR, Greer CA. 2004. Olfactory bulb. In *The Synaptic Organization of the Brain*, ed. GM Shepherd, pp. 165–216. New York: Oxford Univ. Press

Shors TJ, Miesegaes G, Beylin A, Zhao M, Rydel T, Gould E. 2001. Neurogenesis in the adult is involved in the formation of trace memories. *Nature* 410:372–76

Shors TJ, Townsend DA, Zhao M, Kozorovitskiy Y, Gould E. 2002. Neurogenesis may relate to some but not all types of hippocampal-dependent learning. *Hippocampus* 12:578–84

Sidman RL, Miale IL, Feder N. 1959. Cell proliferation and migration in the primitive ependymal zone: an autoradiographic study of histogenesis in the nervous system. *Exp. Neurol.* 1:322–33

Smart I. 1961. The subependymal layer of the mouse brain and its cell production as shown by autography after [H^3]-thymidine injection. *J. Comp. Neurol.* 116:325–27

Snyder JS, Kee N, Wojtowicz JM. 2001. Effects of adult neurogenesis on synaptic plasticity in the rat dentate gyrus. *J. Neurophysiol.* 85:2423–31

Song H, Poo M. 2001. The cell biology of neuronal navigation. *Nat. Cell Biol.* 3:E81–88

Song HJ, Poo MM. 1999. Signal transduction underlying growth cone guidance by diffusible factors. *Curr. Opin. Neurobiol.* 9:355–63

Song H, Stevens CF, Gage FH. 2002a. Astroglia induce neurogenesis from adult neural stem cells. *Nature* 417:39–44

Song HJ, Stevens CF, Gage FH. 2002b. Neural stem cells from adult hippocampus develop essential properties of functional CNS neurons. *Nat. Neurosci.* 5:438–45

Stanfield BB, Trice JE. 1988. Evidence that granule cells generated in the dentate gyrus of adult rats extend axonal projections. *Exp. Brain Res.* 72:399–406

Tattersfield AS, Croon RJ, Liu YW, Kells AP, Faull RL, Connor B. 2004. Neurogenesis in the striatum of the quinolinic acid lesion model of Huntington's disease. *Neuroscience* 127:319–32

Temple S. 2001. The development of neural stem cells. *Nature* 414:112–17

Tropepe V, Craig CG, Morshead CM, van der Kooy D. 1997. Transforming growth factor-alpha null and senescent mice show decreased neural progenitor cell proliferation in the forebrain subependyma. *J. Neurosci.* 17:7850–59

Uchida N, Buck DW, He D, Reitsma MJ, Masek M, et al. 2000. Direct isolation of human central nervous system stem cells. *Proc. Natl. Acad. Sci. USA* 97:14720–25

Ueki T, Tanaka M, Yamashita K, Mikawa S, Qiu Z, et al. 2003. A novel secretory factor, Neurogenesin-1, provides neurogenic environmental cues for neural stem cells in the adult hippocampus. *J. Neurosci.* 23:11732–40

Ullian EM, Christopherson KS, Barres BA. 2004. Role for glia in synaptogenesis. *Glia* 47:209–16

Vallieres L, Campbell IL, Gage FH, Sawchenko PE. 2002. Reduced hippocampal neurogenesis in adult transgenic mice with chronic astrocytic production of interleukin-6. *J. Neurosci.* 22:486–92

van Praag H, Christie BR, Sejnowski TJ, Gage FH. 1999a. Running enhances neurogenesis, learning, and long-term potentiation in mice. *Proc. Natl. Acad. Sci. USA* 96:13427–31

van Praag H, Kempermann G, Gage FH. 1999b. Running increases cell proliferation and neurogenesis in the adult mouse dentate gyrus. *Nat. Neurosci.* 2:266–70

van Praag H, Schinder AF, Christie BR, Toni N, Palmer TD, Gage FH. 2002. Functional neurogenesis in the adult hippocampus. *Nature* 415:1030–34

Walker MC, Ruiz A, Kullmann DM. 2001. Monosynaptic GABAergic signaling from dentate to CA3 with a pharmacological and physiological profile typical of mossy fiber synapses. *Neuron* 29:703–15

Wang S, Scott BW, Wojtowicz JM. 2000. Heterogenous properties of dentate granule neurons in the adult rat. *J. Neurobiol.* 42:248–57

Wu W, Wong K, Chen J, Jiang Z, Dupuis S, et al. 1999. Directional guidance of neuronal migration in the olfactory system by the protein Slit. *Nature* 400:331–36

Yamaguchi M, Saito H, Suzuki M, Mori K. 2000. Visualization of neurogenesis in the central nervous system using nestin promoter-GFP transgenic mice. *Neuroreport* 11:1991–96

Zhao M, Momma S, Delfani K, Carlen M, Cassidy RM, et al. 2003a. Evidence for neurogenesis in the adult mammalian substantia nigra. *Proc. Natl. Acad. Sci. USA* 100:7925–30

Zhao X, Ueba T, Christie BR, Barkho B, McConnell MJ, et al. 2003b. Mice lacking methyl-CpG binding protein 1 have deficits in adult neurogenesis and hippocampal function. *Proc. Natl. Acad. Sci. USA* 100:6777–82

Zupanc GK. 2001. A comparative approach towards the understanding of adult neurogenesis. *Brain Behav. Evol.* 58:246–49

Mechanisms of Vertebrate Synaptogenesis

Clarissa L. Waites,[1] Ann Marie Craig,[2] and Craig C. Garner[1]

[1]Department of Psychiatry and Behavioral Science, Nancy Pritzker Laboratory, Stanford University, Palo Alto, California, 94304-5485; email: clarissa@stanford.edu, cgarner@stanford.edu

[2]Department of Anatomy and Neurobiology, Washington University School of Medicine, St. Louis, Missouri 63110-1093; email: acraig@pcg.wustl.edu

Key Words

synapse, active zone, postsynaptic density, membrane trafficking, cytoskeleton

Abstract

The formation of synapses in the vertebrate central nervous system is a complex process that occurs over a protracted period of development. Recent work has begun to unravel the mysteries of synaptogenesis, demonstrating the existence of multiple molecules that influence not only when and where synapses form but also synaptic specificity and stability. Some of these molecules act at a distance, steering axons to their correct receptive fields and promoting neuronal differentiation and maturation, whereas others act at the time of contact, providing positional information about the appropriateness of targets and/or inductive signals that trigger the cascade of events leading to synapse formation. In addition, correlated synaptic activity provides critical information about the appropriateness of synaptic connections, thereby influencing synapse stability and elimination. Although synapse formation and elimination are hallmarks of early development, these processes are also fundamental to learning, memory, and cognition in the mature brain.

Contents

INTRODUCTION 252
SPECIFICATION AND INDUCTION OF SYNAPSE FORMATION 254
   Diffusible Target-Derived Factors Guiding Synapse Specificity 254
   Cell-Adhesion Molecules Guiding Synapse Specificity 256
   Inducers of Synapse Formation 257
CELLULAR MECHANISMS OF SYNAPSE ASSEMBLY 259
   Membrane Trafficking in Presynaptic Assembly 260
   Membrane Trafficking in Postsynaptic Assembly 261
   Synaptic Maturation 263
ACTIVITY-DEPENDENT REGULATION OF SYNAPTOGENESIS 264
   Synapse Elimination 264
   Ubiquitin Regulation of Synapse Stability 265
CONCLUDING REMARKS 266

Axon: a long, thin neuronal process that carries electrical signals from the cell soma to presynaptic boutons

Dendrite: a tapered neuronal process onto which presynaptic boutons form synapses. Signals from these dendritic synapses are propagated back to the cell soma, summed, and used to trigger an axonal action potential

Synapse: a site of contact between neurons where electrochemical signaling occurs

INTRODUCTION

The human brain is an amazingly complex organ composed of trillions of neurons. Every idea, emotion, and lofty thought we produce is created as a series of electrical and chemical signals transmitted through connected networks of neurons. Neurons transmit these signals to one another at specialized sites of contact called synapses. In the vertebrate nervous system, most neurons communicate via chemical synapses. As the name implies, chemical synapses function by converting electrical signals, in the form of action potentials racing down axons and invading presynaptic boutons, into chemical signals and then back to electrical impulses within the postsynaptic dendrite. Synapses perform this task by releasing neurotransmitters from the presynaptic neuron that bind and activate neurotransmitter-gated ion channels on the postsynaptic cell.

Chemical synapses are asymmetric cellular junctions formed between neurons and their targets, including other neurons, muscles, and glands. Ultrastructurally, synapses are composed of several specialized domains (Gray 1963, Palay 1956) (**Figure 1**). The most prominent is the presynaptic bouton. These small axonal varicosities, ~1 micron in size, stud the length of axons and establish contacts with one or more postsynaptic cells. Each bouton is filled with hundreds to thousands of small ~50-nm clear-centered synaptic vesicles carrying neurotransmitter molecules. A depolarizing action potential invading the bouton causes synaptic vesicles docked at the plasma membrane to fuse and release their neurotransmitter into the synaptic cleft, a small space between the pre- and postsynaptic cells. Synaptic vesicle fusion does not occur randomly at the presynaptic plasma membrane but within another specialized domain called the active zone. Active zones are characterized by the presence of an electron-dense meshwork of proteins, also known as the presynaptic web, and synaptic vesicles that are both embedded within this matrix and docked at the plasma membrane (Burns & Augustine 1995, Hirokawa et al. 1989, Landis 1988, Phillips et al. 2001). Directly apposed to the active zone on the postsynaptic side is a third domain that functions to receive information sent by the presynaptic neuron. Similar to the active zone, this postsynaptic membrane specialization is characterized by the presence of an electron-dense meshwork of proteins that extends across the synaptic cleft to the active zone as well as into the cytoplasm of the postsynaptic cell (Garner et al. 2002, Palay 1956, Sheng 2001). This structure, referred to as the postsynaptic density (PSD), serves to cluster neurotransmitter receptors, voltage-gated ion channels, and various second-messenger signaling molecules at high density directly across from the active zone (Garner et al. 2002, Palay 1956, Sheng 2001). Filamentous proteins extending across the synaptic cleft are

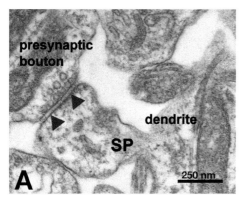 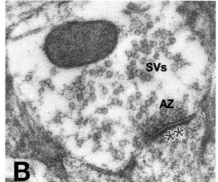

Figure 1

Ultrastructure of excitatory glutamatergic synapses. Electron micrographs of two synapses formed between hippocampal neurons grown for 15 days in culture. The synapse in (*A*) is clearly onto a dendritic spine (SP), indicating that it is excitatory in nature. Docked synaptic vesicles can be seen along a prominent synaptic cleft (*arrowheads*). The synapse in (*B*) has many of the classic features of chemical synapses, including a presynaptic bouton containing ∼50-nm synaptic vesicles (SVs), an active zone (AZ) characterized by an electron-dense meshwork of proteins and clusters of docked synaptic vesicles, and a prominent postsynaptic density (*asterisks*). Micrographs were taken and generously provided by J. Buchanan, Stanford University.

thought to hold the active zone and PSD in register. Although these features are shared by synapses throughout the CNS, variation in the size and organization of presynaptic active zones, as well as in the thickness of the PSD, have been documented and have been correlated with (*a*) synaptic type, e.g., the type of transmitter released by a given bouton; (*b*) synaptic function, e.g., whether the synapse is excitatory, inhibitory, or modulatory; and (*c*) synaptic efficacy, e.g., whether a synapse fires reliably or unreliably, continuously or sporadically. In this review, we restrict our comments to excitatory glutamatergic synapses because most of the insights into the molecular mechanisms of synaptogenesis have come from studies on these synapses.

The formation of synapses in vertebrate organisms occurs over a protracted period of development, beginning in the embryo and extending well into early postnatal life. As discussed below, synapse formation also occurs in adults, where it is thought to contribute significantly to learning and memory. During development, synaptogenesis is tightly coupled to neuronal differentiation and the establishment of neuronal circuitry. For example, shortly after neurons differentiate and extend axonal and dendritic processes, many of the genes encoding synaptic proteins are turned on, resulting in the formation, accumulation, and directional trafficking of vesicles carrying pre- and postsynaptic protein complexes. During this time, the specification of correct neuronal connections is determined, as axons and dendrites make contact and establish initial, often transient, synapses. This courtship involves a myriad of secreted factors, receptors, and signaling molecules that make neurons receptive to form synapses. It also requires interactions between sets of cell-surface adhesion molecules (CAMs) that are involved in cell-cell recognition, as well as inductive signals that trigger the initial stages of synapse formation, including the assembly of pre- and postsynaptic specializations from their component proteins and membranes. Finally, synaptic activity determines whether these synapses will be stabilized or eliminated, both during development and in the mature brain. In this review, we discuss each of these

Active zone: a specialized region of the presynaptic bouton where synaptic vesicles dock and fuse with the plasma membrane

Postsynaptic density: a specialized region of the postsynapse where neurotransmitter receptors and signaling molecules cluster at high density

Synaptogenesis: the complex process by which functional synapses form between neurons

Synaptic activity: correlated or uncorrelated electrochemical signaling between neurons

Glutamate: an amino acid that functions as an excitatory neurotransmitter

Glia: a population of cells in the brain that do not engage in synaptic transmission but are important for neuron survival

issues with a special emphasis on how different classes of molecules appear to guide these events.

SPECIFICATION AND INDUCTION OF SYNAPSE FORMATION

An important aspect of synaptogenesis is target recognition, i.e., the ability of axons from different brain regions to grow into their respective target fields and synapse with the correct cell type. In many cases, axons must navigate across large distances and complex terrains before encountering their target cells. Intriguingly, although they come into contact with a multitude of neurons during their journeys, they refrain from synapsing onto inappropriate target cells. For example, retinal ganglion cell axons traverse long distances from the eye into the lateral geniculate nucleus of the thalamus before synapsing onto thalamic cell dendrites (Shatz 1996, 1997). Similarly, motor neuron axons from the ventral horn of the spinal cord delay synapse formation until they innervate muscle fibers some distance away (Burden 2002, Sanes & Lichtman 2001). In these examples not only is synapse formation delayed until axons reach specific target regions, but even within these target regions there are lag times of days to weeks before synapses form (Lund 1972, Pfrieger & Barres 1996).

What types of factors/cues regulate this temporal and spatial specificity of synaptogenesis? In principle, temporal events can be regulated by intrinsic, genetically encoded programs such as those used in cell-fate determination during development. In the case of neuronal cells, these intrinsic signals could be part of an intracellular clock guiding neuronal growth and differentiation. Studies of dissociated cultures of hippocampal neurons support this idea. In these cultures, axons from E18 rat hippocampal neurons are immediately competent to function as presynaptic partners, whereas dendrites must mature for several days to become competent postsynaptic partners (Fletcher et al. 1994). Alternately, cues for regulating temporal and spatial specificity could be molecules derived from target neurons that act either indirectly, i.e., by activating transcription factors that allow neurons to become synaptogenesis-competent, or directly on axons to induce presynaptic differentiation (Ullian et al. 2004, Umemori et al. 2004).

Although intrinsic signals are a conceptually simple and attractive mechanism for temporally regulating synaptogenesis, very little progress has been made in identifying these important molecules. In contrast, significant progress has been made in the identification of target-derived factors that either accelerate neuronal maturation or directly induce synapse formation. We refer to the former as "priming molecules," because they seem to prime neurons and make them competent to undergo synaptogenesis, and the latter as "inducing molecules," because they appear to trigger synaptogenesis (see **Figure 2**). As discussed below, these molecules can be divided into several classes on the basis of when in the cascade of synaptogenesis they appear to function, and they include diffusible, target-derived factors as well as cell-surface adhesion molecules.

Diffusible Target-Derived Factors Guiding Synapse Specificity

One class of proteins with synaptogenic activity is diffusible factors that are synthesized either by target neurons or the surrounding glia. These molecules have a range of activities, including abilities to guide axonal projections to their correct targets, stimulate local arborization, promote neuronal differentiation and maturation, and create a permissive environment wherein axo-dendritic contact leads to the formation of functional synaptic junctions.

A prominent group of target-derived molecules known to guide axons into reciprocally connected brain areas are those involved in growth cone guidance, including netrins,

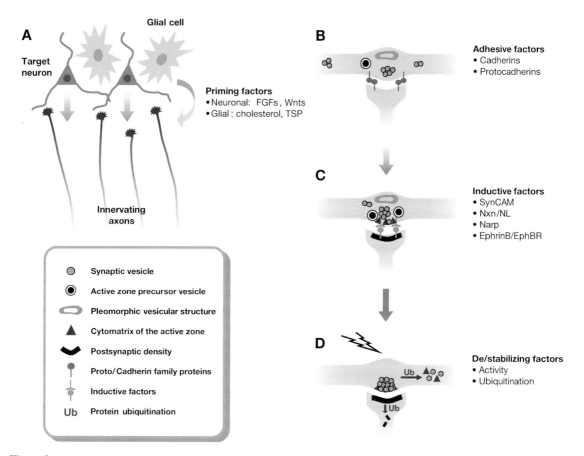

Figure 2

Signaling pathways involved in regulating vertebrate synaptogenesis. Synaptogenesis is a multistep process involving a myriad of signaling molecules. Prior to synapse formation, secreted molecules such as netrins and semaphorins guide axons to their targets. These axons then encounter priming factors secreted by target neurons and the surrounding glia, including fibroblast growth factor (FGFs), Wnts, cholesterol, and thrombospondin (TSP), that act to promote axonal and dendritic maturation and facilitate the ability of these processes to initiate synapse formation (*A*). Recognition that axons are in the correct receptive field is corroborated by CAMs, including members of the cadherin and protocadherin superfamilies, during initial contact between axons and dendrites (*B*). The presence of a second group of CAMs, such as SynCAM and Neuroligin (NL), at these contact sites is then thought to induce the formation of presynaptic active zones (*C*). In this regard, the adhesive CAMs are likely to work in synchrony with the inductive CAMs to stabilize the nascent synaptic junction. The neuroligin binding partner β-neurexin (Nxn) as well as Narp and EphrinB promote the recruitment of glutamate receptors and postsynaptic scaffolding proteins (*C*). Neuronal activity, although not essential for synapse formation, has been strongly implicated in regulating synapse stability (*D*). Intracellular signaling pathways sensitive to the activity states of synapses, including ubiquitin-mediated degradation, not only regulate the turnover of synaptic components but also promote synapse elimination (*D*).

semaphorins, and ephrinA (Bagri & Tessier-Lavigne 2002, Pascual et al. 2004, Tessier-Lavigne 1995). Because these molecules have no demonstrated role in synaptogenesis, we do not discuss them further in this review.

A second group of target-derived molecules include members of the Wnt and fibroblast growth factor (FGF) families. These proteins are secreted by certain subpopulations of neurons and have been shown to induce

Cell-adhesion molecules: cell-surface proteins that mediate close contact between two cells

regional axon arborization and/or accumulation of recycling synaptic vesicles in innervating axons (Scheiffele 2003). Such properties could serve to spatially restrict synaptogenesis. For example, Wnt-3 secreted by motor neuron dendrites in the spinal cord induces the arborization of innervating sensory axons (Krylova et al. 2002), whereas Wnt-7a, secreted by cerebellar granule cells, induces clustering of the synaptic vesicle–associated protein synapsin 1 in innervating mossy fiber terminals (Hall et al. 2000). A second factor secreted by cerebellar granule cells, FGF22, also promotes the formation of presynaptic active zones in innervating mossy fiber axons (Umemori et al. 2004). This action depends on expression of the FGF2 receptor by mossy fiber axons and FGF22 secreted from granule cell neurons. FGF7 and -10 are expressed by other subpopulations of neurons and have similar synaptogenic properties (Umemori et al. 2004). Neurotrophins can also promote neuronal maturation, including regional axon and dendrite arborization. BDNF in particular has been shown to directly regulate the density of synaptic innervation (Alsina et al. 2001), and thus may be considered a synaptogenic "priming molecule."

In addition to neuronally derived synaptogenic factors, several studies indicate that glial-derived factors may also regulate the timing of synapse formation (Ullian et al. 2004). These studies have noted that synaptogenesis in the central nervous system coincides with the birth of astrocytes and that astrocytes or astrocyte-conditioned media dramatically enhance synapse formation in certain populations of neurons (Nagler et al. 2001; Pfrieger & Barres 1996, 1997; Ullian et al. 2001, 2004). Two glial-derived factors shown to promote synapse formation include cholesterol bound to apolipoprotein E (Mauch et al. 2001) and thrombospondin-1 (TSP1) (Ullian et al. 2004). Presumably, these factors act indirectly to facilitate the maturation of both target neurons and incoming axons.

As a final note, it is important to make a distinction between these target-derived synaptogenic molecules and the inductive molecules discussed later. In particular, the target-derived molecules appear to act diffusely from local sources, like the netrins and semaphorins. Further, when added to purified populations of cultured neurons, they cause a global increase in the number of synapses, even between neurons that normally do not form synapses among themselves. These features suggest that these molecules are not acting focally to induce synapse formation, but rather serve to promote neuronal maturation and make neurons competent to proceed with synaptogenesis. Understanding whether they act to change gene transcription and/or protein synthesis would be an excellent way to confirm a role for these proteins in synaptogenic priming.

Cell-Adhesion Molecules Guiding Synapse Specificity

Several classes of CAMs have also been implicated in target recognition and the initial formation of synapses. Candidates include members of the cadherin family of calcium-dependent cell-adhesion molecules, including cadherins and protocadherins (Shapiro & Colman 1999, Takai et al. 2003). With regard to the classical cadherins, there are ~20 members expressed in the CNS (Yagi & Takeichi 2000). Not only are these homotypic CAMs localized to synapses at early stages of synapse formation (Fannon & Colman 1996, Shapiro & Colman 1999, Uchida et al. 1996, Yamagata et al. 1995), but also they exhibit distinct yet complementary expression patterns with respect to subgroups of neurons and their targets. For example, cadherin-6 is expressed in functionally connected groups of neurons involved in audition (Bekirov et al. 2002). Similarly, barrel field pyramidal cells and septal granule cells in the somatosensory cortex, together with their corresponding thalamic inputs, express N-cadherin and cadherin-8, respectively (Gil et al. 2002). As such, the classical cadherins are ideally suited to help guide subclasses of axons to their targets. With regard to synapse formation, individual cadherins not only localize to the pre- and postsynaptic plasma membranes in a variety of synaptic types (Benson et al. 2001, Takai et al. 2003), but also are found at initial axo-dendritic contact sites (Benson & Tanaka 1998). However, cellular expression and reverse genetic studies indicate that classical cadherins are not directly involved in triggering synapse formation. For example, introduction of N-cadherin blocking antibodies in the developing chick optic tectum causes retinal ganglion cell axons to overshoot their targets and to form exuberant synapses but does not inhibit synapse formation per se (Inoue & Sanes 1997). Similarly, axons originating from photoreceptor cells in the *Drosophila* ommaditium are mistargeted when they lack N-cadherin, but synapse formation itself is not disrupted (Lee et al. 2001). Thus, these data support a role for cadherins in target specification and perhaps stabilization of early synaptic contact sites but not in the induction of synapse formation.

A second class of CAMs that may be involved in target recognition and synapse specificity is the protocadherins. These molecules are encoded by a huge family of genes that undergo alternative splicing and have region-specific expression patterns in the developing brain (Hirano et al. 2002; Kohmura et al. 1998; Phillips et al. 2003; Wang et al. 2002a,b; Wu & Maniatis 1999). As such, they are theoretically capable of providing the spatial specificity required for target recognition. Similar to classical cadherins, protocadherin-gammas partially localize to synaptic sites (Phillips et al. 2003), and genetic studies in *Drosophila* indicate that they are involved in target recognition rather than synapse formation (Lee et al. 2003). Studies of protocadherin gamma knockout mice support this conclusion and indicate that these CAMs are not essential for neuronal differentiation or synapse formation but rather for neuronal survival (Wang et al. 2002b).

Inducers of Synapse Formation

In contrast to the cadherins and protocadherins, several classes of molecules are capable of directly inducing various aspects of synapse formation. These include Narp and Ephrin B1, two secreted proteins capable of clustering subsets of postsynaptic proteins, and SynCAM and Neuroligin, two CAMs that can trigger the formation of functional presynaptic boutons (Biederer et al. 2002, Dalva et al. 2000, O'Brien et al. 1999, Scheiffele et al. 2000).

One of the first molecules demonstrated to have synaptogenic activity was Neuronal activity–regulated pentraxin (Narp) (O'Brien et al. 1999). Narp was first identified as an activity-induced transcript in the hippocampus (Tsui et al. 1996). It is a member of the pentraxin family of secreted proteins that not only localizes to synapses, but also binds to the extracellular domains of subunits of the α-amino-3-hydroxy-5-methylisoxazole-4-propionic acid (AMPA)-type glutamate receptor (O'Brien et al. 1999). Intriguingly, when overexpressed in spinal cord neurons, Narp increases the synaptic clustering of AMPA receptors (O'Brien et al. 1999). This activity is blocked by a dominant-negative form of Narp that interferes with its secretion at synapses (O'Brien et al. 2002). Further evidence for Narp's direct role in the synaptic clustering of AMPA receptors came from studies in which Narp was overexpressed in HEK293 cells. This not only induced the clustering of surface AMPA receptors coexpressed with Narp, but also the ectopic clustering of neuronal AMPA receptors on spinal cord neurons cocultured with these Narp-expressing HEK293 cells (O'Brien et al. 2002). These data clearly indicate that Narp has a potent AMPA receptor–clustering activity. Furthermore, Narp promotes the clustering of NMDA as well as AMPA

Hippocampus: a region of the mammalian brain implicated in learning and memory and containing many glutamatergic synapses

FM1-43: a dye that is taken up by recycling synaptic vesicles and used as a marker of functional presynaptic active zones

receptors in certain classes of interneurons (Mi et al. 2002). However, Narp's clustering activity is primarily restricted to glutamatergic synapses forming on inhibitory interneurons and does not influence synapse assembly on pyramidal neurons (Mi et al. 2002).

A second molecule with synaptogenic activity, EphrinB, is a member of the Ephrin family of axonal growth cone guidance molecules. EphrinB family members promote the clustering of subunits of the N-methyl-D-aspartate (NMDA) type of glutamate receptor (Dalva et al. 2000). This activity was initially identified when investigators discovered that the extracellular domain of the EphrinB receptor, EphB, interacts directly with the extracellular domain of the NMDA receptor subunit NR1 (Dalva et al. 2000). Furthermore, EphrinB-mediated aggregation of EphB receptors leads to coaggregation of NMDA receptors. Independent evidence also strongly implicates ephrins and Eph receptors in dendritic spine development (Murai et al. 2003). Activation of postsynaptic EphBs with clustered EphrinB1 promotes spine maturation (Penzes et al. 2003), and triple knockout of EphB1, -2, and -3 results in mice that are viable but have defects in hippocampal spine morphology (Henkemeyer et al. 2003). Interestingly, EphrinB-mediated aggregation of EphB receptors does not induce coaggregation of other postsynaptic components such as PSD-95 family proteins, scaffolding proteins that normally aggregate at synapses prior to or coincident with NMDA receptors. Thus, as with Narp, EphrinB-EphBs act as specialized regulators of certain aspects of postsynaptic differentiation, e.g., the clustering of NMDA receptors and spine morphology. This limited data suggests that the induction of postsynaptic differentiation may involve multiple signaling pathways acting combinatorially. Such a possibility would not be unexpected given the greater molecular heterogeneity of postsynaptic specializations compared with presynaptic specializations (Craig & Boudin 2001).

In contrast to the restricted activities of Narp and Ephrin, molecules that in-

duce presynaptic differentiation cause a more complete functional differentiation of the presynaptic active zone. These synaptogenic molecules fall into two classes: secreted proteins, such as Wnts and FGFs (discussed above), and cell-surface adhesion proteins, such as SynCAM and neuroligin (see below). Whereas the secreted molecules function at a distance, perhaps as synaptogenic "priming molecules," the two CAMs, SynCAM and neuroligin, directly induce presynaptic differentiation through axo-dendritic contact.

The first to be identified was neuroligin, a postsynaptic single-pass transmembrane protein capable of inducing presynaptic differentiation when expressed in HEK293 cells (Scheiffele et al. 2000). The presynaptic receptor for neuroligin is a second single-pass transmembrane protein called β-neurexin (Dean et al. 2003, Scheiffele et al. 2000). The interacting domains of β-neurexin and neuroligin are laminin-G (LG) and acetylcholine esterase (AChE)-like domains, respectively (Dean et al. 2003, Scheiffele et al. 2000). The LG motif was initially characterized in laminin and agrin (Rudenko et al. 1999), molecules important for the differentiation of the neuromuscular junction (Sanes & Lichtman 1999). Remarkably, the binding affinity and biological activity of both agrin and neurexins is regulated by alternative splicing at specific conserved sites within the LG domain (Rudenko et al. 1999). Although structurally similar to AChE, the AChE domain in neuroligin is not catalytically active (Dean et al. 2003). Neuroligins cluster β-neurexin, and this clustering is associated with the formation of functional active zones as assessed by the ability of synaptic vesicles at these sites to recycle the styryl dye FM1-43 (Dean et al. 2003).

A second cell-surface adhesion molecule capable of inducing presynaptic differentiation is a protein called SynCAM (synaptic cell-adhesion molecule). SynCAM is a member of the Ig superfamily of adhesion molecules. It is a homophilic CAM expressed on both sides of the synapse (Biederer et al. 2002, Scheiffele

2003). SynCAM1 overexpression in cultured neurons promoted synapse formation, whereas overexpression in nonneuronal cells rapidly induced the formation of functional presynaptic active zones in axons contacting these cells (Biederer et al. 2002). Furthermore, neurons expressing a dominant-interfering construct against SynCAMs and neurexins exhibit compromised presynaptic differentiation (Biederer et al. 2002). These studies suggest that SynCAM1 is a potent inducer of presynaptic differentiation. Intriguingly, there are three other genes encoding SynCAMs, and heterophilic adhesion between the various SynCAM isoforms can occur (Shingai et al. 2003). However, it is unclear whether this diversity also contributes to synaptic specification.

A very recent study (Graf et al. 2004) reported reciprocal signaling by one of these pairs of inducers. Whereas neuroligin induces presynaptic differentiation in contacting axons, its binding partner β-neurexin induces postsynaptic differentiation in contacting dendrites. β-Neurexin induced local clustering of PSD-95 and NMDA receptors, but not AMPA receptors, by binding to dendritic neuroligins (Graf et al. 2004). Furthermore, β-neurexin induces local dendritic clustering of GABA as well as glutamate postsynaptic proteins, apparently via different neuroligins. This finding highlights the question of how matching of appropriate pre- and postsynaptic specializations is achieved. It remains to be determined whether SynCAMs or Ephrin/EphB also signal bidirectionally.

Obviously, SynCAM, neuroligin, and neurexin are exciting molecules because of their potent hemi-synapse-inducing activities. However, many details about how they function at a molecular level remain to be explored. For example, are they solely involved in triggering pre-/postsynaptic differentiation, or are they also involved in synapse specification via selective cell adhesion? How are their signals transduced within neurons? More globally, how many of these priming, adhesive, and inducing molecules function together to build a given synapse, and how many are essential? Demonstration of the inducing activity of many of these molecules has been performed in cultured neurons. Rigorous analysis of combinatorial conditional knockout mice is warranted to determine the precise in vivo roles of each of these synaptogenic proteins.

Thus, an emerging picture of the specification and induction of synaptogenesis demonstrates that these complex processes involve many molecules and signaling pathways (**Figure 2**). For example, a series of hierarchical interactions between sets of secreted and cell-surface adhesion molecules seems to be required. Some of these, such as Wnts, FGFs, and TSP1, are target- or glia-derived and facilitate neuronal maturation, allowing neurons to undergo synaptogenesis in the correct spatial-temporal window. Others, such as cadherins and protocadherins, may serve to specify appropriate axodendritic connections and stabilize sites of early contact. A third class, including SynCAM, neuroligin/neurexin, Narp, and EphrinB/EphB, seem to trigger various aspects of synapse formation locally. Mechanistically, it is unclear whether pre- and postsynaptic partners engage in bidirectional signaling at the time of contact or in the continuous exchange of factors/signals back and forth between the nascent pre- and postsynaptic compartments. Still less is known about the second messenger signaling pathways that participate in this early inductive stage of synaptogenesis. Much of what we do know has come from genetic screens in worms and flies (for a review see Jin 2002).

CELLULAR MECHANISMS OF SYNAPSE ASSEMBLY

Subsequent to induction and prior to the appearance of a fully functional synapse is the molecular assembly of the synaptic junction and the delivery of pre- and postsynaptic components. How are the inductive signals discussed above translated into the

PSD-95: a well-characterized modular protein of the postsynaptic density involved in receptor anchoring and clustering site-specific recruitment of pre- and postsynaptic molecules, and what cellular mechanisms are responsible for their correct targeting? Of particular interest is whether synapse assembly involves the sequential recruitment of individual synaptic proteins or whether sets of proteins are delivered as preassembled complexes.

Membrane Trafficking in Presynaptic Assembly

Although local recruitment of individual molecules surely contributes to presynaptic assembly, studies coming from a number of laboratories strongly suggest that the vesicular delivery of proteins plays a critical role. For example, during the assembly of presynaptic boutons, clusters of pleiomorphic vesicles are observed at newly forming synapses (Ahmari et al. 2000). These include small, clear-centered vesicles, tubulovesicular structures, and 80-nm dense core vesicles. The exact composition and characteristics of these morphologically distinct vesicle types remain to be investigated, but vesicles within these clusters seem to be somewhat specified. For example, the small clear-centered vesicles appear to be synaptic vesicle precursors, carrying primarily synaptic vesicle proteins (Hannah et al. 1999, Huttner et al. 1995, Matteoli et al. 1992). The 80-nm dense core vesicles contain numerous multidomain scaffold proteins of the active zone, such as Piccolo, Bassoon, and Rab3 interacting molecule (RIM) (Lee et al. 2003, Ohtsuka et al. 2002, Shapira et al. 2003) as well as components of the synaptic vesicle exocytotic machinery including syntaxin, SNAP25, and N-type voltage-gated calcium channels (Shapira et al. 2003, Zhai et al. 2001). Finally, the tubulovesicular structures may represent either post-Golgi membranes of mixed protein composition and/or endosomal intermediates. The former presumably carry newly synthesized proteins and the latter recycled membrane proteins.

The presence of different vesicle types at nascent synapses suggests that active zone formation is driven by the vesicular delivery of proteins. An exciting but unresolved set of questions relates to the hierarchy of protein delivery. Assuming that proteins such as neuroligin and SynCAM are important inducers of presynaptic active zone assembly, it is reasonable to conclude that their delivery to the plasma membrane and clustering at nascent axodendritic contact sites is one of the earliest events. At present, it is unclear whether these molecules are already in the plasma membrane and are simply induced to cluster at contact sites through lateral movement, or whether the contact site itself triggers the directed delivery and fusion of neurexin and SynCAM-containing vesicles.

On the basis of temporal studies, the fusion of 80-nm dense core vesicles carrying structural components of the active zone such as Piccolo, Bassoon, and RIM is likely to occur shortly thereafter (Ziv & Garner 2004). Presumably, delivery of these proteins allows for the rapid establishment of functional synaptic vesicle docking and fusion sites and provides a platform for the subsequent delivery of additional presynaptic proteins and synaptic vesicle precursors (Ziv & Garner 2004). The biogenesis and clustering of mature synaptic vesicles is likely to occur after these initial events. Although this model of active zone formation is consistent with most available data, it is nonetheless quite speculative and cannot account for all observations. For example, recent data on the dynamics of synaptic vesicle recycling demonstrate the existence of functional active zones lacking postsynaptic partners (Takao-Rikitsu et al. 2004). At present, it is unclear whether these so-called "orphan active zones," which retain the ability to recycle synaptic vesicles in an activity-dependent manner, represent an early event in nascent active zone formation, are an artifact of low-density dissociated hippocampal cultures or are remnants of synapses undergoing elimination.

Membrane Trafficking in Postsynaptic Assembly

In contrast to the presynaptic active zone, where delivery of integer numbers of transport vesicles provides sufficient material for synapse formation (Shapira et al. 2003), assembly of the postsynaptic density appears to occur primarily by gradual accumulation of molecules (Bresler et al. 2004, Ziv & Garner 2004) (**Figure 3**). One of the earliest events in postsynaptic differentiation is the recruitment of scaffolding proteins of the PSD-95 family. These molecules are present at synapses in postnatal day 2 hippocampus (Sans et al. 2000) and detectable within 20 min of axodendritic contact in culture (Bresler et al. 2001, Friedman et al. 2000, Okabe et al. 2001). Although some investigators have reported modular transport of recombinant PSD-95 clusters during synaptogenesis (Prange & Murphy 2001), other studies have observed more gradual, nonquantal accumulation of PSD-95 at nascent synapses (Bresler et al. 2001, Marrs et al. 2001). Gradual accumulation of PSD-95 could occur by local trapping of diffuse plasma membrane pools or by sequential local fusion of numerous vesicles, each carrying only small numbers of PSD-95.

Closely following the synaptic recruitment of PSD-95 is recruitment of NMDA-type and AMPA-type glutamate receptors, which are independently regulated. As described above for PSD-95, modular transport of NMDA receptor clusters during synaptogenesis has been reported (Washbourne et al. 2002), but other studies have observed more gradual accumulation at nascent synapses (Bresler et al. 2004). Nonsynaptic clusters of endogenous PSD-95 and NMDA receptors are present early in development (Rao et al. 1998) but are rare and unlikely to represent prefabricated postsynaptic elements. For AMPA receptors, evidence exists to support both local insertion of receptor-containing vesicles near the synapse and insertion over the bulk of the dendritic plasma membrane, followed by diffusion and trapping at the synapse (Borgdorff & Choquet 2002, Passafaro et al. 2001). Interestingly, the mechanisms may differ depending on the AMPA receptor subunit composition, GluR2 being more locally inserted than GluR1 (Passafaro et al. 2001). In the past few years, much effort has focused on how synaptic delivery of AMPA and NMDA receptors is controlled by interacting proteins (Bredt & Nicoll 2003, LeVay et al. 1980, Malinow & Malenka 2002, Wenthold et al. 2003). Both receptor classes interact with different sets of PDZ domain–containing proteins, chaperones, endocytic adaptors, and cytoskeletal proteins. These interactions can regulate exit from the endoplasmic reticulum, transport to the synapse, local receptor stabilization, endocytosis, recycling, and degradation. An interesting example is the linkage to molecular motors. Whereas NR2B-containing NMDA receptors link to the kinesin family motor KIF17 (Setou et al. 2000), GluR2/3-containing AMPA receptors link to the kinesin family motors KIF5 and KIF1A through a different scaffolding complex (Setou et al. 2002, Shin et al. 2003). These NMDA and AMPA receptor complexes lack PSD-95, further suggesting independent synaptic recruitment of PSD-95, NMDA receptors, and AMPA receptors. But there is not universal agreement, as other researchers suggest that NMDA receptors may traffic with PSD-95 family proteins to the synapse (Wenthold et al. 2003).

Accumulation of other postsynaptic components also occurs by individual recruitment. For example, the most intensively studied such component, calcium calmodulin-dependent protein kinase II (CaMKII), accumulates on the cytoplasmic face of the postsynaptic density (Petersen et al. 2003), apparently by regulated trapping of local pools (Shen & Meyer 1999) rather than by active vesicular transport. Regulated synaptic accumulation of the scaffolding proteins Homer 1c and Shank2/3 also occurs by gradual accumulation from a cytosolic pool (Bresler et al. 2004, Okabe et al. 2001). Local protein

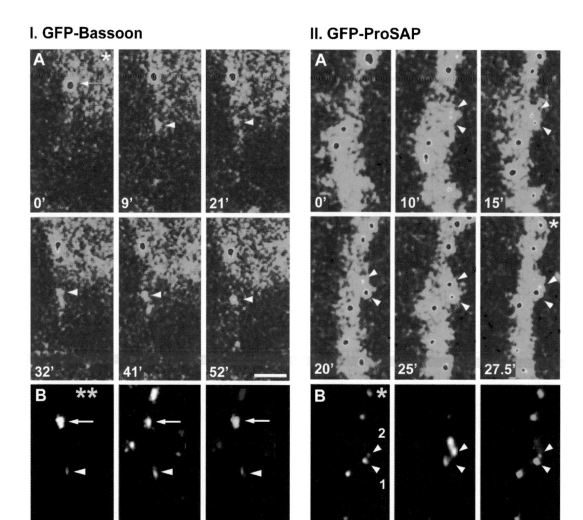

Figure 3

Different mechanisms of protein recruitment at nascent pre- versus postsynaptic sites. Many presynaptic proteins, such as Bassoon (*pictured here in I*), are delivered to nascent active zones via integer numbers of transport vesicles. In contrast, postsynaptic proteins, such as ProSAP1 (*pictured here in II*), appear to be gradually recruited to nascent postsynaptic densities from cytosolic pools. (*IA*) Time-lapse sequence of an axon expressing GFP-Bassoon. At the beginning of the time lapse, a large and static GFP-Bassoon cluster is seen (*arrow*). A new GFP-Bassoon cluster then appears (*arrowhead*) and stabilizes over the course of the next 52 min. (*IB*) FM4-64 labeling at the end of this experiment (*middle panel*) shows that the new GFP-Bassoon cluster (*left panel*) colocalizes with recycling synaptic vesicles (*right panel*), indicating that it is a functional presynaptic active zone. Quantitative measurements of the fluorescence changes at this site indicate that it contains 2–3 unitary amounts of GFP-Bassoon. All times are given in minutes. Scale bar, 5 μm. (*IIA*) Time-lapse sequence of a dendrite expressing GFP-ProSAP1, a PSD protein. The formation of two new GFP-ProSAP1 clusters (*arrowheads*) over the course of 27.5 min is shown. (*IIB*) FM4-64 loading performed at the end of the experiment (*middle panel*) revealed that the new GFP-ProSAP1 clusters (*left panel*) colocalize with functional presynaptic active zones (*right panel*), which suggests that new synapses had formed at these sites. Note the gradual increase in fluorescence intensity of these two clusters. All times are given in minutes. Adapted from figures 6 and 11 in Bresler et al. 2004.

synthesis may also contribute to synapse assembly, particularly in cases where the mRNA is abundant in dendrites, such as for CaMKIIα, Shank, NR1, and GluR1/2 (Bockers et al. 2004, Ju et al. 2004, Steward & Schuman 2001). However, this possibility has not been well explored.

Considering the extensive heterogeneity in postsynaptic composition, even among glutamatergic synapses, and the ability of activity and kinases to separately regulate the density of different components such as PSD-95, NMDA receptors, AMPA receptors, and CaMKII, one might expect the delivery of each of these components to occur independently. Thus, the existence of preassembled mobile postsynaptic transport packets ready to mediate fusion of numerous postsynaptic proteins, as with presynaptic packets, remains an open possibility but one with little experimental support (**Figure 3**) (Bresler et al. 2004). Further biochemical characterization of postsynaptic transport intermediates, high sensitivity ultrastructural localization, and simultaneous time-lapse imaging of multiple tagged postsynaptic proteins may help resolve these issues.

Synaptic Maturation

A general feature of synaptic development is a prolonged maturation phase. During this phase, synapses expand in size; for example, the number of synaptic vesicles per terminal increases two- to threefold over the first month of cortical development (Vaughn 1989). Remarkably, pre- and postsynaptic elements develop in a coordinated fashion, maintaining a correlation among the size of different components: bouton volume, number of total synaptic vesicles, docked vesicles, active zone area, postsynaptic density area, and spine head volume (Harris & Stevens 1989, Pierce & Mendell 1993, Schikorski & Stevens 1997). This finding of correlated size suggests that the cell-adhesion complexes that span the cleft, and perhaps associated secreted factors, signal to each partner to define the area of associated cytomatrix and synapse volume.

Perhaps the most dramatic maturational change in synapses is the change in postsynaptic form. Glutamatergic synapses initially form on filopodia or dendrite shafts that develop over time into dendritic spines. Both are motile actin-based structures (Fischer et al. 1998), filopodia are typically >2 μm long, of thin diameter, and have a half-life of several minutes, whereas spines are typically <2 μm long, have a bulbous head, and have a half-life of days or more (Grutzendler et al. 2002, Sorra & Harris 2000). Mature spines can be of mushroom, branched, thin, or stubby morphology. Live cell-imaging studies suggest that most synapses form on dendritic filopodia and then transform directly to more stable spines (Okabe et al. 2001, Ziv & Smith 1996). Conversion of shaft synapses into spine synapses has also been inferred on the basis of the percentage of synapses of different morphological classes in developing hippocampus (Fiala et al. 1998). Dendritic spine morphogenesis is regulated by numerous mechanisms including cell adhesion via the cadherin, syndecan, and ephrin systems. These in turn signal through proteins that regulate Rho and Ras family GTPases, actin-binding proteins, and calcium regulatory mechanisms (Hering & Sheng 2001, Tashiro & Yuste 2003).

The functional properties of synapses also change with development. As hippocampal synapses mature, the probability of transmitter release decreases (Bolshakov & Siegelbaum 1995, Chavis & Westbrook 2001), and the reserve pool of vesicles increases. Quantal size shows an initial rapid increase and then remains constant (De Simoni et al. 2003, Liu & Tsien 1995, Mohrmann et al. 2003). The kinetics of synaptic responses change concomitantly with developmental changes in receptor subunit composition. For example, expression of the NR2A subunit of the NMDA receptor and its selective incorporation into synaptic NMDA receptors, partially replacing NR2B subunits, mediate a decrease in hippocampal and neocortical

Filopodia: dynamic protrusions found on axons and dendrites, particularly at growth cones

NMDA current duration (Sorra & Harris 2000, Tovar & Westbrook 1999). Many developing brain regions exhibit "silent synapses" characterized by functional NMDA but not AMPA currents (Durand et al. 1996, Isaac et al. 1997). Such silent synapses lack surface AMPA receptors, and great variability in AMPA receptor content of hippocampal synapses is indeed observed (Matsuzaki et al. 2001, Nusser et al. 1998, Takumi et al. 1999). NMDA receptor activation can unsilence these synapses, presumably by recruitment of AMPA receptors to the postsynaptic plasma membrane. Furthermore, receptor content and spine morphology are correlated with larger spines selectively bearing more AMPA receptors. However, other explanations for these silent synapses have been proposed. Alterations in modes of vesicle fusion may activate NMDA receptors while failing to activate AMPA receptors (Choi et al. 2000, Renger et al. 2001). Furthermore, functional AMPA synapses develop normally in genetically targeted neurons lacking functional NMDA receptors (Cottrell et al. 2000, Li et al. 1994). Thus, although not an obligatory step in development, most synapses develop functional NMDA receptors prior to AMPA receptors, and NMDA receptor activity may instruct insertion of AMPA receptors into the plasma membrane.

ACTIVITY-DEPENDENT REGULATION OF SYNAPTOGENESIS

The bulk of synaptogenesis occurs during early postnatal development, but synapses can also form in the mature brain. At both stages, activity sculpts neuronal arbor growth and synapse formation (Knott et al. 2002, Maletic-Savatic et al. 1999, Rajan & Cline 1998, Schmidt et al. 2004, Trachtenberg et al. 2002, Wong & Wong 2001, Hua & Smith 2004). However, several studies have demonstrated that neuronal activity is not required for synapse formation during development (Varoqueaux et al. 2002, Verhage et al. 2000). For example, in hippocampal cultures, synaptogenesis occurs normally in the presence of glutamate receptor blockers that prevent action potentials (Rao & Craig 1997). Moreover, Munc-18 and Munc-13 knockout mice lacking neurotransmitter release exhibit normal synapse morphology and density, although synapses appear less stable over time (Varoqueaux et al. 2002, Verhage et al. 2000). These findings suggest a more nuanced role for activity in synaptogenesis, presumably one involving the regulation of synapse stability and elimination rather than formation per se. Mechanisms by which activity regulates synapse stability are explored in the following sections.

Synapse Elimination

Although synapse formation is the focus of this review, synapse elimination is arguably an equally important developmental process. For example, the initial number of synapses formed in the brain is far greater than the number retained, which suggests that synapse elimination is a crucial step in normal brain development (Hashimoto & Kano 2003, Huttenlocher et al. 1982, Lichtman & Colman 2000, Rakic et al. 1986). Indeed, activity-dependent pruning of synapses appears to underlie many critical aspects of nervous system development, including formation of ocular dominance columns in the visual cortex (LeVay et al. 1980, Shatz & Stryker 1978) and innervation of muscles by neurons originating in the spinal cord (Lichtman & Colman 2000).

In the mature brain, synapse elimination is probably also an important mechanism for removing inappropriate or ineffective connections to fine-tune networks. Evidence that activity regulates synapse elimination, as well as synapse formation, between mature neurons has come from two recent studies of rodent barrel cortex, an area of somatosensory cortex that receives sensory input from the whiskers (Knott et al. 2002, Trachtenberg et al. 2002). In one study, Trachtenberg et al.

(2002) performed in vivo imaging of dendritic arbors and observed high rates of turnover of dendritic protrusions (17% persisting for one day, 23% for only 2–3 days). The authors correlated the appearance of new protrusions with the formation of new synapses and showed that activity affected their stability, as sensory deprivation by whisker trimming increased the ratio of transient-to-stable protrusions. Knott et al. (2002) stimulated a single rodent whisker in a freely moving animal for 24 h then performed serial section electron microsopy to examine synapse density in the region of barrel cortex receiving input from the whisker. Amazingly, after this brief stimulation period, the authors observed a 35% increase in synapse density and a 25% increase in spine density. This increase was largely reversible as excitatory synapse density returned to prestimulation levels several days after stimulation ceased. These studies demonstrate that synapses between mature neurons can be formed and eliminated rapidly and that activity, in the form of whisker stimulation or trimming, regulates these processes.

Ubiquitin Regulation of Synapse Stability

As demonstrated by the studies described above, activity seems to play a fundamental role in synapse formation, stability, and elimination. By which molecular mechanisms does activity regulate these processes? As discussed above, activation of NMDA receptors induces a number of changes, including insertion of AMPA receptors and changes in dendritic spine morphology, that lead to synapse stability and maturation. Activity may also regulate the stability of synaptic proteins via ubiquitination. A recent and intriguing study by Ehlers (2003) demonstrated that large groups of postsynaptic proteins are up- or down regulated together in response to sustained changes in neuronal network activity. Ehlers showed that ubiquitin-mediated protein degradation was responsible for this activity-dependent protein turnover and that only a few postsynaptic density proteins are ubiquitinated, including ProSAP/Shank, GKAP, and AKAP79/150. Thus, activity-dependent ubiquitination of these few proteins could lead to the rapid destabilization and degradation of large postsynaptic protein complexes by the proteosome, providing neurons with a mechanism for regulating synapse stability via activity. Other studies also indicate that ubiquitin-mediated protein degradation is an important mechanism for controlling synapse stability and function. For example, Burbea et al. (2002) showed that expression of the *C. elegans* elegans postsynaptic glutamate receptor GLR-1 is regulated by ubiquitination (Burbea et al. 2002). Mutations that prevent ubiquitination increase GLR-1 abundance at synapses and, concomitantly, synaptic strength, whereas overexpression of ubiquitin decreases GLR-1 levels and the density of GLR-1-containing synapses. In the vertebrate hippocampus, activity-dependent internalization of homologous AMPA receptors was also shown to be regulated by ubiquitination (Colledge et al. 2003). Taken together, these studies indicate that ubiquitination of postsynaptic glutamate receptors can regulate synaptic strength and stability.

Ubiquitin-dependent proteolysis of synaptic proteins also regulates presynaptic function. At the Drosophila neuromuscular junction, application of proteosome inhibitors induced a rapid strengthening of synaptic transmission owing to a 50% increase in the number of synaptic vesicles released (Aravamudan & Broadie 2003, Speese et al. 2003). Along with increased vesicle release, these authors observed a doubling of Dunc-13 (the *Drosophila* homolog of Munc-13), a protein that regulates synaptic vesicle priming/release and is ubiquitinated in vivo (Aravamudan & Broadie 2003). This study suggests that Dunc-13/Munc-13 may be a substrate for the regulation of neurotransmitter release by ubiquitin-mediated protein degradation.

Another example of ubiquitin-mediated regulation of presynaptic assembly and

Ubiquitination: a chemical modification to proteins that targets them to the proteosome for degradation

AChE: acetylcholine esterase

AKAP79/150: A-kinase anchoring protein

AMPA receptor: α-amino-3-hydroxy-5-methylisoxazole-4-propionic acid type glutamate receptor

AZ: active zone

CAMs: cell-adhesion molecules

CaMKII: calcium calmodulin-dependent-protein kinase II

EphB: EphrinB receptor

FGF: fibroblast growth factor

GFP: green fluorescent protein

GKAP: guanylate kinase domain–associated protein

GLR-1: C-elegans postsynaptic glutamate receptor

GluR2, GluR1: subunits of the AMPA receptor

KIF5, KIF17: kinesin super-family

LG: laminin-G

Munc-18: mouse homolog of the *C. elegans* uncoordinated mutant unc-18

Munc-13, Dunc-13: mouse and Drosophila homologs of the *C. elegans* uncoordinated mutant unc-13

NL: neuroligin

Narp: neuronal activity-regulated pentraxin

NMDA: N-methyl-D-aspartate type of glutamate receptor

NR1, NR2A, NR2B: subunits of the NMDA receptor

Nxn: β-neurexin

PDZ: PSD-95/DLG/ZO1 homology domain

ProSAP1: proline rich synapse–associated protein 1

PSD: postsynaptic density

PSD-95: postsynaptic density protein-95

RIM: Rab3 interacting molecule

SP: dendritic spine

SynCAM: synaptic cell-adhesion molecule

SVs: synaptic vesicles

TSP: thrombospondin

function involves the *C. elegans* gene *rpm-1*, the *Drosophila* homolog highwire and the mouse homolog Phr1. In *rpm-1* loss-of-function mutants, synapses either fail to form or are highly disorganized, often lacking active zones and SV clusters (Schaefer et al. 2000, Zhen et al. 2000). *Drosophila highwire* mutants have excessive numbers of synapses that are smaller than normal, whereas mice lacking Phr 1 also exhibit abnormal synaptic morphology (Burgess et al. 2004, Wan et al. 2000). On the basis of the deduced amino acid sequence of rpm-1/highwire/Phr 1, these genes appear to be involved in ubiquitin-mediated protein degradation because they have domains exhibiting sequence homology with E3 ubiquitin ligases (Jin 2002, Wan et al. 2000, Zhen et al. 2000). These data suggest that ubiquitination in general and these proteins in particular have an important role in regulating presynaptic active zone size and organization. Perplexingly, rpm-1 is a periactive zonal protein (Wan et al. 2000, Zhen et al. 2000), which raises some questions about how proteins outside the active zone can influence its formation. One possibility is that these molecules regulate proteins of the periactive zonal plasma membrane that normally serve to delineate active zone size. Thus, a growing body of data indicate that ubiquitination of both pre- and postsynaptic proteins, perhaps in an activity-dependent manner, is an important mechanism for regulating both synapse formation and stability.

CONCLUDING REMARKS

In recent years, considerable progress has been made in understanding the cellular and molecular mechanisms of vertebrate synaptogenesis. By all accounts, it is a complex process that begins before axonal projections reach their targets and involves a series of hierarchical signals. These signals include secreted factors and cell-adhesion molecules that both guide axons to their correct targets and regulate their maturation, insuring that synaptogenesis occurs only once the proper neurons are encountered. These initial priming signals are thought to work in synchrony with subsequent contact-initiated inductive signals to stabilize nascent axodendritic contacts and trigger the assembly of synapses. Assembly of pre- and postsynaptic compartments then occurs through a combination of vesicle trafficking and local recruitment of synaptic proteins. Finally, correlated or uncorrelated synaptic activity leads to synaptic strengthening and stabilization, or weakening and elimination. Although recent studies have begun to unravel the intricacies of synaptogenesis, many important questions remain. For example, what makes some contacts productive for synapse formation and others not? Is there a minimal contact time and/or set of local molecular players needed? What is the distribution of the inductive factors prior to synaptogenesis? How are the inductive signals initiated by axon-dendrite contact translated into the complex, and distinct, processes of pre- and postsynaptic assembly? Specifically, which second-messenger signaling pathways translate the interactions of cell-surface CAMs into the vesicular delivery of proteins on either side of the synapse? What is the temporal sequence of pre- and postsynaptic protein recruitment during synaptogenesis? What are the morphological characteristics and precise molecular compositions of the vesicles that deliver proteins to nascent synapses, and which mechanisms regulate their timely delivery? What are the signals that mediate prolonged maturational changes and synapse-specific features? How stable are synapses during periods of peak synaptogenesis versus in the mature brain? Are the same cellular and molecular mechanisms responsible for synapse formation during development also responsible for this process in the mature brain? By which molecular mechanisms does neuronal activity regulate synapse stability and elimination? These questions are likely to emerge as quintessential issues for developmental neuroscientists in the coming years.

ACKNOWLEDGMENTS

We thank Dr. Martin Meyer for critical reading of the manuscript and Drs. P. Zamorano and N. Ziv for their assistance in designing the figures. We also acknowledge the support of the National Institutes of Health and BSF Grant No. 2000232.

LITERATURE CITED

Ahmari SE, Buchanan J, Smith SJ. 2000. Assembly of presynaptic active zones from cytoplasmic transport packets. *Nat. Neurosci.* 3:445–51

Alsina B, Vu T, Cohen-Cory S. 2001. Visualizing synapse formation in arborizing optic axons in vivo: dynamics and modulation by BDNF. *Nat. Neurosci.* 4:1093–101

Aravamudan B, Broadie K. 2003. Synaptic Drosophila UNC-13 is regulated by antagonistic G-protein pathways via a proteasome-dependent degradation mechanism. *J. Neurobiol.* 54:417–38

Bagri A, Tessier-Lavigne M. 2002. Neuropilins as Semaphorin receptors: in vivo functions in neuronal cell migration and axon guidance. *Adv. Exp. Med. Biol.* 515:13–31

Bekirov IH, Needleman LA, Zhang W, Benson DL. 2002. Identification and localization of multiple classic cadherins in developing rat limbic system. *Neuroscience* 115:213–27

Benson DL, Colman DR, Huntley GW. 2001. Molecules, maps and synapse specificity. *Nat. Rev. Neurosci.* 2:899–909

Benson DL, Tanaka H. 1998. N-cadherin redistribution during synaptogenesis in hippocampal neurons. *J. Neurosci.* 18:6892–904

Biederer T, Sara Y, Mozhayeva M, Atasoy D, Liu X, et al. 2002. SynCAM, a synaptic adhesion molecule that drives synapse assembly. *Science* 297:1525–31

Bockers TM, Segger-Junius M, Iglauer P, Bockmann J, Gundelfinger ED, et al. 2004. Differential expression and dendritic transcript localization of Shank family members: identification of a dendritic targeting element in the 3′ untranslated region of Shank1 mRNA. *Mol. Cell Neurosci.* 26:182–90

Bolshakov VY, Siegelbaum SA. 1995. Regulation of hippocampal transmitter release during development and long-term potentiation. *Science* 269:1730–34

Borgdorff AJ, Choquet D. 2002. Regulation of AMPA receptor lateral movements. *Nature* 417:649–53

Bredt DS, Nicoll RA. 2003. AMPA receptor trafficking at excitatory synapses. *Neuron* 40:361–79

Bresler T, Ramati Y, Zamorano PL, Zhai R, Garner CC, Ziv NE. 2001. The dynamics of SAP90/PSD-95 recruitment to new synaptic junctions. *Mol. Cell Neurosci.* 18:149–67

Bresler T, Shapira M, Boeckers T, Dresbach T, Futter M, et al. 2004. Postsynaptic density assembly is fundamentally different from presynaptic active zone assembly. *J. Neurosci.* 24:1507–20

Burbea M, Dreier L, Dittman JS, Grunwald ME, Kaplan JM. 2002. Ubiquitin and AP180 regulate the abundance of GLR-1 glutamate receptors at postsynaptic elements in C. elegans. *Neuron* 35:107–20

Burden SJ. 2002. Building the vertebrate neuromuscular synapse. *J. Neurobiol.* 53:501–11

Burgess RW, Peterson KA, Johnson MJ, Roix JJ, Welsh IC, O'Brien TP. 2004. Evidence for a conserved function in synapse formation reveals Phr1 as a candidate gene for respiratory failure in newborn mice. *Mol. Cell Biol.* 24:1096–105

Burns ME, Augustine GJ. 1995. Synaptic structure and function: dynamic organization yields architectural precision. *Cell* 83:187–94

Chavis P, Westbrook G. 2001. Integrins mediate functional pre- and postsynaptic maturation at a hippocampal synapse. *Nature* 411:317–21

Choi S, Klingauf J, Tsien RW. 2000. Postfusional regulation of cleft glutamate concentration during LTP at 'silent synapses'. *Nat. Neurosci.* 3:330–36

Colledge M, Snyder EM, Crozier RA, Soderling JA, Jin Y, et al. 2003. Ubiquitination regulates PSD-95 degradation and AMPA receptor surface expression. *Neuron* 40:595–607

Cottrell JR, Dube GR, Egles C, Liu G. 2000. Distribution, density, and clustering of functional glutamate receptors before and after synaptogenesis in hippocampal neurons. *J. Neurophysiol.* 84:1573–87

Craig AM, Boudin H. 2001. Molecular heterogeneity of central synapses: afferent and target regulation. *Nat. Neurosci.* 4:569–78

Dalva MB, Takasu MA, Lin MZ, Shamah SM, Hu L, et al. 2000. EphB receptors interact with NMDA receptors and regulate excitatory synapse formation. *Cell* 103:945–56

Dean C, Scholl FG, Choih J, DeMaria S, Berger J, et al. 2003. Neurexin mediates the assembly of presynaptic terminals. *Nat. Neurosci.* 6:708–16

De Simoni A, Griesinger CB, Edwards FA. 2003. Development of rat CA1 neurones in acute versus organotypic slices: role of experience in synaptic morphology and activity. *J. Physiol.* 550:135–47

Durand GM, Kovalchuk Y, Konnerth A. 1996. Long-term potentiation and functional synapse induction in developing hippocampus. *Nature* 381:71–75

Ehlers MD. 2003. Activity level controls postsynaptic composition and signaling via the ubiquitin-proteasome system. *Nat. Neurosci.* 6:231–42

Fannon AM, Colman DR. 1996. A model for central synaptic junctional complex formation based on the differential adhesive specificities of the cadherins. *Neuron* 17:423–34

Fiala JC, Feinberg M, Popov V, Harris KM. 1998. Synaptogenesis via dendritic filopodia in developing hippocampal area CA1. *J. Neurosci.* 18:8900–11

Fischer M, Kaech S, Knutti D, Matus A. 1998. Rapid actin-based plasticity in dendritic spines. *Neuron* 20:847–54

Fletcher TL, De Camilli P, Banker G. 1994. Synaptogenesis in hippocampal cultures: evidence indicating that axons and dendrites become competent to form synapses at different stages of neuronal development. *J. Neurosci.* 14:6695–706

Friedman HV, Bresler T, Garner CC, Ziv NE. 2000. Assembly of new individual excitatory synapses: time course and temporal order of synaptic molecule recruitment. *Neuron* 27:57–69

Garner CC, Zhai RG, Gundelfinger ED, Ziv NE. 2002. Molecular mechanisms of CNS synaptogenesis. *Trends Neurosci.* 25:243–51

Gil OD, Needleman L, Huntley GW. 2002. Developmental patterns of cadherin expression and localization in relation to compartmentalized thalamocortical terminations in rat barrel cortex. *J. Comp. Neurol.* 453:372–88

Graf ER, Zhang X, Jin SX, Linhoff MW, Craig AM. 2004. Neurexins induce differentiation of GABA and glutamate postsynaptic specializations via neuroligins. *Cell* 119:1013–26

Gray EG. 1963. Electron microscopy of presynaptic organelles of the spinal cord. *J. Anat.* 97:101–6

Grutzendler J, Kasthuri N, Gan WB. 2002. Long-term dendritic spine stability in the adult cortex. *Nature* 420:812–16

Hall AC, Lucas FR, Salinas PC. 2000. Axonal remodeling and synaptic differentiation in the cerebellum is regulated by WNT-7a signaling. *Cell* 100:525–35

Hannah MJ, Schmidt AA, Huttner WB. 1999. Synaptic vesicle biogenesis. *Annu. Rev. Cell Dev. Biol.* 15:733–98

Harris KM, Stevens JK. 1989. Dendritic spines of CA 1 pyramidal cells in the rat hippocampus: serial electron microscopy with reference to their biophysical characteristics. *J. Neurosci.* 9:2982–97

Hashimoto K, Kano M. 2003. Functional differentiation of multiple climbing fiber inputs during synapse elimination in the developing cerebellum. *Neuron* 38:785–96

Henkemeyer M, Itkis OS, Ngo M, Hickmott PW, Ethell IM. 2003. Multiple EphB receptor tyrosine kinases shape dendritic spines in the hippocampus. *J. Cell Biol.* 163:1313–26

Hering H, Sheng M. 2001. Dendritic spines: structure, dynamics and regulation. *Nat. Rev. Neurosci.* 2:880–88

Hirano S, Wang X, Suzuki ST. 2002. Restricted expression of protocadherin 2A in the developing mouse brain. *Brain Res. Mol. Brain Res.* 98:119–23

Hirokawa N, Sobue K, Kanda K, Harada A, Yorifuji H. 1989. The cytoskeletal architecture of the presynaptic terminal and molecular structure of synapsin 1. *J. Cell Biol.* 108:111–26

Hua JY, Smith SJ. 2004. Neural activity and the dynamics of central nervous system development. *Nat. Neurosci.* 7:327–32

Huttenlocher PR, de Courten C, Garey LJ, Van der Loos H. 1982. Synaptogenesis in human visual cortex—evidence for synapse elimination during normal development. *Neurosci. Lett.* 33:247–52

Huttner WB, Ohashi M, Kehlenbach RH, Barr FA, Bauerfeind R, et al. 1995. Biogenesis of neurosecretory vesicles. *Cold Spring Harb. Symp. Quant. Biol.* 60:315–27

Inoue A, Sanes JR. 1997. Lamina-specific connectivity in the brain: regulation by N-cadherin, neurotrophins, and glycoconjugates. *Science* 276:1428–31

Isaac JT, Crair MC, Nicoll RA, Malenka RC. 1997. Silent synapses during development of thalamocortical inputs. *Neuron* 18:269–80

Jin Y. 2002. Synaptogenesis: insights from worm and fly. *Curr. Opin. Neurobiol.* 12:71–79

Ju W, Morishita W, Tsui J, Gaietta G, Deerinck TJ, et al. 2004. Activity-dependent regulation of dendritic synthesis and trafficking of AMPA receptors. *Nat. Neurosci.* 7:244–53

Knott GW, Quairiaux C, Genoud C, Welker E. 2002. Formation of dendritic spines with GABAergic synapses induced by whisker stimulation in adult mice. *Neuron* 34:265–73

Kohmura N, Senzaki K, Hamada S, Kai N, Yasuda R, et al. 1998. Diversity revealed by a novel family of cadherins expressed in neurons at a synaptic complex. *Neuron* 20:1137–51

Krylova O, Herreros J, Cleverley KE, Ehler E, Henriquez JP, et al. 2002. WNT-3, expressed by motoneurons, regulates terminal arborization of neurotrophin-3-responsive spinal sensory neurons. *Neuron* 35:1043–56

Landis DM. 1988. Membrane and cytoplasmic structure at synaptic junctions in the mammalian central nervous system. *J. Electron Microsc. Tech.* 10:129–51

Lee CH, Herman T, Clandinin TR, Lee R, Zipursky SL. 2001. N-cadherin regulates target specificity in the Drosophila visual system. *Neuron* 30:437–50

Lee RC, Clandinin TR, Lee CH, Chen PL, Meinertzhagen IA, Zipursky SL. 2003. The protocadherin Flamingo is required for axon target selection in the Drosophila visual system. *Nat. Neurosci.* 6:557–63

LeVay S, Wiesel TN, Hubel DH. 1980. The development of ocular dominance columns in normal and visually deprived monkeys. *J. Comp. Neurol.* 191:1–51

Li Y, Erzurumlu RS, Chen C, Jhaveri S, Tonegawa S. 1994. Whisker-related neuronal patterns fail to develop in the trigeminal brainstem nuclei of NMDAR1 knockout mice. *Cell* 76:427–37

Lichtman JW, Colman H. 2000. Synapse elimination and indelible memory. *Neuron* 25:269–78

Liu G, Tsien RW. 1995. Properties of synaptic transmission at single hippocampal synaptic boutons. *Nature* 375:404–8

Lund RD. 1972. Anatomic studies on the superior colliculus. *Invest. Ophthalmol.* 11:434–41

Maletic-Savatic M, Malinow R, Svoboda K. 1999. Rapid dendritic morphogenesis in CA1 hippocampal dendrites induced by synaptic activity. *Science* 283:1923–27

Malinow R, Malenka RC. 2002. AMPA receptor trafficking and synaptic plasticity. *Annu. Rev. Neurosci.* 25:103–26

Marrs GS, Green SH, Dailey ME. 2001. Rapid formation and remodeling of postsynaptic densities in developing dendrites. *Nat. Neurosci.* 4:1006–13

Matsuzaki M, Ellis-Davies GC, Nemoto T, Miyashita Y, Iino M, Kasai H. 2001. Dendritic spine geometry is critical for AMPA receptor expression in hippocampal CA1 pyramidal neurons. *Nat. Neurosci.* 4:1086–92

Matteoli M, Takei K, Perin MS, Sudhof TC, De Camilli P. 1992. Exo-endocytotic recycling of synaptic vesicles in developing processes of cultured hippocampal neurons. *J. Cell Biol.* 117:849–61

Mauch DH, Nagler K, Schumacher S, Goritz C, Muller EC, et al. 2001. CNS synaptogenesis promoted by glia-derived cholesterol. *Science* 294:1354–57

Mi R, Tang X, Sutter R, Xu D, Worley P, O'Brien RJ. 2002. Differing mechanisms for glutamate receptor aggregation on dendritic spines and shafts in cultured hippocampal neurons. *J. Neurosci.* 22:7606–16

Mohrmann R, Lessmann V, Gottmann K. 2003. Developmental maturation of synaptic vesicle cycling as a distinctive feature of central glutamatergic synapses. *Neuroscience* 117:7–18

Murai KK, Nguyen LN, Irie F, Yamaguchi Y, Pasquale EB. 2003. Control of hippocampal dendritic spine morphology through ephrin-A3/EphA4 signaling. *Nat. Neurosci.* 6:153–60

Nagler K, Mauch DH, Pfrieger FW. 2001. Glia-derived signals induce synapse formation in neurones of the rat central nervous system. *J. Physiol.* 533:665–79

Nusser Z, Lujan R, Laube G, Roberts JD, Molnar E, Somogyi P. 1998. Cell type and pathway dependence of synaptic AMPA receptor number and variability in the hippocampus. *Neuron* 21:545–59

O'Brien R, Xu D, Mi R, Tang X, Hopf C, Worley P. 2002. Synaptically targeted narp plays an essential role in the aggregation of AMPA receptors at excitatory synapses in cultured spinal neurons. *J. Neurosci.* 22:4487–98

O'Brien RJ, Xu D, Petralia RS, Steward O, Huganir RL, Worley P. 1999. Synaptic clustering of AMPA receptors by the extracellular immediate-early gene product Narp. *Neuron* 23:309–23

Ohtsuka T, Takao-Rikitsu E, Inoue E, Inoue M, Takeuchi M, et al. 2002. Cast: a novel protein of the cytomatrix at the active zone of synapses that forms a ternary complex with RIM1 and munc13-1. *J. Cell Biol.* 158:577–90

Okabe S, Urushido T, Konno D, Okado H, Sobue K. 2001. Rapid redistribution of the postsynaptic density protein PSD-Zip45 (Homer 1c) and its differential regulation by NMDA receptors and calcium channels. *J. Neurosci.* 21:9561–71

Palay SL. 1956. Synapses in the central nervous system. *J. Biophys. Biochem. Cytol.* 2:193–202

Pascual M, Pozas E, Barallobre MJ, Tessier-Lavigne M, Soriano E. 2004. Coordinated functions of Netrin-1 and Class 3 secreted Semaphorins in the guidance of reciprocal septo-hippocampal connections. *Mol. Cell Neurosci.* 26:24–33

Passafaro M, Piech V, Sheng M. 2001. Subunit-specific temporal and spatial patterns of AMPA receptor exocytosis in hippocampal neurons. *Nat. Neurosci.* 4:917–26

Penzes P, Beeser A, Chernoff J, Schiller MR, Eipper BA, et al. 2003. Rapid induction of dendritic spine morphogenesis by trans-synaptic ephrinB-EphB receptor activation of the Rho-GEF kalirin. *Neuron* 37:263–74

Petersen JD, Chen X, Vinade L, Dosemeci A, Lisman JE, Reese TS. 2003. Distribution of postsynaptic density (PSD)-95 and Ca^{2+}/calmodulin-dependent protein kinase II at the PSD. *J. Neurosci.* 23:11270–78

Pfrieger FW, Barres BA. 1996. New views on synapse-glia interactions. *Curr. Opin. Neurobiol.* 6:615–21

Pfrieger FW, Barres BA. 1997. Synaptic efficacy enhanced by glial cells in vitro. *Science* 277:1684–87

Phillips GR, Huang JK, Wang Y, Tanaka H, Shapiro L, et al. 2001. The presynaptic particle web: ultrastructure, composition, dissolution, and reconstitution. *Neuron* 32:63–77

Phillips GR, Tanaka H, Frank M, Elste A, Fidler L, et al. 2003. Gamma-protocadherins are targeted to subsets of synapses and intracellular organelles in neurons. *J. Neurosci.* 23:5096–104

Pierce JP, Mendell LM. 1993. Quantitative ultrastructure of Ia boutons in the ventral horn: scaling and positional relationships. *J. Neurosci.* 13:4748–63

Prange O, Murphy TH. 2001. Modular transport of postsynaptic density-95 clusters and association with stable spine precursors during early development of cortical neurons. *J. Neurosci.* 21:9325–33

Rajan I, Cline HT. 1998. Glutamate receptor activity is required for normal development of tectal cell dendrites in vivo. *J. Neurosci.* 18:7836–46

Rakic P, Bourgeois JP, Eckenhoff MF, Zecevic N, Goldman-Rakic PS. 1986. Concurrent overproduction of synapses in diverse regions of the primate cerebral cortex. *Science* 232:232–35

Rao A, Craig AM. 1997. Activity regulates the synaptic localization of the NMDA receptor in hippocampal neurons. *Neuron* 19:801–12

Rao A, Kim E, Sheng M, Craig AM. 1998. Heterogeneity in the molecular composition of excitatory postsynaptic sites during development of hippocampal neurons in culture. *J. Neurosci.* 18:1217–29

Renger JJ, Egles C, Liu G. 2001. A developmental switch in neurotransmitter flux enhances synaptic efficacy by affecting AMPA receptor activation. *Neuron* 29:469–84

Rudenko G, Nguyen T, Chelliah Y, Sudhof TC, Deisenhofer J. 1999. The structure of the ligand-binding domain of neurexin Ibeta: regulation of LNS domain function by alternative splicing. *Cell* 99:93–101

Sanes JR, Lichtman JW. 1999. Development of the vertebrate neuromuscular junction. *Annu. Rev. Neurosci.* 22:389–442

Sanes JR, Lichtman JW. 2001. Induction, assembly, maturation and maintenance of a postsynaptic apparatus. *Nat. Rev. Neurosci.* 2:791–805

Sans N, Petralia RS, Wang YX, Blahos J Jr, Hell JW, Wenthold RJ. 2000. A developmental change in NMDA receptor-associated proteins at hippocampal synapses. *J. Neurosci.* 20:1260–71

Schaefer AM, Hadwiger GD, Nonet ML. 2000. rpm-1, a conserved neuronal gene that regulates targeting and synaptogenesis in C. elegans. *Neuron* 26:345–56

Scheiffele P. 2003. Cell-cell signaling during synapse formation in the CNS. *Annu. Rev. Neurosci.* 26:485–508

Scheiffele P, Fan J, Choih J, Fetter R, Serafini T. 2000. Neuroligin expressed in nonneuronal cells triggers presynaptic development in contacting axons. *Cell* 101:657–69

Schikorski T, Stevens CF. 1997. Quantitative ultrastructural analysis of hippocampal excitatory synapses. *J. Neurosci.* 17:5858–67

Schmidt JT, Fleming MR, Leu B. 2004. Presynaptic protein kinase C controls maturation and branch dynamics of developing retinotectal arbors: possible role in activity-driven sharpening. *J. Neurobiol.* 58:328–40

Setou M, Nakagawa T, Seog DH, Hirokawa N. 2000. Kinesin superfamily motor protein KIF17 and mLin-10 in NMDA receptor-containing vesicle transport. *Science* 288:1796–802

Setou M, Seog DH, Tanaka Y, Kanai Y, Takei Y, et al. 2002. Glutamate-receptor-interacting protein GRIP1 directly steers kinesin to dendrites. *Nature* 417:83–87

Shapira M, Zhai RG, Dresbach T, Bresler T, Torres VI, et al. 2003. Unitary assembly of presynaptic active zones from Piccolo-Bassoon transport vesicles. *Neuron* 38:237–52

Shapiro L, Colman DR. 1999. The diversity of cadherins and implications for a synaptic adhesive code in the CNS. *Neuron* 23:427–30

Shatz CJ. 1996. Emergence of order in visual system development. *Proc. Natl. Acad. Sci. USA* 93:602–8

Shatz CJ. 1997. Form from function in visual system development. *Harvey Lect.* 93:17–34

Shatz CJ, Stryker MP. 1978. Ocular dominance in layer IV of the cat's visual cortex and the effects of monocular deprivation. *J. Physiol.* 281:267–83

Shen K, Meyer T. 1999. Dynamic control of CaMKII translocation and localization in hippocampal neurons by NMDA receptor stimulation. *Science* 284:162–66

Sheng M. 2001. Molecular organization of the postsynaptic specialization. *Proc. Natl. Acad. Sci. USA* 98:7058–61

Shin H, Wyszynski M, Huh KH, Valtschanoff JG, Lee JR, et al. 2003. Association of the kinesin motor KIF1A with the multimodular protein liprin-alpha. *J. Biol. Chem.* 278:11393–401

Shingai T, Ikeda W, Kakunaga S, Morimoto K, Takekuni K, et al. 2003. Implications of nectin-like molecule-2/IGSF4/RA175/SgIGSF/TSLC1/SynCAM1 in cell-cell adhesion and transmembrane protein localization in epithelial cells. *J. Biol. Chem.* 278:35421–27

Sorra KE, Harris KM. 2000. Overview on the structure, composition, function, development, and plasticity of hippocampal dendritic spines. *Hippocampus* 10:501–11

Speese SD, Trotta N, Rodesch CK, Aravamudan B, Broadie K. 2003. The ubiquitin proteasome system acutely regulates presynaptic protein turnover and synaptic efficacy. *Curr. Biol.* 13:899–910

Steward O, Schuman EM. 2001. Protein synthesis at synaptic sites on dendrites. *Annu. Rev. Neurosci.* 24:299–325

Takai Y, Shimizu K, Ohtsuka T. 2003. The roles of cadherins and nectins in interneuronal synapse formation. *Curr. Opin. Neurobiol.* 13:520–26

Takao-Rikitsu E, Mochida S, Inoue E, Deguchi-Tawarada M, Inoue M, et al. 2004. Physical and functional interaction of the active zone proeins, CAST, RIM1 and Bassoon, in neurotransmitter release. *J. Cell Biology* 164:301–11

Takumi Y, Ramirez-Leon V, Laake P, Rinvik E, Ottersen OP. 1999. Different modes of expression of AMPA and NMDA receptors in hippocampal synapses. *Nat. Neurosci.* 2:618–24

Tashiro A, Yuste R. 2003. Structure and molecular organization of dendritic spines. *Histol. Histopathol.* 18:617–34

Tessier-Lavigne M. 1995. Eph receptor tyrosine kinases, axon repulsion, and the development of topographic maps. *Cell* 82:345–48

Tovar KR, Westbrook GL. 1999. The incorporation of NMDA receptors with a distinct subunit composition at nascent hippocampal synapses in vitro. *J. Neurosci.* 19:4180–88

Trachtenberg JT, Chen BE, Knott GW, Feng G, Sanes JR, et al. 2002. Long-term in vivo imaging of experience-dependent synaptic plasticity in adult cortex. *Nature* 420:788–94

Tsui CC, Copeland NG, Gilbert DJ, Jenkins NA, Barnes C, Worley PF. 1996. Narp, a novel member of the pentraxin family, promotes neurite outgrowth and is dynamically regulated by neuronal activity. *J. Neurosci.* 16:2463–78

Uchida N, Honjo Y, Johnson KR, Wheelock MJ, Takeichi M. 1996. The catenin/cadherin adhesion system is localized in synaptic junctions bordering transmitter release zones. *J. Cell Biol.* 135:767–79

Ullian EM, Christopherson KS, Barres BA. 2004. Role for glia in synaptogenesis. *Glia* 47:209–16

Ullian EM, Sapperstein SK, Christopherson KS, Barres BA. 2001. Control of synapse number by glia. *Science* 291:657–61

Umemori H, Linhoff MW, Ornitz DM, Sanes JR. 2004. FGF22 and its close relatives are presynaptic organizing molecules in the mammalian brain. *Cell* 118:257–70

Varoqueaux F, Sigler A, Rhee JS, Brose N, Enk C, et al. 2002. Total arrest of spontaneous and evoked synaptic transmission but normal synaptogenesis in the absence of Munc13-mediated vesicle priming. *Proc. Natl. Acad. Sci. USA* 99:9037–42

Vaughn JE. 1989. Fine structure of synaptogenesis in the vertebrate central nervous system. *Synapse* 3:255–85

Verhage M, Maia AS, Plomp JJ, Brussaard AB, Heeroma JH, et al. 2000. Synaptic assembly of the brain in the absence of neurotransmitter secretion. *Science* 287:864–69

Wan HI, DiAntonio A, Fetter RD, Bergstrom K, Strauss R, Goodman CS. 2000. Highwire regulates synaptic growth in Drosophila. *Neuron* 26:313–29

Wang X, Su H, Bradley A. 2002a. Molecular mechanisms governing Pcdh-gamma gene expression: evidence for a multiple promoter and cis-alternative splicing model. *Genes Dev.* 16:1890–905

Wang X, Weiner JA, Levi S, Craig AM, Bradley A, Sanes JR. 2002b. Gamma protocadherins are required for survival of spinal interneurons. *Neuron* 36:843–54

Washbourne P, Bennett JE, McAllister AK. 2002. Rapid recruitment of NMDA receptor transport packets to nascent synapses. *Nat. Neurosci.* 5:751–59

Wenthold RJ, Prybylowski K, Standley S, Sans N, Petralia RS. 2003. Trafficking of NMDA receptors. *Annu. Rev. Pharmacol. Toxicol.* 43:335–58

Wong WT, Wong RO. 2001. Changing specificity of neurotransmitter regulation of rapid dendritic remodeling during synaptogenesis. *Nat. Neurosci.* 4:351–52

Wu Q, Maniatis T. 1999. A striking organization of a large family of human neural cadherin-like cell adhesion genes. *Cell* 97:779–90

Yagi T, Takeichi M. 2000. Cadherin superfamily genes: functions, genomic organization, and neurologic diversity. *Genes Dev.* 14:1169–80

Yamagata M, Herman JP, Sanes JR. 1995. Lamina-specific expression of adhesion molecules in developing chick optic tectum. *J. Neurosci.* 15:4556–71

Zhai RG, Vardinon-Friedman H, Cases-Langhoff C, Becker B, Gundelfinger ED, et al. 2001. Assembling the presynaptic active zone: a characterization of an active one precursor vesicle. *Neuron* 29:131–43

Zhen M, Huang X, Bamber B, Jin Y. 2000. Regulation of presynaptic terminal organization by C. elegans RPM-1, a putative guanine nucleotide exchanger with a RING-H2 finger domain. *Neuron* 26:331–43

Ziv NE, Garner CC. 2004. Cellular and molecular mechanisms of presynaptic assembly. *Nat. Rev. Neurosci.* 5:385–99

Ziv NE, Smith SJ. 1996. Evidence for a role of dendritic filopodia in synaptogenesis and spine formation. *Neuron* 17:91–102

Olfactory Memory Formation in *Drosophila*: From Molecular to Systems Neuroscience

Ronald L. Davis

Department of Molecular and Cellular Biology, Department of Psychiatry and Behavioral Sciences, Baylor College of Medicine, Houston, Texas 77030; email: rdavis@bcm.tmc.edu

Key Words

memory trace, cAMP signaling, learning mutants, olfactory learning

Abstract

The olfactory nervous system of insects and mammals exhibits many similarities, which suggests that the mechanisms for olfactory learning may be shared. Molecular genetic investigations of *Drosophila* learning have uncovered numerous genes whose gene products are essential for olfactory memory formation. Recent studies of the products of these genes have continued to expand the range of molecular processes known to underlie memory formation. Recent research has also broadened the neuroanatomical areas thought to mediate olfactory learning to include the antennal lobes in addition to a previously accepted and central role for the mushroom bodies. The roles for neurons extrinsic to the mushroom body neurons are becoming better defined. Finally, the genes identified to participate in *Drosophila* olfactory learning have conserved roles in mammalian organisms, highlighting the value of *Drosophila* for gene discovery.

Contents

INTRODUCTION 276
THE CONDITIONED STIMULUS (CS) PATHWAY IS COMPOSED OF THE INTRINSIC NEURONS IN THE OLFACTORY NERVOUS SYSTEM 278
SEVERAL TYPES OF EXTRINSIC NEURONS PROVIDE INFORMATION REPRESENTING THE UNCONDITIONED STIMULUS TO THE OLFACTORY NERVOUS SYSTEM 279
   Aversive US of Electric Shock 279
   Appetitive US of Sucrose 281
   DPM Neurons 282
THE ANTENNAL LOBES APPEAR TO BE A SITE FOR THE FORMATION OF SHORT-TERM OLFACTORY MEMORIES 283
MUSHROOM BODIES ARE A MAJOR SITE FOR OLFACTORY MEMORY FORMATION 284
OLFACTORY MEMORY FORMATION MEDIATED BY THE MUSHROOM BODY NEURONS EMPLOYS MANY MOLECULAR PLAYERS 287
   Rutabaga 287
   Other Well-Studied Genes 288
   CREB 288
   Notch 290
   Radish 291
   Atypical PKCs 293
   Crammer/Staufen/Pumilio 294
GENES REQUIRED FOR MEMORY FORMATION IN *DROSOPHILA* HAVE CONSERVED BEHAVIORAL FUNCTIONS IN MAMMALS .. 295

INTRODUCTION

Understanding how the brain and the remainder of the nervous system enable animals to learn and remember remains one of the great mysteries in neuroscience. Learning, which is also referred to as memory formation, is often defined as a change in animal behavior in response to experience. The persistence of this behavioral change over time is memory. Many memories that form can be specific to one sensory system, like visual memories, whereas other memories form in response to multimodal sensory experience. Memories can persist for different lengths of time; some memories are short-lived and some are long-lived. In addition, memory formation is a problem that surfaces at several levels of analysis, from issues dealing with the molecules involved in memory formation to the changes that occur with learning in groups of neurons, or at the systems level of neuroscience. Many of the important problems about memory formation deal with the nature, duration, location, and mechanisms of the changes in the nervous system that underlie behavioral memory. These changes are also referred to as memory traces (Yu et al. 2004).

For most of the 30 years that *Drosophila melanogaster* has been used to study olfactory memory formation, investigators have regarded it as an animal model providing relatively rapid, facile, and unbiased access to the genes involved in the process of memory formation (reviewed by Quinn & Greenspan 1984, Dudai 1985, Tully 1987, Heisenberg 1989, Davis 1996, Waddell & Quinn 2001). The molecular genetic prowess of this insect has remained intact, but in the last few years there has been a surprising emphasis on tool development and research at the level of systems neuroscience. Because of these advances, the face of this experimental model has evolved from one suited for gene discovery to one offering the opportunity to integrate the molecular genetics of memory formation with the systems neuroscience of memory formation, with its emphasis on the pathways of information flow, information processing, and information encoding by groups of neurons.

There are two primary laboratory assays for olfactory memory in *Drosophila*. The first, which is an operant olfactory learning assay, was developed by Quinn and colleagues in 1974 (Quinn et al. 1974). In this assay, the flies learn to avoid running into a tube containing an odor after a training session that pairs the odor with electric shock as a negative reinforcer. Flies are allowed to run alternatively into one of two Plexiglas tubes that contain different odorants and receive electric shock only in one of the two tubes and in the presence of only one of the odors. After training, the relative avoidance of the shocked odor versus the control odor is quantitated by allowing the flies the opportunity to run into fresh tubes containing the same odors, but without shock. The selective avoidance behavior is measured and used as a learning, or performance, index.

The second assay, which measures olfactory classical conditioning, was developed by Jellies (1981) and improved upon by Tully & Quinn (1985). In this assay, a group of flies is sequestered in a plastic tube, and two odors are presented to the flies in succession, by pulling odor vapor through the tube with a vacuum. The flies receive multiple electric shock pulses (the unconditioned stimulus, US), in association with one of the odors (the conditioned stimulus, CS+), through an electrified copper grid that comprises the wall of the tube. They are then presented with a counter odor without electric shock (the CS−). After training, the animals are forced to run toward one of the two converging odors presented to them in a T-maze, and their selective avoidance of the shock-associated odor is again calculated into a performance index. Several variations of this assay have been developed, including variations in which the animals are presented with a very short odor presentation along with one shock pulse (Beck et al. 2000). This produces very modest performance gains after one training trial, but multiple training trials performed in succession can then be used to construct acquisition curves and to normalize the initial performance levels between control and mutant animals (Beck et al. 2000, Cheng et al. 2001). Alternatively, multiple training trials, each using multiple shock pulses, can be presented in succession in either a massed or a spaced configuration to produce long-term memory lasting several days (Jellies 1981, Tully et al. 1994). Another recently introduced variation of this assay is the replacement of the electric shock US with an appetitive US: sucrose for a hungry fly (Schwaerzel et al. 2003). Training animals with a sucrose US is performed in a manner similar to the negatively reinforced situation, except that the animals are allowed to feed on a sucrose-saturated filter paper for 30 s in the presence of the CS+ and then are transferred to a tube containing a water-saturated filter paper in the presence of the CS− odor.

Molecular genetic approaches to olfactory memory formation in *Drosophila* and the concepts that have emerged over the past 30 years have been reviewed recently in this series (Waddell & Quinn 2001). A listing of the genes discovered to function in olfactory memory formation is provided in **Table 1** for background, and an abbreviated discussion of some early gene discoveries is included within the text; the emphasis here, however, is on the more recent discoveries and the results that have begun to merge molecular genetic and systems neuroscience approaches. In addition, a critical appraisal of much of the literature is offered to help identify the solid and weak tenets in the field. The focus is also on olfactory operant and classical conditioning. Other types of conditioning that are multimodal but depend partly on olfactory cues such as courtship conditioning (Joiner & Griffith 2000, Siwicki & Ladewski 2003) and olfactory nonassociative learning (Devaud et al. 2003, Cho et al. 2004) are not reviewed. One interesting tangent of the molecular genetic dissection of olfactory memory formation in *Drosophila* is also discussed. This is the surprising functional conservation of the

Conditioned stimulus: any stimulus that becomes predictive of the unconditioned stimulus after behavioral conditioning

Unconditioned stimulus: a stimulus that gives rise to an unconditioned response

TABLE 1 Genes and gene products required for *Drosophila* olfactory memory formation

Mutant or gene	Gene product	Expression pattern
dunce (dnc)	cAMP phosphodiesterase	preferential in mushroom bodies
rutabaga (rut)	calcium/calmodulin-activated adenylyl cyclase	preferential in mushroom bodies
DC0	catalytic subunit of PKA	preferential in mushroom bodies
PKA-RI	regulatory subunit of PKA	preferential in mushroom bodies
fasciclinII (fasII)	fasciclin II	preferential in mushroom bodies
Volado (Vol)	α-integrin	preferential in mushroom bodies
leonardo (leo)	14-3-3	preferential in mushroom bodies
crammer (cre)	cysteine protease inhibitor	preferential in mushroom bodies
pumilio (pum)	ribonucleoprotein translocation/translational repression	preferential in mushroom bodies
oskar (osk)	ribonucleoprotein translocation	preferential in mushroom bodies
eIF-5C	translation initiation factor	preferential in mushroom bodies
staufen (stau)	ribonucleoprotein translocation	—
amnesiac (amn)	neuropeptide(s) related to PACAP	preferential in DPM neurons
radish (rsh)	—	—
CREB	cAMP-response element binding protein	—
neurofibromin (NF1)	neurofibromin	—
aPKC	atypical PKC	—
nebula (nla)	inhibitor of calcineurin	—
Notch (N)	Notch	—

genes involved in *Drosophila* olfactory learning in mammalian behavior.

THE CONDITIONED STIMULUS (CS) PATHWAY IS COMPOSED OF THE INTRINSIC NEURONS IN THE OLFACTORY NERVOUS SYSTEM

The neural system of interest for thinking about and studying olfactory memory is, for the most part, the olfactory nervous system (reviewed by Davis 2004). In *Drosophila*, odors are sensed by ∼60 olfactory receptor proteins, one of which is expressed in each of the ∼1300 olfactory receptor neurons (ORNs) that reside in the sensory bristles on the antennae and maxillary palps on each side of the head (reviewed by Lessing & Carlson 1999, Davis 2004). The ORNs project their axons to the synaptic areas of the antennal lobe, the glomeruli (**Figures 1 & 2**). There are ∼43 glomeruli in each antennal lobe of *Drosophila*, so on average, 30 ORNs project their axons to an individual glomerulus, and ORNs that express the same olfactory receptor protein project to the same glomerulus (Laissue et al. 1999, Gao et al. 2000, Vosshall et al. 2000, Scott et al. 2001). In the antennal lobe, the ORNs form excitatory synapses with the projection neurons (PNs) as well as with local interneurons (LNs). The LNs are primarily GABAergic and ramify extensively within most glomeruli, providing cross-inhibition between glomeruli. Generally, the ∼180 PNs extend dendrites into a single antennal lobe glomerulus and convey olfactory information to the mushroom bodies (MBs) and to the lateral horn (LH) (**Figures 1 & 2**; Jefferis et al. 2001, Marin et al. 2002, Wong et al. 2002). Because there are 43 glomeruli and 180 PNs, each glomerulus is sampled on average by 4–5 PNs. The MBs consist of ∼2000 neurons in each hemisphere and consist of three major cell types: the $\alpha/\beta, \alpha'/\beta'$, and γ MB neurons (Crittenden et al. 1998). The neuropil area in which synaptic contacts are made between the PNs and the MB neurons is called the calyx. The $\alpha/\beta, \alpha'/\beta'$, and γ MB neurons are similar in that each extends a single axon in an anterior direction through a structure called the pedunculus (**Figure 1**), but the γ neurons are different from the α/β and α'/β' neurons by having a single neurite that extends toward the anterior and then turns medially (**Figure 1**). The α/β and α'/β' neurons are branched; one collateral for each type of neuron extends dorsally into neuropil areas denoted as the α and α' lobes, and the other collateral for each type of neuron extends medially into neuropil areas denoted as the β and β' lobes (**Figure 1**). Little is known about the neuronal composition of the LH. Neurons that are postsynaptic to the MBs and the LH also remain poorly characterized (Davis 2004), so the disposition of the olfactory information beyond these brain areas is unclear.

SEVERAL TYPES OF EXTRINSIC NEURONS PROVIDE INFORMATION REPRESENTING THE UNCONDITIONED STIMULUS TO THE OLFACTORY NERVOUS SYSTEM

Aversive US of Electric Shock

For classical conditioning, the pathway of information flow for the CS, the odor, is by definition identical to the olfactory pathway. It is important to define, however, the sites at which US information and neuromodulatory inputs converge on the CS pathway, and how these intersections allow for the formation or modulation of memory formation occurring within the olfactory pathway. For the electric shock US, there is little understanding of the pathways that carry this information, although recent optical imaging results of neural activity in living animals indicate that there is neuronal selectivity for receiving

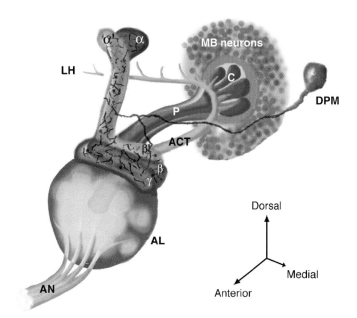

Figure 1

Schematic diagram of the olfactory nervous system in the right hemisphere of the brain of *Drosophila*, as viewed from the anterior, slightly dorsal, and left side of the fly. Olfactory information is passed from olfactory receptor neurons (ORNs) on the antennae via the antennal nerve (AN) to the antennal lobes (AL), which contain ~43 glomeruli (*light green*), each receiving projections from ORNs that express the same olfactory receptor. Odor information is then transmitted via projection neurons from the ALs through the antennal cerebral tract (ACT) to two regions in the dorsal and posterior brain, one containing the mushroom body (MB) neurons. The other is the lateral horn (LH). The PNs synapse on the dendrites of the MB neurons in a neuropil area called the calyx (C). The MB neurons project their axons as a bundle (penduculus, P) to a position just dorsal to the ALs. The axons of the α/β and α'/β' MB neurons split at that location into two collaterals, one which projects dorsally and the other medially. The dorsally projecting processes are termed the α- and α'-collaterals. The medially projecting processes are termed the β- and β'-collaterals. A third class of MB cells, the γ neurons, has only a medially projecting process. These axonal processes reside together in neuropil areas called lobes. A single dorsal paired medial (DPM) neuron resides in each brain hemisphere and extends a process that diffusely innervates the MB lobes. Modified with permission from Davis (2001).

electric shock information (Yu et al. 2004). Using the transgenically supplied optical reporter, synapto-pHluorin, which provides for a fluorescent readout of synaptic activity, Yu and colleagues (2004) expressed the synapto-pHluorin in the three major types of neurons within the antennal lobe and found that PNs, but not ORNs or LNs, release neurotransmitter within the glomeruli in response to electric

Glomeruli: well-defined and distinguishable areas of neuropil that have a high density of synapses

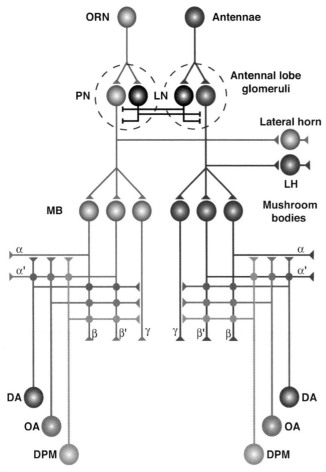

Figure 2

Circuit diagram of the *Drosophila* olfactory nervous system. The antennae contain the olfactory receptor neurons (ORNs), and these synapse on projection neurons (PNs) and local interneurons (LN) of the antennal lobe. The cell bodies of the PNs and LNs are diagrammed for simplicity as being within glomeruli, but in reality, the ORN terminals synapse on PN and LN dendrites within the glomeruli and the cell bodies reside in clusters at the circumference of the antennal lobes. The LNs are generally thought to be inhibitory. A blunt line terminus indicates inhibitory synapses, and a small triangle indicates excitatory synapses. The PNs project to the lateral horn (LH) as well as to the MB neurons. For simplicity, neurons of each glomerulus are depicted as projecting to distinct sets of LH and MB neurons, but PNs are thought to synapse with many different MB neurons. The collaterals of the three types of MB neurons (α/β, α'/β', and γ) are illustrated as receiving inputs from dopaminergic (DA) and octopaminergic (OA) neuromodulatory neurons as well as dorsal paired medial (DPM) neurons. These inputs are likely to be on the MB axons and axon terminals, given that dopamine and octopamine receptors are distributed along these processes (Han et al. 1996, 1998; Kim et al. 2003), but neuromodulatory inputs could also arrive on the MB dendrites (Hammer & Menzel 1998). Modified with permission from Davis (2004).

shock. Thus, PNs receive electric shock US information via some undefined neural pathway, but two other types of neighboring neurons do not. The responses of the PNs to electric shock delivered to the abdomen are fast and robust. Thus the US information being presented to the PNs is likely from presynaptic neurons using a fast, excitatory neurotransmitter rather than from slow, neuromodulatory inputs. There are no other cells in the olfactory nervous system that have been shown, to date, to respond to electric shock pulses to the body, although it is highly likely that other areas such as MBs will also receive shock US information, either indirectly from the PNs or directly from other, presynaptic neurons.

Dopaminergic neurons have long been modeled as potentially mediating or contributing US information (Davis 1993, Han et al. 1996) to the MBs, and recent experimental evidence contributed by Schwaerzel and colleagues (2003) is consistent with the model that dopaminergic neurons may provide part of electric shock US information to the MBs. This intersection is likely at the MB axons or axon terminals because dopaminergic neurons have been shown to innervate broadly areas of the MB lobes (Nassel & Elekes 1992). A transgenic fly line in which the tyrosine hydroxylase promoter was used to drive the expression of yeast GAL4 was used to express *Shibire*[ts] (*Shi*[ts]) in dopaminergic neurons. Tyrosine hydroxylase is the first step in dopamine biosynthesis and therefore is expressed in all dopaminergic neurons. *Shi*[ts] is a dominant-negative and temperature-sensitive variant of dynamin, which is involved in neurotransmitter vesicle recycling. At elevated temperatures, neurotransmitter vesicle recycling is inhibited, producing a block in synaptic transmission (Kitamoto 2001). When tyrosine hydrolyase-GAL4 is used to drive UAS-*Shi*[ts] and the flies trained and tested immediately afterward at elevated temperature, olfactory conditioning using electric shock US is impaired. However, in experiments in which training and testing are separated in time, there is no effect of elevated temperature presented only at the time of retrieval, whereas elevated temperature presented at the time of training severely impairs performance. These data argue, therefore, that dopaminergic synaptic activity is required at the time of conditioning using the electric shock US but not at the time of retrieval. Experiments were not performed to block dopaminergic transmission during times after training but before testing, so it remains unknown whether dopaminergic synaptic transmission has a role only for a very discrete window of time at training or whether dopamine stimulation is also required after training but before testing for memory maintenance. Nevertheless, the available data are consistent with a model that dopamine conveys at least part of the aversive US during electric shock–reinforced, olfactory classical conditioning. However, the data do not discriminate whether dopaminergic synaptic transmission is part of the US signal or whether dopamine is simply a permissive signal for learning to occur.

Appetitive US of Sucrose

Research with the honeybee has provided a reasonable pathway by which a sucrose reward US may intersect the olfactory pathway in *Drosophila*. The ventral unpaired medial (VUM) neurons in the honeybee, which are located in the subesophageal ganglion, are stimulated upon presenting sucrose to the bee's proboscis (Hammer & Menzel 1998). This observation, along with neuron dye-fills, which showed that the VUM neuron extensively innervates multiple areas of the olfactory nervous system, including the antennal lobes, the MBs, and the LH, prompted investigators to test whether electrical stimulation of VUM by itself might be sufficient to replace sucrose application to the proboscis as the US used for conditioning. Honeybees that were presented odor with simultaneous VUM stimulation did indeed exhibit changes in response to conditioning. The muscles that normally drive the proboscis extension response became much more active in response to odors after conditioning compared with before. Because VUM was shown to be octopaminergic, these results indicate that reinforcement using the appetitive US of sucrose is mediated at least in part through the VUM neuron and its intersections with the CS pathway in the antennal lobe, MBs, or LH. An intersection with the calyces of the MBs, as well as the antennal lobes, is suggested by a more recent study showing that injection of octopamine into the calyces or antennal lobes during an odor-conditioning paradigm is sufficient as a US stimulus for proboscis extension (Hammer & Menzel 1998).

Recent studies with *Drosophila* are consistent with the idea that octopaminergic neurons may mediate part of the US signal used for learning associations between odors and appetitive stimuli (Schwaerzel et al. 2003). Mutant flies deficient in tyramine-β-hydroxylase (TβH) are unable to convert tyramine into octopamine. TβH mutant flies show normal olfactory learning when an electric shock US is used, but they are defective when a sucrose US is used. However, the impairment in classical conditioning using sucrose as the US is corrected if the TβH mutant animals are fed on octopamine just prior to training. The fact that this feeding regimen corrects the olfactory learning deficit when feeding occurs just prior to training, but not after training, is consistent with a model that envisions octopaminergic innervation mediating the US used for acquisition, rather than as a reinforcer used for memory stabilization. However, the specific neural pathways that deliver octopamine to the olfactory nervous system in *Drosophila* are unknown, although it is reasonable to expect VUM-like neurons to exist in all insects and to function in similar ways. And as for the involvement of dopamine in electric shock–reinforced olfactory learning, it remains unclear whether octopamine in *Drosophila* is instructive as the US or is simply permissive for olfactory conditioning to occur. In the honeybee, however, the amine appears to be sufficient.

DPM Neurons

The DPM neurons (**Figure 1**) are extremely interesting when considering molecular and systems-level issues of how olfactory memories are formed. Each DPM neuron extends a single axon-like process that broadly ramifies throughout the neuropil areas containing the collaterals of MB cells, and presumably synapses on these neurons, although such synaptic interaction has not yet been demonstrated. The DPM neurons are thought to release acetylcholine as neurotransmitter, evidenced by the activity of the choline acetyltransferase gene promoter in these neurons (Keene et al. 2004). In addition, the DPM neurons strongly express and are believed to release neuropeptide products encoded by the *amnesic* (*amn*) gene (Feany & Quinn 1995, Waddell et al. 2000), neuropeptides that may activate the adenylyl cyclase expressed in the MB neurons and encoded by the *rutabaga* (*rut*) gene (**Table 1**, **Figure 4**). However, no direct evidence shows that the putative neuropeptide products have any activity that regulates MB neurons via the *rut*-encoded adenylyl cyclase, nor is it certain that the *amn* peptides stimulate MB neurons. The DPM neurons may synapse onto other extrinsic neurons in the MB lobes, or the neuropeptides could be released and have an autocrine function on the DPM neurons themselves.

The expression of a wild-type *amn* transgene in these neurons rescues the olfactory memory deficit of the *amn* mutants, and the expression of UAS-*Shibire*^{ts} (UAS-*Shi*^{ts}) at restrictive temperatures in these neurons phenocopies the *amn* mutant (Waddell et al. 2000). The DPM neurons have been conceptualized as providing some of the US input onto MB neurons (**Figure 4**) (Kandel & Abel 1995). If true, this begs the question of where the DPM neurons receive their input. They have no obvious dendritic processes, which indicates that perhaps they receive input onto their cell bodies, that their dendrites are short and fine and therefore difficult to visualize, or that the processes innervating the MB neuropil contain both transmissive and receptive specializations. In this latter scenario, they would be both postsynaptic and presynaptic to the MB neurons. Although a function as a US source is possible, the temporal requirement for DPM synaptic transmission for normal olfactory classical conditioning is limited between 30 and 150 min after training. Keene and colleagues (2004) showed that blocking synaptic transmission from the DPM neurons using UAS-*Shi*^{ts} only at times between 30 and 150 min after training, but not at the time of training or retrieval, impairs memory tested at 3 h, providing evidence for the role of these neurons and their neurotransmitter products in the stabilization of already formed memories, memory consolidation, or for a later temporal phase of memory such as middle-term memory. These observations also indicate that DPM neurons must be spontaneously active during this period or have persistent activity that is induced at the time of conditioning. The temporal requirement for DPM synaptic transmission is not completely consistent with what one would expect from a neuron contributing the US required for acquisition.

The temporal window during which DPM neurons need to be active for the establishment of normal 3-h memories was determined for two odors used for conditioning, methylcyclohexanol and octanol (Keene et al. 2004). However, what is true for some odors does not seem to be true for all. In parallel experiments, the *Shi*^{ts} transgene was used to block DPM synaptic transmission as the animals learned about benzaldehyde. In this case, however, a block in synaptic transmission during conditioning compromised 3-h memory. Why there exists a differential role of DPM synaptic transmission in the formation of memories about different odors remains unclear. However, the reason may be related to the sensory systems that detect the odors. The odors methylcyclohexanol and octanol are detected by the ORNs that reside on the antennae and maxillary palps, as shown by the lack of a behavioral response to these odors when these structures are surgically removed from the flies. In contrast, flies without antennae or palps still respond to benzaldehyde through some other unknown sensory system.

THE ANTENNAL LOBES APPEAR TO BE A SITE FOR THE FORMATION OF SHORT-TERM OLFACTORY MEMORIES

At present, there is no indication that olfactory memory traces are laid down after conditioning in the ORNs. Recent evidence, however, indicates that the initial phases of olfactory memory may be represented by changes in the representation of odors by PNs in the antennal lobe. Yu and associates (2004) utilized transgenic flies that express the synaptic reporter synapto-pHluorin (Ng et al. 2002) in the PNs and classically conditioned the transgenic animals to specific odors using an electric shock US. They quantified synaptic release from the PNs onto other neurons in antennal lobe glomeruli in response to odors before and after conditioning. This is possible because the PN neurites that extend into the glomeruli have both receptive and transmissive specializations (Ng et al. 2002), allowing synaptic release from PNs to be visualized in the glomeruli. As expected, individual odors in unconditioned flies are represented by the stimulation of different subsets of PNs, which are identified by the glomeruli that they innervate. Surprisingly, however, the conditioned odors after training are represented by the same sets of PNs representing the odor in naïve animals along with additional sets of PNs. In other words, conditioning recruits new synaptic activity of PNs that fail to respond to the CS prior to training, which thus indicates that the representation of the CS changes more qualitatively—additional sets of PNs enter the representation—than quantitatively. Furthermore, the memory trace of recruited PN synapses is short-lived, lasting between 5 and 7 min after conditioning. Because flies conditioned and tested behaviorally can remember for days, this memory trace detected by optical imaging can potentially be responsible for conditioned behavior for only the first few minutes after conditioning.

There remain many unknowns surrounding this relatively new observation. Although training recruits the activity of additional PN synapses into the odor CS representation and the activity of these PN synapses is undetectable in the naïve state, the recruited PN synapses may be active prior to conditioning, but this activity is below the sensitivity for detection by optical imaging. Despite this possibility, the recruitment as assayed by optical imaging does appear to occur in a rather large, stepwise fashion, rather than through graded changes in the levels of PN activity. In addition, the memory traces revealed by these studies is at present a behavioral correlate. No direct evidence has confirmed that the trace guides behavior. Another interesting issue concerns the cell biology of this phenomenon. Is the recruitment of PN synapses in the antennal lobe synapse specific, i.e., occurring only on the PN synapses in this region of the olfactory nervous system, or are the synapses that the recruited PNs make in the MBs and lateral horn also recruited? If the memory trace lasts only minutes after training and yet behavioral memories last days, might the memory trace of PN recruitment be transferred to the MBs to be used as substrate to build a more enduring memory trace that could account for the time course of behavioral memory? Finally, the biochemical mechanism by which the presynaptic terminals of PNs become activated within 3 min after training remains completely unexplored.

Despite these unknowns, these recent observations correlate beautifully with now-classic behavioral studies of honeybees performed by Erber and Menzel more than 20 years ago (Erber et al. 1980). These investigators used a miniature cooling probe to inactivate regions of the honeybee brain after olfactory induced proboscis extension learning, and then they tested the memory at 15 min after conditioning. Partial amnesia is produced if the antennal lobe is cooled within a 3-min window after training but not thereafter. These experiments were perhaps the first to indicate that short-lived, memory traces are laid down in the antennal lobe after olfactory classical conditioning.

MUSHROOM BODIES ARE A MAJOR SITE FOR OLFACTORY MEMORY FORMATION

The accumulated evidence supporting a dominant role for the MB neurons in olfactory learning is compelling, and recent studies have continued to support this role. Dubnau and colleagues (2001) and McGuire and colleagues (2001) recently employed the Shi^{ts} transgene to block conditionally synaptic transmission from MB neurons to determine the importance of MB neuron synaptic transmission for memory formation, consolidation, and retrieval. When flies expressing Shi^{ts} in MB neurons are trained and tested at restrictive temperatures, olfactory memory formation is blocked (Dubnau et al. 2001, McGuire et al. 2001). If the block is applied only during training but not testing, no impairment is observed (**Figure 3**), nor is there any impairment when the block is imposed between training and testing. However, when the block is imposed during testing, but not during training or during an interval between training and testing, performance is blocked.

These studies inform us about the role of MBs in olfactory learning in several ways. One conclusion is that synaptic transmission from the MB neurons onto follower neurons is required at the time of testing for normal olfactory memory performance to be observed, which suggests that olfactory memories are formed upstream of the output synapses of MB neurons onto their follower neurons and probably within MB neurons themselves. Second, because olfactory memories are formed normally when synaptic transmission from the MB neurons is blocked during training, synaptic transmission from the MBs at that time is not an essential component of memory formation. Related to this is the conclusion that there is no requirement for follower neurons to provide feedback onto the MB neurons for normal memory formation. The idea that memories form via a feedback loop was a formal possibility prior to these experiments. Finally, in some experiments (McGuire et al. 2001), Shi^{ts} was expressed only in the α/β MB neurons, and this specific expression mimicked the retrieval impairment observed upon expressing Shi^{ts} globally in the MBs. This observation suggests that retrieval of olfactory memories for the odors used for conditioning occurs exclusively through the α/β neurons. It remains possible that olfactory memories are formed in other types of MB neurons and are transferred to the α/β neurons prior to retrieval. The conclusions to these experiments, however, assume that the dominant-negative dynamin has a specific effect on synaptic transmission via the MB neuron output synapses. However, dynamin participates in many aspects of cellular endocytosis (Praefcke & McMahon 2004). So it remains to be proven that the behavioral effects observed are due to blocking the MB output synapses.

Overall, however, these data support the accumulated data already reviewed extensively (Davis 1993, 1996; Roman & Davis 2001; Waddell & Quinn 2001; Dubnau et al. 2003a; Heisenberg 2003), which support the postulate that MB neurons are principle sites for olfactory memory formation. Despite this well-accepted idea, what the MBs contribute specifically to overall behavioral memory formation remains obscure.

Because there exists a short-lived olfactory memory trace in the PNs of the antennal lobe and because localized cooling of the antennal lobes after training honeybees has a defined window of efficacy, an attractive model would show that the MBs form memory traces with a delayed onset relative to the PNs and guide behavior from perhaps several minutes after

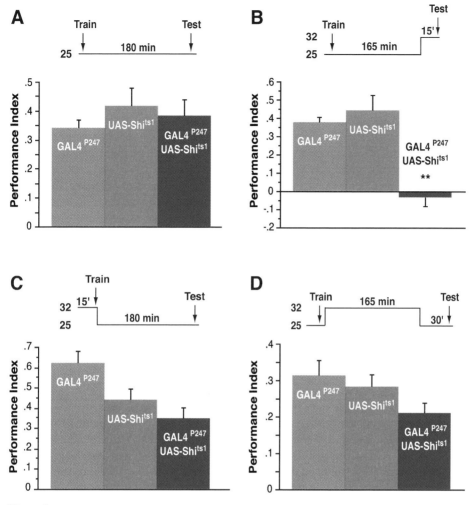

Figure 3

Effects of blocking synaptic transmission from MB neurons with Shi^{ts} on olfactory memory retrieval. The four panels illustrate the 3-h olfactory memory performance scores in flies carrying a GAL4 transposon with high expression specificity to the MBs (GAL4^{P247}), UAS-Shi^{ts}, and both elements (GAL4^{P247}; UAS-Shi^{ts}). (*A*) There is no difference among genotypes after training and testing at 25°. (*B*) Training at 25° and then testing at 32° completely blocks retrieval in the GAL4^{P247}; UAS-Shi^{ts} flies owing to the blockade of synaptic transmission from the MBs. (*C*) There is no significant difference among the genotypes when trained at 32° and then tested at 25°. (*D*) There is no significant difference among the genotypes if the temperature is elevated after training but before testing. These and other similar results have shown that synaptic transmission from the MB neurons is required for retrieval but not for the acquisition or storage of olfactory memories. Adapted from McGuire (2003).

training to hours after training. But where are the long-term memory traces formed that last a day or longer? A recent report claimed that long-term memory formation occurs within the α and α′ collaterals of the α/β and α′/β′ MB neurons, respectively (Pascual & Preat 2001), and if true, offers an intriguing cell-biological dimension to *Drosophila* long-term memory formation, with long-term memories forming in some, but not all, collaterals of the same neurons. The *Drosophila* mutant, *alpha-lobes-absent* (*ala*), is a structural mutant of the MBs that has variable expressivity. Approximately 10% of *ala* flies exhibit normal MBs as analyzed by immunostaining or by reporter gene expression. Despite the mutant's name, 36% of the *ala* flies appear to lack the β and β' collaterals, and 5% lack the α and α' collaterals. The remainder of the *ala* mutants has unilateral defects or other combinations of structural defects. One interpretation of these low-resolution anatomical studies is that the α/β and α'/β' neurons are simply missing their vertical (α and α') or horizontal (β and β') collaterals, although it remains possible that the collaterals are simply misrouted to an alternative area of the brain, perhaps the area occupied by the sister collateral. Studies of single MB neurons in the *ala* mutants are required to resolve these alternatives.

Despite these structural defects in the MBs, the mutant has remarkably normal performance when trained and analyzed as a heterogeneous group. Performance at 1 or 3 h after a single cycle of training and performance at 24 h after 10 cycles of massed or spaced training is not significantly different from the wild-type control. However, when the mutant flies were analyzed for structural defects by immunohistochemistry after testing, separating those that made the correct choice from those that made an incorrect choice, a difference was noted. Mutant flies missing the β/β' collaterals perform indistinguishably from the control flies, but mutant flies missing the α/α' collaterals have no memory after 10 cycles of spaced training. However, the flies missing the α/α' collaterals exhibit normal memory from single-cycle training as well as normal 24-h memory from 10 cycles of massed training. These data are consistent with the possibility that long-term memory formed after spaced training forms specifically in the α/α' collaterals of the α/β and α'/β' neurons, or that the formation or retrieval of long-term memory formed from spaced training requires the neural connections between the α/α' collaterals and their follower neurons. In addition, since short-term memory is normal in these mutants, there may be functional redundancy between the horizontal and vertical collaterals for this process.

Although intriguing, these results suffer from the problem of inadequate sampling. The histological efforts for the study were heroic in terms of the number of flies analyzed, but the low expressivity of the α/α'-absent phenotype softens the conclusion. Only 53 flies missing α/α' collaterals entered the final calculation of the performance index after spaced training. This performance index was zero, which equates to ~27 flies having made the correct choice and 26 the incorrect choice. The wild-type controls after 10 cycles of spaced training had a performance index of 38, which if adjusted to 53 flies, would equate to 37 flies making the correct choice and 16 making the incorrect choice. Thus, the overall conclusion rests on the choice decision made by only 10 flies. Clearly, a mutant with complete expressivity of the α/α'-absent phenotype is needed to confirm the results.

Recent experiments examining long-term memory performance after multiple cycles of spaced training of flies expressing Shi^{ts} are consistent with the possibility that at least some portion of long-term memory forms within the α/β MB neurons (Isabel et al. 2004). The 24-h memory performance of flies expressing Shi^{ts} in the α/β MB neurons is reduced to about half of the control values when such flies are subjected to the restrictive temperature only at the time of testing. This indicates that approximately half of the long-term memory retrieval is blocked by the nonpermissive condition. However, because the block of long-term memory was incomplete, the data leave open the possibility that some of the neural signals employed during retrieval leak from the α collateral owing to an incomplete block in synaptic transmission, or that some long-term memory is formed outside the MBs and can be retrieved independently of MB synaptic transmission.

OLFACTORY MEMORY FORMATION MEDIATED BY THE MUSHROOM BODY NEURONS EMPLOYS MANY MOLECULAR PLAYERS

Genes that impair olfactory memory formation when they are disrupted and have contributed to the conceptualization of the physiological basis for olfactory memory formation are listed in **Table 1**. Numerous reviews contain detailed information about the nature of these genes and their gene products (Davis 1993, 1996; Roman & Davis 2001; Waddell & Quinn 2001; Dubnau et al. 2003a) along with the primary literature references.

The products of most of the genes listed in **Table 1** are thought to participate in the process of olfactory memory formation by acting within the MB neurons, although much of the evidence for selecting the MBs as the site of action is indirect. The major reason for focusing on the MBs is because of the elevated expression of some of these genes within these neurons. For instance, *dunce* (*dnc*), *rutabaga* (*rut*), *DC0*, *PKA-RI*, *leonardo* (*leo*), *Volado* (*Vol*), *fasciclinII* (*fasII*) and *pumilio* (*pum*) have all been directly shown to have preferential expression within these brain neurons. Several other genes have preferential expression in the MBs as inferred from reporter gene expression (*cre*, *osk*, *eIF-5C*). Preferential expression in the MBs alone is a weak argument for site of action of any individual gene product, but the coincidental preferential expression of many different relevant proteins in these neurons has provided a strong basis for this postulate.

Rutabaga

However, for one gene—*rutabaga* (*rut*)—definitive evidence for a function in the MBs has been obtained. Investigators first showed that the expression of a wild-type transgene for *rut* in the MBs of *rut* mutants using the GAL4/UAS system is sufficient to rescue the early- and later-memory impairment of the *rut* mutants (Zars et al. 2000, Schwaerzel et al. 2002). Although this was important, the more critical issue regarding the adenylyl cyclase is whether the enzyme is required at the time of memory formation in adult flies. This issue was critical because the *rut*-type adenylyl cyclases have well-documented roles in brain development (McGuire et al. 2003), and although the GAL4/UAS system offers control over transgene expression in space, it lacks experimenter-controlled expression in time. Thus, the GAL4/UAS rescue experiments failed to delineate whether the behavioral rescue was due to providing wild-type *rut* activity during development, fulfilling a need for the adenylyl cyclase in MB development and subsequently normal adult behavior, or due to a need for wild-type *rut* activity at the time of forming memories in adulthood.

This issue has now been solved by the development of expression systems that offer control over transgene expression in both time and space (McGuire et al. 2004). McGuire and colleagues (2003) rediscovered a forgotten temperature-sensitive repressor of yeast GAL4, called GAL80^{ts}, and fashioned *Drosophila* transgenes expressing GAL80^{ts} to repress GAL4 activity in the MB neurons during development and to derepress this activity with heat pulses during adulthood. This manipulation allowed for the controlled expression of UAS-*rut*$^+$ in a *rut* mutant genetic background. This treatment rescued the learning impairment of *rut* mutants, showing that expression of wild-type *rut* activity in the MB neurons at the time that memories are formed is sufficient for normal memory formation. In a second study, similar results were observed using a protein hybrid of GAL4 and the human progesterone receptor, named Gene-Switch. For this chimera, GAL4 activity is controlled in time by administration of the antiprogestin, RU486. *Rut* mutant flies carrying a transgene that provided Gene-Switch activity in the MB neurons along with UAS-*rut*$^+$ were fed on RU486 as adults, which rescued the *rut* learning impairment (Mao et al. 2004). When RU486 was administered only during developmental periods, no rescue of the adult learning impairment was observed. Thus, these studies have demonstrated conclusively that the *rut*-encoded adenylyl cyclase is required for physiological processes within the MB neurons that underlie memory formation.

Other Well-Studied Genes

The current molecular biological model for olfactory memory formation in *Drosophila*, which emphasizes the role for the MBs, is depicted in **Figure 4** (Davis 2004). One theme gathered from molecular genetic research and emphasized in the model is that cAMP signaling is particularly important for olfactory memory formation, evidenced by the fact that many mutants impaired in olfactory memory formation have mutations in genes that encode components of this signaling pathway (**Table 1**, **Figure 4**). The gene *rutabaga* (*rut*), as mentioned previously, is mutated in the gene for a calcium/calmodulin-stimulated adenylyl cyclase. The mutant, *dunce* (*dnc*), is mutated in the gene encoding a cAMP-specific phosphodiesterase. Genetically reducing the expression of subunits for protein kinase A (PKA) or of CREB, a transcription factor stimulated by PKA, impairs short- and long-term olfactory memory, respectively. And mutants of the *amnesiac* gene (*amn*), which encodes a putative neuropeptide similar to pituitary adenylyl cyclase–activating peptide (PACAP), are deficient in olfactory memory as well. Moreover, mutants of neurofibromin (NF1), which in flies has been reported to activate adenylyl cyclases, are impaired in olfactory memory formation. These findings have combined to produce the model for MB neurons as a site of convergence for the CS and US pathways. The US pathway potentially activates the *rut*-encoded adenylyl cyclase through G protein–coupled receptors for dopamine or octopamine, leading to elevations in intracellular cAMP, activation of PKA, and subsequent protein phosphorylations underlying short-term memory and long-term memory, the latter partly by the activation of the transcription factor, CREB. The importance of the control of protein phosphorylation is underscored by the recent finding that mutants in *nebula*, which encodes an inhibitor of the protein phosphatase calcineurin, are impaired in olfactory conditioning (Chang et al. 2003). A second major theme established from molecular genetic research with *Drosophila* is that cell-adhesion receptors are particularly important for memory formation. The learning mutant *Volado* (*Vol*) encodes an α-integrin, and *fasciclin II* (*fasII*) encodes a member of the immunoglobulin superfamily. In addition, recent evidence discussed below has been reported for the possibility that *Notch* cell-adhesion receptors may mediate long-term memory. The roles for these cell-adhesion receptors in olfactory memory formation may be through their adhesive and/or signaling roles at synapses (**Figure 4**).

One other gene that is preferentially expressed in the MBs has been well studied by Philip and colleagues (2001) for its role in olfactory memory formation, but at present it is more difficult to conceptualize the gene product's specific involvement. *Leonardo* (*leo*) encodes a member of the 14-3-3 family, and careful studies have made its involvement clear: Several different alleles of *leo* have clear impairments in olfactory memory formation, and these impairments can be rescued by expressing the wild-type activity in adulthood using the heat-shock promoter, arguing against a brain developmental defect as underlying the impairment in memory formation (Philip et al. 2001). However, 14-3-3 proteins have extremely diverse roles in cellular physiology, and it remains unclear which specific function of *leo* is required for memory formation.

CREB

CREB was originally implicated in long-term memory using reverse genetic approaches. The *Drosophila CREB2* gene is complex, encoding at least seven isoforms produced via alternative splicing. Yin et al. (1995) reported that the dCREB2-a and dCREB2-b

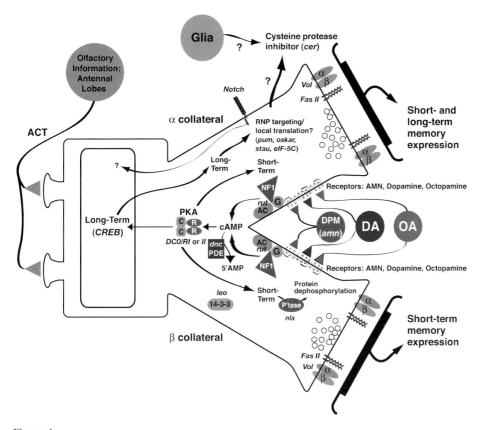

Figure 4

Current molecular model for olfactory memory formation with a focus on the role of the MB neurons. Olfactory information is conveyed by antennal cerebral tract (ACT) (**Figure 1**) to the MB neurons, with unconditioned stimulus (US) information potentially being conveyed, in part, by dopaminergic (DA) and/or octopaminergic (OA) modulatory neurons. The dorsal paired medial (DPM) neurons that express the *amn*-encoded neuropeptides may provide input to the MB neurons for olfactory memory persistence or consolidation. In the model depicted, the DA, OA, and DPM inputs activate the adenylyl cyclase (AC) product of the *rutabaga* (*rut*) gene through G protein (G)–coupled receptors. The product of the *NF1* gene, neurofibromin, is thought to be involved in the activation/maintenance of AC activity. The activation of AC produces elevations in the concentration of intracellular cAMP. The *dunce* (*dnc*)-encoded phosphodiesterase degrades cAMP. It the absence of this enzyme, cAMP is elevated to intolerable levels, which compromises olfactory learning. Cyclic AMP activates the protein kinase A (PKA) tetramer by causing the release of the inhibitory PKA-regulatory (RI or RII) subunits from the catalytic (C) subunits. The *DC0* gene encodes the catalytic subunit of PKA. The activation of PKA leads to either the phosphorylation of a variety of substrates for the establishment of short-term memory or the phosphorylation of CREB for the establishment of long-term memory. The *nebula* (*nla*) gene may be required for normal learning through its control of protein phosphatase activity. The *Volado* (*Vol*)-encoded integrin and *fasII* are cell-adhesion receptors that may mediate signaling or physical alterations of the MB synapses that are important for memory formation. Other genes reported to be involved in olfactory learning potentially by mediating alterations in MB neuron physiology include *leonardo* (*leo*), encoding a 14-3-3 protein, and *Notch*, a cell-adhesion receptor reported to be specifically involved in long-term memory. Recent data have interpreted that long-term memories may form only in the α/α′ collaterals of MB neurons, perhaps in part through the translocation of mRNA in ribonucleoprotein particles (RNP) and the activation of local protein synthesis. The *crammer* gene (*cer*) encodes an inhibitor of cysteine proteases that is required for long-term memory and may be expressed in the MB neurons, in the nearby glia, or in both of these cell types.

isoforms are an activator and blocker of CREB-dependent gene expression, respectively (Yin et al. 1995). Transgenic flies containing a blocking isoform of CREB, dCREB2-b, expressed from a heat-shock promoter, have been generated and used to block dCREB2-activating isoforms in otherwise wild-type flies prior to spaced training experiments, and this produced a specific impairment in long-term memory (Yin et al. 1994). This observation was recently reproduced (Perazzona et al. 2004), confirming that dCREB2-b functions as a suppressor of long-term olfactory memory.

However, overexpression of the dCREB2-a isoform in adult flies using the heat-shock promoter was originally reported to enhance the formation of memory, causing long-term memories to form in flies after a training protocol that produces only short-term memory in wild-type flies. The ability of dCREB2-a to enhance memory formation upon overexpression is now controversial; a recent report failed to reproduce the original behavioral observations using the original transgenic fly lines (Perrazona et al. 2004). Furthermore, the latter study uncovered a nonsense mutation in the dCREB2-a open reading frame of the transgene, which predicts a markedly truncated dCREB2-a gene product from the first ATG. This makes it unlikely that any functional gene product is expressed from the transgene. Correcting this nonsense mutation to the wild-type version also had no effect on long-term memory formation. Indeed, induced overexpression of the full-length dCREB2-a isoform produced lethality. Although the overall model shown in **Figure 4** has not changed with these new observations, they raise serious questions about whether activating forms of dCREB2 can promote memory formation.

Notch

Notch is a well-studied gene in many different organisms and is involved in the specification of cell type during development (Artavanis-Tsakonas et al. 1999). It encodes a single-pass transmembrane domain protein and is the central player in a direct and rapid signaling system from the cell surface to the nucleus. The binding of *Notch* ligands, such as the *Delta* protein, induces the cleavage of a cytoplasmic domain of *Notch*, which then enters the nucleus to regulate the expression of target genes.

The potential role for *Notch* in olfactory learning processes has only recently been examined. Ge and colleagues (2004) reported three different genetic manipulations of *Notch* that alter olfactory memory. First, a temperature-sensitive allele of *Notch*, N^{ts2}, has no effect on early olfactory memory when trained and tested at the nonpermissive temperature, but 24-h memory produced by spaced training is reduced by half. Second, the expression of a dominant-negative transgene at the time of acquisition (under the control of the heat-shock promoter) reduces 24-h memory produced by spaced training to approximately half of that observed in control flies, but expression of the dominant-negative protein has no effect on early memory. The dominant-negative protein is missing the cytoplasmic domain required for nuclear signaling, but it presumably retains the ability to bind ligands. Furthermore, expression of this construct has no effect on 24-h memory produced by massed training. These results therefore suggest that *Notch* functions in long-term memory that is produced by spaced training, which is thought to be a protein synthesis–dependent form. The researchers tested only the effect of inducing the dominant negative just prior to training so that the protein would be present at the time of acquisition. No experiments were performed to determine whether *Notch* is required to maintain memories after formation. Third, the expression of a wild-type transgene of *Notch* at the time of acquisition (under the control of the heat-shock promoter) was reported to facilitate the formation of protein-synthesis dependent, 24-h memory. Training flies with only one cycle of conditioning produces negligible olfactory memory 24-h later, but the overexpression of wild-type *Notch* at acquisition leads to significant levels of memory 1 day later; the enhanced memory is blocked by feeding the flies cycloheximide. Thus, these combined results suggest that *Notch* is required at the time of acquisition for the formation of protein-synthesis-dependent, long-term memory and that its abundance in still-unidentified cells in wild-type animals is rate limiting for the formation of this form of memory.

Independent studies employing a *Notch* RNAi transgene and a GAL4 driver offer a possible focus of *Notch* function in protein synthesis–dependent long-term memory (Presente et al. 2004). Expression of the *Notch* RNAi from a GAL4 driver expressed primarily in MB neurons (GAL4^{c772}) reduces 24-h memory produced by spaced training to about half of the control, with no effect on early memory. However, these results must be regarded as preliminary. The results with only one GAL4 driver were reported; it will be necessary to obtain data with multiple GAL4 drivers expressed in various regions of the brain, and to use multiple GAL4 drivers expressed in the same target region of the brain (i.e., MBs), to rule out nonspecific effects of transgene insertion and expression. A second temperature-sensitive allele of *Notch*, N^{ts1}, was also studied. This allele, like N^{ts2}, is associated with long-term memory impairment at the restrictive temperature.

At present, the strongest conclusion that can be made from these studies is that *Notch* appears to be required for the formation of protein synthesis–dependent long-term memory. Some of the more important remaining questions include: Is *Notch* required only for the formation of protein synthesis–dependent long-term memory, or does it have a role in maintenance? Is *Notch* required in neurons or glia? If in neurons, is the long-term memory role of *Notch* due to a process initiated at the dendrites, the cell bodies, or within the axons of the neurons that require the protein? With which ligand(s) does *Notch* presumably interact? Does the biochemical role for *Notch* in protein synthesis–dependent long-term memory follow its classical mode of action—as a molecule mediating the regulation of gene expression after ligand binding—or does *Notch* have another biochemical role in the adult brain? If it functions in the conventional way, which genes are regulated by *Notch* for long-term memory formation, and how do they confer the state of long-term memory on the neurons?

One of the more curious observations mentioned above is the enhancement of protein synthesis–dependent memory in flies that overexpress *Notch*, when conditioned with only a single cycle. How might we think about this, given the in-depth knowledge of *Notch* function during development? The developmental essence of *Notch* function appears to be in controlling cell fate choices between adjacent cells (Artavanis-Tsakonas et al. 1999). For instance, the specification of the R3 and R4 photoreceptor neurons in the developing eye disc of *Drosophila* occurs through a gradient of an unknown signal, which ultimately causes the neuron closer to the source (R3) to upregulate expression of the ligand *Delta*, which then activates *Notch* in an adjacent precursor, specifying the R4 neuron fate. By analogy, might behavioral conditioning produce a signal that activates *Notch* in perhaps the neurons that represent the conditioned odor, driving the differentiation of the neurons from a state that represents the naïve odor to a state that represents the conditioned odor? This seems plausible and attractive, so long as the newly specified "state" is reversible to allow for forgetting.

Radish

Behavioral studies of rsh^1 flies have revealed several important features (Folkers et al. 1993). First, the mutant impairs memories measured several hours after training more than very early memories. Memory scores measured immediately after olfactory classical conditioning are ~90% of the wild-type scores, but they drop to 0 by 6 h after training.

This mild initial impairment is similar to that produced by the *amn* mutants, although *amn* retains significantly more memory at 6 or more hours after training (Folkers et al. 1993). Second, the mutant is semidominant. Heterozygous $rsh^1/+$ flies exhibit performance scores intermediate between wild-type and rsh^1 homozygous flies (but see Tully et al. 1994, which reports a nonsignificant trend toward semidominance). Third, and most importantly, anesthesia-resistant memory is absent in rsh^1 flies. When wild-type animals are classically conditioned and then anesthetized by cold shock at 2 h after training, memory measured 1 h later (at 3 h after training) is reduced relative to unanesthetized animals (Folkers et al. 1993). This happens because some of the memory at 2 h after training remains susceptible to disruption by anesthesia, whereas some is already consolidated into an anesthesia-resistant form. The rsh^1 mutants have no detectable anesthesia-resistant memory, which suggests that the *rsh* gene product is required for the process of consolidating memories into a form resistant to anesthesia (Folkers et al. 1993, Tully et al. 1994). Clever behavioral experiments performed by Comas and colleagues have suggested that anesthesia-resistant memory may form via *rsh* in a pathway mobilized independently at acquisition of the long-term memory initiated by cAMP signaling because anesthesia-resistant memory forms normally in *rut* mutants (Isabel et al. 2004). Thus, the gene's identification could yield very important insights into how memories are consolidated into a form insensitive to disruption with anesthesia.

Two groups claim to have cloned and identified the *rsh* gene. Unfortunately, the groups have fingered completely different genes as being *rsh*. In one case (Chiang et al. 2004), a GAL4 enhancer detector element named C133 was mapped to a large region of the X-chromosome close to the *rsh* locus (Folkers et al. 1993) and selected as a candidate insertion in *rsh* because of its cytogenetic location and because the element's GAL4 activity drives reporter gene expression in the antennal lobe (Chiang et al. 2004). The authors reported that C133/rsh^1 heterozygotes are impaired in anesthesia-resistant memory, tested by measuring 3-h memory performance after cold anesthesia delivered at 2 h after training. These transheterozygotes have about the same amount of anesthesia-resistant memory as do C133 homozygotes or rsh^1 homozygotes, whereas $rsh^1/+$ and C133/+ have twice these levels of anesthesia-resistant memory as do wild-type flies. Thus, by this genetic criterion, the C133-containing chromosome fails to complement the consolidated memory impairment conferred by the rsh^1 chromosome.

The open reading frame disrupted by the C133 insertion translates conceptually into a protein with significant homology to proteins that have phospholipase A2 (PLA2) activity, which led the authors to conclude that *rsh* encodes a PLA2 enzyme. In addition, the investigators demonstrated that the expression of a UAS-PLA2 transgene driven by the C133-GAL4 element rescues the impairment in performance of the C133 mutants. But the desirable evidence for gene identification that is notably absent in this report includes (*a*) the rescue of the rsh^1 allele with the C133-GAL4/UAS-PLA2 transgene combination; (*b*) the sequence of the *PLA2* gene in the rsh^1 mutant, which presumably might identify the nature of this lesion; and (*c*) olfactory control experiments showing that C133-GAL4 flies are normal for olfaction. The former two lines of evidence are especially critical given that rsh^1 is the prototypic and defining allele of the gene (Folkers et al. 1993, Tully et al. 1994). The latter line of evidence showing normal olfaction in C133 homozygotes and heterozygotes is also critical, given the robust expression of the C133 GAL4 element in the antennal lobe and the semidominance reported for rsh^1. An alternative explanation for the data is that the *PLA2* gene is required for normal olfaction so that C133/+ animals are deficient in the processing of olfactory information. The apparent noncomplementary behavior of the C133 and the rsh^1 chromosomes could be due to inadequate processing of the conditioned stimulus conferred by the C133 chromosome in C133/*rsh¹* flies and the semidominant memory impairment conferred by the *rsh¹* chromosome. A disruption in olfaction in a semidominant memory mutant would produce a more severe, apparent memory deficit and apparent genetic noncomplementation. The reported performance rescue of the C133/*rsh¹* impairment by the PLA2 transgene could be accounted for by a rescue of the deficit in olfaction.

Folkers and colleagues (2004) identified a different gene as *rsh*. The cytogenetic definition of *rsh¹* (Folkers et al. 1993), along with the sequence of the *Drosophila* genome, allowed for a directed search for the *rsh¹* mutation among the genes identified within the cytogenetic interval. One open reading frame in this interval in the *rsh¹* chromosome encoding a novel protein with no obvious sequence motifs contains a nonsense mutation. Antibodies made against this novel protein indicate that this protein is expressed primarily in both the nucleus and cytoplasm of the MBs. In addition, a heat-shock, promoter-driven transgene encoding this novel protein was shown capable of rescuing the *rsh¹* memory impairment in a heat-shock-dependent manner. Thus, this report focused on the *rsh¹* allele as the basis for identifying *rsh*. Therefore, these data support the conclusion that *rsh* encodes a novel protein expressed primarily in the MBs rather than PLA2, which if true provides evidence that the *rsh*-dependent consolidation of memories into an anesthesia-resistant form is a property of the MB neurons.

Atypical PKCs

The atypical PKC isozymes are independent of diacylglycerol and Ca^{2+} for their activity, unlike their conventional sister enzymes. They have an N-terminal regulatory domain containing a pseudosubstrate sequence and a C-terminal catalytic domain. Removal of the N-terminal regulatory domain by proteolysis yields a persistently active kinase, referred to as atypical PKM (aPKM). Drier and colleagues (2002) have intensively studied the role of aPKM in *Drosophila* olfactory memory, starting with leads regarding the enzyme's role in synaptic plasticity. Physiological studies in the hippocampus demonstrated increased aPKM activity during the maintenance phase of long-term potentiation (Sacktor et al. 1993, Osten et al. 1996), decreased aPKM activity during the maintenance phase of long-term depression (Hrabetova and Sacktor 1996), and a requirement for the enzyme's activity for the maintenance of long-term potentiation (Ling et al. 2002).

Expression of a murine aPKM (MaPKM) from a heat-shock-promoter-driven transgene after a single cycle of olfactory classical conditioning enhances long-term memory performance when tested 24 h later. The performance increase depends on when the transgene is induced: Induction of the transgene before training or 2 h or more after training is without effect. Only a heat shock between 30 and 60 min after training enhances 24-h performance. To examine the behavioral specificity of this effect, the investigators induced the aPKM transgene after massed training consisting of 10 back-to-back training cycles or spaced training with the same number of cycles and an intertrial interval of 15 min. Interestingly, when assayed at four days after training, only the memory formed with massed training is enhanced, which led the authors to conclude that aPKM expression specifically enhances memory formed following massed training. These experiments, however, utilized heat-shock induction at 30 min after the completion of the training, and the temporal requirement between elevated enzyme activity and training may not have been met with the spaced training protocol because enzyme induction occurred more than 2.5 h after the beginning of training. Thus, although aPKM induction clearly enhances memory formed from massed training, it might also enhance memory formed from spaced training. In addition, the memory enhancement after a single cycle of training occurs in the background of the *radish* mutant, which indicates that memory enhancement by aPKM functions outside of the *radish*-dependent memory pathway. Furthermore, the expression of a putative aPKM dominant-negative protein after training, or feeding wild-type flies an inhibitor of aPKM, reduced 24-h memory. The latter results argue that aPKM is a component of the endogenous memory machinery, rather than functioning only an exogenous component that can be added to improve memory.

Together, the results argue that aPKM functions in the normal maintenance of memory and that its abundance and/or activity after training is limiting for normal memory formation because the magnitude of memory can be improved by overexpression. The mechanisms for aPKM action and the cellular and subcellular sites of function in memory formation remain unknown, although the role of the enzyme in hippocampal synaptic plasticity provides a sufficient rationale for speculating that the enzyme participates in synaptic plasticity for its behavioral roles.

Crammer/Staufen/Pumilio

Two different groups have recently isolated mutants that reportedly disrupt long-term memory without affecting short-term memory after olfactory classical conditioning. One mutant isolated with such characteristics is *crammer* (*cer*) (Comas et al. 2004). This mutant was isolated from a collection of enhancer detector lines exhibiting preferential expression in the MBs and has normal memory formed from one cycle of conditioning or multiple cycles of massed conditioning, but it is defective in long-term memory formed after multiple cycles of spaced conditioning. The gene disrupted in *crammer* encodes an inhibitor of cysteine proteases, therefore implicating these proteases and their regulation in long-term memory formation. Long-term memory formed from spaced conditioning is also disrupted by overexpression of *crammer* by genomic transgenes. A transgene encoding a *crammer*-GFP protein fusion showed GFP expression in the MBs and in what are potentially a few glial cells in the vicinity of the MB dendrites and axons, which could suggest that normal levels of glial expression or MB expression of the cysteine proteinase inhibitor is required for normal long-term memory. To investigate this, the researchers overexpressed *crammer* in glial cells or MB neurons using glial- or MB neuron-specific drivers and produced an impairment in long-term memory after spaced conditioning only with the glial overexpression. Although this result suggests that glia may be the source for the inhibitor, glial-specific loss-of-function experiments are needed to support the overexpression data. These data together, however, suggest that cysteine proteinase regulation—on the extracellular surfaces of MBs neurons, glial cells around MBs, or in the extracellular space between them—is important for long-term memory formation.

A behavioral screen for new long-term memory mutants was also recently completed (Dubnau et al. 2003b). Mutants in *pumilio*, a protein that participates in microtubule-dependent, subcellular targeting and localized mRNA translation in oocytes, were identified in the screen as having mild effects on short-term memory but a 50% reduction of 24-h memory produced by spaced training. Mutants were also identified in the gene *oskar*, which encodes a protein involved in the translocation of ribonucleoprotein particles to distinct cellular compartments. *Staufen*, which like *oskar* is part of the translocation machinery, is also likely required for normal long-term memory, evidenced by the poor performance 24 h after spaced training of a *staufen* temperature-sensitive mutant at restrictive temperature. Moreover, mutants were also recovered in eIF-5C, which is part of the translation initiation complex. The *oskar* and eIF-5C mutants have a pronounced deficit in long-term memory produced by spaced training but have a relatively normal short-term memory. Evidence was also presented on the basis of reporter gene expression (*pumilio*, *oskar*, and eIF-5C) and immunohistochemistry (*pumilio*) for the preferential expression of these genes in the MBs.

These preliminary results are consistent with the notion that long-term memory is subserved by directed transport of ribonucleoproteins and local translation in specific compartments of the relevant neurons. Given the preferential expression of this ribonucleoprotein transport system in the MB neurons, and the aforementioned evidence suggestive of the localization of long-term memory processes to the α and α′ collaterals of the MB neurons, an attractive model is that long-term memory processes are enabled through the ribonucleoprotein transport system into the α and α′ collaterals for local protein synthesis to aid in the establishment of long-term olfactory memories.

Several other mutants have been isolated and studied for their potential roles in olfactory memory formation, but they have not provided conceptual insights into the process or their characterization remains too preliminary. The *latheo* mutant was originally isolated as a mutant with impaired memory and classed as an acquisition mutant (Tully et al. 1994), but more detailed studies have revealed that the impairment is due to maldevelopment. The mutant *linotte* was also originally classed as an acquisition mutant (Tully et al. 1994) and argued to be an authentic memory formation mutant on the basis of conditional transgenic rescue of the memory impairment with induction during adulthood (Bolwig et al. 1995). However, more detailed studies have revealed that the mutation was originally assigned erroneously to a gene encoding a novel protein rather than to an adjacent gene encoding a receptor tyrosine kinase (Dura et al. 1995, Moreau-Fauvarque et al. 2002), which causes abnormal brain development. Attempts have failed to reproduce the original critical evidence of conditional behavioral rescue using transgenes for the novel protein gene, and the original conclusions have been withdrawn (Bolwig et al. 2002). Therefore, *linotte* now remains as a mutant in the receptor tyrosine kinase gene, which is required for proper brain development. The *nalyot* mutant was also isolated in the same behavioral screen as the *latheo* and *linotte* mutants, has a modest effect on short-term memory, and encodes the Adf1 transcription factor, a Myb-related factor known to regulate the alcohol dehydrogenase gene (DeZazzo et al. 2000). Complete loss of function of Adf1 is lethal. It therefore has clear roles in the development of the organism. Conditional behavioral rescue data using heat shock promoter–based constructs are complicated and unconvincing, but an optimistic interpretation of the behavioral rescue data (DeZazzo et al. 2000) is that Adf1 has roles both during development as well as in adult behavioral plasticity. A preliminary report suggests that mutants in a gene encoding a ribosomal S6 kinase, one kinase target of MAP Kinase, are impaired in olfactory classical conditioning (Putz et al. 2004).

GENES REQUIRED FOR MEMORY FORMATION IN *DROSOPHILA* HAVE CONSERVED BEHAVIORAL FUNCTIONS IN MAMMALS

The homologs for several of the genes listed in **Table 1** have been isolated from mammalian species and studied for their potential roles in mammalian behavior. This has uncovered a remarkable conservation of function in behavioral processes. The first connection made was with the prototypic *Drosophila* memory formation gene, *dunce*. Henkel-Tigges & Davis (1990) first showed that the cAMP phosphodiesterases encoded by the rat homologs of *Drosophila dunce* are inhibited by the antidepressant, rolipram, and this was soon shown true also for the human counterparts of *Drosophila dunce* (Livi et al. 1990). These results provided an intriguing connection and indicated that the mammalian homologs of *dunce* were indeed important for mammalian behavior, in this case, in the regulation of mood.

TABLE 2 Functional conservation of genes required for *Drosophila* memory formation

Mutant or gene	Mouse behavior	Human behavior
dunce (*dnc*)	—	mood
rutabaga (*rut*)	spatial memory	—
PKA-C/R	spatial memory, fear conditioning, conditioned taste aversion	—
CREB	spatial memory, fear conditioning	—
neurofibromin (*NF1*)	spatial memory, contextual discrimination conditioning	learning
N-CAM (related to *fasII*)	spatial memory	—
α-integrins (related to *Volado*)	spatial memory	—
Notch (N)	spatial memory	—

AL: antennal lobe
CS: conditioned stimulus
DPM: dorsal paired medial
GABA: γ-amino butyric acid
LH: lateral horn
LN: local interneuron
MB: mushroom body
ORN: olfactory receptor neuron
PN: projection neurons
US: unconditioned stimulus

A summary of these and related studies employing genetic knockouts in the mouse are presented in **Table 2**. There are two adenylyl cyclase genes in the mouse that encode calcium:calmodulin-activatable enzymes (Wong et al. 1999) and are therefore structurally and functionally homologous to the *Drosophila rutabaga*-encoded enzyme. Storm and colleagues created knockouts lines for both genes (Wu et al. 1995, Wong et al. 1999). The adenylyl cyclase 1 knockout animals have been extensively analyzed and reported to exhibit defective spatial memory in the water maze task (Wu et al. 1995).

The gene families that encode the mammalian PKA subunits include four different regulatory (RIα, RIβ, RIIα, RIIβ) subunit genes and two catalytic subunit genes (Cα, Cβ). A role for this enzyme family in behavior has been uncovered through the use of transgenic animals that express an inhibitory form of the PKA regulatory subunit, R(AB), in the forebrain (Abel et al. 1997). These animals have reduced PKA activity and parallel behavioral deficits in spatial memory and long-term memory of contextual fear conditioning. The RIIβ gene knockouts fail at motor learning on the rotorod (Brandon et al. 1998) and are impaired in long-term memory after conditioned taste aversion (Koh et al. 2003). The mouse Creb1 gene encodes three different CREB isoforms. A knockout that removes all isoforms is homozygous lethal (Rudolph et al. 1998). A different knockout that removes the α and δ forms but produces upregulation of the β isoform was originally reported to be strongly impaired in spatial memory and fear conditioning (Bourtchuladze et al. 1994). Two subsequent reports generally confirmed a role for CREB in fear conditioning (Graves et al. 2002, Gass et al. 1998), but a third report did not (Balschun et al. 2003). Two of the reports generally confirmed poor performance in spatial memory tasks (Balschun et al. 2003, Gass et al. 1998), and a third report did not (Graves et al. 2002), although those confirming the deficit attributed the impairment to increased wall hugging rather than a spatial memory deficit. Thus, although it seems likely that CREB is involved in long-term memory, the specific role of CREB in mammalian memory processes remains controversial.

A more recently defined regulator of calcium:calmodulin-dependent adenylyl cyclases in both *Drosophila* and mice is neurofibromin, the product of the *NF1* gene (Davis 2000). Some mutations in this gene cause neurofibromatosis type I in humans, and about half of the NF1 patients are impaired in learning (Davis 2000). Mice deficient in NF1 exhibit a weak deficiency in spatial memory and contextual discrimination conditioning

(Costa et al. 2001, 2002). Although there exist several possible mechanisms for these deficiencies (Costa et al. 2002), the gene product neurofibromin has been shown to activate G protein–stimulated adenylyl cyclase activity (Tong et al. 2002), so one possibility is that NF1 works partly through adenylyl cyclases (Tong et al. 2002).

The homologs of cell-adhesion receptors of the immunoglobulin superfamily and the integrins, represented by *Drosophila fasII* and the *Volado*-encoded integrins, respectively, are also required for normal mammalian learning. A classic knockout of neural cell-adhesion molecule (N-CAM) impairs spatial memory (Cremer et al. 1994). However, this effect is very modest when conditional knockouts are used (Bukalo et al. 2004), which potentially indicates that the effects observed with the classic knockout are more developmental in etiology. In addition, a genetic reduction in integrin function in the mouse also impairs spatial memory (Chan et al. 2003). Finally, a requirement for the *Notch1* gene in spatial memory of the mouse has been revealed by behavioral analyses of mouse knockouts (Costa et al. 2003).

Overall, these combined studies reveal an amazing conservation of the gene products identified to function in memory formation in *Drosophila*. Continued studies of *Drosophila* memory formation seem highly likely, therefore, to provide insights into the how human memories are formed and how various human diseases disrupt learning and memory. Moreover, the development of systems neuroscience approaches in the fly has provided a unique opportunity to synthesize molecular genetics and systems neuroscience discoveries about memory formation.

ACKNOWLEDGMENTS

Research in the author's laboratory has been supported by grants NS19904 from the NINDS, MH60420 from NIMH, AA13476 from NIAAA, and the Mathers Charitable Trust. The author is the R. P. Doherty-Welch Chair in Science at the Baylor College of Medicine.

LITERATURE CITED

Abel T, Nguyen PV, Barad M, Deuel TA, Kandel ER, Bourtchouladze R. 1997. Genetic demonstration of a role for PKA in the late phase of LTP and in hippocampus-based long-term memory. *Cell* 88:615–26

Artavanis-Tsakonas S, Rand MD, Lake RJ. 1999. Notch signaling: cell fate control and signal integration in development. *Science* 284:770–76

Balschun D, Wolfer DP, Gass P, Mantamadiotis T, Welzl H, et al. 2003. Does cAMP response element-binding protein have a pivotal role in hippocampal synaptic plasticity and hippocampus-dependent memory? *J. Neurosci.* 23:6304–14

Beck CDO, Schroeder B, Davis RL. 2000. Learning performance of *Drosophila* after repeated conditioning trials with discrete stimuli. *J. Neurosci.* 20:2944–53

Bolwig GM, Del Vecchio M, Hannon G, Tully T. 1995. Molecular cloning of *linotte* in *Drosophila*: a novel gene that functions in adults during associative learning. *Neuron* 15:829–42. Erratum. 2002. *Neuron* 34(4):667

Bourtchuladze R, Frenguelli B, Blendy J, Cioffi D, Schutz G, Silva AJ. 1994. Deficient long-term memory in mice with a targeted mutation of the cAMP-responsive element-binding protein. *Cell* 79:59–68

Brandon EP, Logue SF, Adams MR, Qi M, Sullivan SP, et al. 1998. Defective motor behavior and neural gene expression in RIIβ-protein kinase A mutant mice. *J. Neurosci.* 18:3639–49

Bukalo O, Fentrop N, Lee AY, Salmen B, Law JW, et al. 2004. Conditional ablation of the neural cell adhesion molecule reduces precision of spatial learning, long-term potentiation, and depression in the CA1 subfield of mouse hippocampus. *J. Neurosci.* 24:1565–77

Chan CS, Weeber EJ, Kurup S, Sun H, Sweatt JD, Davis RL. 2003. Integrin requirement for hippocampal synaptic plasticity and spatial learning. *J. Neurosci.* 23:7107–16

Chang KT, Shi Y-J, Min K-T. 2003. The Drosophila homolog of Down's syndrome critical region 1 gene regulates learning: implications for mental retardation. *Proc. Natl. Acad. Sci. USA* 100:15794–99

Cheng Y, Endo K, Wu K, Rodan AR, Heberlein U, Davis RL. 2001. *Drosophila* fasciclinII is required for the formation of odor memories and for normal sensitivity to alcohol. *Cell* 105:757–68

Chiang AS, Blum A, Barditch J, Chen YH, Chiu SL, et al. 2004. *radish* encodes a phospholipase-A2 and defines a neural circuit involved in anesthesia-resistant memory. *Curr. Biol.* 14:263–72

Cho W, Heberlein U, Wolf FW. 2004. Habituation of an odorant-induced startle response in *Drosophila*. *Genes Brain Behav.* 3:127–37

Comas D, Petit F, Preat T. 2004. *Drosophila* long-term memory formation involves regulation of cathepsin activity. *Nature* 430:460–63

Costa RM, Federov NB, Kogan JH, Murphy GG, Stern J, et al. 2002. Mechanism for the learning deficits in a mouse model of neurofibromatosis type 1. *Nature* 415:526–30

Costa RM, Honjo T, Silva AJ. 2003. Learning and memory deficits in *Notch* mutant mice. *Curr. Biol.* 13:1348–54

Costa RM, Yang T, Huynh DP, Pulst SM, Viskochil DH, et al. 2001. Learning deficits, but normal development and tumor predisposition, in mice lacking exon 23a of Nf1. *Nat. Genet.* 27:399–405

Cremer H, Lange R, Christoph A, Plomann M, Vopper G, et al. 1994. Inactivation of the N-CAM gene in mice results in size reduction of the olfactory bulb and deficits in spatial learning. *Nature* 367:455–59

Crittenden JR, Skoulakis EMC, Han KA, Kalderon D, Davis RL. 1998. Tripartite mushroom body architecture revealed by antigenic markers. *Learn. Mem.* 5:38–51

Davis RL. 1993. Mushroom bodies and *Drosophila* learning. *Neuron* 11:1–14

Davis RL. 1996. Biochemistry and physiology of *Drosophila* learning mutants. *Physiol. Rev.* 76:299–317

Davis RL. 2000. Neurofibromin progress in the fly. *Nature News Views* 403:846–47

Davis RL. 2001. Mushroom bodies, Ca2+ oscillations, and the memory gene amnesiac. *Neuron* 30:653–56

Davis RL. 2004. Olfactory learning. *Neuron* 44:31–48

Devaud JM, Acebes A, Ramaswami M, Ferrus A. 2003. Structural and functional changes in the olfactory pathway of adult *Drosophila* take place at a critical age. *J. Neurobiol.* 56:13–23

DeZazzo J, Sandstrom D, DeBelle S, Velinzon K, Smith P, et al. 2000. *Nalyot*, a mutation of the *Drosophila* myb-related *Adf1* transcription factor, disrupts synapse formation and olfactory memory. *Neuron* 27:145–58

Drier EA, Tello MK, Cowan M, Wu P, Blace N, et al. 2002. Memory enhancement and formation by atypical PKM activity in *Drosophila* melanogaster. *Nat. Neurosci.* 5:316–24

Dubnau J, Chiang AS, Grady L, Barditch J, Gossweiler S, et al. 2003b. The staufen/pumilio pathway is involved in *Drosophila* long-term memory. *Curr. Biol.* 13:286–96

Dubnau J, Chiang AS, Tully T. 2003a. Neural substrates of memory: from synapse to system. *J. Neurobiol.* 54:238–53

Dubnau J, Grady L, Kitamoto T, Tully T. 2001. Disruption of neurotransmission in *Drosophila* mushroom body blocks retrieval but not acquisition of memory. *Nature* 411:476–80

Dudai Y. 1985. Genes, enzymes and learning in *Drosophila*. *Trends Neurosci.* 8:18–21

Dura JM, Taillebourg E, Preat T. 1995. The Drosophila learning and memory gene linotte encodes a putative receptor tyrosine kinase homologous to the human RYK gene product. *FEBS Lett.* 370:250–54

Erber J, Masuhr T, Menzel R. 1980. Localization of short-term memory in the brain of the bee, Apis mellifera. *Physiol. Entomol.* 5:343–58

Feany MB, Quinn WG. 1995. A neuropeptide gene defined by the *Drosophila* memory mutant amnesiac. *Science* 268:869–73

Folkers E, Drain P, Quinn WG. 1993. Radish, a Drosophila mutant deficient in consolidated memory. *Proc. Natl. Acad. Sci. USA* 90:8123–27

Folkers E, Waddell S, Quinn WG. 2004. Identification of the *radish* memory gene as CG15720. 45th *Annu.* Drosophila *Res. Conf.* Abstr., *Genet. Soc. Am.* San Diego

Gao Q, Yuan Y, Chess A. 2000. Convergent projections of *Drosophila* olfactory neurons to specific glomeruli in the antennal lobe. *Nat. Neurosci.* 8:780–85

Gass P, Wolfer DP, Balschun D, Rudolph D, Frey U, et al. 1998. Deficits in memory tasks of mice with CREB mutations depend on gene dosage. *Learn. Mem.* 5:274–88

Ge X, Hannan F, Xie Z, Feng C, Tully T, et al. 2004. Notch signaling in Drosophila long-term memory formation. *Proc. Natl. Acad. Sci. USA* 101:10172–76

Graves L, Dalvi A, Lucki I, Blendy JA, Abel T. 2002. Behavioral analysis of CREB alphadelta mutation on a B6/129 F1 hybrid background. *Hippocampus* 12:18–26

Hammer M, Menzel R. 1998. Multiple sites of associative odor learning as revealed by local brain microinjections of octopamine in honeybees. *Learn. Mem.* 5:146–56

Han KA, Millar NS, Davis RL. 1998. A novel octopamine receptor with preferential expression in *Drosophila* mushroom bodies. *J. Neurosci.* 18:3650–58

Han KA, Millar NS, Groteweil MS, Davis RL. 1996. DAMB, a novel dopamine receptor expressed specifically in *Drosophila* mushroom bodies. *Neuron* 16:1127–35

Heisenberg M, Borst A, Wagner S, Byers D. 1985. *Drosophila* mushroom body mutants are deficient in olfactory learning. *J. Neurogenet.* 2:1–30

Heisenberg M. 1989. Genetic approaches to learning and memory (mnemogenetics) in *Drosophila melanogaster*. In *Fundamentals of Memory Formation: Neuronal Plasticity and Brain Function*, ed. G Rahmann, *Progress in Zoology* Ser. 37:3–45. Stuttgart/New York: Fischer Verlag

Heisenberg M. 2003. Mushroom body memoir: from maps to models. *Nat. Rev. Neurosci.* 4:266–75

Henkel-Tigges J, Davis RL. 1990. Rat homologs of the *Drosophila dunce* gene code for cyclic AMP phosphodiesterases sensitive to the antidepressant rolipram. *Mol. Pharm.* 37:7–10

Hrabetova S, Sacktor TC. 1996. Bidirectional regulation of protein kinase M zeta in the maintenance of long-term potentiation and long-term depression. *J. Neurosci.* 16:5324–33

Isabel G, Pascual A, Preat T. 2004. Exclusive consolidated memory phases in *Drosophila*. *Science* 304:1024–27

Jefferis GSXE, Marin EC, Stocker RF, Luo L. 2001. Target neuron prespecification in the olfactory map of Drosophila. *Nature* 414:204–8

Jellies JA. 1981. *Associative olfactory conditioning in Drosophila melanogaster and memory retention through metamorphosis*. Master thesis, Illinois State Univ., Normal, IL. 83 pp.

Joiner MA, Griffith LC. 2000. Visual input regulates circuit configuration in courtship conditioning of *Drosophila melanogaster*. *Learn Mem.* 7:32–42

Kandel E, Abel T. 1995. Neuropeptides, adenylyl cyclase, and memory storage. *Science* 268:825–26

Keene AC, Stratmann M, Keller A, Perrat PN, Vosshall LB, Waddell S. 2004. Diverse odor-conditioned memories require uniquely timed dorsal paired medial neuron output. *Neuron* 44:521–33

Kim YC, Lee HG, Seong CS, Han KA. 2003. Expression of a D1 dopamine receptor dDA1/DmDOP1 in the central nervous system of *Drosophila melanogaster*. *Gene Expr. Patterns* 3:237–45

Kitamoto T. 2001. Conditional modification of behavior in *Drosophila* by targeted expression of a temperature-sensitive *shibire* allele in defined neurons. *J. Neurobiol.* 47:81–92

Koh MT, Clarke SN, Spray KJ, Thiele TE, Bernstein IL. 2003. Conditioned taste aversion memory and c-Fos induction are disrupted in RIIβ-protein kinase A mutant mice. *Behav. Brain Res.* 143:57–63

Laissue PP, Reiter C, Hiesinger PR, Halter S, Fischbach KF, Stocker RF. 1999. Three-dimensional reconstruction of the antennal lobe in *Drosophila melanogaster*. *J. Comp. Neurol.* 405:543–52

Lessing D, Carlson JR. 1999. Chemosensory behavior: the path from stimulus to response. *Curr. Opin. Neurobiol.* 9:766–71

Ling DS, Benardo LS, Serrano PA, Blace N, Kelly MT, et al. 2002. Protein kinase Mzeta is necessary and sufficient for LTP maintenance. *Nat. Neurosci.* 54:295–96

Livi GP, Kmetz P, McHale MM, Cieslinski LB, Sathe GM, et al. 1990. Cloning and expression of cDNA for a human low-Km, rolipram-sensitive cyclic AMP phosphodiesterase. *Mol. Cell. Biol.* 10:2678–86

Mao Z, Roman G, Zong L, Davis RL. 2004. Pharmacogenetic rescue in time and space of the *rutabaga* memory impairment using Gene-Switch. *Proc. Natl. Acad. Sci. USA* 101:198–203

Marin EC, Jefferis GS, Komiyama T, Zhu H, Luo L. 2002. Representation of the glomerular olfactory map in the *Drosophila* brain. *Cell* 109:243–55

McGuire SE. 2003. *Temporal studies for the localization of olfactory memory acquisition, storage, and retrieval in Drosophila melanogaster*. PhD thesis. Baylor Coll. Med., Houston. 150 pp.

McGuire SE, Le PT, Davis RL. 2001. The role of *Drosophila* mushroom body signaling in olfactory memory. *Science* 10:1126–29

McGuire SE, Le PT, Osborn AJ, Matsumoto K, Davis RL. 2003. Spatio-temporal rescue of memory dysfunction in *Drosophila*. *Science* 302:1765–68

McGuire SE, Roman G, Davis RL. 2004. Gene expression systems in *Drosophila*: a synthesis of time and space. *Trends Genet.* 20:384–91

Moreau-Fauvarque C, Taillebourg E, Preat T, Dura JM. 2002. Mutation of linotte causes behavioral defects independently of pigeon in Drosophila. *Neuroreport* 13:2309–12

Nassel D, Elekes K. 1992. Aminergic neurons in the brain of blowflies and Drosophila: dopamine- and tyrosine hydroxylase-immunoreactive neurons and their relationship with putative histaminergic neurons. *Cell Tissue Res.* 267:147–67

Ng M, Roorda RD, Lima SQ, Zemelman BV, Morcillo P, Miesenbock G. 2002. Transmission of olfactory information between three populations of neurons in the antennal lobe of the fly. *Neuron* 36:463–74

Osten P, Valsamis L, Harris A, Sacktor TC. 1996. Protein synthesis-dependent formation of protein kinase Mzeta in long-term potentiation. *J. Neurosci.* 168:2444–51

Pascual A, Preat T. 2001. Localization of long-term memory within the *Drosophila* mushroom body. *Science* 294:1115–17

Perazzona B, Isabel G, Preat T, Davis RL. 2004. The role of cAMP response element-binding protein in Drosophila long-term memory. *J. Neurosci.* 24:8823–28

Philip N, Acevedo SF, Skoulakis EM. 2001. Conditional rescue of olfactory learning and memory defects in mutants of the 14–3-3zeta gene *leonardo*. *J. Neurosci.* 21:8417–25

Praefcke GJ, McMahon HT. 2004. The dynamin superfamily: universal membrane tubulation and fission molecules? *Nat. Rev. Mol. Cell. Biol.* 5:133–47

Presente A, Boyles RS, Serway CN, de Belle JS, Andres AJ. 2004. Notch is required for long-term memory in *Drosophila*. *Proc. Natl. Acad. Sci. USA* 101:1764–68

Putz G, Bertolucci F, Raabe T, Zars T, Heisenberg M. 2004. The S6KII (rsk) gene of *Drosophila melanogaster* differentially affects an operant and a classical learning task. *J. Neurosci.* 24:9745–51

Quinn WG, Greenspan RJ. 1984. Learning and courtship in Drosophila: two stories with mutants. *Annu. Rev. Neurosci.* 7:67–93

Quinn WG, Harris WA, Benzer S. 1974. Conditioned behavior in *Drosophila melanogaster*. *Proc. Natl. Acad. Sci. USA* 71:708–12

Roman G, Davis RL. 2001. Molecular biology and anatomy of *Drosophila* olfactory associative learning. *BioEssays* 23:571–81

Rudolph D, Tafuri A, Gass P, Hammerling GJ, Arnold B, Schutz G. 1998. Impaired fetal T cell development and perinatal lethality in mice lacking the cAMP response element binding protein. *Proc. Natl. Acad. Sci. USA* 95:4481–86

Sacktor TC, Osten P, Valsamis H, Jiang X, Naik MU, Sublette E. 1993. Persistent activation of the zeta isoform of protein kinase C in the maintenance of long-term potentiation. *Proc. Natl. Acad. Sci. USA* 90:8342–46

Schwaerzel M, Heisenberg M, Zars T. 2002. Extinction antagonizes olfactory memory at the subcellular level. *Neuron* 35:951–60

Schwaerzel M, Monastirioti M, Scholz H, Friggi-Grelin F, Birman S, Heisenberg M. 2003. Dopamine and octopamine differentiate between aversive and appetitive olfactory memories in *Drosophila*. *J. Neurosci.* 23:10495–502

Scott K, Brady R Jr, Cravchik A, Morozov P, Rzhetsky A, et al. 2001. A chemosensory gene family encoding candidate gustatory and olfactory receptors in *Drosophila*. *Cell* 104:661–73

Siwicki KK, Ladewski L. 2003. Associative learning and memory in *Drosophila*: beyond olfactory conditioning. *Behav. Processes* 64:225–38

Tong J, Hannan F, Zhu Y, Bernards A, Zhong Y. 2002. Neurofibromin regulates G protein-stimulated adenylyl cyclase activity. *Nat. Neurosci.* 52:95–96

Tully T. 1987. *Drosophila* learning and memory revisited. *Trends Neurosci.* 10:330–34

Tully T, Preat T, Boynton SC, Del Vecchio M. 1994. Genetic dissection of consolidated memory in *Drosophila*. *Cell* 79:35–47

Tully T, Quinn WG. 1985. Classical conditioning and retention in normal and mutant *Drosophila melanogaster*. *J. Comp. Physiol.* 157:263–77

Vosshall LB, Wong AM, Axel R. 2000. An olfactory sensory map in the fly brain. *Cell* 102:147–59

Waddell S, Armstrong JD, Kitamoto T, Kaiser K, Quinn WG. 2000. The *amnesiac* gene product is expressed in two neurons in the *Drosophila* brain. *Cell* 103:805–13

Waddell S, Quinn WG. 2001. Flies, genes, and learning. *Annu. Rev. Neurosci.* 24:1283–309

Wong ST, Athos J, Figueroa XA, Pineda VV, Schafer ML, et al. 1999. Calcium-stimulated adenylyl cyclase activity is critical for hippocampus-dependent long-term memory and late phase LTP. *Neuron* 23:787–98

Wong AM, Wang JW, Axel R. 2002. Spatial representation of the glomerular map in the Drosophila protocerebrum. *Cell* 109:229–41

Wu ZL, Thomas SA, Villacres EC, Xia Z, Simmons ML, et al. 1995. Altered behavior and long-term potentiation in type I adenylyl cyclase mutant mice. *Proc. Natl. Acad. Sc. USA* 92:220–24

Yin JCP, Wallach JS, DelVecchio M, Wilder EL, Zhou H, et al. 1994. Induction of a dominant negative CREB transgene specifically blocks long-term memory in Drosophila. *Cell* 79:49–58

Yin JC, Wallach JS, Wilder EL, Klingensmith J, Dang D, et al. 1995. *Drosophila* CREB/CREM homolog encodes multiple isoforms, including a cyclic AMP-dependent protein kinase-responsive transcriptional activator and antagonist. *Mol. Cell. Biol.* 15:5123–30

Yu D, Ponomarev A, Davis RL. 2004. Altered representation of the spatial code for odors after olfactory classical conditioning; memory trace formation by synaptic recruitment. *Neuron* 42:437–49

Zars T, Fischer M, Schulz R, Heisenberg M. 2000. Localization of short-term memory in *Drosophila*. *Science* 288:672–75

The Circuitry of V1 and V2: Integration of Color, Form, and Motion

Lawrence C. Sincich and Jonathan C. Horton

Beckman Vision Center, University of California, San Francisco, California 94143;
email: hortonj@vision.ucsf.edu, sincichl@vision.ucsf.edu

Key Words

striate cortex, extrastriate cortex, lateral geniculate nucleus, cytochrome oxidase, color vision

Abstract

Primary and secondary visual cortex (V1 and V2) form the foundation of the cortical visual system. V1 transforms information received from the lateral geniculate nucleus (LGN) and distributes it to separate domains in V2 for transmission to higher visual areas. During the past 20 years, schemes for the functional organization of V1 and V2 have been based on a tripartite framework developed by Livingstone & Hubel (1988). Since then, new anatomical data have accumulated concerning V1's input, its internal circuitry, and its output to V2. These new data, along with physiological and imaging studies, now make it likely that the visual attributes of color, form, and motion are not neatly segregated by V1 into different stripe compartments in V2. Instead, there are just two main streams, originating from cytochrome oxidase patches and interpatches, that project to V2. Each stream is composed of a mixture of magno, parvo, and konio geniculate signals. Further studies are required to elucidate how the patches and interpatches differ in the output they convey to extrastriate cortex.

Contents

INTRODUCTION 304
RESPONSE ARCHITECTURE
    OF V1 305
        The Geniculate Input 305
        Intracortical Circuitry in V1 307
        Color, Form, and Motion in
            V1 Physiology 310
CONNECTIONS BETWEEN V1
    AND V2 312
        Feedforward Connections 312
        Feedback Connections 315
RESPONSE ARCHITECTURE
    OF V2 316
        Intracortical Circuitry in V2 316
        Physiology of the V2 Stripes 317
RECASTING THE VISUAL
        CORTICAL HIERARCHY 318

Cytochrome oxidase: a mitochondrial enzyme, which can be used to identify particular visual areas (e.g., V1, V2) by its distinct laminar and columnar distribution.

Column: a group of neurons in cortex, clustered radially across at least two laminae, that share similar response properties

Receptive field: a delimited region in visual space for a given neuron, within which a light stimulus elicits a response

INTRODUCTION

In the primate visual system, most of the signals leaving the retina are relayed through the lateral geniculate nucleus (LGN) to V1. Our review starts here in V1 and finishes in V2. We critically assess recent studies that have focused on the organization of these early cortical visual areas. Surveying their function in tandem seems sensible because V1 and V2 are linked intimately on several levels. Both areas are required for the highly evolved sense we commonly think of as "seeing" (Horton & Hoyt 1991). V1 sends most of its cortical output to V2 and in return receives a strong feedback projection. They contain similarly scaled retinotopic maps of the visual field, and both have comparable surface areas. Finally, each area manifests a unique metabolic signature—revealed through cytochrome oxidase (CO) histochemistry—that makes it instantly recognizable. This CO pattern forms the scaffold around which the intra- and intercortical wiring of V1 and V2 is organized. Our goal is to survey the progress made in understanding the function of V1 and V2 since it was last reviewed in this series (Merigan & Maunsell 1993, Callaway 1998). The focus is on observations made in the macaque monkey, because it provides an unsurpassed animal model of the human visual system.

Although our knowledge has certainly grown since the trail blazing work of Hubel and Wiesel, it evokes a sense of humility to survey the progress in our field. For instance, ocular dominance columns were discovered more than 30 years ago but have yet to be invested with a function (Horton & Adams 2005). In biology, function can be notoriously hard to define. The cortex is a fairly uniform tissue adapted, like the skin, for many uses but for no one specific purpose. Neurons in V1 and V2 are not feature detectors, although they can detect features. We shy away from functional assignments and simply describe receptive field properties, recognizing that the most apt stimulus may not yet be known. It is still early in the exploration of the visual cortex and many fundamental premises are open to challenge.

In a key respect, the task of elucidating what Hubel & Wiesel (1962) dubbed the "functional architecture" of visual cortex has been quite fruitful. Originally, in the visual cortex this task involved determining if "there is any tendency for one or more of the [receptive field] characteristics to be shared by neighboring cells" (Hubel & Wiesel 1962, p. 128). Over the years, plentiful evidence has emerged that neurons are grouped within the cortical sheet according to shared response characteristics, although such grouping is not apparent cytoarchitectonically. The continuing empirical problem has been to identify columns of cells by their common features. The task has matured to include understanding how the physiological responses of neurons are sculpted from their inputs, how cells with common response features are interconnected, and how they organize their projections to other cortical areas (Callaway 1998, Lund et al. 2003).

As straightforward as it sounds, mapping the functional architecture of the visual has not been easy, for largely technical reasons. An ideal method would survey the cortex efficiently for the property under investigation and anchor it to an anatomical locus at high spatial resolution (50 μm or less). Traditionally, neuroscientific techniques have relied on "point" methods such as single-cell recording and tracer microinjections. Using such methods to study the organization of columns within a vast expanse of tissue like V1 has obvious limitations. Hubel & Wiesel (1977) called it "a dismaying exercise in tedium, like trying to cut the back lawn with a pair of nail scissors" (p. 28). Single-cell recordings can be correlated with functional architecture by making electrolytic lesions or by depositing fiducial markers along an electrode track. However, accurate alignment of electrode penetrations with individual recording sites in the tissue can be exasperatingly difficult. This problem, and the trend towards experiments in behaving animals, has made histological confirmation of recording sites a vanishing standard. The advent of optical imaging and functional magnetic resonance imaging has overcome the "point" limitation, but these new techniques suffer from poor spatial and temporal resolution, as well as uncertainty regarding the signal source. We emphasize these practical issues because progress in our field has been hampered by methodological hurdles.

RESPONSE ARCHITECTURE OF V1

V1 is the largest single area in the cerebral cortex of the macaque (Felleman & Van Essen 1991). It averages 1343 mm^2, out of a total cortical surface area of ∼10,000 mm^2 (Sincich et al. 2003). According to Livingstone & Hubel (1984a; 1987; 1988), it transforms the three input streams from the LGN into three output streams headed to area V2 (**Figure 1**). This view, however, has begun to erode, undercut by new studies at various levels of the visual pathway that violate the tripartite model of V1 organization. Beginning with the LGN, we examine new anatomical and physiological data that require a fresh consideration of the information flow through V1 and V2.

The Geniculate Input

The LGN contains six major laminae, evident in sections stained for Nissl substance. There are four dorsal parvocellular laminae and two

Optical imaging: a physiological method of mapping the responsiveness of cortex, using reflected light or voltage-sensitive dyes as probes.

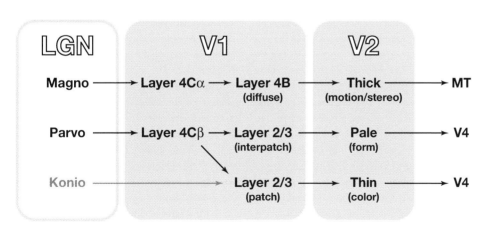

Figure 1

The tripartite model of the visual system. In layer 2/3, parvocellular input splits into two streams, patches and interpatches, segregating color and form signals that are propagated to V2 and subsequently to V4. Magnocellular signals travel via layer 4B to V2 and area MT, conveying information about motion and stereo. Koniocellular input is added to the color stream in layer 2/3 patches. After Livingstone & Hubel (1988) and Van Essen & Gallant (1994).

ventral magnocellular laminae. The parvo laminae receive input from color-opponent midget ganglion cells, whereas the magno laminae are supplied by broadband parasol ganglion cells (Perry et al. 1984). These distinct retinal channels account for the duality of receptive field properties in the LGN. Most parvo cells have color-opponent center-surround receptive fields, e.g., a red on-center and a green off-surround. Magno cells, by comparison, are broadband because their field center and surround receive input from the same mixture of cone types (Wiesel & Hubel 1966, Schiller & Malpeli 1978, Lee et al. 1998, Reid & Shapley 2002). At any given eccentricity, parvo cells have a higher spatial resolution, lower contrast sensitivity, slower conduction velocity, and a more sustained response than do magno cells (Shapley et al. 1981). The output of parvo and magno cells in the LGN is segregated in the primary visual cortex. Parvo cells terminate in layer $4C\beta$ and the upper part of layer 6, whereas magno cells innervate layer $4C\alpha$ and the lower part of layer 6. These distinct anatomical projections persuaded early investigators that parvo and magno channels remain functionally isolated in V1. In fact, as we shall see, they intermingle extensively beyond their input layers.

Livingstone & Hubel (1988) proposed that the parvo and magno systems provide the basis for the segregation of function in the visual system. They pointed out that one's sense of depth is impaired when a colored image is presented against an isoluminant background. Such isoluminant stimuli appear invisible to the magno system's "color blind" cells. Therefore, they reasoned, the loss of depth sensation at isoluminance indicates that magno cells handle stereo perception. In addition, they noted that the sensation of motion dissolves when a moving red/green grating becomes isoluminant, which suggests that motion perception also belongs to the magno channel. This seemed a good choice because magno cells conduct more rapidly than do parvo cells—an advantage perhaps for the perception of motion.

The parvo system was assigned the job of color perception—an easy decision given that only parvo cells have color-opponent receptive fields. This left the problem of form perception. Weighing the evidence, Livingstone and Hubel decided that perceiving form should be a parvo function because parvo cells have the best spatial resolution. However, parvo cells also serve color perception, creating an uncomfortable overlap. At this point, Livingstone and Hubel asserted that although form-perceiving neurons receive input from color-coded parvo geniculate layers, most are not explicitly color coded. They concluded that in the form pathway, "color-coded parvocellular input is pooled in such a way that color contrast can be used to identify borders but that the information about the colors (including black versus white) forming the border is lost" (Livingstone & Hubel 1988, p. 742). Thus, a split was promulgated in the receptive field properties of parvo-derived cortical cells, stripping color coding from those cells involved in the perception of form.

A third, neglected class of cells was later discovered in thin leaflets of tissue intercalated between the classical magno and parvo layers. These additional geniculate laminae were first recognized in the prosimian, where they are better developed than in the macaque. They were called "koniocellular," referring to the small size of the cells that they contain (Kaas et al. 1978), and it is worth noting that they numerically equal the magno population (Blasco et al. 1999). Many cells in the konio layers exhibit strong immunoreactivity for the α-subunit of type II calmodulin-dependent protein kinase (Hendry & Yoshioka 1994). The reason is unclear, but the enzyme provides a handy chemical label to identify the elusive konio layers. By coincidence, a special bistratified blue-on, yellow-off retinal ganglion cell was discovered in the macaque retina just at the time when konio cells were identified firmly as a third class of geniculate cells (Dacey & Lee 1994). This led immediately to speculation that blue-yellow retinal ganglion cells provide input to the konio laminae. In support of this idea, preliminary evidence has emerged that blue-yellow ganglion cells project to the konio layers just ventral to the third and fourth parvocellular layers (Calkins & Hendry 1996). Identifying the properties and connections of konio cells in the LGN has been a struggle because these cells are clustered in thin laminae or in occasional nests of cells embedded within the principal magno and parvo laminae. To complicate matters, konio cells constitute a heterogeneous population of cells, some lacking blue-yellow color opponency (Hendry & Reid 2000) and immunoreactivity for α-subunit of type II calmodulin-dependent protein kinase (Sincich et al. 2004). Certain subpopulations project directly to extrastriate cortex, conveying visual information that bypasses V1 altogether (Yukie & Iwai 1981, Rodman et al. 2001, Sincich et al. 2004). Nonetheless, to many investigators the term konio has become synonymous with the blue-yellow pathway, just as parvo is now equated, too simplistically, with the red-green pathway.

The projections of konio cells are segregated from those of parvo and magno cells in V1. Retrograde tracer studies have shown that konio cells provide the only direct geniculate input to layers 1–3 (Hendry & Yoshioka 1994). The innervation of layer 4A is uncertain. It contains a dense but thin tier of geniculate input organized into a fine, reticular pattern that resembles a honeycomb. Single geniculate afferents either ramify in 4A alone or send collaterals into layer 2/3 (Blasdel & Lund 1983). This implies that the projection to layer 4A is derived from konio cells. Chatterjee & Callaway (2003) have recorded from isolated geniculate afferents in V1 after application of muscimol to silence cortical cells. Sketches of their electrode penetrations and lesions show that exclusively konio afferents are encountered in layer 4A, as well as in the upper layers. However, Yazar et al. (2004) have found that some geniculate fibers terminate in both layers 4Cβ and 4A, implying either a direct parvo input to 4A or a konio input to 4Cβ.

Intracortical Circuitry in V1

CO histochemistry provides valuable information about the organization of V1 in several different ways (**Figure 2**). First, it delineates the cortical layers more crisply than the traditional Nissl stain. Second, CO density in each layer parallels the strength of geniculate input, with greatest activity in layers 6, 4C, 4A, and 2/3 (Fitzpatrick et al. 1983, Horton 1984, Hendry & Yoshioka 1994, Ding & Casagrande 1997). Third, CO reveals a striking array of dark patches (blobs, puffs) present in all layers except 4C and 4A (Hendrickson et al. 1981, Horton & Hubel 1981, Horton 1984, Wong-Riley & Carroll 1984). These patches are separated by paler zones, known logically as interpatches. The transition from patches to interpatches is gradual. Most investigators arbitrarily assign about a third of the cortical surface area to CO patches. The direct konio input to the upper layers coincides perfectly with the patches (Fitzpatrick et al. 1983, Horton 1984, Hendry & Yoshioka 1994).

The discovery that magno, parvo, and konio projections terminate in separate layers has spurred a concerted effort to learn if their signals remain segregated as they filter through the intracortical circuits of V1. One could imagine three isolated, parallel cortical systems operating in V1 to transfer pure magno, parvo, and konio signals to V2. As we shall see, in fact, the organization of cortical circuits in V1 suggests that geniculate channels are combined. Various anatomical approaches allow dissection of cellular networks in the cortex. The traditional Golgi method, or more modern dye-filling techniques, permits reconstruction of single cells along with their dendrites and axonal projections. By studying enough examples of cells in various layers, one can hypothesize about how cortical circuits are put together. Another strategy involves extracellular injection of small amounts of tracer into single layers, with a goal of delineating the connections with other cortical layers. Both these

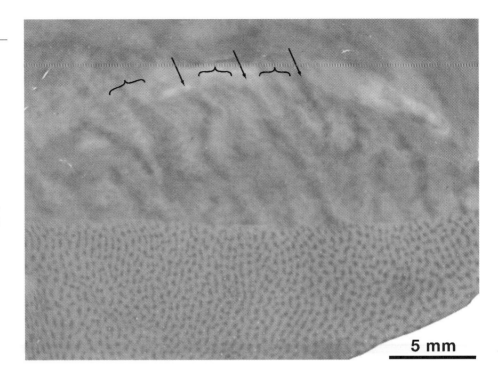

Figure 2
Macaque V1 patches and V2 stripes. A montage prepared from tissue sections cut tangentially to the cortical surface reveals characteristic patterns of endogenous metabolic activity when processed for CO. (*Bottom*) In V1 a fine array of patches is visible. (*Top*) In V2 a more irregular pattern is present, consisting of pale, thin (*arrows*) and thick (*brackets*) stripes arranged in repeating cycles.

approaches suffer from the limitation that they indicate only the potential for synapses to occur wherever axon terminals and dendrites coincide. They do not reveal anything direct about actual cell-to-cell transmission of information through the cortex. Two new methods have been developed to address this latter issue. The first uses transmission of rabies virus across a synapse, followed by immunochemical labeling of the chain of infected cells (Ugolini 1995). The second uses laser photostimulation to release caged glutmate, thereby revealing the inputs from various layers onto a single cell (Callaway & Katz 1993).

New evidence has emerged about the flow of signals within V1 (**Figure 3**). Parvo inputs to the layer $4C\beta$ synapse principally on glutamatergic spiny stellate cells. These cells project in turn to layers 2/3, where about half their synaptic connections are made (Callaway & Wiser 1996, Yabuta & Callaway 1998b). On their way, however, they make numerous synapses in layer $4C\beta$ itself, as well as in layers $4C\alpha$, 4B, and 4A. This implies immediate mixing with magno ($4C\alpha$) and konio (4A) streams, but one cannot be certain because the synapses made in layer 4 actually may be upon the dendrites of cells located in other layers. This uncertainty underscores the difficulty of inferring circuitry from isolated single-cell morphology. There are conflicting data concerning the projections from $4C\beta$ to patches versus interpatches in layer 3. After extracellular biocytin injections, Lachica et al. (1992) found projections to interpatches and patches from $4C\beta$, whereas Yoshioka et al. (1994) found a direct projection only to interpatches. Yabuta & Callaway (1998b) believe that $4C\beta$ projects to both patches and interpatches, but their data are limited to reconstruction of just seven intracellularly filled cells.

Magno inputs terminate in layer $4C\alpha$. Cells in this layer project to all superficial layers, as well as to $4C\alpha$ itself. Most investigators emphasize that layer $4C\alpha$ projects to layer 4B, endowing it with a strong magno bias. However, it actually sends a denser projection to layer 2/3. This projection probably terminates

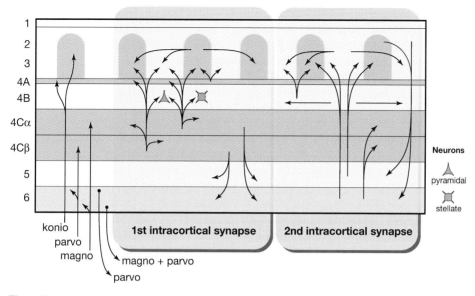

Figure 3

Intracortical circuitry within V1. (*Left*) The three geniculate streams entering V1 terminate in the CO-dense layers and the patches of layer 2/3. (*Middle*) Projections forming the first intracortical synapse yield a blend of the parvo-, magno-, and koniocellular streams. Each geniculate-recipient layer or patch sends axons to interpatches and to other layers or patches. In layer 4B, pyramidal and stellate cells receive different laminar inputs. (*Right*) In the second stage of intracortical projections, the axons continue to mingle the V1 signals, especially between infra- and supragranular layers, with increasing emphasis on horizontal projections. The relative strength of projections is not shown in this schematic diagram, nor is the diversity of cell types and classes comprising the intracortical wiring.

in both patches and interpatches (Yoshioka et al. 1994, Callaway & Wiser 1996), although one study reported that it supplies only patches (Lachica et al. 1992).

Thus, the projection patterns of cells in layer 4C reveal the potential for convergence of all three geniculate channels at their very next tier of synaptic contacts. For example, individual layer 2/3 cells are in a position to receive direct konio input and trans-synaptic parvo and magno input from layer 4C. However, it remains uncertain to what extent single cells actually blend multiple geniculate channels. In principle, cortical neurons might preserve strict segregation through precisely elaborated connections made on a cell-by-cell basis, overcoming the apparent intermingling of parvo, magno, and konio in layers beyond the first cortical synapse. Physiological studies of laminar projections have shown that this is usually not the case.

Callaway and colleagues have recorded from cells in macaque tissue slices, using laser photostimulation to survey the input sources to layers 3B, 4B, 5, and 6 (Sawatari & Callaway 1996, 2000; Briggs & Callaway 2001; Yabuta et al. 2001; Briggs & Callaway 2005). On the whole, these studies demonstrate a wide spectrum of laminar combinations in the input to cells in each of these layers. Of particular interest are layers 4B and 3B because many of their neurons receive input directly from parvo and magno cells in layer 4C. Recordings from four stellate cells in 4B showed significant activity only when stimulation was applied to layer 4Cα, rather than 4Cβ. These limited recordings, which reflect the challenge of acquiring these valuable data, suggest that 4B stellate cells are driven only by magno input (Yabuta et al. 2001). Layer 4B pyramidal cells ($n = 14$), by contrast, have mixed input from 4Cα and 4Cβ, with twofold more

Orientation selectivity: the dependence of a visual neuron's firing rate on the orientation (e.g., horizontal) of a contrast edge or line segment presented in the receptive field.

excitation from 4Cα. Because stellate and pyramidal cells both project to V2 (Rockland 1992), there is little doubt that a combined magno plus parvo signal is conveyed by layer 4B to V2. In layer 3B the cells in patches and interpatches receive input from parvo (4Cβ), magno (4Cα), konio (4A), or mixed (4B) layers, in a range of relative synaptic strengths (Sawatari & Callaway 2000). Most layer 3B cells project locally, almost entirely within layer 2/3, providing a substrate for further mingling of geniculate channels. They also provide a major source of projections to V2.

Infragranular circuits provide further potential for cross talk between geniculate channels. Cells in both 4Cα and 4Cβ project to layers 5 and 6 (Lund & Boothe 1975, Callaway & Wiser 1996). Cells in layers 5 and 6 project up to layer 2/3, which is reciprocally connected back to layers 5 and 6. Cells in layer 6 project back to layer 4C. The function of these reciprocal intracortical loops is not known, but it seems unlikely that the feedback they convey respects the distinction between parvo, magno, and konio. Feedback from layer 6 to the LGN is segregated only partially with respect to magno and parvo, further mixing the geniculate channels (Fitzpatrick et al. 1994).

From these data, it is evident that the intracortical wiring of V1 blurs the distinctiveness of thalamic input by convergence of parvo, magno, and konio signals onto individual cells. Rabies virus provides another means to probe how signals are combined in the visual system by revealing the chain of direct synaptic connections through the cortex. Nassi & Callaway (2004) have injected it into area MT and found infected cells in layers 4B and 4Cα of V1. Virtually no infected cells were located in 4Cβ. These preliminary data indicate that the 4B projection to MT is dominated by the magno geniculate channel.

Color, Form, and Motion in V1 Physiology

The anatomical studies reviewed above imply that magno, parvo, and konio inputs intermingle extensively within V1. Moreover, in layer 2/3 both patches and interpatches receive signals derived from all three geniculate sources. Regardless of the anatomy, the paramount issue is how cells with different receptive field properties are segregated into different functional compartments. Livingstone & Hubel (1988) originally proposed that three main classes of V1 neurons transmit visual signals to V2. Their central hypothesis, in simplest form, was that (*a*) Layer 2/3 patches convey information about color. Most patch cells are unoriented, center-surround, and color-opponent. (*b*) Layer 2/3 interpatches convey information about form. Interpatch cells are orientation tuned but not color coded. (*c*) Layer 4B conveys information about motion and stereo. Its cells are orientation and direction selective but are not tuned for color.

Early studies reporting that color is specifically processed by unoriented cells in CO patches deserve a closer look. These cells are a key feature of the tripartite model because they were described as the origin of a color pathway to V2. Livingstone & Hubel (1984a) made tangential electrode penetrations through the cortex, correlating clusters of unoriented cells with CO patches by making occasional lesions. In these experiments, the color properties of unoriented cells were not addressed. After they had pinned down the association between patches and unoriented cells, they next tested 204 unoriented cells for their color properties. These cells were assumed to be situated in CO patches because they lacked orientation tuning. However, no histological evidence was adduced to show their location. Of the 204 unoriented cells, 133 (65%) were rated as "color coded," establishing the link between CO patches and color. For comparison, of 698 oriented cells, only 148 (21%) were deemed color selective.

More direct evidence implicating patches in color processing was offered subsequently by Ts'o & Gilbert (1988). In their study, clusters of unoriented color cells were identified. The location of these clusters was later compared with the pattern of CO activity.

There was some degree of coincidence between patches and color cells (see their figure 8). Ts'o & Gilbert also made the remarkable observation that some CO patches contain a predominance of red/green cells, whereas others are more richly endowed with blue/yellow cells. This segregation is difficult to reconcile with the fact that all patches get direct blue/yellow konio input and indirect red/green parvo input. The association between unoriented color cells and CO patches has been corroborated by one other study (Yoshioka & Dow 1996). These authors sampled seven cells in patches and found that four were color-coded and unoriented.

Other reports have not confirmed that CO patches are populated by unoriented, color-opponent cells. Leventhal et al. (1995) found no correlation between orientation tuning, color properties, and CO patches. However, corroborative histological data from their electrode tracks were not illustrated. Edwards et al. (1995) and O'Keefe et al. (1998) reported no difference in the orientation tuning of patches and interpatches. In these two studies, color properties were not examined. To date, therefore, the color/patches versus orientation/interpatches dichotomy, derived from the correlation of electrode recordings with anatomy, is not conclusive.

Over the intervening years, studies in anesthetized and awake macaques using cone-isolating stimuli have found that color and orientation are treated as independent features by most cortical neurons. Cells that respond to achromatic, luminance contrast can also respond selectively to color. In addition, orientation-selective cells are frequently color tuned (Thorell et al. 1984, Lennie et al. 1990, Leventhal et al. 1995, Cottaris & DeValois 1998, Vidyasagar et al. 2002, Wachtler et al. 2003, Horwitz et al. 2004). A careful study of color selectivity by layer (Johnson et al. 2001) revealed that cells responsive to isoluminant color (though to varying degrees) are present in all layers, including 4B, which is supposed to be color-blind. The authors found that just 21% of color cells in V1 are unoriented.

This result has been confirmed by Friedman et al. (2003), who reported that only 17% of color-coded units are unoriented. However, Conway (2001) asserts that 80% of color cells are unoriented (Livingstone & Hubel's Class "D" cells). These papers are contradictory, in part because different criteria were used to define orientation and color selectivity.

Imaging studies have also addressed the issue of color and form segregation, subtracting activation due to an isoluminant, chromatic grating from activity evoked by an achromatic, luminance grating. In principle, this differential imaging strategy can isolate color regions in the cortex for subsequent correlation with CO histology. Using optical imaging, Landisman & Ts'o (2002) found zones of high color selectivity in V1 that overlap with CO patches in some instances but not in others. The stimuli were based on isoluminance measures in humans, which are known to differ significantly from those in macaques (Dobkins et al. 2000). Tootell et al. (2004) used a dual-label deoxyglucose technique to show that CO-rich areas of V1 have the strongest uptake of label to color stimuli. The stimuli in this study were tuned to isoluminance by gauging visually evoked potentials (VEPs) to a chromatic grating. Possible limitations of this study include cross talk between the two radioactive labels and difficulty assuring isoluminance with evoked potentials. Collectively, the data from electrode and imaging studies make it difficult to conclude that color properties are the sole province of CO patches in V1.

Livingstone & Hubel (1984a) assigned motion processing to layer 4B because its cells were direction tuned, color nonselective, and apparently magno-dominated. They recorded from 33 "nonblob" cells in layer 4B and reported that two thirds were strongly direction selective. Only five "blob" cells were recorded without any comment on their direction tuning. Subsequent studies have confirmed that direction tuning is prominent in layer 4B, although it is found in other layers as well (Hawken et al. 1988, Ringach et al. 2002, Gur et al. 2005). Some cells in 4B,

Flatmounting: a tissue-dissection technique whereby the cerebral cortex is unfolded and flattened to reveal histological patterns in a plane parallel to the pial surface.

as well as layer 6, exhibit an extremely pronounced direction bias (Livingstone & Hubel 1984a, Hawken et al. 1988). This feature is a striking property of V1 cells that project to MT (Movshon & Newsome 1996) and may be independent of the CO pattern (Leventhal et al. 1995). The projection from layer 4B, which arises from patches and interpatches, probably contributes to the high degree of direction tuning among MT cells. It remains to be proven that cells in layer 4B that project to V2 thick stripes are highly direction biased. Their properties could be different, given that independent populations of cells in layer 4B project to V2 and MT (Sincich & Horton 2003).

CONNECTIONS BETWEEN V1 AND V2

Feedforward Connections

When CO histochemistry was applied to area V2, it yielded a spectacular pattern of coarse, parallel stripes running perpendicular to the V1 border (Horton 1984), divided into repeating cycles of pale-thick-pale-thin (Tootell et al. 1983). Livingstone & Hubel (1984a, 1987), motivated by the idea that areas with comparable levels of CO might be wired together, were first to study the anatomical relationship between patches in V1 and stripes in V2. They found that patches project to thin stripes and that interpatches project to pale stripes. These connections were reported to arise exclusively from layer 2/3. A diffuse projection was described from patches and interpatches in layer 4B to thick stripes. The discovery of three distinct V1 compartments, each providing exclusive input to a type of V2 stripe, provided an enticing clue to the segregation of visual function. Insight into the nature of this functional segregation was furnished by two lines of evidence, which emerged nearly simultaneously. First, investigators showed that thick stripes project to area MT, a region concerned with the perception of motion and stereo, whereas thin and pale stripes project to V4, a region implicated in color and form (DeYoe & Van Essen 1985, Shipp & Zeki 1985). Second, a physiological study of receptive field properties in V2 showed marked segregation according to stripe type (DeYoe & Van Essen 1985, Hubel & Livingstone 1987). Synthesizing these data, Livingstone & Hubel (1988) proposed the following: patch → thin stripe handles color, interpatch → pale stripe processes form, and that layer 4B → thick stripe mediates motion and depth (**Figure 1**). The latter pathway was postulated to be dominated by the magno geniculate input, in contrast to the projections to pale and thin stripes, which were said to be derived principally from the parvo (and later konio) geniculate channel. This tripartite view of the V1-to-V2 pathway has prevailed in visual neuroscience, despite occasional complaints that it is overly reductive (Martin 1988, Merigan & Maunsell 1993).

The strongest piece of evidence in favor of three parallel functional streams (form, color, and stereo/motion) has been Livingstone and Hubel's demonstration of three distinct anatomical projections uniting compartments in V1 and V2. New evidence has emerged that their description of the V1-to-V2 pathway in the macaque was incomplete. Experiments exploiting improvements in tracers and flatmounting techniques have revealed a different pattern of projections between V1 and V2. These findings have rendered the old tripartite model untenable and suggest instead that the V1-to-V2 pathway is organized into a bipartite system (Sincich & Horton 2002a). This new anatomical foundation inclines one to take a more critical look at old physiological data that were interpreted in light of the defunct tripartite model.

Before turning to the physiology, it is worth inquiring why mapping the V1-to-V2 pathway has been such a difficult endeavor. Most of area V2 is buried in the lunate sulcus in the macaque. Only a few millimeters lie exposed on the brain surface, where tracers can be injected under direct visualization. As a result, sections cut tangentially to the pia contain only a sliver of V2. From this fragment of V2 tissue it is often difficult to distinguish between the two CO-dark stripes (thick and thin) or to tell when a transition has occurred between stripes. This problem can be mitigated by dissecting the cortex from the white matter, unfolding it, and flattening it like a sheet (Olavarria & Van Sluyters 1985, Tootell & Silverman 1985). Using this technique, a bird's eye view of V1 and V2 is obtained, facilitating the identification of stripes in V2 (**Figure 2**) (Olavarria & Van Essen 1997, Sincich et al. 2003). Even in such preparations, however, it can be impossible to discriminate thick and thin stripes. For some reason, in macaques the thin and thick stripes are not always clearly defined.

The difficulty of recognizing V2 stripes in some macaques means that data from many injections must be discarded. One can analyze cases only where the identity of a stripe is unequivocal and luck yields a tracer injection perfectly confined to a single stripe. In our reexamination of the V1-to-V2 projections, only 77 of 187 injections met these criteria (Sincich & Horton 2002a). However, they provided a consistent picture of the anatomy (**Figure 4**). The projection to thin stripes arose from patches, most strongly from layer 2/3. However, cells in layers 4A, 4B, and 5/6 also contributed to thin stripes. Cells in the deeper layers tended to be located in patches but were less tightly clustered than the cells in layer 2/3. Tracer injections into pale stripes revealed labeled cells in layer 2/3 interpatches, as expected. In addition, many cells were present in layers 4A, 4B, and 5/6, loosely concentrated in interpatches. Surprisingly, thick stripe injections yielded a pattern of labeling identical to that produced by pale stripe injections.

How does this new description of the V1-to-V2 projections differ from the old account? Previously, according to the tripartite model, each V2 stripe type was believed to receive a different input, derived from a single layer. Instead, it has become clear that multiple layers project to each stripe type and that the pro-

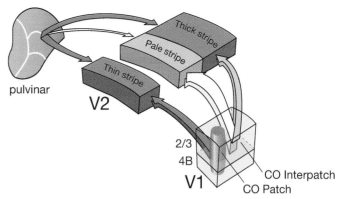

Figure 4

Projections from V1 to V2. Two major pathways originate from separate CO compartments in V1: Neurons in CO patch columns project to V2 thin stripes, and cells in interpatch columns project to pale and thick stripes. The axon terminal fields of these projections are densest in pale stripes. Other projections (*not shown*) arise from layers 4A, 5, and 6. The pulvinar provides the main thalamic input to V2, and the density of its projections is complementary to those from V1.

jections are bipartite, with patches connecting to thin stripes and interpatches connecting to both pale and thick stripes. This result implies that pale and thick stripes receive the same input from V1, rather than different messages concerned with form and stereo/motion, respectively. This notion was tested directly by making paired injections of different tracers into adjacent thick and pale stripes (Sincich & Horton 2002a). About a third of V1 projection neurons were double-labeled, showing that a substantial number of interpatch neurons form a single pathway to both pale and thick stripes. There may be subpopulations within interpatches that carry separate visual signals to pale and thick V2 stripes, but this idea is unproven. It is more likely that many cells remained single-labeled simply because their terminal arbors were smaller than a single V2 stripe (Rockland & Virga 1990). The segregation between CO patch and interpatch streams is nearly perfect, as demonstrated by using different tracers deposited into neighboring pale and thin stripes. In these cases, only a handful of double-labeled neurons was found out of thousands of single-labeled cells (**Figure 5**) (Horton & Sincich 2004).

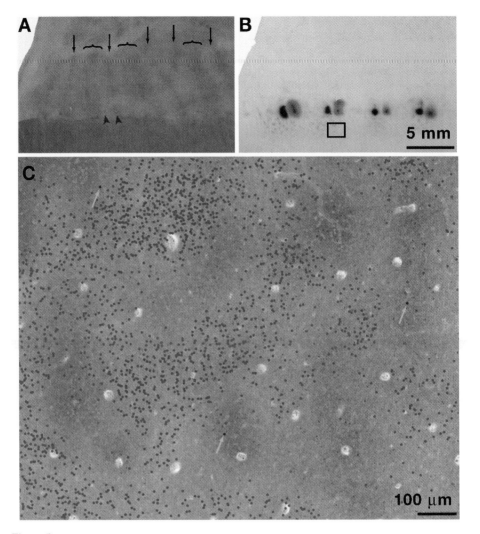

Figure 5

Segregation of V1-to-V2 projections. (*A*) Single CO-stained section from layer 4 in V2, showing the stripe pattern (*brackets, thick stripes; arrows, thin stripes*). One of the thin stripes splits to form a "Y"; such stripe bifurcations occasionally interrupt the regular stripe sequence. Blue arrowheads indicate the location of a CTB-Au injection in a pale stripe (*left*) and a WGA-HRP injection in a thin stripe (*right*). (*B*) A section more superficial to the one shown in (*A*), processed for both tracers. Black box is the area where cells are plotted and shown at higher power below. (*C*) Cells counted in box are superimposed onto the CO pattern from an adjacent section. Neurons projecting to the thin stripe (*green, n = 703*) were located in CO patches, whereas those projecting to the pale stripe (*red, n = 2058*) were situated in the interpatches. Of the 2761 cells in this single section, 3 were double-labeled (*blue, arrows*), demonstrating the high degree of segregation between these two pathways. Adapted from Horton & Sincich (2004).

If pale stripes and thick stripes receive input from the same source in V1, what accounts for their differing CO intensity? One possibility is that thick stripes receive stronger input from V1 than do pale stripes, endowing them with higher metabolic activity. Before the advent of CO histochemistry, investigators observed that V1 projections to V2 terminate in regular clusters (Wong-Riley 1978). These clusters were later shown to coincide with pale stripes (Sincich & Horton 2002b). It runs counter to intuition that the stripes receiving the strongest V1 input should have the weakest metabolic activity.

V2 also receives a major projection from the pulvinar. Its input coincides faithfully with the density of CO staining in V2, perhaps accounting for the increased metabolism of thin and thick stripes (Livingstone & Hubel 1982, Levitt et al. 1995). Pulvinar terminals synapse largely in lower layer 3, whereas V1 input is richer to layer 4 (Rockland & Pandya 1979, Lund et al. 1981, Weller & Kaas 1983, Van Essen et al. 1986, Rockland & Virga 1990). Therefore, the terminal fields of both the pulvinar and V1 are continuous throughout V2, but their densities wax and wane in counterphase and they favor different layers. The dovetailed pulvinar input must exert an influence on the physiological properties of cells in V2, but it has been largely ignored. It provides the first opportunity for the pulvinar to join the flow of information in the cortical visual pathway. The pulvinar is considered a higher-order thalamic relay because it inherits many of its response properties from descending cortical projections, especially from V1, and then projects back to cortex (Sherman & Guillery 1996, Shipp 2001). Therefore, a major source of V1 input to V2 stripes is channeled via the pulvinar. We do not know if this loop originates from distinct CO compartments in V1. Until the nature of the massive pulvinar input to V2 is more clearly defined, it seems premature to assign functions to the CO stripes.

Feedback Connections

Compared with the feedforward V1-to-V2 pathway, the feedback projection has received little attention. Numerically, it is nearly as large. Beneath each square millimeter of cortex, there are an estimated 11,000 feedback neurons in V2 compared with 14,000 feedforward neurons in V1 (Rockland 1997). The feedback neurons reside in layer 2, upper layer 3, and layer 6, situated at least one synapse downstream from V1's input (Tigges et al. 1973, Rockland & Pandya 1981, Weller & Kaas 1983, Kennedy & Bullier 1985). Their axons terminate in layers 1, 2, and 5 of V1, with occasional arbors in layer 3 (Rockland & Virga 1989). A recent study using tritiated amino acids also reported feedback projections to layer 4B (Gattass et al. 1997), although this has not been confirmed by others.

Few studies have asked how the V2 feedback projections are organized with respect to the response architecture of V1. It would be of exceptional interest to know if they differentially target the patches or interpatches. Four studies have reported that the projections form terminal clusters in V1, suggesting a systematic relationship (Wong-Riley 1979, Malach et al. 1994, Angelucci et al. 2002, Shmuel et al. 2005). Comparison was made with CO-stained sections in only one study. Feedback projections from pale and thick stripes were correlated with V1 interpatches as well as with orientation columns (Shmuel et al. 2005). A separate study using an adenoviral anterograde tracer concluded from two pale stripe injections that axons project back diffusely to V1, without clustering in CO patches or columns of the same orientation (Stettler et al. 2002). The distribution of synaptic boutons was not analyzed, making this interpretation problematic. A third injection did reveal a periodic pattern (on a suitable scale of 0.5 mm), but no relationship with orientation columns was observed. The correspondence with CO patches was not examined. Further studies are warranted to probe the organization of V2-to-V1 feedback.

Retrograde tracer injections have shown that V2 gets two thirds of its entire cortical input from V1 (Sincich et al. 2003). Cells in V2 become completely unresponsive after loss of this physiological drive (Schiller & Malpeli 1977, Girard & Bullier 1989). However, the reverse is not true. Withdrawal of V2 feedback by cooling or GABA injections produces surprisingly subtle changes in the responses of V1 neurons. Sandell & Schiller (1982) found no change in orientation tuning and only occasional changes in direction selectivity, although some cells became less responsive. Hupe et al. (2001) report no impact of V2 inactivation on V1 cells' classic receptive fields or on their modulatory surrounds.

RESPONSE ARCHITECTURE OF V2

V2 is the second largest cortical area in the macaque, with a mean area of 1012 mm^2 (Sincich et al. 2003). The representation of visual space in V1 is mirrored across the border in V2 (Allman & Kaas 1974, Gattass et al. 1981, Sereno et al. 1995). Given that V2 is subdivided into 26–34 cycles of stripes that encircle V1 like a corona, it is intriguing to ask whether their presence has any impact on local retinotopic order. At one extreme, each stripe type could represent the visual field independently. In that case, V2 might contain three separate, interleaved visual field maps. Two groups have extensively mapped V2 at high resolution, making electrode recordings that traversed several sets of stripes (Roe & Ts'o 1995, Shipp & Zeki 2002b). Investigators paid particular attention to stripe borders, where a sudden jump in receptive field position might be expected. In addition, evidence was sought that for any given stripe type, retinotopy progresses smoothly from stripe to stripe. The data provide some support for the idea that V2 contains independent maps for each stripe type, but this interpretation is weakened by the receptive field scatter, the gradual transition from one stripe type to another, and the small size of field-position jumps between stripe types. Even if one accepts that V2 stripes contain independent retinotopic maps, that property alone would not warrant dividing V2 into three visual areas.

Intracortical Circuitry in V2

The interlaminar circuitry of V2 has been virtually ignored. Our small store of information is derived entirely from Golgi studies (Valverde 1978, Lund et al. 1981). Neurons in layer 4 project chiefly to layers 3A and 3B. Neurons in 3B, which receive most of the pulvinar input, project to layers 2 and 3A. These layers are the major source of projections to other cortical areas (Rockland 1997). Axons heading to other cortical areas usually have collaterals in layer 5. As in V1, layer 5 neurons form a population of recurrent projections, sending axons to layers 2/3 and 5, as well as to noncortical targets like the pulvinar. Finally, layer 6 neurons appear to differ from those in V1 because they send local projections largely to layer 3 rather than layer 4. The apparent lack of recurrent projections to layer 4 suggests that it may be the only layer that retains response properties reflecting the original V1 input. No studies have examined whether the interlaminar circuitry differs between CO stripe types.

Extensive signal mixing via intralaminar projections occurs across V2 stripes. Within layers 2/3 and 5, horizontal axon projections form periodic terminal clusters, as in V1 (Rockland 1985). Anterograde tracer injections in any individual stripe consistently reveal a set of lateral projections to every stripe type (Levitt et al. 1994b, Malach et al. 1994). Pale stripes project equally to thin and thick stripes as well as to pale stripes. Interestingly, dark CO stripes are more likely to project to other dark stripes, permitting cross talk between thin and thick stripes and, by implication, between pulvinar inputs. The extent of horizontal projections is about 8 mm, twice that of lateral projections in V1, perhaps contributing to the coarser retinotopy of V2. Each terminal cluster is ∼250 μm across, much less than the dimension of a V2 stripe. Columns within V2 stripes have been suggested by physiological and optical imaging studies. The terminal clusters may bear a systematic relationship to purported V2 columns, but this question has not been pursued in the macaque.

The most important insight from these studies of V2 circuitry is that local projections make little effort to confine themselves to the same class of stripe. Therefore, even if it were correct that V1 sends a different signal to each class of V2 stripe, these signals are shared quite freely between V2 stripes. This contrasts with the situation in V1, where CO patches project preferentially to other patches, and interpatches to interpatches (Livingstone & Hubel 1984b, Yoshioka et al. 1996, Yabuta & Callaway 1998a).

Physiology of the V2 Stripes

The physiology of cells in V2 has been studied extensively. We focus explicitly on efforts to correlate receptive field properties and stripe class, setting aside a growing list of interesting studies that address the role of V2 in attention-guided behavior (Ghose et al. 2002) and in the processing of complex stimuli (Kobatake & Tanaka 1994, Ito & Komatsu 2004).

DeYoe & Van Essen (1985) found that color-selective cells were prevalent in both thin and pale stripes, whereas orientation-selective cells were less common in these compartments. Hubel & Livingstone (1987) recorded from 1023 single cells but provided no numerical breakdown of cell properties by stripe class. However, they stated that, with some exceptions, unoriented color-tuned cells were located in thin stripes, oriented cells (which showed no overt color coding) were present in pale stripes, and disparity-tuned cells were concentrated in thick stripes [although "occasional disparity-tuned cells occurred in pale stripes" (p. 3410)]. Their analysis depended on the squirrel monkey because they were unable to distinguish between thick and thin stripes in the macaque. Often stimuli were used selectively in the assessment of receptive fields, injecting a potential bias in their analysis. For example, color responses were not tested systematically in oriented cells, nor disparity tuning in unoriented cells. Nonetheless, their data provided the basis for a link between color and thin stripes, form and pale stripes, and stereo/motion and thick stripes.

Since these original reports, no fewer than 11 studies have reexamined how receptive field properties correlate with different V2 stripe classes (Peterhans & von der Heydt 1993, Levitt et al. 1994a, Roe & Ts'o 1995, Gegenfurtner et al. 1996, Tamura et al. 1996, Yoshioka & Dow 1996, Kiper et al. 1997, Roe & Ts'o 1999, Ts'o et al. 2001, Moutoussis & Zeki 2002, Shipp & Zeki 2002a). These studies are difficult to compare because they differ in their methods, as well as in their criteria for defining a cell as "selective" for any given parameter. No study, with the exception of Shipp & Zeki (2002a), shows electrode tracks marked with lesions in sections containing easily distinguishable cycles of thin-pale-thick-pale CO stripes. It is impossible to say much about the functional specificity of each stripe class without reliable correlation of recording sites with histology. Faced with this difficulty, some investigators have given up trying to use CO to define stripe class. Instead, for example, they use a stimulus thought to activate preferentially color-selective cells, and they define these regions as "thin stripes" (Xiao et al. 2003). It would be preferable to define stripes by their CO appearance because the association between color-selective cells and thin stripes is not yet well established.

The studies mentioned above are quite contradictory; some studies found a high degree of functional segregation by stripe type, and others concluded that little evidence exists to support this idea. Only one property appears in all studies as a robust feature: a higher degree of orientation selectivity in thick and pale stripes. As mentioned above, some studies suggest that color selectivity is more prevalent in thin stripes, but others dissent (Peterhans & von der Heydt 1993, Levitt et al. 1994a, Gegenfurtner et al. 1996, Tamura et al. 1996). The only study containing a laminar analysis of cell properties found that the peak functional "distinctiveness" of the stripes occurs in layer 3 (Shipp & Zeki 2002a). This layer receives the bulk of pulvinar input and also sends the strongest projection to higher visual areas. Ultimately, it is the functional specificity of the output cells within different stripe classes that reflects most meaningfully how V2 segregates the signals it receives from V1.

Optical imaging is an effective technique for the correlation of receptive field properties with anatomical compartments because it allows one to collect signals averaged simultaneously from thousands of cells. In the macaque, however, most of V2 is buried in the lunate suclus, and the small portion situated on the surface lies close to large vessels that produce vascular artifact. In the squirrel monkey, V2 is a more inviting target because it sits in a flat expanse of exposed cortex. In this species, Malach et al. (1994) have shown that orientation columns are prominent in thick and pale stripes but not in thin stripes. This result has been confirmed in the macaque (Vanduffel et al. 2002) and owl monkey (Xu et al. 2004). It is consistent with the verdict from single-cell recordings. Curiously, Xu et al. report that only every other pale stripe has high orientation selectivity.

In the macaque, imaging studies have localized color-selective regions to the thin stripes in V2 (Roe & Ts'o 1995, Xiao et al. 2003, Tootell et al. 2004). In these studies, the response to high-contrast, achromatic gratings was subtracted from the response to an isoluminant, chromatic grating. As mentioned earlier, in monkeys it is difficult to be sure that a stimulus is truly isoluminant. The stimulus can be rendered nearly, but not exactly, isoluminant. Therefore, the comparison may really entail a low-contrast chromatic grating versus a high-contrast black-and-white grating. Many color-selective cells respond well to both stimuli, complicating the interpretation of these experiments. The use of color-exchange stimuli, which equate form and contrast but vary chrominance, are superior for imaging color-specific regions (Wade et al. 2002). Doubt will remain about the localization of color-selective cells until such stimuli are applied to image the stripe compartments in V2.

RECASTING THE VISUAL CORTICAL HIERARCHY

V1 provides the foundation for the visual cortical hierarchy. Although it projects to a number of different cortical areas, most of its output is directed to V2. To understand vision, it is crucial to know what signals V1 conveys to V2. For nearly two decades, our understanding of the V1-to-V2 circuit has rested on the belief that three channels exist, each carrying different information: (*a*) color from layer 2/3 patches to thin stripes, (*b*) form from layer 2/3 interpatches to pale stripes, (*c*) stereo/motion from layer 4B to thick stripes. New findings have called into question this tripartite model of the visual system.

Initially, the discovery that blue-yellow color information is fed directly to patches (and not interpatches) seemed to boost the idea that patches are a color specialization. On the other hand, a direct konio projection to patches exists in the owl monkey, a nocturnal species that lacks color vision (Horton 1984, Casagrande & Kaas 1994, Xu et al. 2004). Perhaps the owl monkey is an exception among primates, making the macaque a more suitable model for humans. In the macaque, however, the konio input to patches appears to be accompanied by a diffuse input to layer 4A. This input to layer 4A has no predilection for patch columns, implying that both patches and interpatches are supplied with ascending konio input from layer 4A. Parvo input from layer $4C\beta$ also appears to be transmitted to both patches and interpatches in layer 2/3. To summarize, there is plenty of evidence that geniculate color information is funneled to both patches and interpatches. This makes it improbable that only patches convey color information to V2.

It has also become clear that inputs to V1, which are stratified by magno, parvo, and konio, become thoroughly intermingled by passage through the elaborate circuitry of V1. As a result, output cells of V1 probably convey a mixed, but transformed, geniculate signal to V2. The old scheme stipulated that only layer 4B projects to thick stripes, carrying a magno signal for stereopsis and motion. This idea has become untenable for several reasons. First, layer 4B gets both parvo and magno input (Yabuta et al. 2001). Second, lesions of magnocellular geniculate laminae have no effect on stereopsis (Schiller et al. 1990). Third, disparity-tuned cells are abundant outside layer 4B (Poggio et al. 1988) and thick stripes (DeYoe & Van Essen 1985, Peterhans & von der Heydt 1993). Fourth, other layers besides 4B project to thick stripes (Sincich & Horton 2002a).

Originally, interpatches were assigned the job of form perception because they were thought to contain oriented cells that lack color tuning. Parenthetically, we point out the flawed logic of assuming that a given V1 compartment constitutes the form pathway merely because it contains cells that are oriented. Cells in 4Cβ are unoriented, but who would argue that they are not part of the form pathway? All cells in V1 contribute to the perception of form, oriented or not. The specious notion that oriented cells are not color selective, and hence serve form but not color, derived from a failure to test oriented cells carefully for their color properties. It also reflected a shrewd bit of guesswork, predicated on the remarkable clinical phenomenon of cerebral achromatopsia. Patients with this rare syndrome perceive the world without color. This rare deficit, produced by a lesion in the fusiform gyrus, proves that perception of form and color eventually becomes divorced in the visual system. However, it is unlikely that their separation occurs as early as V1 and V2.

The pattern of projections from V1 to V2 is actually simpler than proposed by Livingstone & Hubel (1988). Instead of three output channels, there are only two. These two channels are defined by CO compartments. Patches project to thin stripes; interpatches supply pale stripes and thick stripes (**Figure 4**). These projections are columnar because they arise from cells coarsely aligned in layers 2/3, 4A, 4B, 5, and 6. Most of the input to thin stripes is supplied by layer 2/3; pale and thick stripes get strong projections from layers 2/3 and 4B. It should be emphasized that pale stripes and thick stripes receive their input from the same compartment (interpatches) and often from the same cells. This vitiates the proposal that pale stripes get parvo input and thick stripes get magno input.

What functions are dichotomized by patches and interpatches? Embarrassingly, the answer remains elusive, nearly a quarter century after the discovery of CO patches. We must learn if patches are endowed selectively with unoriented, color-opponent cells, as originally described. Do they coincide with orientation singularities ("pinwheels"), where orientation columns seem to converge? For technical reasons, alluded to earlier, a clean verdict has not been forthcoming from single-cell electrode recordings or optical imaging. Perhaps 2-photon confocal imaging of calcium fluxes will furnish the technical breakthrough required to resolve these issues (Ohki et al. 2005). It provides simultaneous information about the physiological responses of hundreds of cells at high spatial resolution. With fluorescent tracers it should be possible to backfill cells in V1, allowing one to focus particular attention on the projection neurons that go to V2. Finally, it may yield data concerning the properties of cells in V2 stripes, where intrinsic signal imaging has been disappointing.

One reason that the tripartite form/color/motion model for the visual system has survived so long is that there is nothing available to replace it. For a neuroscience textbook or an undergraduate class, it provides a compelling story. It would be refreshing, as we conclude, to offer a new, comprehensive picture of the functional organization of V1 and V2. At this point we can offer only a more accurate account of the anatomical projections between these key early visual areas to provide a new foundation for future studies.

CO: cytochrome oxidase

GABA: gamma-aminobutyric acid (inhibitory neurotransmitter)

LGN: lateral geniculate nucleus

MT: middle temporal area (also named cortical area V5)

V1: primary visual cortex, striate cortex

V2: secondary visual cortex

VEP: visual evoked potentials (scalp recordings)

ACKNOWLEDGMENTS

The work was supported by The Larry L. Hillblom Foundation and by National Eye Institute grants to L.C.S., J.C.H., and the Beckman Vision Center. Additional support was provided by a National Institutes of Health core grant to the California Regional Primate Research Center.

LITERATURE CITED

Allman JM, Kaas JH. 1974. The organization of the second visual area (V II) in the owl monkey: a second order transformation of the visual hemifield. *Brain Res.* 76(2):247–65

Angelucci A, Levitt JB, Walton EJ, Hupe JM, Bullier J, Lund JS. 2002. Circuits for local and global signal integration in primary visual cortex. *J. Neurosci.* 22(19):8633–46

Blasco B, Avendano C, Cavada C. 1999. A stereological analysis of the lateral geniculate nucleus in adult *Macaca nemestrina* monkeys. *Vis. Neurosci.* 16:933–41

Blasdel GG, Lund JS. 1983. Termination of afferent axons in macaque striate cortex. *J. Neurosci.* 3(7):1389–413

Briggs F, Callaway EM. 2001. Layer-specific input to distinct cell types in layer 6 of monkey primary visual cortex. *J. Neurosci.* 21(10):3600–8

Briggs F, Callaway EM. 2005. Laminar patterns of local excitatory input to layer 5 neurons in macaque primary visual cortex. *Cereb. Cortex.* In press

Calkins DJ, Hendry SH. 1996. A retinogeniculate pathway expresses the alpha subunit of CaM II kinase in the primate. *Soc. Neurosci. Abstr.* 22:1447

Callaway EM. 1998. Local circuits in primary visual cortex of the macaque monkey. *Annu. Rev. Neurosci.* 21:47–74

Callaway EM, Katz LC.1993. Photostimulation using caged glutamate reveals functional circuitry in living brain slices. *Proc. Natl. Acad. Sci. USA* 90(16):7661–65

Callaway EM, Wiser AK. 1996. Contributions of individual layer 2–5 spiny neurons to local circuits in macaque primary visual cortex. *Vis. Neurosci.* 13(5):907–22

Casagrande VA, Kaas JH. 1994. The afferent, intrinsic, and efferent connections of primary visual cortex in primates. In *Cerebral Cortex*, ed. A Peters, KS Rockland, pp. 201–59. New York: Plenum. Vol. 10

Chatterjee S, Callaway EM. 2003. Parallel colour-opponent pathways to primary visual cortex. *Nature* 426(6967):668–71

Conway BR. 2001. Spatial structure of cone inputs to color cells in alert macaque primary visual cortex (V-1). *J. Neurosci.* 21(8):2768–83

Cottaris NP, DeValois RL. 1998. Temporal dynamics of chromatic tuning in macaque primary visual cortex. *Nature* 395(6705):896–900

Dacey DM, Lee BB. 1994. The 'blue-on' opponent pathway in primate retina originates from a distinct bistratified ganglion cell type. *Nature* 367(6465):731–35

DeYoe EA, VanEssen DC. 1985. Segregation of efferent connections and receptive field properties in visual area V2 of the macaque. *Nature* 317(6032):58–61

Ding Y, Casagrande VA. 1997. The distribution and morphology of LGN K pathway axons within the layers and CO blobs of owl monkey V1. *Vis. Neurosci.* 14(4):691–704

Dobkins KR, Thiele A, Albright TD. 2000. Comparison of red-green equiluminance points in humans and macaques: evidence for different L:M cone ratios between species. *J. Opt. Soc. Am. A Opt. Image Sci. Vis.* 17(3):545–56

Edwards DP, Purpura KP, Kaplan E. 1995. Contrast sensitivity and spatial frequency response of primate cortical neurons in and around the cytochrome oxidase blobs. *Vision Res.* 35(11):1501–23

Felleman DJ, Van Essen DC. 1991. Distributed hierarchical processing in the primate cerebral cortex. *Cereb. Cortex* 1:1–47

Fitzpatrick D, Itoh K, Diamond IT. 1983. The laminar organization of the lateral geniculate body and the striate cortex in the squirrel monkey (Saimiri sciureus). *J. Neurosci.* 3(4):673–702

Fitzpatrick D, Usrey WM, Schofield BR, Einstein G. 1994. The sublaminar organization of corticogeniculate neurons in layer 6 of macaque striate cortex. *Vis. Neurosci.* 11(2):307–15

Friedman HS, Zhou H, von der Heydt R. 2003. The coding of uniform colour figures in monkey visual cortex. *J. Physiol.* 548(Pt. 2):593–613

Gattass R, Gross CG, Sandell JH. 1981. Visual topography of V2 in the macaque. *J. Comp. Neurol.* 201(4):519–39

Gattass R, Sousa AP, Mishkin M, Ungerleider LG. 1997. Cortical projections of area V2 in the macaque. *Cereb. Cortex* 7(2):110–29

Gegenfurtner KR, Kiper DC, Fenstemaker SB. 1996. Processing of color, form, and motion in macaque area V2. *Vis. Neurosci.* 13(1):161–72

Ghose GM, Yang T, Maunsell JH. 2002. Physiological correlates of perceptual learning in monkey V1 and V2. *J. Neurophysiol.* 87(4):1867–88

Girard P, Bullier J. 1989. Visual activity in area V2 during reversible inactivation of area 17 in the macaque monkey. *J. Neurophysiol.* 62(6):1287–302

Gur M, Kagan I, Snodderly DM. 2005. Orientation and direction selectivity of neurons in V1 of alert monkeys: functional relationships and laminar distributions. *Cereb. Cortex*. In press

Hawken MJ, Parker AJ, Lund JS. 1988. Laminar organization and contrast sensitivity of direction-selective cells in the striate cortex of the Old World monkey. *J. Neurosci.* 8(10):3541–48

Hendrickson AE, Hunt SP, Wu JY. 1981. Immunocytochemical localization of glutamic acid decarboxylase in monkey striate cortex. *Nature* 292(5824):605–7

Hendry SH, Reid RC. 2000. The koniocellular pathway in primate vision. *Annu. Rev. Neurosci.* 23:127–53

Hendry SH, Yoshioka T. 1994. A neurochemically distinct third channel in the macaque dorsal lateral geniculate nucleus. *Science* 264(5158):575–77

Horton JC, Adams DL. 2005. The cortical column: a structure without a function? *Philos. Trans. R. Soc. London Ser. B*. In press

Horton JC. 1984. Cytochrome oxidase patches: a new cytoarchitectonic feature of monkey visual cortex. *Philos. Trans. R. Soc. London Ser. B* 304(1119):199–253

Horton JC, Hoyt WF. 1991. Quadrantic visual field defects. A hallmark of lesions in extrastriate (V2/V3) cortex. *Brain* 114(Pt. 4):1703–18

Horton JC, Hubel DH. 1981. Regular patchy distribution of cytochrome oxidase staining in primary visual cortex of macaque monkey. *Nature* 292(5825):762–64

Horton JC, Sincich LC. 2004. How specific is V1 input to V2 thin stripes? *Soc. Neurosci. Abstr.* 34:18.1

Horwitz GD, Chichilnisky EJ, Albright TD. 2004. Spatial opponency and color tuning dynamics in macaque V1. *Soc. Neurosci. Abstr.* 34:370.9

Hubel DH, Livingstone MS. 1987. Segregation of form, color, and stereopsis in primate area 18. *J. Neurosci.* 7(11):3378–415

Hubel DH, Wiesel TN. 1962. Receptive fields, binocular interaction and functional architecture in the cat's visual cortex. *J. Physiol.* 160:106–54

Hubel DH, Wiesel TN. 1977. Ferrier lecture. Functional architecture of macaque monkey visual cortex. *Proc. R. Soc. London Ser. B* 198(1130):1–59

Hupe JM, James AC, Girard P, Bullier J. 2001. Response modulations by static texture surround in area V1 of the macaque monkey do not depend on feedback connections from V2. *J. Neurophysiol.* 85(1):146–63

Ito M, Komatsu H. 2004. Representation of angles embedded within contour stimuli in area V2 of macaque monkeys. *J. Neurosci.* 24(13):3313–24

Johnson EN, Hawken MJ, Shapley R. 2001. The spatial transformation of color in the primary visual cortex of the macaque monkey. *Nat. Neurosci.* 4(4):409–16

Kaas JH, Huerta MF, Weber JT, Harting JK. 1978. Patterns of retinal terminations and laminar organization of the lateral geniculate nucleus of primates. *J. Comp. Neurol.* 182(3):517–53

Kennedy H, Bullier J. 1985. A double-labeling investigation of the afferent connectivity to cortical areas V1 and V2 of the macaque monkey. *J. Neurosci.* 5(10):2815–30

Kiper DC, Fenstemaker SB, Gegenfurtner KR. 1997. Chromatic properties of neurons in macaque area V2. *Vis. Neurosci.* 14(6):1061–72

Kobatake E, Tanaka K. 1994. Neuronal selectivities to complex object features in the ventral visual pathway of the macaque cerebral cortex. *J. Neurophysiol.* 71(3):856–67

Lachica EA, Beck PD, Casagrande VA. 1992. Parallel pathways in macaque monkey striate cortex: anatomically defined columns in layer III. *Proc. Natl. Acad. Sci. USA* 89(8):3566–70

Landisman CE, Ts'o DY. 2002. Color processing in macaque striate cortex: relationships to ocular dominance, cytochrome oxidase, and orientation. *J. Neurophysiol.* 87(6):3126–37

Lee BB, Kremers J, Yeh T. 1998. Receptive fields of primate retinal ganglion cells studied with a novel technique. *Vis. Neurosci.* 15:161–75

Lennie P, Krauskopf J, Sclar G. 1990. Chromatic mechanisms in striate cortex of macaque. *J. Neurosci.* 10(2):649–69

Leventhal AG, Thompson KG, Liu D, Zhou Y, Ault S. 1995. Concomitant sensitivity to orientation, direction, and color of cells in layers 2, 3, and 4 of monkey striate cortex. *J. Neurosci.* 15(3 Pt. 1):1808–18

Levitt JB, Kiper DC, Movshon JA. 1994a. Receptive fields and functional architecture of macaque V2. *J. Neurophysiol.* 71(6):2517–42

Levitt JB, Yoshioka T, Lund JS. 1994b. Intrinsic cortical connections in macaque visual area V2: evidence for interaction between different functional streams. *J. Comp. Neurol.* 342(4):551–70

Levitt JB, Yoshioka T, Lund JS. 1995. Connections between the pulvinar complex and cytochrome oxidase-defined compartments in visual area V2 of macaque monkey. *Exp. Brain Res.* 104(3):419–30

Livingstone M, Hubel D. 1988. Segregation of form, color, movement, and depth: anatomy, physiology, and perception. *Science* 240(4853):740–49

Livingstone MS, Hubel DH. 1982. Thalamic inputs to cytochrome oxidase-rich regions in monkey visual cortex. *Proc. Natl. Acad. Sci. USA* 79(19):6098–101

Livingstone MS, Hubel DH. 1984a. Anatomy and physiology of a color system in the primate visual cortex. *J. Neurosci.* 4(1):309–56

Livingstone MS, Hubel DH. 1984b. Specificity of intrinsic connections in primate primary visual cortex. *J. Neurosci.* 4(11):2830–35

Livingstone MS, Hubel DH. 1987. Connections between layer 4B of area 17 and the thick cytochrome oxidase stripes of area 18 in the squirrel monkey. *J. Neurosci.* 7(11):3371–77

Lund JS, Angelucci A, Bressloff PC. 2003. Anatomical substrates for functional columns in macaque monkey primary visual cortex. *Cereb. Cortex* 13(1):15–24

Lund JS, Boothe RG. 1975. Interlaminar connections and pyramidal neuron organization in the visual cortex, area 17, of the macaque monkey. *J. Comp. Neurol.* 159:305–334

Lund JS, Hendrickson AE, Ogren MP, Tobin EA. 1981. Anatomical organization of primate visual cortex area VII. *J. Comp. Neurol.* 202(1):19–45

Malach R, Tootell RB, Malonek D. 1994. Relationship between orientation domains, cytochrome oxidase stripes, and intrinsic horizontal connections in squirrel monkey area V2. *Cereb. Cortex* 4(2):151–65

Martin KA. 1988. From enzymes to visual perception: a bridge too far? *Trends Neurosci.* 11(9):380–87

Merigan WH, Maunsell JH. 1993. How parallel are the primate visual pathways? *Annu. Rev. Neurosci.* 16:369–402

Moutoussis K, Zeki S. 2002. Responses of spectrally selective cells in macaque area V2 to wavelengths and colors. *J. Neurophysiol.* 87(4):2104–12

Movshon JA, Newsome WT. 1996. Visual response properties of striate cortical neurons projecting to area MT in macaque monkeys. *J. Neurosci.* 16(23):7733–41

Nassi JJ, Callaway EM. 2004. Contributions of magnocellular and parvocellular pathways to neurons in monkey primary visual cortex that provide direct versus indirect input to area MT. *Soc. Neurosci. Abstr.* 34:300.16

Ohki K, Chung S, Ch'ng YH, Kara P, Reid RC. 2005. Functional imaging with cellular resolution reveals precise micro-architecture in visual cortex. *Nature* 433:597–603

O'Keefe LP, Levitt JB, Kiper DC, Shapley RM, Movshon JA. 1998. Functional organization of owl monkey lateral geniculate nucleus and visual cortex. *J. Neurophysiol.* 80(2):594–609

Olavarria J, VanSluyters RC. 1985. Unfolding and flattening the cortex of gyrencephalic brains. *J. Neurosci. Methods* 15(3):191–202

Olavarria JF, VanEssen DC. 1997. The global pattern of cytochrome oxidase stripes in visual area V2 of the macaque monkey. *Cereb. Cortex* 7(5):395–404

Perry VH, Oehler R, Cowey A. 1984. Retinal ganglion cells that project to the dorsal lateral geniculate nucleus in the macaque monkey. *Neuroscience* 12(4):1101–23

Peterhans E, von der Heydt R. 1993. Functional organization of area V2 in the alert macaque. *Eur. J. Neurosci.* 5(5):509–24

Poggio GF, Gonzalez F, Krause F. 1988. Stereoscopic mechanisms in monkey visual cortex: binocular correlation and disparity selectivity. *J. Neurosci.* 8(12):4531–50

Reid RC, Shapley RM. 2002. Space and time maps of cone photoreceptor signals in macaque lateral geniculate nucleus. *J. Neurosci.* 22:6158–75

Ringach DL, Shapley RM, Hawken MJ. 2002. Orientation selectivity in macaque V1: diversity and laminar dependence. *J. Neurosci.* 22:5639–51

Rockland KS. 1985. A reticular pattern of intrinsic connections in primate area V2 (area 18). *J. Comp. Neurol.* 235(4):467–78

Rockland KS. 1992. Laminar distribution of neurons projecting from area V1 to V2 in macaque and squirrel monkeys. *Cereb. Cortex* 2(1):38–47

Rockland KS. 1997. Elements of cortical architecture: hierarchy revisited. In *Cerebral Cortex*, ed. KS Rockland, JH Kaas, A Peters, 12:243–93. New York: Plenum

Rockland KS, Pandya DN. 1979. Laminar origins and terminations of cortical connections of the occipital lobe in the rhesus monkey. *Brain Res.* 179(1):3–20

Rockland KS, Pandya DN. 1981. Cortical connections of the occipital lobe in the rhesus monkey: interconnections between areas 17, 18, 19 and the superior temporal sulcus. *Brain Res.* 212(2):249–70

Rockland KS, Virga A. 1989. Terminal arbors of individual "feedback" axons projecting from area V2 to V1 in the macaque monkey: a study using immunohistochemistry of anterogradely transported Phaseolus vulgaris-leucoagglutinin. *J. Comp. Neurol.* 285(1):54–72

Rockland KS, Virga A. 1990. Organization of individual cortical axons projecting from area V1 (area 17) to V2 (area 18) in the macaque monkey. *Vis. Neurosci.* 4(1):11–28

Rodman HR, Sorenson KM, Shim AJ, Hexter DP. 2001. Calbindin immunoreactivity in the geniculo-extrastriate system of the macaque: implications for heterogeneity in the koniocellular pathway and recovery from cortical damage. *J. Comp. Neurol.* 431(2):168–81

Roe AW, Ts'o DY. 1995. Visual topography in primate V2: multiple representation across functional stripes. *J. Neurosci.* 15(5 Pt. 2):3689–715

Roe AW, Ts'o DY. 1999. Specificity of color connectivity between primate V1 and V2. *J. Neurophysiol.* 82(5):2719–30

Sandell JH, Schiller PH. 1982. Effect of cooling area 18 on striate cortex cells in the squirrel monkey. *J. Neurophysiol.* 48(1):38–48

Sawatari A, Callaway EM. 1996. Convergence of magno- and parvocellular pathways in layer 4B of macaque primary visual cortex. *Nature* 380(6573):442–46

Sawatari A, Callaway EM. 2000. Diversity and cell type specificity of local excitatory connections to neurons in layer 3B of monkey primary visual cortex. *Neuron* 25(2):459–71

Schiller PH, Logothetis NK, Charles ER. 1990. Functions of the colour-opponent and broadband channels of the visual system. *Nature* 343(6253):68–70

Schiller PH, Malpeli JG. 1977. The effect of striate cortex cooling on area 18 cells in the monkey. *Brain Res.* 126(2):366–69

Schiller PH, Malpeli JG. 1978. Functional specificity of lateral geniculate nucleus laminae of the rhesus monkey. *J. Neurophysiol.* 41(3):788–97

Sereno MI, Dale AM, Reppas JB, Kwong KK, Belliveau JW, et al. 1995. Borders of multiple visual areas in humans revealed by functional magnetic resonance imaging. *Science* 268(5212):889–93

Shapley R, Kaplan E, Soodak R. 1981. Spatial summation and contrast sensitivity of X and Y cells in the lateral geniculate nucleus of the macaque. *Nature* 292(5823):543–45

Sherman SM, Guillery RW. 1996. Functional organization of thalamocortical relays. *J. Neurophysiol.* 76(3):1367–95

Shipp S. 2001. Corticopulvinar connections of areas V5, V4, and V3 in the macaque monkey: a dual model of retinal and cortical topographies. *J. Comp. Neurol.* 439(4):469–90

Shipp S, Zeki S. 1985. Segregation of pathways leading from area V2 to areas V4 and V5 of macaque monkey visual cortex. *Nature* 315(6017):322–25

Shipp S, Zeki S. 2002a. The functional organization of area V2, I: specialization across stripes and layers. *Vis. Neurosci.* 19(2):187–210

Shipp S, Zeki S. 2002b. The functional organization of area V2, II: the impact of stripes on visual topography. *Vis. Neurosci.* 19(2):211–31

Shmuel A, Korman M, Sterkin A, Harel M, Ullman S, et al. 2005. Retinotopic axis specificity and selective clustering of feedback projections from V2 to V1 in the owl monkey. *J. Neurosci.* 25:2117–31

Sincich LC, Adams DL, Horton JC. 2003. Complete flatmounting of the macaque cerebral cortex. *Vis. Neurosci.* 20(6):663–86

Sincich LC, Horton JC. 2002a. Divided by cytochrome oxidase: a map of the projections from V1 to V2 in macaques. *Science* 295(5560):1734–37

Sincich LC, Horton JC. 2002b. Pale cytochrome oxidase stripes in V2 receive the richest projection from macaque striate cortex. *J. Comp. Neurol.* 447(1):18–33

Sincich LC, Horton JC. 2003. Independent projection streams from macaque striate cortex to the second visual area and middle temporal area. *J. Neurosci.* 23(13):5684–92

Sincich LC, Park KF, Wohlgemuth MJ, Horton JC. 2004. Bypassing V1: a direct geniculate input to area MT. *Nat. Neurosci.* 7(10):1123–28

Stettler DD, Das A, Bennett J, Gilbert CD. 2002. Lateral connectivity and contextual interactions in macaque primary visual cortex. *Neuron* 36(4):739–50

Tamura H, Sato H, Katsuyama N, Hata Y, Tsumoto T. 1996. Less segregated processing of visual information in V2 than in V1 of the monkey visual cortex. *Eur. J. Neurosci.* 8(2):300–9

Thorell LG, DeValois RL, Albrecht DG. 1984. Spatial mapping of monkey V1 cells with pure color and luminance stimuli. *Vision Res.* 24(7):751–69

Tigges J, Spatz WB, Tigges M. 1973. Reciprocal point-to-point connections between parastriate and striate cortex in the squirrel monkey (*Saimiri*). *J. Comp. Neurol.* 148(4):481–89

Tootell RB, Nelissen K, Vanduffel W, Orban GA. 2004. Search for color 'center(s)' in macaque visual cortex. *Cereb. Cortex* 14(4):353–63

Tootell RB, Silverman MS. 1985. Two methods for flat-mounting cortical tissue. *J. Neurosci. Methods* 15(3):177–90

Tootell RB, Silverman MS, DeValois RL, Jacobs GH. 1983. Functional organization of the second cortical visual area in primates. *Science* 220(4598):737–39

Ts'o DY, Gilbert CD. 1988. The organization of chromatic and spatial interactions in the primate striate cortex. *J. Neurosci.* 8(5):1712–27

Ts'o DY, Roe AW, Gilbert CD. 2001. A hierarchy of the functional organization for color, form and disparity in primate visual area V2. *Vision Res.* 41(10–11):1333–49

Ugolini G. 1995. Specificity of rabies virus as a transneuronal tracer of motor networks: transfer from hypoglossal motoneurons to connected second-order and higher order central nervous system cell groups. *J. Comp. Neurol.* 356(3):457–80

Valverde F. 1978. The organization of area 18 in the monkey. A Golgi study. *Anat. Embryol. (Berlin)* 154(3):305–34

Van Essen DC, Gallant JL. 1994. Neural mechanisms of form and motion processing in the primate visual system. *Neuron* 13(1):1–10

Van Essen DC, Newsome WT, Maunsell JH, Bixby JL. 1986. The projections from striate cortex (V1) to areas V2 and V3 in the macaque monkey: asymmetries, areal boundaries, and patchy connections. *J. Comp. Neurol.* 244(4):451–80

Vanduffel W, Tootell RB, Schoups AA, Orban GA. 2002. The organization of orientation selectivity throughout macaque visual cortex. *Cereb. Cortex* 12(6):647–62

Vidyasagar TR, Kulikowski JJ, Lipnicki DM, Dreher B. 2002. Convergence of parvocellular and magnocellular information channels in the primary visual cortex of the macaque. *Eur. J. Neurosci.* 16(5):945–56

Wachtler T, Sejnowski TJ, Albright TD. 2003. Representation of color stimuli in awake macaque primary visual cortex. *Neuron* 37(4):681–91

Wade AR, Brewer AA, Rieger JW, Wandell BA. 2002. Functional measurements of human ventral occipital cortex: retinotopy and colour. *Philos. Trans. R. Soc. London Ser. B* 357(1424):963–73

Weller RE, Kaas JH. 1983. Retinotopic patterns of connections of area 17 with visual areas V-II and MT in macaque monkeys. *J. Comp. Neurol.* 220(3):253–79

Wiesel TN, Hubel DH. 1966. Spatial and chromatic interactions in the lateral geniculate body of the rhesus monkey. *J. Neurophysiol.* 29(6):1115–56

Wong-Riley M. 1978. Reciprocal connections between striate and prestriate cortex in squirrel monkey as demonstrated by combined peroxidase histochemistry and autoradiography. *Brain Res.* 147(1):159–64

Wong-Riley M. 1979. Columnar cortico-cortical interconnections within the visual system of the squirrel and macaque monkeys. *Brain Res.* 162(2):201–17

Wong-Riley M, Carroll EW. 1984. Effect of impulse blockage on cytochrome oxidase activity in monkey visual system. *Nature* 307(5948):262–64

Xiao Y, Wang Y, Felleman DJ. 2003. A spatially organized representation of colour in macaque cortical area V2. *Nature* 421(6922):535–39

Xu X, Bosking W, Sary G, Stefansic J, Shima D, Casagrande VA. 2004. Functional organization of visual cortex in the owl monkey. *J. Neurosci.* 24(28):6237–47

Yabuta NH, Callaway EM. 1998a. Cytochrome-oxidase blobs and intrinsic horizontal connections of layer 2/3 pyramidal neurons in primate V1. *Vis. Neurosci.* 15(6):1007–27

Yabuta NH, Callaway EM. 1998b. Functional streams and local connections of layer 4C neurons in primary visual cortex of the macaque monkey. *J. Neurosci.* 18(22):9489–99

Yabuta NH, Sawatari A, Callaway EM. 2001. Two functional channels from primary visual cortex to dorsal visual cortical areas. *Science* 292(5515):297–300

Yazar F, Mavity-Hudson J, Ding Y, Oztas E, Casagrande VA. 2004. Layer IIIB-beta of primary visual cortex (V1) and its relationship to the koniocellular (K) pathway in macaque. *Soc. Neurosci. Abstr.* 34:300.17

Yoshioka T, Blasdel GG, Levitt JB, Lund JS. 1996. Relation between patterns of intrinsic lateral connectivity, ocular dominance, and cytochrome oxidase-reactive regions in macaque monkey striate cortex. *Cereb. Cortex* 6(2):297–310

Yoshioka T, Dow BM. 1996. Color, orientation and cytochrome oxidase reactivity in areas V1, V2 and V4 of macaque monkey visual cortex. *Behav. Brain Res.* 76(1–2):71–88

Yoshioka T, Levitt JB, Lund JS. 1994. Independence and merger of thalamocortical channels within macaque monkey primary visual cortex: anatomy of interlaminar projections. *Vis. Neurosci.* 11(3):467–89

Yukie M, Iwai E. 1981. Direct projection from the dorsal lateral geniculate nucleus to the prestriate cortex in macaque monkeys. *J. Comp. Neurol.* 201(1):81–97

Molecular Gradients and Development of Retinotopic Maps

Todd McLaughlin and Dennis D.M. O'Leary

Molecular Neurobiology Laboratory, The Salk Institute, La Jolla, California 92037; email: mclaughlin@salk.edu, doleary@salk.edu

Key Words

axon branching, axon guidance, bidirectional signaling, Ephs, ephrins, visual system development

Abstract

Gradients of axon guidance molecules have long been postulated to control the development of the organization of neural connections into topographic maps. We review progress in identifying molecules required for mapping and the mechanisms by which they act, focusing on the visual system, the predominant model for map development. The Eph family of receptor tyrosine kinases and their ligands, the ephrins, remain the only molecules that meet all criteria for graded topographic guidance molecules, although others fulfill some criteria. Recent reports further define their modes of action and new roles for them, including EphB/ephrin-B control of dorsal-ventral mapping, bidirectional signaling of EphAs/ephrin-As, bifunctional action of ephrins as attractants or repellents in a context-dependent manner, and complex interactions between multiple guidance molecules. In addition, spontaneous patterned neural activity has recently been shown to be required for map refinement during a brief critical period. We speculate on additional activities required for map development and suggest a synthesis of molecular and cellular mechanisms within the context of the complexities of map development.

Contents

- INTRODUCTION 328
  - Toward Discovering Graded Topographic Guidance Molecules 330
- MECHANISMS OF MAP FORMATION 331
  - Axon Extension and Target Overshoot During AP Mapping 333
  - TN Retinotopic Mapping Achieved Through AP-Specific Interstitial Branching 335
  - Mechanisms for AP Branch Specificity 337
  - Parallel AP Gradients of Promoters and Inhibitors of Branching 337
  - Opposing AP Gradients of Branch Inhibitors 338
  - Lateral-Medial Mapping Accomplished by Directed Growth of Interstitial Branches 338
  - EphBs and Ephrin-Bs Control DV Retinotopic Mapping 339
  - Distinctions in Guidance of Primary Axons and Interstitial Branches Require Unique Mechanisms 340
  - Multiple Actions and Models of EphBs and Ephrin-Bs in DV Map Development 342
  - Accounting for Species Differences in Development of Topographic Maps 343
  - Refinement of the Retinotopic Map: Patterned Activity and Axon Repellents 344
  - Additional Activities and Interactions Potentially Required for Map Development 346
  - Genetic Screens: Mapping the Future? 346
- FUTURE DIRECTIONS 347

INTRODUCTION

A critical function of the nervous system is to interpret the environment through the connections of various sensory organs to the brain. To accomplish this task, incoming information must be organized in an efficient manner. Perhaps the most efficient organization is achieved through the use of topographic maps, which are present throughout the brain, to process sensory information (Kaas 1997). In general a topographic map is a projection from one set of neurons to another wherein the receiving set of cells reflects the neighbor relationships of the projecting set. In the nervous system of higher vertebrates topographic maps are common and include sensory maps of the body, tonotopic maps for auditory stimuli, and maps of the visual field. Furthermore, topographic maps persist in some form throughout the circuitry from first-order to higher-order connections.

Map development has been studied in several vertebrate projection systems, including thalamocortical (Dufour et al. 2003, Vanderhaeghen & Polleux 2004), hippocamposeptal (Gao et al. 1996, Yue et al. 2002), olfactory/vomeronasal (Sidebar 1), motor axons to muscles (Feng et al. 2000, Nguyen et al. 2002), and retina to its targets in the brain (see below). However, this latter system, the primary visual projection formed by the axons of retinal ganglion cells (RGCs) to their most prominent midbrain target—the optic tectum (OT) of fish, amphibians, and chick, or the superior colliculus (SC) of mammals—has been far and away the predominant model for studying the development of topographic maps and the gradients of guidance molecules that control their formation. Therefore, we focus on the visual system and primarily on the mechanisms of mapping in the target, with the goal of providing a detailed account of the development of a vertebrate neural map and the molecular mechanisms that control it, though we recognize the importance of growth cone guidance to the target and the intricacies of multiple interacting signaling pathways (see recent reviews Huber et al. 2003, van Horck et al. 2004). We devote most attention to mammals and chickens in which axonal mechanisms of map development require unique actions of topographic guidance molecules in a specific temporal sequence, but we do provide examples of mechanisms and molecules involved in developing maps in lower vertebrates such as frogs and fish.

The representation of the retina onto the OT or SC can be simplified to the mapping of two sets of orthogonally oriented axes: the temporal-nasal (TN) axis of the retina along the anterior-posterior (AP) axis of the OT/SC, and the dorsal-ventral (DV) axis of the retina along the lateral-medial (LM) axis of the OT/SC (corresponding to the ventral-dorsal OT axis in nonmammalian vertebrates). Criteria for a topographic guidance molecule in the retinotectal projection are that it is expressed in a graded or restricted manner in the retina or OT/SC, that RGC axons from different parts of the retina exhibit distinct responses to it, and that it affects RGC mapping in vivo; an additional criterion to provide a stricter definition states that the molecule is required for the development of a proper topographic map, although some molecules can be involved in map development but their role is masked by functional redundancy.

Being well into the molecular era with entire genomes becoming available, one might presume that most major players in topographic guidance are known, although their precise roles and interactions are not. Surprisingly, though, to date, the Eph/ephrin families of receptors and ligands are the only molecules described to meet all criteria for topographic guidance molecules, not only in the retinotectal system but also in other systems in which topographic map development has been studied. A small number of additional molecules, for example, RGM (repulsive guidance molecule) and semaphorins, meet a subset of the criteria.

The diversity and complexity of the expression patterns of Ephs and ephrins,

THE ACCESSORY OLFACTORY SYSTEM

The vomeronasal organ (VNO), located in the ventral nasal cavity, projects to the accessory olfactory bulb (AOB), found posterior-dorsal to the main olfactory bulb using zonal topography. Neurons in apical VNO project to anterior AOB, and neurons in basal VNO project to posterior AOB (**Figure 5**). VNO neurons expressing a given pheromone receptor form multiple glomeruli within each AOB half (Belluscio et al. 1999, Rodriguez et al. 1999). The distributions and known activities of several families of guidance molecules are described in **Figure 5**. As in the visual system, distinct combinations of attractive and repellent molecular activities guide VNO axons to their topographically appropriate zones (Knoll & Drescher 2002). The possibility that neuropilin-2 shapes the gradient of secreted semaphorins is an important concept to be considered in all systems. Furthermore, the apparent use of the GPI-linked ephrin-As as guidance receptors (or part of a receptor complex) has important consequences for models of topographic mapping. The accessory olfactory system, owing to its expression of several molecular guidance families in relatively simple patterns and its zonal topography, is an excellent model system to examine the combinatorial interactions between, and within, families of guidance molecules.

spatially and temporally, and features of their function provide for a vast array of signaling possibilities and therefore guidance activities. The regulation of mRNA and protein localization and the intracellular integration of guidance cues, as well as the influence of the growth substrate, have become major issues (Bassell & Kelic 2004, Huber et al. 2003, van Horck et al. 2004). Many guidance molecules, including Ephs and ephrins, have multiple, often opposite, activities, and the choice of which activity dominates in a given context is critical to the development of topographic maps. Signaling pathways connecting guidance receptors to the cytoskeleton and cell-attachment molecules such as integrins are being defined and provide explanations for the observed functional activities (Davy & Robbins 2000, Gallo & Letourneau 2004, Nakamoto et al. 2004).

Recent system-level progress in the development of retinotopic maps includes the first direct evidence of guidance molecules involved in mapping the DV retinal axis along the LM axis of the OT/SC. In addition, more precise roles for the molecules involved in mapping the TN axis of the retina along the AP axis of the OT/SC have been identified. Also, the refinement of the retinotopic map has been examined through computational models and genetics, lending support to a combinatorial hypothesis involving molecular activities and correlated spontaneous waves of retinal activity.

Toward Discovering Graded Topographic Guidance Molecules

The mechanisms that control the establishment of topographic maps have been intensively studied for many decades, but only in recent years has the molecular control of this process begun to be defined. The chemoaffinity hypothesis, formally proposed by Roger Sperry nearly a half century ago, presaged the dawning of the era of molecular mechanisms of map development (1963). Sperry proposed that molecular tags on projecting axons and their target cells determine the specificity of axonal connections within a neural map. Further, he suggested that these molecular tags might be distributed in complementary gradients that mark corresponding points in both sensory and target structures. Although Sperry based this hypothesis on his studies of regeneration of the retinotectal projection in newts and frogs, it gave direction to the burgeoning field of map development. The basic tenet of the chemoaffinity hypothesis has largely been borne out (Benson et al. 2001), but subsequent mathematical models have substantially refined it to add countergradients of attractants (e.g., Gierer 1983) and graded repellents (e.g., Gierer 1987) and to account more accurately for map development, in particular the sequential phases of complex behaviors exhibited by RGC axons during map development in the OT/SC of higher vertebrates (Yates et al. 2004).

On the basis of the chemoaffinity hypothesis, each point in the OT/SC would have a unique molecular address determined by the graded distribution of topographic guidance molecules along the two tectal axes, and similarly each RGC would have a unique profile of receptors for those molecules that would result in a position-dependent, differential response to them by RGC axons. Over the next several decades, the specificity of the projections of RGC axons to tectal cells was investigated further by analyzing axonal projections in normal animals and following experimental manipulations, first in the regenerating retinotectal system of fish and amphibians, and later in the developing retinotectal/retinocollicular projections when new high-resolution axon-tracing techniques became available. This body of evidence supported the basic tenet of the chemoaffinity hypothesis that the establishment of topographic projections involves RGC axons responding to positional information in the OT/SC.

Searches for molecules with features suggestive of roles in mapping have been carried out by many labs using numerous approaches. Several cell-surface molecules, such as TRAP (McLoon 1991) and TOP$_{AP}$ (Savitt et al. 1995), with graded or restricted patterns in the retina and/or OT/SC consistent with a role in mapping were identified prior to the mid-1990s, but functional studies have yet to show such a role (for review see Roskies et al. 1995). The first description of graded molecules that proved to have properties of topographic guidance molecules came only a decade ago with the cloning of two related genes, ephrin-A2 (originally called Eph ligand family-1, or ELF-1) by Flanagan and colleagues (Cheng et al. 1995, Cheng & Flanagan 1994) and ephrin-A5 by Bonhoeffer, Drescher, and colleagues (originally called repulsive axon guidance signal, or RAGS) (Drescher et al. 1995), both of which are ligands of the receptor tyrosine kinase EphA3 (originally named MEK4), expressed in a graded pattern by RGCs (Cheng et al. 1995). Both ephrin-A2 and ephrin-A5 were thereafter shown to meet functional criteria for graded topographic guidance molecules, including differential responses of RGC axons to them (ephrin-A2: Nakamoto et al. 1996; ephrin-A5: Drescher et al. 1995, Monschau et al. 1997) and their influence on retinotopic mapping in vivo, as well as their requirement for proper map development (ephrin-A2: Nakamoto et al. 1996, Feldheim et al. 2000; ephrin-A5: Frisen et al. 1998).

Prior to the discovery of the ephrins, the most compelling evidence for topographic guidance molecules came from the work of Bonhoeffer's group using in vitro assays, including the membrane stripe and growth cone collapse assays. Using the membrane stripe assay, they showed that chick temporal RGC axons, given a choice between growing on alternating lanes of anterior and posterior tectal membranes, show a strong preference to grow on their topographically appropriate anterior membranes, whereas nasal RGC axons exhibit no preference. One critical finding showed that the growth preference of temporal axons is not due to an attractant or growth-promoting activity associated with anterior tectal membranes but instead to a repellent activity associated with posterior tectal membranes (Walter et al. 1987a,b).

Posterior tectal membranes also preferentially collapse the growth cones of temporal axons, a feature that facilitated biochemically isolating the repellent activity to a 33-kDa, GPI-anchored protein referred to as RGM (Cox et al. 1990, Stahl et al. 1990). RGM is expressed in a graded pattern in the OT similar to ephrin-As, and inactivation of RGM using the CALI (chromophore assisted laser inactivation) technique resulted in a loss of the selective repellent effect of posterior OT membranes on temporal RGC axons (Muller et al. 1996). Chick RGM was recently cloned, and recombinant RGM expressed in 293T cells has a repellent effect on chick RGC axons (Monnier et al. 2002). RGC axons transfected to express neogenin, an RGM receptor, are also repelled by RGM (Rajagopalan et al. 2004). Therefore, it was reasonable to assume that RGM has a required role in retinotopic mapping. Suprisingly, though, mice with targeted deletions of RGMa reportedly have no mapping defects in the retinocollicular projection (Niederkofler et al. 2004), possibly because of a functional redundancy, for example with other RGM family members or with ephrin-As.

MECHANISMS OF MAP FORMATION

Determining the process by which RGCs establish topographic connections is critical for defining the roles of graded guidance molecules in map development and for creating conceptual or computational models of the process. Investigators have used predominantly frogs, fish, chicks, and rodents as vertebrate models for development of retinotopic maps. These species exhibit important differences in the development of the visual system and retinotopic maps, as well as substantial differences in the absolute size of the OT/SC; for example, the AP axis of the chick OT is about 50 times greater than that in frog and fish (**Figure 1**). Although each species has unique features that can be exploited, they also have substantial differences in mechanisms employed by RGC axons to target their correct termination zone (TZ) and therefore the actual roles of topographic guidance molecules in controlling the topographic targeting of RGC axons.

Development of retinotectal topography in chicks (Nakamura & O'Leary 1989, Yates et al. 2001) and rodents (Hindges et al. 2002; Simon & O'Leary 1992a,b,c) is a multistep process that involves axon overshoot and interstitial branching. Detailed quantitative analyses have indicated that this is the exclusive mechanism for map development and have begun to define the relative importance of directed axon extension and branching and the roles of guidance molecules in controlling them (Hindges et al. 2002, McLaughlin et al.

2003b, Yates et al. 2001). Initially, the primary growth cones of RGC axons enter the OT/SC and extend posteriorly past the location of their future TZ (Simon & O'Leary 1992a,c; Yates et al. 2001) (**Figure 1**). RGC axons from a given DV location have a broad distribution along the LM tectal axis, with a peak centered on the location of the future TZ, mirroring their coarse ordering within the optic tract (Simon & O'Leary 1991, 1992b; Yates et al. 2001). In rodents, these two features result in RGC axons originating from a focal source in the retina covering virtually the entire SC at perinatal ages and covering a sizeable fraction of the chick OT at E10 to E13.

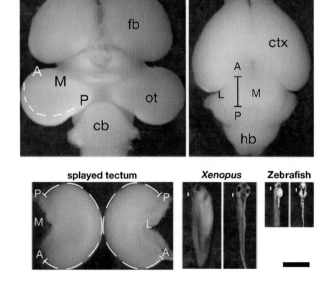

Figure 1

Development of the retinotopic projection and relative scale of the tectum in primary model species. (*A, Top*) In mouse and chick, RGC axons enter the OT/SC and initially extend well posterior to the location of their future termination zone (TZ) (*circle*). Interstitial branches form along the axon shaft in a distribution biased for the AP location of the TZ and subsequently exhibit bidirectional growth along the LM axis toward their correct TZ. Upon reaching their TZ, branches elaborate complex arbors and the initial axon overshoot is eliminated. All arbors are formed by interstitial branches. (*A, Bottom*) In frog (*Xenopus laevis*) and zebrafish, the tectum and retina expand throughout the development of the retinal projection. During retinotopic map development the tectum is much smaller in relation to a typical growth cone in frog and fish than in chick and mouse. RGC axons extend into the tectum and elaborate many small branches from the base of the growth cone. Arbors elaborate from these backbranches and the thinned growth cone. The TZ becomes dense and refines as the tectum enlarges. (*A, Background*) The two ovals in the background represent the relative sizes of the chick tectum (*large oval*) and frog or zebrafish tectum (*small oval*). (*B*) The photographs are at the same scale. The chick OT rotates during development such that the posterior pole (P) is near the midline. The OT is cut along the AP axis at the LM midline (*dashed line*) and splayed. The distance from the anterior to posterior pole along the cut edge is 1 cm (*dashed lines in the splayed tectum*). The mouse SC is about 2 mm along the AP pole at the LM midline (*bar*). For frog and zebrafish the entire animal is shown in lateral and dorsal views. The white bar on the left of each panel represents the approximate AP position and size of the tectum. The tecta for these organisms are approximately 200 μm along the AP axis. cb, cerebellum; ctx, cortex; fb, forebrain; hb, hindbrain. Scale bar = 2 mm. Adapted from McLaughlin et al. 2003a.

A period of interstitial branch formation begins and is the first indication of appropriate topography. Branches form de novo from the axon shaft hundreds of microns or even millimeters behind the growth cone. Interstitial branching exhibits a significant degree of topographic specificity along the AP axis; the highest percentage of branches are found at the AP location of the future TZ (Yates et al. 2001). Interstitial branches form roughly perpendicular to the primary axon and preferentially extend along the LM axis toward their future TZ (Hindges et al. 2002, McLaughlin et al. 2003b, Nakamura et al. 1989). The branches arborize at the appropriate LM and AP location of their TZ and are the exclusive means by which RGCs form permanent, topographically ordered synaptic connections (Yates et al. 2001). Although RGC axons originating from the same site along the DV retinal axis are broadly distributed across the LM extent of the OT/SC, with most being well outside the LM position of their appropriate TZ, their distribution does not change even though their number declines as the map undergoes considerable refinement coincident with the death of a substantial proportion of RGCs (Hindges et al. 2002, Simon & O'Leary 1992b). Therefore, the position of an RGC axon along the LM axis relative to its TZ does not bias its ability to make a connection to the TZ and to be maintained.

In frogs and fish, initial DV mapping along the LM axis is much more accurate than in chicks and rodents. In addition, RGC axons extend along the AP axis directly to the correct location of their TZ (**Figure 1**). As the growth cone of the primary RGC axon reaches the location of its future TZ, it stops and exhibits a phenomenon termed backbranching. Backbranching is characterized by the formation of short terminal branches at or near the base of the growth cone, which itself often acquires a branch-like morphology, and together they locally elaborate a terminal arborization of the distal part of the primary axon (Harris et al. 1987, Kaethner & Stuermer 1992, O'Rourke et al. 1994). Thus, backbranching, as originally defined in frogs and fish, is a phenomenon distinct in scale, location, and purpose from interstitial branching in chicks and rodents.

The size of an individual arbor in frog and fish is much larger in relation to the OT than in chick and mouse, in part owing to the fact that RGC axons reach the OT and arborize at relatively early stages of tectal neurogenesis when the OT is very small. Therefore, though RGC axons are not described to overshoot their TZ in frogs or fish, early on their RGC axonal arbors are disproportionately large compared with the OT, particularly along its AP axis, and cover a greater percentage of its surface area than at later stages. In frogs and fish, arbors cover progressively less of the AP axis over the period of map development because the OT expands substantially more than the arbors, and some arbor refinement occurs (Debski & Cline 2002). In contrast, the surface area of the OT/SC of chick and rodents expands relatively little over the period of map development.

Axon Extension and Target Overshoot During AP Mapping

RGC axons enter their midbrain target at its anterior edge and extend posteriorly parallel to the AP axis of the OT/SC. In amphibians and fish the primary RGC axonal growth cone halts its posterior extension at, or just past, its TZ (Harris et al. 1987). However, in higher vertebrates, essentially all RGC axons extend well posterior to the topographically appropriate location of their future TZ (Nakamura & O'Leary 1989; Simon & O'Leary 1992a,c; Yates et al. 2001). Thus, the growth cones of RGC axons in birds and mammals do not target their future TZ but instead extend a millimeter or more posterior to it.

The extent of the posterior overshoot in higher vertebrates and targeted posterior extension in lower vertebrates is controlled, in large part, by repellent EphA/ephrin-A interactions (**Figure 2**) (Ciossek et al. 1998; Drescher et al. 1995; Feldheim et al. 2000,

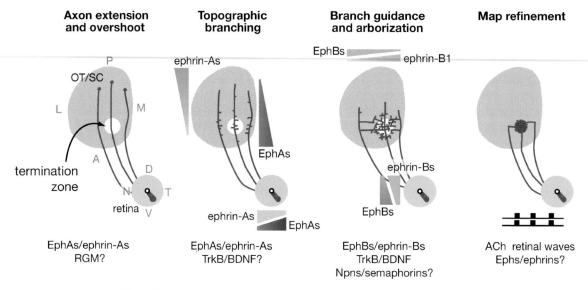

Figure 2

Mechanisms and molecules controlling retinotopic mapping in chicks and rodents. The names and/or distributions of molecules known, or potentially able, to control the dominant mechanisms at each stage are listed. The gradients represent the consensus distribution for a combination of related molecules (i.e., ephrin-A's), which are not listed individually owing to distinctions in the individual members expressed and the precise distributions between species. Molecules other than those listed are likely to participate.

2004; Nakamoto et al. 1996). EphAs are expressed in an overall low-to-high NT gradient in RGCs, and ephrin-As are expressed in an overall low-to-high AP gradient in the OT/SC, though the individual receptors and ligands and their exact distributions vary between species (**Figure 3**). Thus, in lower vertebrates, which lack a posterior overshoot of the TZ, this single repellent gradient can guide RGC axons to the appropriate location of their TZ (**Figure 4A**). Furthermore, in species with an extended posterior overshoot, the shape of the ephrin-A gradient in the OT/SC predicts the extent of the overshoot. For example, in chick, temporal RGC axons extend a greater distance past their future TZ than do nasal RGC axons, a feature predicted by the relatively shallow slope of ephrin-As in anterior and central OT and the steep slope of ephrin-As in posterior OT (Yates et al. 2001). These correlations are consistent with in vitro (Rosentreter et al. 1998) and in vivo (Brown et al. 2000) data indicating that the incremental change in ligand concentration is a critical factor and that absolute concentration (at least, apparently, in physiological ranges) is not.

Considerable evidence indicates that ephrin-As are repellents acting through EphAs to control the advance of the primary axon's growth cone, dependent on the amount of EphA present on the growth cone and the shape of the ephrin-A gradient it encounters. Reducing signaling through EphAs expressed by RGCs results in a decrease in the repellent response of temporal RGC axons to ephrin-As in vitro (Ciossek et al. 1998; Feldheim et al. 2000, 2004; Frisen et al. 1998) and an increase in the extent of the posterior overshoot in vivo (Frisen et al. 1998, Sakurai et al. 2002). Blocking EphA/ephrin-A interactions in vitro results in a decreased repellent response (Ciossek et al. 1998; Feldheim et al. 2000, 2004). Complementing these findings, increasing signaling through EphAs by overexpression (Brown et al. 2000), or ectopic expression of EphAs (Feldheim et al. 2004) or ephrin-As

(Nakamoto et al. 1996), results in an increase in the repellent response of EphA expressing axons to ephrin-As in vitro (Feldheim et al. 2004, Nakamoto et al. 1996) and a decreased extension of RGC axons along the AP axis in vivo (Brown et al. 2000, Nakamoto et al. 1996).

In zebrafish and frog, the growth cones of RGC axons directly target their appropriate AP location in the OT. The in vitro action of two ephrin-A homologs in zebrafish and their expression patterns, one in a low-to-high AP gradient and another in a dense band posterior to the OT, suggest that ephrin-As act as a molecular barrier to prevent RGC axons from exiting the posterior end of the OT (Brennan et al. 1997). Furthermore, the in vitro action and in vivo distribution of sema3A in posterior OT and neuropilin-1 in RGCs in frog suggest that it may be involved in controlling the posterior extension (and terminal branching, described below) of RGC axons (Campbell et al. 2001). In mice ephrin-As also act as a molecular barrier to prevent RGC axons from extending posteriorly from the SC into the inferior colliculus (Feldheim et al. 2004, Frisen et al. 1998).

TN Retinotopic Mapping Achieved Through AP-Specific Interstitial Branching

In frogs and fish RGC axon extension along the AP axis is determined in part by EphAs and ephrin-As; RGC axonal growth cones stop at or very near the appropriate topographic location and undergo terminal arborization, in part via backbranching (Harris

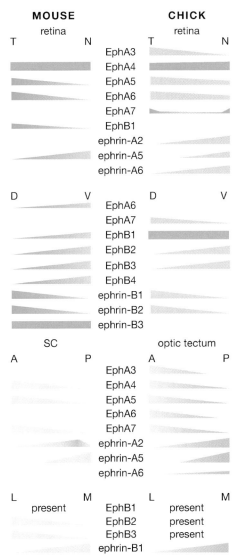

Figure 3

Expression of Ephs and ephrins in the retinocollicular/retinotectal systems of mouse and chick. The table represents our view of the consensus expression patterns for individual Ephs and ephrins in the retina (and likely RGCs) and OT/SC (in positions likely to affect mapping) during the primary molecular-dependent events in topographic map formation (about E15-P7 in mouse and E6-E14 in chick). The list may not be complete, and exclusion from this chart does not necessarily signify absence of expression in vivo. The sizes and shapes of the gradients are generalized, and relative expression levels should not be inferred (for an example of relative expression levels of Eph receptors see Reber et al. 2004). We have included only Eph family members with published expression domains determined by in situ hybridization with antisense riboprobes or specific antibodies in preparations where the listed pattern is evident at an appropriate age. A, anterior; D, dorsal; L, lateral; M, medial; N, nasal; P, posterior; present, receptor is expressed but the pattern is unclear; T, temporal; V, ventral.

et al. 1987, Kaethner & Stuermer 1992). In vitro studies suggest that backbranching may be causally linked to growth cone collapse or to the halting of axonal extension: For example, a neuropilin-1-mediated collapse of the growth cones of frog RGC axons in response to Sema3A leads to an increase in backbranching around the collapsed growth cone (Campbell et al. 2001). RGC axon arborization via terminal branching is also likely controlled, at least in part, by TrkB/BDNF interactions (Alsina et al. 2001, Cohen-Cory 2002, Cohen-Cory & Fraser 1995).

In mammals and birds, critical events in retinotopic mapping are the topographically selective addition and stabilization of interstitial branches, which subsequently form all RGC axonal arbors after they directionally extend to their topographically correct site (Simon & O'Leary 1992a,b,c; Yates et al. 2001). In these species, growth cones of RGC axons typically grow well over a millimeter posterior to the location of their future TZ before they halt their extension. The majority of branches extending from the axon shaft do so at or near the AP location of the nascent TZ, with a paucity of branches anterior and posterior to it (Simon & O'Leary 1992a,b,c; Yates et al. 2001). At least two distinct activities

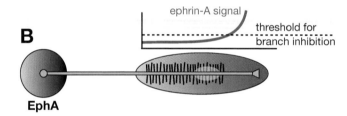

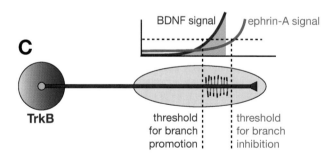

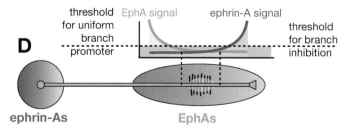

Figure 4

Requirement for two molecular activities to control branch distribution along the AP axis. (*A*) A gradient of repulsive guidance molecules, such as ephrin-As, is in principle sufficient to guide RGC axonal growth cones topographically to their appropriate TZ. This is the mapping mechanism in lower vertebrates. (*B*) However, a single repulsive gradient cannot result in the topographic branching observed in chicks and rodents. If the inhibitory signal allows for branching at the TZ, it must also allow for branching anterior to the TZ at the same, or higher, levels. This is not observed in vivo. (*C*) Gradients of molecules with branch-promoting activities could act in concert with inhibitory activities to result in branching restricted to the topographically appropriate domain. TrkB, in a similar distribution to EphAs in the retina (and/or if TrkB were graded along each RGC axon), and brain-derived neurotrophic factor (BDNF), in the OT/SC, have the appropriate activities to act with EphAs/ephrin-As in mapping. (*D*) Gradients of branch-inhibiting molecules in gradients opposite the EphA/ephrin-A gradients depicted in (*A*) could also result in topographic branching. Dual inhibitory gradients would require either a branching mechanism intrinsic to RGC axons or a branch-promoting molecule (*blue shading*) distributed to allow branching at the low point of the combined inhibitory gradients. The distributions of ephrin-As in the retina and EphAs in the OT/SC fulfill these requirements.

are required to center the distribution of branches around the future TZ: one activity to limit branching posterior to the TZ and one activity to limit branching anterior to the TZ (O'Leary et al. 1999; Yates et al. 2001, 2004). A single graded activity, whether negative or positive, cannot do both, though a single molecule with two activities could, in theory.

In addition to controlling the posterior overshoot, in vitro branching assays show that ephrin-As also inhibit branch formation along RGC axons (Roskies & O'Leary 1994, Yates et al. 2001). This conclusion is supported by in vivo studies showing enhanced RGC axon branching in the OT coincident with a local inactivation of ephrin-As using CALI (Sakurai et al. 2002). Consistent with these data, temporal axons form ectopic TZs in aberrantly posterior locations in ephrin-A-deficient mice (Feldheim et al. 2000, Frisen et al. 1998). These findings indicate that the low-to-high AP gradient of ephrin-As in the OT/SC exposes RGC axons posterior to their correct TZ to levels of ephrin-As that inhibits their branching and thereby helps generate the topographic bias in branching along the AP axis of the OT/SC observed in vivo (**Figure 2**) (Simon & O'Leary 1992a, Yates et al. 2001). This mechanism of branch inhibition is a primary role for ephrin-As in retinotopic map development (Yates et al. 2001).

Mechanisms for AP Branch Specificity

Though an activity that limits branching along the portion of RGC axons anterior to their TZ remains to be identified experimentally, one must exist, and investigators have suggested some candidates (O'Leary et al. 1999; Yates et al. 2001, 2004). The "activity" could be an intrinsic bias for RGC axons to branch distally due, for example, to a proximal-to-distal gradient in the maturity of the axonal cytoskeleton and the polymerization dynamics of actin and tubulin into neurofilaments and microtubules, possibly coupled with a trigger, for example, collapse of the growth cone of the primary axon in response to ephrin-As as suggested by in vitro work (Davenport et al. 1999). Alternatively, the activity may be due to activation of branch-promoting and/or branch-inhibiting signaling pathways by exogenous ligands binding to their receptors along the axon shaft. Straightforward examples would include a low-to-high AP gradient of a signal that promotes branching along each RGC axon or a high-to-low proximal AP gradient of a signal that inhibits branching along each RGC axon (**Figure 4**). These alternative models are not mutually exclusive and could cooperate to develop AP-specific branching.

Parallel AP Gradients of Promoters and Inhibitors of Branching

In this model, the graded branch-promoting activity is at a level sufficient to overcome the graded ephrin-A branch inhibitory activity and thereby to promote branching only near the TZ; posterior to the TZ its level is insufficient to overcome the ephrin-A branch-inhibitory activity, and anterior to the TZ its level is below a threshold required to promote branching (**Figure 4**) (O'Leary et al. 1999, Yates et al. 2001). In chick, BDNF/TrkB signaling is a good candidate for this activity on the basis of expression patterns and the finding that focal application of BDNF to the shaft of RGC axons selectively induces, via activation of TrkB receptors, the formation of primary branches in vitro (Choi et al. 1998, Choi & O'Leary 2000). Other potential candidates include the ephrin-As, -A2 and -A5, if they acted bifunctionally as branch promoters in addition to their demonstrated role as branch inhibitors (Yates et al. 2001).

Recent findings are consistent with this proposed bifunctional action for ephrin-As. Several guidance molecules, including semaphorins, netrin, and ephrin-B1 (see below), can have both attractive and repulsive functions, depending on developmental context, substrate, and intracellular concentrations of second messengers. In vitro, soluble ephrin-A5 can act as either an attractant or a repellent for frog RGC axons, dependent on the substrate (Weinl et al. 2003), and in vivo ephrin-A5 can have positive or inhibitory effects on distinct subsets of EphA4-expressing motor neurons (Eberhart et al. 2004). More directly relevant is a recent in vitro study concluding that ephrin-A2 can have an adhesive, attractive, or growth-promoting effect on RGC axons at concentrations below those that result in its previously defined repellent effect (Hansen et al. 2004).

Opposing AP Gradients of Branch Inhibitors

An alternative to parallel gradients of branch promoters and inhibitors is a set of opposing gradients along the AP axis, each of which inhibits branching (Yates et al. 2001, 2004). In this scenario, one gradient is the low-to-high AP gradient of ephrin-A2 and -A5, which inhibits branching along RGC axons posterior to their TZ; opposing it is a high-to-low AP signaling gradient that inhibits branching along RGC axons anterior to their TZ (**Figure 4**).

Several ephrin-As and EphAs are expressed by RGCs and the OT/SC in opposing expression gradients that complement the high-to-low graded TN expression of EphAs by RGCs and low-to-high graded AP expression of ephrin-As in the OT/SC (Brennan et al. 1997, Connor et al. 1998, Hornberger et al. 1999, Marcus et al. 1996); in addition to EphAs, ephrin-As are present along RGC axons and exhibit a low-to-high graded TN distribution (Hornberger et al. 1999). The opposing gradients of ephrin-As in the retina sharpen the graded distribution of functional EphA receptors (Hornberger et al. 1999). Others investigators have proposed that the opposing gradients in the retina and OT/SC act as opposing gradients of branch inhibition through EphA-ephrin-A reverse signaling (Yates et al. 2001, 2004). Some ephrin-As and EphAs are expressed in ways that suggest that they act in map development predominantly via reverse signaling. For example, in chick, ephrin-A6 is expressed in a high-to-low NT gradient by RGCs but is sparsely expressed in the OT (Menzel et al. 2001).

EphB-ephrin-B binding is well established to initiate both forward and reverse signaling (Bruckner et al. 1997, Henkemeyer et al. 1996, Holland et al. 1996). Reverse signaling into ephrin-A-expressing cells upon binding EphAs has been implicated in topographic mapping in the accessory olfactory system, although in this system, axonal ephrin-As act as attractant receptors for EphAs in the target (Davy et al. 1999, Knoll et al. 2001) (See sidebar on the Accessory Olfactory System and **Figure 5**). Computational modeling of retinotopic mapping shows that opposing gradients of EphAs and ephrin-As can act as branch inhibitors through bidirectional signaling and generate the major phases of map development in chick and mouse, including progressive increases in the topographic specificity of AP branching exhibited by RGC axons originating from all TN positions, and can recapitulate the phenotypes reported for ephrin-A knockout and EphA knock-in (KI) mice (Yates et al. 2004).

Lateral-Medial Mapping Accomplished by Directed Growth of Interstitial Branches

In zebrafish and frog, the growth cones of RGC axons target directly to the appropriate LM location of their future TZ. However, in rodents and chicks, RGC axons from the same retinal location enter and grow across the OT/SC with a broad distribution over its LM axis, though biased for the LM location of their future TZ (Hindges et al. 2002; Simon & O'Leary 1991, 1992a,b,c). Topographic specificity along the LM axis emerges through the bidirectional guidance of branches that form along RGC axons (Hindges et al. 2002, McLaughlin et al. 2003b, Nakamura & O'Leary 1989) with an AP bias as described above. Branches that extend from RGC axons located lateral to their future TZ grow medially, branches that extend from RGC axons located medial to their future TZ grow laterally, and branches extending from RGC axons located within the LM extent of the future TZ exhibit no directional bias (Hindges et al. 2002, McLaughlin et al. 2003b, Nakamura & O'Leary 1989). Branches that reach the area of the nascent TZ selectively form complex arbors. In summary, in mammals and chicks, the bidirectional guidance of interstitial branches is the critical feature in retinotopic mapping along the LM axis of the OT/SC, analogous to the importance of topographic specificity in branch formation in AP mapping.

EphBs and Ephrin-Bs Control DV Retinotopic Mapping

Characterization of the molecular control of DV mapping along the LM axis of the OT/SC has lagged behind that of TN mapping along the AP axis of the OT/SC, in part because in vitro assays that reveal strong TN responses to endogenous AP target tissues fail to reveal differential DV responses from RGC axons. Only in the past few years have defined molecules been shown to control DV mapping; the reports are few and demonstrate roles for EphBs and ephrin-Bs but implicate both bidirectional signaling and bifunctional action.

In retina, EphB receptors are expressed by RGCs during map development in an overall low-to-high DV gradient, complemented by an overall high-to-low DV gradient of ephrin-Bs (McLaughlin et al. 2003a). In both chick OT and mouse SC, ephrin-B1 is expressed in a low-to-high LM gradient (Braisted et al. 1997, Hindges et al. 2002), complemented by an overall high-to-low LM EphB gradient (Hindges et al. 2002). Analyses of EphB2 and EphB3 mutant mice, with and without reverse signaling intact, show aberrant LM mapping due to defects in the guidance of interstitial branches; these findings show that ephrin-B1 acts as a branch attractant via EphB2/B3

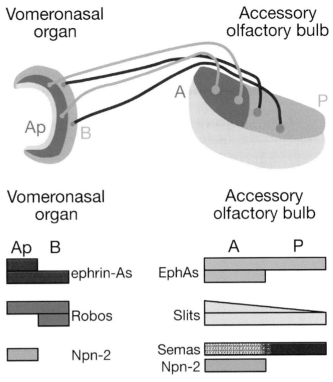

Figure 5

Representation of the accessory olfactory system in cross section. Apical (Ap, *blue*) vomeronasal (VNO) neurons (*light blue*) extend axons into the anterior (A, *blue*) accessory olfactory bulb (AOB) and form glomeruli. Basal (B, *green*) VNO neurons (*dark green*) extend axons into posterior (P, *green*) AOB and form glomeruli. This zonal topography is controlled by the guidance molecules charted below the illustration in representative form. Neuropilin-2 (Npn-2) in the AOB is present primarily in the anterior external plexiform layer of the AOB (*gray*) and may act to sequester secreted semaphorins (Semas), thus converting the uniform expression of secreted semaphorins in the AOB into a functional distribution of semaphorin protein (*purple stippled area*) highest in posterior AOB by reducing the availability of ligand in anterior AOB. Apical VNO neurons are guided to anterior AOB by attractive ephrin-A/EphA interactions and repellent Npn-2/sema interactions, whereas basal VNO neurons are guided to posterior AOB by repellent Robo/Slit interactions (Cloutier et al. 2002, 2004; Knoll et al. 2001, 2003; Walz et al. 2002). Other activities may play a role as well, and the described activities likely act in concert with pheromone receptors and other guidance cues (such as MHC class molecules; Loconto et al. 2003) to control the formation of glomeruli (Belluscio et al. 1999, Rodriguez et al. 1999). The gray area in the VNO represents the lumen.

forward signaling; modeling of these data indicates that bidirectional branch extension requires a branch repellent in a distribution paralleling ephrin-B1 (Hindges et al. 2002). This model is supported by the demonstration that high levels of ephrin-B1 repel interstitial branches in a selective manner (McLaughlin et al. 2003b). Taken together, these studies show that in mice and chicks ephrin-B1 acts through EphB forward signaling as both an attractant and repellent: A branch located lateral to its nascent TZ is attracted up the gradient of ephrin-B1 toward its future TZ, whereas a branch located medial to its nascent TZ is repelled down the ephrin-B1 gradient toward its future TZ (**Figure 6**) (Hindges et al. 2002, McLaughlin et al. 2003b). Importantly, the trajectories of primary RGC axons are not changed in wild-type OT/SC nor in the SC of EphB mutant mice or when encountering domains of ectopic ephrin-B1 expression in chick OT (Hindges et al. 2002, McLaughlin et al. 2003b). Therefore, the response of RGC axons to ephrin-B1 is specific to their interstitial branches (and later their arborizations; McLaughlin et al. 2003b) and is context dependent—the location of the branch on the ephrin-B1 gradient in relation to the location of its future TZ and its EphB level determine its response.

In frog, ephrin-B reverse signaling has been implicated in retinotopic mapping (Mann et al. 2002). Increasing ephrin-B expression in retina results in defects in RGC axon targeting, consistent with an attractant response for retinally expressed ephrin-Bs being activated on RGC axons by EphBs expressed in the OT (Mann et al. 2003). Expression of a dominant-negative construct in retina to interfere with this interaction results in a response also consistent with an attractant effect of ephrin-Bs acting by reverse signaling (Mann et al. 2002). It remains to be determined if reverse signaling has a role in mapping in mice and chicks and that forward signaling has a role in mapping in frog. In zebrafish, ephrin-B2a expressed in the OT has a repellent effect on RGC axons via forward signaling through EphB receptors (Wagle et al. 2004). In zebrafish, DV retinotopic mapping is also likely controlled, in part, by Sema3D, which is expressed primarily in ventral (lateral) OT and repels ventral RGC axons that map to dorsal (medial) OT (Liu et al. 2004).

Distinctions in Guidance of Primary Axons and Interstitial Branches Require Unique Mechanisms

In frog and zebrafish, EphBs/ephrin-Bs affect the primary axon growth cone in vivo and in vitro (Mann et al. 2002, 2003; Wagle et al. 2004). Conversely, in mice and chicks, EphBs/ephrin-B1 do not influence the trajectories of primary RGC axons in the OT/SC but do affect the guidance of interstitial branches (Hindges et al. 2002, McLaughlin et al. 2003b). A priori, a potential explanation is that RGC axons extend parallel to the ephrin-B1 gradient and thus do not encounter a gradient along their primary direction of extension, whereas their interstitial branches extend perpendicular to the gradient and therefore extend either directly up or directly down the ephrin-B1 gradient. However, primary RGC axons do not respond even to very high levels of graded ephrin-B1 achieved by electroporation of ephrin-B1 retroviral expression vectors (McLaughlin et al. 2003b), whereas they do stop their posterior extension across the OT when they confront an ectopic domain of ephrin-A2 created by retroviral infection (Nakamoto et al. 1996). A potential explanation is that growth cones of primary RGC axons lack sufficient EphB receptors and signaling to respond to ephrin-B1 in the OT/SC. An intriguing possibility is that the transport of EphB mRNA and its subsequent translation, or the export of EphB receptors, may be preferentially targeted to interstitial branches as they form rather than to the primary RGC axon growth cones; such differential mRNA transport, local translation, and protein export to selected parts of the axon have been described for other proteins and RNAs (Brittis at al. 2002, Campbell & Holt 2001).

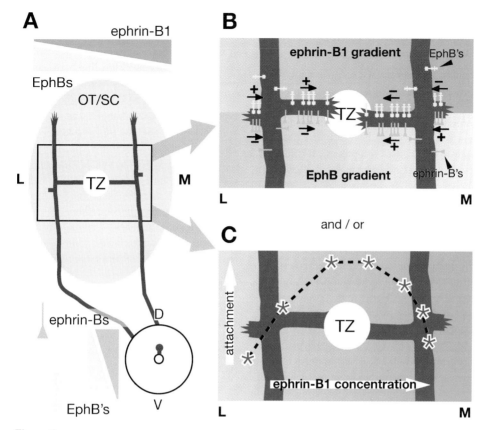

Figure 6

Potential bidirectional and bifunctional interactions resulting in topographic branch guidance along the LM axis of the OT/SC. (*A*) Representation of the projection from two RGCs in the same retinal location. One RGC axon has extended lateral (L) to the TZ and preferentially extends branches medially, toward the TZ. The other RGC axon has extended medial (M) to the TZ and preferentially extends branches laterally, toward the TZ. (*B*) EphBs and ephrin-Bs can cooperate via bifunctional and bidirectional signaling to guide branches appropriately both medially and laterally. The top half of this panel summarizes in vivo data demonstrating branch guidance by EphBs (*yellow icons*) on RGC axons. Lateral to the TZ (*left axon*), EphB receptors encounter an ephrin-B1 level (*green gradient*) lower than that at their TZ (*circle*) and, at that relatively low level of ephrin-B1, branches are attracted up the ephrin-B1 gradient. Branches initially probing down the ephrin-B1 gradient from this axon have no incentive to do so and do not extend. Medial to the TZ (*right axon*), EphB receptors encounter a relatively high level of ephrin-B1 that repels branches down the ephrin-B1 gradient. The bottom half of this panel illustrates potential reverse signaling events. Lateral to the TZ, ephrin-Bs (*green icons*) on RGC axons encounter a higher level of EphBs (*yellow gradient*) than that at the TZ and thus are repulsed down the gradient. Medial to the TZ, ephrin-Bs along RGC axons encounter a relatively low level of EphBs, and branches are attracted up the EphB gradient laterally toward the TZ. (*C*) Branches may also be guided to the TZ by forward signaling if EphBs act as ligand density sensors. Overlaid on the schematic are data from Huynh-do et al. (1999) showing EphB-expressing cell attachment (*y-axis*) on substrates containing different concentrations of ephrin-B1 (*x-axis*). At low concentrations (i.e., lateral to the TZ), attachment is favored at a higher concentration (i.e., medial positions). At high concentrations (i.e., medial to the TZ) increased attachment is favored at a lower concentration (i.e., lateral positions). RGC axonal branches may be guided by a similar principle. The point of maximal attachment for each RGC axon is centered on the TZ and is dependent on its EphB concentration.

Multiple Actions and Models of EphBs and Ephrin-Bs in DV Map Development

EphBs and ephrin-Bs control map development along the LM axis of the OT/SC likely by acting both bifunctionally (i.e., one molecule acting as an attractant and repellent) and bidirectionally (i.e., both EphBs and ephrin-Bs acting as receptors or ligands). Although several guidance molecules have been shown to be bifunctional (van Horck et al. 2004), and EphB/ephrin-Bs have long been known to signal bidirectionally (Bruckner et al. 1997, Henkemeyer et al. 1996, Holland et al. 1996), an individual RGC axon has the unique ability to exhibit a response to all of these signaling possibilities simultaneously (**Figure 6**). For example, two neighboring RGCs may extend axons with multiple EphBs and ephrin-Bs on their membranes and may encounter multiple ephrin-Bs and EphBs in the OT/SC. The responses to these cues, being transmitted by forward signaling through EphBs and reverse signaling through ephrin-Bs, are dependent on the location of each RGC axon in relation to its future TZ. One RGC axon may be located medial to its future TZ and extend branches laterally toward its future TZ through a combination of a repellent response of EphBs binding ephrin-B1 in the OT/SC and an attractant response of ephrin-Bs binding EphBs in the OT/SC (**Figure 6**). Its neighboring RGC, which may have extended lateral to their future TZ, will respond in the exact opposite manner, despite expressing an identical complement of EphBs/ephrin-Bs and responding to identical guidance cues, though at different concentrations, reflecting its different location on the LM axis and therefore on the gradients of EphBs and ephrin-B1.

The bifunctional action of ephrin-B1 through EphBs present along RGC axons may be due to the balance of distinct responses through each receptor type (i.e., EphB2 signaling results in attraction and EphB1 signaling in repulsion) or, alternatively, to a combinatorial thresholding mechanism in which the combined signaling through all EphBs results in either attraction or repulsion, controlled by a transition of EphB signaling between attraction and repulsion to ephrin-B1 that is balanced at the TZ, with lower signaling levels occurring lateral to the TZ and resulting in branch attraction and higher levels occurring medial to the TZ and resulting in branch repulsion (Hindges et al. 2002, McLaughlin et al. 2003b). Trans-endocytosis of EphBs and ephrin-Bs may be responsible for the switch from attraction to repulsion (Marston et al. 2003, Zimmer et al. 2003). At signaling levels above threshold, endocytosis, which initiates repulsion, is favored, whereas at low signaling levels attraction is favored (Marston et al. 2003, Zimmer et al. 2003).

A third model, and in particular an alternative mechanism to a receptor threshold model, is that EphB receptor signaling may act as a ligand-density sensor to control DV retinotopic mapping (**Figure 6**) (McLaughlin et al. 2003b). This model is based on studies showing that EphB1-induced attachment of cell lines to a substrate of extracellular matrix molecules is dependent on the concentration of ephrin-B1 in the substrate (Huynh-Do et al. 1999). Within a critical concentration range, cells attach to their substrate in an integrin-dependent manner at a much higher density; if ephrin-B1 concentration is either above or below this optimal level, cell attachment is decreased. Furthermore, as described above, the point of maximal attachment may depend on the point at which trans-endocytosis of EphB/ephrin-Bs is favored (Marston et al. 2003, Zimmer et al. 2003).

Such a ligand-density sensor model could account for the bidirectional targeted extension of interstitial branches observed in LM mapping. An interstitial branch senses the ephrin-B1 concentration (i.e., ligand density) through EphB receptors and is directed toward its TZ. The amount of EphB receptors expressed by the parent RGC is determined by its DV location and sets the most favorable concentration of ephrin-B1 for its interstitial branches that by definition is found at the appropriate LM position of the TZ. A branch located either medial or lateral to the TZ would encounter a gradient of increasingly favored attachment centered on its TZ (**Figure 6**). EphB-ephrin-B1 signaling on the branch would control the density of other molecules (e.g., integrins, cell adhesion molecules, etc.) that mediate its attachment to ECM components and cells in the OT/SC as well as cytoskeletal changes required for axon growth to promote branch extension toward the optimal ephrin-B1 concentration found at the future TZ.

Accounting for Species Differences in Development of Topographic Maps

Analyses of topographic map development in frogs (O'Rourke & Fraser 1990), fish (Kaethner & Stuermer 1992, Yoda et al. 2004), chick (Nakamura & O'Leary 1989, Yates et al. 2001), rat (Simon & O'Leary 1992a,b), mouse (Hindges et al. 2002), ferret (Chalupa et al. 1996, Chalupa & Snider 1998), and wallaby (Ding & Marrote 1997) show that the initial topographic targeting of RGC axons differs substantially across species. Differences in the initial DV ordering of RGC axons along the LM axis of the OT/SC are a direct reflection of their ordering in the optic tract as they enter the OT/SC from its anterior margin. The broad LM distribution of axons arising from neighboring RGCs in chicks and mammals relative to the tight, topographically ordered distribution in frog and fish requires an additional mechanism to achieve DV retinotopic mapping, such as the directed growth of interstitial branches. Differences in the initial targeting of RGC axons over the AP axis of the OT/SC are likely due to differences in the slope of gradients of guidance molecules, the length of the axis over which the gradient is distributed, and the sensitivity of RGC axons to these molecules. Clear differences do exist between species in the expression of Ephs and ephrins (**Figure 3**, see sidebar on EPH Family of Receptor Tyrosine Kinases).

EPH FAMILY OF RECEPTOR TYROSINE KINASES

The Eph family of receptor tyrosine kinases (RTKs) is the largest known family of RTKs, composed of 14 Ephs and 8 ephrins in mouse and 15 Ephs and 9 ephrins in chick. Signaling through Ephs and ephrins has been implicated in a wide variety of processes including axon guidance, cell migration, vascular development, synapse development, structure and plasticity, and midline fusion events (Himanen & Nikolov 2003, Murai & Pasquale 2003, Palmer & Klein 2003, Poliakov et al. 2004, Rossant & Hirashima 2003, Surawska et al. 2004). Ephs and ephrins are separated into two subclasses on the basis of homology, the EphA/ephrin-As and EphB/ephrin-Bs, within which receptor-ligand binding and activation are promiscuous. In addition, some cross talk occurs between subclasses and may be functionally relevant in some systems. Ephs and ephrins are membrane bound, Ephs and ephrin-Bs are transmembrane proteins, and ephrin-As are GPI-linked to the plasma membrane, allowing for complex and precise patterns of expression within a single tissue or organ and between a projecting set of neurons and their targets (e.g., the retinocollicular/retinotectal projection; **Figure 3**). Eph-ephrin binding initiates signal transduction cascades by both Ephs and ephrins, resulting in bidirectional signaling. Additionally, Eph/ephrin signaling can be bifunctional, resulting in opposing responses (i.e., attraction or repulsion) to the same cues, depending on context (e.g., level of signaling). These features combine to make the Ephs and ephrins major players in many intricate problems of development, including the development of topographic maps described here.

The shape of topographic guidance molecule gradients is an integral part in forming theories of topographic mapping and axon guidance. Early studies indicated that a ~1% difference in concentration of a repellent guidance molecule across the growth cone is required to halt the forward extension of a growth cone (Baier & Bonhoeffer 1992). Calculations based on this estimate and other parameters including growth cone size and effective concentration range of a guidance molecule indicate that a single, graded guidance activity could function over a maximum distance of 5–10 mm (Goodhill & Baier

1998), which intriguingly approximates the 10-mm length of the AP axis of the chick OT during map development (**Figure 1**). However, recent technological advances suggest a more sensitive mechanism that allows for directed growth along much more shallow gradients, implying that relatively shallow gradients of a single class of molecule can be effective for axon guidance, theoretically allowing for guidance over a longer distance than the 5–10 mm previously calculated (Rosoff et al. 2004). Nonetheless, intuitively, target size should be a variable in determining the slope of the gradient of a topographic guidance molecule across a target axis and therefore the distance over which the molecule is effective. For example, if the same concentration range of ephrin-As is distributed along the AP tectal axis in zebrafish as in chick, because the AP axis of the zebrafish OT is only ~2% of the length of the chick OT axis during map development, the slope of the gradient of the full range of effective concentration could be much steeper in the smaller zebrafish OT. If the threshold of growth cone response to the ephrin-A repellent is conserved, a steeper gradient should result in a greater degree of topographic precision in growth cone targeting, as is observed in zebrafish compared with chick. This proposal is supported by the correlation between the position-dependent differential growth cone overshoot and the gradients of ephrin-As in the chick OT (Yates et al. 2001): The greater overshoot exhibited by temporal axons within anterior OT than by nasal axons in posterior OT correlates with the shallow slope of the graded distribution of ephrin-As in anterior OT (solely due to ephrin-A2) and the steep ephrin-A slope in posterior OT (due to ephrin-A2 and -A5 combined).

Refinement of the Retinotopic Map: Patterned Activity and Axon Repellents

As described above, in mice and chicks, all arbors are formed by interstitial branches that preferentially arborize at or in the vicinity of the topographically appropriate TZ. In frog, refinement of individual arbors is a dynamic process involving the addition and subtraction of higher-order branches (Alsina et al. 2001, Ruthazer et al. 2003); further, the continued but disparate growth of the retina and OT requires a continuous small-scale remodeling throughout life (Cline 1998). These processes shape and refine developing arbors and are dependent on TrkB/BNDF interactions and neural activity (Alsina et al. 2001, Ruthazer et al. 2003). Map refinement in fish and frogs is a precise shaping of arbors rather than the large-scale remodeling of a topographically diffuse projection that occurs in rodents and chicks.

In chicks and mice, the initial collection of arbors is loosely organized around the topographically appropriate position of the future TZ and requires a substantial degree of remodeling to develop the precise connections evident in the mature retinotopic map. This large-scale remodeling requires the elimination of overshooting portions of RGC axons as well as the removal of inappropriately located branches and even entire ectopic arbors. In mice, the retinotopic map resembles its mature form by P8, days before the onset of visually evoked activity and the opening of the eyes (Tian & Copenhagen 2003). However, the large-scale map remodeling is coincident with a period of correlated waves of spontaneous neural activity that propagate across the retina (Galli & Maffei 1998, Meister et al. 1991, Wong et al. 1993). These waves are mediated by a network of cholinergic amacrine cells and correlate the activity of neighboring RGCs, thereby relating an RGC's position to its pattern of activity (Wong 1999). Correlated activity has long been postulated to refine topographic connections by strengthening coordinated inputs and/or weakening uncorrelated inputs (Butts 2002, Debski & Cline 2002, Hebb 1949, Stent 1973). Pharmacological activity blockade in mice and chicks suggested a small but evident role for neural activity in map remodeling (Kobayahsi et al. 1990, Simon et al. 1992).

However, a recent analysis of mice lacking cholinergic-mediated retinal waves indicates that correlated patterns of RGC activity are required for the large-scale remodeling of the retinocollicular projection into a refined map (McLaughlin et al. 2003c). Mice lacking the $\beta 2$ subunit of the nicotinic acetylcholine receptor maintain spontaneous activity, but the correlation evident in nearby RGCs in wild-type retina is lost (McLaughlin et al. 2003c). The topographic projection in $\beta 2$ mutant mice is aberrant in that RGC axons from a given retinal location do not form a dense TZ but rather maintain a loose collection of diffuse arborizations around the appropriate location of their TZ (McLaughlin et al. 2003c). In $\beta 2$ mutant mice, correlated activity does resume during the second postnatal week through a glutamatergic process, and visually evoked activity begins soon thereafter, but neither process leads to proper map refinement, indicating a brief early critical period for retinotopic map remodeling in mice (McLaughlin et al. 2003c).

In addition to the required role for correlated RGC activity in retinotopic map refinement, the graded expression of guidance molecules may play a role as well (Yates et al. 2004). Indeed, in the $\beta 2$ mutant mice some remodeling does occur; for example, the portions of the primary RGC axons that overshoot the aberrantly diffuse TZ are eliminated (McLaughlin et al. 2003c). Although the mechanisms responsible for the elimination of these overshooting axon segments are unknown, this finding and other evidence suggest that is due to graded repellents that are also involved in generating the initial retinotopic map (Yates et al. 2004). The opposing gradients of EphAs/ephrin-As in the retina result in EphAs predominantly along temporal RGC axons and ephrin-As predominantly along nasal RGC axons. These distributions create imposed countergradients across the OT/SC, which presumably add to the EphA/ephrin-A countergradients expressed by OT/SC cells, but which progressively increase over time as axons branch and arborize (Yates et al. 2001, 2004). Thus, as development proceeds, the level of ephrin-As in posterior OT/SC increases as nasal RGC axons elaborate branches, and the level of EphAs in anterior OT/SC increases as temporal RGCs branch. These changes in EphA/ephrin-A over time amplify the endogenous OT/SC gradients and could be responsible, in part, for the restriction in branching patterns around the TZ, observed over the course of map development, as well as the elimination of the initial RGC axon overshoot (Yates et al. 2001, 2004).

This dynamic type of axon-axon interaction could also provide a partial explanation for aspects of phenotypes observed in mutant mice. One example comes from the analysis of the Isl2-IRES-EphA3 knock-in mouse (EphA3 KI), in which EphA3 is misexpressed by approximately half of the RGCs (RGCs expressing the LIM homeodomain transcription factor Isl2) (Brown et al. 2000). RGCs in homozygous EphA3 KI mice form two topographic maps in the SC: One map comprised essentially entirely from Isl2-positive cells (which therefore misexpress EphA3) forms in anterior SC, and the second map is composed of Isl2-negative RGCs (which have wild-type levels of EphAs) and forms in posterior SC (Brown et al. 2000). Thus, temporal RGCs with wild-type levels of EphAs project to mid-SC, rather than their topographically appropriate TZ in anterior SC, indicating that absolute signaling of EphAs is not the dominant determinant of mapping (Brown et al. 2000, Reber et al. 2004). Crossing EphA3 KI mice with an EphA knockout allows for many combinations and levels of EphA signaling. Results from such crosses confirm that absolute EphA levels are not the absolute determinant of mapping but rather that secondary interactions must take place to form a continuous map (Reber et al. 2004). These studies (Brown et al. 2000, Reber et al. 2004) suggest that competition-exclusion rules (Prestige & Willshaw 1975) may be acting in retinotopic mapping. Thus, another potential way gradients shape topographic maps is by being dynamic over the period of map development.

Additional Activities and Interactions Potentially Required for Map Development

Some mutant phenotypes are not easily explained by the reported actions and distributions of known guidance molecules. For example, an appropriately located TZ (in addition to ectopic TZs) is found in all mice deficient for the Ephs or ephrins required for retinotopic mapping in the SC (Feldheim et al 2000, 2004; Frisen et al. 1998; Hindges et al. 2002), which indicates the action of additional guidance activities along both the AP and LM axes. The dramatic LM mapping defects observed in mice deficient for the homeodomain protein Vax2 also suggest the action of DV guidance molecules other than EphBs and ephrin-Bs. Vax2 is expressed in a tilted gradient in the developing retina, being highest in nasal-ventral RGCs and lowest in temporal-dorsal RGCs (Mui et al. 2002). Targeted deletion of Vax2 in mice results in flattened or diminished gradients of retinal EphBs and ephrin-Bs and a complete shift in the TZs of temporal-ventral RGC axons from antero-medial SC to antero-lateral SC (Mui et al. 2002; also see Schulte et al. 1999), a phenotype much more dramatic than seen in EphB2/B3 double mutants (Hindges et al. 2002). Finally, the functional interactions of Ephs and ephrins are still being detailed and new ones uncovered. For example, EphB2 and ephrin-A5, guidance molecules critical for LM and AP mapping, respectively (Frisen et al. 1998, Hindges et al. 2002), reportedly bind one another, leading to activation of EphB2 signaling pathways (Himanen et al. 2004). Such interactions could potentially influence retinotopic mapping along both axes of the OT/SC.

Though guidance molecules play critical roles in map formation, other interactions, e.g., axon-axon interactions, are likely critical for mapping. Mice deficient for the cell adhesion molecule, L1, which is transiently expressed on RGC axons during pathfinding and mapping, reportedly have defects in both AP and LM mapping in the SC (Demyanenko & Maness 2003). Why L1 is required for proper retinotopic mapping is not known; but considering its roles in other systems, the investigators suggest that it modulates RGC axon-axon interactions required for mapping or the appropriate function of Ephs and ephrins (Itoh et al. 2004, Suh et al. 2004).

Other interactions suggested to influence mapping include competitive interactions for limiting diffusible factors (such as BDNF) or synaptic sites, as well as the interplay between neural activity, response to guidance molecules, and branch dynamics (Alsina et al. 2001, Ruthazer et al. 2003). One example of evidence for this type of secondary interaction comes from analysis of EphA3 KI mice described above. In these mice, Isl2-negative temporal RGCs form TZs in aberrant locations, despite having wild-type levels of topographic guidance molecules, e.g., Ephs and ephrins (Brown et al. 2000). One explanation for this result is that the TZs of Isl2-positive, EphA3 KI RGCs are limited to anterior SC, owing to their enhanced sensitivity to the low-to-high AP gradient of ephrin-A repellents, and exclude Isl2-negative RGC TZs through axon-axon interactions and/or competitive interactions resulting in their orderly, ectopic mapping in posterior SC (Brown et al. 2000).

Genetic Screens: Mapping the Future?

To date, only certain members of the Eph and ephrin families fulfill all criteria for topographic guidance molecules, with a handful of other molecules being studied. Several groups, though, have carried out forward- and reverse-genetic screens to identify additional candidate genes involved in retinotopic mapping or screens that could produce candidate genes as a by-product. The most prominent such undertaking has been the large-scale Tubingen genetic screen in mutagenized zebrafish designed in part to identify genes involved in RGC axon pathfinding and mapping (Baier et al. 1996, Karlstrom et al. 1996, Trowe et al. 1996). This near-saturation screen resulted in the identification of about a hundred mutants, representing scores of genes; a subset of these mutants have defects in DV (LM) or AP mapping in the retinotectal projection, some of which appear to be primary defects (Hutson & Chien 2002, Karlstrom et al. 1997). A subsequent behavioral screen of mutants previously obtained from broad-based screens of mutagenized zebrafish has identified a subset as defective in visual behavior and therefore may provide additional means to identify genes required for retinotopic mapping (Neuhauss 2003, Neuhauss et al. 1999).

A recent microarray screen has identified many known and unknown genes expressed in gradients or restricted patterns along the TN or DV axes of the developing mouse retina; a common pattern is differential gene expression along the DV axis (Diaz et al. 2003). On the basis of their expression pattern and functional class, a subset of these genes will likely prove to be involved in retinotopic mapping.

FUTURE DIRECTIONS

Future work will be focused most likely on small-scale events within the growth cone and axon shaft as well as on population-wide effects, such as the dynamics in gradients of topographic guidance moelcules and the effects of the progressive elaboration of the projection itself on its own development. Furthermore, given the complexity and array of guidance molecules, along with their complementary expression both in the projecting neurons and in the target, conditional mutants in which a gene can be deleted from a defined structure or neuronal population at a defined time point will be fruitful. Additionally, systems in which reciprocal expression of ligands and receptors is asymmetric (i.e., a receptor present in the projecting population but not in the target population) will prove useful in sorting through the actions of multiple related genes expressed in complex expression patterns.

Among the topics most amenable to examination are the roles of ephrin-As in the retina and their potential as axon guidance receptors in the visual system. Identification of conditional mutants in which Ephs can be selectively removed from the SC, or in which ephrin-As can be removed from the retina or from the SC, would be informative. Furthermore, determining the molecular cascade involved with ephrin-A reverse signaling is of fundamental importance (Davy et al. 1999). In addition to sorting out the potential bifunctional and bidirectional actions of Ephs and ephrins, interactions between various families of guidance molecules may provide evidence of new molecular mechanisms to control topographic map formation and are likely to indicate a level of control not yet fully appreciated. For example, the Eph/ephrins have been linked to multiple guidance molecules including slits, Trk receptors, L1, and laminin (Kong et al. 2001, Suh et al. 2004, Wong et al. 2004). In addition, phosphatases clearly have a critical role in axon guidance, and their roles require further examination (Ensslen-Craig & Brady-Kalnay 2004, Johnson et al. 2001, Palmer et al. 2002).

Lastly, we find that the role a system plays in its own development is an intriguing facet of mapping. In many systems the projecting axons themselves likely affect the development of the projection. These potential interactions are not well defined but may include competition for synaptic space or diffusible factors, direct axon-axon interactions, or the addition of guidance molecules present along projecting axons to the target, to name a few. For example, RGC axons may alter the landscape of guidance molecules as they elaborate arbors in the target, owing in part to the presence of guidance molecules on RGC axons, such as ephrin-As on nasal RGC axons (Hornberger et al. 1999). As nasal RGC axons branch in posterior OT/SC, the level

AOB: accessory olfactory bulb

AP: anterior-posterior

DV: dorsal-ventral

IRES: internal ribosome entry site

KI: knock-in

LM: lateral-medial

NT: nasal-temporal

OB: olfactory bulb

OT: optic tectum

RGC: retinal ganglion cell

SC: superior colliculus

TZ: termination zone

VNO: vomeronasal organ

of ephrin-As should increase in that domain, adding to the expression of ephrin-As by posterior tectal or collicular cells. This alteration of the ephrin-A gradient will likely affect topographic mapping (Yates et al. 2004). Thus, guidance information is unlikely to be static, and therefore the sum total of analyses of small numbers of axons may not yield data that seem able to account fully for topographic mapping. However, in the dynamic developing topographic map the same information may, in fact, be sufficient.

ACKNOWLEDGMENTS

Work in the authors' lab on the topic of this review is supported by National Eye Institute grant R01 EY 07025 (DO'L). We thank current and former members of the O'Leary lab (especially Carol Armstrong, Octavio Choi, Glenn Friedman, Robert Hindges, Winnie Pak, Adina Roskies, Alicia Santiago, David Simon, and Paul Yates) who have worked on this topic and have made valued contributions to discussions of concepts of mapping. We also thank our collaborators, in particular Mariano Barbacid, Jonas Frisen, Mark Henkemeyer, and Greg Lemke, for sharing mutant mice for our analyses of molecular mechanisms of retinotopic mapping.

LITERATURE CITED

Alsina B, Vu T, Cohen-Cory S. 2001. Visualizing synapse formation in arborizing optic axons in vivo: dynamics and modulation by BDNF. *Nat. Neurosci.* 4:1093–101

Baier H, Bonhoeffer F. 1992. Axon guidance by gradients of a target-derived component. *Science* 255:472–75

Baier H, Klostermann S, Trowe T, Karlstrom RO, Nusslein-Volhard C, Bonhoeffer F. 1996. Genetic dissection of the retinotectal projection. *Development* 123:415–25

Bassell GJ, Kelic S. 2004. Binding proteins for mRNA localization and local translation, and their dysfunction in genetic neurological disease. *Curr. Opin. Neurobiol.* 14:574–81

Belluscio L, Koentges G, Axel R, Dulac C. 1999. A map of pheromone receptor activation in the mammalian brain. *Cell* 97:209–20

Benson DL, Colman DR, Huntley GW. 2001. Molecules, maps and synapse specificity. *Nat. Rev. Neurosci.* 2:899–909

Braisted JE, McLaughlin T, Wang HU, Friedman GC, Anderson DJ, O'Leary DDM. 1997. Graded and lamina-specific distributions of ligands of EphB receptor tyrosine kinases in the developing retinotectal system. *Dev. Biol.* 191:14–28

Brennan C, Monschau B, Lindberg R, Guthrie B, Drescher U, et al. 1997. Two Eph receptor tyrosine kinase ligands control axon growth and may be involved in the creation of the retinotectal map in the zebrafish. *Development* 124:655–64

Brittis PA, Lu Q, Flanagan JG. 2002. Axonal protein synthesis provides a mechanism for localized regulation at an intermediate target. *Cell* 110:223–35

Brown A, Yates PA, Burrola P, Ortuño D, Vaidya A, et al. 2000. Topographic mapping from the retina to the midbrain is controlled by relative but not absolute levels of EphA receptor signaling. *Cell* 102:77–88

Bruckner K, Pasquale EB, Klein R. 1997. Tyrosine phosphorylation of transmembrane ligands for Eph receptors. *Science* 275:1640–43

Butts DA. 2002. Retinal waves: implications for synaptic learning rules during development. *Neuroscientist* 8:243–53

Campbell DS, Holt CE. 2001. Chemotropic responses of retinal growth cones mediated by rapid local protein synthesis and degradation. *Neuron* 32:1013–26

Campbell DS, Regan AG, Lopez JS, Tannahill D, Harris WA, Holt CE. 2001. Semaphorin 3A elicits stage-dependent collapse, turning, and branching in *Xenopus* retinal growth cones. *J. Neurosci.* 21:8538–47

Chalupa LM, Snider CJ. 1998. Topographic specificity in the retinocollicular projection of the developing ferret: an anterograde tracing study. *J. Comp. Neurol.* 392:35–47

Chalupa LM, Snider CJ, Kirby MA. 1996. Topographic organization in the retinocollicular pathway of the fetal cat demonstrated by retrograde labeling of ganglion cells. *J. Comp. Neurol.* 368:295–303

Cheng HJ, Flanagan JG. 1994. Identification and cloning of ELF-1, a developmentally expressed ligand for the Mek4 and Sek receptor tyrosine kinases. *Cell* 79:157–68

Cheng HJ, Nakamoto M, Bergemann AD, Flanagan JG. 1995. Complementary gradients in expression and binding of ELF-1 and Mek4 in development of the topographic retinotectal projection map. *Cell* 82:371–81

Choi O, O'Leary DDM. 2000. Potential roles for BDNF and TrkB in developing topographic retinotectal projections. *Soc. Neurosci. Abstr.* 26 (Abstr.)

Choi SB, Yates PA, O'Leary DDM. 1998. Localized BDNF application induces branch-like structures along retinal axons. *Soc. Neurosci. Abstr.* 24 (Abstr.)

Ciossek T, Monschau B, Kremoser C, Loschinger J, Lang S, et al. 1998. Eph receptor-ligand interactions are necessary for guidance of retinal ganglion cell axons in vitro. *Eur. J. Neurosci.* 10:1574–80

Cline HT. 1998. Topographic maps: developing roles of synaptic plasticity. *Curr. Biol.* 8:R836–39

Cloutier JF, Giger RJ, Koentges G, Dulac C, Kolodkin AL, Ginty DD. 2002. Neuropilin-2 mediates axonal fasciculation, zonal segregation, but not axonal convergence, of primary accessory olfactory neurons. *Neuron* 33:877–92

Cloutier JF, Sahay A, Chang EC, Tessier-Lavigne M, Dulac C, et al. 2004. Differential requirements for semaphorin 3F and Slit-1 in axonal targeting, fasciculation, and segregation of olfactory sensory neuron projections. *J. Neurosci.* 24:9087–96

Cohen-Cory S. 2002. The developing synapse: construction and modulation of synaptic structures and circuits. *Science* 298:770–76

Cohen-Cory S, Fraser SE. 1995. Effects of brain-derived neurotrophic factor on optic axon branching and remodelling in vivo. *Nature* 378:192–96

Connor RJ, Menzel P, Pasquale EB. 1998. Expression and tyrosine phosphorylation of Eph receptors suggest multiple mechanisms in patterning of the visual system. *Dev. Biol.* 193:21–35

Cox EC, Muller B, Bonhoeffer F. 1990. Axonal guidance in the chick visual system: posterior tectal membranes induce collapse of growth cones from the temporal retina. *Neuron* 4:31–37

Davenport RW, Thies E, Cohen ML. 1999. Neuronal growth cone collapse triggers lateral extensions along trailing axons. *Nat. Neurosci.* 2:254–59

Davy A, Gale NW, Murray EW, Klinghoffer RA, Soriano P, et al. 1999. Compartmentalized signaling by GPI-anchored ephrin-A5 requires the Fyn tyrosine kinase to regulate cellular adhesion. *Genes Dev.* 13:3125–35

Davy A, Robbins SM. 2000. Ephrin-A5 modulates cell adhesion and morphology in an integrin-dependent manner. *EMBO J.* 19:5396–405

Debski EA, Cline HT. 2002. Activity-dependent mapping in the retinotectal projection. *Curr. Opin. Neurobiol.* 12:93–99

Demyanenko GP, Maness PF. 2003. The L1 cell adhesion molecule is essential for topographic mapping of retinal axons. *J. Neurosci.* 23:530–38

Diaz E, Yang YH, Ferreira T, Loh KC, Okazaki Y, et al. 2003. Analysis of gene expression in the developing mouse retina. *Proc. Natl. Acad. Sci. USA* 100:5491–96

Ding Y, Marotte LR. 1997. The initial stages of development of the retinocollicular projection in the wallaby (Macropus eugenii): distribution of ganglion cells in the retina and their axons in the superior colliculus. *Anat. Embryol.* 194:301–17

Drescher U, Kremoser C, Handwerker C, Loschinger J, Noda M, Bonhoeffer F. 1995. In vitro guidance of retinal ganglion cell axons by RAGS, a 25 kDa tectal protein related to ligands for Eph receptor tyrosine kinases. *Cell* 82:359–70

Dufour A, Seibt J, Passante L, Depaepe V, Ciossek T, et al. 2003. Area specificity and topography of thalamocortical projections are controlled by ephrin/Eph genes. *Neuron* 39:453–65

Eberhart J, Barr J, O'Connell S, Flagg A, Swartz ME, et al. 2004. Ephrin-A5 exerts positive or inhibitory effects on distinct subsets of EphA4-positive motor neurons. *J. Neurosci.* 24:1070–78

Ensslen-Craig SE, Brady-Kalnay SM. 2004. Receptor protein tyrosine phosphatases regulate neural development and axon guidance. *Dev. Biol.* 275:12–22

Feldheim DA, Kim YI, Bergemann AD, Frisen J, Barbacid M, Flanagan JG. 2000. Genetic analysis of ephrin-A2 and ephrin-A5 shows their requirement in multiple aspects of retinocollicular mapping. *Neuron* 25:563–74

Feldheim DA, Nakamoto M, Osterfield M, Gale NW, DeChiara TM, et al. 2004. Loss-of-function analysis of EphA receptors in retinotectal mapping. *J. Neurosci.* 24:2542–50

Feng G, Laskowski MB, Feldheim DA, Wang H, Lewis R, et al. 2000. Roles for ephrins in positionally selective synaptogenesis between motor neurons and muscle fibers. *Neuron* 25:295–306

Frisen J, Yates PA, McLaughlin T, Friedman GC, O'Leary DDM, Barbacid M. 1998. Ephrin-A5 (AL-1/RAGS) is essential for proper retinal axon guidance and topographic mapping in the mammalian visual system. *Neuron* 20:235–43

Galli L, Maffei L. 1988. Spontaneous impulse activity of rat retinal ganglion cells in prenatal life. *Science* 242:90–91

Gallo G, Letourneau PC. 2004. Regulation of growth cone actin filaments by guidance cues. *J. Neurobiol.* 58:92–102

Gao PP, Zhang JH, Yokoyama M, Racey B, Dreyfus CF, et al. 1996. Regulation of topographic projection in the brain: Elf-1 in the hippocamposeptal system. *Proc. Natl. Acad. Sci. USA* 93:11161–66

Gierer A. 1983. Model for the retino-tectal projection. *Proc. R. Soc. London B. Biol. Sci.* 218:77–93

Gierer A. 1987. Directional cues for growing axons forming the retinotectal projection. *Development* 101:479–89

Goodhill GJ, Baier H. 1998. Axon guidance: stretching gradients to the limit. *Neural Comput.* 10:521–27

Hansen MJ, Dallal GE, Flanagan JG. 2004. Retinal axon response to ephrin-As shows a graded, concentration-dependent transition from growth promotion to inhibition. *Neuron* 42:717–30

Harris WA, Holt CE, Bonhoeffer F. 1987. Retinal axons with and without their somata, growing to and arborizing in the tectum of *Xenopus* embryos: a time-lapse video study of single fibres in vivo. *Development* 101:123–33

Hebb DO. 1949. *The Organization of Behavior: A Neuropsychological Theory*. New York: Wiley

Henkemeyer M, Orioli D, Henderson JT, Saxton TM, Roder J, et al. 1996. Nuk controls pathfinding of commissural axons in the mammalian central nervous system. *Cell* 86:35–46

Himanen J-P, Chumley MJ, Lackmann M, Li C, Barton WA, et al. 2004. Repelling class discrimination: ephrin-A5 binds to and activates EphB2 receptor signaling. *Nat. Neurosci.* 7:501–9

Himanen JP, Nikolov DB. 2003. Eph signaling: a structural view. *Trends Neurosci.* 26:46–51

Hindges R, McLaughlin T, Genoud N, Henkemeyer M, O'Leary DDM. 2002. EphB forward signaling controls directional branch extension and arborization required for dorsal ventral retinotopic mapping. *Neuron* 35:475–87

Holland SJ, Gale NW, Mbamalu G, Yancopoulos GD, Henkemeyer M, Pawson T. 1996. Bidirectional signalling through the EPH-family receptor Nuk and its transmembrane ligands. *Nature* 383:722–25

Hornberger MR, Dutting D, Ciossek T, Yamada T, Handwerker C, et al. 1999. Modulation of EphA receptor function by coexpressed ephrinA ligands on retinal ganglion cell axons. *Neuron* 22:731–42

Huber AB, Kolodkin AL, Ginty DD, Cloutier JF. 2003. Signaling at the growth cone: ligand-receptor complexes and the control of axon growth and guidance. *Annu. Rev. Neurosci.* 26:509–63

Hutson LD, Chien CB. 2002. Wiring the zebrafish: axon guidance and synaptogenesis. *Curr. Opin. Neurobiol.* 12:87–92

Huynh-Do U, Stein E, Lane AA, Liu H, Cerretti DP, Daniel TO. 1999. Surface densities of ephrin-B1 determine EphB1-coupled activation of cell attachment through alphavbeta3 and alpha5beta1 integrins. *EMBO J.* 18:2165–73

Itoh K, Cheng L, Kamei Y, Fushiki S, Kamiguchi H, et al. 2004. Brain development in mice lacking L1-L1 homophilic adhesion. *J. Cell Biol.* 165:145–54

Johnson KG, McKinnell IW, Stoker AW, Holt CE. 2001. Receptor protein tyrosine phosphatases regulate retinal ganglion cell axon outgrowth in the developing Xenopus visual system. *J. Neurobiol.* 49:99–117

Kaas JH. 1997. Topographic maps are fundamental to sensory processing. *Brain Res. Bull.* 44:107–12

Kaethner RJ, Stuermer CA. 1992. Dynamics of terminal arbor formation and target approach of retinotectal axons in living zebrafish embryos: a time-lapse study of single axons. *J. Neurosci.* 12:3257–71

Karlstrom RO, Trowe T, Bonhoeffer F. 1997. Genetic analysis of axon guidance and mapping in the zebrafish. *Trends Neurosci.* 20:3–8

Karlstrom RO, Trowe T, Klostermann S, Baier H, Brand M, et al. 1996. Zebrafish mutations affecting retinotectal axon pathfinding. *Development* 123:427–38

Knoll B, Drescher U. 2002. Ephrin-As as receptors in topographic projections. *Trends Neurosci.* 25:145–49

Knoll B, Schmidt H, Andrews W, Guthrie S, Pini A, et al. 2003. On the topographic targeting of basal vomeronasal axons through Slit-mediated chemorepulsion. *Development* 130:5073–82

Knoll B, Zarbalis K, Wurst W, Drescher U. 2001. A role for the EphA family in the topographic targeting of vomeronasal axons. *Development* 128:895–906

Kobayashi T, Nakamura H, Yasuda M. 1990. Disturbance of refinement of retinotectal projection in chick embryos by TTX and grayanotoxin. *Dev. Brain Res.* 57:29–35

Kong H, Boulter J, Weber JL, Lai C, Chao MV. 2001. An evolutionarily conserved transmembrane protein that is a novel downstream target of neurotrophin and ephrin receptors. *J. Neurosci.* 21:176–85

Liu Y, Berndt J, Su F, Tawarayama H, Shoji W, et al. 2004. Semaphorin3D guides retinal axons along the dorsoventral axis of the tectum. *J. Neurosci.* 24:310–18

Loconto J, Papes F, Chang E, Stowers L, Jones EP, et al. 2003. Functional expression of murine V2R pheromone receptors involves selective association with the M10 and M1 families of MHC class Ib molecules. *Cell* 112:607–18

Mann F, Miranda E, Weinl C, Harmer E, Holt CE. 2003. B-type Eph receptors and ephrins induce growth cone collapse through distinct intracellular pathways. *J. Neurobiol.* 57:323–36

Mann F, Ray S, Harris WA, Holt CE. 2002. Topographic mapping in dorsoventral axis of the *Xenopus* retinotectal system depends on signaling through ephrin-B ligands. *Neuron* 35:461–73

Marcus RC, Gale NW, Morrison ME, Mason CA, Yancopoulos GD. 1996. Eph family receptors and their ligands distribute in opposing gradients in the developing mouse retina. *Dev. Biol.* 180:786–89

Marston DJ, Dickinson S, Nobes CD. 2003. Rac-dependent trans-endocytosis of ephrinBs regulates Eph-ephrin contact repulsion. *Nat. Cell Biol.* 5:879–88

McLaughlin T, Hindges R, O'Leary DDM. 2003a. Regulation of axial patterning of the retina and its topographic mapping in the brain. *Curr. Opin. Neurobiol.* 13:57–69

McLaughlin T, Hindges R, Yates PA, O'Leary DDM. 2003b. Bifunctional action of ephrin-B1 as a repellent and attractant to control bidirectional branch extension in dorsal-ventral retinotopic mapping. *Development* 130:2407–18

McLaughlin T, Torborg CL, Feller MB, O'Leary DDM. 2003c. Retinotopic map refinement requires spontaneous retinal waves during a brief critical period of development. *Neuron* 40:1147–60

McLoon SC. 1991. A monoclonal antibody that distinguishes between temporal and nasal retinal axons. *J. Neurosci.* 11:1470–77

Meister M, Wong RO, Baylor DA, Shatz CJ. 1991. Synchronous bursts of action potentials in ganglion cells of the developing mammalian retina. *Science* 252:939–43

Menzel P, Valencia F, Godement P, Dodelet VC, Pasquale EB. 2001. Ephrin-A6, a new ligand for EphA receptors in the developing visual system. *Dev. Biol.* 230:74–88

Monnier PP, Sierra A, Macchi P, Deitinghoff L, Andersen JS, et al. 2002. RGM is a repulsive guidance molecule for retinal axons. *Nature* 419:392–95

Monschau B, Kremoser C, Ohta K, Tanaka H, Kaneko T, et al. 1997. Shared and distinct functions of RAGS and ELF-1 in guiding retinal axons. *EMBO J.* 16:1258–67

Mui SH, Hindges R, O'Leary DD, Lemke G, Bertuzzi S. 2002. The homeodomain protein Vax2 patterns the dorsoventral and nasotemporal axes of the eye. *Development* 129:797–804

Muller BK, Jay DG, Bonhoeffer F. 1996. Chromophore-assisted laser inactivation of a repulsive axonal guidance molecule. *Curr. Biol.* 6:1497–502

Murai KK, Pasquale EB. 2003. 'Eph'ective signaling: forward, reverse and crosstalk. *J. Cell Sci.* 116:2823–32

Nakamoto M, Cheng HJ, Friedman GC, McLaughlin T, Hansen MJ, et al. 1996. Topographically specific effects of ELF-1 on retinal axon guidance in vitro and retinal axon mapping in vivo. *Cell* 86:755–66

Nakamoto T, Kain KH, Ginsberg MH. 2004. Neurobiology: new connections between integrins and axon guidance. *Curr. Biol.* 14:121–23

Nakamura H, O'Leary DDM. 1989. Inaccuracies in initial growth and arborization of chick retinotectal axons followed by course corrections and axon remodeling to develop topographic order. *J. Neurosci.* 9:3776–95

Neuhauss SC. 2003. Behavioral genetic approaches to visual system development and function in zebrafish. *J. Neurobiol.* 54:148–60

Neuhauss SC, Biehlmaier O, Seeliger MW, Das T, Kohler K, et al. 1999. Genetic disorders of vision revealed by a behavioral screen of 400 essential loci in zebrafish. *J. Neurosci.* 19:8603–15

Nguyen QT, Sanes JR, Lichtman JW. 2002. Pre-existing pathways promote precise projection patterns. *Nat. Neurosci.* 5:861–67

Niederkofler V, Salie R, Sigrist M, Arber S. 2004. Repulsive guidance molecule (RGM) gene function is required for neural tube closure but not retinal topography in the mouse visual system. *J. Neurosci.* 24:808–18

O'Leary DDM, Yates P, McLaughlin T. 1999. Mapping sights and smells in the brain: distinct mechanisms to achieve a common goal. *Cell* 96:255–69

O'Rourke NA, Cline HT, Fraser SE. 1994. Rapid remodeling of retinal arbors in the tectum with and without blockade of synaptic transmission. *Neuron* 12:921–34

O'Rourke NA, Fraser SE. 1990. Dynamic changes in optic fiber terminal arbors lead to retinotopic map formation: an in vivo confocal microscopic study. *Neuron* 5:159–71

Palmer A, Klein R. 2003. Multiple roles of ephrins in morphogenesis, neuronal networking, and brain function. *Genes Dev.* 17:1429–50

Palmer A, Zimmer M, Erdmann KS, Eulenburg V, Porthin A, et al. 2002. EphrinB phosphorylation and reverse signaling: regulation by Src kinases and PTP-BL phosphatase. *Mol. Cell* 9:725–37

Poliakov A, Cotrina M, Wilkinson DG. 2004. Diverse roles of eph receptors and ephrins in the regulation of cell migration and tissue assembly. *Dev. Cell* 7:465–80

Prestige MC, Willshaw DJ. 1975. On a role for competition in the formation of patterned neural connexions. *Proc. R. Soc. London B Biol. Sci.* 190:77–98

Rajagopalan S, Deitinghoff L, Davis D, Conrad S, Skutella T, et al. 2004. Neogenin mediates the action of repulsive guidance molecule. *Nat. Cell Biol.* 6:756–62

Reber M, Burrola P, Lemke G. 2004. A relative signalling model for the formation of a topographic neural map. *Nature* 431:847–53

Rodriguez I, Feinstein P, Mombaerts P. 1999. Variable patterns of axonal projections of sensory neurons in the mouse vomeronasal system. *Cell* 97:199–208

Rosentreter SM, Davenport RW, Loschinger J, Huf J, Jung J, Bonhoeffer F. 1998. Response of retinal ganglion cell axons to striped linear gradients of repellent guidance molecules. *J. Neurobiol.* 37:541–62

Roskies A, Friedman G, O'Leary DDM. 1995. Molecules and mechanisms in the development of retinotopic maps. *Perspect. Dev. Neurobiol.* 3:63–75

Roskies AL, O'Leary DDM. 1994. Control of topographic retinal axon branching by inhibitory membrane-bound molecules. *Science* 265:799–803

Rosoff WJ, Urbach JS, Esrick MA, McAllister RG, Richards LJ, Goodhill GJ. 2004. A new chemotaxis assay shows the extreme sensitivity of axons to molecular gradients. *Nat. Neurosci.* 7:678–82

Rossant J, Hirashima M. 2003. Vascular development and patterning: making the right choices. *Curr. Opin. Genet. Dev.* 13:408–12

Ruthazer ES, Akerman CJ, Cline HT. 2003. Control of axon branch dynamics by correlated activity in vivo. *Science* 301:66–70

Sakurai T, Wong E, Drescher U, Tanaka H, Jay DG. 2002. Ephrin-A5 restricts topographically specific arborization in the chick retinotectal projection in vivo. *Proc. Natl. Acad. Sci. USA* 99:10795–800

Savitt JM, Trisler D, Hilt DC. 1995. Molecular cloning of TOP_{AP}: a topographically graded protein in the developing chick visual system. *Neuron* 14:253–61

Schulte D, Furukawa T, Peters MA, Kozak CA, Cepko CL. 1999. Misexpression of the Emx-related homeobox genes cVax and mVax2 ventralizes the retina and perturbs the retinotectal map. *Neuron* 24:541–53

Simon DK, O'Leary DDM. 1991. Relationship of retinotopic ordering of axons in the optic pathway to the formation of visual maps in central targets. *J. Comp. Neurol.* 307:393–404

Simon DK, O'Leary DDM. 1992a. Development of topographic order in the mammalian retinocollicular projection. *J. Neurosci.* 12:1212–32

Simon DK, O'Leary DDM. 1992b. Influence of position along the medial-lateral axis of the superior colliculus on the topographic targeting and survival of retinal axons. *Dev. Brain Res.* 69:167–72

Simon DK, O'Leary DDM. 1992c. Responses of retinal axons in vivo and in vitro to position-encoding molecules in the embryonic superior colliculus. *Neuron* 9:977–89

Simon DK, Prusky GT, O' Leary DDM, Constantine-Paton M. 1992. N-methyl-D-aspartate receptor antagonists disrupt the formation of a mammalian neural map. *Proc. Natl. Acad. Sci. USA* 89:10593–97

Sperry R. 1963. Chemoaffinity in the orderly growth of nerve fiber patterns and connections. *Proc. Natl. Acad. Sci. USA* 50:703–10

Stahl B, Muller B, von Boxberg Y, Cox EC, Bonhoeffer F. 1990. Biochemical characterization of a putative axonal guidance molecule of the chick visual system. *Neuron* 5:735–43

Stent GS. 1973. A physiological mechanism for Hebb's postulate of learning. *Proc. Natl. Acad. Sci. USA* 70:997–1001

Suh LH, Oster SF, Soehrman SS, Grenningloh G, Sretavan DW. 2004. L1/Laminin modulation of growth cone response to EphB triggers growth pauses and regulates the microtubule destabilizing protein SCG10. *J. Neurosci.* 24:1976–86

Surawska H, Ma PC, Salgia R. 2004. The role of ephrins and Eph receptors in cancer. *Cytokine Growth Factor Rev.* 15:419–33

Tian N, Copenhagen DR. 2003. Visual stimulation is required for refinement of ON and OFF pathways in postnatal retina. *Neuron* 39:85–96

Trowe T, Klostermann S, Baier H, Granato M, Crawford AD, et al. 1996. Mutations disrupting the ordering and topographic mapping of axons in the retinotectal projection of the zebrafish, Danio rerio. *Development* 123:439–50

van Horck FP, Weinl C, Holt CE. 2004. Retinal axon guidance: novel mechanisms for steering. *Curr. Opin. Neurobiol.* 14:61–66

Vanderhaeghen P, Polleux F. 2004. Developmental mechanisms patterning thalamocortical projections: intrinsic, extrinsic and in between. *Trends Neurosci.* 27:384–91

Wagle M, Grunewald B, Subburaju S, Barzaghi C, Le Guyader S, et al. 2004. EphrinB2a in the zebrafish retinotectal system. *J. Neurobiol.* 59:57–65

Walter J, Henke-Fahle S, Bonhoeffer F. 1987b. Avoidance of posterior tectal membranes by temporal retinal axons. *Development* 101:909–13

Walter J, Kern-Veits B, Huf J, Stolze B, Bonhoeffer F. 1987a. Recognition of position-specific properties of tectal cell membranes by retinal axons in vitro. *Development* 101:685–96

Walz A, Rodriguez I, Mombaerts P. 2002. Aberrant sensory innervation of the olfactory bulb in neuropilin-2 mutant mice. *J. Neurosci.* 22:4025–35

Weinl C, Drescher U, Lang S, Bonhoeffer F, Loschinger J. 2003. On the turning of *Xenopus* retinal axons induced by ephrin-A5. *Development* 130:1635–43

Wong EV, Kerner JA, Jay DG. 2004. Convergent and divergent signaling mechanisms of growth cone collapse by ephrinA5 and slit2. *J. Neurobiol.* 59:66–81

Wong RO, Meister M, Shatz CJ. 1993. Transient period of correlated bursting activity during development of the mammalian retina. *Neuron* 11:923–38

Wong RO. 1999. Retinal waves and visual system development. *Annu. Rev. Neurosci.* 22:29–47

Yates PA, Holub AD, McLaughlin T, Sejnowski TJ, O' Leary DD. 2004. Computational modeling of retinotopic map development to define contributions of EphA-ephrinA gradients, axon-axon interactions, and patterned activity. *J. Neurobiol.* 59:95–113

Yates PA, Roskies AR, McLaughlin T, O' Leary DDM. 2001. Topographic specific axon branching controlled by ephrin-As is the critical event in retinotectal map development. *J. Neurosci.* 21:8548–63

Yoda H, Hirose Y, Yasuoka A, Sasado T, Morinaga C, et al. 2004. Mutations affecting retinotectal axonal pathfinding in Medaka, Oryzias latipes. *Mech. Dev.* 121:715–28

Yue Y, Chen ZY, Gale NW, Blair-Flynn J, Hu TJ, et al. 2002. Mistargeting hippocampal axons by expression of a truncated Eph receptor. *Proc. Natl. Acad. Sci. USA* 99:10777–82

Zimmer M, Palmer A, Kohler J, Klein R. 2003. EphB-ephrinB bi-directional endocytosis terminates adhesion allowing contact mediated repulsion. *Nat. Cell Biol.* 5:869–78

Neural Network Dynamics

Tim P. Vogels, Kanaka Rajan, and L.F. Abbott

Volen Center for Complex Systems and Department of Biology, Brandeis University, Waltham, Massachusetts 02454-9110; email: vogels@brandeis.edu

Key Words

balance, memory, signal propagation, states, sustained activity

Abstract

Neural network modeling is often concerned with stimulus-driven responses, but most of the activity in the brain is internally generated. Here, we review network models of internally generated activity, focusing on three types of network dynamics: (*a*) sustained responses to transient stimuli, which provide a model of working memory; (*b*) oscillatory network activity; and (*c*) chaotic activity, which models complex patterns of background spiking in cortical and other circuits. We also review propagation of stimulus-driven activity through spontaneously active networks. Exploring these aspects of neural network dynamics is critical for understanding how neural circuits produce cognitive function.

Contents

INTRODUCTION 358
FIRING-RATE AND SPIKING
    NETWORK MODELS 359
    Firing-Rate Networks 360
    Integrate-and-Fire Networks 360
FORMS OF NETWORK
    DYNAMICS 362
    Sustained Activity 364
    Oscillations 365
    Chaotic Spiking Networks 366
SIGNAL PROPAGATION 368
    Avalanche Model 368
    Synfire Chains 369
    Propagation of Firing Rates 372
DISCUSSION 372

INTRODUCTION

We generate most of our thoughts and behaviors internally, but our actions can be modified drastically by small changes in our perception of the external environment. Stated another way, the neural circuits of the brain perpetually generate complex patterns of activity with an extraordinarily rich spatial and temporal structure, yet they remain highly sensitive to sensory input. The majority of modeling in neuroscience is concerned with activity that is driven by a stimulus. Such models are constructed to account for sensitivity, selectivity, and other features of neuronal responses to sensory input (reviewed, for example, in Dayan & Abbott 2001). In the absence of that input, neurons in these models are typically silent. Although this approach has had considerable success in accounting for response properties in primary sensory areas, such as the early visual system, it clearly cannot account for the majority of activity in the brain, which is internally generated. This review is devoted to modeling work at the other extreme: models that produce their own activity, even in the absence of external input.

Understanding how neural circuitry generates complex patterns of activity is challenging, and it is even more difficult to build models of this type that remain sensitive to sensory input. In mathematical terms, we need to understand how a system can reconcile a rich internal state structure with a high degree of sensitivity to external variables. This problem is far from solved, but here we review progress that has been made in recent years. Rather than surveying a large number of models and applications, we illustrate the existing issues and the progress made using two basic models: a network model described in terms of neuronal firing rates that exhibits sustained and oscillatory activity and a network of spiking model neurons that displays chaotic activity.

We begin the review with a discussion of sustained responses to transient stimuli. Neuronal activity evoked by a transient stimulus often continues beyond the period of stimulus presentation and, in cases where short-term memory of the stimulus is required for a task, such sustained activity can last for tens of seconds (Wang & Goldman-Rakic 2004). Neuronal firing at a constant rate is a form of internally generated activity known as fixed-point behavior. This time-independent activity is too simple to address the issue of how complex patterns of activity are generated. On the other hand, these models provide an excellent example of the problem of making internally generated activity sensitive to external input because, to be of any use in a memory task, self-sustained neural activity must be sensitive to those aspects of the stimulus that are being remembered (Compte et al. 2000, Seung et al. 2000).

From sustained activity, we move on to a discussion of oscillations. Oscillations are a widespread feature of neural systems, and oscillating neural network models have been studied extensively (Marder & Calabrese 1996, Buzsaki & Draguhn 2004). We illustrate this form of network activity by modifying a model with self-sustained activity.

Our next topic covers large networks of spiking model neurons that display complex chaotic activity. In these networks, excitation and inhibition are balanced so that they nearly cancel, and neuronal firing is driven by fluctuations that transiently spoil this cancellation (Shadlen & Newsome 1994, Tsodyks & Sejnowski 1995, Troyer & Miller 1997). Neuronal responses recorded in vivo are highly variable (Burns & Webb 1976, Dean 1981, Softky & Koch 1993, Holt et al. 1996, Anderson et al. 2000), and it has long been recognized that some form of "noise" has to be included if models of such responses are to match the data (see, for example, Usher et al. 1994). In models, this noise is often added from a random external source, such as a random number generator, which does not match what happens in real neural circuits. Although neurons are subject to thermal fluctuations that act like an external source of noise, it appears that most of the variability in cortical circuits comes from activity generated within the neural circuits themselves (Arieli et al. 1996). Sparsely connected networks of spiking model neurons can generate what looks like random, noisy activity without the need for any external source of randomness (van Vreeswijk & Sompolinsky 1996, 1998; Amit & Brunel 1997; Brunel 2000; Mehring et al. 2003; Lerchner et al. 2004). This is a significant achievement toward the goal of understanding the dynamics of complex neuronal activity.

Spiking network models go a long way toward solving the problem of producing complex, self-sustained patterns of activity, but they fail at accounting for the input sensitivity of biological networks. Returning to this problem, we conclude this review by examining studies of signal propagation in neural networks. For a network to be sensitive to external input, the activity generated by that input must propagate through the network. There has been considerable discussion about the different ways that information might be encoded by neural activity. To be viable, a coding scheme must support propagation of information from one brain region to another (Diesmann et al. 1999, van Rossum et al. 2002). Propagation of signals across neural networks is difficult to achieve because of two sources of instability. First, signals tend to either grow or shrink in amplitude as they propagate from one group of neurons to another. Rather precise tuning is required to prevent signals from either blowing up or decaying away before they reach their targets. This problem is well illustrated in a simple avalanche model of propagation that we review (Harris 1963, Zapperi et al. 1995, de Carvalho & Prado 2000, Beggs & Plenz 2003). Second, signal propagation in neural networks can lead to unrealistic large-scale synchronization of neuronal firing (Marsalek et al. 1997, Golomb 1998, Burkitt & Clark 1999, van Rossum et al. 2002, Reyes 2003, Litvak et al. 2003). Avoiding this problem requires the introduction of noise, which leads us back to sparsely coupled networks of spiking neurons that can generate the required noise internally. As discussed below, signal propagation and sensitivity to input remain significant challenges to our understanding of neural network dynamics.

FIRING-RATE AND SPIKING NETWORK MODELS

The power of present-day computers permits simulation of large networks, even in cases when the individual neurons are modeled in considerable detail. Of course, there is a trade-off between the amount of detail that can be devoted to modeling each individual neuron (or each synapse, which is even more costly owing to their larger numbers) and the size and complexity of the network that can be simulated. A common compromise is to use a relatively simple spiking model, the integrate-and-fire model, to describe each neuron. This allows simulations of networks with tens or even hundreds of thousands of neurons.

Such networks are complex dynamical systems involving the numerical integration of many thousands of coupled differential equations. In computer simulations, as opposed to experiments, any variable in any neuron or synapse of the network can be monitored and manipulated. Nevertheless, characterizing and understanding what is going on at the network level can be difficult. Furthermore, time constraints on the simulation of these systems makes it difficult to survey the complete parameter space (which typically has a high dimension) adequately (Prinz et al. 2004). Spiking models are difficult to analyze mathematically, so modeling networks often involves a second approach that uses firing rates, rather than action potential sequences, to characterize neuronal responses (Wilson & Cowan 1972, Shriki et al. 2003).

Firing-Rate Networks

In a firing-rate network, each neuron is described at time t by a firing rate $r_i(t)$ for neuron i, where $i = 1, 2, \ldots N$ labels the N neurons of the network. Each firing rate relaxes with a time constant τ_r to a steady-state value given by a function F that describes the relationship between firing rate and input current for the neuron. The input current for neuron i is the sum of the input from sources outside the network such as sensory input, denoted by $I_i(t)$, and a term describing input from other neurons within the network. The resulting dynamic equation is

$$\tau_r \frac{dr_i}{dt} = -r_i(t) + F\left(I_i(t) + \sum_{j=1}^{N} \mathcal{J}_{ij} r_j(t) + \Theta\right),$$

1.

where $\mathcal{J}_{ij}$ describes the strength and type (excitatory if $\mathcal{J}_{ij} > 0$ and inhibitory if $\mathcal{J}_{ij} < 0$) of the synapse from presynaptic neuron j to postsynaptic neuron i. The constant Θ acts as a current bias that can induce spontaneous firing if it is positive or suppress firing if it is negative. The time constant τ_r determines how quickly the firing rate can change. For the network shown in **Figures 1** and **2** (see Sustained Activity, below), $\tau_r = 10$ ms.

The assumption behind Equation 1 is that, on average, the input from a given presynaptic neuron is proportional to its firing rate (the factor r_j in Equation 1) and that the total synaptic input is obtained by summing the contributions of all the presynaptic neurons (the sum over j in Equation 1). A number of different forms can be used for the firing-rate function F, but we restrict our discussion to one simple form,

$$F(x) = \begin{cases} x & \text{if } x \geq 0 \\ 0 & \text{if } x < 0, \end{cases}$$

2.

which assumes a linear relationship between firing rate and current but accounts for rectification, i.e., the fact that firing rates cannot be negative.

To examine the activity generated internally by the model of Equation 1, we set $I_i = 0$ for all i. This spontaneous activity is then determined by the values of the synaptic weights $\mathcal{J}_{ij}$ for all i, j pairs; the constant Θ; and, in some cases, the initial state of the network. We consider different forms for the synaptic weights that generate different types of internally generated activity. To probe the input sensitivity of these networks, we can transiently activate the external inputs and examine what happens.

Integrate-and-Fire Networks

In addition to network models described by firing rates, we discuss networks constructed from a simple model of a spiking neuron, the integrate-and-fire model. In the integrate-and-fire approach, network neuron i is described by a membrane potential V_i. Although the model generates action potentials, it contains no biophysical spike-generating mechanism. Instead, action potentials are generated by a simple rule: An action potential occurs whenever the membrane potential reaches a threshold value V_{th}, and immediately thereafter the membrane potential is reset to a value V_{reset}. Refractoriness can be imposed by holding the membrane potential to this value for a time, the refractory period, following the spike. The network model shown in **Figure 3** (see Chaotic Spiking Networks, below), for example, uses a refractory period of 5 ms.

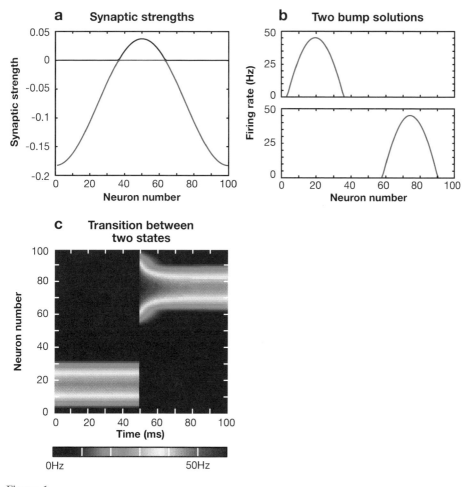

Figure 1

Sustained network activity in a bump configuration. (*a*) The distribution of excitation (*blue line*) and inhibition (*red line*) for a network neuron. These lines indicate the strength and sign of the synaptic weight linking neuron 50 to the other neurons in the network indicated by the value on the horizontal axis. (*b*) Two bumps centered around different neuronal populations. Both plots indicate the firing rate for 100 network neurons. The top panel shows a bump of activity centered around neuron 20, and the bottom a similar bump centered around neuron 75. (*c*) Activity of 100 network neurons as a function of time. A bump of activity centered around neuron 20 was perturbed at time 50 ms to shift it to a bump centered around neuron 75, after a transient.

When an action potential occurs, the time of the spike, denoted by t_i^a, is recorded. The superscript *a* refers to which particular action potential fired by neuron *i* occurred at time t_i^a. In other words, the sequence of action potentials generated by neuron *i* is described by the firing times t_i^a for $a = 1, 2, \ldots$.

The membrane potential in the subthreshold range of the integrate-and-fire model is described by a simple resistor-capacitor circuit or, equivalently, by the equation

$$\tau_m \frac{dV_i}{dt} = V_{\text{rest}} - V_i(t) + \Theta + I_i(t)$$

$$+ \sum_{j=1}^{N} \mathcal{J}_{ij} \sum_{t_j^a < t} f\left(t - t_j^a\right) \qquad 3.$$

for neuron *i*. Here, τ_m is the membrane time constant, V_{rest} is the resting membrane

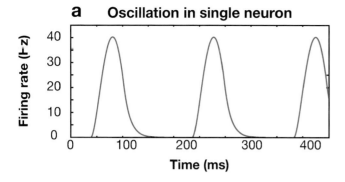

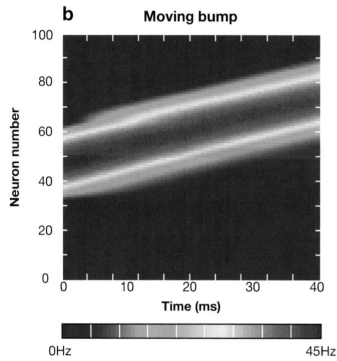

Figure 2
Oscillatory network activity produced by a traveling bump of activity. (*a*) The firing rate of one neuron as a function of time. (*b*) The activity of the entire population of 100 neurons as a function of time. The angled stripe indicates the traveling bump.

potential of the neuron, and Θ is a bias current similar to what appears in Equation 1. As in the case of a firing-rate network, the total input current, given by the last two terms in Equation 3, consists of an external input, $I_i(t)$, and input coming from the other neurons within the network. For the networks shown in **Figure 3**, $\tau_m = 20$ ms, $V_{rest} = V_{reset} = -60$ mV, $V_{th} = -50$ mV, and $\Theta = 15$ mV. For excitatory connections, $\mathcal{J}_{ij} = 1.6$ mV and for inhibitory connections, $\mathcal{J}_{ij} = -8.7$ mV.

The interaction of neurons through chemical synapses arises when presynaptic action potentials produce transient changes in the conductance of the postsynaptic neuron. This can be duplicated in an integrate-and-fire network. However, in the models we review a simplification is made: The postsynaptic effect of a presynaptic action potential is modeled as current injection into the neuron, rather than as a change in its conductance. The synaptic current generated in this way depends on the timing of the presynaptic action potentials. The second sum in the double sum within Equation 3 adds up the contributions from all the action potentials fired by neuron j prior to the time t when the membrane potential of neuron i is being evaluated. The factor $\mathcal{J}_{ij}$ describes, as in the case of firing-rate networks, the strength and type of the synapse from neuron j to neuron i. The function f describes the time course of the postsynaptic current evoked by a presynaptic action potential and, in the examples we show, it takes the form

$$f(t) = \begin{cases} \exp(-t/\tau_s) & \text{if } t \geq 0 \\ 0 & \text{if } t < 0. \end{cases} \quad 4.$$

For the network in **Figure 3**, $\tau_s = 5$ ms for excitatory synapses and $\tau_s = 10$ ms for inhibitory synapses. As in the case of firing-rate networks, the internally generated activity of the integrate-and-fire network depends on the values of the synaptic weights used in the model and on Θ.

FORMS OF NETWORK DYNAMICS

The long-term behavior of dynamical systems is typically classified into four categories: fixed, periodic, quasi-periodic, or chaotic (Strogatz 1994). Fixed or, more properly, fixed-point dynamics means that the system is in a state in which the variables do not change over time. Periodic behavior refers to a time-varying asymptotic state over a particular interval that repeats indefinitely. Quasi-periodic behavior is nonrepeating because it is composed of two or more periodic patterns of activity with incommensurate frequencies.

Incommensurate means that the ratio of the frequencies is an irrational number, which implies that the phase relationship between the different periodic patterns changes on every cycle forever. Finally, chaotic activity is nonrepeating and is characterized by extreme sensitivity to initial conditions. Stable fixed, oscillatory, or chaotic states are often called attractors because nearby states are drawn to them over time.

The dynamic states we have outlined can be linked to activity patterns of obvious importance to neuroscience. The sustained activity characteristic of short-term memory (Wang & Goldman-Rakic 2004) resembles fixed-point behavior, but with an important twist. A system with a single fixed-point attractor, meaning that all initial states end up with the same time-independent pattern of activity, is useless as a memory device. Memory requires the final state of the system to retain some trace of its initial state; this is how sensitivity to the remembered stimulus is expressed in the final state of the system. Thus, memory models require multiple fixed points, each one used to retain a different memory. If a continuous parameter related to the stimulus, such as its position or size, is to be remembered, the memory model must contain a continuum of

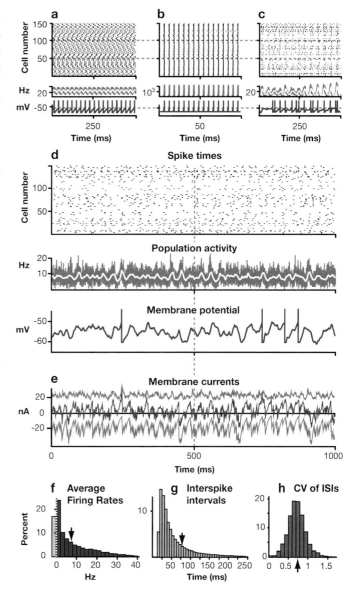

Figure 3

Different forms of activity in a network of spiking model neurons. In panels *a–d*, the top plot is a rastor showing spikes produced over time by 150 of the 10,000 network neurons, the middle plot shows the average firing rate of the entire population (green for 0.1 ms binning and white for 5 ms binning), and the bottom plot shows the voltage trace of a single representative neuron. (*a*) Asynchronous regular activity. The individual neurons fire regularly, but the population rate is roughly constant. (*b*) Synchronous regular activity. Both the individual neurons and the population rate oscillate. (*c*) Synchronous irregular activity. Individual neurons fire irregularly, but the population activity is oscillatory. (*d*) Asynchronous irregular activity. The individual neurons fire irregularly, and the population rate is roughly constant. (*e*) Plots of the excitatory (*green curve*) and inhibitory (*red curve*) currents into the neuron shown above (plotted against time) illustrate that the total current (*blue curve*) is made up of roughly canceling components. (*f*) A histogram of the time-average firing rates of the network neurons. The average firing rate for the entire population, indicated by the arrow, is 8 Hz. (*g*) A histogram of the interspike intervals (ISIs) of the network neurons with the average indicated by the arrow. (*h*) A histogram of the coefficients of variation (the standard deviation of the ISIs over their mean) for the network neurons, with the arrow indicating the average.

fixed points configured along what is called a line attractor, which is just a line of fixed-point attractors. We illustrate a model of this general type in the following section.

Periodic dynamics is obviously connected to the many oscillatory states seen in neural recordings (Marder & Calabrese 1996, Buzsaki & Draguhn 2004). A stable periodic state is also called a limit cycle because it involves cyclic activity that describes the limiting behavior, in time, of many nearby initial states. A model with a continuous line of fixed-point attractors that loops back on itself can easily be turned into an oscillator simply by making the system move around on this loop, a situation we illustrate below.

Although periodic oscillations are frequently seen in cortical recordings (Buzsaki & Draguhn 2004), the overall activity is extremely complex and nonrepeating. The fact that this activity appears to involve superimposed oscillations at many different frequencies (Penttonen & Buzsaki 2003, Leopold et al. 2003) might suggest that the total activity is quasi-periodic, constructed from many different incommensurate oscillations. However, nonrepeating dynamical systems constructed from large numbers of oscillatory elements tend to be chaotic rather than quasi-periodic (Ott 2002). Therefore, the overall activity is more likely to be modeled by a chaotic system, and we review a spiking model of this type below.

Sustained Activity

To sustain their own activity, a group of neurons must feed back enough excitation to each other to maintain the firing that is the source of that excitation. At the same time, sufficient inhibition must be present to prevent the excitatory feedback from producing runaway activity. It might appear that precise parameter adjustment would be required to balance the runaway effects of excitation and the suppressive effects of inhibition, but this is not necessarily the case. In firing-rate models with short-range excitation and long-range inhibition (a so-called Mexican hat configuration), the balance needed for stability can arise automatically. Consider what happens when a group of neurons that excite each other locally are activated. If the excitation is strong enough, more neurons will be recruited into this active group and its size will grow. The number of active excitatory inputs received by a single neuron in this group will grow initially as more neurons become excited, but then it will saturate when all of the neurons within the range of its excitatory connections are already active. On the other hand, the number of active inhibitory inputs will continue to grow at this point because of the longer range of the inhibitory connections. At some point, as the group of active neurons grows, the amount of inhibition will catch up to and then surpass the amount of excitation (provided the inhibition is strong enough) and, at this point, the active group will stop growing and stabilize. In this way, the growth of the active group to an equilibrium size adjusts the balance of excitation and inhibition automatically without requiring fine tuning of parameters (Hansel & Mato 1993; Ben-Yishai et al. 1995, 1997; Hahnloser et al. 2003).

Such a model can be constructed using the network model of Equation 1 by choosing the synaptic weights appropriately (Ben-Yishai et al. 1995, 1997; Hansel & Sompolinsky 2000). For this purpose, we set the weight for the synapse from neuron j to neuron i to

$$\mathcal{J}_{ij} = -\mathcal{J}_0 + \mathcal{J}_2 \cos\left(\frac{2\pi(i-j)}{N}\right), \quad 5.$$

where $\mathcal{J}_0$ and $\mathcal{J}_2$ are constants. For the network shown in **Figures 1** and **2**, $N = 100$, $\mathcal{J}_0 = 0.073$, $\mathcal{J}_2 = 0.11$, and $\Theta = 20$ Hz. The patterns of excitation for one particular neuron (given by the positive part of $\mathcal{J}_{ij}$ for $i = 50$ and the full range of j values) and inhibition for the same neuron (given by the negative part of $\mathcal{J}_{ij}$) are plotted in **Figure 1a**. The pattern of short-range excitation and long-range inhibition discussed in the previous paragraph is readily apparent. As a consequence of this configuration, the model has a time-independent steady-state solution consisting of a "bump" of activity, as seen in **Figure 1b,c**.

An important feature of this model is that it has not one, but a multitude of bump configurations. The synaptic weights given in Equation 5 do not change if we make the transformations $i \to i + c$ and $j \to j + c$, where c is any integer. This symmetry implies that if the model has one bump solution (which it does), it must have a whole family of such solutions. Two of these are shown in **Figure 1b,c**. For any given bump, there is one neuron that fires faster than all the others, and any one of the N neurons in the network can play this role. Thus, there are at least N different bump states.

The existence of multiple bump states allows the network to retain a memory of its initial pattern of activity, as shown in **Figure 1c**. Here, the population of neurons started out in a state with activity centered around neuron 20. Halfway through the simulation, we changed the state of the system and created a new pattern of activity centered around neuron 75. Thus, in this case, the system maintains a memory of one pattern of activity for half the simulation and then of another pattern of activity thereafter.

The model shown in **Figure 1** provides a useful description of short-term memory because it combines a simple form of self-sustained activity with sensitivity to external input. However, there is a price for this success. The model maintains sensitivity to input through the imposition of a symmetry, and anything that breaks this symmetry, even slightly, will spoil that sensitivity. Various remedies have been proposed for the sensitivity of these models to small symmetry-breaking effects (Camperi & Wang 1998, Compte et al. 2000, Seung et al. 2000, Tegner et al. 2002, Renart et al. 2003), but the general issue of reconciling self-sustained activity and sensitivity to input is not completely resolved even in the simple case of fixed patterns of activity.

Oscillations

Oscillations and synchrony in neural networks have received an enormous amount of attention (Ermentrout & Cowan 1979, Marder & Calabrese 1996, Gray 1994, Laurent 1996, Rinzel & Ermentrout 1998, Traub et al. 1999). We do not attempt to review the vast literature on this subject, but instead treat oscillations as a step toward understanding more complex, nonperiodic activity. To illustrate one mechanism through which network oscillations can arise, we modify the model of sustained activity introduced in the previous section to make it oscillate.

Network oscillations often arise from a dynamic interplay of excitatory and inhibitory populations of cells (Wilson & Cowan 1972), with inhibition playing a particularly important role (Traub et al. 1989, 1997; Golomb & Rinzel 1993; Wang & Buzsaki 1996; White et al. 1998; Brunel & Hakim 1999; Whittington et al. 2000). In the model we use to illustrate network oscillations, a periodic pattern of activity is set up in each neuron by turning the steady-state bump of the previous section into a traveling wave. One way to do this is to introduce adaptation into the neurons of the model (Hansel & Sompolinsky 2000). As the active neurons adapt, the bump of activity moves to recruit neurons not previously active. Another way to produce a moving bump of activity is to modify the synaptic weights of Equation 5, replacing them with

$$\mathcal{J}_{ij} = -\mathcal{J}_0 + \mathcal{J}_2 \left(\cos\left(\frac{2\pi(i-j)}{N}\right) - \omega\tau_r \sin\left(\frac{2\pi(i-j)}{N}\right) \right). \quad 6.$$

The parameter ω, which is set to $\omega = 40$ radians/s for **Figure 2**, determines the speed of propagation of the moving bump or, equivalently, the oscillation frequency of the individual neurons in the network. Note that these synaptic weights retain the symmetry that existed for the model of sustained activity discussed in the previous section.

The resulting pattern of activity is shown in **Figure 2**. **Figure 2a** shows the firing rate of a single neuron plotted as a function of time, indicating the periodic, oscillatory nature of the activity. Each neuron in the network has an oscillating firing rate, but with a different phase depending on when the moving bump sweeps past it. The entire moving bump is shown in **Figure 2b**, with the firing rate plotted against time for all of the neurons in the network.

Chaotic Spiking Networks

The large number of synaptic connections received by cortical neurons does not, at first sight, appear to be consistent with the high degree of variability in cortical responses (Softky & Koch 1993). Naively, one would expect a neuron with n synaptic inputs, each of strength g, to receive a total synaptic input of order gn. If we think of the strength factor g as the probability that a given presynaptic action potential evokes a postsynaptic response, we must require the total synaptic input (gn) to be of order 1, otherwise the postsynaptic neuron would fire much more rapidly than its presynaptic partners within the network. In a recurrently connected network, such a situation is inconsistent because each neuron plays both presynaptic and postsynaptic roles. This condition requires g to be of order $1/n$. For synaptic inputs that fluctuate independently, the variance of the total synaptic input is proportional to $g^2 n$, which for $g \sim 1/n$ is of order $1/n$. With n being approximately 10,000, this argument would suggest that synapses should be quite weak ($g \approx 0.0001$) and that the variance of the total synaptic input should be quite small.

The argument of the preceding paragraph is clearly incorrect for cortical circuits. Where they have been measured, synapses between cortical neurons have been found to be much stronger than the estimate obtained above (Song et al. 2004), and cortical neurons are highly variable in their responses, suggesting that the input variance is much larger than of order $1/n$. Instead, cortical synapses seem to have a strength of order $g \sim \sqrt{1/n}$, which makes the input variance, $g^2 n$, of order 1 (van Vreeswijk & Sompolinsky 1996, 1998). The reason that cortical neurons do not experience runaway firing is that the total synaptic input is not of order $gn \sim \sqrt{n}$ as suggest above, but of order 1 because of a balance between excitatory and inhibitory inputs. The order-of-magnitude estimate given above did not take into account such a cancelation, but this appears to be what happens in cortical circuits (Shadlen & Newsome 1994, Tsodyks & Sejnowski 1995, Troyer & Miller 1997). This realization has important implications that we now explore, and it represents an important advance in our understanding of cortical dynamics.

In the previous sections, we considered model networks in which the individual neurons are described by firing rates. We now discuss networks of spiking neurons. The activity of spiking networks has been divided into four classes depending on how the individual neurons fire and how the activities of the different neurons relate to each other (Brunel 2000). Individual neurons are classified as firing in either a regular or an irregular pattern. At the network level, the neurons may either synchronize their firing or fire asynchronously. **Figure 3** shows examples of each of the four possible combinations of these attributes for a network of 10,000 integrate-and-fire model neurons, 80% of which are excitatory and 20% inhibitory (meaning that they inject positive and negative exponentially decaying pulses of current into their target postsynaptic neurons, respectively).

Individual neurons in the example of **Figure 3a** fire at a steady rate in a periodic pattern, but, in this case, they fire asynchronously because they are uncoupled and were started with random initial conditions. This is a rather trivial example of an asynchronous regular state; more interesting cases also exist (Brunel 2000). Weakly coupled networks can also exhibit such asynchronous regular activity.

In **Figure 3b**, excitatory synapses between regularly firing neurons have caused the spikes to synchronize, producing synchronous regular network activity. As the excitatory and inhibitory currents are brought into a more balanced configuration, the individual neurons start to fire in irregular patterns, as seen in panels c and d of **Figure 3**. The difference between these panels is that there is partial synchrony in the case of **Figure 3c**, as can be seen by the oscillations in the average firing rate of the population of neurons shown in the middle trace of this panel, whereas the population activity in the case of **Figure 3d** is asynchronous. A transition between synchronous and asynchronous states with irregular firing occurs, for example, when Θ is decreased or inhibition is increased (Brunel 2000).

Asynchronous states have received considerable attention as candidates for the background activity seen in cortical and other neural circuits (Abbott & van Vreeswijk 1993, Destexhe 1994, Brunel & Hakim 1999, Fusi & Mattia 1999, Gerstner 2000, van Vreeswijk 2000, Hansel & Mato 2002). The key features that permit the existence of an asynchronous network state with irregular firing of the individual neurons that are illustrated in panels d–h of **Figure 3** are (*a*) a balance of excitation and inhibition, as indicated by **Figure 3e**, and (*b*) sparse connectivity (van Vreeswijk & Sompolinsky 1996, 1998; Amit & Brunel 1997; Brunel 2000; Lerchner et al. 2004). For the example shown in **Figure 3**, neurons were connected randomly with a connection probability of 1.5%. Asynchronous irregular activity can also arise in sparsely connected networks with local patterns of connectivity (Mehring et al. 2003). In all cases, the sparseness of the connectivity means that large numbers of neurons are required to achieve this type of activity. The example of **Figure 3** is based on synapses that inject current into their postsynaptic targets, but asynchronous irregular activity can also arise from conductance-based synapses, although the firing of individual neurons tends to be considerably more burst-like in this case.

The asynchronous irregular state of network activity is quite remarkable. Note that the highly irregular voltage trace for an individual neuron of the network shown in the bottom panel of **Figure 3d** arises in this model without the addition of any external source of randomness (i.e., no random number generator is used in the simulation). As shown in panel *f* of **Figure 3**, the network neurons display a roughly exponentially distributed range of firing rates. The spiking statistics of the network show an exponentially distributed range of interspike intervals (except for short intervals forbidden by the refractory period imposed on the neurons) and a range of values of the coefficients of variation (standard deviation divided by the mean) of interspike intervals for the individual neurons (**Figures 3g** and **3h**, respectively).

It is possible to investigate the asynchronous irregular state through analytic calculations, not merely by simulation (van Vreeswijk & Sompolonsky 1996, 1998; Amit & Brunel 1997; Brunel 2000). This analysis is based on self-consistent mean-field calculations. The idea underlying these calculations is that neurons in a closed network receive inputs from other neurons with firing statistics similar to their own. The self-consistent calculation involves determining the firing statistics of a neuron receiving a total synaptic input characterized by a particular mean and variance (Ricciardi 1977, Tuckwell 1988, Brunel & Sergi 1998). Self-consistency is then imposed by demanding that the assumed mean and variance match what would be obtained by summing synaptic inputs from a set of presynaptic neurons with the computed postsynaptic firing statistics. These calculations provide an accurate description of what happens in computer simulations of model networks, such as that shown in **Figure 3**.

Given that cortical connectivity is sparse and that cortical background activity is of the asynchronous irregular form, the fact that sparsely connected model networks with balanced excitation and inhibition produce such a pattern of activity lends further support to the idea that excitation and inhibition are in a balanced configuration within cortical circuits. Balancing two opposing influences in this way is an odd approach from a computational perspective. A standard warning in numerical calculations is to avoid subtracting two large numbers that are close to being equal because the answer obtained in such a case is sensitive to small errors in the calculation of either of the large numbers. Yet this appears to be exactly how cortical circuits operate and, indeed, the result is highly variable responses. Why would the system have evolved to operate in this way? One proposed answer is that networks in a balanced configuration can respond more rapidly to inputs than those with unbalanced excitation and inhibition (Tsodyks & Sejnowski 1995, van Vreeswijk & Sompolinsky 1998). Another suggestion is that the large input variance produced by a balanced network is not merely a source of noise. Instead, having comparable levels of mean and variance for synaptic inputs permits two types of signals to be transmitted simultaneously within these circuits, one (the mean) that drives neuronal responses and the other (the variance) that serves to modify neuronal gain (Chance et al. 2002) and can evoke rapid responses (Silberberg et al. 2004). Finally, as we discuss in the following section, high variance is important for supporting stable signal propagation.

SIGNAL PROPAGATION

Cognitive processing requires that signals propagate reliably through multiple regions of the brain. Achieving stable signal propagation in neural networks is a difficult problem. After introducing some of the problems associated with signal propagation, we discuss two proposed modes of propagation, synfire chains and firing-rate propagation.

As activity propagates through a network, action potentials tend to synchronize, and avoiding network-wide synchronization of activity requires noise (Marsalek et al. 1997, Golomb 1998, Burkitt & Clark 1999, Diesmann et al. 1999, van Rossum et al. 2002, Reyes 2003, Litvak et al. 2003). In early models, this noise was introduced from an external source, such as a random-number generator (Diesmann et al. 1999, van Rossum et al. 2002), but more recent studies have used the asynchronous irregular state discussed in the previous section to provide the variability needed to prevent large-scale synchronization (Mehring et al. 2003, Aviel et al. 2003, Vogels & Abbott 2004). However, noise, whether externally or internally generated, does not necessarily remove all synchronization. Forms of signal propagation can be distinguished by whether or not they involved synchrony. Synchrony plays a critical role in synfire propagation, whereas for firing-rate propagation synchronization destroys the signal.

Avalanche Model

We begin our discussion of signal propagation by discussing a highly simplified model that is, nevertheless, useful for illustrating basic issues relevant to signal propagation in neural networks. This model, known as the avalanche model (Harris 1963, Zapperi et al. 1995, de Carvalho & Prado 2000), is defined by the following rule: When a neuron fires an action potential, a spike is evoked in each of its n postsynaptic target neurons with probability p. The same rule is then applied to any of the postsynaptic neurons that fired owing to the first application of the rule. Each firing is treated as an independent event (certainly an unrealistic assumption, especially for late stages of the propagation chain). By applying this rule to a single initial neuron, then to all the neurons that the initial neuron caused to fire, and then to all the neurons that subsequently fire, the model can be iterated sequentially, describing the propagation of a neuronal signal.

The average number of neurons that fire at the second stage, after the single initial neuron has fired, is pn. At the next stage, the average number of firing neurons is $(pn)^2$ because each of the pn neurons firing at the second stage induces an average of pn neurons to fire at the third stage. At stage s, the average number of neurons that fire is $(pn)^s$. This simple calculation illustrates one major issue in signal propagation. Signals tend either to decay, which happens in the avalanche model if $pn < 1$, or to blow up, which happens if $pn > 1$. To maintain stable average propagation over multiple stages, the probability of evoking an action potential must be adjusted so that pn is close to 1. This feature of propagation in the avalanche model is illustrated in **Figure 4b** for subcritical ($pn < 1$), critical ($pn = 1$), and supracritical ($pn > 1$) cases.

Although choosing $pn = 1$ stabilizes the average level of propagation from stage to stage, there are large fluctuations from trial to trial. The red sequence in **Figure 4a** shows a case where one neuron firing at stage 1 evokes one neuron firing at stage 5. Although this represents the average behavior, it is actually quite rare. The blue sequence in **Figure 4a** is an example of a propagation failure, and the green sequence shows a propagation explosion. A failure can occur, for example, if the first neuron does not activate any other neurons, which makes the second stage silent. Such failures obviously make the wave of propagating activity die out. If the number of activated neurons at the second stage is larger than average, the system becomes prone to sequential increases, and the propagating activity tends to explode. **Figure 4c** shows the percentage of successful propagations (meaning that at least one neuron fired) at various stages of a critical avalanche. Note that beyond 3 stages, propagation fails more than 50% of the time. **Figure 4d** shows the number of activity explosions. This is highly sensitive to the value of pn, but even for the critical case, there are a fair number of explosions beyond layer 5.

The large fluctuations that we have discussed cause the number of neurons activated by a critical avalanche to have a power-law distribution (Harris 1963, Zapperi et al. 1995, de Carvalho & Prado 2000), which agrees with what is seen in multielectrode data from organotypic cultures or slices (Beggs & Plenz 2003). Although the avalanche model is an oversimplified description, it highlights two basic points. First, tuning is required so that, on average, propagation neither dies out nor explodes. Second, even if this critical condition is met, large fluctuations can cause frequent failures and occasional explosions.

Synfire Chains

We have already mentioned that the spiking of different neurons tends to synchronize as signals propagate through a network. As long as this synchronization can be kept from spreading across the entire network, it can provide an effective method for transmitting signals (Salinas & Sejnowski 2002). This idea is the basis of propagation along synfire chains (Abeles 1991), which are groups of neurons coupled in a feedforward manner that support synchronous signal propagation (Herrmann et al. 1995). **Figure 5a** shows an example of a feedforward chain in which every neuron of a given layer makes synapses onto every neuron of the next layer. Signal propagation along such synfire chains has been studied extensively in network models (Aertsen et al. 1996, Diesmann et al. 1999, Cateau & Fukai 2001).

By signal propagation, we mean the transmission of activity along specific pathways across a neural network, not the activation of an entire network. Noise is essential in a network that supports synfire activity to prevent synchronous activation from spreading beyond the specified synfire chain. Stable propagation in a network receiving noisy input from an external source is illustrated in **Figure 5b**. This propagation is subject to the same types of fluctuations seen in the avalanche model, leading, for example, to failures of propagation as in **Figure 5c**. In these networks, synfire signals can travel through a number of embedded feedforward layers, provided they are large enough (Diesmann et al. 1999). Propagation requires a critical level of activity in the initial layer-1 pulse packet seen in **Figure 5b**, and during propagation the level of synchrony stays at a constant value determined by the level of noise.

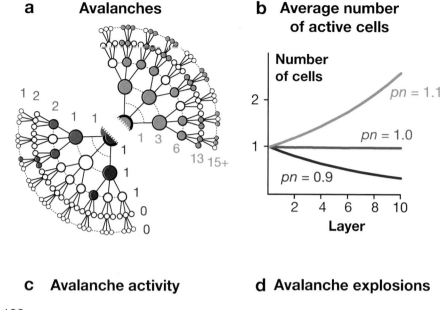
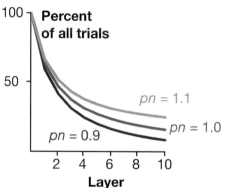
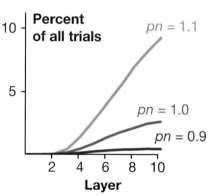

Figure 4

The avalanche model. (*a*) Avalanches develop from the center of the graphic and travel outward. Systems with $pn > 1$ (*green*) are likely to show sharply increasing numbers of active cells in consecutive layers. When $pn = 1$ (*red*), the system is more likely to propagate activity without explosive multiplication of active cells in higher layers. $pn < 1$ frequently leads to an eventual extinction of the activated wave (*blue*). (*b*) Average number of activated cells per layer in 10^4 independently activated avalanches. Avalanches with $pn > 1$ show an exponential increase in the number of active cells in higher layers. When $pn = 1$, the average number of active cells is constant, and with $pn < 1$ it declines to zero exponentially. (*c*) Distribution of avalanche run lengths. The run length is the number of layers before the avalanche stops. Higher values of pn lead to longer-surviving avalanches. (*d*) Occurrence of events with more than 10 active cells in a single layer for the same trials as in *c*. Higher values of pn increase the number of such "explosions."

As mentioned above, the synfire propagation seen in **Figures 5b** and **5c** occurs in a network that receives noise input from an external source. A more realistic approach is to generate noise within the network through the mechanism reviewed in the previous section. This has been done by embedding synfire chains (i.e., constructing specific feedforward synaptic pathways) in large, sparsely connected networks of integrate-and-fire neurons (Mehring et al. 2003, Aviel et al. 2003). Signal propagation has been achieved in such networks when they are large (**Figure 5d**), but only for very specific sets of parameters. Consistent, stable propagation is problematic. These difficulties arise because, unlike the situation with external noise, synfire activity interacts with the chaotic background activity in these networks, and this interaction can be destructive to both the propagating signal and the background activity. Similar to what can occur in the avalanche model, synfire propagation can set off large-scale synchronization within the network, as seen in **Figure 5e**. Furthermore, through the over-excitation of the inhibitory cell population these "synfire explosions" can subsequently silence the network completely

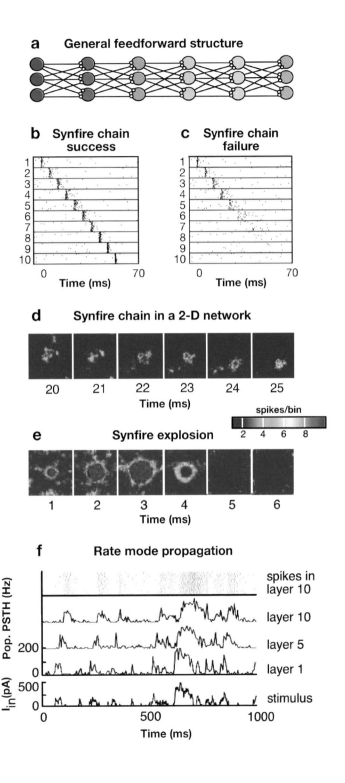

Figure 5

Signal propagation. (*a*) General layout of a synfire chain with all-to-all coupling in the feedforward direction between neighboring layers of neurons. (*b*) Propagation of a synfire wave through a feedforward network with added noise. A group of cells is activated in layer 1. This evokes activity that propagates through all layers with a time lag of approximately 5 ms per layer. (*c*) A propagation failure. The synfire wave dissolves into background activity and fails to propagate. (*d*) Synfire propagation in a two-dimensional, locally coupled, sparse, balanced network with an embedded synfire chain. Six frames show the activity of the network at different times indicated below the plots. Activity propagates from the center of the network to the lower right corner. The rest of the network is quiescent. (*e*) Activity in the same network for different initial conditions. Now the synfire event evokes a large shock wave, affects the majority of cells, and ultimately silences all network activity. (*f*) Firing-rate propagation in a feedforward network with added noise. The lowest plot shows the current injected into layer 1 of the network to evoke activity. The plots above this show the firing rates in various layers of the network, and the top plot is a raster of the spikes produced in the 20 cells of layer 10. The network accurately propagates the activity from layer 1 to 10. Panels were adapted from previously published figures: *b* and *c* (Diesmann et al. 1999), *d* and *e* (Mehring et al. 2003), and *f* (van Rossum et al. 2002).

(Gutkin et al. 2001, Mehring et al. 2003). To prevent this, the perturbation in network activity caused by the synfire chain must be diluted either by embedding it in a very large network or by introducing some form of canceling inhibition (Aviel et al. 2003).

Propagation of Firing Rates

Signals can also propagate through networks in the form of transiently elevated or depressed firing rates, rather than as waves of synchronized firing. This requires a feedforward structure similar to a synfire chain (**Figure 5a**), but more noise to prevent synchronization even within the groups of neurons carrying the signal (Litvak et al. 2003). As shown in **Figure 5f**, firing-rate propagation has been demonstrated in feedforward networks receiving external noise (van Rossum et al. 2002). The network used for **Figure 5f** (van Rossum et al. 2002) consists of 200 integrate-and-fire neurons organized in 10 layers, in which every neuron of one layer synapses on to every neuron of the next. Signal propagation through all 10 layers is observed when a sufficiently strong input is fed into layer 1.

There are actually two different modes of propagation within this network, depending on the level of externally applied noise. Without external noise, the network shows an all-or-none response. The otherwise silent input layer fires only when stimuli succeed in driving the membrane potential above threshold. All neurons of that layer then fire simultaneously and their activity evokes a traveling wave through the layers of the network similar to a synfire chain (Abeles 1991). When noise is introduced, network behavior changes. Noise was adjusted to maintain an average firing rate of 5 Hz. At optimal noise levels, all neurons desynchronize, and their membrane potentials hover slightly below threshold. Both small and large stimuli are now transmitted because the number of neurons driven to fire is proportional to the amplitude of the stimulus. Graded signals are reproduced accurately and in an approximately linear manner across 10 layers with a minimum of 20 cells in each layer.

As in the case of synfire chains, it is important to study the propagation of firing rates in networks with internally generated noise. Preliminary indications are that firing rates can indeed be propagated along embedded feedforward chains in large networks of sparsely connected integrate-and-fire neurons of the type reviewed in the previous section (Vogels & Abbott 2004).

DISCUSSION

To understand neural network activity in behaving animals we must account for both internally generated activity and activity evoked by external stimuli. The history of neuroscience is full of examples in which stimulus-evoked activity has been successfully modeled. This review has covered models that describe three forms of internally generated activity: persistent, oscillatory, and asynchronous irregular. Networks that self-sustain activity without being sensitive to external stimuli are useless in shaping behavior in response to environmental cues. On the other hand, networks in which stimulus-driven responses are not accompanied by internally generated activity cannot sustain useful forms of signal propagation. Although there is much to be done in the separate modeling of both internally and externally generated activity, the bigger challenge is to reconcile these two forms of activity and construct models in which they coexist synergistically.

ACKNOWLEDGMENTS

Research was supported by the National Science Foundation (IBN-0235463), an NIH Pioneer Award, and the Swartz Foundation.

LITERATURE CITED

Abbott LF, van Vreeswijk C. 1993. Asynchronous states in networks of pulse-coupled oscillators. *Phys. Rev. E* 48:1483–90

Abeles M. 1991. *Corticonics: Neural Circuits of the Cerebral Cortex*. Cambridge, UK: Cambridge Univ. Press. 280 pp.

Aertsen A, Diesmann M, Gewaltig MO. 1996. Propagation of synchronous spiking activity in feedforward neural networks. *J. Physiol. Paris* 90:243–47

Amit DJ, Brunel N. 1997. Model of global spontaneous activity and local structured activity during delay periods in the cerebral cortex. *Cereb. Cortex* 7:237–52

Anderson JS, Lampl I, Gillespie DC, Ferster D. 2000. The contribution of noise to contrast invariance of orientation tuning in cat visual cortex. *Science* 290:1968–72

Arieli A, Sterkin A, Grinvald A, Aertsen A. 1996. Dynamics of ongoing activity: explanation of the large variability in evoked cortical responses. *Science* 273:1868–71

Aviel Y, Mehring C, Abeles M, Horn D. 2003. On embedding synfire chains in a balanced network. *Neural Comput.* 15:1321–40

Beggs JM, Plenz D. 2003. Neuronal avalanches in neocortical circuits. *J. Neurosci.* 23:11167–77

Beggs JM, Plenz D. 2004. Neuronal avalanches are diverse and precise activity patterns that are stable for many hours in cortical slice cultures. *J. Neurosci.* 24:5216–29

Ben-Yishai R, Bar-Or RL, Sompolinsky H. 1995. Theory of orientation tuning in visual cortex. *Proc. Natl. Acad. Sci. USA* 92:3844–48

Ben-Yishai R, Hansel D, Sompolinsky H. 1997. Traveling waves and the processing of weakly tuned inputs in a cortical network module. *J. Comput. Neurosci.* 4:57–77

Brunel N. 2000. Dynamics of networks of randomly connected excitatory and inhibitory spiking neurons. *J. Physiol. Paris* 94:445–63

Brunel N, Hakim D. 1999. Fast global oscillations in networks of integrate-and-fire neurons with low firing rates. *Neural Comput.* 11:1621–71

Brunel N, Sergi S. 1998. Firing frequency of leaky intergrate-and-fire neurons with synaptic current dynamics. *J. Theor. Biol.* 195:87–95

Burkitt AN, Clark GM. 1999. Analysis of integrate-and-fire neurons: synchronization of synaptic input and output spikes. *Neural Comput.* 11:871–901

Burns BD, Webb AC. 1976. The spontaneous activity of neurones in the cat's visual cortex. *Proc. R. Soc. London B Biol. Sci.* 194:211–23

Buzsaki G, Draguhn A. 2004. Neuronal oscillations in cortical networks. *Science* 304:1926–29

Camperi M, Wang X-J. 1998. A model of visuospatial short-term memory in prefrontal cortex: recurrent network and cellular bistability. *J. Comput. Neurosci.* 5:383–405

Cateau H, Fukai T. 2001. Fokker-Planck approach to the pulse packet propagation in synfire chain. *Neural Netw.* 14:675–85

Chance FS, Abbott LF, Reyes AD. 2002. Gain modulation through background synaptic input. *Neuron* 35:773–82

Compte A, Brunel N, Goldman-Rakic PS, Wang X-J. 2000. Synaptic mechanisms and network dynamics underlying spatial working memory in a cortical network model. *Cereb. Cortex* 10:910–23

Dayan P, Abbott LF. 2001. *Theoretical Neuroscience: Computational and Mathematical Modeling of Neural Systems*. Cambridge, MA: MIT Press. 460 pp.

de Carvalho JX, Prado CP. 2000. Self-organized criticality in the olami-feder-christensen model. *Phys. Rev. Lett.* 84:4006–9

Dean AF. 1981. The variability of discharge of simple cells in the cat striat cortex. *Exp. Brain Res.* 44:437–40

Destexhe A. 1994. Oscillations, complex spatiotemporal behavior, and information transport in networks of excitatory and inhibitory neurons. *Phys. Rev. E* 50:1594–606

Diesmann M, Gewaltig MO, Aertsen A. 1999. Stable propagation of synchronous spiking in cortical neural networks. *Nature* 402:529–33

Ermentrout GB, Cowan JD. 1979. Temporal oscillations in neuronal nets. *J. Math. Biol.* 7:265–80

Fusi S, Mattia M. 1999. Collective behavior of networks with linear (VLSI) integrate and fire neurons. *Neural Comput.* 11:633–52

Gerstner W. 2000. Population dynamics of spiking neurons: fast transients, asynchronous state, and locking. *Neural Comput.* 12:43–89

Golomb D. 1998. Models of neuronal transient synchrony during propagation of activity through neocortical circuitry. *J. Neurophysiol.* 79:1–12

Golomb D, Rinzel J. 1993. Dynamics of globally coupled inhibitory neurons with hereogeneity. *Phys. Rev. E* 48:4810–14

Gray CM. 1994. Synchronous oscillations in neuronal systems: mechanisms and functions. *J. Comput. Neurosci.* 1:11–38

Gutkin BS, Laing CR, Colby CL, Chow CC, Ermentrout GB. 2001. Turning on and off with excitation: the role of spike-timing asynchrony and synchrony in sustained neural activity. *J. Comput. Neurosci.* 11:121–34

Hahnloser RHR, Seung HS, Slotine JJ. 2003. Permitted and forbidden sets in symmetric threshold-linear networks. *Neural Comput.* 15:621–38

Hansel D, Mato G. 1993. Existence and stability of persistent states in large neuronal networks. *Phys. Rev. Lett.* 86:4175–78

Hansel D, Mato G. 2002. Asynchronous states and the emergence of synchrony in large networks of interacting excitatory and inhibitory neurons. *Neural Comput.* 15:1–56

Hansel D, Sompolinsky H. 2000. Modeling feature selectivity in local cortical circuits. In *Methods in Neuronal Modeling: From Synapses to Networks*, ed. C Koch, I Segev, pp. 499–567. Cambridge, MA: MIT Press

Harris TE. 1963. *The Theory of Branching Processes*. Berlin: Springer. 229 pp.

Herrmann M, Hertz J, Pruegel-Bennett A. 1995. Analysis of synfire chains. *Netw.: Comput. Neural Syst.* 6:403–14

Holt GR, Softky WR, Koch C, Douglas RJ. 1996. Comparison of discharge variability in vitro and in vivo in cat visual cortex neurons. *J. Neurophysiol.* 75:1806–14

Laurent G. 1996. Dynamical representation of odors by oscillating and evolving neural assemblies. *Trends Neurosci.* 19:489–96

Leopold DA, Murayama Y, Logothetis NK. 2003. Very slow activity fluctuations in monkey visual cortex: Implications for functional brain imaging. *Cereb. Cortex* 13:422–433

Lerchner A, Ahmadi M, Hertz J. 2004. High-conductance states in a mean-field cortical network model. *Neurocomputing* 58–60:935–40

Litvak V, Sompolinsky H, Segev I, Abeles M. 2003. On the transmission of rate code in long feedforward networks with excitatory-inhibitory balance. *J. Neurosci.* 23:3006–15

Marder E, Calabrese RL. 1996. Principles of rhythmic motor pattern generation. *Physiol. Rev.* 76:687–17

Marsalek PR, Koch C, Maunsell J. 1997. On the relationship between synaptic input and spike output jitter in individual neurons. *Proc. Natl. Acad. Sci. USA* 94:735–40

Mehring C, Hehl U, Kubo M, Diesmann M, Aertsen A. 2003. Activity dynamics and propagation of synchronous spiking in locally connected random networks *Biol. Cybern.* 88:395–408

Ott E. 2002. *Chaos in Dynamical Systems*. Cambridge, UK: Cambridge Univ. Press. 478 pp.

Penttonen M, Buzsaki G. 2003. Natural logarithmic relationship between brain oscillators. *Thalamus Relat. Syst.* 2:145–52

Prinz AA, Bucher D, Marder E. 2004. Similar network activity from widely disparate combinations of intrinsic properties and synaptic strengths. *Nat. Neurosci.* 12:1345–52

Renart A, Song P, Wang X-J. 2003. Robust spatial working memory through homeostatic synaptic scaling in heterogeneous cortical networks. *Neuron* 38:473–85

Reyes AD. 2003. Synchrony-dependent propagation of firing rate in iteratively constructed networks in vitro. *Nat. Neurosci.* 6:593–99

Ricciardi LM. 1977. *Diffusion Processes and Related Topics in Biology*. Berlin: Springer-Verlag. 200 pp.

Rinzel J, Ermentrout GB. 1998. Analysis of neural excitability and oscillations. In *Methods in Neuronal Modeling*, ed. C Koch, I Segev, pp. 251–91. Cambridge, MA: MIT Press

Salinas E, Sejnowski T. 2002. Integrate-and-fire neurons driven by correlated stochastic input. *Neural Comput.* 14:2111–55

Seung HS, Lee DD, Reis BY, Tank DW. 2000. Stability of the memory of eye position in a recurrent network of conductance-based model neurons. *Neuron* 26:259–71

Shadlen MN, Newsome WT. 1994. Noise, neural codes and cortical organization. *Curr. Opin. Neurobiol.* 4:569–79

Shriki O, Hansel D, Sompolinsky H. 2003. Rate models for conductance-based cortical neuronal networks *Neural Comput.* 15:1809–41

Silberberg G, Bethge M, Markram H, Pawelzik K, Tsodyks M. 2004. Dynamics of population rate codes in ensembles of neocortical neurons. *J. Neurophysiol.* 91:704–9

Softky WR, Koch C. 1993. The highly irregular firing of cortical cells is inconsistent with temporal integration of random EPSPs. *J. Neurosci.* 13:334–50

Song S, Sjöström PJ, Reigl M, Nelson SB, Chklovskii DB. 2005. Highly non-random features of synaptic connectivity in local cortical circuits. In press

Strogatz SH. 1994. *Nonlinear Dynamics and Chaos: With Applications to Physics, Biology, Chemistry, and Engineering*. Reading, MA: Perseus Books. 498 pp.

Tegner J, Compte A, Wang X-J. 2002. The dynamical stability of reverberatory neural circuits. *Biol. Cybern.* 87:471–81

Traub R, Jefferys JGR, Whittington MA. 1997. Simulation of gamma rhythms in networks of interneurons and pyramidal cells. *J. Comput. Neurosci.* 4:141–50

Traub R, Jefferys JGR, Whittington MA. 1999. *Fast Oscillations in Cortical Circuits*. Cambridge, MA: MIT Press. 308 pp.

Traub RD, Miles R, Wong RKS. 1989. Model of the origin of rhythmic population oscillations in the hippocampal slice. *Science* 243:1319–25

Troyer TW, Miller KD. 1997. Physiological gain leads to high ISI variability in a simple model of a cortical regular spiking cell. *Neural Comput.* 9:971–83

Tsodyks M, Sejnowski TJ. 1995. Rapid switching in balanced cortical network models. *Network* 6:1–14

Tuckwell HC. 1988. *Introduction to Theoretical Neurobiology*, Vol. 2. Cambridge, UK: Cambridge Univ. Press. 265 pp.

Usher M, Stemmler M, Koch C. 1994. Network amplification of local fluctuations causes high spike rate variability, fractal patterns and oscillatory local field potentials. *Neural Comput.* 6:795–836

van Rossum MC, Turrigiano GG, Nelson SB. 2002. Fast propagation of firing rates through layered networks of noisy neurons. *J. Neurosci.* 22:1956–66

van Vreeswijk C. 2000. Analysis of the asynchronous state in networks of strongly coupled oscillators. *Phys. Rev. Lett.* 84:5110–13

van Vreeswijk C, Sompolinsky H. 1996. Chaos in neuronal networks with balanced excitatory and inhibitory activity *Science* 274:1724–26

van Vreeswijk C, Sompolinsky H. 1998. Chaotic balanced state in a model of cortical circuits. *Neural Comput.* 10:1321–71

Vogels TP, Abbott LF. 2004. Signal propagation in large networks of integrate-and-fire neurons. *Soc. Neurosci.* 970.7 (Abstr.)

Wang X-J, Buzsaki G. 1996. Gamma oscillations by synaptic inhibition in a hippocampal interneuronal network model. *J. Neurosci.* 19:9587–603

Wang X-J, Goldman-Rakic, P, eds. 2004. Special issue: Persistent neural activity: experiments and theory. *Cereb. Cortex* 13:1123–269

White JA, Chow CC, Ritt J, Soto-Trovino C, Kopell N. 1998. Synchronization and oscillatory dynamics in heterogeneous, mutually inhibitory neurons. *J. Comput. Neurosci.* 5:5–16

Whittington MA, Traub RD, Kopell N, Ermentrout GB, Buhl EH. 2000. Inhibition-based rhythms: experiments and mathematical observations on network dynamics. *Int. J. Psychophysiol.* 38:315–36

Wilson HR, Cowan JD. 1972. Excitatory and inhibitory interactions in localized populations of model neurons. *Biophys. J.* 12:1–24

Zapperi S, Baekgaard Lauritsen K, Stanley HE. 1995. Self-organized branching processes: mean-field theory for avalanches. *Phys. Rev. Lett.* 75:4071–74

The Plastic Human Brain Cortex

Alvaro Pascual-Leone, Amir Amedi, Felipe Fregni, and Lotfi B. Merabet

Center for Non-Invasive Brain Stimulation, Department of Neurology, Beth Israel Deaconess Medical Center, Harvard Medical School, Boston, Massachusetts 02215; email: apleone@bidmc.harvard.edu

Key Words

stroke, blindness, neurorehabilitation, neuromodulation, crossmodal plasticity, cortical stimulation, functional neuroimaging

Abstract

Plasticity is an intrinsic property of the human brain and represents evolution's invention to enable the nervous system to escape the restrictions of its own genome and thus adapt to environmental pressures, physiologic changes, and experiences. Dynamic shifts in the strength of preexisting connections across distributed neural networks, changes in task-related cortico-cortical and cortico-subcortical coherence and modifications of the mapping between behavior and neural activity take place in response to changes in afferent input or efferent demand. Such rapid, ongoing changes may be followed by the establishment of new connections through dendritic growth and arborization. However, they harbor the danger that the evolving pattern of neural activation may in itself lead to abnormal behavior. Plasticity is the mechanism for development and learning, as much as a cause of pathology. The challenge we face is to learn enough about the mechanisms of plasticity to modulate them to achieve the best behavioral outcome for a given subject.

Contents

THE CONCEPT OF PLASTICITY 378
TWO-STEP CHANGES 379
THE RAPIDLY SHIFTING MAPPING BETWEEN BRAIN ACTIVITY AND BEHAVIOR 382
THE RISK OF CHANGE AND THE OPPORTUNITY FOR INTERVENTION 383
PLASTICITY IN THE SETTING OF BRAIN INJURY 384
THE OCCIPITAL CORTEX IN THE BLIND 387
UNMASKING CONNECTIONS: THE BLINDFOLD EXPERIMENT 390
ESTABLISHING NEW CONNECTIONS: OCCIPITAL ACTIVATION IN HIGH-LEVEL COGNITIVE TASKS 392
DRAWING CONCLUSIONS FROM THE BLINDFOLDED AND THE BLIND 392
SUMMARY 395

THE CONCEPT OF PLASTICITY

"Plastic" is derived from the Greek word πλαστόσ (plastos), which means molded. According to the Oxford English Dictionary, being plastic refers to the ability to undergo a change in shape. William James (1890) in *The Principles of Psychology* was the first to introduce the term plasticity to the neurosciences in reference to the susceptibility of human behavior to modification.

> Plasticity [...] means the possession of a structure weak enough to yield to an influence, but strong enough not to yield all at once. Each relatively stable phase of equilibrium in such a structure is marked by what we may call a new set of habits. Organic matter, especially nervous tissue, seems endowed with a very extraordinary degree of plasticity of this sort; so that we may without hesitation lay down as our first proposition the following, that the phenomena of habit in living beings are due to the plasticity. (p. 68)

Some years later, Santiago Ramón y Cajal (1904) in the *Textura del Sistema Nervioso* argued that behavioral modifiability must have an anatomical basis in the brain and thus extended the notion of plasticity to the neural substrate. Considering the acquisition of new skills, Cajal wrote

> La labor de un pianista [...] es inaccesible para el hombre ineducado ya que la adquisición de nuevas habilidades requiere muchos años de práctica mental y física. Para entender plenamente este complejo fenómeno se hace necesario admitir, además del refuerzo de vias orgánicas pre-establecidas, la formación de vias nuevas por ramificación y crecimiento progresivo de la arborización dendrítica y terminales nerviosas.[1] (p. 296)

We argue that plasticity is an intrinsic property of the nervous system retained throughout a lifespan and that it is not possible to understand normal psychological function or the manifestations or consequences of disease without invoking the concept of brain plasticity. The brain, as the source of human behavior, is by design molded by environmental changes and pressures, physiologic modifications, and experiences. This is the mechanism for learning and for growth and development—changes in the input of any neural system, or in the targets or demands of its efferent connections, lead to system

---

[1] The labor of a pianist [...] is inaccessible for the uneducated man as the acquisition of new skill requires many years of mental and physical practice. In order to fully understand this complex phenomenon it becomes necessary to admit, in addition to the reinforcement of pre-established organic pathways, the formation of new pathways through ramification and progressive growth of the dendritic arborization and the nervous terminals.

reorganization that might be demonstrable at the level of behavior, anatomy, and physiology and down to the cellular and molecular levels.

Therefore, plasticity is not an occasional state of the nervous system; instead, it is the normal ongoing state of the nervous system throughout the life span. A full, coherent account of any sensory or cognitive theory has to build into its framework the fact that the nervous system, and particularly the brain, undergoes continuous changes in response to modifications in its input afferents and output targets. Implicit to the commonly held notion of plasticity is the concept that there is a definable starting point after which one may be able to record and measure change. In fact, there is no such beginning point because any event falls upon a moving target, i.e., a brain undergoing constant change triggered by previous events or resulting from intrinsic remodeling activity. We should not therefore conceive of the brain as a stationary object capable of activating a cascade of changes that we call plasticity, nor as an orderly stream of events driven by plasticity. Instead we should think of the nervous system as a continuously changing structure of which plasticity is an integral property and the obligatory consequence of each sensory input, motor act, association, reward signal, action plan, or awareness. In this framework, notions such as psychological processes as distinct from organic-based functions or dysfunctions cease to be informative. Behavior will lead to changes in brain circuitry, just as changes in brain circuitry will lead to behavioral modifications.

The mapping between behavioral modifiability (James 1890) and brain plasticity (Cajal 1904) is not one to one. Therefore, depending on the circumstances, neural plasticity can confer no perceptible change in the behavioral output of the brain, can lead to changes demonstrated only under special testing conditions, or can cause behavioral changes that may force the patient to seek medical attention. There may be loss of a previously acquired behavioral capacity, release of behaviors normally suppressed in the uninjured brain, takeover of lost function by neighboring systems (albeit perhaps incompletely or via different strategies and computations), or the emergence of new behaviors that may prove adaptive or maladaptive for the individual. Plasticity at the neural level does not speak to the question of behavioral change and certainly does not necessarily imply functional recovery or even functional change. The challenge we face is to learn enough about the mechanisms of plasticity and the mapping relations between brain activity and behavior to be able to guide it, suppressing changes that may lead to undesirable behaviors while accelerating or enhancing those that result in a behavioral benefit for the subject or patient.

In this review we first discuss mechanisms of plasticity and strategies for its modulation in the motor system during the acquisition of motor skills and the recovery of function after a stroke. Then we focus on crossmodal plasticity following sensory loss, i.e., blindness, to illustrate the fundamental nature of plasticity and emphasize the principles that are applicable across systems.

TWO-STEP CHANGES

Cajal (1904) predicted that with the acquisition of new skills the brain would change through rapid reinforcement of preestablished organic pathways and later formation of new pathways. We hypothesize that the first of these processes is in fact a necessary requirement for the development of the second. Formation of new pathways is possible only following initial reinforcement of preexistent connections. Therefore, the scope of possible plastic changes is defined by existing connections, which are the result of genetically controlled neural development and are ultimately different across individuals. Reinforcement of existing connections, on the other hand, is the consequence of environmental influences, afferent input, and efferent demand.

These two steps of plasticity are illustrated by the following experiment (Pascual-Leone 1996, Pascual-Leone et al. 1995). Normal subjects were taught to perform with one hand a five-finger exercise on a piano keyboard connected via computer musical interface. The subjects were instructed to perform the sequence of finger movements fluently, without pauses, and without skipping any keys, while paying particular attention to keeping the interval between the individual key presses constant and the duration of each key press the same. A metronome gave a tempo of 60 beats per minutes for which the subjects were asked to aim, as they performed the exercise under auditory feedback. Subjects were studied on five consecutive days, and each day they had a two-hour practice session followed by a test. The test consisted of the execution of 20 repetitions of the five-finger exercise. The number of sequence errors decreased, and the duration, accuracy, and variability of the intervals between key pushes (as marked by the metronome beats) improved significantly over the course of the five days. Before the first practice session on the first day of the experiment and daily thereafter, we used focal transcranial magnetic stimulation (TMS) to map the motor cortical areas targeting long finger flexor and extensor muscles bilaterally. As the subjects' performance improved, the threshold for TMS activation of the finger flexor and extensor muscles decreased steadily. Even considering this change in threshold, the size of the cortical representation for both muscle groups increased significantly (**Figure 1A**). However, this increase could be demonstrated only when the cortical mapping studies were conducted following a 20- to 30-min rest period after the practice (and test) session (Pascual-Leone 1996). No such modulation in the cortical output maps was noted when maps were obtained before each daily practice session (**Figure 1A**).

Remarkably, mental practice resulted in a similar reorganization of the motor outputs to the one observed in the group of subjects that physically practiced the movements (**Figure 1C**). Mental simulation of movements activates some of the same central neural structures required for the performance of the actual movements (Roland et al. 1987, Decety & Ingvar 1990). In doing so, mental practice alone may be sufficient to promote the plastic modulation of neural circuits placing the subjects at an advantage for faster skill learning with minimal physical practice, presumably by making the reinforcement of existing connections easier and perhaps speeding up the process of subsequent sprouting and consolidating of memories.

Once near-perfect level of performance was reached at the end of a week of daily practice, subjects were randomized into two groups (**Figure 1B**). Group 1 continued daily practice of the same piano exercise during the following four weeks. Group 2 stopped practicing. During the four weeks of follow-up, cortical output maps for finger flexor and extensor muscles were obtained in all subjects on Mondays (before the first practice session of that week in group 1), and on Fridays (after the last practice session for the week in group 1). In the group that continued practicing (group 1), the cortical output maps obtained on Fridays showed an initial peak and eventually a slow decrease in size despite continued performance improvement. On the other hand, the maps obtained on Mondays, before the practice session and following the weekend rest, showed a small change from baseline with a tendency to increase in size over the course of the study. In group 2, the maps returned to baseline after the first week of follow-up and remained stable thereafter.

This experiment reveals that acquisition of the necessary motor skills to perform a five-finger movement exercise correctly is associated with reorganization in the cortical motor outputs to the muscles involved in the task. The rapid time course in the initial modulation of the motor outputs, by which a certain region of motor cortex can reversibly increase its influence on a motoneuron pool, is most compatible with the unmasking of previously existing connections (Jacobs & Donoghue 1991, Sanes et al. 1992). Supporting this notion, the initial changes are quite transient: demonstrable after practice, but

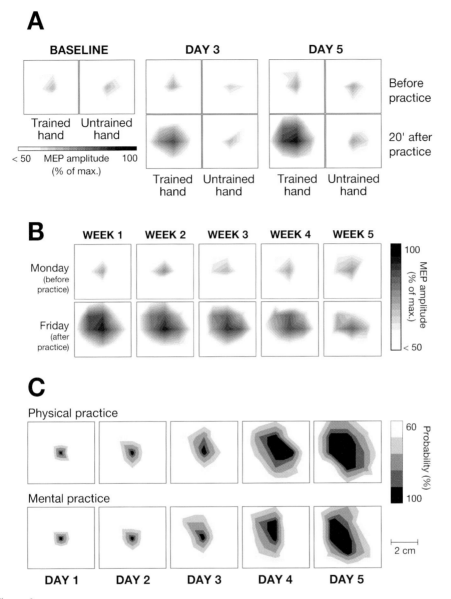

Figure 1

Changes in cortical output maps associated with learning a five-finger exercise on the piano (modified from Pascual-Leone 1996, Pascual-Leone et al. 1995). *A*: Cortical output maps for the finger flexors of the trained and the untrained hands of a representative subject (see text and Pascual-Leone et al. 1995 for details on mapping method). Note the marked changes of the output maps for the trained hand following practice and the lack of changes for the untrained hand. Note further the significant difference in cortical output maps for the trained hand after the practice sessions on days 3–5. *B*: Serial cortical output maps to finger flexors in a representative subject during five weeks of daily (Monday to Friday) practice of the five-finger exercise on the piano. Note that there are two distinct processes in action, one accounting for the rapid modulation of the maps from Mondays to Fridays and the other responsible for the slow and more discrete changes in Monday maps over time. *C*: Average cortical output maps for the finger flexors of the trained hand in subjects undergoing daily physical versus mental practice. Note the similarity in output maps with either form of practice.

returning to baseline after a weekend rest. As the task becomes overlearned over the course of five weeks, the pattern of cortical activation for optimal task performance changes as other neural structures take a more leading role in task performance. We suggest that flexible, short-term modulation of existing pathways represents a first and necessary step leading up to longer-term structural changes in the intracortical and subcortical networks as skills become overlearned and automatic.

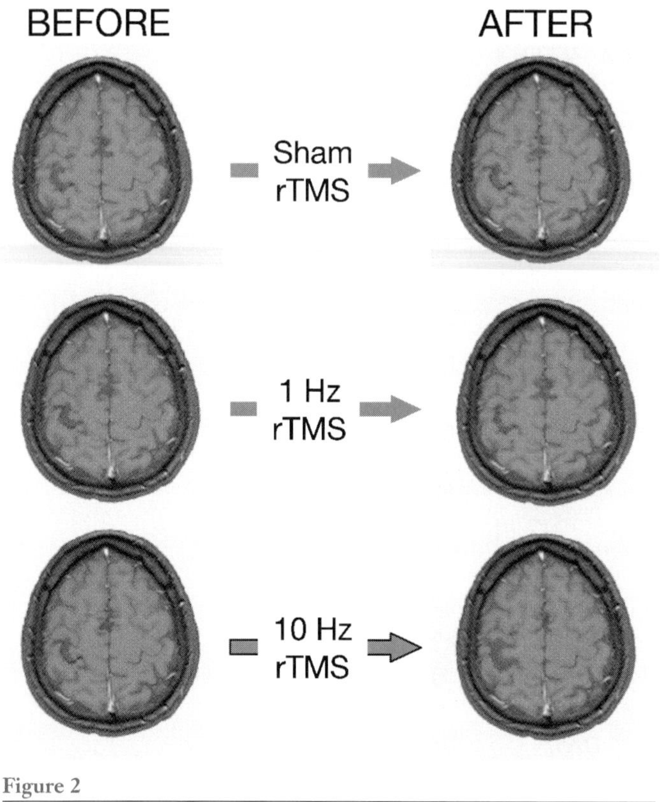

A growing number of neuroimaging studies have suggested a similar two-step process (Seitz et al. 1990; Grafton et al. 1992; Jenkins et al. 1994; Karni et al. 1995, 1998).

THE RAPIDLY SHIFTING MAPPING BETWEEN BRAIN ACTIVITY AND BEHAVIOR

Behavior is the manifestation of the coordinated workings of the entire nervous system. As long as an output pathway to manifest the behavior is preserved (even if alternate pathways need to be unmasked or facilitated), changes in the activity across a distributed neural network may be able to establish new patterns of brain activation and sustain function. These notions are illustrated by the following experiment (**Figure 2**). We asked normal subjects to open and close their fist deliberately at a self-paced rhythm of approximately one movement every second while lying in an fMRI scanner. As compared with during rest, during movement there was a significant activation of the motor cortex (M1) contralateral to the moving hand and of the rostral supplementary motor cortex (SMA). If motor cortex activity is modified by repetitive TMS, the pattern of brain activation changes as behavioral integrity is maintained. Application of slow, repetitive TMS to the contralateral M1 (presumed to suppress neuronal firing; Walsh & Pascual-Leone 2003) results in increased activation of the rostral SMA and of M1 ipsilateral to the moving hand. Conversely, increasing excitability in the contralateral M1 (by application of fast, repetitive TMS) leads to a decrease in activation of rostral SMA.

In a very elegant study, Lee et al. (2003) combining TMS and positron emission tomography (PET) have provided supporting evidence to these notions and critically extended it by revealing the shifts in cortico-cortical and cortico-subcortical connectivity underlying the changes in cortical activation patterns (**Figure 3**). Following rTMS, task-dependent increases in rCBF were seen

Figure 2

Brain activation in fMRI while subjects performed the same rhythmic hand movement (under careful kinematic control) before and after repetitive transcranial magnetic stimulation (rTMS) of the contralateral motor cortex. Following sham rTMS (*top row*) there is no change in the significant activation of the motor cortex (M1) contralateral to the moving hand and of the rostral supplementary motor cortex (SMA). After M1 activity is suppressed using 1Hz rTMS (1600 stimuli, 90% of motor threshold intensity; *middle row*), there is an increased activation of the rostral SMA and of M1 ipsilateral to the moving hand. Increasing excitability in the contralateral M1 using high-frequency rTMS (20 Hz, 90% of motor threshold intensity, 1600 stimuli; *bottom row*) results in a decrease in activation of rostral SMA.

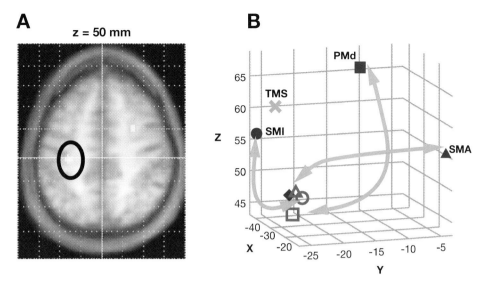

Figure 3

Areas of the brain showing differential movement-related responses and coupling after rTMS. Modified from Lee et al. 2003 (copyright 2003 by the Society for Neuroscience). *A*: Increased movement-related activity after 1Hz rTMS to the motor cortex. Results are displayed on an axial section of averaged anatomical MRI scans ($p < 0.001$, uncorrected). *B*: Circle, square, and triangle symbols indicating sites in primary motor cortex (*open symbols*) that are more strongly coupled to activity in sensorimotor cortex (SM1), dorsal premotor cortex (PMd), and supplementary motor cortex (SMA) after rTMS. The solid diamond indicates the position of the SM1 site circled in **Figure 3A**. X marks the site of stimulation with 1Hz rTMS.

during movement in the directly stimulated primary motor cortex and the dorsal premotor cortex in the unstimulated hemisphere, whereas motor performance remained unchanged. Analyses of effective connectivity showed that after rTMS there is a remodeling of the motor system, with increased movement-related connectivity from the SMA and premotor cortex to sites in primary sensorimotor cortex.

Both of these experiments demonstrate that in the face of a change in motor cortex activity (in these cases transient disruption induced by rTMS; Walsh & Pascual-Leone 2003) performance of a relatively simple movement task can be maintained by rapid operational remapping of motor representations, recruitment of additional motor areas, and task-related changes in cortico-cortical and cortico-muscular coherence (Strens et al. 2002, Chen et al. 2003, Lee et al. 2003, Oliviero et al. 2003).

THE RISK OF CHANGE AND THE OPPORTUNITY FOR INTERVENTION

A system capable of such flexible reorganization harbors the risk of unwanted change. Increased demand of sensorimotor integration poses such a risk. Faulty practice or excessive demand in the presence of certain predisposing factors (for example, genetic) may result in unwanted cortical rearrangement and lead to disease. Focal hand dystonia in musicians (Chamagne 2003) is such an example of pathological consequences of plasticity.

We examined five guitarists using fMRI during dystonic symptom provocation by means of an adapted guitar inside the magnet (Pujol et al. 2000). As reference we used the activation pattern obtained in the same subjects during other hand movements and in matched guitar players without dystonia during the execution of the same guitar-playing

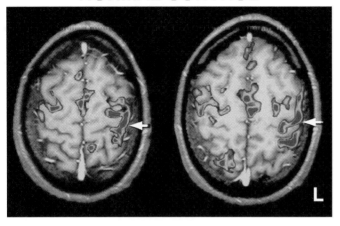

NORMAL CONTROL

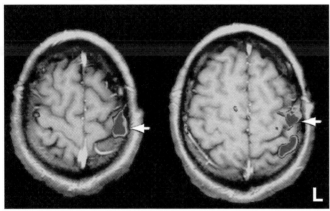

DYSTONIA PATIENT

Figure 4
BOLD fMRI images of a normal and a dystonic guitar player executing right hand arpeggios in the scanner. Note the greater activation of the sensorimotor cortex (*arrows*) contralateral to the performing hand and the lack of activation of premotor and supplementary motor cortices in the dystonic patient. Modified from Pujol et al. 2000.

exercises. Dystonic musicians compared with both control situations showed a significantly larger activation of the contralateral primary sensorimotor cortex that contrasted with a conspicuous bilateral underactivation of premotor areas (**Figure 4**). Our results coincide with studies of other dystonia types because they show an abnormal recruitment of cortical areas involved in the control of voluntary movement. Our study demonstrates that the primary sensorimotor cortex in patients with focal dystonia is overactive when they are tested during full expression of the task-induced movement disorder. The implication is that the established mapping between brain activity and behavior is inadequate and ultimately maladaptive, giving rise to symptoms of pathology. A sensory disturbance or sensorimotor mismatch may play a crucial role in contributing to the establishment of such an undesirable pattern of cortical activation (Hallett 1995, Bara-Jimenez et al. 1998, Elbert et al. 1998, Pantev et al. 2001).

Regardless of the role that sensory dysfunction may play, suppression of the task-specific excessive activation of the motor cortex has beneficial behavioral consequences for the symptoms of dystonia. For example, application of slow, repetitive TMS to the contralateral motor cortex suppresses cortical excitability by increasing intracortical inhibition and leads to a transient but significant improvement of dystonic symptoms in some patients (Siebner et al. 1999). Sensorimotor retuning (Candia et al. 1999, 2002), where motor activity is constrained using an individually designed splint so as to prevent dystonic posturing, is likely to induce its remarkable beneficial effects on dystonia by a similar mechanism of reducing task-specific motor cortical activation, thus promoting the establishment of a more adaptive mapping between brain activity and behavior (Candia et al. 2003). It is thus possible to induce shifts in brain activity either by guiding (and constraining) behavior or by directly modulating neuronal firing, for example through cortical stimulation. In either case, the plastic property of the nervous system is utilized to induce a behaviorally desirable outcome.

PLASTICITY IN THE SETTING OF BRAIN INJURY

That plasticity is a capacity of the brain that can be activated in response to an insult to promote functional recovery or compensate for lost function is a misconception. Rather, plasticity is always activated. Following brain injury, behavior (regardless of whether normal or manifesting injury-related deficits) remains the consequence of the functioning of the entire brain, and thus the consequence of a plastic nervous system. Symptoms are not the manifestation of the injured brain region, but rather the expression of plastic changes in the rest of the brain. After a lesion, just like after an rTMS-induced shift in activity in the primary motor cortex (see above), parallel motor circuits might be activated to establish some alternative input to the spinal motoneurons. These parallel circuits may originate from the contralateral, undamaged primary motor area (M1), bilateral premotor areas (PMA), bilateral supplementary motor areas (SMA), bilateral somatosensory areas, cerebellum, basal ganglia, etc. As long as efferent, cortico-spinal output pathways exist, cortico-cortical and cortico-subcortico-cortical interactions will shift weights across the involved functional network, aiming to establish a suitable brain activation map for a desired behavioral result. Conceptually it might be worth thinking of processes occurring after brain injury and leading to restoration of function as fitting different mechanisms that may proceed partly in parallel but which have variable time frames. Initial plastic changes aim to minimize damage. Rapid functional improvement is likely to occur as dysfunctional, but not damaged, neuronal elements recover from the postinjury shock and penumbra processes resolve. Partially damaged neural elements may be able to be repaired relatively quickly after the insult as well, thus contributing to early functional improvement. Subsequent processes, once the final damage has been established, involve relearning (rather than recovery) and will follow the two steps discussed above: initial unmasking and strengthening of existing neural pathways, and eventually the establishment of new structural changes.

These concepts can be illustrated by examining the role of the ipsilateral motor cortex in the recovery of hand motor function following stroke. After stroke, there is an increase in the excitability of the unaffected hemisphere, presumably owing to reduced transcallosal inhibition from the damaged hemisphere and increased use of the intact hemisphere. Several studies have demonstrated the increased cortical excitability in the unaffected hemisphere after a stroke. For example, in patients with acute cortical stroke, intracortical inhibition is decreased and intracortical facilitation increased in the unaffected hemisphere (Liepert et al. 2000). Furthermore, the interhemispheric inhibitory drive from the unaffected to the affected motor cortex in the process of voluntary movement generation is abnormal (Murase et al. 2004). Interestingly, the duration of stroke is inversely correlated to the imbalance of the excitability between the hemispheres. A disease duration of more than four months after stroke onset results in a tendency to normalization of the intracortical facilitation (ICF) of the unaffected hemisphere (Shimizu et al. 2002).

Acutely after a stroke, increased inhibitory input from the undamaged to the damaged hemisphere makes conceptual sense if one considers it a manifestation of a neural attempt to control perilesional activity, reduce oxygen and glucose demands in the penumbra of the stroke, and thus limit the extension of the lesion (**Figure 5**). However, after an acute phase, and once the injury is stable, input to the perilesional area would seem to be best as excitatory in nature to maximize the capability of the preserved neurons in the injured tissue to drive behavioral output. If so, following the acute phase, we might expect a shift of interhemispheric (and many intrahemispheric) interactions, from inhibitory to excitatory. Should such a shift fail to take place, the resulting functional outcome may be undesirable, with limited behavioral restoration, in part owing to persistent inhibitory inputs from the intact to the damaged hemisphere (**Figure 5**). In fact, some neuroimaging studies demonstrate that long-term, persistent activation of the ipsilateral cortex during motor tasks is associated with poor motor outcomes, whereas a good motor recovery is associated with a decrease in activity in the unaffected and an increase in the affected primary sensorimotor cortex activity (Carey et al. 2002, Rossini & Dal Forno 2004). Furthermore, the pattern of activation

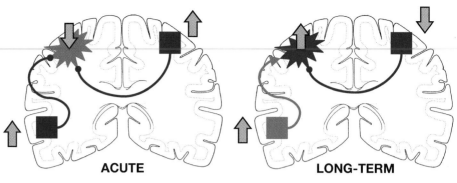

Figure 5

Schematic illustration showing that in the acute phase after a stroke, increased inhibitory input (within or across the hemispheres) may limit the extension of the lesion. Increased excitability (increased glutamatergic activity and reduced GABAergic activity) and postischemic LTP harbor an otherwise increased risk for further damage. However, after the acute phase, and once the injury is stable (long-term), excitatory input increases excitability and may further increase the efferent (e.g., motor) output. In contrast, inhibitory input at chronic stages is a maladaptive strategy, and the resulting functional outcome may be undesirable, with limited behavioral restoration. Note that the sources (intra- or interhemispheric) of such inputs may differ across neural systems and across individuals. Nevertheless, this provides a road map for neuromodulatory interventions whose aims differ in the acute and long-term stages (*block arrows indicate a desirable increase or decrease in excitability*).

in well-recovered patients is similar to healthy subjects (Ward et al. 2003). More longitudinal studies of patients following a stroke and correlation of interhemispheric interactions with functional measures are needed to explore these issues further. If correct, neuromodulatory approaches targeting the intact hemisphere may be useful to limit injury and promote recovery after a stroke.

For instance, suppression of the ipsilateral motor cortex through slow rTMS (Pascual-Leone et al. 1998, Maeda et al. 2000) may enhance motor performance in patients stable following the acute phase of a stroke. In patients 1–2 months after a stroke, Mansur et al. (2005) applied 0.5 Hz rTMS for 10 min to the unaffected hemisphere to suppress cortical activity and thus release the damaged hemisphere from potentially excessive transcallosal inhibition. The results of this pilot study support the notion that the overactivity of the unaffected hemisphere (ipsilateral hemisphere) may hinder hand function recovery, and neuromodulation can be an interventional tool to accelerate this recovery. The findings are consistent with results in normal subjects, where ipsilateral motor cortex activation on functional MRI during unilateral hand movements is indeed related primarily to interhemispheric interactions (Kobayashi et al. 2003), and disruption of the activity of one hemisphere reduces transcallosal inhibition to the contralateral hemisphere and can indeed improve ipsilateral motor function (Kobayashi et al. 2004). However, Werhahn et al. (2003) conducted a similar study to evaluate the modulation effects of 1Hz rTMS of the unaffected hemisphere on the paretic hand and found different results. In that study, 1 Hz rTMS of the unaffected hemisphere did not affect the finger tapping in the paretic hand in a small sample of five patients more than one year after a stroke. The time since the brain insult is likely to be a critical variable to consider. Studies with larger samples of patients are needed to investigate this question further.

Behavioral motor therapy may also shift cortical excitability balance between hemispheres and thus influence outcome. For example, the beneficial effects of constraint-induced therapy on motor function (Mark & Taub 2004, Grotta et al. 2004) are achieved through immobilization of the unaffected arm, which results in a reduction of the excitability of the contralateral (undamaged) motor cortex owing to the decreased efferent demand and afferent input (Liepert et al. 2001). The reduced activity of the undamaged motor cortex may decrease transcallosal inhibition of the damaged motor cortex and thus promote recovery, ultimately by mechanisms similar to those recruited by suppressing cortical excitability through slow rTMS.

Of course, the alternative neuromodulatory approach, directly aimed to enhance excitability of the damaged hemisphere perilesionally, can also be entertained. Results of a pilot study in primates support the feasibility of using a therapy approach, combining peri-infarct electrical stimulation with rehabilitative training to alleviate chronic motor deficits and promote recovery from cortical ischemic injury (Plautz et al. 2003). Very early experiments with invasive cortical stimulation in humans reveal similarly encouraging results (Brown et al. 2003). However, noninvasive rTMS at appropriate parameters can also be applied to enhance cortical excitability (Pascual-Leone et al. 1998, Maeda et al. 2000, Huang et al. 2005) and thus may exert similar beneficial effects, particularly if coupled with physical therapy. In this setting, functional neuroimaging might be useful, among other things, to identify the perilesional areas to be targeted (Baron et al. 2004), and EEG or fMRI may allow investigators to define precisely and optimize the physiologic effects of TMS (Bestmann et al. 2004).

Similar principles of neuromodulation can be applied to the recovery of nonmotor strokes and other focal brain lesions as illustrated by studies on the effects of cortical stimulation on neglect (Hilgetag et al. 2001, Oliveri et al. 2001, Brighina et al. 2003) or aphasia (Knecht et al. 2002; Martin et al. 2004; Naeser et al. 2005a,b).

Therefore, functional recovery after a focal brain injury, e.g., a stroke, is essentially learning with a partially disrupted neural network. A main neural mechanism underlying relearning of skills and preservation of behavior involves shifts of distributed contributions across a specific neural network (fundamentally, the network engaged in learning the same skills in the healthy brain). Intra- and particularly interhemispheric interactions may shift from being initially inhibitory (to minimize damage) to later excitatory (to promote functional recovery). Changes in the time course of such connectivity shifts may result in the establishment of dead-end strategies and limit functional recovery. Ultimately, activation of brain areas that are not normally recruited in normal subjects may represent a nonadaptative strategy resulting in a poor prognosis.

THE OCCIPITAL CORTEX IN THE BLIND

The core principles of neural function that apply to the motor system should also apply to sensory systems. We now switch from the motor to the visual system and briefly discuss the impact of visual loss on occipital cortical function to illustrate the fundamental nature of plasticity.

We live in a society that relies heavily on vision. Therefore, blind individuals have to make striking adjustments to their loss of sight to interact effectively with their environment. One may thus imagine that blind individuals need to develop superior abilities in the use of their remaining senses (compensatory hypothesis). However, blindness could also be the cause for maladjustments (general-loss hypothesis). For example, the loss of sight could be detrimental to sensory perception/spatial information processing mediated by the remaining senses because of our strong reliance on vision for the acquisition and construction of spatial and form representations. Against the general-loss hypothesis is evidence that blind individuals show normal and often superior skills in tasks implicating touch and hearing as compared with the average sighted population (Rauschecker 1995; Hollins & Kelley 1988; Lessard et al. 1998; Van Boven et al.

2000; Gougoux et al. 2004, 2005; Voss et al. 2004; Doucet et al. 2005). Growing experimental evidence suggests that in blind persons brain areas commonly associated with the processing of visual information are recruited in a compensatory cross-modal manner that may account for these superior nonvisual capabilities (Merabet et al. 2005, Theoret et al. 2004).

Phelps et al. (1981) and Wanet-Defalque et al. (1988) were among the first to suggest that the occipital cortex is active in the blind and furthermore, that puberty may represent an important developmental milestone for this activation. Using event-related electroencephalograph (EEG) recordings, Uhl et al. (1991) [and later confirmed by a follow-up study using single photon emission computerized tomography (SPECT) imaging; Uhl et al. 1993] provided early support for the notion of task-related (tactile) occipital cortex activation in blind subjects. Sadato et al. (1996, 1998) employed PET imaging and demonstrated that the primary visual cortex is activated in early-blind subjects performing a Braille reading task (**Figure 6A**). Specifically, they observed bilateral activation in medial occipital cortex (area 17) with concomitant activity in extrastriate areas. Activation of the primary visual cortex was also evident in non-Braille tactile discrimination tasks (e.g., discrimination of angle, width, and Roman-embossed characters encoded in Braille cells), though to a lesser extent. However, passive sweeping of the finger (without responding) over a homogeneous pattern of Braille dots (i.e., meaningless Braille symbols) did not result in activation of the primary visual cortex. Subsequent investigators have further refined and extended these early findings addressing the role of imagery, the differences between early and late blind, and the role of tactile versus verbal/linguistic aspects of the task (Büchel et al. 1998, Melzer et al. 2001, Burton et al. 2002a, Sadato et al. 2002, Amedi et al. 2003; for review see Merabet et al. 2005).

One must realize that functional neuroimaging at best establishes an association between activity in a given region or network with task performance. Therefore, the observation of activity in visual cortical areas in the blind fails to prove that this activity is necessary for the sensory processing. In support of a causal link between occiptal function and the ability to read Braille is the remarkable patient reported by Hamilton et al. (2000, **Figure 6B**). This congenitally blind woman (from retinopathy of prematurity) was once a highly proficient Braille reader (learning at the age of six and able to read at a rate of 120–150 symbols per minute). Following bilateral posterior cerebral artery strokes, she was rendered unable to read Braille despite the fact that her somatosensory sensation, peripheral motor, and sensory nerve functions were all intact. Even though she was well aware of the presence of the dot elements contained in the Braille text, she was "unable to extract enough information" to determine which letters and words were written. Despite her profound inability to read, she had no difficulty in performing simple discrimination tactile tasks, such as identifying the roughness of a surface, distinguishing between different coins, or identifying her house key from a given set. However, she was not able to judge distance between Braille dots or read Braille (Hamilton et al. 2000, Merabet et al. 2004b). This serendipitous experiment of nature (and tragic event for our patient) provides strong clinical evidence that a functioning occipital cortex is needed to carry out the task of Braille reading.

In an experimental setting, Cohen and colleagues (1997) used TMS to induce transient disruption of cortical function during a Braille identification task. Identification of Braille characters or embossed Roman letters was impaired following TMS to the occipital cortex in early-onset blind subjects but not in sighted subjects. In blind subjects, occipital stimulation with TMS not only induced errors in Braille identification, but also distorted tactile perceptions. Subjects knew that they were touching Braille symbols but were unable to identify them, reporting instead

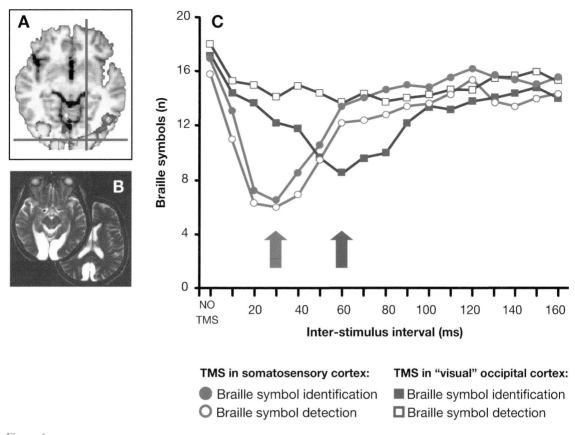

Figure 6
Braille activation in the visual cortex of the blind and its functional relevance. *A*: Brain imaging (PET) of occipital activation during Braille reading in an early-blind subject (modified from Sadato et al. 1996). *B*: Braille alexia following a bilateral occipital stroke. T2-weighted structural MRI of a congenitally blind woman who was rendered unable to read Braille following a bilateral occipital stroke (modified from Hamilton et al. 2000). *C*: Effects of transient disruption of the occipital cortex using single-pulse TMS on tactile Braille character recognition in congenitally blind subjects. The TMS pulse was delivered at different times (interstimulus interval) after delivery of a tactile stimulus to the index finger pad. The graph displays the number of tactile stimuli detected (*open symbols*) and correctly identified (*filled symbols*). Disruption of somatosensory cortex leads to a decrease in the number of detected and identified letters at an interstimulus interval of 30 ms (*red arrow*). Disruption of occipital cortex does not affect detection but leads to a decrease in the number of identified letters at a later interstimulus interval of approximately 60 ms (*blue arrow*) (modified from Hamilton et al. 1998).

that the Braille dots felt "different," "flatter," "less sharp and less well-defined" (1997). Occasionally, some subjects even reported feeling additional ("phantom") dots in the Braille cell. The functional significance of the occipital activation during Braille reading in the early blind has been further evaluated using single-pulse TMS to obtain information about the timing (chronometry) of information processing (Pascual-Leone et al. 2000, Walsh & Pascual-Leone 2003). A disruptive TMS pulse was delivered to the occipital or the somatosensory cortex (contralateral to the reading hand) at a variable interval after a peripheral stimulus was applied to the pad of the subject's index finger (Hamilton & Pascual-Leone 1998). In normal sighted subjects, stimuli to the occipital cortex had no effect, but TMS delivered to the somatosensory cortex ~20–30 ms after a tactile stimulus to a contralateral finger interfered with the detection of the peripheral somatosensory stimulus (presumably by disrupting the arrival of the thalamo-cortical volley into the primary sensory cortex; Cohen et al. 1991, Pascual-Leone et al. 1994). In congenitally blind subjects, TMS to the left somatosensory cortex disrupted detection of Braille stimuli presented to their right index finger at interstimulus intervals of 20–40 ms (**Figure 6C**). Similar to the findings in the sighted, in some cases the subjects did not realize that a peripheral stimulus had been presented to their finger. When they did realize it, they were able to identify correctly which Braille symbol was presented. On the other hand, TMS to the striate cortex disrupted the processing of the peripheral stimuli at interstimulus intervals of 50–80 ms. Contrary to the findings after sensorimotor TMS, the subjects generally knew whether a peripheral stimulus had been presented. However, they could not discriminate which Braille symbol had been presented. These results suggest that in early-blind subjects, the somatosensory cortex appears engaged in detection, whereas the occipital cortex contributes to the perception of tactile stimuli.

UNMASKING CONNECTIONS: THE BLINDFOLD EXPERIMENT

Complete and transient visual deprivation in sighted subjects (i.e., five days of blindfolding) seems to be sufficient to lead to recruitment of the primary visual cortex for tactile and auditory processing (Pascual-Leone & Hamilton 2001). The speed of these functional changes is such that it is highly improbable that new cortical connections are established in these sighted individuals. Therefore, somatosensory and auditory connections to the occipital cortex must already be present and are unmasked under our experimental conditions (**Figure 7A**). These could be cortico-cortical connections, linking Heschl gyrus or postcentral cortex and striate cortex directly, via cortical multisensory areas, through thalamic or other subcortical relay nuclei. Ultimately, the occipital cortex recruitment mechanisms in tactile processing in the blind and under blindfolded conditions are not likely to

Figure 7

Changes in connectivity between somatosensory and visual areas in blindfolded sighted and early-blind subjects. *A*: BOLD fMRI activation focused on the calcarine sulcus (area V1) of a sighted subject undergoing complete visual deprivation for five days through blindfolding. During this period the patient was scanned conducting different tasks including tactile object recognition versus corresponding sensorimotor controls. This contrast results in negligible V1 activation during baseline (day 1 of blindfolding; *not shown*). Following five days of blindfolding robust V1 activation is evident in V1 (*left*). The V1 region of interest shows significant activation for tactile object recognition with either the left or the right hand (*red and purple time course, respectively*) but none during low-level sensorimotor control (imitating object palpation movements in the air with right or left hand; *orange and cyan colors, respectively*). This activation is dramatically reduced only hours following the removal of the blindfold (*right*). The speed of these functional changes is such that it is highly improbable that new cortical connections are established in these sighted individuals. Therefore, somatosensory connections to the occipital cortex, perhaps via the ventral pathway and the lateral occipital (LO) region, must already be present and are unmasked under our experimental conditions. *B*: Combined PET and rTMS study to probe the connectivity between primary somatosensory cortex (S1) and early visual cortex (V1 and neighboring areas). The figure presents the pattern of PET activation in early- versus late-blind subjects contrasting the effects of real versus sham rTMS to S1. Note significant activation in parietal, occipital, and occipito-temporal areas. Most striking, activation was found also in V1, which suggests that cortico-cortical connection between S1 and V1 is stronger in early-blind subjects, possibly supporting enhanced tactile information processing (modified from Wittenberg et al. 2004).

be identical. Fast changes during blindfolding reveal the capacity of the plastic brain to change in response to environmental changes (in this case, visual deafferentation) and maintain functional behavior (in this case, perceptual capturing of the world, for example, by enhancing auditory and tactile processing). Such shifts in connectivity are rapidly reversed with visual input restoration (**Figure 7A**). However, sustained and early sight loss may result in lasting structural brain changes with the establishment of new pathways following the initial reinforcement of preestablished connections (just as Cajal predicted). Indeed, in early-blind subjects connectivity between primary somatosensory cortex (S1) and early visual cortex is changed, as recently shown in an elegant study combining rTMS with PET. Wittenberg et al. (2004) showed that rTMS over S1 evoked activation of peri-striate cortex in early-blind but not in late-blind or sighted individuals (**Figure 7B**).

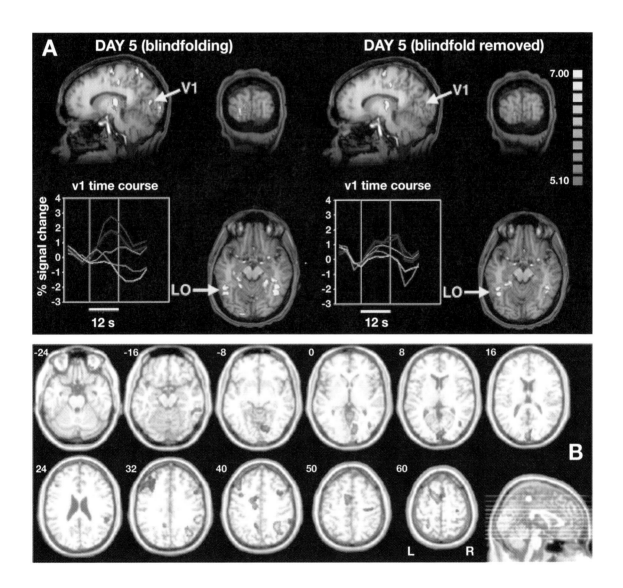

ESTABLISHING NEW CONNECTIONS: OCCIPITAL ACTIVATION IN HIGH-LEVEL COGNITIVE TASKS

Recent fMRI studies in the blind have demonstrated occipital cortex activation (including V1) during tasks requiring auditory verb-generation and similar linguistic tasks (Burton et al. 2002b, Amedi et al. 2003), semantic judgment tasks (Burton 2003, Noppeney et al. 2003), and speech processing (Röder et al. 2002). In a comparative analysis of brain activation in early and late blind during a verb-generation task, Burton et al. (2002b) instructed subjects to generate covertly a verb in response to reading a noun cue presented in Braille (e.g., reading the word "cake" would generate "bake") or using auditory words. They found that occipital cortex activation (including primary visual cortex) was much more prominent in early- than in late-blind subjects (**Figure 8**).

Amedi et al. (2003) observed robust left-lateralized V1 activation for a verbal-memory task requiring the retrieval of abstract words from long-term memory. The striking finding of this report is that contrary to previous studies, the observed occipital activation was demonstrated without introducing any tactile or auditory sensory input. Notably, blind subjects showed superior verbal memory capabilities compared not only with age-matched, sighted controls, but also with reported population averages (using the Wechsler verbal memory test). Furthermore, investigators found a strong positive correlation between the magnitude of V1 activation and the verbal memory capability in that the degree of activation increased with increasing word-recall ability (**Figure 9**).

The functional relevance of these findings was demonstrated with rTMS. When activity in the left calcarine sulcus or left occipito-temporal cortex was disrupted by rTMS, performance in a verb-generation task was impaired (the error rate increased) (Amedi et al. 2004). An analysis of error types revealed that the most common error produced after rTMS was semantic (e.g., apple ⇒ jump, instead of eat, one possible correct response). Phonological errors and interference with motor execution or articulation (stuttering and slurring of the responses) were rare. Thus, in blind subjects, a transient virtual lesion of the left occipital cortex can interfere with high-level verbal processing, and not only with the processing of tactile stimuli and Braille reading. This finding suggests that beyond changes in connectivity across sensory systems, in early blind the visually deafferented occipital cortex becomes engaged in higher-order cognitive functions, presumably through establishment of new connections.

DRAWING CONCLUSIONS FROM THE BLINDFOLDED AND THE BLIND

The functional and structural identity of the occipital cortex may change from processing visual information to processing information related to another sensory modality or even supramodal high-level cognitive functions (**Figure 10**). In comparison, the occipital cortex may inherently possess the computational machinery necessary for nonvisual information processing (Pascual-Leone & Hamilton 2001). Under specific conditions (such as blindness or prolonged blindfolding) this potential could be revealed. Burton (2003) suggested the definition of two distinct mechanisms: (*a*) "cross-modal plasticity de novo" in response to visual deprivation, and (*b*) "expression of normal physiology" that is normally inhibited or masked when sight is present. However, as discussed above, in the context of the intrinsically plastic brain, these two mechanisms are inextricably linked. The unmasking of preexisting connections and shifts in connectivity represent rapid, early plastic changes, which can lead, if sustained and reinforced, to slower developing but more permanent structural changes, with dendritic arborization, sprouting, and growth. This hypothesis can account for the findings in

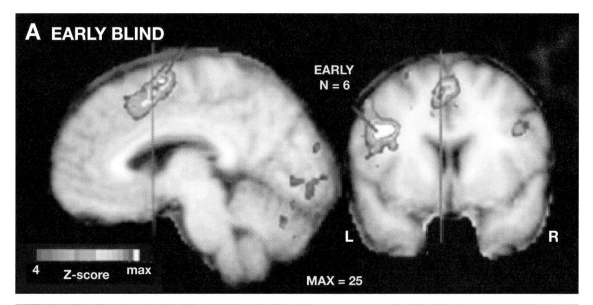

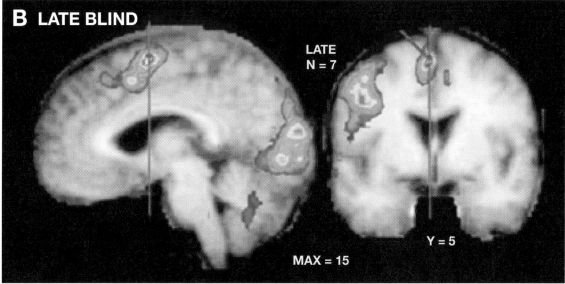

Figure 8
BOLD fMRI activation within occipital cortex in early- and late-blind Braille readers. Note that the magnitude of activation is greater in early-blind individuals. Scale denotes the Z-scores for BOLD responses (Z-max early-blind 25; Z-Max late-blind: 15; modified from Burton et al. 2002).

blindfolded subjects and for the magnitude difference of the reorganization between early and late blind. This hypothesis also results in the strong prediction that careful task choice and experimental design will reveal the nonvisual roles of the occipital cortex in the sighted. Indeed, Amedi & colleagues (2001, 2002) reported convergence of visual and tactile object recognition in the ventral visual stream in an occipito-temporal area termed the lateral-occipital tactile visual area (LOtv). The defining feature of this region is that

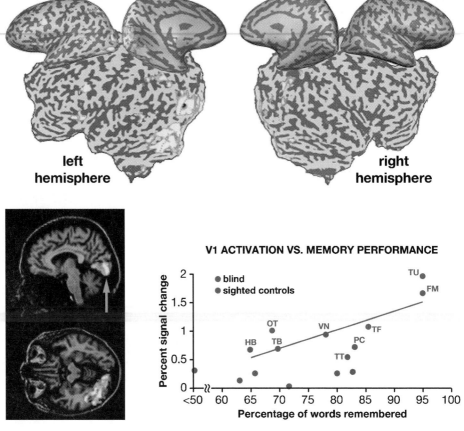

Figure 9

Verbal memory fMRI activation in early visual cortex of congenitally blind subjects correlates with their superior verbal memory abilities. Congenitally blind subjects show robust activation in the left visual cortex during a verbal memory task of abstract word retrieval, which involves no sensory stimulation. The left-lateralized activity stretches from V1, via extrastriate retinotopic areas, to nonretinotopic areas such as the lateral occipital complex (LOC; *top panel*). This activation was correlated with the subjects' verbal memory abilities (*lower panel*). Subjects were tested on the percentage of words they remembered six months after the scan. In general, blind subjects remembered more words and showed greater V1 activation than did the sighted controls. The correlation between V1 activity and performance was significant only in blind subjects (modified from Amedi et al. 2003).

it is activated preferentially by object shape rather than by texture and scrambled images of the object. Similarly, TMS studies have revealed that the visual cortex of the sighted is functionally involved in tactile processing of orientation (Zangaladze et al. 1999) and judging of distance between Braille dots (Merabet et al. 2004a). Recent fMRI work even suggests that as compared with phonemic word generation, semantic word generation involves (in addition to a series of other brain regions) activation of bilateral occipital cortices in sighted subjects (Press et al. 2004).

Therefore, in the visual system in response to loss of visual input, we encounter a similar situation to the one discussed in the context of the motor system: In a first step the nervous system is molded rapidly by shifts of strength in existing connections. In a second step new structural connections are established giving rise to new capacities such as improved verbal memory through recruitment of the occipital cortex (Amedi et al. 2003); the remarkably high incidence of absolute pitch in early-blind subjects in the absence of the expected changes in planum temporal asymmetry (Hamilton et al. 2004); or the superior auditory localization ability of blind subjects, which correlates with the amount of striate cortex activation (Gougoux et al. 2005).

However, just as in the motor system, plasticity risks becoming the cause of pathology. For example, acutely after visual deafferentation, just as after a focal lesion of the occipital cortex, altered cortical excitability and rapid changes in cortico-cortical connectivity frequently lead to visual hallucinations and phantom vision (Merabet et al. 2004a). Such hallucinations can be suppressed by reducing cortical excitability through slow rTMS (Merabet et al. 2003). Eventually, hallucinations tend to subside, perhaps correlating with the long-lasting, cross-modal plasticity changes and the recruitment of the occipital cortex for high-order cognitive tasks. However, such plasticity, although aiding in the adaptation to blindness, poses difficult challenges to vision restoration. Appropriate delivery of electrical stimulation to the retina or the occipital cortex can evoke patterned sensations of light even in those who have been blind for many years. However, success in developing functional visual prostheses requires an understanding of how to communicate effectively with the plastically changed, visually deprived brain to merge what is perceived visually with what is generated electrically (Merabet et al. 2005). Similarly challenging is the situation for patients who recover sight through surgical approaches (for example, cataract removal or corneal stem-cell transplant) after long-term visual deprivation (Fine et al. 2003).

SUMMARY

Plasticity in the motor system may be primarily driven by efferent demand, whereas in the visually deprived visual system it is the consequence of afferent input changes. Nevertheless, across systems the fundamental aspects of plasticity remain the same: Plasticity is an obligatory consequence of all neural activity (even mental practice), and environmental

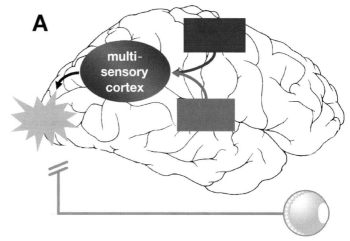
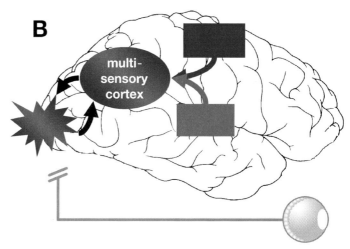

Figure 10

Schematic representation of changes affecting the occipital cortex after visual deafferentation. *A*: Following visual deafferentation, inputs from other sensory processing areas (*red and blue*) reach the occipital cortex via connections through multisensory cortical areas (and possibly through direct connections). *B*: The unmasking and strengthening of such connections (*thicker lines*) lead to enhanced multisensory processing at the level of the occipital cortex and may eventually lead to the establishment of new connections and functional roles.

ICF: intracortical facilitation

M1: motor cortex

LOtv: lateral-occipital tactile visual area

PMA: premotor area

PMd: dorsal premotor cortex

S1: somatosensory cortex

SMA: supplementary motor area

SPECT: single photon emission computerized tomography

TMS: transcranial magnetic stimulation

pressures, functional significance, and experience are critical factors. For example, the role of the occipital cortex for tactile processing in the blind is likely to be fundamentally different for those who learn Braille early and those who do not.

Two steps of plasticity can be identified: unmasking existing connections that may be followed by establishment of new ones. The former provides an opportunity to learn about core aspects of normal physiology (for example, the role of cross-modal interactions in visual perception). The latter can give rise to unexpected capacities (for example, supranormal auditory abilities or verbal memory in the blind).

Plasticity is an intrinsic property of the human nervous system, and plastic changes may not necessarily represent a behavioral gain for a given subject. Plasticity is the mechanism for development and learning, as much as a cause of pathology and the cause of clinical disorders. Our challenge is to modulate neural plasticity for optimal behavioral gain, which is possible, for example, through behavioral modification and through invasive and noninvasive cortical stimulation.

So, what makes our dynamic plastic cortex possible? Experience and behavior correspond to the activity of all relevant neurons throughout the brain. Neuronal networks provide a most energy efficient, spatially compact, and precise means to process input signals and generate responses (Laughlin & Sejnowski 2003). Nodes in such networks, specific brain regions, are conceptualized as operators that contribute a given computation independent of the input (Pascual-Leone & Hamilton 2001). Inputs shift depending on the integration of a region in a distributed neural network, and the layered and reticular structure of the cortex with rich reafferent loops provides the substrate for rapid modulation of the engaged network nodes. Ultimately, plasticity is a most efficient way to utilize the brain's limited resources.

ACKNOWLEDGMENTS

Work on this review was supported by K24 RR018875, RO1-EY12091, RO1-DC05672, RO1-NS 47,754, RO1-NS 20,068, and R01-EB 00,5047. F.F. is supported by a grant from the Harvard Medical School Scholars in Clinical Sciences Program (NIH K30 HL004095-03). L.M. is supported by a K 23 mentored career development award from the National Eye Institute (K23 EY016131-01) and A.A. by a Human Frontiers Science Program award. The authors thank Mark Thivierge for invaluable administrative support.

LITERATURE CITED

Amedi A, Floel A, Knecht S, Zohary E, Cohen LG. 2004. Transcranial magnetic stimulation of the occipital pole interferes with verbal processing in blind subjects. *Nat. Neurosci.* 7:1266–70

Amedi A, Jacobson G, Hendler T, Malach R, Zohary E. 2002. Convergence of visual and tactile shape processing in the human lateral occipital complex. *Cereb. Cortex* 11:1202–12

Amedi A, Malach R, Hendler T, Peled S, Zohary E. 2001. Visuo-haptic object-related activation in the ventral visual pathway. *Nat. Neurosci.* 3:324–30

Amedi A, Raz N, Pianka P, Malach R, Zohary E. 2003. Early 'visual' cortex activation correlates with superior verbal memory performance in the blind. *Nat. Neurosci.* 6:758–66

Bara-Jimenez W, Catalan MJ, Hallett M, Gerloff C. 1998. Abnormal somatosensory homunculus in dystonia of the hand. *Ann. Neurol.* 44:828–31

Baron JC, Cohen LG, Cramer SC, Dobkin BH, Johansen-Berg H, et al. 2004. Neuroimaging in stroke recovery: a position paper from the First International Workshop on Neuroimaging and Stroke Recovery. *Cerebrovasc. Dis.* 18:260–67

Bestmann S, Baudewig J, Siebner HR, Rothwell JC, Frahm J. 2004. Functional MRI of the immediate impact of transcranial magnetic stimulation on cortical and subcortical motor circuits. *Eur. J. Neurosci.* 19:1950–62

Brighina F, Bisiach E, Oliveri M, Piazza A, La Bua V, et al. 2003. 1 Hz repetitive transcranial magnetic stimulation of the unaffected hemisphere ameliorates contralesional visuospatial neglect in humans. *Neurosci. Lett.* 336:131–33

Brown JA, Lutsep H, Cramer SC, Weinand M. 2003. Motor cortex stimulation for enhancement of recovery after stroke: case report. *Neurol. Res.* 25:815–18

Büchel C, Price C, Frackowiak RSJ, et al. 1998. Different activation patterns in the visual cortex of late and congenitally blind subjects. *Brain* 121:409–19

Burton H. 2003. Visual cortex activity in early and late blind people. *J. Neurosci.* 23:4005–11

Burton H, Snyder AZ, Conturo TE, Akbudak E, Ollinger JM, Raichle ME. 2002a. Adaptive changes in early and late blind: a fMRI study of Braille reading. *J. Neurophysiol.* 87:589–607

Burton H, Snyder AZ, Diamond JB, Raichle ME. 2002b. Adaptive changes in early and late blind: a FMRI study of verb generation to heard nouns. *J. Neurophysiol.* 88:3359–71

Candia V, Elbert T, Altenmuller E, Rau H, Schafer T, Taub E. 1999. Constraint-induced movement therapy for focal hand dystonia in musicians. *Lancet* 353:42

Candia V, Schafer T, Taub E, Rau H, Altenmuller E, et al. 2002. Sensory motor retuning: a behavioral treatment for focal hand dystonia of pianists and guitarists. *Arch. Phys. Med. Rehabil.* 83:1342–48

Candia V, Wienbruch C, Elbert T, Rockstroh B, Ray W. 2003. Effective behavioral treatment of focal hand dystonia in musicians alters somatosensory cortical organization. *Proc. Natl. Acad. Sci. USA* 100:7942–46

Carey JR, Kimberley TJ, Lewis SM, Auerbach EJ, Dorsey L, et al. 2002. Analysis of fMRI and finger tracking training in subjects with chronic stroke. *Brain* 125:773–88

Chamagne P. 2003. Functional dystonia in musicians: rehabilitation. *Hand Clin.* 19:309–16

Chen WH, Mima T, Siebner HR, Oga T, Hara H, et al. 2003. Low-frequency rTMS over lateral premotor cortex induces lasting changes in regional activation and functional coupling of cortical motor areas. *Clin. Neurophysiol.* 114:1628–37

Cohen LG, Bandinelli S, Sato S, Kufta C, Hallett M. 1991. Attenuation in detection of somatosensory stimuli by transcranial magnetic stimulation. *Electroencephalogr. Clin. Neurophysiol.* 81:366–76

Cohen LG, Celnik P, Pascual-Leone A, Corwell B, Falz L, et al. 1997. Functional relevance of cross-modal plasticity in blind humans. *Nature* 389:180–83

Decety J, Ingvar DH. 1990. Brain structures participating in mental simulation of motor behavior: a neuropsychological interpretation. *Acta Psychol. (Amst.)* 73:13–34

Doucet ME, Guillemot JP, Lassonde M, Gagne JP, Leclerc C, Lepore F. 2005. Blind subjects process auditory spectral cues more efficiently than sighted individuals. *Exp. Brain Res.* 160:194–202

Elbert T, Candia V, Altenmuller E, Rau H, Sterr A, et al. 1998. Alteration of digital representations in somatosensory cortex in focal hand dystonia. *Neuroreport* 9:3571–75

Fine I, Wade AR, Brewer AA, May MG, Goodman DF, et al. 2003. Long-term deprivation affects visual perception and cortex. *Nat. Neurosci.* 6(9):915–16

Gougoux F, Lepore F, Lassonde M, Voss P, Zatorre RJ, Belin P. 2004. Neuropsychology: pitch discrimination in the early blind. *Nature* 430:309

Gougoux F, Zatorre RJ, Lassonde M, Voss P, Lepore F. 2005. A functional neuroimaging study of sound localization: visual cortex activity performance in early-blind individuals. *PLoS Biol.* 3:e27

Grafton ST, Mazziota JC, Presty S, Friston KJ, Frackowiak RSJ, Phleps ME. 1992. Functional anatomy of human procedural learning determined with regional cerebral blood flow and PET. *J. Neurosci.* 12:2542–48

Grotta JC, Noser EA, Ro T, Boake C, Levin H, et al. 2004. Constraint-induced movement therapy. *Stroke* 35:2699–701

Hallett M. 1995. Is dystonia a sensory disorder? *Ann. Neurol.* 38:139–40

Hamilton R, Keenan JP, Catala M, Pascual-Leone A. 2000. Alexia for Braille following bilateral occipital stroke in an early blind woman. *Neuroreport* 11:237–40

Hamilton RH, Pascual-Leone A. 1998. Cortical plasticity associated with Braille learning. *Trends Cogn. Sci.* 2:168–74

Hamilton RH, Pascual-Leone A, Schlaug G. 2004. Absolute pitch in blind musicians. *Neuroreport* 15:803–6

Hilgetag CC, Theoret H, Pascual-Leone A. 2001. Enhanced visual spatial attention ipsilateral to rTMS-induced 'virtual lesions' of human parietal cortex. *Nat. Neurosci.* 4:953–57

Hollins M, Kelley EK. 1988. Spatial updating in blind and sighted people. *Percept. Psychophys.* 43:380–88

Huang YZ, Edwards MJ, Rounis E, Bhatia KP, Rothwell JC. 2005. Theta burst stimulation of the human motor cortex. *Neuron* 45:201–6

Jacobs KM, Donoghue JP. 1991. Reshaping the cortical motor map by unmasking latent intracortical connections. *Science* 251:944–47

Jenkins IH, Brooks DJ, Nixon PD, Frackowiak RS, Passingham RE. 1994. Motor sequence learning: a study with positron emission tomography. *J. Neurosci.* 14:3775–90

Karni A, Meyer G, Jezzard P, Adams MM, Turner R, Ungerleider LG. 1995. Functional MRI evidence for adult motor cortex plasticity during motor skill learning. *Nature* 377:155–58

Karni A, Meyer G, Rey-Hipolito C, Jezzard P, Adams MM, et al. 1998. The acquisition of skilled motor performance: fast and slow experience-driven changes in primary motor cortex. *Proc. Natl. Acad. Sci. USA* 95:861–68

Knecht S, Floel A, Drager B, Breitenstein C, Sommer J, et al. 2002. Degree of language lateralization determines susceptibility to unilateral brain lesions. *Nat. Neurosci.* 5:695–99

Kobayashi M, Hutchinson S, Schlaug G, Pascual-Leone A. 2003. Ipsilateral motor cortex activation on functional magnetic resonance imaging during unilateral hand movements is related to interhemispheric interactions. *Neuroimage* 20:2259–70

Kobayashi M, Hutchinson S, Theoret H, Schlaug G, Pascual-Leone A. 2004. Repetitive TMS of the motor cortex improves ipsilateral sequential simple finger movements. *Neurology* 62:91–98

Laughlin SB, Sejnowski TJ. 2003. Communication in neuronal networks. *Science* 301:1870–73

Lee L, Siebner HR, Rowe JB, Rizzo V, Rothwell JC, et al. 2003. Acute remapping within the motor system induced by low-frequency repetitive transcranial magnetic stimulation. *J. Neurosci.* 23:5308–18

Lessard N, Pare M, Lepore F, Lassonde M. 1998. Early-blind human subjects localize sound sources better than sighted subjects. *Nature* 395:278–80

Liepert J, Storch P, Fritsch A, Weiller C. 2000. Motor cortex disinhibition in acute stroke. *Clin. Neurophysiol.* 111:671–76

Liepert J, Uhde I, Graf S, Leidner O, Weiller C. 2001. Motor cortex plasticity during forced-use therapy in stroke patients: a preliminary study. *J. Neurol.* 248:315–21

Maeda F, Keenan JP, Tormos JM, Topka H, Pascual-Leone A. 2000. Modulation of corticospinal excitability by repetitive transcranial magnetic stimulation. *Clin. Neurophysiol.* 111:800–5

Mansur CG, Fregni F, Boggio PS, Riberto M, Gallucci-Neto J, et al. 2005. A sham-stimulation controlled trial of rTMS of the unaffected hemisphere on hand motor function after stroke. *Neurology*. In press

Mark VW, Taub E. 2004. Constraint-induced movement therapy for chronic stroke hemiparesis and other disabilities. *Restor. Neurol. Neurosci.* 22:317–36

Martin PI, Naeser MA, Theoret H, Tormos JM, Nicholas M, et al. 2004. Transcranial magnetic stimulation as a complementary treatment for aphasia. *Semin. Speech Lang.* 25:181–91

Melzer P, Morgan VL, Pickens DR, Price RR, Wall RS, Ebner FF. 2001. Cortical activation during Braille reading is influenced by early visual experience in subjects with severe visual disability: a correlational fMRI study. *Hum. Brain Mapp.* 14:186–95

Merabet L, Amedi A, Pascual-Leone A. 2005. Activation of the visual cortex by Braille reading in blind subjects. In *Reprogramming Cerebral Cortex: Plasticity Following Central and Peripheral Lesions*, ed. S Lomber, JJ Eggermont. Oxford, UK: Oxford Univ. Press. In press

Merabet L, Kobayashi M, Barton J, Pascual-Leone A. 2003. Suppression of complex visual hallucinatory experiences by occipital transcranial magnetic stimulation: a case report. *Neurocase* 9(5):436–40

Merabet L, Maguire D, Warde A, Altruescu K, Stickold R, Pascual-Leone A. 2004a. Visual hallucinations during prolonged blindfolding in sighted subjects. *J. Neuro-Ophthalmol.* 24:109–13

Merabet L, Rizzo J, Amedi A, Somers D, Pascual-Leone A. 2005. What blindness can tell us about seeing again: Merging neuroplasticity and neuroprostheses. *Nat. Rev. Neurosci.* 6:71–77

Merabet L, Thut G, Murray B, Andrews J, Hsiao S, Pascual-Leone A. 2004b. Feeling by sight or seeing by touch? *Neuron* 42:173–79

Murase N, Duque J, Mazzocchio R, Cohen LG. 2004. Influence of interhemispheric interactions on motor function in chronic stroke. *Ann. Neurol.* 55:400–9

Naeser MS, Martin PI, Nicholas M, Baker EH, Seekins H, et al. 2005a. Improved naming after TMS treatments in a chronic, global Aphasia patient—case report. *Neurocase*. In press

Naeser MS, Martin PI, Nicholas M, Baker EH, Seekins H, et al. 2005b. Improved picture naming in chronic Aphasia after TMS to part of right Broca's area, an open-protocol study. *Brain Lang.* In press

Noppeney U, Friston KJ, Price CJ. 2003. Effects of visual deprivation on the organization of the semantic system. *Brain* 126:1620–27

Oliveri M, Bisiach E, Brighina F, Piazza A, La Bua V, et al. 2001. rTMS of the unaffected hemisphere transiently reduces contralesional visuospatial hemineglect. *Neurology* 57:1338–40

Oliviero A, Strens LH, Di Lazzaro V, Tonali PA, Brown P. 2003. Persistent effects of high frequency repetitive TMS on the coupling between motor areas in the human. *Exp. Brain Res.* 149:107–13

Pantev C, Engelien A, Candia V, Elbert T. 2001. Representational cortex in musicians. Plastic alterations in response to musical practice. *Ann. N.Y. Acad. Sci.* 930:300–14

Pascual-Leone A. 1996. Reorganization of cortical motor outputs in the acquisition of new motor skills. In *Recent Advances in Clinical Neurophysiology*, ed. J Kinura, H Shibasaki, pp. 304–8. Amsterdam: Elsevier Sci.

Pascual-Leone A, Cohen LG, Brasil-Neto JP, Valls-Sole J, Hallett M. 1994. Differentiation of sensorimotor neuronal structures responsible for induction of motor evoked potentials, attenuation in detection of somatosensory stimuli, and induction of sensation of movement by mapping of optimal current directions. *Electroencephalogr. Clin. Neurophysiol.* 93:230–36

Pascual-Leone A, Hamilton RH. 2001. The metamodal organization of the brain. *Prog. Brain Res.* 134:427–45

Pascual-Leone A, Nguyet D, Cohen LG, Brasil-Neto JP, Cammarota A, Hallett M. 1995. Modulation of muscle responses evoked by transcranial magnetic stimulation during the acquisition of new fine motor skills. *J. Neurophysiol.* 74:1037–45

Pascual-Leone A, Tormos JM, Keenan J, Tarazona F, Canete C, Catala MD. 1998. Study and modulation of human cortical excitability with transcranial magnetic stimulation. *J. Clin. Neurophysiol.* 15:333–43

Pascual-Leone A, Walsh V, Rothwell J. 2000. Transcranial magnetic stimulation in cognitive neuroscience—virtual lesion, chronometry, and functional connectivity. *Curr. Opin. Neurobiol.* 10:232–37

Phelps ME, Mazziotta JC, Kuhl DE, Nuwer M, Packwood J, et al. 1981. Tomographic mapping of human cerebral metabolism visual stimulation and deprivation. *Neurology* 31:517–29

Plautz EJ, Barbay S, Frost SB, Friel KM, Dancause N, et al. 2003. Post-infarct cortical plasticity and behavioral recovery using concurrent cortical stimulation and rehabilitative training: a feasibility study in primates. *Neurol. Res.* 25:801–10

Press DZ, Casement MD, Moo LR, Alsop DC. 2004. Imaging phonological and semantic networks with fMRI. *Soc. Neurosci. Abstr. Viewer/Itinerary Planner*. No. 80.13

Pujol J, Roset-Llobet J, Rosines-Cubells D, Deus J, Narberhaus B, et al. 2000. Brain cortical activation during guitar-induced hand dystonia studied by functional MRI. *Neuroimage* 12:257–67

Rauschecker JP. 1995. Developmental plasticity and memory. *Behav. Brain Res.* 66:7–12

Roder B, Stock O, Bien S, Neville H, Rosler F. 2002. Speech processing activates visual cortex in congenitally blind humans. *Eur. J. Neurosci.* 16:930–36

Roland PE, Eriksson L, Stone-Elander S, Widen L. 1987. Does mental activity change the oxidative metabolism of the brain? *J. Neurosci.* 7:2373–89

Rossini PM, Dal Forno G. 2004. Neuronal post-stroke plasticity in the adult. *Restor. Neurol. Neurosci.* 22:193–206

Sadato N, Okada T, Honda M, Yonekura Y. 2002. Critical period for cross-modal plasticity in blind humans: a functional MRI study. *Neuroimage* 16:389–400

Sadato N, Pascual-Leone A, Grafman J, Deiber MP, Ibanez V, Hallett M. 1998. Neural networks for Braille reading by the blind. *Brain* 121(Pt. 7):1213–29

Sadato N, Pascual-Leone A, Grafman J, Ibanez V, Deiber MP, et al. 1996. Activation of the primary visual cortex by Braille reading in blind subjects. *Nature* 380:526–28

Sanes JN, Wang J, Donoghue JP. 1992. Immediate and delayed changes of rat motor cortical output representation with new forelimb configurations. *Cereb. Cortex* 2:141–52

Seitz RJ, Roland E, Bohm C, Greitz T, Stone ES. 1990. Motor learning in man: a positron emission tomographic study. *Neuroreport* 1:57–60

Shimizu T, Hosaki A, Hino T, Sato M, Komori T, et al. 2002. Motor cortical disinhibition in the unaffected hemisphere after unilateral cortical stroke. *Brain* 125:1896–907

Siebner HR, Tormos JM, Ceballos-Baumann AO, Auer C, Catala MD, et al. 1999. Low-frequency repetitive transcranial magnetic stimulation of the motor cortex in writer's cramp. *Neurology* 52:529–37

Strens LH, Oliviero A, Bloem BR, Gerschlager W, Rothwell JC, Brown P. 2002. The effects of subthreshold 1 Hz repetitive TMS on cortico-cortical and interhemispheric coherence. *Clin. Neurophysiol.* 113:1279–85

Theoret H, Merabet L, Pascual-Leone A. 2004. Behavioral and neuroplastic changes in the blind: evidence for functionally relevant cross-modal interactions. *J. Physiol. (Paris)* 98(1–3):221–33

Uhl F, Franzen P, Lindinger G, et al. 1991. On the functionality of the visually deprived occipital cortex in early blind person. *Neurosci. Lett.* 124:256–59

Uhl F, Franzen P, Podreka I, et al. 1993. Increased regional cerebral blood flow in inferior occipital cortex and the cerebellum of early blind humans. *Neurosci. Lett.* 150:162–64

van Boven R, Hamilton R, Kaufman T, Keenan JP, Pascual-Leone A. 2000. Tactile spatial resolution in blind Braille readers. *Neurology* 54:2030–46

Voss P, Lassonde M, Gougoux F, Fortin M, Guillemot JP, Lepore F. 2004. Early- and late-onset blind individuals show supra-normal auditory abilities in far-space. *Curr. Biol.* 14:1734–38

Walsh V, Pascual-Leone A. 2003. *Neurochronometrics of Mind: TMS in Cognitive Science*. Cambridge, MA: MIT Press

Wanet-Defalque MC, Veraart C, De Volder A, Metz R, Michel C, et al. 1988. High metabolic activity in the visual cortex of early blind human subjects. *Brain Res.* 446:369–73

Ward NS, Brown MM, Thompson AJ, Frackowiak RS. 2003. Neural correlates of motor recovery after stroke: a longitudinal fMRI study. *Brain* 126:2476–96

Werhahn KJ, Conforto AB, Kadom N, Hallett M, Cohen LG. 2003. Contribution of the ipsilateral motor cortex to recovery after chronic stroke. *Ann. Neurol.* 54:464–72

Wittenberg GF, Werhahn KJ, Wassermann EM, Herscovitch P, Cohen LG. 2004. Functional connectivity between somatosensory and visual cortex in early blind humans. *Eur. J. Neurosci.* 20(7):1923–27

Zangaladze A, Epstein CM, Grafton ST, Sathian K. 1999. Involvement of visual cortex in tactile discrimination of orientation. *Nature* 401:587–90

An Integrative Theory of Locus Coeruleus-Norepinephrine Function: Adaptive Gain and Optimal Performance

Gary Aston-Jones[1,*] and Jonathan D. Cohen[2,*]

[1]Laboratory of Neuromodulation and Behavior, Department of Psychiatry, University of Pennsylvania, Philadelphia, Pennsylvania 19104; email: gaj@mail.med.upenn.edu

[2]Department of Psychology, Center for the Study of Brain, Mind and Behavior, Princeton University, Princeton, New Jersey 08540, and Department of Psychiatry Western Psychiatric Institute and Clinic University of Pittsburgh, Pittsburgh, Pennsylvania 15260; email: jdc@princeton.edu

*The order of authorship is alphabetical. The authors contributed equally to the research reported in this review.

Key Words

neuromodulation, decision making, utility, optimization, orbitofrontal cortex, anterior cingulate cortex

Abstract

Historically, the locus coeruleus-norepinephrine (LC-NE) system has been implicated in arousal, but recent findings suggest that this system plays a more complex and specific role in the control of behavior than investigators previously thought. We review neurophysiological and modeling studies in monkey that support a new theory of LC-NE function. LC neurons exhibit two modes of activity, phasic and tonic. Phasic LC activation is driven by the outcome of task-related decision processes and is proposed to facilitate ensuing behaviors and to help optimize task performance (exploitation). When utility in the task wanes, LC neurons exhibit a tonic activity mode, associated with disengagement from the current task and a search for alternative behaviors (exploration). Monkey LC receives prominent, direct inputs from the anterior cingulate (ACC) and orbitofrontal cortices (OFC), both of which are thought to monitor task-related utility. We propose that these frontal areas produce the above patterns of LC activity to optimize utility on both short and long timescales.

Contents

- INTRODUCTION AND OVERVIEW 404
  - Neuromodulatory Systems and the Regulation of Behavior: A Historical Perspective 405
  - A Modern View of the LC-NE System: Optimization of Performance 407
- EMPIRICAL AND MODELING STUDIES REVEAL ROLES FOR THE LC-NE SYSTEM IN COGNITIVE PROCESSES 408
  - Classical Findings Concerning the Neurobiology of the LC-NE System Suggested a General Role in Regulating Neural Processing and Behavior 408
  - Recent Findings Concerning the Physiology of LC-NE Neurons Reveal Two Modes of Operation 410
  - Neural Network Models of LC-NE Function Relate Physiological Mechanisms to Behavioral Effects 415
- AN INTEGRATIVE MODEL OF LC FUNCTION: THE LC PRODUCES ADAPTIVE ADJUSTMENTS IN GAIN THAT SERVE TO OPTIMIZE PERFORMANCE 417
  - A Simple Mathematical Model Can Be Used to Describe Decision Processes and Analyze Them for Optimality 417
  - The LC Phasic Mode Produces Adaptive Ajustments of Gain that Optimize Performance Within a Task (Temporal Filtering) 418
  - The LC Tonic Mode Produces Adaptive Adjustments of Gain that Optimize Performance Across Tasks (Exploration Versus Exploitation) 420
  - Utility Assessment in Frontal Cortex Regulates LC Mode 423
- GENERAL DISCUSSION.......... 430
  - Descending Influences Play an Important Role in LC-NE Function...................... 431
  - LC Activity is Plastic and May Play a Role in Learning............. 431
  - Interactions Between the LC-NE and DA Systems are Important for Normal and Disordered Cognition.................... 432
  - How do Specific Effects Arise from a "Nonspecific" System? 433
  - Optimization is Critical in a Competitive Context 433
  - Relation to Earlier Studies of the LC-NE System 434
- SUMMARY AND CONCLUSIONS................ 438

INTRODUCTION AND OVERVIEW

Adaptive behavior in a diverse and changing world requires a trade-off between exploiting known sources of reward and exploring the environment for other, potentially more valuable or stable opportunities. The capacity to support such adaptive behavior introduces another trade-off, between complexity of mechanisms required to sustain a broad and flexible repertoire of behaviors and the efficiency of function that comes with simpler designs. Both of these trade-offs—between exploitation and exploration, and between complexity and efficiency—are well recognized by engineers and computer scientists. The evolution of the brain has likely faced similar pressures. In this review, we propose that the LC-NE system serves to adjudicate these trade-offs and thereby contributes to the optimization of behavioral performance. This proposal contrasts with traditional views of the LC-NE system.

Neuromodulatory Systems and the Regulation of Behavior: A Historical Perspective

The LC-NE system is one of several brainstem neuromodulatory nuclei with widely distributed, ascending projections to the neocortex (see **Figure 1**); others include the dopaminergic, serotonergic, and cholinergic systems. These neurons play critical roles in regulating cortical function, and disturbances in these systems are central to major psychiatric disorders, such as schizophrenia, depression, and bipolar disorder. Traditionally investigators have assumed that these neurotransmitters serve relatively simple and basic functions. For example, many have thought that dopamine (DA) release signals reward or motivation, and NE mediates arousal (Berridge & Waterhouse 2003, Jouvet 1969, Robinson & Berridge 1993, Wise & Rompre 1989). Such functions seemed to accord well with the characteristic anatomy of these systems (widely distributed projections throughout the forebrain), and it is easy to understand how disturbances in such basic and pervasive functions would have profound disruptive effects on cognition, emotion, and behavior such as those associated with psychiatric disorders. Furthermore, although NE, DA, serotonin, and acetylcholine are sometimes referred to as "classical neurotransmitters" (presumably because of their early discovery and their effects in peripheral systems), an equally and perhaps more important function of these substances in cortex is neuromodulation. That is, rather than producing direct excitatory or inhibitory effects on postsynaptic neurons, they modulate such effects produced by other neurotransmitters such as glutamate and gamma amino butyric acid (GABA). Neuromodulatory actions,

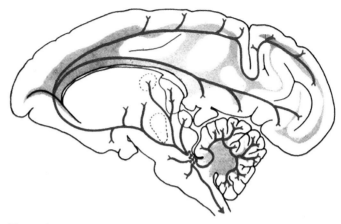

Figure 1

Illustration of projections of the LC system. Saggital view of a monkey brain showing LC neurons located in the pons with efferent projections throughout the central nervous system. Note that only few areas do not receive LC innervation (e.g., hypothalamus and caudate-putamen).

especially when they are distributed over a wide area, seemed well suited to basic, pervasive functions such as the signaling of reward and the mediation of arousal.

Whereas functions such as reward and arousal have intuitive appeal, they have often escaped precise characterization, at both the neural and the computational levels. Recently, however, this has begun to change. For example, considerable progress has been made in developing a formal theory of the role of DA in reinforcement learning. According to this theory, DA does not signal reward per se but rather mediates a learning signal that allows the system to predict better when rewards are likely to occur and thereby contribute to the optimization of reward-seeking behaviors (Montague et al. 1996, 2004). This represents a significant refinement in understanding of the relationship of DA to reward and the role that this neuromodulatory system plays in the regulation of cognition and behavior. In this review, we propose a theory that offers a similar refinement to our understanding of LC-NE function and its relationship to arousal, and how this in turn relates to the optimization of reward-seeking behaviors.

Arousal reflects a fundamental property of behavior that has proven difficult to define

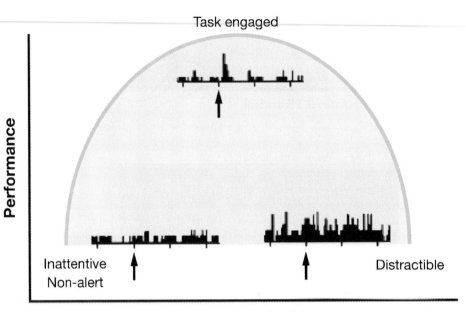

Figure 2

Inverted-U relationship between LC activity and performance on tasks that require focused attention. Performance is poor at very low levels of LC tonic discharge because animals are drowsy and nonalert. Performance is optimal with moderate LC tonic activity and prominent phasic LC activation following goal-relevant stimuli (phasic LC mode). Performance is poor at high levels of tonic LC activity (tonic mode, lacking phasic LC activity). This resembles the classical Yerkes-Dodson relationship between arousal and performance. From Aston-Jones et al. 1999.

or to explain precisely with neurobiological mechanisms. The importance of arousal is undeniable: It is closely related to other phenomena such as sleep, attention, anxiety, stress, and motivation. Dampened arousal leads to drowsiness and, in the limit, sleep. Heightened arousal (brought on by the sudden appearance of an environmentally salient event or a strongly motivating memory) can facilitate behavior but in the limit can also lead to distractibility and anxiety. Traditional theories of LC-NE function, which have tied this structure to arousal, have not described specific mechanisms by which this system produces changes in arousal and have left important unanswered questions about the relationship between arousal and behavior. For example, performance on most tasks is best with an intermediate level of arousal and is worse with too little or too much arousal. This inverted U-shaped relationship is described by the classic Yerkes-Dodson curve (see **Figure 2**). As we discuss below, a similar relationship has been observed between performance and LC-NE activity. This relationship could be interpreted as consistent with the view that the LC-NE system mediates arousal. In this review, however, we propose a theory of LC-NE function that, rather than addressing arousal per se, specifies a role for the LC-NE system in optimizing behavioral performance, which in turn may explain effects conventionally interpreted in terms of arousal.

A Modern View of the LC-NE System: Optimization of Performance

Some theories of LC function suggested that the LC-NE system has its primary effects on sensory processing and in so doing serves to regulate arousal (Berridge & Waterhouse 2003). This was motivated largely by the consistent observation that highly salient and arousing stimuli elicit a phasic activation of LC neurons (Aston-Jones & Bloom 1981b, Grant et al. 1988, Herve-Minvielle & Sara 1995, Rasmussen et al. 1986) and concomitant NE release (Abercrombie et al. 1988, Brun et al. 1993). In addition, NE was found to augment the throughput of signals in sensory brain areas (Devilbiss & Waterhouse, 2000, 2004; Hurley et al. 2004; Waterhouse & Woodward 1980; Waterhouse et al. 1980, 1998). This observation led some to think of LC as the brain's analog of the adrenal gland, orienting the system to and augmenting the processing of motivationally relevant stimuli. Over the past decade, however, neuronal recordings from the primate LC during performance of simple decision-making tasks, coupled with new anatomic studies, have suggested a revision of traditional views of LC-NE function (Aston-Jones et al. 1994, 1997; Clayton et al. 2004; Rajkowski et al. 2004; Usher et al. 1999). Specifically, these recordings indicate that in the waking state there are at least two distinguishable modes of LC function. In a phasic mode, bursts of LC activity are observed in association with the outcome of task-related decision processes and are closely coupled with behavioral responses that are generally highly accurate. In a tonic mode, LC baseline activity is elevated but phasic bursts of activity are absent, and behavior is more distractible. Moreover, strong projections to the LC found from the OFC and ACC (Aston-Jones et al. 2002; M. Iba, W. Lu, Y. Zhu, J. Rajkowski, R. Morecraft & G. Aston-Jones, manuscript in preparation; Rajkowski et al. 2000), and the functions of these frontal areas in evaluating rewards and costs, suggest that these regions are important in generating these patterns of LC activity.

Here we review these findings and describe a theory of LC-NE function that seeks to integrate them with an emerging understanding of the neural mechanisms underlying performance in simple decision-making tasks. We propose that within the context of a given task, phasic activity of the LC-NE system facilitates behavioral responses to the outcome of task-specific decision processes, filtering responses to irrelevant events.[1] By selectively facilitating responses to task-relevant processes, the LC-NE phasic response serves to optimize the trade-off between system complexity (which can support a broad range of functions) and efficiency of function (optimizing performance in the current task). We further propose that the LC-NE system is responsive to ongoing evaluations of task utility (that is, the costs and benefits associated with performance), provided by input from frontal structures. When utility persistently wanes, changes in LC-NE tonic activity withdraw support for task performance, facilitating other forms of behaviors that serve to explore alternative sources of reward. These functions are accomplished by the neuromodulatory effects of NE release at cortical target sites, modulating the gain (responsivity) of processing in cortical circuits responsible for task performance. The different modes of LC activity adaptively adjust the gain of these cortical circuits, both phasically and tonically, facilitating or disengaging task-specific

---

[1] By decision processes, we mean those processes responsible for mapping task-relevant stimuli onto the corresponding response. As we discuss further below, there is growing evidence that, for simple tasks, such processes may be implemented relatively early in the processing stream, distinct from and preceding those responsible for response execution by as much as 100–200 ms. Furthermore, whereas in this review we focus on tasks involving motoric responses (which are most readily accessible to measurement and therefore have yielded the most data), our theory is intended to apply equally to tasks involving internal "responses" such as the encoding of information into long-term memory.

processes. This adjustment of modes serves to optimize the trade-off between exploitation and exploration of opportunities for reward and thereby maximizes utility. This adaptive gain theory integrates and explains new findings concerning LC physiology and its relationship to behavior, as well as newly discovered projections to LC from key frontal structures involved in utility assessment. It also suggests specific ways in which LC-NE function interacts with the proposed role that DA plays in reinforcement learning, providing a formal framework within which to explore dynamic interactions between these systems.

EMPIRICAL AND MODELING STUDIES REVEAL ROLES FOR THE LC-NE SYSTEM IN COGNITIVE PROCESSES

We begin with a brief summary of the basic neurobiology of the LC-NE system (for more detailed reviews, see Foote et al. 1983, Berridge & Waterhouse 2003, and Moore & Bloom 1979). We follow with a consideration of recent neurophysiological findings that suggest the need for revising how we think about this system. We then review computational models developed to explain these findings. We conclude by presenting our theory for a role of the LC-NE system in optimizing task performance, which relates the functioning of this system to cortical mechanisms involved in the evaluation of costs and benefits associated with task performance, and the trade-off between exploiting task-related sources of reward and exploring other possible rewards.

Classical Findings Concerning the Neurobiology of the LC-NE System Suggested a General Role in Regulating Neural Processing and Behavior

"Locus coeruleus" means blue spot in Latin, reflecting the pigmented nature of LC neurons in human. The LC nucleus is a small collection of noradrenergic neurons (about 16,000 per hemisphere in the human), located just behind the periaqueductal gray in the dorsorostral pons. Although few in number, LC-NE neurons give rise to highly divergent and extensive efferents in rats, monkeys, and humans (Dahlstrom & Fuxe 1964, Morrison et al. 1982). These cells provide the bulk of the brain's NE and are the sole source of NE to the cerebral, cerebellar, and hippocampal cortices (Aston-Jones 2004, Aston-Jones et al. 1984, Moore & Bloom 1979). LC projections are unmyelinated and therefore slowly conducting (typically <1 m/s; Aston-Jones et al. 1985). Early studies also reported that LC terminals have nonsynaptic release sites that may provide a paracrine-type of neurotransmission (Beaudet & Descarries 1978, Seguela et al. 1990).

NE can have different effects on target neurons, depending on the receptor that is activated (reviewed in Berridge & Waterhouse 2003 and Foote et al. 1983). Thus, alpha1 adrenoceptor activation is often associated with excitation, and alpha2 adrenoceptor activation (the dominant type within LC itself) is associated with inhibition (Rogawski & Aghajanian 1982, Williams et al. 1985). However, modulatory effects that do not evoke simple excitatory or inhibitory effects are also frequently described. For example, NE increased the ratio of synaptically evoked activity to spontaneous activity in target neurons in early studies (Foote et al. 1975, Segal & Bloom 1976). Later studies found that in many target areas NE augments evoked responses (either excitatory or inhibitory) while decreasing spontaneous activity of the same neuron (Waterhouse et al. 1980, 1984; Waterhouse & Woodward 1980). Thus, modulation of neuronal responses to other inputs is a prominent effect of NE actions on target cells.

This modulatory action was captured in an early computational model of NE effects as an increase in the gain of the activation function of neural network units (**Figure 3**), which was shown to mimic many of the physiologic effects of NE and could explain patterns of

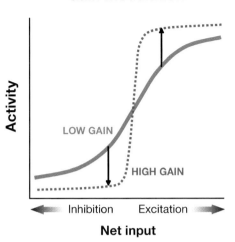

Figure 3

Effect of gain modulation on nonlinear activation function. The activation (or transfer) function relates the net input of a unit to its activity state (e.g., the firing rate of a single neuron or the mean firing rate of a population). The function illustrated here is given by

$$\text{activation} = \frac{1}{1 + e^{-(gain * net\ input)}}.$$

An increase in gain (*dotted line*) increases the activity of units receiving excitatory input (*upward arrow on right*) and decreases the activity of units receiving inhibitory input (*downward arrow on left*), thus increasing the contrast between activated and inhibited units and driving them toward more binary function. Adapted from Servan-Schreiber et al. 1990.

behavior associated with manipulations of NE (Servan-Schreiber et al. 1990). This computational model of NE's modulatory effects set the stage for further studies using more elaborate models involving LC neurons and their targets, as described in more detail below.

The above properties—widespread slowly conducting projections and neuromodulatory action—suggested that LC may play a general role in regulating neural processing and behavior. Commensurate with this view, tonic impulse activity of LC-NE neurons strongly covaries with stages of the sleep-waking cycle. These neurons fire most rapidly during waking, slowly during drowsiness and slow-wave/non-REM sleep, and become virtually silent during REM/paradoxical sleep (Aston-Jones & Bloom 1981a, Hobson et al. 1975, Rajkowski et al. 1998, Rasmussen et al. 1986). LC activity may in fact be a primary factor that differentiates REM sleep (when other systems, including the neocortex, exhibit signs of heightened arousal) from wakefulness (Steriade et al. 1993). These and related findings support the view that low levels of LC activity facilitate sleep and disengagement from the environment.

Further supporting the view that the LC-NE system plays a role in general arousal and environmental responsiveness, LC neurons in rats and monkeys activate robustly following salient stimuli in many modalities that elicit behavioral responses (Aston-Jones & Bloom 1981b, Foote et al. 1980, Grant et al. 1988). For example, tapping the cage door around feeding time elicits LC activation accompanied by a behavioral orienting response and increased physiological signs of arousal. Conversely, stimuli that elicit no behavioral response typically do not evoke an LC response.

The classical observations described above suggest that the LC-NE system has a relatively broad, nonspecific effect on cortical information processing. However, other findings indicate that substantial specificity exists in the LC-NE system in several domains. For example, although they are widespread, LC projections exhibit substantial regional and laminar specificity (Morrison et al. 1982). Notably, brain areas thought to be involved in attentional processing (e.g., parietal cortex, pulvinar nucleus, superior colliculus) as well as motor responding (e.g., primary motor cortex) receive a particularly dense LC-NE innervation (Foote & Morrison 1987). Also, LC terminals make conventional synapse-like appositions with postsynaptic specializations on target neurons (Olschowka et al. 1981; Papadopoulos et al. 1987, Papadopoulos et al. 1989), in addition to having possible nonsynaptic release sites.

Recent neurophysiological findings also indicate that LC may play a specific role in information processing and that it may interact closely with top-down influences from cortical systems. These findings have led to the development of mechanistically explicit computational models that describe the physiological mechanisms governing LC-NE function and their interaction with cortical mechanisms responsible for the execution and evaluation of behavior. Below, we review the recent findings concerning LC-NE physiology and anatomy in monkey that suggest a more specific role in information processing, and computational models of LC-NE function that provide a formal description of the proposed mechanisms involved.

Recent Findings Concerning the Physiology of LC-NE Neurons Reveal Two Modes of Operation

LC phasic mode involves phasic activation of LC neurons following task-relevant processes. A number of studies have revealed that during accurate task performance reflecting focused attentiveness, LC neurons fire tonically at a moderate rate and respond phasically and selectively shortly following task-relevant target stimuli, but not after distractors that may differ subtly from targets. In one series of experiments, LC activity was recorded while monkeys performed a simple signal-detection task in which they were required to respond by releasing a lever immediately following a specific visual target (e.g., a small vertical bar of light—target cue, 20% of trials) but to withhold responding for another similar cue (e.g., a horizontal bar of light—distractor, 80% of trials). Correct responses were rewarded by the delivery of a small quantity of juice, whereas incorrect responses (target misses and false alarms to the distractor) were punished by a brief time-out. Monkeys performed this task with high accuracy, typically greater than 90%. **Figure 4** shows a representative recording of LC neurons, demonstrating substantial phasic activation shortly following target stimuli but only a weak (if any) response following distractors (Aston-Jones et al. 1994). Systematic examination of the LC phasic response following targets indicated that it is not specific to particular sensory attributes. Also, LC does not respond phasically to distractors even if they

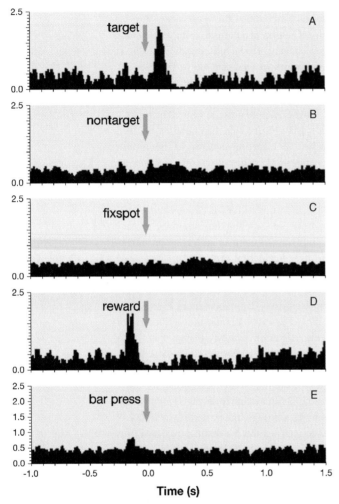

Figure 4
Phasic activation of monkey LC neurons in a signal-detection task. Peri-event time histograms (PETHs) for a typical individual LC neuron in response to various events during performance of the signal-detection task. PETHs are each accumulated for 100 sweeps of activity in this neuron synchronized with (*A*) target stimuli, (*B*) nontarget stimuli, (*C*) fix spot presentation, (*D*) juice solenoid activation, or (*E*) bar press and release performed outside of the task, as indicated. Note the selective activation following target stimuli (*panel A*). The small tendency for a response in (*C*) may reflect activation after target stimuli that occur at short but somewhat variable times after fix spots. Similarly, the activation seen before reward presentation (*D*) is due to activation following target cues. From Aston-Jones et al. 1994.

are infrequent, and (in a forced-choice task) LC responses occur even when targets are presented on every trial. The LC response is also not linked to a specific reward because similar responses are observed for different juice rewards or for water reward in fluid-restricted subjects. Furthermore, in reversal experiments in which the distractor becomes the target and vice versa, LC phasic responses are quickly acquired to the new target and extinguished for the new distractor. This reversal in LC response precedes stable behavioral reversal within a single testing session (Aston-Jones et al. 1997). These findings indicate that the LC response is highly plastic and that it is not rigidly linked to specific sensory attributes of a stimulus but rather responds to events in a task-sensitive manner.

The timing of LC phasic responses is also informative and contrasts with traditional concepts of a slowly acting, nonspecific system. The latency of LC phasic activation following targets is surprisingly short (~100 ms onset) and precedes lever-release responses by about 200 ms. The conduction latency for monkey LC impulses to reach the frontal cortex is ~60–70 ms (Aston-Jones et al. 1985), making it possible for NE release to occur at about the time that neural activity in motor cortex associated with the behavioral response begins to develop (about 150 ms before the manual response; Mountcastle et al. 1972). Thus, although the conduction velocity of LC impulses is slow, the timing of impulse arrival in cortical targets makes it possible for the LC phasic response and NE release to influence the behavioral response on the same trial. Consistent with this possibility, the latency of LC neuron response and lever release are significantly correlated over trials; shorter LC response is associated with shorter behavioral response to the same cue (Aston-Jones et al. 1994).

LC recordings in monkeys performing two-alternative forced choice (2AFC) tasks strongly suggest that these phasic responses are associated with decision processes. 2AFC tasks have been used in a growing number of studies examining the neural mechanisms involved in simple forms of decision making (e.g., Gold & Shadlen 2000; Hanes & Schall 1996; Schall & Thompson 1999; Shadlen & Newsome 1996, 2001). In one such experiment with LC recordings, monkeys were rewarded for responding with the left lever for one stimulus and the right for another (Clayton et al. 2004). As in previous experiments, LC phasic responses were observed shortly following task cues and preceding lever responses. More detailed analysis revealed that LC activation was more tightly time-locked to the behavioral response than to presentation of the stimulus. This result is shown in **Figure 5** with peri-event time histograms that tabulate LC activity with respect to either the sensory stimulus or the behavioral response. Such analyses showed that LC responses precede behavioral responses by about 230 ms regardless of trial type or response time (RT). Note in particular that, in the stimulus-locked histograms, LC responses are greater for correct trials than for error trials. In contrast, in the response-locked histograms, LC activity is comparable for correct and error trials. This can be explained by the additional observation that RT variability was significantly greater for error than for correct trials. Because LC activity is more tightly coupled to the behavioral response than to stimulus onset, and RT is more variable for error trials, stimulus-locked distributions of LC activity are also more variable (relative to response-locked distributions) for error trials. LC activation did not occur on trials in which the animal made no response despite viewing the cue, and there was no LC response associated with spurious lever responses that occasionally occurred between trials when no stimulus was present. Finally, note that the LC phasic response appears to be closely coupled with, and precedes, the task-related behavioral responses. These observations have been confirmed in a signal-detection task in which trial difficulty was manipulated to produce variable RTs. Once again, LC phasic activity

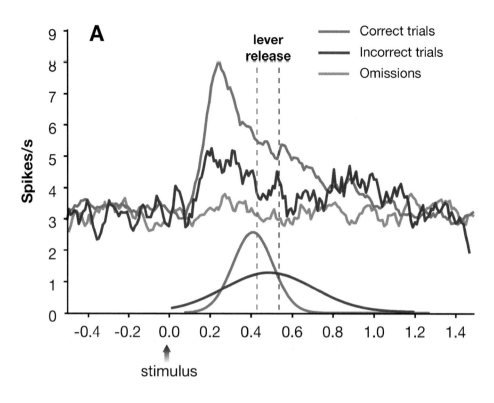
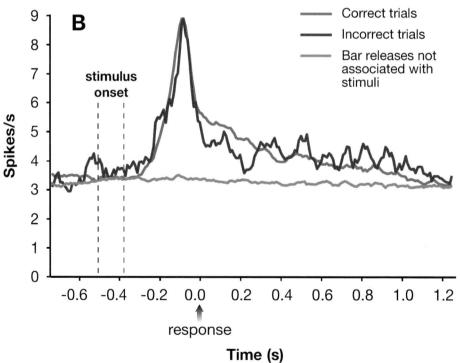

was more tightly linked to the RT than to the sensory stimulus and preceded lever responses by ∼200 ms (Rajkowski et al. 2004).

Similar results were obtained in a recent study that recorded LC neurons in behaving rats (Bouret & Sara 2004). Here, the animal was presented with conditioned odor stimuli that instructed it when to respond to obtain food reward. As found in the monkey, LC neurons exhibited phasic responses to the conditioned stimulus but not to stimuli (nontargets) that were not associated with food availability; these responses preceded the behavioral response and were linked more tightly to the behavioral response than to the sensory stimulus. Moreover, in reversal conditioning the rat LC responses tracked the significance of the stimuli rather than stimulus identity, and newly acquired LC responses preceded those observed in behavior by several trials. These findings all closely parallel those for the monkey LC described above (Aston-Jones et al. 1997, Clayton et al. 2004).

The pattern of results described above precludes the possibility that LC phasic activation is driven strictly by stimulus onset, response generation, or reward. As discussed in greater detail below, these results have led us to hypothesize that LC phasic activity is driven by the outcome of internal decision processes that may vary in duration from trial to trial (accounting for RT variability) but precede response generation with a regular latency. Along these lines, an important observation is that LC phasic responses are largest and most consistent when the animal is performing the task well. During epochs of poor performance, LC phasic responses are considerably diminished or absent. These observations are consistent with our hypothesis that the LC phasic response plays a role in facilitating task-relevant behavioral responses.

As the phasic LC response occurs only for identification of a task-relevant stimulus, and not task-irrelevant stimuli, it can be thought of as an attentional filter that selects for the occurrence (i.e., timing) of task-relevant stimuli. This filter is temporally specific but, given the broad projections of LC neurons, spatially global. In addition, the link to decision outcome, rather than to stimulus presentation, indicates that this LC response primarily modulates specific behaviors rather than sensory processing. Therefore, we propose that the LC phasic response provides a temporal attentional filter that selectively facilitates task-relevant behaviors. This conclusion is supported by the results of neural network–modeling studies discussed later.

LC tonic mode involves increased baseline activity and diminished phasic responses of LC neurons. In addition to LC phasic responses, levels of LC tonic (baseline) activity vary significantly in relation to measures of task performance. For example, during performance of a signal-detection task, periods of elevated LC tonic activity were

Figure 5
Phasic activation of monkey LC neurons in a two-alternative forced choice (2AFC) task. Stimulus- and response-locked population PETHs showing LC responses for trials yielding correct and incorrect behavioral responses. (*A*) Stimulus-locked population PETHs showing LC response to cues (presented at time 0) for trials yielding correct or incorrect behavioral responses. Note that the LC response peaks sooner and is less prolonged on correct compared with incorrect trials in this analysis (17,533 and 1362 trials, respectively). No LC activation was detected on omission trials (*orange line*, 1128 trials). Vertical dashed lines indicate the mean behavioral RTs. Curves represent the normalized RT distributions for correct and incorrect trials. (*B*) The difference in the phasic LC response between correct and incorrect trials was not evident in response-locked population PETHs. In addition, no LC activation occurred prior to or following lever releases not associated with stimulus presentation (*orange line*, 3381 trials). Dashed vertical lines indicate the mean stimulus onset times. From Clayton et al. 2004.

consistently accompanied by more frequent false-alarm errors (Aston-Jones et al. 1996, Kubiak et al. 1992, Usher et al. 1999). Analyses using standard signal-detection measures revealed that, during periods of elevated tonic LC activity, the animal's ability to discriminate targets from distractors (D-prime) and its threshold for responding to stimuli (beta) both decreased (Aston-Jones et al. 1994). RT distributions were also wider. Furthermore, the experimental paradigm required the animal to foveate the center of the computer display prior to stimulus presentation (as an indicator of task preparedness). Such foveations were less frequent during periods of elevated LC tonic activity, resulting in a significantly greater number of aborted trials (Aston-Jones et al. 1996, 1998). Collectively, these findings indicate that when baseline LC activity is increased, the animal is less effectively engaged in task performance, displaying increased distractibility with a greater tendency to respond to nontarget stimuli (lower response threshold). Such periods are also consistently associated with a diminution or absence of the LC phasic responses seen during periods of best performance. These observations are consistent with the hypothesis that LC phasic activity facilitates behavioral responses engaged by task-related decision processes. However, it begs the question of whether and how the LC tonic mode is adaptive and what information-processing function it may serve. As described below, we propose that although this mode is disadvantageous for performance on a specific task, it may be important for sampling alternative behaviors and adaptively pursuing other tasks in a changing environment.

The above results indicate an association of LC activity with task performance. However, they do not establish whether alterations in LC activity are causative of, correlated with, or result from other mechanisms responsible for changes in performance. Preliminary evidence using microinfusions into the monkey LC supports the view that the LC plays a causal role in influencing performance. In a recent study, the alpha2 adrenoceptor agonist clonidine was used to decrease tonic LC activity, and the muscarinic cholinergic agonist pilocarpine was used to stimulate tonic LC discharge. Direct microinfusion of clonidine into the LC of a monkey exhibiting an unusual degree of distractibility (hyperactivity) and poor performance on the signal-detection task significantly decreased tonic LC activity, increased LC phasic responses to target stimuli, and improved performance by decreasing false alarm and omission errors. In contrast, during error-free performance in other monkeys, local microinjection of the muscarinic cholinergic agonist pilocarpine caused tonic activation of LC neurons, decreased phasic responsiveness to task stimuli, and interfered with task performance (Ivanova et al. 1997).

Summary. Overall, these results indicate that two modes of LC activity correspond to different patterns of performance. (*a*) In the phasic mode, LC cells exhibit phasic activation in response to the processing of task-relevant stimuli but display only a moderate level of tonic discharge. This mode of function is consistently associated with high levels of task performance. (*b*) In the tonic mode, LC cells fail to respond phasically to task events but exhibit higher levels of ongoing tonic activity. This mode is associated with poor performance on tasks that require focused attention and corresponds to apparent increases in distractibility. We should note here that, whereas we have described the phasic and tonic modes as distinct, they likely represent the extremes of a continuum of function. This idea is consistent with findings from the computational models reviewed in the following section, which suggest transitions between these modes may be regulated by simple, continuous physiological variables. Nevertheless, for expository purposes we continue to refer to the phasic and tonic modes as distinct in the remainder of this review.

Neural Network Models of LC-NE Function Relate Physiological Mechanisms to Behavioral Effects

The data described above pose important questions about LC activity and its relationship to behavioral performance: Which physiological mechanisms underlie the phasic and tonic modes of LC activity and transitions between them, and how do these mechanisms interact with cortical mechanisms responsible for task performance to produce the patterns of behavior associated with each mode of LC function? Computational modeling has recently begun to address these questions. In an initial effort, Usher et al. (1999) constructed a model composed of two components (see **Figure 6**): (*a*) a detailed, population-level model of LC; and (*b*) a more abstract connectionist network that was the simplest network capable of simulating performance in the signal-detection task. This model revealed that alterations in electrotonic coupling among LC neurons can produce the two modes of LC activity. The model also revealed how the corresponding alterations in gain of cortical units receiving LC inputs can either facilitate task performance (phasic mode) or produce more distractible, less-task-focused responding (tonic mode).

The behavioral network was composed of three layers: an input and a decision layer (each with two units, representing the target and distractor stimuli), and a response layer (with a single unit corresponding to the behavioral response).[2] Connections from units in one layer to the next were excitatory (information flow), whereas connections between units in the decision layer were inhibitory (competition for representation). A response was recorded when activity of the response unit exceeded a specified threshold.

The LC component of the model was composed of 250 integrate-and-fire units, con-

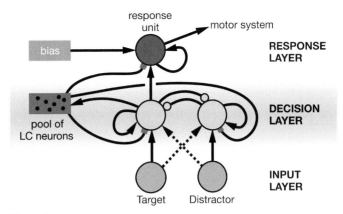

Figure 6

Architecture of the neural network model of LC function in the signal-detection task. Arrows represent excitatory links, small circles depict inhibition, and squares (from LC projections) represent modulation of gain (see **Figure 3**). There is a moderate positive bias on the response unit, which captures the observation that monkeys in this task make many false alarms but very few misses (Aston-Jones et al. 1994). Note the projection from the decision-layer target unit to the LC, which captures the observation that in the well-trained animal LC neurons are selectively activated following target stimuli. From Usher et al. 1999.

nected by mutually inhibitory noradrenergic collaterals as well as weak electrotonic coupling, both of which have been observed empirically (Aghajanian et al. 1977; Christie et al. 1989; Christie & Jelinek 1993; Egan et al. 1983; Ennis & Aston-Jones 1986; Ishimatsu & Williams 1996; Travagli et al. 1995, 1996). LC units received afferent connections from the decision unit representing the target stimulus in the behavioral network (consistent with decision-driven LC activity described above)[3] and sent projections back to all units in the behavioral network (consistent with the known broad efferent projections of LC neurons). All units in the model were subject to noise in their input, producing a baseline level of activity in the LC and the possibility for spurious responses in the behavioral network.

The effect of NE release in the behavioral network was simulated as an increase in the responsivity (i.e., gain of the activation

---

[2] Units in this network were assumed to represent populations of recurrently connected cortical neurons contributing to the representation of a given piece of information (e.g., Amit 1989).

[3] These are assumed to have been learned for this task but in other circumstances could represent "hardwired" connections for evolutionarily important signals such as highly salient sensory events.

function) of these units (Servan-Schreiber et al. 1990). This mimicked the modulatory influence of NE (discussed above) and augmented the activity of units in the behavioral network that were already activated (e.g., because of noise), while further suppressing the activity of units that were already being inhibited. A systematic examination of these simulated NE effects revealed that their greatest influence on performance was achieved by their impact on the response unit, which produced behavioral changes that closely matched those observed empirically. This point becomes important in our discussion below about the role of the LC-NE system in optimizing task performance.

An important finding from this model was that modest changes in the strength of electronic coupling among LC units reproduced the entire set of neurophysiological and behavioral results described above concerning modes of LC activity and their relationship to task performance. Within LC, increased coupling facilitated phasic activation of LC neurons in response to activation of the target decision unit in the behavioral network by allowing summation of concurrent responses across electrotonic links. At the same time, coupling reduced spontaneous firing by averaging the effects of uncorrelated noise across LC units. Conversely, reduced coupling accurately simulated the effects associated with the LC tonic mode by producing a modest rise in baseline LC activity but a diminished phasic response to input from the target decision unit in the behavioral network.

Within the behavioral network, LC phasic responses (associated with high coupling and generated by activation of the target decision unit) produced NE release and corresponding increases in gain within the response unit. This occurred within the same time frame that feed-forward activation was being received from the target decision unit, facilitating activation of the response unit. This, in turn, reproduced the empirical observation of improved target detection performance associated with LC phasic responses. In this respect, the LC phasic response can be thought of as an attentional filter that selects for the occurrence (i.e., timing) of task-relevant events and facilitates responses to these events. The existence of such a temporal filter, and its association with noradrenergic function, is consistent with several recent psychophysical and psychopharmacological studies (e.g., Coull & Nobre 1998, Coull et al. 2001). Conversely, during low coupling the attenuated LC phasic response to target detection produced somewhat slower and more variable behavioral responses, whereas increased baseline NE release (as a result of increased LC tonic activity) rendered the response unit more susceptible to the effects of noise and therefore to the production of spurious (false alarm) responses. Although this degrades performance in the present task, Usher et al. (1999) suggest that the general increase in responsiveness associated with the LC tonic mode facilitates exploration of alternative behaviors, which is adaptive in a changing environment or when current rewards lose their value (as discussed in more detail below).

Recent modeling work has extended these basic results by refining the original population model of LC using more biophysically detailed, phase oscillator units for each LC neuron (Brown et al. 2004). The individual units of this model have been matched closely with detailed physiological properties empirically observed for LC neurons, such as firing rate, variance in interspike interval, calcium and voltage-dependent potassium currents, electrotonic coupling, and collateral inhibitory connections. One important finding that has emerged from this work is that the phasic and tonic modes of LC activity (described above) can be produced also by changes in baseline excitatory drive to LC cells (in the absence of any change in coupling). Decreases in baseline afferent drive promote the phasic mode of LC activity, whereas increases favor the tonic mode. These observations are consistent with other recent modeling studies (Alvarez et al. 2002) and suggest that changes in baseline activity provide another simple mechanism by which LC mode can be regulated. Additional studies are needed to determine if altered baseline drive or electronic coupling, or both, is responsible for the different LC firing modes.

Summary. In the models described above, the outcome of processing at the decision layer drives an LC phasic response, increasing the gain of units throughout the behavioral network. This lowers the threshold for the response unit, thus facilitating a response to the outcome of the decision process. The timing of the LC phasic response suggests that it can be thought of as a temporal attentional filter, much as cortical attentional mechanisms act as a content filter, facilitating responses to task-relevant events. At the same time, by increasing the gain of cortical representations, the LC phasic response may also enhance the effects of attentional selection by content within the neocortex (e.g., Robertson et al. 1998). Together, these effects allow the LC phasic response to facilitate selectively responses to task-relevant stimuli. These mechanisms provide an account of how LC phasic responses may contribute to enhanced task performance associated with the LC phasic mode of function, and how the LC tonic mode may lead to degraded task performance. At the same time, the models have identified specific physiological parameters—electrotonic coupling and baseline firing rate—that can drive transitions between the LC tonic and phasic modes of function. However, these models leave several important questions unaddressed: What function does the LC tonic mode serve, and what drives transistions between the phasic and tonic modes? In the sections that follow, we consider both LC phasic and tonic function in the context of recent theories concerning the neural mechanisms underlying simple decision processes and mathematical analyses concerning the optimization of such processes. A consideration of LC function within this context suggests how both modes of LC activity may contribute to the overall optimization of performance.

AN INTEGRATIVE MODEL OF LC FUNCTION: THE LC PRODUCES ADAPTIVE ADJUSTMENTS IN GAIN THAT SERVE TO OPTIMIZE PERFORMANCE

A Simple Mathematical Model Can Be Used to Describe Decision Processes and Analyze Them for Optimality

To consider how LC may play a role in optimization of task performance, we must first more precisely define what we mean by optimization. This, in turn, requires a more formal characterization of the mechanisms that underlie task performance. We can think of these as being composed of a set of decision processes, which may involve perception ("Was that a ball or a strike?"), memory ("Was the count level or full?"), evaluation ("Was the last call fair or unfair?"), and/or action ("Should I swing high or low?"). Cognitive and neuroscientific studies have made considerable progress in identifying and characterizing mechanisms associated with the simplest decision processes involved in 2AFC tasks (e.g., Gold & Shadlen 2000; Hanes & Schall 1996; Schall & Thompson 1999; Shadlen & Newsome 1996, 2001). There is a remarkable convergence of views that agree that the dynamics of both neural activity and behavioral performance observed in such tasks can be described accurately by a simple mathematical model, often referred to as the drift diffusion model (DDM). This, in turn, provides a useful framework for defining and evaluating the optimization of performance.

The DDM describes decision processes in terms of simple accumulators that integrate signals favoring each of the two choices and respond when the difference between these signals exceeds a threshold value. The DDM offers a mathematically precise characterization of the dynamics and outcome of decision making in such tasks (Laming 1968; Ratcliff 1978, 2004; Stone 1960), on the basis of the

Originally introduced in discrete form as the sequential probability ratio test (SPRT, also known as the random walk model), investigators proved that this is the optimal procedure for making a binary decision under uncertainty (Barnard 1946, Wald 1947). That is, for a specified level of accuracy it is the fastest method to reach a decision; or, conversely, for a specified time in which to make the decision (i.e., deadline) it is the most accurate. Accordingly, this procedure was used by Turing to decipher the Enigma code used by the German navy in World War II. Furthermore, researchers have recently proven that for a given set of task variables (e.g., stimulus strength and intertrial interval), there is a single decision threshold that maximizes reward rate; that is, there is an optimal trade-off between speed and accuracy that maximizes reward rate.[4] Recent empirical work indicates that performers can approximate this maximum (R. Bogacz, E.T. Brown, J. Moehlis, P. Hu, P. Holmes & J.D. Cohen, manuscript under review).

analytically tractable simplification of single-layered neural network models that simulate performance in such tasks (R. Bogacz, E.T. Brown, J. Moehlis, P. Hu, P. Holmes & J.D. Cohen, manuscript under review; Usher & McClelland 2001).[5]

The LC Phasic Mode Produces Adaptive Ajustments of Gain that Optimize Performance Within a Task (Temporal Filtering)

The DDM is appealing in the present context because it provides a framework within which we can formally define optimal performance.[6] The DDM itself is, in fact, the optimal process for 2AFC decision making (i.e., it is the most accurate for a given speed of decision following three assumptions: (*a*) the decision-making process is stochastic (that is, it is subject to random fluctuations in accumulation of evidence in favor of each alternative); (*b*) evidence favoring each alternative is accumulated over time; and (*c*) the decision is made when sufficient evidence (exceeding a threshold) has accumulated in favor of one alternative over the other. What is remarkable about this model is that, although originally developed to describe behavior (e.g., reaction-time distributions and error rates), it also provides a detailed and accurate account of neuronal responses in such tasks and their relationship to performance (Gold & Shadlen 2002, Ratcliff et al. 2003, Schall & Thompson 1999). Furthermore, the DDM provides an

---

[4]In principle, we can imagine utility functions that differentially weigh types of outcomes under varying circumstances (e.g., speed over accuracy, or vice versa; Bohil & Maddox 2003, Mozer et al. 2002, Wald 1947). However, it seems reasonable to assume that under many (if not most) circumstances, a critical objective is to maximize the rewards accrued by performance. Reward rate can be formalized as a function of decision time, interstimulus interval, and error rate and may be used as an objective function in evaluating maximization of utility (R. Bogacz, E.T. Brown, J. Moehlis, P. Hu, P. Holmes & J.D. Cohen under review; Busemeyer & Myung 1992, Gold & Shadlen 2002).

---

[5]Typically, such models involve pairs of mutually inhibiting (i.e., competing) leaky accumulators, the activity of which represents the amount of evidence favoring each alternative in the choice. A response is produced when the activity of one of the units crosses a specified threshold. This is precisely the mechanism used in the decision layer of the LC models described above, and it has been used to simulate a wealth of findings from other 2AFC tasks (e.g., Botvinick et al. 2001; Cohen et al. 1990, 1992; Usher & McClelland 2001) as well as the dynamics of neural activity in response-selective brain areas associated with the performance of such tasks (e.g., Gold & Shadlen 2002, Shadlen & Newsome 2001). Mathematical analyses of these models suggest that, under certain assumptions, they can be reduced to the DDM without significant loss of their ability to describe behavioral and neurophysiological data accurately (R. Bogacz, E.T. Brown, J. Moehlis, P. Hu, P. Holmes & J.D. Cohen, manuscript under review). In this respect, the DDM provides a mathematically precise, theoretical bridge from neural mechanisms to behavioral performance.

[6]In discussing optimization of performance, we are not asserting people always behave perfectly optimally. In many instances, people do not behave optimally (e.g., Herrnstein 1961, 1997; Kahneman & Tversky 1984; Loewenstein & Thaler 1989). Nevertheless, the assumption of optimality is useful because it provides theoretical traction: It allows us to define formally and precisely the goal of an adaptive system, even when this goal may not be fully achieved in practice. This definition can be used to generate testable predictions and provides a valuable reference against which actual behavior can be compared and understood, even when it turns out to be suboptimal. This approach has been used effectively in a wide variety of disciplines, including neuroscience and psychology (Anderson 1990, Barlow 1981, Bialek & Owen 1990, Gallistel et al. 2001, Mozer et al. 2002).

making, and the fastest for a given level of accuracy; see sidebar). Furthermore, because the DDM provides a good description of simple, single-layered neural networks that implement 2AFC decision processing, we can infer that such single-layered networks can approximate optimal performance in 2AFC decision processing. We know, however, that real neural architectures, in fact, involve many layers (e.g., Schall 2003, Reddi 2001), presumably because different tasks require decision processes that integrate information of different types, at varying levels of analysis, and from a variety of sources. Although all of these decision processes may involve fundamentally similar mechanisms—possibly well described by the DDM—they are implemented by different parts of ("layers" within) the full neural architecture. This presents a challenge for optimal performance. On the one hand, as noted above, a single-layered network implements the optimal decision process. On the other hand, the layer implementing this process for a given task may be several layers away from the response mechanism. It would be inefficient if a decision process in the layer integrating information relevant to the current task crossed threshold but then had to drive a subsequent series of repeated accumulator processes—each of which introduces additional noise and requires addition integration time—before a behavioral response could be elicited. This problem reflects the fundamental trade-off between the complexity of a multilayered system that can support a wide range of decision processes (and the flexibility of behavior that this affords) and the efficiency of a simpler, single-layered system (that is, the optimality of function that this affords).

The inefficiency of multilayered integration can be ameliorated if, at the time a unit in the task-relevant decision layer crosses threshold, a signal is issued ensuring that this information rapidly and directly influences the behavioral response. The LC phasic signal accomplishes precisely this effect in the models described above. The LC phasic response is trigged when sufficient activity accumulates in one of the units in the decision layer of the behavioral network. The resulting LC phasic response increases the gain of all units in the behavioral network, which drives units toward binary responding (see **Figure 3**), in effect eliminating further integration in any subsequent layers. Because this occurs at a time when the relevant decision unit has just crossed threshold and is therefore highly active, all units "downstream" will assume states that are heavily determined by this particular input. Thus, we can think of the sudden increase in gain as an adaptive sampling bias (this is the sense in which the LC acts as a temporal filter), favoring the selection of states of the entire system that are most heavily influenced by the activity of the units in the decision layer, and thereby allowing that layer efficiently to determine the behavioral response. Thus, the effect of an LC phasic response driven by the threshold crossing in the task-relevant decision layer can be thought of as collapsing the multilayered network into a single-layered network, thus approximating the optimal decision-making process implemented by the task-relevant decision layer. In this way, the LC phasic response resolves a fundamental trade-off between the flexibility of a complex, multilayered system (that can support a wide variety of decision processes responsive to information from different sources and different levels of analysis) and the optimality of a single-layered decision mechanism. From this perspective, the decision-driven LC phasic activation provides a mechanism for optimizing task performance by a multilayered system.

In the specific models previously discussed, there are only two layers: the decision layer and the response layer. Nevertheless, even in such a two-layered system, formal analysis reveals that the adaptive gain-adjustment mechanism implemented by LC produces reliable improvements in performance that more closely approximate the optimal performance of a single-layered network (Gilzenrat et al. 2004). These improvements are expected to be considerably greater for more realistic, multilayered networks.[7] We should note, however, that the function of existing models relies on the detection of discrete threshold-crossing events (e.g., by LC, for generating its phasic response; and by the cortical network, for generating an overt behavioral response). This may reflect intrinsic nonlinearities of processing units involved; the specifics of these mechanisms remain to be described in further research.

The LC Tonic Mode Produces Adaptive Adjustments of Gain that Optimize Performance Across Tasks (Exploration Versus Exploitation)

In considering optimal performance thus far, our focus has been on performance within a single task, in which the LC phasic response produces adaptive adjustments in gain that serve to optimize performance of that task. In contrast, the LC tonic mode produces a persistent increase in gain (i.e., responsivity of widespread LC target neurons) that renders the system more sensitive to task-irrelevant stimuli. With respect to the current task, this is clearly disadvantageous. However, this tonic increase in gain may be adaptive by facilitating a change in behavior if either the current task is no longer remunerative or if the environment has changed and more valuable opportunities for reward or new behavioral imperatives have appeared. That is,

---

[7]A variant on this model allows the integration process to be distributed over several tightly coupled, interacting layers of the network (rather than a single layer) but assumes that there are two thresholds for the integrators: one for driving the LC phasic response, and a much higher one for driving the behavioral response. The latter insures that spurious activity of task-unrelated units will not drive a behavioral response. When the threshold for activating the LC is crossed by task-related (decision) units, the LC phasic response increases gain throughout the network, driving the most currently activated units toward the behavioral response threshold. This mechanism allows a more distributed implementation of decision processing, while still exploiting the LC phasic response as a "temporal filter." Formal analyses suggest this distributed case also exhibits improvements in performance with adaptive gain adjustment (Gilzenrat et al. 2004).

in addition to pursuing optimal performance within the specific task at hand, organisms face the broader and equally important challenge of deciding whether and for how long it is best to continue performing the current task. From this perspective, optimization involves not only determining how to best perform the current task, but also considering its utility against alternative courses of action and pursuing these if they are more valuable. This is, of course, a more complex and less well-defined problem, which presents significant challenges to formal analysis. Reinforcement learning models represent one approach to this problem. Such models describe mechanisms that seek to exploit opportunities optimally for reward by sampling a wide range of behaviors and strengthening actions leading to states with the highest value (Montague et al. 2004, Sutton & Barto 1981).

A conundrum faced by reinforcement learning models is how to sample the values of a large number of different states. Early sampling will reveal some states that are more valuable than others. Then the agent must decide whether to spend most of its time engaged in behaviors associated with the most valuable ones that it has already discovered (that is, exploit these known states) or to seek new behaviors by continuing to sample the environment (that is, explore a broader range of states) in search of novel and potentially more valuable opportunities than those already discovered. This conundrum is often referred to as the trade-off between exploitation and exploration (e.g., Kaelbling et al. 1996). How it is handled can have a profound effect on reward accrued. If the agent favors exploitation too heavily before the environment has been adequately explored, then it risks missing valuable opportunities (in the terminology of thermodynamics, it may get caught in a local minimum). However, if it favors exploration too heavily, then it may fail to exploit known sources of reward adequately (i.e., it will be "unstable").

In standard reinforcement learning models, the trade-off between exploitation and exploration is handled using a procedure that regulates the amount of noise (randomness) in the agent's behavior. This is akin to annealing in thermodynamic systems. Annealing is a procedure in which a molten metal is slowly cooled so that the molecules can move around thoroughly during recrystallization and achieve thermodynamic equilibrium producing the most uniform (optimal) lattice structure. This process prevents gaps in the lattice, i.e., local minima in thermodynamic terms. During initial exposure to an environment when exploration is more valuable than exploitation, noise is set at a high level, encouraging exploration (similar to a free flow of molecules, or melting).[8] However, as increasingly valuable states are identified, noise is gradually diminished (akin to slow cooling and annealing, restricting flow and solidifying the lattice) so that the agent can most effectively exploit the environment by focusing on behaviors tied to the most valuable states. Such annealing procedures help optimize the agent's ability to extract reward from the environment. However, typically these procedures are introduced *deus ex machina* (that is, using predefined schedules imposed by the modeler). Furthermore, such schedules typically cannot accommodate changes in the environment. For example, what happens when a previously identified source of reward becomes unavailable or less valuable as the agent becomes sated (e.g., the defining structure of the lattice changes)? Under such conditions, the optimal strategy is to resume exploring the environment (melting), sampling different behaviors until new sources of reward are discovered. This is exactly the role played by increases in tonic LC activity and attenuation of the phasic LC response in the adaptive gain theory: Increased baseline release of NE increases the gain of units in the network indiscriminately, making them more responsive to any stimulus. This uniform increase in responsivity is tantamount to increasing noise and favoring exploration. The broad efferent network of LC projections is well suited for this role because it applies the tonic gain increase across global targets and circuits, thereby allowing a broad scan of possible new reward sources.

At least two sources of recent evidence support this theory concerning the function of the LC tonic mode. The first is provided by the target reversal experiment described earlier (Aston-Jones et al. 1997). Following reversal of reward contingencies, LC phasic responses to the former target rapidly diminished, while baseline (tonic) LC firing increased. This was maintained until phasic responses appeared for the new target and disappeared for the old one. That is, LC transitioned from a phasic to a tonic mode and then reversed as the new target was acquired. These findings are what would be predicted if LC implemented the annealing procedure associated with reinforcement learning in the task. Of course, this requires that LC has the relevant information to determine when to transition between phasic and tonic modes, an important question that we address shortly.

The second line of evidence comes from studies of human performance, and measurements of pupil diameter as an indirect index of LC activity. As shown in **Figure 7**, pupil diameter correlates remarkably well with LC tonic activity in the monkey (Rajkowski et al. 1993) and shows the same relationship to behavioral performance as LC tonic activity. The latter finding has recently been corroborated in the human (Gilzenrat et al. 2003). Furthermore, numerous studies have shown that task processing is accompanied by rapid and dramatic pupil dilation, consistent with the occurrence of an LC phasic response to task-relevant events (Beatty 1982a,b; Richer

---

[8]An increase in noise promotes the simplest form of exploration, which is random search. However, more sophisticated agents may, at least under some circumstances, use more structured, model-based forms of search (involving heuristics or explicit algorithms). Even in these cases, however, a transient increase in noise may serve to disengage the current behavioral set, facilitating the discovery and pursuit of a new one.

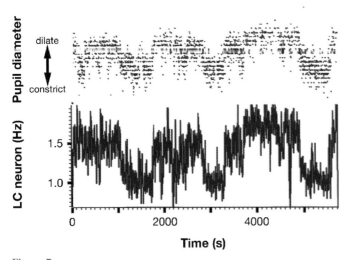

Figure 7

Relationship between tonic pupil diameter and baseline firing rate of an LC neuron in monkey. Pupil diameter measurements were taken by remote eye-tracking camera at each instant in time when the monkey achieved fixation of a visual spot during the signal-detection task (described in text). Note the close direct relationship between the pupil diameter and the rate of LC activity.

& Beatty 1987).[9] We recently measured both baseline- and stimulus-related pupil diameter in a task involving diminishing utility, designed to test our hypothesis concerning the function of tonic LC activity. Human subjects performed a series of tone discriminations of progressively increasing difficulty with rewards for correct performance that increased in value with increasing task difficulty. Initially, the increases in reward value outpaced increases in difficulty (and associated increases in errors) so that subjects remained engaged in the task. However, after several trials, the increases in difficulty led to sufficient numbers of errors as to reduce reward rate even in the face of the increasing value of correct responses. At the beginning of every trial, participants were allowed to press a reset button, which would start a new series of discriminations, beginning again with low difficulty and low reward value. Participants behaved optimally on average, choosing to reset when the success (expected utility) of the discriminations began to decline. Early in each trial series there were large phasic pupil dilations for each discrimination. As would be predicted for LC phasic responses, these dilations declined in amplitude, and baseline (tonic) pupil diameter rose as the task became more difficult and expected utility began to decline. Baseline pupil diameter was greatest at the point at which subjects chose to abandon the current series, consistent with the hypothesis that this was mediated by an increase in LC tonic activity.

Although these findings are consistent with the hypothesis that LC tonic activity supports optimal performance by favoring exploration, this remains to be tested using more direct measurements. This dimension of our theory also has yet to be expressed in formal terms. Existing models of LC function provide a mechanism by which this could occur (i.e., tonically increasing gain throughout the network) but have not directly established that such tonic activity optimizes performance when opportunities exist for sources of reward outside the current task. A recent line of work, however, has begun to make strides in this direction and converges on an interpretation of the function of NE release similar to the one we propose here. A. Yu and P. Dayan (manuscript under review) have used a Bayesian framework to develop a model of how an agent can optimally determine whether a failure of prediction (and therefore performance) reflects variability inherent in the current task (which they term "expected uncertainty") or an underlying change in the environment (termed "unexpected uncertainty"). They propose that estimates of expected (task-related) uncertainty are mediated by acetylcholine, whereas estimates of unexpected uncertainty—which promote a revision of expectations—are mediated by NE. The latter corresponds closely to our theory that LC tonic activity (and corresponding

---

[9]Despite the close relationship between pupil diameter and LC activity, the mechanisms underlying this relationship are not yet understood. There are presently no known anatomic pathways that could mediate this relationship directly, which suggests that these effects may reflect parallel downstream influences of a common source mechanism.

increases in tonic NE release) favors exploration. Note, however, that both our hypothesis and that of Yu and Dayan assume the operation of evaluative mechanisms that identify violations of expectation or, more generally, decreases in utility that favor behavioral changes. Sustained increases in gain mediated by the LC tonic mode, which lead to shifts in behavior, are adaptive in the sense that they are responsive to such evaluative information. Below we review evidence that leads us to propose that projections to LC from the frontal cortex may provide this evaluative information.

Summary. The findings reviewed above support the theory that the LC-NE system helps optimize performance by adaptively adjusting gain (responsivity) of target sites in two distinct ways. LC phasic responses produce a transient, system-wide increase in gain driven by task-related decision processes, insuring that the outcome of such processes is efficiently expressed in behavior (recall from above that the timing of phasic LC activation allows it to facilitate task-related behavior). This serves to optimize performance within a given task. In contrast, the LC tonic mode produces a more enduring and less discriminative increase in gain. Although this degrades performance within the current task, it facilitates the disengagement of performance from this task and thereby the sampling of others. This action provides a mechanism by which the system can optimize performance in a broader sense—by identifying and pursuing more remunerative forms of behavior when utility associated with the current task begins to diminish. Note that this may transiently accelerate a current reduction in utility by further degrading performance in the current task. This degradation of performance accounts for the far-right end of the Yerkes-Dodson curve (see **Figure 2**), where performance decreases as "arousal" (associated with tonic NE release) increases. According to our theory, this effect reflects the local consequences (for the current task) of a mechanism that is serving to optimize performance on a broader scale. From this perspective, although the right end of the Yerkes-Dodson curve may appear to be maladaptive, in fact it reflects the operation of an important mechanism of longer term adaptation.[10]

Utility Assessment in Frontal Cortex Regulates LC Mode

Our theory proposes that the LC phasic mode supports optimization of current task performance as long as task-related utility remains sufficiently high, whereas the LC tonic mode supports optimization on a broader scale, favoring exploration when current task-related utility falls below an acceptable value (or that which may be available from other tasks). This begs a critical question: What determines when LC should transition between phasic and tonic modes? That is, how does LC know when current task utility exceeds or has fallen below an acceptable value? If LC is to respond in a truly adaptive fashion, then it must have access to information about rewards and costs. As briefly reviewed below, a growing body of evidence suggests that two frontal structures, the OFC and ACC, play critical roles in evaluating rewards and costs, respectively. Furthermore, recent studies—motivated by our theory of LC—reveal that the most prominent descending cortical projections to LC come from these two frontal structures. We review this evidence in the sections that follow, concluding with a simple formal theory about how top-down evaluative information from the frontal cortex may be combined to regulate LC function.

The frontal cortex plays an important role in the evaluation of utility. The OFC and ACC have been the focus of increasingly intense study, using both direct neuronal

---

[10]Although LC tonic activity may be adaptive, overly persistent LC tonic activity (or high arousal) can, in the limit, be maladaptive, as may be the case for clinical syndromes such as anxiety disorders and attention deficit disorder.

recording techniques in nonhuman primates and neuroimaging methods in humans. The results of these studies consistently indicate a role for these structures in the evaluation of utility.

OFC plays a role in the evaluation of reward. The OFC receives input from all modalities of high-order sensory cortices, in particular areas processing information with strong appetitive significance, such as taste and olfaction, as well as primary limbic structures such as the ventral striatum and amygdala (Baylis et al. 1995; Carmichael et al. 1994; Carmichael & Price 1995a,b; Ongur & Price 2000; Rolls et al. 1990). Neurons in monkey OFC are activated by rewarding stimuli in various modalities but not by stimulus identification alone nor by response preparation (Roesch & Olson 2004, Rolls 2004). Furthermore, OFC responses vary in magnitude in proportion to the relative reward value of the corresponding stimulus (Tremblay & Schultz 1999; Wallis & Miller 2003), and reward-specific responses diminish as the animal becomes sated for that particular reward (Critchley & Rolls 1996, Rolls et al. 1989). OFC neurons in monkey also are sensitive to the anticipation and delivery of reward (Schultz et al. 2000, Hollerman et al. 2000), and recent evidence suggests that OFC responses may be able to integrate the ongoing rate of reward over relatively extended periods (Sugrue et al. 2004). Many of these findings have been corroborated in human neuroimaging studies involving a variety of rewards, including money, food, and drugs of abuse (Breiter et al. 1997, 2001; Knutson et al. 2000; O'Doherty et al. 2002; Small et al. 2001; Thut et al. 1997; McClure et al. 2004). Collectively, these findings provide strong evidence that OFC plays an important role in the evaluation of reward.

ACC plays a role in the evaluation of cost. Like OFC, ACC receives convergent inputs from a broad range of neocortical and subcortical structures, including somatosensory areas and limbic structures such as insular cortex, amygdala, and ventral striatum (Devinsky et al. 1995, Mesulam 1981). ACC is known to be directly responsive to aversive interoceptive and somatosensory stimuli, and to pain, in particular (e.g., Peyron et al. 2000). More recently, neurophysiological studies in the monkey as well as human electrophysiological and neuroimaging studies have consistently demonstrated that ACC is also responsive to negatively valenced information of a more abstract nature, such as errors in performance, negative feedback, monetary loss, and even social exclusion (Eisenberger et al. 2003; Falkenstein et al. 1991; Gehring et al. 1993; Holroyd et al. 2003, 2004a,b; Ito et al. 2003; Kiehl et al. 2000; Miltner et al. 1997; Yeung et al. 2005; Yeung & Sanfey 2004). In addition to explicitly negative information, ACC responds robustly and reliably to task difficulty and conflicts in processing (e.g., Barch et al. 1997; Botvinick et al. 1999, 2001, 2004; Carter et al. 1998; Duncan & Owen 2000; Ullsperger & von Crammon 2001). Conflicts occur when simultaneously active processes compete for the expression of incompatible alternatives, a factor that may be directly (and possibly causally) related to task difficulty and has been formalized in neural network models of task performance (Botvinick et al. 2001, Yeung et al. 2004). Thus, converging evidence suggests that ACC is responsive to a variety of negatively valenced signals—from pain to internal states that predict degraded performance—all of which may serve as indices of performance-related cost.

OFC and ACC send strong convergent projections to LC. Historically, anatomic studies of afferents to LC have focused almost entirely on subcortical structures; there have been very few published reports of studies examining possible inputs from cortical areas (for a review, see Aston-Jones 2004). This may reflect the emphasis placed by traditional theories on the roles for LC in bottom-up processes such as sensory encoding and arousal (as discussed in the Introduction section). Importantly, most studies have also been conducted in nonprimate species in which cortical structures (and the frontal cortex in particular) are substantially less well developed. Although Jodo and Aston-Jones observed that stimulation of rat prefrontal cortex can activate LC neurons (Jodo & Aston-Jones 1997, Jodo et al. 1998), and others reported inhibitory responses (Sara & Herve-Minvielle 1995), anatomical studies showed that prefrontal projections in rat terminate adjacent to, but not within, the LC nucleus (Aston-Jones 2004). Until recently, therefore, little was known about the extent to which primate LC neurons receive direct descending projections from the neocortex. The adaptive gain theory of LC-NE function outlined above, however, predicts that LC should receive information about task-related utility from high-level structures. Motivated directly by this prediction, a series of anatomic studies was undertaken in monkey to determine the extent to which LC receives top-down cortical projections. These studies have revealed a consistent and striking pattern of cortical projections in the primate LC, the preponderance of which come from OFC and ACC.

As illustrated in **Figures 8** and **9**, focal injections of retrograde tracer into the monkey LC reveal a large number of labeled neurons in OFC and ACC (Aston-Jones et al. 2002; Rajkowski et al. 2000; Zhu et al. 2004; M. Iba, W. Lu, Y. Zhu, J. Rajkowski, R. Morecraft & G. Aston-Jones, manuscript in preparation). These retrograde results have been confirmed by injection of anterograde tracers into OFC or ACC, which yielded prominent fiber and terminal labeling in the monkey LC nucleus and peri-LC dendritic area (**Figure 8**) (M. Iba, W. Lu, Y. Zhu, J. Rajkowski, R. Morecraft & G. Aston-Jones, manuscript in preparation; Zhu et al. 2004). Importantly, these OFC and ACC projections appear to be the major cortical inputs to LC; relatively few neurons in other cortical areas were retrogradely labeled from LC injections. For example, very few neurons were retrogradely labeled in area 46, and anterograde tracing from area 46 produced labeling nearby, but not within, the LC, consistent with a previous report (Arnsten & Goldman 1984). Interestingly, the bulk of OFC inputs to LC appears to originate in the caudolateral OFC, the same area that receives strong direct olfactory and taste primary reinforcer inputs (as noted above). Numerous retrogradely labeled neurons also extend caudally from the OFC into the anterior insular cortex. ACC neurons that innervate LC are located in both dorsal and ventral ACC subdivisions (including areas 24, 25, and 32) and densely populate layer 5/6 throughout the rostral ACC. These results indicate that OFC and ACC provide prominent direct input to LC in monkey and that these projections are the major cortical influences on LC neurons. These studies also indicate that OFC and ACC inputs to LC in the monkey are stronger than those in the rat, where prefrontal fibers terminate nearly exclusively in the peri-LC dendritic zone and do not appreciably enter the LC nucleus proper (described above).

OFC and ACC may regulate LC function. The evidence reviewed above indicates that OFC and ACC each play an important role in assessing utility and that both of these structures project directly to LC. These findings suggest that these frontal areas could influence LC function on the basis of assessments of utility, consistent with the adaptive gain theory of LC function. There are two ways in which this could occur: OFC and ACC could drive LC phasic activity directly, and they could modulate LC mode of function.

OFC and ACC may drive LC phasic activation. If LC phasic responses are driven by the outcome of decision processes, an important question is, what cortical regions convey this information to LC? The obvious candidates are regions that house the neural accumulators associated with the decision processes themselves, which have typically been localized to cortical sensorimotor integration areas (e.g., to lateral intraparietal cortex and frontal eye fields for visual tasks requiring an

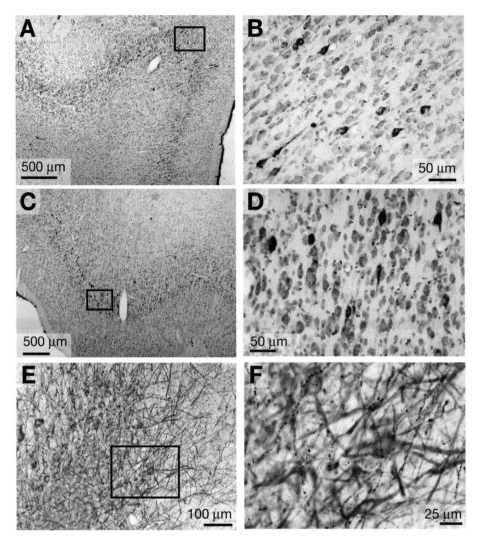

Figure 8

Projections to the LC from the anterior cingulate cortex (ACC) and the orbitofrontal cortex (OFC) in monkey. (*A*) Low-power photomicrograph of a frontal section through the ACC showing retrogradely labeled neurons in area 24b/c. Area shown is just ventral to the cingulate sulcus. (*B*) High-power photomicrograph showing retrogradely labeled neurons in the ACC (*corresponding to rectangle in panel A*). Note labeled cells located in deep layer V/superficial layer VI. (*C*) Low-power photomicrograph of a frontal section through the OFC showing retrogradely labeled neurons in area 12. Lateral orbital sulcus is at star. (*D*) High-power photomicrograph showing retrogradely labeled neurons in the OFC (*corresponding to rectangle in panel A*). Note cells located in deep layer V/superficial layer VI. Neutral red counterstain used for sections in panels *A–D*. Medial is at right, and dorsal is at top. (*E, F*) Low (*E*)- and high-power (*F*) photomicrographs of a frontal section through the LC and peri-LC showing fibers and terminals labeled from an injection of the anterograde tracer biotinylated dextran amine in the ipsilateral OFC (area 12) of an African green monkey. Noradrenergic neurons and processes are stained brown with an antibody against tyrosine hydroxylase. Note close juxtaposition of OFC fibers and terminals with noradrenergic somata and dendrites. Lateral is at right, and dorsal is at top.

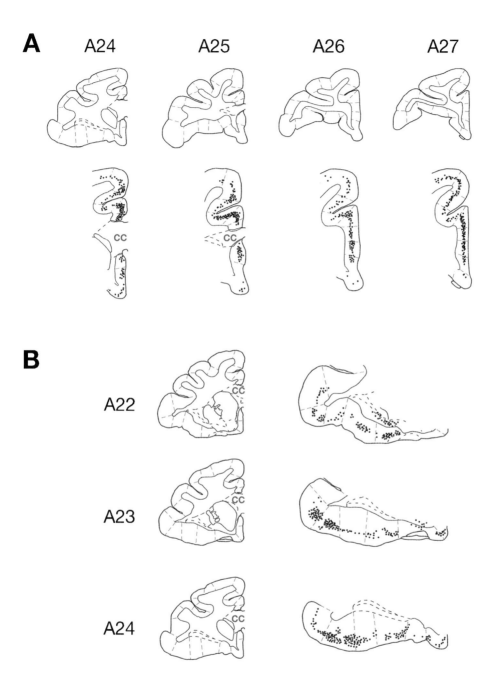

Figure 9
Plots of retrogradely labeled neurons in ACC and OFC after injections of CTb into monkey LC. (*A*) ACC neurons labeled from the monkey LC. Lower sections are high-power views containing plotted cells; upper sections are low-power views to give orientation. (*B*) OFC neurons labeled from the monkey LC. Sections at right are high-power views containing plotted cells; sections at left are low-power views to give orientation. For both panels, A22–A27 refer to distances in mm from the interaural line. Plots were composed on atlas sections from Paxinos et al. 2000.

oculomotor response). However, as reviewed above, these areas do not provide direct projections to LC. Alternatively, the outcome of decision processes may be relayed to LC indirectly via OFC, ACC, or both, which receive inputs from a wide array of sensorimotor areas (Baylis et al. 1995, Carmichael & Price 1995b, Morecraft et al. 1992). Such relays may weight decision-related signals by motivational significance—that is, perceived utility of the current task. This possibility is supported by the pattern of task-specific responses commonly observed within OFC and ACC that are also closely related to the motivational significance of the eliciting event. This, in turn, is consistent with the fact that LC phasic responses are limited to goal-related events (e.g., target but not distractor stimuli in a signal-detection task), with an amplitude modulated by the motivational significance of the stimulus (e.g., the reward associated with appropriate performance) (Aston-Jones et al. 1994, Rajkowski et al. 2004). Thus, OFC and ACC may relay the outcome of task-related decision processes, modulated by their assessed utility, driving the LC phasic response. Studies are presently underway to test this hypothesis more directly.

OFC and ACC may regulate LC mode. The above considerations address the possibility that OFC and ACC drive LC phasic activation in response to individual events (e.g., decision outcome within the trial of a task). The adaptive gain theory suggests that these frontal structures also influence LC function by driving transitions between phasic and tonic modes to regulate the balance between exploitation and exploration: When evaluations in OFC or ACC indicate the current task is providing adequate utility, they drive LC toward the phasic mode (by increasing electrotonic coupling, reducing baseline drive, or both), favoring exploitation of that task for associated rewards. However, when utility diminishes sufficiently over prolonged durations, they drive it toward the tonic mode, favoring exploration. This process requires that utility be evaluated over both short and longer time frames.

For example, consider a case in which the current task is associated with high utility (e.g., the animal is thirsty and correct performance provides juice). In this situation, it is advantageous to optimize performance on the task and maximally exploit the utility it provides. The adaptive gain theory states that this is promoted by the LC phasic mode, during which event-related phasic LC activity facilitates task-appropriate behavioral responses. Furthermore, if performance should temporarily flag (e.g., owing to a momentary lapse of attention), the LC phasic mode should be augmented to restore performance.

Compensatory adjustments following lapses in performance have repeatedly been observed (e.g., Botvinick et al. 1999; Gratton et al. 1992; Jones et al. 2002; Laming 1968, 1979; Rabbitt 1966). Furthermore, there is strong evidence that they are mediated by an evaluative function in frontal cortex, consistent with the mounting evidence discussed above that monitoring mechanisms within ACC detect lapses in performance and signal the need to augment top-down control for the current task (Botvinick et al. 2001). However, previous theories have assumed ACC-based monitoring mechanisms act directly on prefrontal systems responsible for top-down control (e.g., Botvinick et al. 2001, Jones et al. 2002). The adaptive gain theory of LC suggests an additional, more general mechanism for improving performance. Signals from ACC to LC (indicating an adverse outcome), possibly complemented by signals from OFC to LC (indicating absence of an expected reward), may augment the LC phasic mode (by further increasing electronic coupling, reducing LC baseline drive, or both). This, in turn, would improve performance on subsequent trials by enhancing the LC phasic response and thereby augmenting the gain of units responsible for task execution. Increased phasic release of NE may also have direct enhancing effects on task-specific control representations in prefrontal cortex (PFC) (Arnsten et al. 1996, Cohen et al. 2004). This effect could further contribute to the compensatory increase in control following a transient decrease in performance and/or reward. One appeal of this hypothesis is that it provides a general mechanism by which the detection of momentary reductions in utility can augment task control without requiring the monitoring mechanism to have special knowledge about the nature of control required for every possible task. That is, a global signal (LC-mediated NE release) that adaptively adjusts gain throughout the processing system can interact with task-specific control mechanisms (e.g., in PFC) to produce requisite improvements in performance.

The foregoing account addresses circumstances in which overall task-related utility remains high, and momentary lapses in performance (and utility) can be rectified by enhancement of control. However, what happens if there is a persistent decline in utility? In such circumstances, augmenting control associated with the current task may no longer be advantageous. For example, if performance fails to improve despite compensatory adjustments, or if task-related utility progressively declines for other reasons (e.g., satiety, depletion of the task-related source of reward, or an increase in the costs associated with its procurement), then the relationship between utility and task investment should reverse: Further decreases in utility should promote task disengagement rather than attempts to restore performance. That is, they should favor exploration over exploitation. The adaptive gain theory proposes that this is mediated by a transition to the LC tonic mode. Importantly, the determination of when to promote exploration over exploitation requires that evaluative mechanisms take account of both short- and long-term changes in utility. There are many ways of doing so. The following equation describes one simple means (shown graphically in **Figure 10**):

Engagement in current task
$$= [1 - \text{logistic}(\text{short term utility})]$$
$$* [\text{logistic}(\text{long-term utility})],$$
(Equation 1)

where *logistic* refers to the sigmoid function $1/(1 + e^{-\text{utility}})$, and high values of the equation favor the LC phasic mode, whereas low values favor the tonic mode.

We assume that evaluations of utility are computed within both OFC and ACC and then integrated (averaged) over relatively short (e.g., seconds) and longer (e.g., minutes) timescales. How estimations in OFC and ACC are combined and then averaged over different timescales remains a matter for fur-

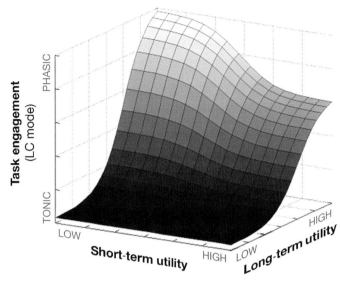

Figure 10

Plot of the relationship between engagement in the current task and task-related utility integrated over relatively brief (e.g., seconds) and longer (e.g., minutes) timescales given by Equation 1 (see text). The adaptive gain theory of LC-NE function proposes that high values of this equation favor the LC phasic mode, whereas low values favor the tonic mode. Accordingly, low values of long-term utility favor the LC tonic mode (exploration), whereas high values favor the LC phasic mode (exploitation). Note that when long-term utility is low, changes in short-term utility have little impact. However, when long-term utility is high, a decrease in short-term utility augments the LC phasic mode, implementing an adaptive adjustment that serves to restore performance.

ther research. However, allowing that such computations take place, Equation 1 provides a simple means by which OFC and ACC may regulate engagement in the current task by controlling a single, or small number of, simple physiological parameters in LC. High values of this function favor task engagement (by driving a transition to LC phasic mode through increases in electrotonic coupling, decreases in baseline drive, or both), discounted by enduring declines in utility that favor task disengagement (LC tonic mode). When long-term utility is high, decreases in short-term utility augment the LC phasic mode. Thus, when overall utility is high, momentary reductions favor improvements in task performance. In contrast, these effects are diminished when long-term utility is low: Persistent declines in utility drive LC toward the tonic mode, favoring task disengagement. Note that this proposed modulation of LC mode occurs simultaneously and independently of the phasic excitation of LC proposed also to originate in these two structures in response to the outcome of individual decisions.

Summary. The findings reviewed above indicate that neurons in OFC and ACC show task-selective responses that represent the present value of task-related events. These structures provide substantial direct projections to LC in monkey. As illustrated in **Figure 11**, the adaptive gain theory of LC function proposes that these projections provide the information necessary to drive LC phasic responses directly, as well as to drive transitions between its phasic and tonic modes of function. Specifically, the theory proposes that the outcome of decision processes associated with high potential utility are represented in frontal structures (OFC and ACC) that drive LC phasic responses. Furthermore, assessments of utility by these frontal structures are integrated over different timescales and used to regulate LC mode. Brief lapses in performance, in the context of otherwise high utility, augment the LC phasic mode, improving task performance. In contrast, enduring decreases in utility drive transitions to the tonic mode, promoting disengagement from the current task and facilitating exploration of behavioral alternatives. Taken together, these mechanisms constitute a self-regulating system by which LC—informed by evaluations of utility in frontal structures—can control behavioral strategy through adaptive adjustments of gain in its global efferent targets.

GENERAL DISCUSSION

The adaptive gain theory suggests that the LC-NE system plays a more important and specific role in the control of performance than has traditionally been thought. This new theory is an important evolution of an LC theory that proposed roles for this system in vigilance and response initiation (Aston-Jones 1985, 1991b) by further specifying the mechanisms that might be involved in executing such functions. The presently proposed framework has a wide range of implications for understanding the neural mechanisms underlying both normal behavior and its impairment in clinical disorders associated with disturbances of decision making and control. At the same time, the theory raises several questions that remain to be addressed. In the reminder of this review, we consider some of these questions and the broader implications of our theory.

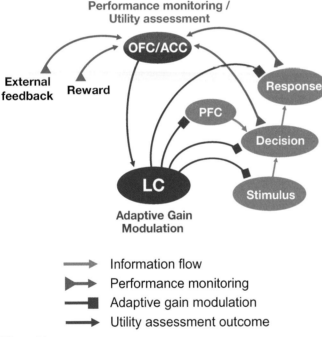

Figure 11

Integrated neural system for the adaptive regulation of performance. Components in green represent neural pathways responsible for task execution, including top-down control from prefrontal cortex (PFC). Components in red represent monitoring and evaluative mechanisms in OFC and ACC. These components assess task-related utility based on indicators of performance (including internal conflict and external feedback). This information is used both to drive LC activity (in the phasic mode) and to regulate LC mode of function (phasic versus tonic). These influences on LC, in turn, regulate performance through release of NE throughout the processing system, either phasically in response to task-relevant events (phasic mode) or in a more sustained manner (tonic mode).

Descending Influences Play an Important Role in LC-NE Function

Many previous analyses of LC function have focused on its role in mediating ascending ("bottom-up") influences on cortical processing. For example, a major focus of research has been on inputs to LC from subcortical mechanisms involved in the sleep-waking cycle, supporting the view that LC translates these mechanisms into modulatory changes in neocortex producing alertness or sleep (Aston-Jones et al. 2005, Berridge & Waterhouse 2003, Jones 1991, Saper et al. 2001). Similarly, theories about the role of LC in sensory processing have generally assumed that it receives inputs from low-level processing mechanisms, mediating the alerting effects of highly salient events via its widespread ascending projections to the neocortex (Devilbiss & Waterhouse 2000, 2004; Hurley et al. 2004). Our theory allows a more precise understanding of LC's role in bottom-up processes such as sleep and sensory processing. For example, descending regulation of LC suggests a mechanism for volitional control of waking in the face of fatigue, and for increased sensory focus with increased task difficulty. Critcally, however, our theory highlights the importance of descending cortical influences on LC and the refined role that the LC-NE system plays in regulating cortically based mechanisms for decision making and control. There is reason to believe that such influences are more highly developed in the primate brain, commensurate with the expansion of frontal structures and their top-down influence on processing and behavior. We have highlighted the roles of orbital and anterior cingulate regions of frontal cortex in particular, both of which send strong projections directly to LC in monkeys (projections that appear to be substantially more developed than in the rat) (Aston-Jones et al. 2002, Rajkowski et al. 2000, Zhu et al. 2004). These areas have been consistently implicated in evaluative functions that are commensurate with the role we propose for LC in regulating behavioral performance.

LC Activity is Plastic and May Play a Role in Learning

A critical component of the adaptive gain theory is that the LC phasic response is driven by the outcome of task-relevant decision processes. This hypothesis implies that the LC phasic activation must be plastic to adapt to changes in task demands (e.g., the relevance of different decision processes). This implication is consistent with the physiological evidence. Most strikingly, reversal experiments demonstrate that LC phasic responses reliably track changes in the target stimulus. However, this plasticity must reflect the response characteristics of the systems that drive LC. This too is consistent with the evidence, given our hypothesis that the LC phasic response is driven by frontal structures, including OFC and ACC. As reviewed above, these structures demonstrate strong task-selective responses. Furthermore, previous studies indicate that OFC exhibits marked plasticity in its response to reward-related stimuli during reversal experiments when the valence of stimuli is abruptly altered. For example, Rolls has shown that in behaving monkeys OFC responses change quickly to track changes in the motivational relevance of stimuli (Rolls et al. 1996, Thorpe et al. 1983; see also Wallis & Miller 2003). These observations support the idea that both the selectivity and plasticity of the LC phasic response reflect afferent drive by frontal structures. At the same time, LC itself may contribute to plasticity within these structures. A role for the LC-NE system in learning has frequently been suggested in the literature (Amaral & Foss 1975, Anlezark et al. 1973, Archer et al. 1984, Bouret & Sara 2004, Cirelli & Tononi 2004, Harley 1991, Harris & Fitzgerald 1991, Koob et al. 1978, Velley et al. 1991). The adaptive gain theory suggests at least one way in which this influence on learning might occur. As discussed above, changes in reward contingency should drive LC into the tonic mode. This shift, in turn, should promote exploration and facilitate the discovery of new reward contingencies that in turn provide a new source of drive for the LC phasic response. As discussed earlier, this hypothesis is supported by the observation that reversals in reward contingency (requiring new learning) precipitate a shift to the LC tonic mode, followed by shifts back to the phasic mode once the new target has been acquired. This hypothesis is also consistent with interactions between the LC-NE and DA systems suggested by the adaptive gain theory, as discussed in the following section. However, a direct test of this hypothesis will require more detailed studies that characterize the relative timing of frontal and LC phasic activity in response to task-relevant stimuli and its evolution over the course of reversal conditioning.[11]

Interactions Between the LC-NE and DA Systems are Important for Normal and Disordered Cognition

There are many similarities between the LC-NE and DA systems. Both NE and DA are neuromodulatory neurotransmitters that have similar physiological effects on target systems (e.g., modulation of gain; Nicola et al. 2000; Servan-Schreiber et al. 1990; Waterhouse et al. 1980, 1984); both are responsive to motivationally salient events (e.g., reward predictors); and disturbances of both have been implicated in highly overlapping sets of clinical disorders (such as schizophrenia, depression, and attention deficit disorder). Despite these similarities, the relationships between these systems and how they interact has remained unclear. In part this ambiguity has been due to the lack of formal theories about the function of either system. Recently, however, Montague et al. (1996) have proposed a sophisticated theory of DA function that suggests it implements the learning signal associated with a reinforcement learning mechanism. This theory affords a direct point of contact with the adaptive gain theory of LC-NE function.

As previously discussed, reinforcement learning requires an annealing procedure, favoring exploration during learning in new (or changing) environments and promoting exploitation when reliable sources of reward are discovered. The adaptive gain theory proposes that the LC-NE system serves this function, implementing an annealing mechanism that is adaptive to ongoing estimates of current utility. Thus, early in learning, when utility is low, LC remains in the tonic mode, favoring exploration. However, as sources of reward are discovered DA-dependent reinforcement learning strengthens behaviors that produce these rewards. This strengthening increases current utility, driving LC into the phasic mode, which further stabilizes and exploits the utility associated with DA-reinforced behaviors. This process continues until the current source of reward is either no longer valued or available. As utility declines, LC is driven back into the tonic mode, promoting exploration and learning of new behaviors. In this way, the proposed functions of the LC-NE and DA systems may interact synergistically to implement an auto-annealing reinforcement learning mechanism that is adaptive both to the needs of the organism and changes in the environment. Although these theories of the LC-NE and DA systems are both early in their development, together they potentially offer a powerful new account of how these systems interact, which may provide conceptual traction in understanding how disruptions in these systems may impact one another in producing the complex patterns of disturbance observed in clinical disorders.

---

[11]The form of plasticity discussed here involves learning over many trials. However, behavioral flexibility can also be exhibited in more rapid form from trial to trial. Such flexibility is thought to rely on PFC mechanisms responsible for cognitive control (e.g., Duncan 1986, Shallice 1988). According to one recent theory, PFC supports such behavioral flexibility by providing top-down modulation of processing along task-relevant pathways (Miller & Cohen 2001). In addition to influencing the flow of activity along cortical pathways, such mechanisms could also dynamically modulate which cortical circuits (e.g., within OFC or ACC) most effectively drive LC activity, providing another mechanism by which LC phasic responses are driven by the outcome of task-relevant processes.

How do Specific Effects Arise from a "Nonspecific" System?

The broad efferent anatomy, slow conduction speed, and modulatory postsynaptic effects of LC neurons have traditionally been interpreted as evidence that the LC-NE system plays a relatively nonspecific role in state setting and arousal (as discussed above). Given these properties, it is natural to wonder how this system could support the more precise form of regulation proposed by the adaptive gain theory (e.g., involving, in the phasic mode, real-time responses that have within-trial effects on performance). The findings we have reviewed suggest two responses to this concern. First, direct recording studies from LC in animals performing complex conditioned tasks have revealed detailed, task-specific patterns of LC response that were not previously observed in nonconditioned subjects. This finding indicates that recordings of these neurons during conditioned behavior may be critical to observing these properties of the LC-NE system. This point is further underlined by the finding that OFC and ACC send strong projections to LC in monkey that are not apparent in subprimate species. A prominent role for these prefrontal structures in the regulation of LC function indicates that studies of the LC system in primates may be critical to uncovering more fully the precise and subtle roles played in cognitive processing.

Second, the work we have reviewed also highlights the importance of formally explicit theoretical models. With respect to the LC-NE system, such models have demonstrated how a mechanism with low spatial (efferent) specificity and modulatory effects can play an important role in regulating, and even mediating, high-level cortical function. In particular, modeling work has characterized the dynamics of the mechanisms involved and has verified that these mechanisms are plausible given the observed temporal properties of LC activity and NE release. For example, modeling work was essential in showing that LC phasic activity at the appropriate time (typically within 150 ms of the stimulus and immediately postdecision) can increase gain in target areas and have an impact on processing of the behavioral response. Modeling also suggested an important function for the global projections of the LC-NE system: These projections allow the system to facilitate a broad range of possible behaviors favoring exploration in the tonic mode. At the same time, modeling work has shown that these widespread projections do not compromise the more precise function performed by the LC phasic mode (given appropriate timing of the phasic response). Indeed, and perhaps most important, these broad projections allow LC to carry out its function without the need for special knowledge of where the changes in gain are needed for a particular task. Finally, modeling work has demonstrated how these effects can arise from the modulation of simple low-level physiological variables within LC (electronic coupling or baseline afferent drive) that are sufficient to regulate transitions between the LC phasic and tonic modes. Thus, modeling work has shown how a distributed system with subtle modulatory effects, driven by a small number of simple parameters, can have precise and profound effects on higher-level cognitive processes.

Optimization is Critical in a Competitive Context

The adaptive gain theory proposes that the LC-NE system helps to optimize performance within and across tasks. Recent modeling analyses indicate that adaptive adjustments of gain in a two-layer network improve reward rate by approximately 10%–20% (Gilzenrat et al. 2004). Although this is a conservative estimate (analyses of multilayered systems may reveal greater benefits), the benefits noted thus far appear to be modest. However, a modest increase in performance in an isolated context may translate into a highly significant effect in a competitive environment, where resources are scarce and can be lost to another agent working with a similar motivation and strategy. Thus, increases in response speed or accuracy that have a modest impact on performance in the isolated environment of a laboratory task may make the difference between survival and extinction in a context where food is limited and competition from other agents exists. This hypothesis may provide some insight into the adaptive advantage of a system such as the LC-NE and may give a reason for its evolution in vertebrates. Future modeling studies employing a game-theory approach will be of interest to examine the benefits of adaptive gain adjustment in a competitive environment.

Relation to Earlier Studies of the LC-NE System

Our review has focused primarily on recent findings from studies of LC in monkeys. Although limited in number, such studies provide the most detailed and elaborate data addressing the role of the LC-NE system in behavior. Nevertheless, a theory of LC-NE function should also be consistent with findings using other methods, including lesion and pharmacological manipulations of the LC-NE system, recording of LC activity and NE release in nonprimate species, and electrophysiological findings related to LC function in humans.

Previous lesion and pharmacological studies implicated the LC in cognitive functions. This literature is vast, and page limitations permit only a selective review of studies of the role of LC in performance and cognitive function (for more extensive reviews see Berridge & Waterhouse 2003, Coull 1994, Robbins & Everitt 1995). Many such studies are compatible with the view that the LC-NE system is important for supporting task-focused behavior. For example, Roberts and colleagues (1976) and others (Oke & Adams 1978) reported that rats with lesions of the ascending LC pathway were more disrupted by distractors during discriminative learning than were intact animals. Robbins and colleagues (Carli et al. 1983, Cole & Robbins 1992) found that lesions of ascending LC projections in rats produced deficits in a continuous performance task that required sustained monitoring for visual target stimuli that could occur in multiple locations. Lesions impaired target discrimination performance and prolonged reaction times when distracting stimuli were presented just before targets or when target presentation was unpredictable. This group also found that lesions of LC projections in rats increased the effectiveness of conditioning to contextual stimuli but decreased conditioning to explicit cues (Selden et al. 1990a,b). Although these findings are consistent with the idea that the LC-NE system is important in task-focused performance, the mechanisms underlying behavioral changes following these lesions have been difficult to define. Moreover, not all lesion studies of LC produced similar results (e.g., Pisa & Fibiger 1983, Pisa et al. 1988). The conflicting results of some studies may be due at least in part to plasticity following the intervention, known to be substantial for LC (Fritschy & Grzanna 1992, Haring et al. 1986, Levin et al. 1985). In addition, the adaptive gain theory suggests that experimental manipulations should interact critically with LC mode of function (phasic vs. tonic), a factor that has not been considered in the design or interpretation of most previous studies. When this is taken into account, previous lesion studies can often be seen in a new light that is compatible with the adaptive gain theory.

For example, a number of plastic responses have been observed following LC lesions, including increases in NE receptor number or sensitivity (Dooley et al. 1987, Harik et al. 1981, U'Prichard et al. 1980) and in tyrosine hydroxylase (Acheson et al. 1980) and firing rate (Chiodo et al. 1983) in the remaining LC neurons, as well as compensatory changes in other non-LC systems (Carboni & Silvagni 2004, Harik et al. 1981, Martin et al. 1994, Valentini et al. 2004). Such lesion-induced plasticity could effectively upregulate LC-NE function tonically. At the same time, LC lesions would eliminate any temporally specific (e.g., decision-driven) phasic NE release in LC target areas; that is, compensatory changes would not be able to mimic the temporal properties of LC neuronal discharge. Thus, the net effect of LC lesions may have been to produce a persistent state similar to the LC tonic mode. This idea would be consistent with increased responsiveness to distractors and decreased task-focused performance, as observed in several lesion studies (discussed above). This postlesion tonic function of LC could also produce decreased conditioning with explicit sensory cues (which require more focus on the task at hand, supported by the phasic LC mode) and increased performance and conditioning with contextual cues (facilitated by increased responsivity to a broader set of events or exploration associated with the tonic LC mode). Although length constraints prohibit a detailed treatment here, similar considerations are applicable to prior studies of LC function in attention using pharmacological approaches.

Several pharmacological studies have also investigated the effects of LC-NE function on memory (see Berridge & Waterhouse 2003 for a review). Our focus in this review has been on optimization of performance in simple tasks involving sensory stimuli and motor responses. However, as noted earlier, the adaptive gain theory applies equally to tasks involving other types of processes, such as the retrieval and storage of information in long-term memory. For example, in one study, LC was stimulated by injection of the alpha2 autoreceptor antagonist idazoxan (Sara & Devauges 1989). This agent blocks the local inhibitory effects of NE within the LC, which may have potentiated LC phasic responses. When animals were trained for several weeks on a set of associations, idazoxan administration just prior to a retention test produced fewer errors than in nontreated animals. This result is consistent with the possibility that potentiation of LC phasic responses facilitated the outcome of memory-driven decision processes, akin to its effects on stimulus-driven decision processes that have been the focus of this review.

Studies by Arnsten and colleagues indicate that manipulations of LC-NE function also have an impact on working-memory function. They found that systemic injections of alpha2 NE autoreceptor agonists, such as clonidine or guanficine, facilitated working-memory performance in aged monkeys (Arnsten et al. 1996). The doses used are thought to preferentially activate alpha2 receptors located postsynaptically on LC target neurons and are assumed to augment NE effects. It is difficult to interpret these results in the context of the adaptive gain theory without a more precise understanding of the dynamics of these pharmacologic manipulations (e.g., the extent to which they impact phasic versus tonic NE release). Nevertheless, they could be consistent with a relationship of LC-NE function to the Yerkes-Dobson curve (see **Figure 2**) if it is assumed that aged animals have tonically diminished LC-NE projections and receptor function (consistent with prior data from Burnett et al. 1990, Eriksdotter Jonhagen et al. 1995, Iversen et al. 1983, and Tejani-Butt & Ordway 1992). Thus, injection of NE autoreceptor agonists into aged animals could improve function by increasing stimulation of deficient postsynaptic NE receptors and preferentially restoring the phasic mode of LC activity. (The latter is suggested by a previous study showing that stimulation of NE autoreceptors on LC neurons preferentially reduces baseline activity while leaving glutamate-driven responses intact, in effect emulating an increased phasic mode; Aston-Jones et al. 1991a). The same manipulation in normal animals, where clonidine may preferentially activate intact cortical NE receptors, could maladaptively mimick a tonic LC mode.

Recordings of LC activity or NE release in nonprimate species. Although relatively few studies involving nonprimate species have recorded impulse activity of LC neurons in behaving animals, the overall results are consistent with the adaptive gain theory. Early studies reported that NE-LC neurons in rats and cats (Aston-Jones & Bloom 1981b, Rasmussen et al. 1986), as in monkeys (Grant et al. 1988), were responsive to salient unconditioned stimuli in many modalities. These LC responses occurred in close relation to orienting behaviors and were much smaller or absent when stimuli produced no overt behavioral response. These results are consistent with other studies in rat, which indicates that LC neurons are activated following novel stimuli (McQuade et al. 1999, Sara et al. 1995, Sara & Herve-Minvielle 1995). This set of findings can be understood in terms of the adaptive gain theory by noting that orienting and novelty responses are rudimentary decisions to act in a nonconditioned manner to highly salient events likely to have motivational significance (e.g., a loud cracking sound likely signifies the danger of a falling branch or an approaching predator). Thus, the associated LC activity may represent an evolutionarily hard-wired response that has facilitating effects on behavior analogous to the learned LC phasic responses in the 2AFC tasks studied in the laboratory (described above).

Finally, the adaptive gain theory also predicts that LC activity should increase (with a shift to the tonic mode) as unexpected and prolonged changes in reward contingencies occur. Results of a recent study by Dalley et al. (2001) are consistent with this prediction. Using microdialysis, they found that cortical release of NE (for which LC is the sole source) was increased in a sustained manner when an instrumental task was abruptly changed to noncontingent delivery of reward that was otherwise equal in overall amount. This is consistent with the predicted shift to LC tonic mode and associated increase in sustained NE release.

Human electrophysiological findings and LC function. One of the most robust findings from scalp recordings of event-related potentials (ERPs) in humans is the P3. This is a positive potential with a broad scalp distribution that typically occurs ~300 ms following task-relevant stimuli. The literature on the P3 is vast and well beyond our ability to review here. However, early in the study of this phenomenon investigators proposed that this potential may reflect neuromodulatory function (Desmedt & Debecker 1979) and implicated the LC-NE system in particular (Pineda et al. 1989). Elsewhere, we have reviewed the literature on the P3 with regard to this hypothesis and the adaptive gain theory of LC-NE function (Niewenhuis et al. 2005). There we suggest that the adaptive gain theory of LC-NE function can explain a wide range of seemingly disparate findings concerning the P3. Both the LC phasic response and the P3 appear shortly following target stimuli in oddball paradigms (e.g., the signal-detection task described above), and both depend on the motivational significance of, as well as attention paid to, eliciting stimuli. Furthermore, similar to the LC phasic response, P3 amplitude is greater for hits than for false alarms or misses in a signal-detection task; and factors that increase the duration of the decision process (e.g., stimulus degradation, reduced stimulus intensity) have generally been found to increase P3 latency and RT by a similar amount, paralleling the consistent relationship between LC phasic response and RT. Finally, the P3 is widely considered to be associated with completion of stimulus categorization, consistent with our hypothesis that the LC phasic response is driven by the outcome of task-related decision processes. These considerations have led us to hypothesize that the P3 reflects the phasic enhancement of gain in the cerebral cortex induced by LC-mediated release of NE (Niewenhuis et al. 2005). This hypothesis awaits direct testing in nonhuman primates using coordinated LC and P3 recordings in the same subjects. However, if validated, it will provide an important link between neurophysiological studies in nonhuman species and the study of LC function in humans, including the vast literature that already exists concerning the P3 both in normal and disordered cognition (Duncan 2003).

The adaptive gain hypothesis also suggests an explanation for the close relationship of the P3 to the attentional blink. The attentional blink refers to the failure to process the second of two sequentially presented targets when they are embedded in a series of rapidly presented stimuli (e.g., every 100 ms; Raymond et al. 1992). Models implementing the adaptive gain theory (S. Nieuwenhuis, M.S. Gilzenrat, B.D. Holmes & J.D. Cohen, manuscript under review) have successfully simulated this effect by attributing it to the empirically observed postactivation inhibition of LC neurons (Aghajanian et al. 1977, Andrade & Aghajanian 1984, Williams et al. 1984), which is presumed to render these cells unresponsive to the second target.

Limitations and future directions. Many elements in the adaptive gain theory require testing, further analysis, and additional elaboration. Nevertheless, even at this early stage of development, the theory makes several specific predictions that may stimulate new research. Below we consider some of the areas that call for additional work and that indicate predictions to be tested.

One limitation of the theory is that it is based heavily on animal studies. It will be important, therefore, to develop methods to measure or selectively manipulate LC activity in humans. As discussed in this review, pupilometry and scalp electrophysiology represent two promising avenues. Another is the use of neuroimaging methods, such as functional magnetic resonance imaging and positron emission tomography. Reports have begun to appear about the use of these methods to study brainstem nuclei, including the LC (e.g., Raizda & Poldrack 2003). Although the results of these studies require independent anatomic verification, methods for doing so are under development.

A key proposal that requires empirical confirmation is that LC phasic activation reflects decision outcome. One way to test this hypothesis is to analyze LC activity during a decision task simultaneously with activity of a brain region that has been strongly linked to decision processes in prior studies. Regions to consider in this regard include the lateral interoparietal cortex (LIP) and frontal eye fields (FEF), areas whose neurons meet many criteria for neural integrators of information in models of simple decision making (Gold & Shadlen 2001, Hanes & Schall 1996, Mazurek et al. 2003, Schall & Thompson 1999, Shadlen & Newsome 1996). A related component of our theory is that inputs from OFC and ACC drive LC phasic responses to decision outcome. This idea is suggested by the fact that decision-related cortical areas such as LIP and FEF do not strongly innervate the monkey LC (M. Iba, J. Rajkowski & G. Aston-Jones, unpublished observations) and that LC phasic responses are modulated by motivational significance. This idea can be tested by manipulating these prefrontal areas while recording LC neurons during decision tasks. The theory predicts, for example, that inactivation of ipsilateral OFC should diminish decision-driven LC phasic responses on positively motivated tasks.

The adaptive gain theory also states that phasic activation of LC facilitates behaviors (e.g., speeds correct responses) associated with task-related decisions. This idea can be tested by direct phasic electrical stimulation of LC during decision task performance and by combining this with modeling work that predicts the outcome of stimulation at different times in the task. Conversely, blockade of phasic LC activation (e.g., via local infusions of glutamate antagonists) should decrease the speed of decision-driven behavioral responses and increase the probability that no response will be elicited.

Finally, a central prediction of the adaptive gain theory is that the tonic LC mode facilitates transitions in behavioral focus so that tasks offering greater reward than the current one can be identified and pursued. Additional recordings of LC activity under conditions in which task-related utility is manipulated (such

2AFC: two-alternative forced choice

ACC: anterior cingulate cortex

DDM: drift diffusion model

LC: locus coeruleus

NE: norepinephrine

OFC: orbitofrontal cortex

REM: rapid eye movement

RT: response/reaction time

PFC: prefrontal cortex

as the diminishing utility task described earlier) will be important for testing and more precisely formulating this component of the theory. Moreover, manipulations of LC (e.g., tonic stimulation or inhibition of baseline excitatory inputs that drive the tonic mode), as well as manipulations of OFC and ACC that are proposed to regulate transitions between LC phasic and tonic modes, will be important for testing the causal role that these transitions play in behavioral changes.

SUMMARY AND CONCLUSIONS

We have reviewed findings indicating that during the awake state the LC-NE system has two distinguishable modes of activity (phasic and tonic) and have described an adaptive gain theory of the LC-NE system, which proposes specific functions for each of these modes. In the phasic mode, a burst of LC activity, driven by the outcome of task-related decision processes, produces a widespread but temporally specific release of NE, increasing gain of cortical processing units and facilitating task-appropriate behavior. In this mode, the event-locked nature of the LC phasic response acts as a temporal attentional filter, facilitating task-relevant processes relative to distracting events, thereby augmenting performance of the current task. The LC phasic response serves to optimize the trade-off between the flexibility of a complex, multilayered system and the efficiency of a single-layered decision-making mechanism. The theory further proposes that computations regarding decision and utility in OFC and ACC drive these LC phasic responses. We also propose that utility computations in the OFC and ACC produce the transitions between phasic and tonic modes in LC. Such transitions occur by the regulation of simple physiological variables within LC, such as electronic coupling, baseline drive, or both. High utility associated with performance of the current task favors the LC phasic mode. This mode is further augmented in response to momentary lapses in performance to exploit maximally the utility associated with the current task. However, persistent declines in utility drive a transition to the LC tonic mode. In the tonic mode, a lasting increase in baseline NE release augments responsivity of target neurons to a broader class of events, while a concomitant attenuation of the LC phasic response degrades processing of events related to the current task. This indiscriminate release of NE promotes disengagement from the current behavioral routine while facilitating the sampling of others that may provide greater utility. These different LC modes serve to optimize the trade-off between exploitation of stable sources of reward and exploration of other, potentially more remunerative, opportunities in a changing environment.

The role proposed here for the LC-NE system indicates that it would interact with many other brain circuits. In addition to its interaction with cortical systems, we have considered what this theory implies about the relationship between the functions of the LC-NE and DA systems, which may have particular relevance to psychiatric disorders. We propose that these systems work in synergy, the LC-NE system regulating the balance between exploitation and exploration—a factor that is central to reinforcement learning mechanisms of the sort thought to be implemented by the DA system. Improved understanding of these systems has substantial potential for understanding not only normal function, but also disturbances of function in a variety of clinical disorders. Several decades of research have made it clear that disturbances of NE and DA are involved in most of the major psychiatric illnesses, including schizophrenia, depression, and anxiety disorders. Some disorders have been associated more closely with NE and others more closely with DA. However, until recently, research has focused primarily on relatively simple hypotheses concerning static excesses or deficits of activity in these systems and has given virtually no consideration to interactions between them. The simplicity of these hypotheses has reflected a general lack of knowledge about the more complex dynamics that characterize the functioning of the NE and DA systems individually as well as their interaction. A more sophisticated understanding of these dynamics, and their relationship to cognition and behavior, promises to open up new avenues of inquiry. Realizing this potential, in turn, will afford greater understanding of how disruptions in the LC-NE and DA systems contribute to the complex patterns of behavior associated with psychiatric illness and how appropriate and effective interventions can be designed.

ACKNOWLEDGMENTS

The authors acknowledge the following individuals for the important contributions they have made to the development of the ideas and conduct of the research reviewed in this chapter: Rafal Bogacz, Eric Brown, Edwin Clayton, Mark Gilzenrat, Phil Holmes, Michiyo Iba, Sander Nieuwenhuis, Janusz Rajkowski, David Servan-Schreiber, and Marius Usher. The authors also express their profound appreciation to the National Institute of Mental Health for its enduring support of this work, which has included the following grants: P50 MH62196, R01 MH55309, and R01 MH58480. Work reported here was also supported by the Human Frontiers Science Program and the Air Force Office of Scientific Research. We also thank Mark Gilzenrat, Josh Gold, and Samuel McClure for valuable comments on the manuscript, and Stephanie Aston-Jones and Yan Zhu for assistance with illustrations.

LITERATURE CITED

Abercrombie ED, Keller RJ, Zigmond MJ. 1988. Characterization of hippocampal norepinephrine release as measured by microdialysis perfusion: pharmacological and behavioral studies. *Neuroscience* 27:897–904

Acheson AL, Zigmond MJ, Stricker EM. 1980. Compensatory increase in tyrosine hydroxylase activity in rat brain after intraventricular injections of 6-hydroxydopamine. *Science* 207:537–40

Aghajanian GK, Cedarbaum JM, Wang RY. 1977. Evidence for norepinephrine-mediated collateral inhibition of locus coeruleus neurons. *Brain Res.* 136:570–77

Alvarez VA, Chow CC, Van Bockstaele EJ, Williams JT. 2002. Frequency-dependent synchrony in locus coeruleus: role of electrotonic coupling. *Proc. Natl. Acad. Sci. USA* 99:4032–36

Amaral DG, Foss JA. 1975. Locus coeruleus lesions and learning. *Science* 188:377–78

Amit DJ. 1989. *Modeling Brain Function*. Cambridge, UK: Cambridge Univ. Press

Anderson JR. 1990. The adaptive character of thought. Hillsdale, NJ: Lawrence Erlbaum

Andrade R, Aghajanian GK. 1984. Intrinsic regulation of locus coeruleus neurons: electrophysiological evidence indicating a predominant role for autoinhibition. *Brain Res.* 310:401–6

Anlezark GM, Crow JT, Greenway AP. 1973. Impaired learning and decreased norepinephrine after bilateral locus coeruleus lesions. *Science* 181:682–84

Archer T, Jonsson G, Ross SB. 1984. A parametric study of the effects of the noradrenaline neurotoxin DSP4 on avoidance acquisition and noradrenaline neurones in the CNS of the rat. *Br. J. Pharmacol.* 82:249–57

Arnsten AF, Goldman RP. 1984. Selective prefrontal cortical projections to the region of the locus coeruleus and raphe nuclei in the rhesus monkey. *Brain Res.* 306:9–18

Arnsten AF, Steere JC, Hunt RD. 1996. The contribution of alpha 2-noradrenergic mechanisms of prefrontal cortical cognitive function. Potential significance for attention-deficit hyperactivity disorder. *Arch. Gen. Psychiatry* 53:448–55

Aston-Jones G. 1985. Behavioral functions of locus coeruleus derived from cellular attributes. *Physiol. Psych.* 13:118–26

Aston-Jones G. 2004. Locus coeruleus, A5 and A7 noradrenergic cell groups. In *The Rat Nervous System*, ed. G Paxinos, pp. 259–94. San Diego: Elsevier Academic

Aston-Jones G, Akaoka H, Charlety P, Chouvet G. 1991a. Serotonin selectively attenuates glutamate-evoked activation of locus coeruleus neurons in vivo. *J. Neurosci.* 11:760–69

Aston-Jones G, Bloom FE. 1981a. Activity of norepinephrine-containing locus coeruleus neurons in behaving rats anticipates fluctuations in the sleep-waking cycle. *J. Neurosci.* 1:876–86

Aston-Jones G, Bloom FE. 1981b. Norepinephrine-containing locus coeruleus neurons in behaving rats exhibit pronounced responses to non-noxious environmental stimuli. *J. Neurosci.* 1:887–900

Aston-Jones G, Chiang C, Alexinsky T. 1991b. Discharge of noradrenergic locus coeruleus neurons in behaving rats and monkeys suggests a role in vigilance. *Prog. Brain Res.* 88:501–20

Aston-Jones G, Foote SL, Bloom FE. 1984. Anatomy and physiology of locus coeruleus neurons: functional implications. In *Norepinephrine: Frontiers of Clinical Neuroscience*, ed. M Ziegler, CR Lake, pp. 92–116. Baltimore: Williams and Wilkins. Vol. 2

Aston-Jones G, Foote SL, Segal M. 1985. Impulse conduction properties of noradrenergic locus coeruleus axons projecting to monkey cerebrocortex. *Neuroscience* 15:765–77

Aston-Jones G, Gonzalez MM, Doran SM. 2005. Role of the locus coeruleus-norepinephrine system in arousal and circadian regulation of the sleep-waking cycle. In *Norepinephrine: Neurobiology and Therapeutics for the 21st Century*, ed. GA Ordway, M Schwartz, A Frazer. Cambridge UK: Cambridge Univ. Press. In press

Aston-Jones G, Rajkowski J, Cohen J. 1999. Role of locus coeruleus in attention and behavioral flexibility. *Biol. Psychiatry* 46:1309–20

Aston-Jones G, Rajkowski J, Kubiak P. 1997. Conditioned responses of monkey locus coeruleus neurons anticipate acquisition of discriminative behavior in a vigilance task. *Neuroscience* 80:697–715

Aston-Jones G, Rajkowski J, Kubiak P, Alexinsky T. 1994. Locus coeruleus neurons in the monkey are selectively activated by attended stimuli in a vigilance task. *J. Neurosci.* 14:4467–80

Aston-Jones G, Rajkowski J, Kubiak P, Ivanova S, Usher M, Cohen J. 1998. Neuromodulation and cognitive performance: recent studies of noradrenergic locus coeruleus neurons in behaving monkeys. In *Catecholamines: Bridging Basic Science with Clinical Medicine*, ed. D Goldstein, G Eisenhofer, R McCarty, pp. 755–59. New York: Academic

Aston-Jones G, Rajkowski J, Kubiak P, Valentino R, Shipley M. 1996. Role of the locus coeruleus in emotional activation. *Prog. Brain Res.* 107:379–402

Aston-Jones G, Rajkowski J, Lu W, Zhu Y, Cohen JD, Morecraft RJ. 2002. Prominent projections from the orbital prefrontal cortex to the locus coeruleus in monkey. *Soc. Neurosci. Abstr.* 28:86–9

Barch DM, Braver TS, Nystrom LE, Forman SD, Noll DC, Cohen JD. 1997. Dissociating working memory from task difficulty in human prefrontal cortex. *Neuropsychologia* 35:1373–80

Barlow HB. 1981. Critical limiting factors in the design of the eye and visual cortex. *Proc. R. Soc. London B* 212:1–34

Barnard GA. 1946. Sequential tests in industrial statistics. *J. R. Stat. Soc. Suppl.* 8:1–26

Baylis LL, Rolls ET, Baylis GC. 1995. Afferent connections of the caudolateral orbitofrontal cortex taste area of the primate. *Neuroscience* 64:801–12

Beatty J. 1982a. Phasic not tonic pupillary responses vary with auditory vigilance performance. *Psychophysiology* 19:167–72

Beatty J. 1982b. Task-evoked pupillary responses, processing load, and structure of processing resources. *Psychol. Bull.* 91:276–92

Beaudet A, Descarries L. 1978. The monoamine innervation of rat cerebral cortex: synaptic and nonsynaptic axons terminals. *Neuroscience* 3:851–60

Berridge CW, Waterhouse BD. 2003. The locus coeruleus-noradrenergic system: modulation of behavioral state and state-dependent cognitive processes. *Brain Res. Brain Res. Rev.* 42:33–84

Bialek W, Owen WG. 1990. Temporal filtering in retinal bipolar cells: elements of an optimal computation? *Biophys. J.* 58:1227–33

Bohil CJ, Maddox WT. 2003. On the generality of optimal versus objective classifer feedback effects on decision criterion learning in perceptual categorization. *Mem. Cognit.* 31(2):181–98

Botvinick MM, Braver TS, Barch DM, Carter CS, Cohen JD. 2001. Conflict monitoring and cognitive control. *Psychol. Rev.* 108(3):624–52

Botvinick MM, Cohen JD, Carter CS. 2004. Conflict monitoring and anterior cingulate cortex: an update. *Trends Cogn. Sci.* 8(12):539–46

Botvinick MM, Nystrom LE, Fissell K, Carter CS, Cohen JD. 1999. Conflict monitoring versus selection-for-action in anterior cingulate cortex. *Nature* 402:179–81

Bouret S, Sara SJ. 2004. Reward expectation, orientation of attention and locus coeruleus-medial frontal cortex interplay during learning. *Eur. J. Neurosci.* 20:791–802

Breiter HC, Aharon I, Kahneman D, Dale A, Shizgal P. 2001. Functional imaging of neural responses to expectancy and experience of monetary gains and losses. *Neuron* 30:619–39

Breiter HC, Gollub RL, Weisskoff RM, Kennedy DN, Makris N, et al. 1997. Acute effects of cocaine on human brain activity and emotion. *Neuron* 19(3):591–611

Brown E, Moehlis J, Holmes P, Clayton E, Rajkowski J, Aston-Jones G. 2004. The influence of spike rate and stimulus duration on noradrenergic neurons. *J. Comput. Neurosci.* 17:5–21

Brun P, Suaud-Chagny MF, Gonon F, Buda M. 1993. In vivo noradrenaline release evoked in the anteroventral thalamic nucleus by locus coeruleus activation: an electrochemical study. *Neuroscience* 52:961–72

Burnett DM, Bowyer JF, Masserano JM, Zahniser NR. 1990. Effect of aging on alpha-1 adrenergic stimulation of phosphoinositide hydrolysis in various regions of rat brain. *J. Pharmacol. Exp. Ther.* 255:1265–70

Busemeyer JR, Myung IJ. 1992. An adaptive approach to human decision making: learning theory, decision theory, and human performance. *J. Exp. Psychol. Gen.* 121(2):177–94

Carboni E, Silvagni A. 2004. Dopamine reuptake by norepinephrine neurons: exception or rule? *Crit. Rev. Neurobiol.* 16:121–28

Carli M, Robbins TW, Evenden JL, Everitt BJ. 1983. Effects of lesions to ascending noradrenergic neurones on performance of a 5-choice serial reaction task in rats; implications for theories of dorsal noradrenergic bundle function based on selective attention and arousal. *Behav. Brain Res.* 9:361–80

Carmichael ST, Clugnet MC, Price JL. 1994. Central olfactory connections in the macaque monkey. *J. Comp. Neurol.* 346:403–34

Carmichael ST, Price JL. 1995a. Limbic connections of the orbital and medial prefrontal cortex in macaque monkeys. *J. Comp. Neurol.* 363:615–41

Carmichael ST, Price JL. 1995b. Sensory and premotor connections of the orbital and medial prefrontal cortex of macaque monkeys. *J. Comp. Neurol.* 363:642–64

Carter CS, Braver TS, Barch DM, Botvinick MM, Noll DC, Cohen JD. 1998. Anterior cingulate cortex, error detection and the on-line monitoring of performance. *Science* 280:747–49

Chiodo LA, Acheson AL, Zigmond MJ, Stricker EM. 1983. Subtotal destruction of central noradrenergic projections increases the firing rate of locus coeruleus cells. *Brain Res.* 264:123–26

Christie MJ, Jelinek HF. 1993. Dye-coupling among neurons of the rat locus coeruleus during postnatal development. *Neuroscience* 56:129–37

Christie MJ, Williams JT, North RA. 1989. Electrical coupling synchronizes subthreshold activity in locus coeruleus neurons in vitro from neonatal rats. *J. Neurosci.* 9:3584–89

Cirelli C, Tononi G. 2004. Locus coeruleus control of state-dependent gene expression. *J. Neurosci.* 24:5410–19

Clayton EC, Rajkowski J, Cohen JD, Aston-Jones G. 2004. Phasic activation of monkey locus coeruleus neurons by simple decisions in a forced choice task. *J. Neurosci.* 24:9914–20

Cohen JD, Aston-Jones G, Gilzenrat MS. 2004. A systems-level perspective on attention and cognitive control: guided activation, adaptive gating, conflict monitoring, and exploitation vs. exploration. In *Cognitive Neuroscience of Attention*, ed. MI Posner, pp. 71–90. New York: Guilford

Cohen JD, Dunbar K, McClelland JL. 1990. On the control of automatic processes: a parallel distributed processing model of the Stroop effect. *Psychol. Rev.* 97(3):332–61

Cohen JD, Servan-Schreiber D, McClelland JL. 1992. A parallel distributed processing approach to automaticity. *Am. J. Psychol.* 105:239–69

Cole BJ, Robbins TW. 1992. Forebrain norepinephrine: role in controlled information processing in the rat. *Neuropsychopharmacology* 7:129–41

Coull JT. 1994. Pharmacological manipulations of the alpha 2-noradrenergic system. Effects on cognition. *Drugs Aging* 5:116–26

Coull JT, Nobre AC. 1998. Where and when to pay attention: the neural systems for directing attention to spatial locations and to time intervals as revealed by both PET and fMRI. *J. Neurosci.* 18:7426–35

Coull JT, Nobre AC, Frith CD. 2001. The noradrenergic alpha2 agonist clonidine modulates behavioural and neuroanatomical correlates of human attentional orienting and alerting. *Cereb. Cortex* 11:73–84

Critchley HD, Rolls ET. 1996. Hunger and satiety modify the responses of olfactory and visual neurons in the primate orbitofrontal cortex. *J. Neurophysiol.* 75:1673–86

Dahlstrom A, Fuxe K. 1964. Evidence for the existence of monoamine-containing neurons in the central nervous system. I. Demonstration of monoamines in the cell bodies of brain stem neurons. *Acta Physiol. Scand.* 62:5–55

Dalley JW, McGaughy J, O'Connell MT, Cardinal RN, Levita L, Robbins TW. 2001. Distinct changes in cortical acetylcholine and noradrenaline efflux during contingent and noncontingent performance of a visual attentional task. *J. Neurosci.* 21:4908–14

Desmedt JE, Debecker J. 1979. Wave form and neural mechanism of the decision P350 elicited without pre-stimulus CNV or readiness potential in random sequences of near-threshold auditory clicks and finger stimuli. *Electroencephalogr. Clin. Neurophysiol.* 47:648–70

Devilbiss DM, Waterhouse BD. 2000. Norepinephrine exhibits two distinct profiles of action on sensory cortical neuron responses to excitatory synaptic stimuli. *Synapse* 37:273–82

Devilbiss DM, Waterhouse BD. 2004. The effects of tonic locus coeruleus output on sensory-evoked responses of ventral posterior medial thalamic and barrel field cortical neurons in the awake rat. *J. Neurosci.* 24:10773–85

Devinsky O, Morrell MJ, Vogt BA. 1995. Contributions of anterior cingulate cortex to behaviour. *Brain* 118:279–306

Dooley DJ, Jones GH, Robbins TW. 1987. Noradrenaline- and time-dependent changes in neocortical alpha 2- and beta 1-adrenoceptor binding in dorsal noradrenergic bundle-lesioned rats. *Brain Res.* 420:152–56

Duncan CC. 2003. Brain potentials in normal and disordered attention: findings in search of a theory. Presidential address presented at the *Annu. Meet. Soc. Psychophysiol. Res.*, Chicago, IL

Duncan J. 1986. Disorganisation of behaviour after frontal lobe damage. *Cogn. Neuropsychol.* 3:271–90

Duncan J, Owen AM. 2000. Common regions of the human frontal lobe recruited by diverse cognitive demands. *Trends Neurosci.* 23:475–83

Egan TM, Henderson G, North RA, Williams JT. 1983. Noradrenaline-mediated synaptic inhibition in rat locus coeruleus neurones. *J. Physiol. (London)* 345:477–88

Eisenberger NI, Lieberman MD, Williams KD. 2003. Does rejection hurt? An FMRI study of social exclusion. *Science* 302:290–92

Ennis M, Aston-Jones G. 1986. Evidence for self- and neighbor-mediated postactivation inhibition of locus coeruleus neurons. *Brain Res.* 374:299–305

Eriksdotter Jonhagen M, Hoffer B, Luthman J. 1995. Alterations in alpha-adrenoceptors in aging intraocular hippocampal grafts. *Neurobiol. Aging* 16:633–38

Falkenstein M, Hohnsbein J, Hoorman J, Blanke L. 1991. Effects of crossmodal divided attention on late ERP components: II. Error processing in choice reaction tasks. *Electroencephalogr. Clin. Neurophysiol.* 78:447–55

Foote SL, Aston-Jones G, Bloom FE. 1980. Impulse activity of locus coeruleus neurons in awake rats and monkeys is a function of sensory stimulation and arousal. *Proc. Natl. Acad. Sci. USA* 77:3033–37

Foote SL, Bloom FE, Aston-Jones G. 1983. Nucleus locus coeruleus: new evidence of anatomical and physiological specificity. *Physiol. Rev.* 63:844–914

Foote SL, Freedman R, Oliver AP. 1975. Effects of putative neurotransmitters on neuronal activity in monkey auditory cortex. *Brain Res.* 86:229–42

Foote SL, Morrison JH. 1987. Extrathalamic modulation of cortical function. *Annu. Rev. Neurosci.* 10:67–95

Fritschy JM, Grzanna R. 1992. Restoration of ascending noradrenergic projections by residual locus coeruleus neurons: compensatory response to neurotoxin-induced cell death in the adult rat brain. *J. Comp. Neurol.* 321:421–41

Gallistel CR, Mark TA, King AP, Latham PE. 2001. The rat approximates an ideal detector of changes in rates of reward: implications for the law of effect. *J. Exp. Psychol. Anim. Behav.* 27:354–72

Gehring WJ, Goss B, Coles MGH, Meyer DE, Donchin E. 1993. A neural system for error detection and compensation. *Psychol. Sci.* 4:385–90

Gilzenrat MS, Cohen JD, Rajkowski J, Aston-Jones G. 2003. Pupil dynamics predict changes in task engagement mediated by locus coeruleus. *Soc. Neurosci. Abstr.* No. 515.19

Gilzenrat MS, Brown ET, Aston-Jones G, Cohen JD. 2004. Locus coeruleus, adaptive gain, and the optimization of decision tasks. *Soc. Neurosci. Abstr.* No. 899.6

Gold JI, Shadlen MN. 2000. Representation of a perceptual decision in developing oculomotor commands. *Nature* 404:390–94

Gold JI, Shadlen MN. 2001. Neural computations that underlie decisions about sensory stimuli. *Trends Cogn. Sci.* 5:10–16

Gold JI, Shadlen MN. 2002. Banburismus and the brain: decoding the relationship between sensory stimuli, decisions, and reward. *Neuron* 36:299–308

Grant SJ, Aston-Jones G, Redmond DEJ. 1988. Responses of primate locus coeruleus neurons to simple and complex sensory stimuli. *Brain Res. Bull.* 21:401–10

Gratton G, Coles MGH, Donchin E. 1992. Optimizing the use of information: strategic control of activation and responses. *J. Exp. Psychol. Gen.* 4:480–506

Hanes DP, Schall JD. 1996. Neural control of voluntary movement initiation. *Science* 18:427–30

Harik SI, Duckrow RB, LaManna JC, Rosenthal M, Sharma VK, Banerjee SP. 1981. Cerebral compensation for chronic noradrenergic denervation induced by locus ceruleus lesion: recovery of receptor binding, isoproterenol-induced adenylate cyclase activity, and oxidative metabolism. *J. Neurosci.* 1:641–49

Haring JH, Miller GD, Davis JN. 1986. Changes in the noradrenergic innervation of the area dentata after axotomy of coeruleohippocampal projections or unilateral lesion of the locus coeruleus. *Brain Res.* 368:233–38

Harley C. 1991. Noradrenergic and locus coeruleus modulation of the perforant path-evoked potential in rat dentate gyrus supports a role for the locus coeruleus in attentional and memorial processes. *Prog. Brain Res.* 88:307–21

Harris GC, Fitzgerald RD. 1991. Locus coeruleus involvement in the learning of classically conditioned bradycardia. *J. Neurosci.* 11:2314–20

Herrnstein RJ. 1961. Stereotypy and intermittent reinforcement. *Science* 133:2067–69

Herrnstein RJ. 1997. *The Matching Law*, ed. H Rachlin, DI Laibson. Cambridge, MA: Harvard Univ. Press

Herve-Minvielle A, Sara SJ. 1995. Rapid habituation of auditory responses of locus coeruleus cells in anaesthetized and awake rats. *Neuroreport* 6:1363–68

Hobson JA, McCarley RW, Wyzinski PW. 1975. Sleep cycle oscillation: reciprocal discharge by two brainstem neuronal groups. *Science* 189:55–58

Hollerman JR, Tremblay L, Schultz W. 2000. Involvement of basal ganglia and orbitofrontal cortex in goal-directed behavior. *Prog. Brain Res.* 126:193–215

Holroyd CB, Larsen JT, Cohen JD. 2004a. Context dependence of the event-related brain potential associated with reward and punishment. *Psychophysiology* 41(2):245–53

Holroyd CB, Nieuwenhuis S, Yeung N, Cohen JD. 2003. Errors in reward prediction are reflected in the event-related brain potential. *Neuroreport* 14:2481–84

Holroyd CB, Nieuwenhuis S, Yeung N, Nystrom L, Mars R, et al. 2004b. Dorsal anterior cingulate cortex is sensitive to internal and external sources of error information: a functional magnetic resonance imaging study. *Nat. Neurosci.* 7:497–98

Hurley LM, Devilbiss DM, Waterhouse BD. 2004. A matter of focus: monoaminergic modulation of stimulus coding in mammalian sensory networks. *Curr. Opin. Neurobiol.* 14:488–95

Ishimatsu M, Williams JT. 1996. Synchronous activity in locus coeruleus results from dendritic interactions in pericoerulear regions. *J. Neurosci.* 16:5196–204

Ito S, Stuphorn V, Brown JW, Schall JD. 2003. Performance monitoring by the anterior cingulate cortex during saccade countermanding. *Science* 302:120–22

Ivanova S, Rajkowski J, Silakov V, Watanabe T, Aston-Jones G. 1997. Local chemomanipulations of locus coeruleus (LC) activity in monkeys alter cortical event-related potentials (ERPs) and task performance. *Soc. Neurosci. Abstr.* 23:1587

Iversen LL, Rossor MN, Reynolds GP, Hills R, Roth M, et al. 1983. Loss of pigmented dopamine-beta-hydroxylase positive cells from locus coeruleus in senile dementia of Alzheimer's type. *Neurosci. Lett.* 39:95–100

Jodo E, Aston-Jones G. 1997. Activation of locus coeruleus by prefrontal cortex is mediated by excitatory amino acid inputs. *Brain Res.* 768:327–32

Jodo E, Chiang C, Aston-Jones G. 1998. Potent excitatory influence of prefrontal cortex activity on noradrenergic locus coeruleus neurons. *Neuroscience* 83:63–80

Jones AD, Cho RY, Nystrom LE, Cohen JD, Braver TS. 2002. A computational model of anterior cingulate function in speeded response tasks: effects of frequency, sequence, and conflict. *Cogn. Affect. Behav. Neurosci.* 2:300–17

Jones BE. 1991. The role of noradrenergic locus coeruleus neurons and neighboring cholinergic neurons of the pontomesencephalic tegmentum in sleep-wake states. *Prog. Brain Res.* 88:533–43

Jouvet M. 1969. Biogenic amines and the states of sleep. *Science* 163:32–41

Kaelbling LP, Littman ML, Moore AW. 1996. Reinforcement learning: a survey. *J. Artif. Intell. Res.* 4:237–85

Kahneman D, Tversky A. 1984. Choices, values, and frames. *Am. Psychol.* 39:341–50

Kiehl KA, Liddle PF, Hopfinger JB. 2000. Error processing and the rostral anterior cingulate: an event-related fMRI study. *Psychophysiology* 37:216–23

Koob GF, Kelley AE, Mason ST. 1978. Locus coeruleus lesions: learning and extinction. *Physiol. Behav.* 20:709–16

Knutson B, Westdorp A, Kaiser E, Hommer D. 2000. FMRI visualization of brain activity during a monetary incentive delay task. *Neuroimage* 12(1):20–27

Kubiak P, Rajkowski J, Aston-Jones G. 1992. Behavioral performance and sensory responsiveness of LC neurons in a vigilance task varies with tonic LC discharge rate. *Soc. Neurosci. Abstr.* 18:538

Laming DRJ. 1968. *Information Theory of Choice-Reaction Times*. New York: Academic

Laming DRJ. 1979. Choice reaction performance following an error. *Acta-Psychologica* 43:199–224

Levin BE, Battisti WP, Murray M. 1985. Axonal transport of beta-receptors during the response to axonal injury and repair in locus coeruleus neurons. *Brain Res.* 359:215–23

Loewenstein G, Thaler R. 1989. Anomalies: intertemporal choice. *J. Econ. Perspect.* 3:181–93

Martin P, Ohno M, Southerland SB, Mailman RB, Suzuki K. 1994. Heterotypic sprouting of serotonergic forebrain fibers in the brindled mottled mutant mouse. *Brain Res. Dev. Brain Res.* 77:215–25

Mazurek ME, Roitman JD, Ditterich J, Shadlen MN. 2003. A role for neural integrators in perceptual decision making. *Cereb. Cortex* 13:1257–69

McClure SM, Laibson DI, Loewenstein G, Cohen JD. 2004. Separate neural systems value immediate and delayed monetary rewards. *Science* 306:503–7

McQuade R, Creton D, Stanford SC. 1999. Effect of novel environmental stimuli on rat behaviour and central noradrenaline function measured by in vivo microdialysis. *Psychopharmacology (Berlin)* 145:393–400

Mesulam M-M. 1981. A cortical network for directed attention and unilateral neglect. *Ann. Neurol.* 10:309–25

Miller EK, Cohen JD. 2001. An integrative theory of prefrontal cortex function. *Annu. Rev. Neurosci.* 24:167–202

Miltner WHR, Braun CH, Coles MGH. 1997. Event-related potentials following incorrect feedback in a time-estimation task: evidence for a 'generic' neural system for error detection. *J. Cogn. Neurosci.* 9(6):788–98

Montague PR, Dayan P, Sejnowski TJ. 1996. A framework for mesencephalic dopamine systems based on predictive Hebbian learning. *J. Neurosci.* 16:1936–47

Montague PR, Hyman SE, Cohen JD. 2004. Computational roles for dopamine in behavioural control. *Nature* 431(7010):760–67

Moore RY, Bloom FE. 1979. Central catecholamine neuron systems: anatomy and physiology of the norepinephrine and epinephrine systems. *Annu. Rev. Neurosci.* 2:113–68

Morecraft RJ, Geula C, Mesulam MM. 1992. Cytoarchitecture and neural afferents of orbitofrontal cortex in the brain of the monkey. *J. Comp. Neurol.* 323:341–58

Morrison JH, Foote SL, O'Connor D, Bloom FE. 1982. Laminar, tangential and regional organization of the noradrenergic innervation of the monkey cortex: dopamine-beta-hydroxylase immunohistochemistry. *Brain Res. Bull.* 9:309–19

Mountcastle VB, LaMotte RH, Carli G. 1972. Detection thresholds for stimuli in humans and monkeys: comparison with threshold events in mechanoreceptive afferent nerve fibers innervating the monkey hand. *J. Neurophysiol.* 35:122–36

Mozer MC, Colagrosso MD, Huber DE. 2002. A rational analysis of cognitive control in a speeded discrimination task. In *Advances in Neural Information Processing Systems*, ed. T Dietterich, S Becker, Z Ghahramani, pp. 51–57. Cambridge, MA: MIT Press. Vol. XI

Nicola SM, Surmeier J, Malenka RC. 2000. Dopaminergic modulation of neuronal excitability in the striatum and nucleus accumbens. *Annu. Rev. Neurosci.* 23:185–215

Niewenhuis S, Aston-Jones G, Cohen JD. 2005. Decision making, the P3, and the locus coeruleus-norepinephrine system. *Psych. Bull.* In press

O'Doherty JP, Deichmann R, Critchley HD, Dolan RJ. 2002. Neural responses during anticipation of a primary taste reward. *Neuron* 33(5):815–26

Oke AF, Adams RN. 1978. Selective attention dysfunctions in adult rats neonatally treated with 6-hydoxydopamine. *Pharmacol. Biochem. Behav.* 9:429–32

Olschowka JA, O' Donohue T, Jacobowitz DM. 1981. The distribution of bovine pancreatic polypeptide-like immunoreactive neurons in rat brain. *Peptides* 2:309–31

Ongur D, Price JL. 2000. The organization of networks within the orbital and medial prefrontal cortex of rats, monkeys and humans. *Cereb. Cortex* 10:206–19

Papadopoulos GC, Parnavelas JG, Buijs R. 1987. Monoaminergic fibers form conventional synapses in the cerebral cortex. *Neurosci. Lett.* 76:275–79

Papadopoulos GC, Parnavelas JG, Buijs RM. 1989. Light and electron microscopic immunocytochemical analysis of the noradrenaline innervation of the rat visual cortex. *J. Neurocytol.* 18:1–10

Paus T, Koski L, Caramanos Z, Westbury C. 1998. Regional differences in the effects of task difficulty and motor output on blood flow response in the human anterior cingulate cortex: a review of 107 PET activation studies. *Neuroreport* 9(9):R37–47

Paxinos G, Huang X-F, Toga AW. 2000. *The Rhesus Monkey Brain in Stereotaxic Coordinates*. San Diego: Academic

Peyron R, Laurent B, Garcia-Larrea L. 2000. Functional imaging of brain responses to pain. A review and meta-analysis. *Neurophysiol. Clin.–Clin. Neurophysiol.* 30:263–88

Pineda JA, Foote SL, Neville HJ. 1989. Effects of locus coeruleus lesions on auditory, long-latency, event-related potentials in monkey. *J. Neurosci.* 9:81–93

Pisa M, Fibiger HC. 1983. Evidence against a role of the rat's dorsal noradrenergic bundle in selective attention and place memory. *Brain Res.* 272:319–29

Pisa M, Martin-Iverson MT, Fibiger HC. 1988. On the role of the dorsal noradrenergic bundle in learning and habituation to novelty. *Pharmacol. Biochem. Behav.* 30:835–45

Rabbitt PMA. 1966. Errors and error-correction in choice-response tasks. *J. Exp. Psychol.* 71:264–72

Raizada RD, Poldrack RA. 2003. Difficult, unpredictable trials coactivate noradrenergic and frontal attentional systems. *Soc. Neurosci. Abstr.* 401:4

Rajkowski J, Kubiak P, Aston-Jones G. 1993. Correlations between locus coeruleus (LC) neural activity, pupil diameter and behavior in monkey support a role of LC in attention. *Soc. Neurosc. Abstr.* 19:974

Rajkowski J, Kubiak P, Ivanova S, Aston-Jones G. 1998. State related activity and reactivity of locus coeruleus neurons in behaving monkeys. In *Catecholamines: Bridging Basic Science with Clinical Medicine*, ed. D Goldstein, G Eisenhofer, R McCarty, pp. 740–44. New York: Academic

Rajkowski J, Lu W, Zhu Y, Cohen J, Aston-Jones G. 2000. Prominent projections from the anterior cingulate cortex to the locus coeruleus in Rhesus monkey. *Soc. Neurosci. Abstr.* 26:838–15

Rajkowski J, Majczynski H, Clayton E, Aston-Jones G. 2004. Activation of monkey locus coeruleus neurons varies with difficulty and behavioral performance in a target detection task. *J. Neurophysiol.* 92:361–71

Rasmussen K, Morilak DA, Jacobs BL. 1986. Single unit activity of locus coeruleus neurons in the freely moving cat. I. During naturalistic behaviors and in response to simple and complex stimuli. *Brain Res.* 371:324–34

Ratcliff R. 1978. A theory of memory retrieval. *Psychol. Rev.* 85:59–108

Ratcliff R, Cherian A, Segraves MJ. 2003. A comparison of macaque behavior and superior colliculus neuronal activity to predictions from models of two choice decisions. *J. Neurophysiol.* 90:1392–407

Ratcliff R. Smith PL. 2004. A comparison of sequential sampling models for two-choice reaction time. *Psychol. Rev.* 111:333–67

Raymond JE, Shapiro KL, Arnell KM. 1992. Temporary suppression of visual processing in an RSVP task: an attentional blink? *J. Exp. Psychol. Hum. Percept. Perform.* 18:849–60

Reddi B. 2001. Decision making: two stages of neural judgement. *Curr. Biol.* 11(15):R603–6

Richer F, Beatty J. 1987. Contrasting effects of response uncertainty on the task-evoked pupillary response and reaction time. *Psychophysiology* 24:258–62

Robbins TW, Everitt BJ. 1995. Central norepinephrine neurons and behavior. In *Psychopharmacology: The Fourth Generation of Progress*, ed. DJ Kupfer, FE Bloom, pp. 363–72. New York: Raven

Roberts DCS, Price MTC, Fibiger HC. 1976. The dorsal tegmental noradrenergic bundle: an analysis of its role in maze learning. *J. Comp. Physiol. Psychol.* 90:363–72

Robertson IH, Mattingley JB, Rorden C, Driver J. 1998. Phasic alerting of neglect patients overcomes their spatial deficit in visual awareness. *Nature* 395:169–72

Robinson TE, Berridge KC. 1993. The neural basis of drug craving: an incentive-sensitization theory of addiction. *Brain Res. Rev.* 18:247–91

Roesch MR, Olson CR. 2004. Neuronal activity related to reward value and motivation in primate frontal cortex. *Science* 304:307–10

Rogawski MA, Aghajanian GK. 1982. Activation of lateral geniculate neurons by locus coeruleus or dorsal noradrenergic bundle stimulation: selective blockade by the alpha 1-adrenoceptor antagonist prazosin. *Brain Res.* 250:31–39

Rolls ET. 2004. The functions of the orbitofrontal cortex. *Brain Cogn.* 55:11–29

Rolls ET, Critchley HD, Mason R, Wakeman EA. 1996. Orbitofrontal cortex neurons: role in olfactory and visual association learning. *J. Neurophysiol.* 75:1970–81

Rolls ET, Sienkiewicz ZJ, Yaxley S. 1989. Hunger modulates the responses to gustatory stimuli of single neurons in the caudolateral orbitofrontal cortex of the macaque monkey. *Eur. J. Neurosci.* 1:53–60

Rolls ET, Yaxley S, Sienkiewicz ZJ. 1990. Gustatory responses of single neurons in the caudolateral orbitofrontal cortex of the macaque monkey. *J. Neurophysiol.* 64:1055–66

Saper CB, Chou TC, Scammell TE. 2001. The sleep switch: hypothalamic control of sleep and wakefulness. *Trends Neurosci.* 24:726–31

Sara SJ, Devauges V. 1989. Idazoxan, an alpha-2 antagonist, facilitates memory retrieval in the rat. *Behav. Neural Biol.* 51:401–11

Sara SJ, Dyon-Laurent C, Herve A. 1995. Novelty seeking behavior in the rat is dependent upon the integrity of the noradrenergic system. *Brain Res. Cogn. Brain Res.* 2:181–87

Sara SJ, Herve-Minvielle A. 1995. Inhibitory influence of frontal cortex on locus coeruleus neurons. *Proc. Natl. Acad. Sci. USA* 92:6032–36

Schall J. 2003. Neural correlates of decision processes: neural and mental chronometry. *Curr. Opin. Neurobiol.* 13(2):182–86

Schall JD, Thompson KG. 1999. Neural selection and control of visually guided eye movements. *Annu. Rev. Neurosci.* 22:241–59

Schultz W, Tremblay L, Hollerman JR. 2000. Reward processing in primate orbitofrontal cortex and basal ganglia. *Cereb. Cortex* 10:272–84

Segal M, Bloom FE. 1976. The action of norepinephrine in the rat hippocampus. IV. The effects of locus coeruleus stimulation on evoked hippocampal unit activity. *Brain Res.* 107:513–25

Seguela P, Watkins KC, Geffard M, Descarries L. 1990. Noradrenaline axon terminals in adult rat neocortex: an immunocytochemical analysis in serial thin sections. *Neuroscience* 35:249–64

Selden NR, Cole BJ, Everitt BJ, Robbins TW. 1990a. Damage to coeruleo-cortical noradrenergic projections impairs locally cued but enhances spatially cued water maze acquisition. *Behav. Brain Res.* 39:29–51

Selden NR, Robbins TW, Everitt BJ. 1990b. Enhanced behavioral conditioning to context and impaired behavioral and neuroendocrine responses to conditioned stimuli following coeruleocortical noradrenergic lesions: support for an attentional hypothesis of central noradrenergic function. *J. Neurosci.* 10:531–39

Servan-Schreiber D, Printz H, Cohen JD. 1990. A network model of catecholamine effects gain signal to noise ratio and behavior. *Science* 249:892–95

Shadlen MN, Newsome WT. 1996. Motion perception: seeing and deciding. *Proc. Natl. Acad. Sci. USA* 93:628–33

Shadlen MN, Newsome WT. 2001. Neural basis of a perceptual decision in the parietal cortex (area LIP) of the rhesus monkey. *J. Neurophysiol.* 86:1916–36

Shallice T. 1988. *From Neuropsychology to Mental Structure*. Cambridge, UK: Cambridge Univ. Press

Small DM, Zatorre RJ, Dagher A, Evans AC, Jones-Gotman M. 2001. Changes in brain activity related to eating chocolate: from pleasure to aversion. *Brain* 124(Pt. 9):1720–33

Steriade M, McCormick DA, Sejnowski TJ. 1993. Thalamocortical oscillations in the sleeping and aroused brain. *Science* 262:679–85

Stone M. 1960. Models for choice reaction time. *Psychometrika* 25:251–60

Sugrue LP, Corrado GC, Newsome WT. 2004. Neural correlates of value in orbitofrontal cortex of the rhesus monkey. *Soc. Neurosci. Abstr.* 30:671–78

Sutton RS, Barto AG. 1981. Toward a modern theory of adaptive networks: expectation and prediction. *Psychol. Rev.* 88:135–70

Tejani-Butt SM, Ordway GA. 1992. Effect of age on [3H]nisoxetine binding to uptake sites for norepinephrine in the locus coeruleus of humans. *Brain Res.* 583:312–15

Thorpe SJ, Rolls ET, Maddison S. 1983. The orbitofrontal cortex: neuronal activity in the behaving monkey. *Exp. Brain Res.* 49:93–115

Thut G, Schultz W, Roelcke U, Nienhusmeier M, Missimer J, et al. 1997. Activation of the human brain by monetary reward. *Neuroreport* 8:1225–28

Travagli RA, Dunwiddie TV, Williams JT. 1995. Opioid inhibition in locus coeruleus. *J. Neurophysiol.* 74:519–28

Travagli RA, Wessendorf M, Williams JT. 1996. Dendritic arbor of locus coeruleus neurons contributes to opioid inhibition. *J. Neurophysiol.* 75:2029–35

Tremblay L, Schultz W. 1999. Relative reward preference in primate orbitofrontal cortex. *Nature* 398:704–8

Ullsperger M, von Cramon DY. 2001. Subprocesses of performance monitoring: a dissociation of error processing and response competition revealed by event-related fMRI and ERPs. *NeuroImage* 14:1387–401

U'Prichard DC, Reisine TD, Mason ST, Fibiger HC, Yamamura HI. 1980. Modulation of rat brain alpha- and beta-adrenergic receptor populations by lesion of the dorsal noradrenergic bundle. *Brain Res.* 187:143–54

Usher M, Cohen JD, Servan-Schreiber D, Rajkowski J, Aston-Jones G. 1999. The role of locus coeruleus in the regulation of cognitive performance. *Science* 283:549–54

Usher M, McClelland JL. 2001. On the time course of perceptual choice: the leaky competing accumulator model. *Psychol. Rev.* 108:550–92

Valentini V, Frau R, Di Chiara G. 2004. Noradrenaline transporter blockers raise extracellular dopamine in medial prefrontal but not parietal and occipital cortex: differences with mianserin and clozapine. *J. Neurochem.* 88:917–27

Velley L, Cardo B, Kempf E, Mormede P, Nassif-Caudarella S, Velly J. 1991. Facilitation of learning consecutive to electrical stimulation of the locus coeruleus: cognitive alteration or stress-reduction? *Prog. Brain Res.* 88:555–69

Wald A. 1947. *Sequential Analysis*. New York: Wiley

Wallis JD, Miller EK. 2003. Neuronal activity in primate dorsolateral and orbital prefrontal cortex during performance of a reward preference task. *Eur. J. Neurosci.* 18(7):2069–81

Waterhouse BD, Moises HC, Woodward DJ. 1980. Noradrenergic modulation of somatosensory cortical neuronal responses to iontophoretically applied putative neurotransmitters. *Exp. Neurol.* 69:30–49

Waterhouse BD, Moises HC, Woodward DJ. 1998. Phasic activation of the locus coeruleus enhances responses of primary sensory cortical neurons to peripheral receptive field stimulation. *Brain Res.* 790:33–44

Waterhouse BD, Moises HC, Yeh HH, Geller HM, Woodward DJ. 1984. Comparison of norepinephrine- and benzodiazepine-induced augmentation of Purkinje cell responses to gamma-aminobutyric acid (GABA). *J. Pharmacol. Exp. Ther.* 228:257–67

Waterhouse BD, Woodward DJ. 1980. Interaction of norepinephrine with cerebrocortical activity evoked by stimulation of somatosensory afferent pathways in the rat. *Exp. Neurol.* 67:11–34

Williams JT, Henderson G, North RA. 1985. Characterization of alpha 2-adrenoceptors which increase potassium conductance in rat locus coeruleus neurones. *Neuroscience* 14:95–101

Williams JT, North RA, Shefner SA, Nishi S, Egan TM. 1984. Membrane properties of rat locus coeruleus neurones. *Neuroscience* 13:137–56

Wise RA, Rompre PP. 1989. Brain dopamine and reward. *Annu. Rev. Psychol.* 40:191–225

Yeung N, Botvinick MM, Cohen JD. 2004. The neural basis of error detection: conflict monitoring and the error-related negativity. *Psychol. Rev.* 111(4):931–59

Yeung N, Holroyd CB, Cohen JD. 2005. ERP correlates of feedback and reward processing in the presence and absence of response choice. *Cereb. Cortex*. In press

Yeung NP, Sanfey AG. 2004. Independent coding of reward magnitude and valence in the human brain. *J. Neurosci.* 24:6258–64

Zhu Y, Iba M, Rajkowski J, Aston-Jones G. 2004. Projection from the orbitofrontal cortex to the locus coeruleus in monkeys revealed by anterograde tracing. *Soc. Neurosci. Abstr.* 30:211.3

Neuronal Substrates of Complex Behaviors in *C. elegans*

Mario de Bono[1] and Andres Villu Maricq[2]

[1]MRC Laboratory of Molecular Biology, Hills Road, Cambridge CB2 2QH, United Kingdom; email: debono@mrc-lmb.cam.ac.uk

[2]Department of Biology, University of Utah, Salt Lake City, Utah 84112-0840; email: maricq@biology.utah.edu

Key Words

Caenorhabditis, sensory processing, signal transduction, neural circuits, plasticity, genetics, learning, mating, foraging, Ca^{2+} imaging

Abstract

A current challenge in neuroscience is to bridge the gaps between genes, proteins, neurons, neural circuits, and behavior in a single animal model. The nematode *Caenorhabditis elegans* has unique features that facilitate this synthesis. Its nervous system includes exactly 302 neurons, and their pattern of synaptic connectivity is known. With only five olfactory neurons, *C. elegans* can dynamically respond to dozens of attractive and repellant odors. Thermosensory neurons enable the nematode to remember its cultivation temperature and to track narrow isotherms. Polymodal sensory neurons detect a wide range of nociceptive cues and signal robust escape responses. Pairing of sensory stimuli leads to long-lived changes in behavior consistent with associative learning. Worms exhibit social behaviors and complex ultradian rhythms driven by Ca^{2+} oscillators with clock-like properties. Genetic analysis has identified gene products required for nervous system function and elucidated the molecular and neural bases of behaviors.

Contents

INTRODUCTION 452
   Wiring of the *C. Elegans* Nervous
      System 453
WORM LOCOMOTION 454
   Gatekeepers for Forward or
      Backward Movement 455
   Nociception by the ASH
      Polymodal Neuron 457
   ASH-Mediated Changes in
      Locomotion: Cell Biology
      of a Model Synapse 461
   Area-Restricted Search: A Strategy
      to Locate Resources that is
      Regulated by Glutamate and
      Dopamine Signaling 462
   Food-Induced Slowing Responses . 463
   Chemotaxis Toward Water-Soluble
      Attractants 464
   Olfaction 465
   Thermosensation 468
CALCIUM OSCILLATIONS AND
RHYTHMIC BEHAVIOR: THE
DEFECATION MOTOR
PROGRAM 471
   Defecation Motor Program 472
   Genetic Analysis of the Defecation
      Motor Program 473
   Control of the Defecation Cycle:
      A Complex Clock 474
   Ca^{2+} Oscillations and Rhythm
      Generation: Future Directions . 476
COMPLEX CONSEQUENCES
OF SIMPLE VARIATION:
AGGREGATION AND
AEROTAXIS 477
   Differential Activation of the Two
      Natural NPR-1 Isoforms by
      FLP-18 and FLP-21 Peptides .. 479
   How Do the Two Alleles of *npr-1*
      Lead to Such Dramatic
      Differences in Behavior? 479
   The Nociceptive Neurons
      ASH and ADL Promote
      Aggregation 480
   Neuroendocrine Regulation
      of Aggregation 481
ANALYSIS OF MALE MATING
BEHAVIOR 481
   Detection of Hermaphrodites:
      Male Search Behavior 482
   Neuronal Control of Male
      Mating Behavior 483
   Gene Products that Contribute
      to Male Mating Behavior 484
WORM LEARNING AND
SYNAPTIC PLASTICITY 486
   Glutamatergic Signaling
      Modulates Some Forms of
      Plasticity 486
   Other Molecules Required
      for Learning 487
FUTURE PROSPECTS 489

INTRODUCTION

One millimeter long, transparent, and with a life cycle of 3.5 days at 20°C, the nematode *Caenorhabditis elegans* is a favored model for behavioral studies because of its compact nervous system and experimental tractability. These advantages allow behavior to be dissected at the level of genes, individual neurons, and neural circuits. The natural environment of *C. elegans* is the soil, where it feeds on bacteria and other microbes. It moves by propagating bends along its body, much like a snake, navigating its environment using mechanical, thermal, and most especially chemical cues. The behavioral repertoire of this simple organism is surprisingly rich, and because many of its behaviors can easily be quantified, one can rapidly identify and categorize informative genetic mutants.

C. elegans occurs in two highly dimorphic sexes, males and self-fertilizing hermaphrodites. Most behavioral studies have focused on the hermaphrodite, the exception being studies of mating. Adults of both sexes are composed of a precise number of cells: Hermaphrodites have 959 somatic nuclei, and males have 1031 (Sulston et al. 1983, Sulston & Horvitz 1977). These cells make up hypodermis, muscle, the digestive tract, gonad, and the nervous system. In hermaphrodites the nervous system consists of 302 neurons and 56 glial and support cells, whereas males have 381 neurons and 92 glial and support cells. About half the neuronal cell bodies are located in the head, surrounding a central neuropil called the nerve ring. The remainder is found along the ventral cord and in tail ganglia. Male-specific neurons are located mostly in the copulatory tail. In both sexes each neuron is uniquely recognizable in different individuals by its characteristic position and morphology (Albertson & Thomson 1976, Sulston et al. 1983, Sulston & Horvitz 1977, White et al. 1986).

Wiring of the *C. elegans* Nervous System

The complete structure of the hermaphrodite nervous system has been reconstructed from serial section electron micrographs (EM) so that the morphology of each neuron and its chemical synapses and gap junctions are known (White et al. 1986; http://www.wormatlas.org). The 302 hermaphrodite neurons can be grouped by anatomical criteria into 118 classes. These include 39 classes of predicted sensory neurons, of which 21 have specialized sensory endings at the tip of the head (Ward et al. 1975, Ware et al. 1975). Many sensory neurons have been assigned some function by assessing the behavioral deficit of animals in which a neuron class is killed with a laser microbeam (Bargmann & Avery 1995, Bargmann & Kaplan 1998, Sulston & White 1980). Another 27 classes are motoneurons, and the remainder are classed as interneurons (White et al. 1986). To some extent this classification is a simplification: Many sensory neurons receive significant synaptic input and probably also act as interneurons; similarly, some motoneurons likely have sensory and/or interneuronal functions (White et al. 1986).

Most classes of sensory neurons and interneurons consist of two left/right homologs with similar synaptic connectivity. For simplicity, we indicate only the neuronal class in the text, with the understanding that this shorthand refers to the pair(s) of neurons. Many of these pairs are connected by a gap junction and probably act cooperatively. The processes of neurons travel in bundles and make *en passant* synapses. A neuron can have anywhere from 1 to 30 synaptic partners; on average it synapses with only ~15% of its neighbors. The number of synapses between two partners can vary from 1 to 19 synapses; typically it is ~5 synapses. The synaptic connections made by a particular neuron are ~75% reproducible across different animals (Durbin 1987). Together, the 302 neurons make approximately 5000 chemical synapses, 600 gap junctions, and 2000 neuromuscular junctions (White et al. 1986). For comparison, the number of chemical synapses in *C. elegans* is approximately equivalent to that made by any single hippocampal pyramidal neuron.

C. elegans contains many of the classic neurotransmitters found in vertebrates, including acetylcholine, glutamate, gamma-aminobutyric acid (GABA), serotonin, and dopamine. Neurotransmitter assignments have been made for many *C. elegans* neurons (Lee et al. 1999, Rand & Nonet 1997), but whether specific synapses are inhibitory or excitatory is usually unknown. In addition to classic neurotransmitter vesicles, many *C. elegans* neurons also have dense core vesicles, characteristic of catecholamine- and neuropeptide-containing vesicles in other animals (White et al. 1986). The *C. elegans* genome sequence reveals numerous neuropeptides. For example, the Phe-Met-Arg-Phe-amide FMRFamide family of neuropeptides is represented by at least 23 genes that can encode 60 peptides (Kim & Li 2004, Li et al. 1999b).

C. elegans behavioral studies have been driven by the isolation and analysis of mutants. Unlike in mice and to a lesser extent in flies, phenotypic variation due to genetic background differences is rarely a problem in C. elegans: Almost all studies are carried out in the same wild strain, N2 (Bristol), which is homozygous for one allele (defined as wild type) at all gene loci. Moreover strains can be stored as frozen stocks, minimizing genetic drift. A further advantage is that most of the C. elegans nervous system is dispensable for viability in the lab: Hermaphrodites are self-fertile, so even completely paralyzed worms can have progeny. Mutations have been isolated that disrupt genes required for sensory transduction, development of specific neuron classes, and neurotransmitter production or release; in many cases these genes provide entry points to signaling pathways and neural circuits underpinning a behavior.

The C. elegans genome sequence encodes ~20,000 genes; these include homologs of many neurally expressed mammalian genes (Bargmann 1998). Reverse genetic methods make it possible to isolate knockouts in these genes (Zwaal et al. 1993). The development of a genome-wide RNA interference (RNAi) library has also facilitated candidate gene analysis because the function of genes can easily and systematically be "knocked down" by feeding or injecting RNA (Kamath et al. 2003). However, many neural genes are resistant to RNAi (Kennedy et al. 2004), and mutants are usually required for detailed studies of gene function. One limitation is that anatomically targeted and temporally controlled reduction of gene function is currently inefficient, although new strategies are being developed (Tavernarakis et al. 2000).

Because C. elegans is transparent, each cell in the animal can be visualized by differential interference contrast (DIC) and fluorescence microscopy. Green fluorescent protein (GFP) tags permit rapid identification of the cellular and subcellular location of proteins and the differential labeling of neuronal compartments, e.g., sensory cilia and synapses (Chalfie et al. 1994, Nonet 1999, Rongo et al. 1998, Sengupta et al. 1996). A caveat of GFP protein-tagging experiments is that transgenes can be significantly overexpressed compared with endogenous levels. A growing number of promoters are available to target transgene expression to specific neurons. In the near future, we anticipate the availability of a library of promoters that allow each class of C. elegans neurons to be selectively tagged or modified. Often a gene product important for a specific behavior is expressed in many neurons. These new reagents allow one to define the neurons where a particular gene functions to regulate a behavior (Coates & de Bono 2002, de Bono et al. 2002).

Until recently, the way genetically defined signaling pathways altered electrical activity and synaptic release could only be inferred by analogy with physiological studies in other systems. This limitation is gradually being overcome: Several C. elegans electrophysiology preparations have been developed (Brockie et al. 2001b, Francis 2003, Goodman et al. 1998, Raizen & Avery 1994, Richmond et al. 1999), and the genetically encoded sensors cameleon and synaptopHluorin have been used to monitor calcium influx and synaptic release (Kerr et al. 2000, Samuel et al. 2003).

WORM LOCOMOTION

Most behavioral studies have focused on sensory processing of environmental stimuli that elicit changes in locomotory pattern. Worms move by undulatory propulsion in which a train of waves is passed along the body. C. elegans lay on either their left or right side, so these waves are formed by the successive contraction and relaxation of dorsal and ventral longitudinal body wall muscles. Three forces make movement on a solid surface possible: the hydrostatic skeleton of the animal, against which muscles act; surface tension, which pushes the worm against its substrate;

and friction between the worm and its substrate, which allows body waves to travel along the animal's length without side-slip, thereby generating forward propulsion (Niebur & Erdos 1993).

The neural circuitry responsible for locomotion must explain three features of *C. elegans* movement: How is the waveform able to propagate smoothly along an animal? How is the direction of propagation determined and controlled? How is the rate of wave propagation regulated? The locomotory musculature consists of 95 body-wall muscle cells that receive excitatory inputs at cholinergic neuromuscular junctions (NMJs) and inhibitory inputs at GABAergic NMJs (Chalfie & White 1988). Different classes of motor neurons distributed along the length of the animal innervate a local region of musculature. One set of excitatory and inhibitory dorsal (DB, DD) and ventral (VB, VD) neurons controls forward movement, and a second set (DA, DD and VA, VD) controls backward movement. The connectivity of these neurons means that if a particular region of the ventral musculature contracts owing to excitation by cholinergic motor neurons, e.g., VB or VA, the opposing dorsal musculature relaxes because of inhibition by GABAergic motor neurons (DD) (**Figure 1**). In this manner, the body is bent at a particular place and time. This reciprocal inhibition is especially important during rapid escape movements. For example, mechanostimulation of mutants defective in GABAergic signaling leads to a shrinking phenotype rather than a coordinated escape movement (McIntire et al. 1993a).

In *C. elegans*, no classical central pattern generators have been identified that drive body bends. Instead, wave propagation may depend on two features: coupling local bending to activation of proprioceptive elements in motorneurons of adjacent body segments (Chalfie & White 1988), and relaying excitation between neighboring motor neurons via gap junctions. Stretch activation of motorneurons may occur via tension-activation of DEGENERIN/ENaC ion channels as suggested by the locomotory defects observed in *unc-8 degenerin* mutants (Tavernarakis et al. 1997). Contraction should directly lead to activation of the next neuron in the series, and one can envision propagation of an electrical signal along the length of the worm. In this simple model, interneurons set the direction of wave propagation by providing excitatory and/or inhibitory input to motor neuron circuits that intrinsically coordinate either forward or backward wave propagation (Niebur & Erdos 1993).

Gatekeepers for Forward or Backward Movement

The interneurons that set forward or backward movement were identified by combining information from the EM reconstruction of the nervous system with studies of the neural control of touch avoidance (for two recent and comprehensive reviews of mechanosensation, see Ernstrom & Chalfie 2002 and Goodman & Schwarz 2003). Specialized sensory neurons detect touch to the anterior (AVM, ALM) or posterior (PVM, PLM) parts of the *C. elegans* body (**Figure 1B**). These neurons make electrical gap junctions and chemical synapses with a prominent set of interneurons. The nature of the chemical synapses is not known, but PLM and ALM neurons make gap junction contacts with PVC and AVD, respectively. AVD and a second interneuron called AVA innervate the VA and DA motor neurons, which coordinate backward movement. PVC and the AVB interneurons innervate the VB and DB motor neurons, which coordinate forward movement (Chalfie et al. 1985, White et al. 1986) (**Figure 1A**). Although the function of these four interneurons appears distributed and intertwined (all four neurons are either directly synaptic to each other or via one additional neuron), the strong prediction from the anatomical connections is that these interneurons mediate the withdrawal response to body touch and control the choice between forward and backward movement (Chalfie et al. 1985, White et al. 1986).

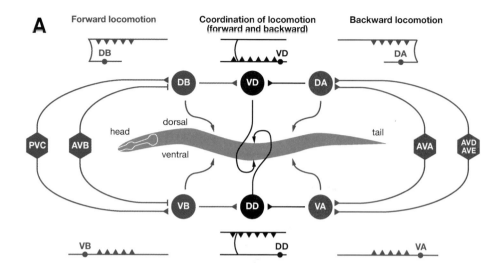
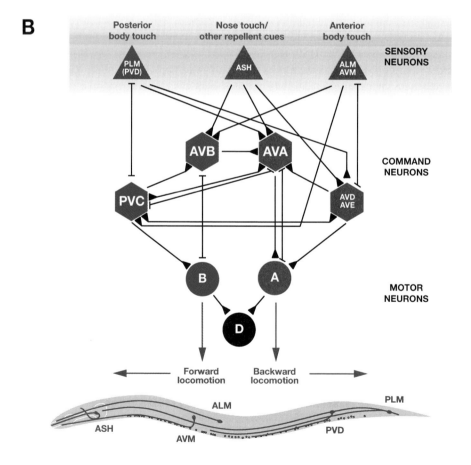

Laser ablation studies support this hypothesis: Ablation of AVA and AVD abolishes anterior touch avoidance and backward movement, whereas ablation of AVB and PVC abolishes posterior touch avoidance and forward movement (Chalfie et al. 1985, White et al. 1986). Because of their importance to forward and backward movement, AVA, AVB, AVD, and PVC are termed command interneurons.

The conclusion that the command interneurons are required for locomotion was buttressed and extended by two complementary sets of genetic perturbation studies in transgenic worms. When a transgene causing necrotic cell death is expressed in the command interneurons (as well as additional neurons) all touch sensitivity is lost, and backward and forward locomotion is severely disrupted (Maricq et al. 1995). On the other hand, depolarization of these neurons by an ionotropic glutamate receptor subunit with a gain-of-function mutation [originally identified in the *lurcher* mouse mutant (Zuo et al. 1997)] causes a dramatic change in locomotory behavior (Zheng et al. 1999). Wild-type worms spend most of their time moving in a forward direction, interrupted by infrequent and brief backward movements usually associated with a subsequent change in the direction of forward movement (Croll 1975). In contrast, the transgenic worms (called lurcher worms) rapidly switch between short-duration forward and backward movements and are unable to move an appreciable net distance, even though motor movement itself is not disrupted (Zheng et al. 1999).

These genetic perturbations further suggest that the command interneurons do not coordinate forward or backward locomotion per se, but rather gate whether a worm goes forward or backward. Movement and the decision to go forward or to reverse and change direction are central to almost all worm behavior. How then is sensory and interneuron processing of environmental cues relayed to the command interneurons, and which molecules subserve the information processing?

Nociception by the ASH Polymodal Neuron

C. elegans avoids a variety of noxious stimuli including high osmolarity (Culotti & Russell 1978), touch to the nose (Kaplan & Horvitz 1993), some odors (Troemel et al. 1995), heavy metals such as Cu^{2+} and Cd^{2+} (Bargmann et al. 1990, Sambongi et al. 1999), low pH (Sambongi et al. 2000), alkaloids such as quinine (Hilliard et al. 2004), and detergents (Bargmann et al. 1990, Hilliard et al. 2002). These cues all elicit an escape response that involves the animal reversing rapidly and then resuming forward movement, usually in a different direction. The sensory neurons mediating these avoidance responses have been defined by laser ablation (**Table 1**). Remarkably, despite their physical and chemical

---

Figure 1

Neural control of forward and backward locomotion. (*A*) Connectivity between the command interneurons (*hexagons*) and motor neurons (*circles*) is shown. Triangles indicate chemical synapses, and bars indicate gap junctions. The upper- and lowermost panels represent cell bodies (*small circles*), neural processes (*lines*) and synapses (*triangles*) of the motor neurons. All cell bodies lie on the ventral side of the worm. The DB, DD, and DA neurons send neural processes to the dorsal cord where they innervate dorsal body-wall muscles. The VB, VD, and VA neurons innervate ventral body-wall muscles. Figure is adapted from Chalfie et al. 1985. (*B*) The touch-response circuit controls forward and backward escape responses. *Upper diagram*: connectivity between sensory neurons (*triangles*), command interneurons, and motor neurons. PVD neurons are required for the response to harsh body touch, whereas PLM, ALM, and AVM are required for the response to light body touch. *Lower diagram*: an adult hermaphrodite (*anterior to the left and ventral down*) showing the morphology of the sensory neurons shown in the upper diagram. Blue and red represent the neurons primarily involved in forward and backward locomotion, respectively.

TABLE 1 Sensory neurons mediating aversive responses in *C. elegans*

Noxious stimulus	Sensory neuron
High osmolarity	ASH
Nose touch	ASH, OLQ, FLP
Odors	ASH, ADL, AWB
Heavy metals	ASH, ADL, ASE
Protons	ASH, ADF, ASE, ASK
Alkaloids	ASH, ASK
Detergents	ASH, ASK, PHA, PHB

diversity, these stimuli can all be transduced by the ciliated head neuron ASH, implicating it as a polymodal nociceptive neuron.

All ASH-mediated avoidance responses require the transient receptor potential vanilloid (TRPV)-related channel proteins OSM-9 and OCR-2, which promote each other's localization to the ASH sensory cilia (located at the tip of the worm's nose; see **Figure 2A,B**) and appear to be the sensory transduction channels of these neurons (Colbert et al. 1997, Tobin et al. 2002). In vivo imaging shows that noxious stimuli evoke Ca^{2+} influxes in ASH, but these are abolished in *osm-9* mutants (Hilliard et al. 2005). OSM-9 is a *C. elegans* ortholog of the mammalian TRPV4 channel, and the osmotic avoidance and nose-touch response defects of *osm-9* mutants can be rescued by expressing rat TRPV4 in the ASH neurons (Liedtke et al. 2003). In contrast, expressing the rat TRPV1 channel in ASH does not restore native avoidance responses but is sufficient to induce an ectopic avoidance response to capsaicin, a TRPV1 channel agonist (Tobin et al. 2002). ASH sensory responses are also dependent on the Gα-protein ODR-3 (Hilliard et al. 2005, Roayaie et al. 1998) and on biosynthetic enzymes that make polyunsaturated fatty acids (PUFA) (Kahn-Kirby et al. 2004). Mutations in these genes do not disrupt capsaicin avoidance in transgenic animals that express rat TRPV1 in ASH, which suggests they are not required for signal transduction downstream of TRP channel activation. One possibility is that transduction of noxious stimuli involves a G-protein cascade that activates OSM-9/OCR-2 by generating PUFA-containing lipids. Interestingly, the ASH neurons express at least eight other Gα protein subunits apart from ODR-3 (Jansen et al. 1999). At least one of these Gα subunits, GPA-3, is important for some nociceptive responses (Hilliard et al. 2004, Jansen et al. 1999).

The pathways by which different noxious stimuli are detected by ASH and coupled to gating of OCR-2/OSM-9 channels are still mysterious. Several pieces of data suggest that distinct processes detect different noxious stimuli. First, several genes have been identified that are required for transduction of some noxious stimuli detected by ASH but not others. For example, animals lacking the novel cytosolic protein OSM-10 fail to respond to hyperosmolarity but avoid nose touch, quinine, and 1-octanol (Hart et al. 1999, Hilliard et al. 2004). Conversely, *qui-1*, a novel protein with WD-40 repeats, is required for avoiding quinine but not hyperosmolarity (Hilliard et al. 2004). Loss of *grk-2*, G protein–coupled receptor kinase-2, attenuates response to hyperosmolarity, octanol, and quinine, but not avoidance of nose touch (Fukuto et al. 2004). Second, behavioral and imaging studies provide evidence for repellent-specific adaptation. For example animals subjected to repeated nose touch adapt to this stimulus but continue to avoid high osmolarity and the volatile repellent 1-octanol (Hart et al. 1999, Hilliard et al. 2005). Also, prolonged exposure to Cu^{2+} ions causes adaptation to this stimulus but not to hyperosmolarity (Hilliard et al. 2005). A key component for ASH adaptation is *gpc-1*, one of two G-protein gamma subunits encoded in the *C. elegans* genome (Hilliard et al. 2005; Jansen et al. 1999, 2002). Mutants in *gpc-1* resemble wild type in their initial response to most repellents detected by ASH but adapt poorly to repeated noxious stimulation. The Ca^{2+}-dependent phosphatase calcineurin may also be important to attenuate avoidance responses: *tax-6* mutants, which are defective in the calcineurin A

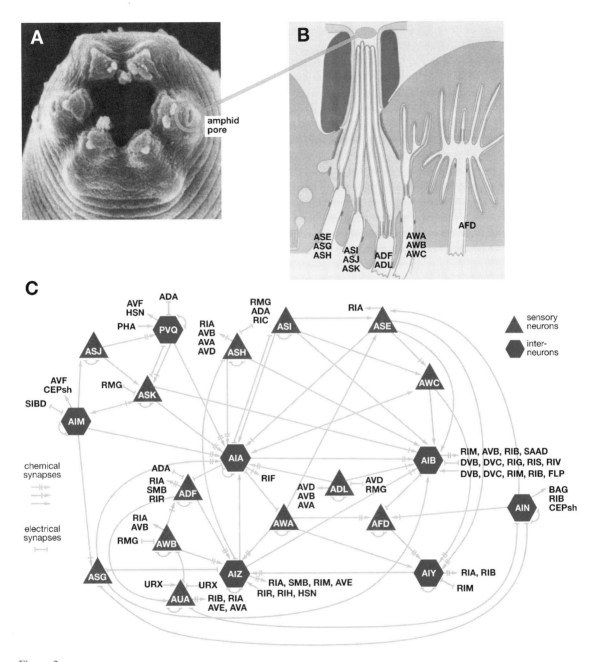

Figure 2

Anatomy of chemosensory and thermosensory neurons. (*A*) Scanning electron micrograph of the head of the wild-type N2 adult showing the amphid pores. (*B*) Schematic diagram of the amphid, the major sensory sensillum in *C. elegans*. Six single ciliated and two dually ciliated chemosensory neurons enter the amphid channel. Three olfactory neurons, AWA, AWB, and AWC, have endings embedded in the sheath cell (*shown in gray*). The sensory ending of the AFD thermosensory neuron is characterized by multiple finger-like projections that are also embedded in the sheath cell. (*C*) The neural circuitry associated with amphid neurons. Panel *A* is reprinted, with permission, from Albert et al. 1981. Panel *B* is reprinted, with permission from Perkins et al. 1986. Panel *C* is reprinted, with permission, from White et al. 1986.

subunit, take longer to cross a hyperosmotic barrier (Kuhara et al. 2002).

The Ca^{2+} influx evoked by different noxious stimuli varies in size depending on the type and intensity of the noxious stimulus (Hilliard et al. 2005, Kahn-Kirby et al. 2004). For example, the Ca^{2+} influx in ASH cell bodies evoked by osmotic shock is greater than that evoked by 10-mM quinine or nose touch, and stimulation by 10-mM Cu^{2+} for increasing lengths of time evokes longer-lasting increases in Ca^{2+} concentration. Genetic studies suggest that differential activation of ASH results in differential release of neurotransmitter and distinct patterns of downstream signaling. The ASH neurons are glutamatergic: They express the EAT-4 glutamate vesicular transporter (VGLUT), which is required for ASH nociceptive responses (Berger et al. 1998, Lee et al. 1999). Differential release of glutamate from ASH may activate different types of glutamate receptors on command interneurons that are postsynaptic to ASH and that mediate the nociceptive escape response (Hart et al. 1995, Maricq et al. 1995, Mellem et al. 2002) (**Figure 1B**).

Weak activation, such as that elicited by nose touch, activates channels containing GLR-1 (*glr*, glutamate receptor) and/or GLR-2 non-NMDA ionotropic glutamate receptor (iGluR) subunits, and mutations in these genes disrupt nose-touch avoidance (Hart et al. 1995, Maricq et al. 1995, Mellem et al. 2002). Stimuli that evoke higher levels of Ca^{2+} release, such as hyperosmolarity (Hilliard et al. 2005), can signal avoidance through not only GLR-1/GLR-2 channels but also through NMDA iGluR–containing subunits (NMR) encoded by the *nmr-1* and *nmr-2* genes (Brockie et al. 2001a, Mellem et al. 2002). Mutations in either *glr-1* or *glr-2* attenuate but do not abolish avoidance of hyperosmolarity, whereas loss of both GLR and NMR channels additively disrupts hyperosmotic avoidance. Thus differential activation of postsynaptic glutamate receptors serves to decode the intensity and type of aversive stimuli (Hart et al. 1995, Maricq et al. 1995, Mellem et al. 2002). Consistent with this interpretation, *nmr-1* mutants are not defective for avoidance of nose touch, which suggests that tactile stimuli do not release sufficient glutamate to activate NMDA receptors, which may be located more perisynaptically (Brockie et al. 2001b).

Worms spontaneously interrupt their forward movement with brief backward movement, even in the absence of aversive stimuli. This may be due to basal activation of ASH during the worm's exploration of its environment. Support for this model is provided by the observation that the time spent going forward is increased in *glr-1* mutants (Juo & Kaplan 2004, Mellem et al. 2002). *nmr-1* mutants are also defective in the regulation of forward movement (Brockie et al. 2001b).

ASH avoidance responses are regulated by neuropeptides (Kass et al. 2001). The *egl-3* gene encodes a *C. elegans* ortholog of the mammalian proprotein convertase type 2 (PC2) that helps process neuropeptide proproteins into active peptides. Mutations in *egl-3* suppress the ASH defects of *glr-1* mutants, i.e., they restore nose-touch sensation and decrease the time taken to respond to hyperosmotic stimuli (Kass et al. 2001, Mellem et al. 2002). This restoration depends on glutamate release: It requires the EAT-4 glutamate vesicular transporter (VGLUT) and the NMR-1 NMDA receptor subunit. Electrophysiological and cell-specific rescue studies suggest that neuropeptides processed by EGL-3 in the command interneurons downregulate glutamate release from ASH (Kass et al. 2001, Mellem et al. 2002). In this model, synaptic transmission in *glr-1* mutants is partly rescued because the increased release of glutamate in *egl-3* mutants is sufficient to activate NMDA receptors.

C. elegans avoidance responses can also be modulated by food. ASH-mediated avoidance of nose touch and 1-octanol is stronger when food is present (Chao et al. 2004). The effect of food can be mimicked by serotonin (5-HT), which signals food abundance for many *C. elegans* behaviors (Avery & Horvitz

1990, Horvitz et al. 1982, Sawin et al. 2000, Segalat et al. 1995). 5-HT can modify ASH itself: It potentiates Ca^{2+} influx in ASH in response to nose touch (Hilliard et al. 2005). However, its effect is selective: Exogenous 5-HT does not stimulate Ca^{2+} influx evoked by Cu^{2+} or osmotic shock. 5-HT stimulation of octanol avoidance is probably mediated by metabotropic receptors because it requires the Gα protein GPA-11, which is expressed in ASH and ADL (Chao et al. 2004, Jansen et al. 1999).

ASH-Mediated Changes in Locomotion: Cell Biology of a Model Synapse

Because noxious avoidance depends on GLR-1, one might expect that mutations that disrupt GLR-1 stability, function, or synaptic localization would result in avoidance defects. Mutations in the *lin-10* gene, which encodes a PDZ-domain protein, are defective for ASH-mediated avoidance behaviors (Rongo et al. 1998). In vertebrates, PDZ-domain proteins interact with select NMDA and non-NMDA subunits and are required for the proper localization and stabilization of iGluRs at the postsynaptic density (Kim & Sheng 2004). Mutations in *lin-10* change the normally punctate (synaptic) distribution of GLR-1 to one that appears more uniformly distributed. The punctate distribution of GLR-1 is also dependent on *unc-43*, which encodes calcium and calmodulin-dependent protein kinase II (CaMKII) (Rongo & Kaplan 1999). An *unc-43* loss-of-function mutation disrupts trafficking, leading to an accumulation of GLR-1 in cell bodies and a reduction in the density of GLR-1 synapses.

Recently, Shim et al. (2004) implicated the unfolded protein response (UPR) in the export of *C. elegans* iGluR subunits from the endoplasmic reticulum (ER). Homologs of key proteins of the vertebrate UPR, IRE1 and XBP1, are found in *C. elegans*. Interestingly, mutations in *ire-1* and *xbp-1* prevent GLR-1 export from the ER (Shim et al. 2004). In addition to the UPR, GLR-1 exit from the ER is regulated by a quality-control mechanism that prevents defective receptors from reaching the synapse. Thus, mutations in GLR-1 predicted to disrupt glutamate-binding or modify ion permeability increase GLR-1 retention in the ER (Grunwald & Kaplan 2003) or influence neuronal viability (Aronoff et al. 2004).

Synaptic iGluRs are not static but rapidly cycle between a subcellular pool and the postsynaptic membrane (Bredt & Nicoll 2003, Ehlers 2003). *unc-11* encodes an ortholog of the AP180 clathrin adaptin protein, mutations in which cause endocytotic defects (Nonet et al. 1999). Either disrupting *unc-11* or blocking ubiquitination of GLR-1 leads to increased levels of GLR-1 at ventral cord puncta (Burbea et al. 2002). The anaphase-promoting complex (APC) also regulates the abundance of GLR-1 at synapses (Juo & Kaplan 2004). The APC is a multisubunit ubiquitin ligase that assembles ubiquitin chains on substrate molecules thereby regulating their degradation (Harper et al. 2002, Peters 2002). Mutations in genes that encode subunits of the APC result in an accumulation of GLR-1 at ventral cord synapses (Juo & Kaplan 2004). The accumulation of synaptic GLR-1 results in a predicted change in worm behavior, partly phenocopying the locomotion defects observed in *glr-1* gain-of-function mutants (Juo & Kaplan 2004, Zheng et al. 1999). Together, the data suggest that both ubiquitination of GLR-1 and the APC modulate the strength of glutamatergic signaling. Recently, another pathway regulating GLR-1 was identified. LIN-23, a subunit of a SCF (Skp1/Cullin/F-Box) ubiquitin ligase, modulates the β-catenin/Wnt pathway to alter postsynaptic GLR-1 levels (Dreier et al. 2005).

Homeostatic regulation of iGluR abundance at synapses, termed synaptic scaling, is a proposed compensatory mechanism by which neurons maintain a normal level of synaptic input (Turrigiano & Nelson 2004). In *C. elegans* a mutation in either *unc-2*, which encodes the only worm non-L-type high-threshold voltage–gated calcium channel (Schafer & Kenyon 1995), or the vesicular glutamate transporter *eat-4* is predicted to suppress activity at glutamatergic synapses. Both *unc-2* and *eat-4* mutants display an increased abundance of synaptic GLR-1 that parallels an increase in the amplitude of glutamate-gated currents in response to exogenously applied glutamate (Grunwald et al. 2004). *unc-11; eat-4* double mutants have the same abundance of GLR-1 at synapses as *eat-4* mutants, which suggests that synaptic scaling in *C. elegans* depends on clathrin-mediated endocytosis.

Once localized to the appropriate postsynaptic sites, iGluRs must be competent to gate open in response to glutamate-binding. Are additional gene products required for iGluR synaptic function? The SOL-1 CUB-domain transmembrane protein was identified in a genetic screen for modifiers of a constitutively active variant of GLR-1 (Zheng et al. 2004). SOL-1 is required for GLR-1-dependent glutamate-gated currents and for behaviors dependent on GLR-1-mediated synaptic transmission. Although the data suggest that SOL-1 is required for either the gating open of GLR-1 receptors or their rate of desensitization once open, the mechanism of SOL-1 function remains to be determined. Nonetheless, the identification of this novel protein in *C. elegans* shows that iGluRs are not simply stand-alone molecules and suggests that synaptic signaling in other organisms may be similarly dependent on SOL-1-like proteins.

Area-Restricted Search: A Strategy to Locate Resources that is Regulated by Glutamate and Dopamine Signaling

Area-restricted search (ARS) is a well-described universal foraging strategy where animals turn more frequently following an encounter with food to locally restrict their search (Bell 1991). As the period of time since the last food encounter increases, animals turn less frequently and begin to move to new areas. This makes ecological sense: Animals learn that food is not clumped nearby and so move on to seek greener pastures. ARS is also observed in *C. elegans* (Hills et al. 2004). Thus, if a feeding worm is transferred to an environment with no food, it initially makes short-duration forward movements and high-angle turns (Gray et al. 2005, Hills et al. 2004, Shingai 2000, Tsalik & Hobert 2003, Wakabayashi et al. 2004). If food is nearby, this behavior increases the likelihood that the worm will again find food. If no food is nearby, over a period of 30 min, the behavior of the worm changes such that the duration of forward movement increases, and the likelihood of reversals or high-angled turns decreases (Hills et al. 2004). Ablation of dopaminergic neurons, or mutations in either the GLR-1/GLR-2 iGluR or the EAT-4 VGLUT, disrupts ARS. Together, the data suggest that dopamine, released when the worms encounter food, modulates glutamatergic signaling to the command interneurons, resulting in an increased turn frequency thus improving the probability of staying in the vicinity of food (Hills et al. 2004).

Chemosensory neurons also regulate the duration of forward movement. Ablation of ASK and AWC causes an increase, and ablation of salt-sensing ASE and ADF, a decrease in the duration of forward locomotion (Gray et al. 2005, Wakabayashi et al. 2004). The neural anatomy suggests that these sensory neurons send their outputs to a small set of sensory interneurons, including AIA, AIB, AIY, and AIZ (**Figure 2C**) (White et al. 1986). Ablation of these neurons causes dramatic changes in movement. Ablation of AIB or AIZ leads to large increases in the duration of forward movement, whereas ablation of AIY and, less dramatically, AIA shortens the duration of forward movement (Gray et al. 2005, Tsalik & Hobert 2003, Wakabayashi et al. 2004). How these four pairs of interneurons integrate sensory input and provide outputs to the command interneurons is not currently understood. One possibility is that when a worm leaves food, this event is detected by

ASK, which via its synaptic output depolarizes AIB/AIZ, leading to shortened forward movements and high-angled turns, which would reorient the worm and increase the probability of returning to the food (Wakabayashi et al. 2004).

In the presence of food, worms also switch between high- and low-activity states (Fujiwara et al. 2002). These states have been called roaming, where a worm moves at high speed with little turning, and dwelling, characterized by slower movements and more turning. Roaming depends on sensory input—worms with defective chemosensory cilia (*che-2* mutants) move normally if provoked but otherwise remain in the dwell state for long periods. Mutations in the EGL-4 cGMP-dependent protein kinase suppress the roaming defects of *che-2* mutants, and *egl-4* mutants alone have increased roaming compared with wild-type. EGL-4 is expressed in sensory neurons, and selective expression of EGL-4 in the sensory neurons of transgenic *egl-4* mutants is sufficient for behavioral rescue. The genetic data suggest that EGL-4 suppresses roaming and that sensory input functions to downregulate EGL-4. Currently we do not know which sensory inputs regulate EGL-4. For ARS, lack of food sensory input eventually makes the worms roam (Hills et al. 2004). The *che-2* loss-of-function phenotype indicates that other sensory inputs are required for roaming to proceed in the presence of food (Fujiwara et al. 2002).

Food-Induced Slowing Responses

When worms encounter food they slow down (de Bono & Bargmann 1998, Sawin et al. 2000). This slowing response is partly dependent on dopamine and dopaminergic neurons that detect mechanical properties of bacteria (Sawin et al. 2000). Mutations that disrupt this response have been identified, including two genes that encode G protein–coupled D1-like and D2-like dopamine receptors (Chase et al. 2004). Dopamine's effects appear to be humorally mediated—the relevant dopamine receptors are expressed in cholinergic motor neurons, which do not receive synaptic inputs from known dopaminergic neurons. In mammalian neurons, D1 and D2 receptors can act antagonistically. A similar scenario is seen in *C. elegans*. In the worm, knockout of the D2 receptor (*dop-3*) makes worms resistant to exogenous dopamine, and these mutants fail to slow when they encounter food. Mutations in the D1 receptor (*dop-1*) do not on their own affect the slowing response (Sanyal et al. 2004), but they suppress the *dop-3* mutant phenotype: *dop-1*; *dop-3* animals slow down in response to food. D2-mediated signaling involves activation of the G-protein $G\alpha_o$, which reduces synaptic transmission via downstream effects on diacylglycerol (DAG) and the DAG-binding protein UNC-13 (Miller et al. 1999, Nurrish et al. 1999). D1 signaling involves activation of $G\alpha_q$, which increases synaptic transmission by stimulating DAG production (Lackner et al. 1999). In one model, in the absence of food, when little dopamine is present, the activity of D1 predominates (presumably D1 has a higher affinity for dopamine, but other mechanisms may be in place as well), and cholinergic synaptic transmission to muscle cells is promoted. Food leads to increased dopamine release, which binds to D2 as well as D1 receptors. The net signal from activation of these two antagonistic receptors is the inhibition of synaptic transmission and the slowing of the worm (Chase et al. 2004).

In contrast to well-fed worms, when worms encounter food after 30 min of food deprivation, two things happen: Worms are more likely to initiate high-angle turns (Hills et al. 2004) and their rate of movement is dramatically slowed, more so than the normal slowing observed with food contact. The slowing following starvation is called the enhanced slowing response (Sawin et al. 2000). Ablating the NSM serotonergic neurons or mutations in genes required for serotonin biosynthesis disrupts the enhanced slowing response. This response can also be blocked by exogenous application of serotonin antagonists and potentiated by the selective serotonin reuptake inhibitor (SSRI) fluoxetine (Prozac). More of this signaling pathway has been revealed by genetic screens for mutants defective in the slowing response or serotonin transport. Mutations in *mod-1*, which encodes a serotonin-gated chloride channel (Ranganathan et al. 2000), and *goa-1*, which encodes a G-protein alpha subunit (Segalat et al. 1995), disrupt the slowing response (Sawin et al. 2000), whereas mutations in *mod-5*, which encodes a fluoxetine-sensitive serotonin reuptake transporter (Ranganathan et al. 2001), enhance the slowing response. However, the neural circuit that leads to the slowing response has not yet been identified. MOD-1 appears to be expressed in a few neurons and not in muscle cells; the identity of these neurons has not been established.

Chemotaxis Toward Water-Soluble Attractants

C. elegans is attracted to a variety of water-soluble chemicals in the millimolar and submillimolar range. These include cations such as Na^+, K^+, and NH_4^+, anions such as Cl^- and SO_4^{2-}, and small organic molecules such as cAMP, cGMP, lysine, biotin, and glucose (Bargmann & Horvitz 1991, Dusenbery 1974, Jansen et al. 1999, Ward 1973). The behavioral mechanism for chemotaxis has been quantitatively described and modelled by tracking individual animals in defined chemical gradients (Pierce-Shimomura et al. 1999). To migrate up an attractant gradient *C. elegans* employs a course-correction mechanism. When the rate of change of concentration of an attractant, dC/dt, is <0, animals increase the probability of a sharp turn. Conversely, when dC/dt is >0, animals suppress turning and maintain runs. Sharp turns, also called pirouettes, are composed of reversals and omega bends and serve to reorient the worm. The size of the turn is weakly correlated with how far the animal was off course before the pirouette. Thus animals that are running down an attractant gradient tend to change course by close to 180° after a pirouette, and this orients them up the gradient. Variations of this course-correction mechanism mediate all other *C. elegans* taxes studied thus far (see Olfaction and Thermotaxis below).

Laser ablation studies have identified a set of head neurons with exposed sensory cilia that mediate chemotaxis to water-soluble attractants (Bargmann & Horvitz 1991). The principal *C. elegans* gustatory neurons appear to be the ASEL (ASE Left) and ASER (ASE Right) neurons; additional neurons in the amphid play subsidiary roles for chemotaxis to specific subsets of cues (Bargmann & Horvitz 1991). Chemosensory receptors have not been identified, but sensory transduction in ASE involves a cGMP pathway: Mutants lacking the TAX-2 and TAX-4 cGMP-gated ion channel subunits fail to migrate to ASE-sensed cues, and these channels are coexpressed in ASE neurons (Coburn & Bargmann 1996; Dusenbery et al. 1975; Komatsu et al. 1996, 1999). Mutations in a transmembrane guanylate cyclase, DAF-11, also disrupt chemotaxis to water-soluble chemoattractants (Vowels & Thomas 1994), but it is unclear if this molecule is expressed in ASE (Birnby et al. 2000). Although the ASEL and ASER neurons appear to be homologous at the structural level, they are bilaterally asymmetric at the level of gene expression and function (Pierce-Shimomura et al. 2001, Yu et al. 1997). This left-right asymmetry is set up by a gene-regulatory cascade that involves asymmetric microRNA regulation of transcription factor activity (Chang et al. 2003, 2004; Chang et al. 2003; Johnston & Hobert 2003; Pierce-Shimomura et al. 2001). At the molecular level, the asymmetry can be visualized using GFP reporters for receptor guanylate cyclases: Whereas *gcy-5* is expressed in ASER, *gcy-6* and *gcy-7* are expressed in ASEL (Yu et al. 1997). Functionally, ASEL and ASER make different contributions to chemotaxis toward different chemical cues (Pierce-Shimomura et al. 2001). For example ASEL is more important for chemotaxis toward sodium acetate than ASER

(Chang et al. 2004, Pierce-Shimomura et al. 2001). Asymmetry between the ASE neurons is also important for discrimination between different water-soluble attractants. *C. elegans* can migrate to the peak concentration of one water-soluble cue in a saturating concentration of another. Animals having only one type of ASE neuron exhibit poor discrimination for some combinations of attractant cues (Pierce-Shimomura et al. 2001, Ward 1973).

Continuous exposure to water-soluble attractants results in decreased chemotaxis to that stimulus (Dusenbery 1980, Jansen et al. 2002, Ward 1973). This adaptation is reversible and is partly stimulus-specific (Jansen et al. 2002). Genetic studies have identified components that regulate adaptation to water-soluble attractants, including the G-protein gamma subunit GPC-1 and the TRPV-related channel protein OSM-9 (Jansen et al. 2002). Null mutants in *gpc-1* adapt poorly to all water-soluble attractants tested, whereas *osm-9* mutants have more circumscribed deficits, adapting normally to some attractants but not to others (Jansen et al. 2002). Both *gpc-1* and *osm-9* mutants exhibit normal responses when initially encountering water-soluble attractants. Both molecules have been implicated in sensory adaptation in other neurons: OSM-9 in adaptation to odors sensed by the AWC neurons (Colbert et al. 1997), and *gpc-1* in adaptation to nociceptive cues (Hilliard et al. 2005). The neurons in which these molecules exert their function on taste adaptation have not been identified. However, GPC-1 does not appear to be expressed in ASE, but is expressed in other chemosensory neurons.

Chemotaxis to water-soluble cues can be modified by experience. *C. elegans* starved in the presence of NaCl suppress chemotaxis toward NaCl (Saeki et al. 2001). This suppression does not occur if animals are starved in the absence of NaCl or conditioned with NaCl in the presence of food, which indicates it is neither a nonspecific effect of starvation nor is it due to sensory adaptation. As well as suppressing chemotaxis to NaCl, animals starved in the presence of NaCl suppress chemotaxis to other water-soluble chemoattractants, although they spare olfactory responses. This suggests that animals modify the activity of a gustatory cell or group of cells to alter behavior. The suppression of chemotaxis occurs gradually over a period of 3–4 hours and is rapidly reversed if animals are fed (within 10 min), or if starving animals are placed in media that do not contain NaCl (within 1 h). Serotonin, which appears to encode food abundance in *C. elegans* (Avery & Horvitz 1990, Horvitz et al. 1982, Sawin et al. 2000, Segalat et al. 1995), can also partially restore NaCl chemotaxis to conditioned animals. Together, these data suggest a form of associative learning, in which *C. elegans* associates an unconditioned stimulus (starvation) with a conditioned stimulus (NaCl). The mechanistic basis for this plasticity is as yet unclear but requires HEN-1, a secretory protein with an LDL receptor motif that is similar to the *Drosophila* signaling molecule *Jelly belly* (Jeb) (Ishihara et al. 2002, Lee et al. 2003, Weiss et al. 2001) (see below, Learning and Memory).

Olfaction

As well as responding to water-soluble cues, *C. elegans* is highly sensitive and broadly tuned to volatile chemicals. More than half of 121 volatiles tested either attract or repel worms, with sensitivity to some odorants retained at dilutions of 10^{-5} (Bargmann et al. 1993). As in other animals, odors are detected by G protein–coupled receptors. The worm genome contains ~1000 genes that encode G protein–coupled receptors. On the basis of sequence and/or expression patterns most of these genes are predicted to be chemosensory receptors for volatile and nonvolatile cues (Bargmann 1998, Robertson 2001, Troemel 1999, Troemel et al. 1995). The first putative chemosensory receptor to be definitively linked to sensation of a specific odor was discovered in *C. elegans*. *odr-10*, the founding member of one of the largest *C. elegans* chemosensory receptor superfamilies, encodes a high-affinity receptor for diacetyl. *odr-10* null mutants display about a 100-fold weaker attraction to diacetyl than do wild-type animals (Sengupta et al. 1996).

C. elegans olfaction is mediated by 5 pairs of ciliated amphid (head) neurons that express multiple chemosensory receptors and detect multiple odors (Bargmann et al. 1993, Troemel 1999, Troemel et al. 1995) (**Figure 2**). The sensory cilia of these neurons are associated with a pore to the outside of the animal and bear olfactory receptors as well as sensory transduction components (Dwyer et al. 1998, Sengupta et al. 1996, Tobin et al. 2002, Ward et al. 1975, Ware et al. 1975). Two pairs of neurons, AWA and AWC, mediate attractive odor responses (Bargmann et al. 1993), whereas three pairs of neurons, ASH, ADL, and AWB, mediate repulsive ones (Troemel et al. 1995, 1997). The ASH and ADL neurons also mediate avoidance of other stimuli (see above). Whether a particular odorant is attractive or repellent depends on the neuron in which its cognate receptor is expressed and not on the odorant receptor itself. Thus the ODR-10 diacetyl receptor mediates attraction to diacetyl when expressed in the AWA neurons but induces avoidance of this odorant when ectopically expressed in the AWB avoidance neurons (Troemel et al. 1997).

All odor responses mediated by a particular neuron appear to be transduced by a common signaling pathway. Genetic studies have outlined these transduction pathways for all five odor-sensing neurons. In the AWA, ASH, and ADL neurons, G-protein pathways are proposed to regulate a TRPV-related channel encoded by the *ocr-2* and *osm-9* genes; the exact coupling mechanism is unknown (Colbert et al. 1997, Tobin et al. 2002). In AWA and ASH, the principle Gα protein involved is ODR-3, but other Gα subunits are expressed in these neurons and can have regulatory roles (Chao et al. 2004, Jansen et al. 1999, Lans et al. 2004, Roayaie et al. 1998). ODR-3 also appears to mediate olfactory receptor signaling in the AWB and AWC neurons through a cGMP signaling pathway (Roayaie et al. 1998, Troemel et al. 1997). By an as yet unknown mechanism, activated ODR-3 [and also GPA-3 and for some odors, GPA-2 Gα (Lans et al. 2004)] regulates the transmembrane guanylate cyclases ODR-1 and DAF-11 (Birnby et al. 2000, L'Etoile & Bargmann 2000). Odor-induced changes in cGMP levels control the activity of a cGMP-gated cation channel encoded by the *tax-2* and *tax-4* genes to direct the AWB and AWC olfactory responses (Coburn & Bargmann 1996; Komatsu et al. 1996, 1999). Thus attraction to and repulsion from an odor can be mediated by the same signaling pathway: the ODR-3–DAF-11/ODR-1–TAX-2/TAX-4 pathway mediates odor attraction in AWC and odor repulsion in AWB; similarly, the ODR-3–OSM-9/OCR-2 pathway mediates chemoattraction in AWA but chemorepulsion in ASH (**Figure 3**). Why do the AWA and AWC neurons mediate attractive responses and yet AWB, ASH, and ADL mediate repellant ones? The answer probably lies in the properties of the synaptic connections made by these neurons. However, the neural circuitry by which olfactory information is relayed is only just beginning to be dissected (Tsalik & Hobert 2003).

The AWA and AWC olfactory neurons have complex sensory properties. Each neuron mediates responses to multiple odors across a broad range of concentrations (Bargmann et al. 1993, Sengupta et al. 1996). By analogy with chemotaxis toward water-soluble chemicals (Pierce-Shimomura et al. 1999), a wide dynamic range in *C. elegans* olfactory response is probably achieved by animals monitoring the rate of change of odor concentration (dC/dt), rather than absolute odor concentration. This implies that the neurons adapt over short timescales (of seconds or less) to increasing odor concentration, but molecular mechanisms for this rapid desensitization are unknown. One potential clue is that odor stimulation rapidly induces (<10 s) MAP kinase activity (Hirotsu et al. 2000), which could direct desensitization by an as yet unknown mechanism. Arrestin, which in mammals cooperates with G protein–coupled receptor kinase to mediate receptor desensitization, is not required for *C. elegans* chemosensation, although loss of *C. elegans* G protein–coupled receptor kinase-2 (*grk-2*) leads to profound defects in both AWA- and AWC-mediated olfactory responses (Fukuto et al. 2004).

C. elegans olfactory responses also adapt over longer timescales of minutes to hours, selectively turning off an odor response after prolonged exposure to that odor. This reversible adaptation is not due to neuron fatigue: It spares responses to other odors detected by the same olfactory neuron (Colbert & Bargmann 1995). Interestingly, olfactory adaptation is regulated by context. Food attenuates adaptation of odor responses, whereas starved animals adapt more quickly than recently fed ones (Colbert & Bargmann 1997, Nuttley et al. 2002, Saeki et al. 2001). This plasticity is reminiscent of associative learning, in which the presence or absence of food acts as unconditioned stimuli that sustain or inhibit chemotaxis to odor, the conditioned stimulus. The mechanisms for this modulation are unclear but involve serotonin (Colbert & Bargmann 1997, Nuttley et al. 2002).

Most studies of olfactory adaptation have focused on responses mediated by the AWC neurons (Colbert & Bargmann 1995, 1997; Kuhara et al. 2002; L'Etoile et al. 2002; Nuttley et al. 2002). Multiple adaptation pathways operate in this neuron, mediated by Ca^{2+}, the TRP channel OSM-9, the cGMP-dependent protein kinase EGL-4, and the as yet unconed *adp-1* (adaptation–1) gene. These pathways regulate subsets of odor responses mediated by AWC in an arrangement that appears thus far to be complex and idiosyncratic. Most information is known about adaptation mediated by the cGMP-dependent protein kinase EGL-4 (L'Etoile et al. 2002). EGL-4 is proposed to contribute to an early phase of olfactory adaptation by phosphorylating the cGMP-gated ion

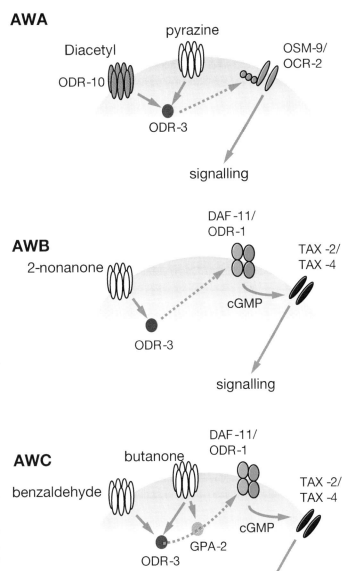

Figure 3

Signal transduction pathways in olfactory neurons. The AWA and AWC neurons mediate chemotaxis toward attractive odors, whereas the AWB neurons mediate avoidance of repellent odors. ODR-10, high affinity diacetyl receptor; ODR-3 and GPA-2, G protein alpha subunits; OSM-9 and OCR-2, TRPV-related channel proteins; DAF-11 and ODR-1, transmembrane guanylate cyclases; TAX-2 and TAX-4, cGMP-gated cation channel subunits.

channel subunit TAX-2, thereby downregulating the responsiveness of the TAX-2/TAX-4 channel to cGMP. EGL-4 also contributes to later phases of adaptation induced by prolonged odor exposure; this function requires an intact nuclear localization signal (L'Etoile et al. 2002).

Investigators have also discovered a pathway that represses AWC adaptation to odors, mediated by the Ca^{2+}-activated protein phosphatase calcineurin (Kuhara et al. 2002). Animals lacking the calcineurin A subunit TAX-6 adapt more rapidly. Having several positive and negative pathways that regulate adaptation in different time windows may provide flexibility to olfactory responses in a complex and dynamic environment.

By discriminating between different smells, C. elegans can also chemotax to one odor in a uniform saturating concentration of another odor (Bargmann et al. 1993). Odor discrimination occurs rapidly, within a few minutes. One way to discriminate odors would be to segregate their perception to different neurons that can act independently to direct odortaxis. In this way, C. elegans distinguishes odors detected by the AWA neurons from those detected by AWC. Moreover, the animal programs the two AWC neurons to express different odor receptors; thus, they mediate partly nonoverlapping odor responses that can be discriminated (Troemel et al. 1999, Wes & Bargmann 2001). Asymmetry between the two AWC neurons is achieved by a stochastic, calcium-dependent intercellular signaling mechanism (Troemel et al. 1999).

Segregating odor receptors to three neuron types should permit discrimination of three classes of odors, but C. elegans can discriminate at least seven odor classes (Bargmann et al. 1993). The mechanisms underlying the additional discrimination of odors sensed by the same neuron are unclear. One possibility is that signaling complexes for different odors are segregated from one another in complexes that can be independently downregulated (L'Etoile & Bargmann 2000,

Montell 1999). Alternatively, different odor receptors could recruit distinct regulators that selectively turn off the cognate olfactory response.

Most studies of C. elegans olfactory behavior have focused on events that occur in the AWA and AWC olfactory neurons themselves. Little is known about how signaling from olfactory neurons orchestrates the activity of downstream interneurons to generate chemotaxis behavior. One study suggests that odor sensitivity can be enhanced by expressing a hyperactive UNC-86 POU transcription factor in the AIZ interneurons that are downstream targets of olfactory neurons, which suggests that olfaction may also be modulated at the interneuron level (Sze & Ruvkun 2003).

The richness of C. elegans olfactory behavior makes it particularly attractive for study with electrophysiology or calcium imaging methods. Thus far both these approaches have been difficult because of technical limitations, but linking the extensive genetic and molecular studies of olfaction with physiology remains an important goal.

Thermosensation

C. elegans can grow and reproduce between approximately 12°C and 27°C. High temperatures (>33°C) induce a reverse-turn-and-run escape response, the molecular and neural basis of which is poorly understood (Wittenburg & Baumeister 1999). Within its viable thermal range well-fed C. elegans seeks its most recent cultivation temperature; once at this temperature the animal tracks isotherms (Hedgecock & Russell 1975, Mori & Ohshima 1995). Thermotaxis is part of a food-searching strategy. Feeding animals at a new temperature for 2–4 h resets the preferred temperature to that now associated with food. By contrast, depriving animals of food for 4–6 h induces avoidance of the starvation-associated temperature (Hedgecock & Russell 1975, Mori 1999). In behavioral terms, C. elegans migrates toward its preferred temperature by modulating turning rate and run length as a function of temperature change (Ryu & Samuel 2002, Zariwala et al. 2003). Once within 3°C of its preferred temperature *C. elegans* switches to a different locomotory pattern, tracking isotherms that can be less than 0.05°C wide by constantly reorienting head movement (Hedgecock & Russell 1975, Mori & Ohshima 1995, Ryu & Samuel 2002). How does *C. elegans* sense ambient temperature? How does it store and update information about past cultivation temperature? How is thermal information associated with information about favorable (food present) or unfavorable (food absent) growth conditions to determine the response to a temperature gradient?

The main *C. elegans* thermosensors are a pair of head neurons called AFD, which have ciliated sensory endings located at the tip of the nose (Mori & Ohshima 1995) (**Figure 2B**). Animals lacking AFD neurons are either athermotactic or cryophilic (they migrate to cold parts of a thermal gradient irrespective of cultivation temperature) and fail to track isotherms. The residual cryophilia of AFD(−) animals suggests that *C. elegans* has at least one other thermosensory neuron besides AFD, but its identity is unknown (Cassata et al. 2000, Hobert et al. 1997, Mori & Ohshima 1995). A characteristic of AFD and thermosensory neurons in other nematodes is sensory endings with elaborated membrane development (Li et al. 2000, Lopez et al. 2000, Ward et al. 1975, Ware et al. 1975). In the case of AFD, the sensory ending is composed of numerous finger-like projections. Mutations that disrupt formation of these microvilli also disrupt normal thermosensation (Cassata et al. 2000, Perkins et al. 1986, Satterlee et al. 2001).

The temperature sensor(s) in AFD neurons is unknown, but thermal cues may be transduced by a cGMP pathway: AFD neurons express a cGMP-gated cation channel encoded by the *tax-2* and *tax-4* genes that is required for thermosensation (Coburn & Bargmann 1996; Hedgecock & Russell 1975; Komatsu et al. 1996, 1999). Like vertebrate cyclic nucleotide-gated channels, TAX-2/TAX-4 channels are permeable to both Na^+ and Ca^{2+} ions (Komatsu et al. 1999). In vivo imaging studies using cameleon show that Ca^{2+} levels transiently rise in the cell body of AFD neurons in response to warming above a threshold temperature (Kimura et al. 2004). This threshold is approximately the same as the cultivation temperature. Warming above the threshold—but not cooling—induces a rise in Ca^{2+} in the AFD cell body. As for the thermotaxis behavioral response itself, the threshold temperature for the Ca^{2+} response can be altered by cultivation temperature experience. The Ca^{2+} increase requires *tax-4* function; its transience suggests that it is evoked by temperature change rather than by absolute temperature (Kimura et al. 2004). Thus the AFD neurons encode at least one bit of information: They sense when animals are moving up temperature gradients that exceed the cultivation temperature. Moreover the "memory" of cultivation temperature is encoded, at least in part, either in the properties of AFD or in the properties of neurons presynaptic to AFD (Kimura et al. 2004).

Synaptic release from AFD neurons in animals kept at fixed temperatures has also been studied by examining fluorescence recovery after photobleaching (FRAP) of AFD synapses labeled with synaptobrevin::pHluorin (Samuel et al. 2003). pHluorin is more fluorescent and more easily photobleached at the cell surface than inside vesicles because the low pH inside vesicles inhibits photon absorption. The rate of FRAP therefore provides a measure of synaptobrevin::pHluorin released to the AFD cell surface. Measured by this method, synaptic release from AFD appears to be low when ambient temperature is equal to the most recent cultivation temperature, but high when the two temperatures are different (Samuel et al. 2003). These results suggest that AFD can decode the relative difference between ambient and cultivation temperatures. Although measurements of AFD neural activity using cameleon and synaptopfluorin yield different results, these likely reflect the different conditions under which cells were imaged. The cameleon experiments monitored responses to transient changes in ambient temperature, whereas the synaptopHluorin studies measured synaptic release at constant temperatures. The complex thermosensory properties of AFD elucidated by the two methods raise the possibilities that this neuron has multiple thermosensory pathways or that feedback from other thermosensory neurons contributes to AFD function.

How is recent cultivation temperature encoded? How is it altered by thermal experience? One tantalizing possibility is that different cultivation temperatures elicit activity-dependent changes in gene expression in AFD (Satterlee et al. 2004). Supporting this possibility, expression of some genes in AFD neurons is regulated by the TAX-4 cGMP-gated ion channel and the calcium/calmodulin–dependent protein kinase I (*cmk-1*) in a temperature-dependent manner (Satterlee et al. 2004). Several transcription factors that are expressed in AFD and that could be involved in reprogramming AFD according to cultivation temperature have been identified (Cassata et al. 2000; Satterlee et al. 2001, 2004). The interdependence of thermosensation and the resetting of the memory of cultivation temperature in AFD suggests it may be difficult to tease apart the role of genes in the two processes using behavioral phenotypes alone.

Appropriate AFD function requires the calcium-dependent phosphatase calcineurin, whose A subunit is encoded by the *tax-6* gene (Kuhara et al. 2002). Lack of *tax-6* in AFD leads to constitutive migration to high temperatures irrespective of cultivation temperature (thermophilic behavior), whereas constitutive calcineurin activity in AFD results in athermotactic or cryophilic behavior, as if AFD is inactive (Kuhara et al. 2002). These phenotypes implicate calcineurin in either thermal transduction or in setting cultivation temperature memory or both. Moni-

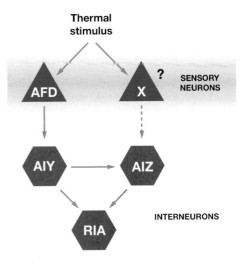

Figure 4

Simplified circuit for thermosensory responses. Arrows signify chemical synapses defined by the EM reconstruction of the *C. elegans* nervous system (White et al. 1986) and first implicated in thermotaxis behavior by laser ablation studies (Mori & Ohshima 1995). The thermosensory neuron X is predicted to exist because animals lacking AFD neurons retain the ability to migrate to cold parts of a thermal gradient.

toring neuronal activity, for example using cameleon, may help tease apart these alternatives (Kimura et al. 2004).

The AFD sensory neurons appear to mediate thermotaxis by regulating a neural circuit that includes the interneurons AIY and AIZ (Mori & Ohshima 1995) (**Figure 4**). AIY is the major postsynaptic target of AFD and makes chemical synapses onto AIZ (**Figure 2C**) (White et al. 1986). Animals in which AIY has been ablated, either genetically or using a laser microbeam, are strongly cryophilic and have defects in isothermal tracking (Gomez et al. 2001, Hobert et al. 1997, Mori & Ohshima 1995). In contrast, animals lacking AIZ are strongly thermophilic (Hobert et al. 1998, Mori & Ohshima 1995). One simple model that elegantly explains these data is that AIY and AIZ mediate opposing drives and that *C. elegans* accumulates at the balance point of these drives (Mori & Ohshima 1995, Mori 1999) (**Figure 4**).

How do AFD and its associated neural circuitry regulate locomotion? *C. elegans* increases reversal frequency when it experiences a rise in temperature above cultivation temperature (Ryu & Samuel 2002, Satterlee et al. 2004), and the cameleon studies suggest that such a temperature rise activates Ca^{2+} influx in AFD (Kimura et al. 2004). Laser or genetic ablation of AFD and AIY neurons has opposite effects on reversal frequency: Loss of AFD suppresses spontaneous reversals, whereas ablation of AIY increases spontaneous reversals (Tsalik & Hobert 2003). The hyporeversal phenotype of AFD(−) animals appears to be mediated through AIY because it is not observed in AFD(−); AIY(−) animals (Tsalik & Hobert 2003). Together, the ablation and cameleon data suggest a model in which activation of the AFD neurons by warming above a threshold temperature can induce *C. elegans* to reverse by inhibiting AIY. Less is known about the mechanism by which *C. elegans* responds to cooling below the ambient temperature. However, the hyper-reversal phenotype of AIY-ablated animals is temperature dependent: It is observed at 25°C but not at lower temperatures (Tsalik & Hobert 2003). Because AFD(−); AIY(−) animals also show the hyper-reversal, this regulation may be mediated by another temperature-sensitive process, which further supports the notion that *C. elegans* has another thermosensory neuron apart from AFD (**Figure 4**).

Ablating AIZ neurons, the major postsynaptic target of AIY (**Figure 2C**), also suppresses spontaneous reversals (Tsalik & Hobert 2003, Wakabayashi et al. 2004). Thus, ablations of AIY and AIZ cause opposite behavioral phenotypes. One interpretation is that the synapses made by AIY onto AIZ are inhibitory. Alternatively information from AIY and AIZ could be integrated in a downstream neuron. Consistent with this possibility AIY and AIZ both make synapses onto RIA, and animals lacking RIA are either cryophilic or athermotactic (Mori & Ohshima 1995).

Finally, how does feeding state modulate the thermotaxis circuitry so that well-fed animals seek cultivation temperature, whereas starved ones avoid it? The mechanism for this plasticity remains mysterious, but shutting off thermotaxis to the cultivation temperature in response to starvation requires the secreted LDL receptor motif–containing protein HEN-1, which is expressed in the ASE and AIY neurons (Ishihara et al. 2002). Moreover, exogenous serotonin induced thermotaxis to the cultivation temperature in food-deprived animals, whereas exogenous octopamine induced well-fed animals to avoid cultivation temperature (Mori 1999). These two neuromodulators have been postulated to signal the well-fed and starved state, respectively, in *C. elegans* (Avery & Horvitz 1990, Horvitz et al. 1982, Sawin et al. 2000, Segalat et al. 1995).

CALCIUM OSCILLATIONS AND RHYTHMIC BEHAVIOR: THE DEFECATION MOTOR PROGRAM

Rhythmic activities are ubiquitous biological phenomena and can be observed in cells, tissues, and the behavior of most organisms. Perhaps the best known of these are circadian rhythms, which are entrained by the 24-h light-dark cycle. A recent report suggests that worm locomotory speed (Saigusa et al. 2002) cycles on an approximately 24-h time course [see also cycling of resistance to stress (Kippert et al. 2002)]. These circadian behaviors could be studied using genetic analysis, especially since relatives of known timekeeper genes are present in *C. elegans*.

Other behaviors occur rhythmically with periods of seconds to hours (known as ultradian rhythms). Many of these are driven by oscillations in free intracellular Ca^{2+} concentration (Berridge et al. 2000). Three well-characterized rhythmic behaviors in *C. elegans* are pharyngeal pumping (which controls feeding), defecation, and gonadal sheath cell contractions (which controls egg laying). The periodicities of the rhythms range from subsecond (pharynx) (Avery & Thomas 1997), to seconds (gonadal sheath) (Yin et al. 2004), to minutes (detection) (Liu & Thomas 1994). The molecular mechanisms underlying all these rhythmic behaviors are being intensively studied, but for reasons of space we can review only studies of defecation behavior. Excellent reviews of other rhythmic behaviors can be found in a new online book on *C. elegans* (*WormBook*; **http://www.wormbase.org/**).

Defecation Motor Program

The *C. elegans* hermaphrodite defecates approximately every 50 s while it is eating (Avery & Thomas 1997, Croll & Smith 1978, Liu & Thomas 1994, Thomas 1990). The defecation motor program (DMP) is highly stereotyped and is composed of three invariant steps (**Figure 5**): posterior body contraction (pBoc), anterior body contraction (aBoc), and enteric muscle contraction (Emc) that results in expulsion (Exp) (Thomas 1990). The DMP begins with the simultaneous contraction of dorsal and ventral posterior body-wall muscles (pBoc). This contraction pressurizes the intestine and drives its contents forward. Approximately 2 s later, the dorsal and ventral anterior body-wall muscles simultaneously contract (aBoc) and the pharynx is driven backward, also pressurizing the gut. Shortly after the aBoc, the anal musculature opens the anus (Emc), releasing the gut contents (Exp). The entire motor program of contractions lasts about 5 s. Remarkably, the standard deviation for the cycle length (pBoc-pBoc) of feeding worms is only about 3 s (Liu & Thomas 1994). How is this behavior so tightly regulated?

The periodicity of the DMP is mostly independent of pharyngeal pumping (feeding rate), which suggests that these two rhythmic behaviors are regulated by different oscillators (Liu & Thomas 1994, Thomas 1990). Worms with mutations in genes such as *eat-2* or *unc-89* have slow irregular pharyngeal pumping and poor food intake but have a near normal defecation cycle, which suggests that defecation is not triggered simply by stretching of the intestines. However, in certain situations food intake does affect the duration of the cycle. For example, cycle duration increased almost twofold when the concentration of food is decreased 20-fold (Liu & Thomas 1994), which indicates that some food-metering process influences the periodicity of the DMP. Interestingly, the phase of the DMP is maintained even when the worm stops feeding. If a worm leaves the lawn of bacterial food, pumping soon stops and the defecation motor program is not activated. However, when the worm returns to food the DMP is usually reactivated in phase with its previous DMP cycle (Liu & Thomas 1994). Lightly prodding the worm provides a signal that resets the DMP to the time of mechanosensory stimulation. Surprisingly, the duration of the defecation cycle is unaffected by perturbations, such as mutations or laser ablation of cells that disrupt the expulsion step. Such animals become constipated with bloated intestines because they do not follow each pBoc with an Exp, but retain rather normal pBoc-pBoc intervals.

An additional property of the defecation oscillator is that it is stable between 19°C

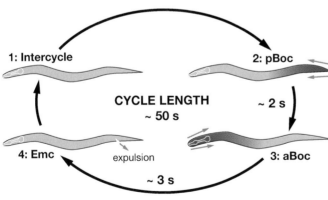

Figure 5

The defecation motor program. After the intercycle period (1), the program is initiated with contraction of posterior body-wall muscles (2, pBoc) followed by an anterior body-wall muscle contraction (3, aBoc) and subsequent enteric muscle contraction (4, Emc) that results in the expulsion of intestinal content. aBoc to Emc takes about 5 s, whereas the cycle duration from pBoc to pBoc is approximately 50 s. Figure adapted from Thomas 1990.

and 30°C—below 19°C the period dramatically increases with lowered temperature (Liu & Thomas 1994). These data, obtained from animals raised at one temperature, suggest that a temperature-compensated oscillator (or "clock") controls the DMP. However, a different conclusion is reached when animals are grown at one temperature and tested at another. For example, if animals are raised at 20°C and then shifted to 25°C, the cycle length is decreased by 13 s (Branicky et al. 2001), suggesting an interaction between temperature and oscillator frequency. In this study, temperature compensation was observed only in a clk-1 mutant background (see below). Regardless of whether a clock is regulating the cycle, or whether regulation is by a stable oscillator that can be modified by experience, the locus of the oscillator, its molecular make-up, and the signaling pathways upstream and downstream of the oscillator are fascinating topics of study.

Genetic Analysis of the Defecation Motor Program

Mutations that affect the DMP can be conveniently divided into those that affect defecation (disrupt pBoc, aBoc, or Exp) and those that alter the period of the clock (generally defined as the pBoc-pBoc interval) (Iwasaki et al. 1995, Thomas 1990). Relatively few mutations specifically affect pBoc, whereas mutations that disrupt aBoc or Exp, or both aBoc and Exp (*aex* genes), are more numerous (Thomas 1990).

The expulsion step is dependent on the neurotransmitter GABA. Mutations in *unc-25* (glutamic acid decarboxylase) or *unc-47* (vesicular GABA transporter), which are respectively required for production and packaging of GABA, cause a constipated phenotype because of the failure of enteric muscle contractions. The enteric muscles are innervated by the AVL and DVB GABAergic neurons (McIntire et al. 1993b, White et al. 1986). Like loss of GABAergic transmission, laser ablation of AVL and DVB causes severe constipation owing to defective expulsion (Liu & Thomas 1994). These data suggest an unexpected excitatory function for GABA either directly via ionotropic channels gated by GABA or indirectly via G protein–coupled receptors.

Worms with a mutation in the *exp-1* gene have a specific defect in the expulsion step, which suggests that *exp-1* may encode a receptor for GABA (Thomas 1990). Indeed, cloning of *exp-1* revealed that it encodes a member of the ligand-gated superfamily of ionotropic receptors (Beg & Jorgensen 2003). Expression of EXP-1 in *Xenopus* oocytes demonstrated that EXP-1 functions as a GABA-activated cation-conducting ion channel. This was the first demonstration that GABA can have an excitatory role by activating a cation-selective channel. The EXP-1 channel is expressed in the enteric musculature and concentrated at the neuromuscular junctions from AVL and DVB. These results nicely explain how GABA release by AVL and DVB promotes the depolarization of the postsynaptic membrane and contraction of the enteric muscles (Beg & Jorgensen 2003).

The enteric contractions mediated by EXP-1 can be modulated by the neurotransmitter serotonin. Exposing worms to exogenous serotonin (5-HT) can inhibit enteric muscle contractions. This inhibition is dependent on the $G_o\alpha$ encoded by *goa-1* (Segalat et al. 1995), although defecation itself is normal in *goa-1* mutants. Because GOA-1 is widely expressed in neurons and muscles (including the enteric muscles), it is unclear whether 5-HT directly inhibits enteric muscle or acts in the neurons that provide synaptic input to the muscle.

One might predict that mutations in genes that disrupt synaptic transmission would also disrupt steps in the defecation cycle. Indeed many gene products with roles in synaptic communication are involved in some aspect of defecation (Thomas 1990). These include genes that may be involved in synapse formation (Ginzburg et al. 2002), Ca^{2+} entry (Mathews et al. 2003), and synaptic vesicle release (Mee et al. 2004, Nonet et al. 1993). One important example is *unc-43*, the gene that encodes the only calcium/calmodulin-dependent serine/threonine kinase type II (CamKII) in *C. elegans* (Reiner et al. 1999, Rongo & Kaplan 1999). Both loss-of-function (*lf*) and gain-of-function (*gf*) alleles of *unc-43* have been identified, and these have opposite effects on the DMP. Thus, the Exp step is missing in 85% of defecation cycles in *unc-43(gf)* mutants, and the cycle length is longer and more variable. In contrast, in *unc-43(lf)* mutants the cycle length is shorter in duration (Reiner et al. 1999). In this study, one likely target of *unc-43* was identified. *unc-103* encodes a potassium channel of the *eag* family. Members of this family regulate cardiac excitability in vertebrates. Loss of *unc-103* does not cause defecation defect, but does suppress the lack of enteric contractions in *unc-43(gf)* mutants. Thus, *unc-103; unc-43(gf)* double mutants have a nearly normal Exp frequency, which suggests that UNC-103 may be a direct target of UNC-43 CamKII.

Screens for mutations that disrupt defecation (constipated phenotype) have also identified new gene products that are generally required for synaptic transmission. *aex-3* mutants have normal pBoc but frequently lack both aBoc and Exp steps (Aex phenotype) (Thomas 1990). *aex-3* encodes a large predicted protein related to the DENN protein in humans and to rat Rab3 GEF (Iwasaki et al. 1997, Iwasaki & Toyonaga 2000). Synaptic transmission in the pharynx and at body-wall muscles is impaired in *aex-3* mutants, presumably a consequence of the observed mislocalization of RAB-3. RAB-3 is a small GTPase that may regulate recruitment of synaptic vesicles to the active zone (Nonet et al. 1997). Surprisingly, however, *rab-3* mutants do not have an Aex phenotype, which suggests that AEX-3 and RAB-3 may function in parallel pathways. In support of this hypothesis, AEX-3 was found to bind to a novel protein CAB-1, which is expressed in neurons (Iwasaki & Toyonaga 2000). *cab-1* mutants have aBoc and Exp defects, and a *cab-1; rab-3* double mutant has approximately the same overall phenotype as that observed in *aex-3* mutants. Thus, although *aex-3* may mediate aBoc and Exp via its effects on synaptic transmission, it may have more general effects on signal transduction.

The cloning of another *aex* gene identified a novel pathway for retrograde regulation of synaptic transmission. *aex-1* mutants have defects in defecation as well as disrupted neuromuscular synaptic transmission (Doi & Iwasaki 2002). *aex-1* encodes a protein that is distantly related to the family of UNC-13/Munc13 proteins. These proteins are presynaptic and regulate the fusion of synaptic vesicles in response to depolarization (Martin 2002). In contrast, AEX-1 is expressed in muscle and intestinal cells. Surprisingly, selective expression of AEX-1 in intestinal cells or muscle cells rescues the defecation and neuromuscular transmission defects, respectively. These results suggest the existence of a new signaling mechanism, where *aex-1* is required for retrograde release of vesicles. A mutation in *aex-5* has defecation defects similar to *aex-1*. *aex-5* encodes a furin-like prohormone convertase (Thacker & Rose 2000) and may function in a pathway with *aex-1* to enable activity-dependent retrograde peptidergic signaling (Doi & Iwasaki 2002). Clearly, the identities of the peptidergic signal and its receptor are of intense interest.

Control of the Defecation Cycle: A Complex Clock

How is defecation periodicity maintained? Genetic screens for altered periodicities revealed mutants with either shorter or longer cycles (Iwasaki et al. 1995). In the former class was the mutant *flr-1*, originally identified in a screen for mutants resistant to extracellular fluoride. Compared with wild-type worms with a mean cycle length of 47 ± 6 s, *flr-1* mutants have a shorter and more variable cycle of 29 ± 18 s (Take-Uchi et al. 1998). *flr-1* encodes an ion channel of the DEG/ENaC superfamily; many of these channels are thought to be mechanosensitive and permeable to

Na$^+$. When open, they would therefore tend to depolarize the cell (Kellenberger & Schild 2002). FLR-1 expression appears limited to intestinal cells, which suggests that mechanical deformation of intestinal cells might contribute to the defecation cycle.

A fundamental insight into the molecular regulation of the defecation cycle came from the analysis of the *dec-4/itr-1* mutant. *dec-4* mutants have a long defecation cycle (Iwasaki et al. 1995). A stronger allele of *dec-4* was subsequently identified and cloned (Dal Santo et al. 1999). Because the gene was shown to encode an inositol trisphosphate receptor, it was renamed *itr-1*. A genetic analysis of available *itr-1* alleles revealed a key role for ITR-1 in the defecation cycle. A strong allele (*n2559*) eliminated the cycle such that no defecations were observed in a 10–20 min window. A weaker allele (*sa73*) had a cycle length nearly twofold increased in duration, whereas overexpression of the wild-type ITR-1 in transgenic worms decreased the cycle length. (See, however, Jee et al. 2004. In this study, gain-of-function *itr-1* alleles had normal cycle lengths.) In accordance with an earlier study implicating the intestine in control of cycle length (Take-Uchi et al. 1998), careful mosaic analysis revealed that ITR-1 functions in intestinal cells to control the defecation cycle (Dal Santo et al. 1999). Furthermore, using the fluorescent indicator fura-2 to monitor Ca^{2+} oscillations, the authors were able to measure in vivo Ca^{2+} transients in intestinal cells from wild-type and mutant worms. A rise in cytoplasmic Ca^{2+} occurred immediately prior to the onset of pBoc. In wild-type worms, the Ca^{2+} transients occurred every ∼50 s, whereas the rate was slower in *itr-1(sa73)* mutants and nonexistent in the *itr-1(n2559)* mutants. These important results demonstrated that ITR-1 is a central component of the timekeeper for defecation, and the cellular locus of the timekeeper is in the intestine.

A subsequent study supported key roles for IP$_3$ and ITR-1 in the control of the defecation cycle. IP$_3$ levels were manipulated in transgenic worms by using an inducible promoter to express the IP$_3$ binding region of the ITR-1 receptor (Walker et al. 2002a). Expression of the "sponge" increased the mean cycle duration but also increased the variability of the cycle. Thus, decreasing IP$_3$ levels led to a dramatically decreased rhythmicity of the defecation cycle. This contrasts what was found in *itr-1(sa73)* mutants. Here, the mean cycle length was increased, but variability was not affected (Dal Santo et al. 1999).

Activation of cell-surface receptors coupled to phospholipase C (PLC) causes the production of IP$_3$ (Berridge et al. 2000, 2003). In *C. elegans*, the *egl-8* gene encodes a PLCβ, and mutations in this gene cause defects in synaptic transmission as well as in the defecation cycle (Lackner et al. 1999, Miller et al. 1999, Thomas 1990). The defecation defect is relatively specific to the pBoc step. *egl-8* mutants have a weak or absent pBoc but have a relatively normal cycle duration (Lackner et al. 1999, Thomas 1990). The EGL-8 PLCβ protein is expressed in many neurons, explaining the synaptic defects observed in *egl-8* mutants. Interestingly, most synaptic mutants do not affect the strength of pBoc. However, EGL-8 PLCβ is also expressed in posterior intestinal cells, where it may signal neighboring body-wall muscles to initiate the pBoc step of the DMP.

Gene products that directly interact with ITR-1 are also important for the defecation cycle. A yeast two-hybrid screen identified IP$_3$ receptor interacting protein (IRI-1), a novel protein related only to LIN-15B, a regulator of vulval growth (Walker et al. 2004). IRI-1 is coexpressed with ITR-1 in many tissues, including the intestines. Knockdown of IRI-1 levels using RNAi moderately increases the mean cycle duration but dramatically increases the variability in cycle length. Moreover, *iri-1* RNAi treatment of *itr-1(sa73)* mutants causes a dramatic increase in both the variability and duration of the defecation cycle (Walker et al. 2004). How IRI-1 modulates the ITR-1 receptor is unclear. Because IRI-1 has proline-rich domains that may bind to

SH3 domains, IRI-1 may play a role in localizing ITR-1 or organizing other components of the signaling pathway.

In the postsynaptic membranes of vertebrate neurons, members of the Shank family of scaffolding proteins are involved in the assembly and localizaton of proteins such as ligand-gated receptors (Sheng & Kim 2000). In *C. elegans*, a homolog of Shank, SHN-1, is expressed in many tissues, including the intestines (Jee et al. 2004). RNAi of *shn-1* does not modify the timing of the defecation cycle in wild-type worms but significantly increases the length and variability of cycles in *itr-1(sa73)* mutants. *shn-1* RNAi has no effect on the cycle properties of *itr-1* gain-of-function mutants. By immuno-gold criteria, SHN-1 appears localized to vesicular structures in intestinal cells. Perhaps, SHN-1 has a role in optimally localizing the ITR-1 receptor or accessory proteins. ITR-1 may need to be localized to maximize binding to the IP_3 ligand or to localize Ca^{2+} efflux to specific downstream signaling pathways.

Another interesting gene required for normal-length defecation cycles is *clk-1*. This gene encodes a mitochondrial protein required for the synthesis of ubiquinone, a lipid that functions in the respiratory chain as an electron transporter. All known alleles of *clk-1* eliminate ubiquinone but have differential effects on the length of the defecation cycle (Branicky et al. 2001, Wong et al. 1995). *clk-1* mutants have a longer defecation cycle than do wild type and notably do not shift their cycle length when temperature is shifted (Branicky et al. 2001). How *clk-1* participates in the timing of the defecation cycle is unknown. One strategy to identify other genes in this signaling pathway is to identify genetic suppressors of the *clk-1* phenotype. This strategy shows promise, and to date seven defecation suppressor of *clk-1* (*dsc*) mutants have been identified (Branicky et al. 2001). The cloning and molecular characterization of these *dsc* genes may provide important insights into how the defecation cycle is regulated.

Ca^{2+} Oscillations and Rhythm Generation: Future Directions

Signaling by oscillations, rather than by shifting the level of stationary Ca^{2+}, is believed to confer certain physiological advantages, including increased signal detection and prevention of desensitization. In addition, the amplitude and frequency of the oscillations can be controlled and tuned to specific downstream signaling pathways, leading to increased signaling specificity (Schuster et al. 2002). For example, the frequency of intracellular Ca^{2+} concentration differentially activates transcription factors (Dolmetsch et al. 1998). Thus, it is of great interest to understand the molecular requirements for the establishment of Ca^{2+} oscillations.

The genetic strategies and tools available to *C. elegans* researchers have allowed fundamental new insights into the regulation of rhythmic behaviors. Two important observations should provide the focus for future studies. First, the ITR-1 IP_3 receptor is important for the control of the duration and periodicity of most rhythmic behaviors in *C. elegans*, including pharyngeal pumping (Walker et al. 2002a,b, 2004) and gonadal sheath cell contractions (Yin et al. 2004). In particular, ITR-1 has a key role in the defecation cycle. Second, ITR's key role in the control of the cycle takes place in intestinal cells. Thus, future studies will likely focus on what controls the oscillations in intestinal cells and the nature of the signaling pathway that translates oscillatory changes in intestinal Ca^{2+} concentration to body-wall muscle contraction and the depolarization of neurons.

Much remains to be known about how the numerous Ca^{2+} conductances in most cells contribute to oscillatory changes in Ca^{2+} concentration. Although the ITR-1 receptor certainly has an important role, other conductances may have equally important roles. Thus, most cells have store-operated Ca^{2+} channels (SOCC), which allow Ca^{2+} to enter the cytoplasm and in some cases contribute to Ca^{2+} oscillations.

Second-messenger-operated Ca^{2+} channels (SMOCC) also contribute to this process. A necessary first step toward understanding the regulation of Ca^{2+} oscillations in intestinal cells is to study the properties of the ionic conductances present in these cells. In a recent study, patch-clamp recordings of currents from identified cultured *C. elegans* intestinal cells revealed two different Ca^{2+} conductances (Estevez et al. 2003). One current (I_{ORCa}) was blocked by intracellular Mg^{2+} but was not activated by depletion of intracellular Ca^{2+} stores. A second current (I_{SOC}) had many of the characteristics of store-operated Ca^{2+} conductances, including activation by intracellular IP_3. The molecular identity of the proteins that contribute to I_{SOC} has not yet been determined in any system, although certain TRP channels may be candidates. This important study lays the groundwork for further studies that should link gene products to intestinal cell electrophysiological properties and the behavior of the organism.

Whether signals that modulate cycle length (dilute food, temperature shift) change Ca^{2+} oscillations in intestinal cells is currently unknown. Presumably, the application of cameleon-based fluorescence measurement techniques (Kerr et al. 2000) to the study of Ca^{2+} transients in the intestines of wild-type and mutant *C. elegans* should provide a powerful means of studying the dynamics of the Ca^{2+} changes. For example, does the Ca^{2+} oscillation begin in the posterior intestinal cells and spread anteriorly through gap junctions? These imaging techniques should also help address the question of whether the ENaC channel encoded by *flr-1* is required for normal Ca^{2+} oscillations and whether other signaling pathways modulate the size or frequency of Ca^{2+} transients.

The related question of how Ca^{2+} transients in intestinal cells are coupled to activation of neighboring neurons and muscles, is a major challenge. Again, using RNAi to test whether candidate genes disrupt pBoc or cycle duration should lead to much progress in the near future.

COMPLEX CONSEQUENCES OF SIMPLE VARIATION: AGGREGATION AND AEROTAXIS

Isolates of *C. elegans* have been collected from many different parts of the world (Hodgkin & Doniach 1997). Differences between these strains provide an opportunity to investigate at a molecular level the basis for natural phenotypic variation. One striking behavioral difference between wild *C. elegans* strains is their response to bacterial food (**Figure 6A,B**) (de Bono & Bargmann 1998, Hodgkin & Doniach 1997). A subset of wild races, including the standard laboratory strain N2, isolated in Bristol, England, respond to bacteria by reducing locomotory activity and dispersing over a bacterial food lawn to feed as individuals. In contrast a different set of strains, exemplified by an isolate from Hawaii called CB4856, continue moving rapidly on encountering food, strongly accumulate where bacteria grow thickest (a behavior termed bordering), and burrow. Upon encountering one another these animals dramatically change their locomotory behavior to form groups that can contain from two to hundreds of individuals.

The constellation of behavioral differences between solitary and social *C. elegans* is associated with two natural alleles of a neuropeptide Y-like G protein–coupled receptor called NPR-1. The two alleles differ at a single amino acid position: Aggregating strains encode phenylalanine (F) at position 215 of the receptor, whereas dispersing strains encode valine (V) (de Bono & Bargmann 1998) (**Figure 6C**). Animals heterozygous for the two natural alleles are solitary feeders, indicating that the *npr-1 215V* allele is genetically dominant to *npr-1 215F*. Chemically generated null mutations of *npr-1* induce solitary wild strains to exhibit the behavioral phenotypes associated with social wild strains, which suggests that this neuropeptide receptor represses aggregation and associated behaviors. In evolutionary terms *npr-1 215F*

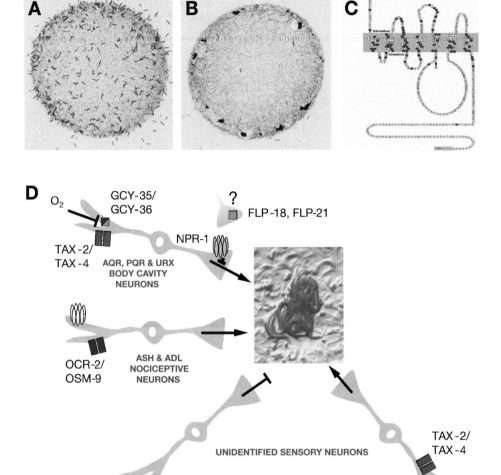

Figure 6

Aggregation behavior. Animals from "solitary" wild isolates of *C. elegans* disperse to feed alone (*A*), whereas animals from "social" strains aggregate into groups and accumulate preferentially where food is thickest, on the border of the bacterial lawn (*B*). Panels *A* and *B* show N2 (Bristol) and *npr-1* animals, respectively. (*C*) Natural variation at residue 215 of NPR-1 is associated with solitary and social feeding. Aggregating wild strains encode NPR-1 215F, whereas dispersing strains encode NPR-1 215V. The residues highlighted in red are conserved in the human Y4 NPY receptor. The -TLV motif at the C-terminus has the potential to bind PDZ-containing proteins. (*D*) Shown are neurons and signaling pathways that regulate aggregation behavior.

appears to be the ancestral *npr-1* allele: Only this allele has been found in the related nematode species *C. briggsae*, *C. remanei*, and *C. sp* CB5161 (Rogers et al. 2003). Moreover all *C. elegans* wild strains that bear the *npr-1 215V* allele are closely related (Denver et al. 2003), which suggests that this allele arose once in *C. elegans* as a gain-of-function allele that spread to different parts of the world (Rogers et al. 2003).

Differential Activation of the Two Natural NPR-1 Isoforms by FLP-18 and FLP-21 Peptides

How do the two natural NPR-1 isoforms differ in their signaling properties? Residue 215 lies at the N-terminal region of transmembrane domain 5 of NPR-1, at a position implicated in G protein coupling in other receptors (**Figure 6C**). A clue to how variation at this residue affects NPR-1 signaling came from studying the signaling properties of these receptors in heterologous systems (Kubiak et al. 2003, Rogers et al. 2003). Although NPR-1 bears resemblance to neuropeptide Y (NPY) receptors, the *C. elegans* genome does not appear to encode NPY-like sequences. However, *C. elegans* encodes at least 60 peptides that belong to the FMRFamide family of neuropeptides; these are encoded by 23 *flp* (FMRFamide-like peptide) genes (Kim & Li 2004, Li et al. 1999a). Synthesized versions of these peptides have been tested for their ability to activate NPR-1 in three preparations: one in *Xenopus* oocytes, using K$^+$ currents carried by G protein–coupled inward rectifying (GIRK) channels as a readout; one in *C. elegans* pharyngeal muscle, where the readout was frequency of action potentials; and a third in Chinese hamster ovary cells (CHO cells) looking at ligand-stimulated recruitment of GTPγS to the membrane (Kubiak et al. 2003, Rogers et al. 2003). All three assays identified a peptide encoded by the *flp-21* gene as a cognate ligand for NPR-1. The *Xenopus* oocyte and pharyngeal assays also identified peptides encoded by *flp-18* as NPR-1 ligands. Both in *Xenopus* oocytes and in CHO cells, the FLP-21 peptide was more potent at the genetically dominant NPR-1 215V receptor associated with solitary feeding than at the NPR-1 215F receptor. Moreover, whereas FLP-18 peptides potently activated the NPR-1 215V receptor, they did not activate the NPR-1 215F receptor in *Xenopus* oocytes. Thus, natural variation at residue 215 can dramatically alter NPR-1 signaling and ligand coupling. FLP-18 peptides can signal through both NPR-1 215F and NPR-1 215V receptors in the pharyngeal assay, which suggests that signaling differences depend on the cellular context of the receptor.

Both loss-of-function and gain-of-function studies confirm that FLP-21 acts in vivo as an NPR-1 ligand that regulates aggregation and bordering behaviors (Rogers et al. 2003). Transgenic overexpression of *flp-21* can act through the NPR-1 215F receptor to suppress these behaviors, and deleting *flp-21* enhances aggregation of *npr-1* 215F animals. However, loss of *flp-21* only weakly increases aggregation and bordering in animals bearing the *npr-1* 215V allele, which suggests that a second ligand can activate this receptor to inhibit social feeding. The in vitro data suggest this second ligand could be encoded by *flp-18*.

How Do the Two Alleles of *npr-1* Lead to Such Dramatic Differences in Behavior?

NPR-1 215V acts in several neural circuits to reconfigure sensory responses. One of these circuits includes the body cavity neurons AQR, PQR, and URX. Targeted expression of NPR-1 215V or of a constitutively active K$^+$ channel, EGL-2(gf) (Weinshenker et al. 1999) in AQR, PQR, and URX represses aggregation and bordering (Coates & de Bono 2002). NPR-1 215V acts by antagonizing another signaling pathway in AQR, PQR, and URX that promotes social feeding (**Figure 6D**). This pathway is mediated by a cGMP-gated ion channel encoded by the *tax-2* and *tax-4* genes (Coates & de Bono 2002). AQR, PQR, and URX express five soluble guanylate cyclase subunits called GCY-32, GCY-34, GCY-35, GCY-36, and GCY-37 (Cheung et al. 2004, Gray et al. 2004, Yu et al. 1997). Two of these, GCY-35 and GCY-36, are required for both aggregation and bordering behaviors and act as an α/β soluble guanylate cyclase to activate the TAX-2/TAX-4 channel and induce social feeding (Cheung et al. 2004, Gray et al. 2004).

Mammalian soluble guanylate cyclases are heme-binding proteins activated by gaseous ligands, most prominently by nitric oxide (NO) (Koesling 1999). *C. elegans* appears to lack nitric oxide synthase (Morton et al. 1999), raising the question, what regulates the activity of its soluble guanylate cyclases? Several pieces of data suggest that GCY-35/GCY-36 cyclase activity is regulated by dioxygen (O_2). First, the heme-binding domain of GCY-35 binds oxygen, a property that sets it apart from mammalian soluble guanylate cyclases (Gray et al. 2004). Second, *gcy-35* mutations disrupt *C. elegans* aerotaxis behavior. Wild-type *C. elegans* aerotax to a preferred ambient oxygen concentration of 5%–11%; in contrast *gcy-35* mutants tend to accumulate at higher oxygen concentrations (Gray et al. 2004). Third, biochemical experiments show that high oxygen levels repress the cyclase activity of *Drosophila* homologs of *C. elegans*–soluble guanylate cyclases (Morton 2004). Interestingly, both aggregation and bordering behaviors are repressed in a quantitative manner when ambient oxygen levels are reduced. Moreover, NPR-1 215V, which strongly inhibits social feeding, blunts *C. elegans* aerotaxis to 5%–12% oxygen when food is present, whereas NPR-1 215F does not (Gray et al. 2004). Together, these data suggest a model in which reducing oxygen derepresses GCY-35/GCY-36, leading to a rise in cGMP, which gates the TAX-2/TAX-4 cGMP channel (**Figure 6D**). By a mechanism that is as yet unclear, this leads to avoidance of high ambient oxygen, promoting aggregation and accumulation where oxygen-consuming bacteria grow most thickly.

GCY-35 and GCY-36 function in the URX neurons alone is sufficient for strong social feeding (Cheung et al. 2004). A major postsynaptic target of URX is AUA (White et al. 1986). NPR-1 215V functions in both the body cavity neurons and in AUA to repress aggregation and bordering behaviors, which suggests that it may act both presynaptically and postsynaptically to regulate these behaviors (Coates & de Bono 2002).

The *gcy-35/gcy-36* aerotaxis pathway is not always required for aggregation behavior (see below), which suggests that it is one of several sensory circuits that contribute to this behavior (Cheung et al. 2004). Supporting this idea, the cGMP-gated ion channel subunit TAX-4 is required in other neurons besides AQR, PQR, and URX for aggregation and bordering behaviors (Coates & de Bono 2002), although this neuron(s) has not yet been identified.

The Nociceptive Neurons ASH and ADL Promote Aggregation

The aggregation and bordering behaviors of *npr-1* animals are dependent on signaling from the nociceptive neurons ASH and ADL (de Bono et al. 2002). Ablation of either ASH or ADL does not reduce aggregation (de Bono et al. 2002). However, simultaneous ablation of both pairs of neurons transforms aggregating animals into solitary feeders. TRP-related OCR-2/OSM-9 channels are required for all known avoidance responses mediated by the ASH and ADL neurons (see above). Mutations in *ocr-2* or *osm-9* suppress social feeding, and, consistent with the ablation data, expression of OCR-2 in either ASH or ADL is sufficient to restore social feeding to *ocr-2* mutants (de Bono et al. 2002). The *odr-4* gene also acts in nociceptive neurons to promote aggregation and bordering (de Bono et al. 2002). *odr-4* encodes a novel transmembrane protein that localizes odorant receptors to sensory cilia (Dwyer et al. 1998). The requirement of ODR-4 suggests that nociceptive neurons signal through chemosensory receptors to promote aggregation and bordering behaviors.

Signaling from the ASH and ADL neurons is not absolutely required for *C. elegans* to aggregate into groups and accumulate where food is most abundant (de Bono et al. 2002). Instead, these neurons may serve to antagonize a signal from an unidentified neuron(s) that suppresses these behaviors. Such a scheme of balanced signaling may ensure that aggregation and bordering are exhibited only in the appropriate environmental contexts. Two pieces of data support this scenario. First, mutations in *osm-3*, a kinesin required for appropriate development of sensory cilia in ASH, ADL, and nine other sensory neurons, do not suppress social feeding (de Bono et al. 2002), despite disrupting nociceptive responses (Culotti & Russell 1978; Hilliard et al. 2004, 2005; Sambongi et al. 1999, 2000). Second, *osm-3* mutations restore social feeding behavior to *ocr-2; npr-1* and *odr-4; npr-1* mutants (de Bono et al. 2002). One interpretation of these data is that *osm-3* mutations disrupt both the ASH/ADL signal required for aggregation and bordering and a counterbalancing signal from another neuron that suppresses social feeding.

Neuroendocrine Regulation of Aggregation

Aggregation and bordering behaviors are also negatively regulated by the *daf-7* TGF-β neuroendocrine pathway. Loss-of-function mutations in the *daf-1* TGF-β receptor, *daf-7* TGF-β, and *daf-8* or *daf-14* SMAD-related genes induce N2 animals to aggregate and border on food, although less strongly than do *npr-1* mutations (Thomas et al. 1993). The DAF-7 neuroendocrine pathway acts in parallel to NPR-1 to suppress aggregation and bordering because *daf-7; npr-1(null)* double mutants aggregate more strongly than does either single mutant alone (de Bono et al. 2002). Moreover, whereas loss-of-function mutations in *daf-3* SMAD or *daf-5* Ski completely suppress aggregation of *daf-7* animals, they do not suppress aggregation of *npr-1* mutant animals (de Bono et al. 2002, Thomas et al. 1993). Finally, aggregation of *daf-7* mutants is not dependent on the oxygen-regulated soluble guanylate cyclases *gcy-35* and *gcy-36* (Cheung et al. 2004). Perhaps DAF-7 and NPR-1 downregulate different neural circuits that promote aggregation.

DAF-7 is expressed in the ASI head chemosensory neurons, where its transcription is regulated by a constitutively synthesized dauer pheromone that acts as a measure of population density (Ren et al. 1996, Schackwitz et al. 1996). High levels of dauer pheromone repress *daf-7* activity, leading to dauer formation. DAF-7 expression and reproductive growth are restored when pheromone is removed and animals receive food signals. The *daf-7* pathway may play a similar role in regulating aggregation and bordering behaviors, as in controlling dauer formation, by downregulating these behaviors when population density is low and food is available. Interestingly, dauer pheromone and *daf-7* signaling regulate expression of chemoreceptors (Nolan et al. 2002, Peckol et al. 2001), suggesting a possible mechanism for altering behavioral responses.

ANALYSIS OF MALE MATING BEHAVIOR

The most complex and sophisticated patterns of locomotion are observed during male mating behavior. *C. elegans* is highly sexually dimorphic: More than 30% of adult somatic nuclei are sexually specialized in each sex. Males can be easily distinguished from hermaprodites by differences in their external appearance. The most prominent feature of adult males is a specialized sexual organ that is part of the male tail (**Figure 7***A*). This genital structure contains sensory specializations for detection of the hermaprodite and the hermaphrodite vulva and discrete structures for mating and sperm transfer (Emmons & Sternberg 1997, Sulston et al. 1980). The nervous system of the male is substantially more complicated than that of the hermaphrodite, and only part of it has been reconstructed from serial electron micrographs. Compared with the 302 neurons in the hermaphrodite nervous system, males have 381 neurons of which 87 are sex-specific. The majority of these are sensory or motor neurons; only a handful of the neurons appear to be interneurons.

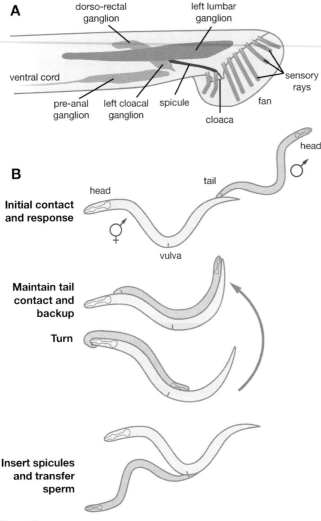

Figure 7

The *C. elegans* male tail and mating behavior. (*A*) A lateral view of the male tail showing the ganglia, sensory rays, and spicule on the left side of the worm. Figure adapted from Sulston et al. 1980. (*B*) Male mating behavior is characterized by a series of steps beginning with contact between the male tail and the hermaphrodite and a halt in forward movement. The male then initiates backward tracking movements, during which the tail is pressed firmly against the hermaphrodite. Tracking continues along both the ventral and dorsal sides until the vulva is located, after which the spicules are inserted allowing sperm transfer. Figure adapted from Liu & Sternberg 1995.

from detection of the hermaphrodite to successful sperm transfer (Emmons & Sternberg 1997, Liu & Sternberg 1995, Loer & Kenyon 1993). Genetics and cell ablation analysis have led to the identification of gene products and neurons required for these discrete steps (Emmons & Lipton 2003). To date, major advances have occurred in our understanding of the development of male-specific anatomy, the sensory regulation of male mating behavior, and components of the motor program. The tractability of this sex-specific behavior to genetic and experimental analyses offers the real possibility of gaining a cellular- and molecular-based understanding of a complex behavior.

Detection of Hermaphrodites: Male Search Behavior

Males preferentially seek and mate with hermaphrodites (Simon & Sternberg 2002). Most organisms use odorant cues to detect potential mates. Reversal frequency in males has been used as a behavioral readout to investigate whether hermaphrodite worms secrete an attractant cue. As males encounter regions where hermaphrodites have previously resided they increase their reversals. The net result of this increased reversal frequency is that males are much more likely to stay near the spot that previously contained hermaphrodites. The odorant cue was found to work at a distance as well. Males that venture away from a conditioned spot are more likely to return to the spot compared with control spots. Males do not increase their reversal frequency in response to male-conditioned spots, and hermaphrodites are not attracted to either hermaphrodite- or male-conditioned spots. Thus, the attractant cue is produced by hermaphrodites and specifically attracts males. The tissues that contribute to the production of this cue have not been determined, although the vulva is known not to be required. The neurons that detect the attractant are also unknown. However, males that harbor mutations in either *osm-5* or *osm-6*, genes

How do these additional neurons contribute to the dramatic behavioral differences between males and hermaphrodites?

Mating behavior itself can be broken down into a series of defined steps (**Figure 7B**),

required for proper formation of sensory cilia (Perkins et al. 1986), are unable to detect the hermaphrodite-specific cue (Simon & Sternberg 2002).

Males not only actively respond to the presence of hermaphrodites, but in the absence of hermaphrodites appear to initiate a search for suitable mating partners. Whereas hermaphrodites and immature males usually do not leave a bacterial food lawn, adult males do (Lipton et al. 2004). This leaving behavior is suppressed by the presence of adult hermaphrodites. Interestingly, food trumps sex. In the absence of food, males do not linger in close proximity to hermaphrodites but quickly leave to explore other environments (perhaps a version of male area restricted search? See above). Adult males in which the gonad or germ line are ablated, or which contain mutations in genes that block germ-line proliferation or differentiation, are far less likely to leave the food (Lipton et al. 2004). The signaling pathways and neural circuitry that control food-leaving behavior are poorly understood. However, *tph-1* mutant males that are serotonin deficient have a decreased rate of leaving, as do males mutant for the *daf-2* insulin receptor. DAF-2 is known to suppress the activity of the DAF-16 transcription factor. Thus, one prediction is that mutations in *daf-16* should restore normal leaving behavior in *daf-2* mutants. Indeed, the *daf-16* and *daf-2; daf-16* double mutants have close to wild-type leaving rates (Lipton et al. 2004).

A combination of biochemical techniques to purify attractant factors and genetics to isolate mutants that either fail to produce or respond to attractant should help to reveal, at a mechanistic level, how males are attracted to their mates. *C. elegans* males are more attracted to females of the related dioecious strain *C. remanei* than they are to *C. elegans* hermaphrodites, even though these interspecies crosses do not result in viable progeny (Chasnov & Chow 2002). This finding suggests that biochemical efforts to purify attractant factors may profit from analysis of *C. remanei* females.

Neuronal Control of Male Mating Behavior

Successful foraging for mates leads to males physically contacting hermaphrodites. Contact initiates the following series of five behavioral steps leading to copulation and sperm transfer (**Figure 7B**) (Liu & Sternberg 1995, Loer & Kenyon 1993): (*a*) Contact between the male tail and the hermaphrodite is followed by immediate cessation of forward movement, the placement of the ventral side of the tail against the hermaphrodite, and backward movement with the tail pressed firmly against the body of the hermaphrodite. The backward-searching movement ceases with successful location of vulva, in which case the worm may proceed with the steps of copulation. Often however, the backward search is initiated on the dorsal surface of the hermaphrodite, or the vulva is not found immediately, and the worm turns around and tries again. (*b*) To turn, the male makes a deep ventral flexure in its tail while maintaining contact with the hermaphrodite and, by continuing its backward movement, propels itself around 180° to the other side of the worm. Males successfully time this coordinated flexure and turn so that they usually do not lose contact with the hermaphrodite. This pattern of backing and turning is repeated until the male successfully locates the vulva. (*c*) When the vulva is detected, the male extrudes its spicules and makes small corrective movements to find its precise location. (*d*) Spicule insertion into the vulva commences after localization. This appears to be a difficult step because in a majority of mating events spicule insertion is not successful and the male disengages from the hermaphrodite. (*e*) Successful insertion of the spicules is followed by sperm release from the seminal vesicle and its passage via the vas deferens and cloaca into the vulva.

Each step in mating can be thought of as a behavioral module, and the behavioral modules link together to bring about successful copulation. This behavior sequence is not a fixed or stereotyped program set in ballistic motion by sensory contact. For example, turning behavior need not occur if the male immediately locates the vulva. Moreover, the male can execute its complicated mating maneuvers even as the hermaphrodite changes the speed and direction of its own movement, which suggests that the male constantly tracks and matches the hermaphrodite's movements as it mates. Distinct sets of neurons controlling each behavioral module have been identified by laser ablation studies (Liu & Sternberg 1995). Less clear is how each behavioral module is coupled to the next.

The male tail is elaborated into a complex genital structure containing a cuticular fan from which protrude nine pairs of finger-like specialized sensillae known as rays (**Figure 7A**) (Emmons & Sternberg 1997, Lints et al. 2004, Sulston et al. 1980). The fan and rays are used to contact the hermaphrodite. Each ray contains two ciliated sensory neurons (RnA, RnB), which have processes that extend the length of the ray to a sensory opening (Sulston et al. 1980, Sulston & Horvitz 1977). Ablation of individual rays has assigned specific roles for sensory rays in detecting contact with the hermaphrodite and for successful backward tracking and turning. Location of the vulva is dependent on the hook sensilla (innervated by sensory neurons HOA and HOB), bilateral postcloacal sensilla (each containing sensory neurons PCA, PCB, and PCC), and bilateral spicules (each containing sensory neurons SPD and SPV and motor neuron SPC). Spicule insertion is required for anchoring the male to the hermaphrodite and is a requirement for successful sperm transfer via the cloaca. Spicule insertion is dependent on the spicule sensory neurons and a spicule motor neuron. The SPV spicule neuron helps regulate sperm transfer (Liu & Sternberg 1995).

Gene Products that Contribute to Male Mating Behavior

Genetics and pharmacology have begun to provide insights into the molecular basis for male mating behavior (Hodgkin 1983, Liu & Sternberg 1995, Loer & Kenyon 1993). An early clue that led to the identification of signaling pathways that contribute to mating was the observation that male, but not hermaphrodite, worms exposed to the neurotransmitter serotonin (5-HT) exhibit a strong ventral curl of the tail that phenocopies the ventral flexure observed during the turning phase of mating (Loer & Kenyon 1993). This effect of serotonin probably mimics the action of the male-specific CP serotonergic motor neurons, which are located in the ventral cord and synapse onto male-specific diagonal muscles that mediate curling of the tail. Mutations in genes required for serotonin biosynthesis or packaging, including *cat-1* (vesicular monoamine transporter) (Duerr et al. 1999), *cat-4* (cyclohydrolase), and *bas-1* (aromatic amino acid decarboxylase) (Nass & Blakely 2003), reduce 5-HT immunoreactivity and tail curling and moderately disrupt mating behavior. Interestingly, the behavior of mutant males that are serotonin deficient can be rescued simply by exposure to exogenous 5-HT (Loer & Kenyon 1993), which suggests that these mutants do not affect the development of neural circuitry.

A forward genetic screen for mutants defective in vulval location identified *lov-1* (location of vulva), which encodes a predicted transmembrane protein with homology to the polycystin proteins PC1 and PC2 encoded by PKD1 and PKD2 (Barr & Sternberg 1999). In mammals, mutations in PKD1 or PKD2 are associated with polycystic kidney disease (Anyatonwu & Ehrlich 2004, Delmas et al. 2004), and PC1 and PC2 physically interact to form a cation channel. The LOV-1 protein is expressed in ray processes as well as the HOB sensory neurons that innervate the hook structure. A *C. elegans* homolog of PKD2 (*pkd-2*) is localized to the same male-specific sensory neurons, which suggests that PKD-2 and LOV-1 may function together. Consistent with this idea, the *pkd-2; lov-1* double mutant is no worse at vulval localization than is either single mutant (Barr et al. 2001,

Kaletta et al. 2003). A mutation in an RNA-binding protein expressed in many neurons, including those of the male rays, also moderately disrupts male turning and vulva location (Yoda et al. 2000). No ultrastructural defects are observed in the above mutants, which suggests that these genes have a primary role in signaling rather than establishing morphology.

The precise roles of LOV-1 and PKD-2 have not yet been established. In mammals, PC1 and PC2 are thought to participate in mechanosensation by the apical primary cilia of renal epithelial cells. In this model, PC1 detects mechanical stress induced by fluid flow and activates both the PC2 channel and G protein–coupled pathways, although the mechanisms are not yet clear. PC2 activation permits entry of extracellular Ca^{2+} and may also participate in a secondary amplification of the Ca^{2+} signal. This amplification may occur directly if PC2 also has a role in the ER as a Ca^{2+}-activated Ca^{2+}-release channel, or indirectly via the second messenger inositol trisphosphate (IP_3) or ryanodine receptors (Nauli & Zhou 2004). The molecular mechanisms of fluid detection by PC1 or PC2 activation have not been elucidated. Unraveling the PC1/PC2 complex signaling pathway could be aided by genetic analysis in *C. elegans*.

After the male locates the vulva and cloacal contact is established, the male copulatory spicules are inserted. Spicule insertion has two separable components: prodding behavior and insertion. Following cloacal contact, the spicule protractor muscles contract at approximately 7 Hz generating a rhythmic prodding of the vulva. Spicule penetration into the vulva is associated with a much stronger and more extended contraction of the protractors that lasts until sperm transfer is complete (Garcia et al. 2001). This insertion behavior is dependent on the SPC motor neuron as well as unknown factors from the male gonad. SPC and the sensory neurons that detect vulva location express a vesicular acetylcholine (ACh) transporter, which suggests that ACh release leads to protractor contraction. A role for ACh in spicule protraction was further supported by pharmacological and genetic experiments designed to interfere with cholinergic neurotransmission.

ACh appears to work redundantly through G protein–coupled and ionotropic receptor pathways to effect spicule protraction. Muscle contraction itself depends on a rise in intracellular Ca^{2+}, either because of influx from the extracellular space mediated by voltage-gated Ca^{2+} channels or release from intracellular stores mediated by ryanodine or IP_3 receptors. Mutations that affect these two different pathways have differential effects on spicule protraction. Thus, *unc-68* mutants (ryanodine receptor) are defective for prodding behavior but are still able to insert their spicules fully and mate. In contrast, *egl-19* mutants (L-type voltage-gated Ca^{2+} channel) have a mild decrease in prodding frequency but are unable to insert their spicules fully and cannot sire cross-progeny (Garcia et al. 2001). Interestingly, the behaviors were restored in transgenic mutants that expressed a wild-type copy of *egl-19* under control of a muscle-specific promoter.

Mating-independent spicule protraction is observed in *unc-103* mutants, which have defects in the homolog of the h-erg-encoded delayed inward rectifying K^+ channel in humans (Garcia & Sternberg 2003). UNC-103 probably has several roles in mating behavior, including suppressing spicule protraction until mating and regulating the frequency of prodding behavior. UNC-103 is expressed in many neurons, including the SPC neurons; however, expression is not observed in muscle cells. Spontaneous spicule protraction is suppressed in *unc-103; cha-1* double mutants. *cha-1* encodes choline acetyltransferase (required for ACh synthesis), which suggests that UNC-103 regulates cholinergic neurotransmission. In addition, no protraction is observed in *unc-103; egl-19* double mutants, as would be expected from the finding that *egl-19* is required in muscle for spicule protraction.

These studies have revealed an unanticipated complexity in the neural control of spicule insertion. One possibility is that the rhythmic prodding behavior is set in motion by neural input but that the control of prodding is then regulated by the dynamics of the ryanodine receptor, which may function as part of a pacemaker pathway in the muscle. However, the switch from rhythmic prodding to spicule insertion, which depends on multiple classes of AChRs and differential recruitment of ryanodine receptors, is poorly understood (Garcia et al. 2001).

Many of the numerous mutations that affect male mating were described in a survey of known mutants by Hodgkin (1983) or were identified in more recent work that examined the development of the male tail (reviewed by Emmons & Lipton 2003). More recently, forward screens have identified genes that are important for the differentiation of the ciliated sensory neurons that help locate the vulva (Yu et al. 2003). The identification of the sensory neurons that are required for various parts of mating behavior is the first step toward the larger goal of understanding how the nervous system controls a coordinated, complicated behavior. In the future, studies using imaging and electrophysiological strategies in conjunction with genetics should provide fundamental insights into how simple neural circuits control the remarkable tracking and mating capabilities of *C. elegans* males.

WORM LEARNING AND SYNAPTIC PLASTICITY

A consideration of the question "Do worms learn?" depends entirely on how one defines learning. A simple definition states that learning is a change of behavior that follows experience. This broad definition encompasses both nonassociative learning, e.g., habituation or adaptation to sensory inputs, and associative learning, e.g., changes in the behavioral response to a conditioning stimulus (CS) following its pairing with an unconditioned stimulus (US). Worms clearly can adapt to odorants (Colbert & Bargmann 1995), habituate to mechanical stimuli (Rankin & Wicks 2000, Wicks & Rankin 1995), and change their behavior according to their feeding state (Hills et al. 2004, Sawin et al. 2000). In addition, what emerges from studies on thermotaxis, olfaction, chemotaxis, and mechanosensation is that worms are capable of associative learning. In all these paradigms, animals make a positive association between food (US) and a second stimulus (CS), with training requiring several hours. Conversely, animals can make a negative association between aversive stimuli, typically lack of food or starvation, and an initially neutral or positive stimulus (Saeki et al. 2001, Wen et al. 1997). In previous sections, we have described these behaviors and how they relate to certain forms of learning. Here we focus primarily on examples of associative learning and dissection of their genetic basis.

Glutamatergic Signaling Modulates Some Forms of Plasticity

In vertebrates, synaptic plasticity and learning are critically dependent on experience-dependent changes at glutamatergic synapses (Bredt & Nicoll 2003, Malenka & Bear 2004, Seidenman et al. 2003). Two different paradigms, olfactory associative learning and mechanosensory habituation, show that glutamatergic neurotransmission is also important for behavioral plasticity in *C. elegans*.

Worms demonstrate associative learning in classical conditioning paradigms. They chemotax to NaCl if it was previously associated with food but not if it was associated with starvation (Saeki et al. 2001, Wen et al. 1997). In a variant of the original Pavlovian conditioning protocol the attractant odorant diacetyl was paired with the aversive stimulus acetic acid. Worms normally strongly chemotax to diacetyl. Following pairing of diacetyl with acetic acid, wild-type worms are far less likely to track diacetyl. In contrast, *glr-1* mutants show no evidence of a learned association between diacetyl and acetic acid and maintain their tracking to diacetyl (Morrison & van der Kooy 2001). The lack of association cannot be simply explained by a primary sensory defect because *glr-1* mutants track normally toward diacetyl and avoid acetic acid. Interestingly, although acetic acid is sensed by ASH, *glr-1* mutants avoid it (Morrison & van der Kooy 2001), which suggests that NMDA-type glutamate receptors may be activated by acetic acid stimulation of ASH, as is observed for hyperosmotic stimuli (Mellem et al. 2002). These results suggest that GLR-1 receptors function in an as yet unspecified interneuron(s) to mediate the association between these disparate sensory stimuli and modify tracking behavior following conditioning.

When a Petri plate is tapped, causing vibrations, the worms crawling on the agar surface stop their forward movement and initiate a backing response. With continued taps, the worms habituate and back less frequently (Rankin & Wicks 2000, Rose et al. 2002, 2003). Habituation to tap probably does not occur at the level of the sensory neurons because the receptor potential in touch neurons shows no evidence of decay with repeated taps (O'Hagan et al. 2005). The habituation lasts a long time and can be paired with environmental context, which suggests that there is a form of memory for habituation (Rankin 2000). This habituation can be thought of as a form of experience-dependent nonassociative learning. Any student that has crammed for an exam will attest that spaced or interval training leads to greater retention than massed training. Memory for habituation shows the same effect, and retention is dependent on the number of training sessions (Rose et al. 2002). Worms with a mutation in EAT-4, a *C. elegans* vesicular glutamate transporter (Bellocchio et al. 2000, Lee et al. 1999, Takamori et al. 2000), display more rapid habituation to tap than observed for wild-type controls (Rose et al. 2002). How *eat-4* mutation affects habituation is not known, and there may be additional genes encoding vesicular glutamate transporters, so glutamatergic neurotransmission might not be completely abolished in these mutants. However, and importantly, long-term memory of habituation (tested 24 h later) was not observed in *eat-4* or *glr-1* mutants (Rose et al. 2002), which suggests that glutamatergic transmission, acting elsewhere in the nervous system, is required for the memory of past habituation.

Other Molecules Required for Learning

In a choice paradigm, naive worms track slightly better to a defined concentration of NaCl than to a separate defined dilution of isoamylalcohol. In contrast, worms subjected to aversive conditioning by first being placed for 4 h on a plate that contained NaCl but no food (starvation) greatly preferred the isoamylalcohol upon subsequent testing (Saeki et al. 2001) (**Figure 8***B*). Two mutants, JN603 and JN683, defective in this learning assay were isolated in a forward genetic screen, but we do not currently know the identity of the mutated genes in the two strains.

A limited genetic screen for animals defective in a taste-associative learning paradigm identified 2 mutants, *lrn-1* and *lrn-2* (Wen et al. 1997). These mutants also showed defects in an olfactory associative learning paradigm (Morrison et al. 1999). Tests of chemotaxis, habituation, and dishabituation failed to show any defects for these mutants, which suggests they specifically disrupt associative learning. However, the molecular identities of *lrn-1* and *lrn-2*, and whether the alleles identified are full or partial loss of function, are unknown. *lrn-2* mutants are also defective in an qualitatively different model of associative learning (Law et al. 2004). Worms typically adapt to odorants with prolonged exposure. This adaptation appears to be context dependent in that adaptation in one

adp: sensory *ad*aptation abnormal

aex: *a*boc, *ex*pulsion defective

bas: *b*iogenic *a*mine *s*ynthesis related

cat: abnormal *cat*echolamine distribution

cha: *ch*oline *a*cetyl transferase

clk: *cl*ock abnormal

cmk: *C*am *k*inase

daf: *da*uer larva *f*ormation abnormal

dec: *de*fecation *c*ycle abnormal

dop: *dop*amine receptor

eat: *eat*ing abnormal

egl: *e*gg-*l*aying defective

exp: *exp*ulsion defective

flp: *F*MRFamide-*l*ike *p*eptide

flr: *fl*uoride *r*esistant

gcy: *g*uanylyl *cy*clase

glr: *gl*utamate *r*eceptor class

goa: *G*-protein class *O* *a*lpha subunit

gpa: *G*-*p*rotein *a*lpha subunit

grk: *G* protein–coupled *r*eceptor *k*inase

hen-1: *h*esitation in crossing an aversive barrier

itr: *i*nositol *tr*iphosphate receptor

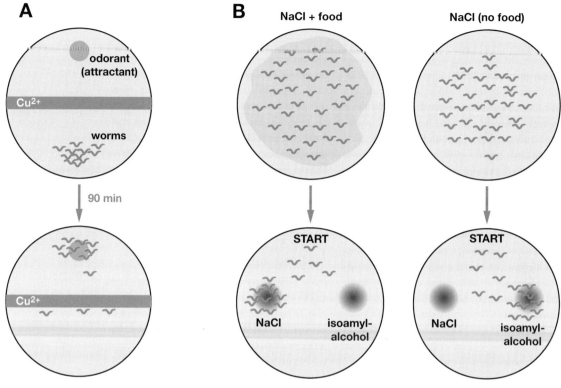

Figure 8

Learning paradigms in *C. elegans*. (*A*) An assay that requires worms to integrate two conflicting cues—aversive Cu^{2+} and an attractive odorant. Worms are placed at one pole of an agar plate opposite an attractive odorant. To reach the odorant the worms must cross a repellent barrier of Cu^{2+}. Figure adapted from Ishihara et al. 2002. (*B*) Worms learn to associate an aversive stimulus (starvation) with an attractive cue (NaCl). When NaCl is paired with food, worms preferentially track toward a spot of NaCl. Conversely, worms starved for 4 h in the presence of NaCl prefer isoamylalcohol (*right*). Figure adapted from Saeki et al. 2001.

lov: *lo*cation of *v*ulva defective

lrn: *l*ea*rn*ing abnormal

nmr: *N*MDA-class glutamate *r*eceptor

npr: *n*euro*p*eptide *r*eceptor

ocr: *o*sm-9 and *c*apsaicin *r*eceptor related

environment (agar containing the salt sodium acetate) is lost when the worms are tested for adaptation in the presence of a novel environment (agar containing the salt ammonium chloride). State-dependent learning has also been demonstrated for olfactory adaptation in the presence of food (Nuttley et al. 2002) and the drug ethanol (Bettinger & McIntire 2004).

Perhaps the most interesting gene product thus far implicated in learning is HEN-1 (Ishihara et al. 2002). *hen-1* mutants were identified in an assay that required worms to cross an aversive rubicon of Cu^{2+} ions to reach an attractive odorant cue (diacetyl) (**Figure 8***A*). Whether to cross over the Cu^{2+} depends on integrating the two conflicting cues—the decision to cross can be manipulated by increasing diacetyl or decreasing Cu^{2+} concentrations. *hen-1* mutants are far less likely to cross the rubicon, but in separate experiments, both alleles of *hen-1* showed normal aversion to Cu^{2+} and normal attraction to diacetyl. Thus, only the integrated response to the two stimuli is disrupted in *hen-1* mutants. An explanation for this finding is that diacetyl inhibits the response to Cu^{2+} and vice versa. The failure of *hen-1* mutants can be explained by a lack of cross-inhibition.

hen-1 mutants were also defective in two associative learning protocols that tested the effects of pairing food deprivation with chemotaxis to NaCl and thermotaxis to cultivation temperature.

HEN-1 encodes a secretory protein with an LDL receptor motif that is similar to the *Drosophila* signaling molecule *Jelly belly* (Jeb) (Ishihara et al. 2002, Lee et al. 2003, Weiss et al. 2001). The Jeb receptor is the *Drosophila* homolog of anaplastic lymphoma kinase (Alk), a receptor tyrosine kinase of the insulin receptor superfamily (Lee et al. 2003). *hen-1* mutants poorly suppress chemotaxis to NaCl when starved in its presence. *hen-1* is expressed in the ASE and AIY interneurons, and targeted *hen-1* expression in either of these neurons is sufficient to rescue the *hen-1* mutant phenotype. AIY is a major postsynaptic target of the ASE salt-sensing neurons, and mutants defective in AIY development exhibit a *hen-1*-like phenotype (Ishihara et al. 2002). HEN-1 does not act cell-autonomously. Ectopic expression of HEN-1 in the AWB and AWC sensory neurons is sufficient for behavioral rescue in transgenic *hen-1* mutants. Importantly, heat shock–induced expression of HEN-1 protein in adult *hen-1* mutants is sufficient for behavioral rescue, which strongly suggests that HEN-1 has a modulatory role in nervous system function rather than being required for the development, maintenance, or modeling of neural circuits. The study of plasticity and learning in *C. elegans* is still in its infancy, but the long-term outlook is promising for a molecular-based understanding of short- and long-term plastic changes in nervous system function. What bodes well for future studies is the development of new assays to explore behavior and a firm context, in terms of neural circuitry, to help analyze the function of gene products that modify plasticity.

FUTURE PROSPECTS

The study of behavior in *C. elegans* has been aided by the wealth of analytical tools and structural information available to researchers in the field and has led to fundamental advances in our understanding of neuronal function, the encoding of sensory information, and the identification of gene products that contribute to synaptic and neuronal function. Each of the behaviors we have discussed is also observed in more complex organisms, but in *C. elegans* behaviors are generally controlled by only a handful of identified cells, providing hope that coordinated genetic, biochemical, and physiological studies will provide fundamental new insights into the neuronal control of behavior. Lessons learned from the study of *C. elegans* will almost certainly aid in the study of more complex nervous systems. Odorant discriminators, thermometers, sophisticated tracking capability, and Ca^{2+} oscillators driving rhythmic behaviors all localized to identified classes of neurons that are now slowly revealing their secrets as a result of improved electrophysiology and microscopy. In the future, genetic analysis will continue to serve as a particularly sharp scalpel with which to dissect the molecular complexity underlying behavior.

odr: *od*orant *r*esponse abnormal

osm: *osm*otic avoidance abnormal

qui: *qui*nine non-avoider

sol: *s*uppressor *o*f *l*urcher movement defect

tax: abnormal chemo*tax*is

tph: *t*ryptophan *h*ydroxylase

unc: *unc*oordinated

ACKNOWLEDGMENTS

We thank Josh Kaplan, Oliver Hobert, William Schafer, and members of the de Bono and Maricq laboratories, for helpful comments on the manuscript, and Penelope Brockie and Candida Rogers, for assistance with the figures. M.d.B is supported by the Medical Research Council of the United Kingdom. A.V.M. is supported by grants from the National Institutes of Health and the National Science Foundation, and is an independent investigator of the National Alliance for Research on Schizophrenia and Depression.

LITERATURE CITED

Albert PS, Brown SJ, Riddle DL. 1981. Sensory control of dauer larva formation in *Caenorhabditis elegans*. *J. Comp. Neurol.* 198:435–51

Albertson DG, Thomson JN. 1976. The pharynx of *Caenorhabditis elegans*. *Philos. Trans. R. Soc. London B. Biol. Sci.* 275:299–325

Anyatonwu GI, Ehrlich BE. 2004. Calcium signaling and polycystin-2. *Biochem. Biophys. Res. Commun.* 322:1364–73

Aronoff R, Mellem JE, Maricq AV, Sprengel R, Seeburg PH. 2004. Neuronal toxicity in *Caenorhabditis elegans* from an editing site mutant in glutamate receptor channels. *J. Neurosci.* 24:8135–40

Avery L, Horvitz HR. 1990. Effects of starvation and neuroactive drugs on feeding in *Caenorhabditis elegans*. *J. Exp. Zool.* 253:263–70

Avery L, Thomas JH. 1997. Feeding and defecation. See Riddle et al. 1997, pp. 679–716

Bargmann CI. 1998. Neurobiology of the *Caenorhabditis elegans* genome. *Science* 282:2028–33

Bargmann CI, Avery L. 1995. Laser killing of cells in *Caenorhabditis elegans*. In *Methods in Cell Biology*, ed. HF Epstein, DC Shakes, pp. 225–50. New York: Academic

Bargmann CI, Hartwieg E, Horvitz HR. 1993. Odorant-selective genes and neurons mediate olfaction in *C. elegans*. *Cell* 74:515–27

Bargmann CI, Horvitz HR. 1991. Chemosensory neurons with overlapping functions direct chemotaxis to multiple chemicals in *C. elegans*. *Neuron* 7:729–42

Bargmann CI, Kaplan JM. 1998. Signal transduction in the *Caenorhabditis elegans* nervous system. *Annu. Rev. Neurosci.* 21:279–308

Bargmann CI, Thomas JH, Horvitz HR. 1990. Chemosensory cell function in the behavior and development of *Caenorhabditis elegans*. *Cold Spring Harb. Symp. Quant. Biol.* 55:529–38

Barr MM, DeModena J, Braun D, Nguyen CQ, Hall DH, Sternberg PW. 2001. The *Caenorhabditis elegans* autosomal dominant polycystic kidney disease gene homologs *lov-1* and *pkd-2* act in the same pathway. *Curr. Biol.* 11:1341–46

Barr MM, Sternberg PW. 1999. A polycystic kidney-disease gene homologue required for male mating behaviour in *C. elegans*. *Nature* 401:386–89

Beg AA, Jorgensen EM. 2003. EXP-1 is an excitatory GABA-gated cation channel. *Nat. Neurosci.* 6:1145–52

Bell WJ. 1991. *Searching Behavior: The Behavioral Ecology of Finding Resources*. New York: Chapman and Hall

Bellocchio EE, Reimer RJ, Fremeau RT Jr, Edwards RH. 2000. Uptake of glutamate into synaptic vesicles by an inorganic phosphate transporter. *Science* 289:957–60

Berger AJ, Hart AC, Kaplan JM. 1998. G alphas-induced neurodegeneration in *Caenorhabditis elegans*. *J. Neurosci.* 18:2871–80

Berridge MJ, Bootman MD, Roderick HL. 2003. Calcium signalling: dynamics, homeostasis and remodelling. *Nat. Rev. Mol. Cell Biol.* 4:517–29

Berridge MJ, Lipp P, Bootman MD. 2000. The versatility and universality of calcium signalling. *Nat. Rev. Mol. Cell Biol.* 1:11–21

Bettinger JC, McIntire SL. 2004. State-dependency in *C. elegans*. *Genes Brain Behav.* 3:266–72

Birnby DA, Link EM, Vowels JJ, Tian H, Colacurcio PL, Thomas JH. 2000. A transmembrane guanylyl cyclase (DAF-11) and Hsp90 (DAF-21) regulate a common set of chemosensory behaviors in *Caenorhabditis elegans*. *Genetics* 155:85–104

Branicky R, Shibata Y, Feng J, Hekimi S. 2001. Phenotypic and suppressor analysis of defecation in *clk-1* mutants reveals that reaction to changes in temperature is an active process in *Caenorhabditis elegans*. *Genetics* 159:997–1006

Bredt DS, Nicoll RA. 2003. AMPA receptor trafficking at excitatory synapses. *Neuron* 40:361–79

Brockie PJ, Madsen DM, Zheng Y, Mellem J, Maricq AV. 2001a. Differential expression of glutamate receptor subunits in the nervous system of *Caenorhabditis elegans* and their regulation by the homeodomain protein UNC-42. *J. Neurosci.* 21:1510–22

Brockie PJ, Mellem JE, Hills T, Madsen DM, Maricq AV. 2001b. The *C. elegans* glutamate receptor subunit NMR-1 is required for slow NMDA-activated currents that regulate reversal frequency during locomotion. *Neuron* 31:617–30

Burbea M, Dreier L, Dittman JS, Grunwald ME, Kaplan JM. 2002. Ubiquitin and AP180 regulate the abundance of GLR-1 glutamate receptors at postsynaptic elements in C. elegans. *Neuron* 35:107–20

Cassata G, Kagoshima H, Andachi Y, Kohara Y, Durrenberger MB, et al. 2000. The LIM homeobox gene *ceh-14* confers thermosensory function to the AFD neurons in *Caenorhabditis elegans*. *Neuron* 25:587–97

Chalfie M, Sulston JE, White JG, Southgate E, Thomson JN, Brenner S. 1985. The neural circuit for touch sensitivity in *Caenorhabditis elegans*. *J. Neurosci.* 5:956–64

Chalfie M, Tu Y, Euskirchen G, Ward WW, Prasher DC. 1994. Green fluorescent protein as a marker for gene expression. *Science* 263:802–5

Chalfie M, White JG. 1988. The nervous system. In *The Nematode Caenorhabditis elegans*, ed. WB Wood. Cold Spring Harbor, NY: Cold Spring Harbor Lab. Press

Chang S, Johnston RJ Jr, Frokjaer-Jensen C, Lockery S, Hobert O. 2004. MicroRNAs act sequentially and asymmetrically to control chemosensory laterality in the nematode. *Nature* 430:785–89

Chang S, Johnston RJ Jr, Hobert O. 2003. A transcriptional regulatory cascade that controls left/right asymmetry in chemosensory neurons of *C. elegans*. *Genes Dev.* 17:2123–37

Chao MY, Komatsu H, Fukuto HS, Dionne HM, Hart AC. 2004. Feeding status and serotonin rapidly and reversibly modulate a *Caenorhabditis elegans* chemosensory circuit. *Proc. Natl. Acad. Sci. USA* 101:15512–17

Chase DL, Pepper JS, Koelle MR. 2004. Mechanism of extrasynaptic dopamine signaling in *Caenorhabditis elegans*. *Nat. Neurosci.* 7:1096–103

Chasnov JR, Chow KL. 2002. Why are there males in the hermaphroditic species *Caenorhabditis elegans*? *Genetics* 160:983–94

Cheung BH, Arellano-Carbajal F, Rybicki I, De Bono M. 2004. Soluble guanylate cyclases act in neurons exposed to the body fluid to promote *C. elegans* aggregation behavior. *Curr. Biol.* 14:1105–11

Coates J, de Bono M. 2002. Antagonistic pathways in neurons exposed to body fluid regulate social feeding in *Caenorhabditis elegans*. *Nature* 419:925–29

Coburn CM, Bargmann CI. 1996. A putative cyclic nucleotide-gated channel is required for sensory development and function in *C. elegans*. *Neuron* 17:695–706

Colbert HA, Bargmann CI. 1995. Odorant-specific adaptation pathways generate olfactory plasticity in *C. elegans*. *Neuron* 14:803–12

Colbert HA, Bargmann CI. 1997. Environmental signals modulate olfactory acuity, discrimination, and memory in *Caenorhabditis elegans*. *Learn. Mem.* 4:179–91

Colbert HA, Smith TL, Bargmann CI. 1997. OSM-9, a novel protein with structural similarity to channels, is required for olfaction, mechanosensation, and olfactory adaptation in *Caenorhabditis elegans*. *J. Neurosci.* 17:8259–69

Croll NA. 1975. Components and patterns in the behaviour of the nematode *Caenorhabditis elegans*. *J. Zool. London* 176:159–76

Croll NA, Smith JM. 1978. Integrated behaviour in the feeding phase of *Caenorhabditis elegans* (Nematoda). *J. Zool. London* 184:507–17

Culotti JG, Russell RL. 1978. Osmotic avoidance defective mutants of the nematode *Caenorhabditis elegans*. *Genetics* 90:243–56

Dal Santo P, Logan MA, Chisholm AD, Jorgensen EM. 1999. The inositol trisphosphate receptor regulates a 50-second behavioral rhythm in *C. elegans*. *Cell* 98:757–67

de Bono M, Bargmann CI. 1998. Natural variation in a neuropeptide Y receptor homolog modifies social behavior and food response in *C. elegans*. *Cell* 94:679–89

de Bono M, Tobin D, Davis MW, Avery L, Bargmann C. 2002. Social feeding in *Caenorhabditis elegans* is induced by neurons that detect aversive stimuli. *Nature* 419:899–903

Delmas P, Padilla F, Osorio N, Coste B, Raoux M, Crest M. 2004. Polycystins, calcium signaling, and human diseases. *Biochem. Biophys. Res. Commun.* 322:1374–83

Denver DR, Morris K, Thomas WK. 2003. Phylogenetics in *Caenorhabditis elegans*: an analysis of divergence and outcrossing. *Mol. Biol. Evol.* 20:393–400

Doi M, Iwasaki K. 2002. Regulation of retrograde signaling at neuromuscular junctions by the novel C2 domain protein AEX-1. *Neuron* 33:249–59

Dolmetsch RE, Xu K, Lewis RS. 1998. Calcium oscillations increase the efficiency and specificity of gene expression. *Nature* 392:933–36

Dreier L, Burbea M, Kaplan JM. 2005. LIN-23-mediated degradation of β-catenin regulates the abundance of GLR-1 glutamate receptors in the ventral nerve cord of C. elegans. *Neuron*. In press

Duerr JS, Frisby DL, Gaskin J, Duke A, Asermely K, et al. 1999. The *cat-1* gene of *Caenorhabditis elegans* encodes a vesicular monoamine transporter required for specific monoamine-dependent behaviors. *J. Neurosci.* 19:72–84

Durbin R. 1987. *Studies on the Development and Organisation of the Nervous System of* Caenorhabditis elegans. Cambridge, UK: Univ. Cambridge Press

Dusenbery DB. 1974. Analysis of chemotaxis in the nematode *Caenorhabditis elegans* by countercurrent separation. *J. Exp. Zool.* 188:41–47

Dusenbery DB. 1980. Responses of the nematode *Caenorhabditis elegans* to controlled chemical stimulation. *J. Comp. Physiol.* 136:327–31

Dusenbery DB, Sheridan RE, Russell RL. 1975. Chemotaxis-defective mutants of the nematode *Caenorhabditis elegans*. *Genetics* 80:297–309

Dwyer ND, Troemel ER, Sengupta P, Bargmann CI. 1998. Odorant receptor localization to olfactory cilia is mediated by ODR-4, a novel membrane-associated protein. *Cell* 93:455–66

Ehlers MD. 2003. Activity level controls postsynaptic composition and signaling via the ubiquitin-proteasome system. *Nat. Neurosci.* 6:231–42

Emmons SW, Lipton J. 2003. Genetic basis of male sexual behavior. *J. Neurobiol.* 54:93–110

Emmons SW, Sternberg PW. 1997. Male development and mating behavior. See Riddle et al. 1997, pp. 295–334

Ernstrom GG, Chalfie M. 2002. Genetics of sensory mechanotransduction. *Annu. Rev. Genet.* 36:411–53

Estevez AY, Roberts RK, Strange K. 2003. Identification of store-independent and store-operated Ca^{2+} conductances in *Caenorhabditis elegans* intestinal epithelial cells. *J. Gen. Physiol.* 122:207–23

Fujiwara M, Sengupta P, McIntire SL. 2002. Regulation of body size and behavioral state of *C. elegans* by sensory perception and the EGL-4 cGMP-dependent protein kinase. *Neuron* 36:1091–102

Fukuto HS, Ferkey DM, Apicella AJ, Lans H, Sharmeen T, et al. 2004. G protein-coupled receptor kinase function is essential for chemosensation in C. elegans. *Neuron* 42:581–93

Garcia LR, Mehta P, Sternberg PW. 2001. Regulation of distinct muscle behaviors controls the *C. elegans* male's copulatory spicules during mating. *Cell* 107:777–88

Garcia LR, Sternberg PW. 2003. *Caenorhabditis elegans* UNC-103 ERG-like potassium channel regulates contractile behaviors of sex muscles in males before and during mating. *J. Neurosci.* 23:2696–705

Ginzburg VE, Roy PJ, Culotti JG. 2002. Semaphorin 1a and semaphorin 1b are required for correct epidermal cell positioning and adhesion during morphogenesis in *C. elegans*. *Development* 129:2065–78

Gomez M, De Castro E, Guarin E, Sasakura H, Kuhara A, et al. 2001. Ca2+ signaling via the neuronal calcium sensor-1 regulates associative learning and memory in *C. elegans*. *Neuron* 30:241–48

Goodman MB, Hall DH, Avery L, Lockery SR. 1998. Active currents regulate sensitivity and dynamic range in *C. elegans* neurons. *Neuron* 20:763–72

Goodman MB, Schwarz EM. 2003. Transducing touch in *Caenorhabditis elegans*. *Annu. Rev. Physiol.* 65:429–52

Gray JM, Hill JJ, Bargmann CI. 2005. A circuit for navigation in *Caenorhabditis elegans*. *Proc. Natl. Acad. Sci. USA*. In press

Gray JM, Karow DS, Lu H, Chang AJ, Chang JS, et al. 2004. Oxygen sensation and social feeding mediated by a *C. elegans* guanylate cyclase homologue. *Nature* 430:317–22

Grunwald ME, Kaplan JM. 2003. Mutations in the ligand-binding and pore domains control exit of glutamate receptors from the endoplasmic reticulum in *C. elegans*. *Neuropharmacology* 45:768–76

Grunwald ME, Mellem JE, Strutz N, Maricq AV, Kaplan JM. 2004. Clathrin-mediated endocytosis is required for compensatory regulation of GLR-1 glutamate receptors after activity blockade. *Proc. Natl. Acad. Sci. USA* 101:3190–95

Harper JW, Burton JL, Solomon MJ. 2002. The anaphase-promoting complex: It's not just for mitosis anymore. *Genes Dev.* 16:2179–206

Hart AC, Kass J, Shapiro JE, Kaplan JM. 1999. Distinct signaling pathways mediate touch and osmosensory responses in a polymodal sensory neuron. *J. Neurosci.* 19:1952–58

Hart AC, Sims S, Kaplan JM. 1995. Synaptic code for sensory modalities revealed by *C. elegans* GLR-1 glutamate receptor. *Nature* 378:82–85

Hedgecock EM, Russell RL. 1975. Normal and mutant thermotaxis in the nematode *Caenorhabditis elegans*. *Proc. Natl. Acad. Sci. USA* 72:4061–65

Hilliard MA, Apicella AJ, Kerr R, Suzuki H, Bazzicalupo P, Schafer WR. 2005. In vivo imaging of *C. elegans* ASH neurons: cellular response and adaptation to chemical repellents. *EMBO J.* 24:63–72

Hilliard MA, Bargmann CI, Bazzicalupo P. 2002. *C. elegans* responds to chemical repellents by integrating sensory inputs from the head and the tail. *Curr. Biol.* 12:730–34

Hilliard MA, Bergamasco C, Arbucci S, Plasterk RH, Bazzicalupo P. 2004. Worms taste bitter: ASH neurons, QUI-1, GPA-3 and ODR-3 mediate quinine avoidance in *Caenorhabditis elegans*. *EMBO J.* 23:1101–11

Hills T, Brockie PJ, Maricq AV. 2004. Dopamine and glutamate control area-restricted search behavior in *Caenorhabditis elegans*. *J. Neurosci.* 24:1217–25

Hirotsu T, Saeki S, Yamamoto M, Iino Y. 2000. The Ras-MAPK pathway is important for olfaction in *Caenorhabditis elegans*. *Nature* 404:289–93

Hobert O, D'Alberti T, Liu Y, Ruvkun G. 1998. Control of neural development and function in a thermoregulatory network by the LIM homeobox gene *lin-11*. *J. Neurosci.* 18:2084–96

Hobert O, Mori I, Yamashita Y, Honda H, Ohshima Y, et al. 1997. Regulation of interneuron function in the *C. elegans* thermoregulatory pathway by the *ttx-3* LIM homeobox gene. *Neuron* 19:345–57

Hodgkin J. 1983. Male phenotypes and mating efficiency in *Caenorhabditis elegans*. *Genetics* 103:43–64

Hodgkin J, Doniach T. 1997. Natural variation and copulatory plug formation in *Caenorhabditis elegans*. *Genetics* 146:149–64

Horvitz HR, Chalfie M, Trent C, Sulston JE, Evans PD. 1982. Serotonin and octopamine in the nematode *Caenorhabditis elegans*. *Science* 216:1012–14

Ishihara T, Iino Y, Mohri A, Mori I, Gengyo-Ando K, et al. 2002. HEN-1, a secretory protein with an LDL receptor motif, regulates sensory integration and learning in *Caenorhabditis elegans*. *Cell* 109:639–49

Iwasaki K, Liu DW, Thomas JH. 1995. Genes that control a temperature-compensated ultradian clock in *Caenorhabditis elegans*. *Proc. Natl. Acad. Sci. USA* 92:10317–21

Iwasaki K, Staunton J, Saifee O, Nonet M, Thomas JH. 1997. *aex-3* encodes a novel regulator of presynaptic activity in *C. elegans*. *Neuron* 18:613–22

Iwasaki K, Toyonaga R. 2000. The Rab3 GDP/GTP exchange factor homolog AEX-3 has a dual function in synaptic transmission. *EMBO J.* 19:4806–16

Jansen G, Thijssen KL, Werner P, van der Horst M, Hazendonk E, Plasterk RH. 1999. The complete family of genes encoding G proteins of *Caenorhabditis elegans*. *Nat. Genet.* 21:414–19

Jansen G, Weinkove D, Plasterk RH. 2002. The G-protein gamma subunit *gpc-1* of the nematode *C. elegans* is involved in taste adaptation. *EMBO J.* 21:986–94

Jee C, Lee J, Lee JI, Lee WH, Park BJ, et al. 2004. SHN-1, a Shank homologue in *C. elegans*, affects defecation rhythm via the inositol-1,4,5-trisphosphate receptor. *FEBS Lett.* 561:29–36

Johnston RJ, Hobert O. 2003. A microRNA controlling left/right neuronal asymmetry in *Caenorhabditis elegans*. *Nature* 426:845–49

Juo P, Kaplan JM. 2004. The anaphase-promoting complex regulates the abundance of GLR-1 glutamate receptors in the ventral nerve cord of *C. elegans*. *Curr. Biol.* 14:2057–62

Kahn-Kirby AH, Dantzker JL, Apicella AJ, Schafer WR, Browse J, et al. 2004. Specific polyunsaturated fatty acids drive TRPV-dependent sensory signaling in vivo. *Cell* 119:889–900

Kaletta T, Van der Craen M, Van Geel A, Dewulf N, Bogaert T, et al. 2003. Towards understanding the polycystins. *Nephron. Exp. Nephrol.* 93:e9–17

Kamath RS, Fraser AG, Dong Y, Poulin G, Durbin R, et al. 2003. Systematic functional analysis of the *Caenorhabditis elegans* genome using RNAi. *Nature* 421:231–37

Kaplan JM, Horvitz HR. 1993. A dual mechanosensory and chemosensory neuron in *Caenorhabditis elegans*. *Proc. Natl. Acad. Sci. USA* 90:2227–31

Kass J, Jacob TC, Kim P, Kaplan JM. 2001. The EGL-3 proprotein convertase regulates mechanosensory responses of *Caenorhabditis elegans*. *J. Neurosci.* 21:9265–72

Kellenberger S, Schild L. 2002. Epithelial sodium channel/degenerin family of ion channels: a variety of functions for a shared structure. *Physiol. Rev.* 82:735–67

Kennedy S, Wang D, Ruvkun G. 2004. A conserved siRNA-degrading RNase negatively regulates RNA interference in *C. elegans*. *Nature* 427:645–49

Kerr R, Lev-Ram V, Baird G, Vincent P, Tsien RY, Schafer WR. 2000. Optical imaging of calcium transients in neurons and pharyngeal muscle of *C. elegans*. *Neuron* 26:583–94

Kim E, Sheng M. 2004. PDZ domain proteins of synapses. *Nat. Rev. Neurosci.* 5:771–81

Kim K, Li C. 2004. Expression and regulation of an FMRFamide-related neuropeptide gene family in *Caenorhabditis elegans*. *J. Comp. Neurol.* 475:540–50

Kimura KD, Miyawaki A, Matsumoto K, Mori I. 2004. The *C. elegans* thermosensory neuron AFD responds to warming. *Curr. Biol.* 14:1291–95

Kippert F, Saunders DS, Blaxter ML. 2002. *Caenorhabditis elegans* has a circadian clock. *Curr. Biol.* 12:R47–49

Koesling D. 1999. Studying the structure and regulation of soluble guanylyl cyclase. *Methods* 19:485–93

Komatsu H, Jin YH, L'Etoile N, Mori I, Bargmann CI, et al. 1999. Functional reconstitution of a heteromeric cyclic nucleotide-gated channel of *Caenorhabditis elegans* in cultured cells. *Brain Res.* 821:160–68

Komatsu H, Mori I, Rhee JS, Akaike N, Ohshima Y. 1996. Mutations in a cyclic nucleotide-gated channel lead to abnormal thermosensation and chemosensation in *C. elegans*. *Neuron* 17:707–18

Kubiak TM, Larsen MJ, Nulf SC, Zantello MR, Burton KJ, et al. 2003. Differential activation of "social" and "solitary" variants of the *Caenorhabditis elegans* G protein-coupled receptor NPR-1 by its cognate ligand AF9. *J. Biol. Chem.* 278:33724–29

Kuhara A, Inada H, Katsura I, Mori I. 2002. Negative regulation and gain control of sensory neurons by the *C. elegans* calcineurin TAX-6. *Neuron* 33:751–63

L'Etoile ND, Bargmann CI. 2000. Olfaction and odor discrimination are mediated by the *C. elegans* guanylyl cyclase ODR-1. *Neuron* 25:575–86

L'Etoile ND, Coburn CM, Eastham J, Kistler A, Gallegos G, Bargmann CI. 2002. The cyclic GMP-dependent protein kinase EGL-4 regulates olfactory adaptation in *C. elegans*. *Neuron* 36:1079–89

Lackner MR, Nurrish SJ, Kaplan JM. 1999. Facilitation of synaptic transmission by EGL-30 Gqalpha and EGL-8 PLCbeta: DAG binding to UNC-13 is required to stimulate acetylcholine release. *Neuron* 24:335–46

Lans H, Rademakers S, Jansen G. 2004. A network of stimulatory and inhibitory Galpha-subunits regulates olfaction in *Caenorhabditis elegans*. *Genetics* 167:1677–87

Law E, Nuttley WM, van der Kooy D. 2004. Contextual taste cues modulate olfactory learning in *C. elegans* by an occasion-setting mechanism. *Curr. Biol.* 14:1303–8

Lee HH, Norris A, Weiss JB, Frasch M. 2003. *Jelly belly* protein activates the receptor tyrosine kinase Alk to specify visceral muscle pioneers. *Nature* 425:507–12

Lee RY, Sawin ER, Chalfie M, Horvitz HR, Avery L. 1999. EAT-4, a homolog of a mammalian sodium-dependent inorganic phosphate cotransporter, is necessary for glutamatergic neurotransmission in *Caenorhabditis elegans*. *J. Neurosci.* 19:159–67

Li C, Kim K, Nelson LS. 1999a. FMRFamide-related neuropeptide gene family in *Caenorhabditis elegans*. *Brain Res.* 848:26–34

Li C, Nelson L, Kim K, Nathoo A, Hart AC. 1999b. Neuropeptide gene families in the nematode *Caenorhabditis elegans*. *Ann. N.Y. Acad. Sci.* 897:239–52

Li J, Zhu X, Boston R, Ashton FT, Gamble HR, Schad GA. 2000. Thermotaxis and thermosensory neurons in infective larvae of *Haemonchus contortus*, a passively ingested nematode parasite. *J. Comp. Neurol.* 424:58–73

Liedtke W, Tobin DM, Bargmann CI, Friedman JM. 2003. Mammalian TRPV4 (VR-OAC) directs behavioral responses to osmotic and mechanical stimuli in *Caenorhabditis elegans*. *Proc. Natl. Acad. Sci. USA* 100(Suppl. 2):14531–36

Lints R, Jia L, Kim K, Li C, Emmons SW. 2004. Axial patterning of *C. elegans* male sensilla identities by selector genes. *Dev. Biol.* 269:137–51

Lipton J, Kleemann G, Ghosh R, Lints R, Emmons SW. 2004. Mate searching in *Caenorhabditis elegans*: a genetic model for sex drive in a simple invertebrate. *J. Neurosci.* 24:7427–34

Liu DW, Thomas JH. 1994. Regulation of a periodic motor program in *C. elegans*. *J. Neurosci.* 14:1953–62

Liu KS, Sternberg PW. 1995. Sensory regulation of male mating behavior in *Caenorhabditis elegans*. *Neuron* 14:79–89

Loer CM, Kenyon CJ. 1993. Serotonin-deficient mutants and male mating behavior in the nematode *Caenorhabditis elegans*. *J. Neurosci.* 13:5407–17

Lopez PM, Boston R, Ashton FT, Schad GA. 2000. The neurons of class ALD mediate thermotaxis in the parasitic nematode, *Strongyloides stercoralis*. *Int. J. Parasitol.* 30:1115–21

Malenka RC, Bear MF. 2004. LTP and LTD: an embarrassment of riches. *Neuron* 44:5–21

Maricq AV, Peckol E, Driscoll M, Bargmann CI. 1995. Mechanosensory signalling in *C. elegans* mediated by the GLR-1 glutamate receptor. *Nature* 378:78–81

Martin TF. 2002. Prime movers of synaptic vesicle exocytosis. *Neuron* 34:9–12

Mathews EA, Garcia E, Santi CM, Mullen GP, Thacker C, et al. 2003. Critical residues of the *Caenorhabditis elegans unc-2* voltage-gated calcium channel that affect behavioral and physiological properties. *J. Neurosci.* 23:6537–45

McIntire SL, Jorgensen E, Horvitz HR. 1993a. Genes required for GABA function in *Caenorhabditis elegans*. *Nature* 364:334–37

McIntire SL, Jorgensen E, Kaplan J, Horvitz HR. 1993b. The GABAergic nervous system of *Caenorhabditis elegans*. *Nature* 364:337–41

Mee CJ, Tomlinson SR, Perestenko PV, De Pomerai D, Duce IR, et al. 2004. Latrophilin is required for toxicity of black widow spider venom in *Caenorhabditis elegans*. *Biochem. J.* 378:185–91

Mellem JE, Brockie PJ, Zheng Y, Madsen DM, Maricq AV. 2002. Decoding of polymodal sensory stimuli by postsynaptic glutamate receptors in *C. elegans*. *Neuron* 36:933–44

Miller KG, Emerson MD, Rand JB. 1999. Goalpha and diacylglycerol kinase negatively regulate the Gqalpha pathway in *C. elegans*. *Neuron* 24:323–33

Montell C. 1999. Visual transduction in *Drosophila*. *Annu. Rev. Cell Dev. Biol.* 15:231–68

Mori I. 1999. Genetics of chemotaxis and thermotaxis in the nematode *Caenorhabditis elegans*. *Annu. Rev. Genet.* 33:399–422

Mori I, Ohshima Y. 1995. Neural regulation of thermotaxis in *Caenorhabditis elegans*. *Nature* 376:344–48

Morrison GE, van der Kooy D. 2001. A mutation in the AMPA-type glutamate receptor, *glr-1*, blocks olfactory associative and nonassociative learning in *Caenorhabditis elegans*. *Behav. Neurosci.* 115:640–49

Morrison GE, Wen JY, Runciman S, van der Kooy D. 1999. Olfactory associative learning in *Caenorhabditis elegans* is impaired in *lrn-1* and *lrn-2* mutants. *Behav. Neurosci.* 113:358–67

Morton DB. 2005. Atypical soluble guanylate cyclases in *Drosophila* can function as molecular oxygen sensors. *J. Biol. Chem.* In press

Morton DB, Hudson ML, Waters E, O'Shea M. 1999. Soluble guanylyl cyclases in *Caenorhabditis elegans*: NO is not the answer. *Curr. Biol.* 9:R546–47

Nass R, Blakely RD. 2003. The *Caenorhabditis elegans* dopaminergic system: opportunities for insights into dopamine transport and neurodegeneration. *Annu. Rev. Pharmacol. Toxicol.* 43:521–44

Nauli SM, Zhou J. 2004. Polycystins and mechanosensation in renal and nodal cilia. *Bioessays* 26:844–56

Niebur E, Erdos P. 1993. Theory of the locomotion of nematodes: control of the somatic motor neurons by interneurons. *Math Biosci.* 118:51–82

Nolan KM, Sarafi-Reinach TR, Horne JG, Saffer AM, Sengupta P. 2002. The DAF-7 TGF-beta signaling pathway regulates chemosensory receptor gene expression in *C. elegans*. *Genes Dev.* 16:3061–73

Nonet ML. 1999. Visualization of synaptic specializations in live *C. elegans* with synaptic vesicle protein-GFP fusions. *J. Neurosci. Methods* 89:33–40

Nonet ML, Grundahl K, Meyer BJ, Rand JB. 1993. Synaptic function is impaired but not eliminated in *C. elegans* mutants lacking synaptotagmin. *Cell* 73:1291–305

Nonet ML, Holgado AM, Brewer F, Serpe CJ, Norbeck BA, et al. 1999. UNC-11, a *Caenorhabditis elegans* AP180 homologue, regulates the size and protein composition of synaptic vesicles. *Mol. Biol. Cell.* 10:2343–60

Nonet ML, Staunton JE, Kilgard MP, Fergestad T, Hartwieg E, et al. 1997. *Caenorhabditis elegans rab-3* mutant synapses exhibit impaired function and are partially depleted of vesicles. *J. Neurosci.* 17:8061–73

Nurrish S, Segalat L, Kaplan JM. 1999. Serotonin inhibition of synaptic transmission: Galpha(0) decreases the abundance of UNC-13 at release sites. *Neuron* 24:231–42

Nuttley WM, Atkinson-Leadbeater KP, Van Der Kooy D. 2002. Serotonin mediates food-odor associative learning in the nematode *Caenorhabditis elegans*. *Proc. Natl. Acad. Sci. USA* 99:12449–54

O'Hagan R, Chalfie M, Goodman MB. 2005. The MEC-4 DEG/ENaC channel of *Caenorhabditis elegans* touch receptor neurons transduces mechanical signals. *Nat. Neurosci.* 8:43–50

Peckol EL, Troemel ER, Bargmann CI. 2001. Sensory experience and sensory activity regulate chemosensory receptor gene expression in *Caenorhabditis elegans*. *Proc. Natl. Acad. Sci. USA* 98:11032–38

Perkins LA, Hedgecock EM, Thomson JN, Culotti JG. 1986. Mutant sensory cilia in the nematode *Caenorhabditis elegans*. *Dev. Biol.* 117:456–87

Peters JM. 2002. The anaphase-promoting complex: proteolysis in mitosis and beyond. *Mol. Cell* 9:931–43

Pierce-Shimomura JT, Faumont S, Gaston MR, Pearson BJ, Lockery SR. 2001. The homeobox gene *lim-6* is required for distinct chemosensory representations in *C. elegans*. *Nature* 410:694–98

Pierce-Shimomura JT, Morse TM, Lockery SR. 1999. The fundamental role of pirouettes in *Caenorhabditis elegans* chemotaxis. *J. Neurosci.* 19:9557–69

Raizen DM, Avery L. 1994. Electrical activity and behavior in the pharynx of *Caenorhabditis elegans*. *Neuron* 12:483–95

Rand JB, Nonet ML. 1997. Neurotransmitter assignments for specific neurons. See Riddle et al. 1997, pp. 1049–52

Ranganathan R, Cannon SC, Horvitz HR. 2000. MOD-1 is a serotonin-gated chloride channel that modulates locomotory behaviour in *C. elegans*. *Nature* 408:470–75

Ranganathan R, Sawin ER, Trent C, Horvitz HR. 2001. Mutations in the *Caenorhabditis elegans* serotonin reuptake transporter MOD-5 reveal serotonin-dependent and -independent activities of fluoxetine. *J. Neurosci.* 21:5871–84

Rankin CH. 2000. Context conditioning in habituation in the nematode *Caenorhabditis elegans*. *Behav. Neurosci.* 114:496–505

Rankin CH, Wicks SR. 2000. Mutations of the *Caenorhabditis elegans* brain-specific inorganic phosphate transporter *eat-4* affect habituation of the tap-withdrawal response without affecting the response itself. *J. Neurosci.* 20:4337–44

Reiner DJ, Newton EM, Tian H, Thomas JH. 1999. Diverse behavioural defects caused by mutations in *Caenorhabditis elegans unc-43* CaM kinase II. *Nature* 402:199–203

Ren P, Lim CS, Johnsen R, Albert PS, Pilgrim D, Riddle DL. 1996. Control of *C. elegans* larval development by neuronal expression of a TGF-β homolog. *Science* 274:1389–91

Richmond JE, Davis WS, Jorgensen EM. 1999. UNC-13 is required for synaptic vesicle fusion in *C. elegans*. *Nat. Neurosci.* 2:959–64

Riddle DL, Blumenthal T, Meyer BJ, Priess JR, eds. 1997. *C. elegans II*. Cold Spring Harbor, NY: Cold Spring Harb. Lab. Press

Roayaie K, Crump JG, Sagasti A, Bargmann CI. 1998. The G alpha protein ODR-3 mediates olfactory and nociceptive function and controls cilium morphogenesis in *C. elegans* olfactory neurons. *Neuron* 20:55–67

Robertson HM. 2001. Updating the *str* and *srj* (*stl*) families of chemoreceptors in *Caenorhabditis* nematodes reveals frequent gene movement within and between chromosomes. *Chem. Senses* 26:151–59

Rogers C, Reale V, Kim K, Chatwin H, Li C, et al. 2003. Inhibition of *Caenorhabditis elegans* social feeding by FMRFamide-related peptide activation of NPR-1. *Nat. Neurosci.* 6:1178–85

Rongo C, Kaplan JM. 1999. CaMKII regulates the density of central glutamatergic synapses in vivo. *Nature* 402:195–99

Rongo C, Whitfield CW, Rodal A, Kim SK, Kaplan JM. 1998. LIN-10 is a shared component of the polarized protein localization pathways in neurons and epithelia. *Cell* 94:751–59

Rose JK, Kaun KR, Chen SH, Rankin CH. 2003. GLR-1, a non-NMDA glutamate receptor homolog, is critical for long-term memory in *Caenorhabditis elegans*. *J. Neurosci.* 23:9595–99

Rose JK, Kaun KR, Rankin CH. 2002. A new group-training procedure for habituation demonstrates that presynaptic glutamate release contributes to long-term memory in *Caenorhabditis elegans*. *Learn. Mem.* 9:130–37

Ryu WS, Samuel AD. 2002. Thermotaxis in *Caenorhabditis elegans* analyzed by measuring responses to defined Thermal stimuli. *J. Neurosci.* 22:5727–33

Saeki S, Yamamoto M, Iino Y. 2001. Plasticity of chemotaxis revealed by paired presentation of a chemoattractant and starvation in the nematode *Caenorhabditis elegans*. *J. Exp. Biol.* 204:1757–64

Saigusa T, Ishizaki S, Watabiki S, Ishii N, Tanakadate A, et al. 2002. Circadian behavioural rhythm in *Caenorhabditis elegans*. *Curr. Biol.* 12:R46–47

Sambongi Y, Nagae T, Liu Y, Yoshimizu T, Takeda K, et al. 1999. Sensing of cadmium and copper ions by externally exposed ADL, ASE, and ASH neurons elicits avoidance response in *Caenorhabditis elegans*. *Neuroreport* 10:753–57

Sambongi Y, Takeda K, Wakabayashi T, Ueda I, Wada Y, Futai M. 2000. *Caenorhabditis elegans* senses protons through amphid chemosensory neurons: proton signals elicit avoidance behavior. *Neuroreport* 11:2229–32

Samuel AD, Silva RA, Murthy VN. 2003. Synaptic activity of the AFD neuron in *Caenorhabditis elegans* correlates with thermotactic memory. *J. Neurosci.* 23:373–76

Sanyal S, Wintle RF, Kindt KS, Nuttley WM, Arvan R, et al. 2004. Dopamine modulates the plasticity of mechanosensory responses in *Caenorhabditis elegans*. *EMBO J.* 23:473–82

Satterlee JS, Ryu WS, Sengupta P. 2004. The CMK-1 CaMKI and the TAX-4 cyclic nucleotide-gated channel regulate thermosensory neuron gene expression and function in *C. elegans*. *Curr. Biol.* 14:62–68

Satterlee JS, Sasakura H, Kuhara A, Berkeley M, Mori I, Sengupta P. 2001. Specification of thermosensory neuron fate in *C. elegans* requires *ttx-1*, a homolog of otd/Otx. *Neuron* 31:943–56

Sawin ER, Ranganathan R, Horvitz HR. 2000. *C. elegans* locomotory rate is modulated by the environment through a dopaminergic pathway and by experience through a serotonergic pathway. *Neuron* 26:619–31

Schackwitz WS, Inoue T, Thomas JH. 1996. Chemosensory neurons function in parallel to mediate a pheromone response in *C. elegans*. *Neuron* 17:719–28

Schafer WR, Kenyon CJ. 1995. A calcium-channel homologue required for adaptation to dopamine and serotonin in *Caenorhabditis elegans*. *Nature* 375:73–78

Schuster S, Marhl M, Hofer T. 2002. Modelling of simple and complex calcium oscillations. From single-cell responses to intercellular signalling. *Eur. J. Biochem.* 269:1333–55

Segalat L, Elkes DA, Kaplan JM. 1995. Modulation of serotonin-controlled behaviors by Go in *Caenorhabditis elegans*. *Science* 267:1648–51

Seidenman KJ, Steinberg JP, Huganir R, Malinow R. 2003. Glutamate receptor subunit 2 Serine 880 phosphorylation modulates synaptic transmission and mediates plasticity in CA1 pyramidal cells. *J. Neurosci.* 23:9220–28

Sengupta P, Chou JH, Bargmann CI. 1996. *odr-10* encodes a seven transmembrane domain olfactory receptor required for responses to the odorant diacetyl. *Cell* 84:899–909

Sheng M, Kim E. 2000. The Shank family of scaffold proteins. *J. Cell. Sci.* 113(Pt. 11):1851–56

Shim J, Umemura T, Nothstein E, Rongo C. 2004. The unfolded protein response regulates glutamate receptor export from the endoplasmic reticulum. *Mol. Biol. Cell.* 15:4818–28

Shingai R. 2000. Durations and frequencies of free locomotion in wild type and GABAergic mutants of *Caenorhabditis elegans*. *Neurosci. Res.* 38:71–83

Simon JM, Sternberg PW. 2002. Evidence of a mate-finding cue in the hermaphrodite nematode *Caenorhabditis elegans*. *Proc. Natl. Acad. Sci. USA* 99:1598–603

Sulston JE, Albertson DG, Thomson JN. 1980. The *Caenorhabditis elegans* male: postembryonic development of nongonadal structures. *Dev. Biol.* 78:542–76

Sulston JE, Horvitz HR. 1977. Post-embryonic cell lineages of the nematode, *Caenorhabditis elegans*. *Dev. Biol.* 56:110–56

Sulston JE, Schierenberg E, White JG, Thomson JN. 1983. The embryonic cell lineage of the nematode *Caenorhabditis elegans*. *Dev. Biol.* 100:64–119

Sulston JE, White JG. 1980. Regulation and cell autonomy during postembryonic development of *Caenorhabditis elegans*. *Dev. Biol.* 78:577–97

Sze JY, Ruvkun G. 2003. Activity of the *Caenorhabditis elegans* UNC-86 POU transcription factor modulates olfactory sensitivity. *Proc. Natl. Acad. Sci. USA* 100:9560–65

Takamori S, Rhee JS, Rosenmund C, Jahn R. 2000. Identification of a vesicular glutamate transporter that defines a glutamatergic phenotype in neurons. *Nature* 407:189–94

Take-Uchi M, Kawakami M, Ishihara T, Amano T, Kondo K, Katsura I. 1998. An ion channel of the degenerin/epithelial sodium channel superfamily controls the defecation rhythm in *Caenorhabditis elegans*. *Proc. Natl. Acad. Sci. USA* 95:11775–80

Tavernarakis N, Shreffler W, Wang S, Driscoll M. 1997. *unc-8*, a DEG/ENaC family member, encodes a subunit of a candidate mechanically gated channel that modulates *C. elegans* locomotion. *Neuron* 18:107–19

Tavernarakis N, Wang SL, Dorovkov M, Ryazanov A, Driscoll M. 2000. Heritable and inducible genetic interference by double-stranded RNA encoded by transgenes. *Nat. Genet.* 24:180–83

Thacker C, Rose AM. 2000. A look at the *Caenorhabditis elegans* Kex2/Subtilisin-like proprotein convertase family. *Bioessays* 22:545–53

Thomas JH. 1990. Genetic analysis of defecation in *Caenorhabditis elegans*. *Genetics* 124:855–72

Thomas JH, Birnby DA, Vowels JJ. 1993. Evidence for parallel processing of sensory information controlling dauer formation in *Caenorhabditis elegans*. *Genetics* 134:1105–17

Tobin D, Madsen D, Kahn-Kirby A, Peckol E, Moulder G, et al. 2002. Combinatorial expression of TRPV channel proteins defines their sensory functions and subcellular localization in *C. elegans* neurons. *Neuron* 35:307–18

Troemel ER. 1999. *Chemosensory Receptors in Caenorhabditis Elegans*. Ph.D. Thesis, Univ. Calif. San Francisco

Troemel ER, Chou JH, Dwyer ND, Colbert HA, Bargmann CI. 1995. Divergent seven transmembrane receptors are candidate chemosensory receptors in *C. elegans*. *Cell* 83:207–18

Troemel ER, Kimmel BE, Bargmann CI. 1997. Reprogramming chemotaxis responses: sensory neurons define olfactory preferences in *C. elegans*. *Cell* 91:161–69

Troemel ER, Sagasti A, Bargmann CI. 1999. Lateral signaling mediated by axon contact and calcium entry regulates asymmetric odorant receptor expression in *C. elegans*. *Cell* 99:387–98

Tsalik EL, Hobert O. 2003. Functional mapping of neurons that control locomotory behavior in *Caenorhabditis elegans*. *J. Neurobiol.* 56:178–97

Turrigiano GG, Nelson SB. 2004. Homeostatic plasticity in the developing nervous system. *Nat. Rev. Neurosci.* 5:97–107

Vowels JJ, Thomas JH. 1994. Multiple chemosensory defects in *daf-11* and *daf-21* mutants of *Caenorhabditis elegans*. *Genetics* 138:303–16

Wakabayashi T, Kitagawa I, Shingai R. 2004. Neurons regulating the duration of forward locomotion in *Caenorhabditis elegans*. *Neurosci. Res.* 50:103–11

Walker DS, Gower NJ, Ly S, Bradley GL, Baylis HA. 2002a. Regulated disruption of inositol 1,4,5-trisphosphate signaling in *Caenorhabditis elegans* reveals new functions in feeding and embryogenesis. *Mol. Biol. Cell.* 13:1329–37

Walker DS, Ly S, Gower NJ, Baylis HA. 2004. IRI-1, a LIN-15B homologue, interacts with inositol-1,4,5-triphosphate receptors and regulates gonadogenesis, defecation, and pharyngeal pumping in *Caenorhabditis elegans*. *Mol. Biol. Cell.* 15:3073–82

Walker DS, Ly S, Lockwood KC, Baylis HA. 2002b. A direct interaction between IP(3) receptors and myosin II regulates IP(3) signaling in *C. elegans*. *Curr. Biol.* 12:951–56

Ward S. 1973. Chemotaxis by the nematode *Caenorhabditis elegans*: identification of attractants and analysis of the response by use of mutants. *Proc. Natl. Acad. Sci. USA* 70:817–21

Ward S, Thomson N, White JG, Brenner S. 1975. Electron microscopical reconstruction of the anterior sensory anatomy of the nematode *Caenorhabditis elegans*. *J. Comp. Neurol.* 160:313–37

Ware RW, Clark D, Crossland K, Russell RL. 1975. The nerve ring of the nematode *Caenorhabditis elegans*: sensory input and motor output. *J. Comp. Neurol.* 162:71–110

Weinshenker D, Wei A, Salkoff L, Thomas JH. 1999. Block of an *ether-a-go-go*-like K(+) channel by imipramine rescues *egl-2* excitation defects in *Caenorhabditis elegans*. *J. Neurosci.* 19:9831–40

Weiss JB, Suyama KL, Lee HH, Scott MP. 2001. *Jelly belly*: a *Drosophila* LDL receptor repeat-containing signal required for mesoderm migration and differentiation. *Cell* 107:387–98

Wen JY, Kumar N, Morrison G, Rambaldini G, Runciman S, et al. 1997. Mutations that prevent associative learning in *C. elegans*. *Behav. Neurosci.* 111:354–68

Wes PD, Bargmann CI. 2001. *C. elegans* odour discrimination requires asymmetric diversity in olfactory neurons. *Nature* 410:698–701

White JG, Southgate E, Thomson JN, Brenner S. 1986. The structure of the nervous system of the nematode *Caenorhabditis elegans*. *Phil. Trans. Roy. Soc. (London) B* 314:1–340

Wicks SR, Rankin CH. 1995. Integration of mechanosensory stimuli in *Caenorhabditis elegans*. *J. Neurosci.* 15:2434–44

Wittenburg N, Baumeister R. 1999. Thermal avoidance in *Caenorhabditis elegans*: an approach to the study of nociception. *Proc. Natl. Acad. Sci. USA* 96:10477–82

Wong A, Boutis P, Hekimi S. 1995. Mutations in the *clk-1* gene of *Caenorhabditis elegans* affect developmental and behavioral timing. *Genetics* 139:1247–59

Yin X, Gower NJ, Baylis HA, Strange K. 2004. Inositol 1,4,5-trisphosphate signaling regulates rhythmic contractile activity of myoepithelial sheath cells in *Caenorhabditis elegan*s. *Mol. Biol. Cell.* 15:3938–49

Yoda A, Sawa H, Okano H. 2000. MSI-1, a neural RNA-binding protein, is involved in male mating behaviour in *Caenorhabditis elegans*. *Genes Cells* 5:885–95

Yu H, Pretot RF, Burglin TR, Sternberg PW. 2003. Distinct roles of transcription factors EGL-46 and DAF-19 in specifying the functionality of a polycystin-expressing sensory neuron necessary for *C. elegans* male vulva location behavior. *Development* 130:5217–27

Yu S, Avery L, Baude E, Garbers DL. 1997. Guanylyl cyclase expression in specific sensory neurons: a new family of chemosensory receptors. *Proc. Natl. Acad. Sci. USA* 94:3384–87

Zariwala HA, Miller AC, Faumont S, Lockery SR. 2003. Step response analysis of thermotaxis in *Caenorhabditis elegans*. *J. Neurosci.* 23:4369–77

Zheng Y, Brockie PJ, Mellem JE, Madsen DM, Maricq AV. 1999. Neuronal control of locomotion in *C. elegans* is modified by a dominant mutation in the GLR-1 ionotropic glutamate receptor. *Neuron* 24:347–61

Zheng Y, Mellem JE, Brockie PJ, Madsen DM, Maricq AV. 2004. SOL-1 is a CUB-domain protein required for GLR-1 glutamate receptor function in *C. elegans*. *Nature* 427:451–57

Zuo J, De Jager PL, Takahashi KA, Jiang W, Linden DJ, Heintz N. 1997. Neurodegeneration in Lurcher mice caused by mutation in delta2 glutamate receptor gene. *Nature* 388:769–73

Zwaal RR, Broeks A, van Meurs J, Groenen JT, Plasterk RH. 1993. Target-selected gene inactivation in *Caenorhabditis elegans* by using a frozen transposon insertion mutant bank. *Proc. Natl. Acad. Sci. USA* 90:7431–35

Dendritic Computation

Michael London and Michael Häusser

Wolfson Institute for Biomedical Research and Department of Physiology, University College London, London WC1E 6BT; email: m.london@ucl.ac.uk, m.hausser@ucl.ac.uk

Key Words

dendrites, coding, synaptic integration, spikes, ion channels

Abstract

One of the central questions in neuroscience is how particular tasks, or computations, are implemented by neural networks to generate behavior. The prevailing view has been that information processing in neural networks results primarily from the properties of synapses and the connectivity of neurons within the network, with the intrinsic excitability of single neurons playing a lesser role. As a consequence, the contribution of single neurons to computation in the brain has long been underestimated. Here we review recent work showing that neuronal dendrites exhibit a range of linear and nonlinear mechanisms that allow them to implement elementary computations. We discuss why these dendritic properties may be essential for the computations performed by the neuron and the network and provide theoretical and experimental examples to support this view.

Contents

INTRODUCTION 504
THE DENDRITIC "TOOLKIT" ... 505
  Computations in Passive Dendrites 505
  Computations in Active Dendrites . 509
EXAMPLES OF REAL-WORLD
   DENDRITIC COMPUTATION 516
  Directional Selectivity 517
  Coincidence Detection in Auditory
    Neurons 519
  Temporal Integration Over Long
    Timescales 520
  Image Processing in Dendrites of
    Fly Neurons 520
  Looming Sensitive Neurons
    in the Locust 522
  Forward Masking of Cricket Songs 522
  Dendritic Mechanisms and
    Behavior 524
CHALLENGES FOR THE
   FUTURE: A WISH LIST FOR
   DENDRITIC COMPUTATION 524
  For Molecular Biologists:
    Designer Dendrites 524
  For Neurophysiologists: Putting
    Dendrites Back into the
    Network 526
  For the Theorist: Proving the
    Benefits of Dendritic
    Computation 526
CONCLUSIONS 527

INTRODUCTION

Brains compute. This means that they process information, creating abstract representations of physical entities and performing operations on this information in order to execute tasks. One of the main goals of computational neuroscience is to describe these transformations as a sequence of simple elementary steps organized in an algorithmic way. The mechanistic substrate for these computations has long been debated. Traditionally, relatively simple computational properties have been attributed to the individual neuron, with the complex computations that are the hallmark of brains being performed by the network of these simple elements. In this framework, the neuron (often called a "Perceptron," "Spin," or "Unit") sums up the synaptic input and, by comparing this sum against a threshold, "decides" whether to initiate an action potential. In computational terms this process includes only one nonlinear term (thresholding), which is usually counted as a single operation. Thus, the neuron operates as a device where analog computations are at some decision point transformed into a digital output signal. Such a design forms the backbone of many artificial neuronal networks, starting from the original work of McCullough & Pitts (1943) to the present day. In this review we argue that this model is oversimplified in view of the properties of real neurons and the computations they perform. Rather, additional linear and nonlinear mechanisms in the dendritic tree are likely to serve as computational building blocks, which combined together play a key role in the overall computation performed by the neuron.

In the first part of this review we describe the dendritic computational toolkit, i.e., the biophysical mechanisms in dendrites, which endow them with potential computational powers. We focus on the most recent findings and group them according to the type of computations being carried out. In the second part of the review, we present several examples where the role of dendritic mechanisms in computation has strong circumstantial support, such as directional selectivity in retinal neurons or coincidence detection in the auditory system. We conclude with a discussion of how we can ultimately prove the role of dendrites in neural computation and outline a process for achieving this goal. Throughout the review, we focus on the *online* aspects of computation. Naturally, the results of such computations must ultimately be read out and stored within the network. Given that dendrites are also the site where synaptic plasticity takes place, their properties are likely to affect the induction and expression of plasticity. These issues have been discussed in several recent reviews (Häusser & Mel 2003, Linden 1999, Mainen 1999, Chklovskii et al. 2004), and we therefore do not address this aspect in detail here.

THE DENDRITIC "TOOLKIT"

Why discuss what a neuron does in terms of computation? We use a simple example to illustrate the problem. When one watches a movie, each frame is presented for about 50 ms, during which time we have to process any changes from the last frame before the next frame appears. To a first approximation the timescale of the computational cycle of a neuron involved in this computation is thus on the order of milliseconds to tens of milliseconds. Because a typical neuron in the brain receives thousands of synaptic inputs, a neuron involved in processing the visual input thus receives hundreds to thousands of 50-ms input spike trains. But the neuron has only one axon, which provides the output signal, and thus the final conversion represents a compression into a much smaller amount of information. Because there are so many inputs, an essential feature of this transformation is the amount of input information that must be discarded by the neuron. This process is analogous to what some mathematical functions are doing: projecting a huge space onto a narrow one. Thus, the neuron throws away the irrelevant information and selects only the information relevant to its function. The essence of computing a function in computer science is usually considered to be implementing an algorithm, that is, a sequence of simple computational steps organized in a flowchart that leads from the input to the output (**Figure 1**). Are there ways to deconstruct into such simpler building blocks the very complex mapping done by a neuron? Can we identify the crucial steps that occur during this process? Are decisions taken only at the final step of transforming the somatic voltage into an action potential, or are there decision points on the way?

Although the flowchart representation may not be the most appropriate and comprehensive way to describe what neurons are doing, it can nevertheless be very instructive to define and explore the computational building blocks (i.e., the boxes on the flowchart) that neurons can perform. During the past decade a rapidly increasing number of studies investigating signaling mechanisms in dendrites have appeared, and several recent reviews have discussed these findings in depth (Euler & Denk 2001, Häusser et al. 2000, Koch & Segev 2000, Magee 2000, Segev & London 2000, Williams & Stuart 2003). Rather than listing all the known facts about dendritic signaling, in this section we focus on identifying the unique dendritic mechanisms that can act as computational building blocks. We start with a brief discussion of how passive dendrites transform their inputs and implement nonlinear interactions between synaptic inputs. Then we review the role of dendritic voltage-dependent channels and conclude by evaluating the possible role of dendrites in chemical computation.

Computations in Passive Dendrites

It is important to recognize that the passive properties of the dendritic tree provide the backbone for the electrical signaling in dendrites, even when they are richly endowed with voltage-dependent ionic currents. For example, the threshold for initiation of a dendritic spike is determined in part by the availability of sodium channels, but perhaps even more by the passive load of the surrounding dendrites, which dictates how much of the input current will depolarize the membrane and how much will flow axially toward other dendritic regions (Segev & London 1999). Thus, understanding the passive properties of dendritic trees remains crucial for understanding computation in dendritic trees. Aside from their role in regulating the conditions for active dendritic signaling, the passive properties

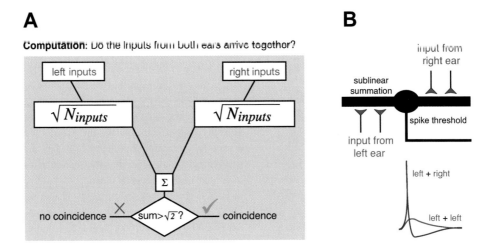

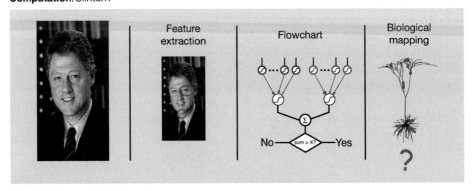

Figure 1

Dendritic computation. The task of a brainstem auditory neuron performing coincidence detection in the sound localization system of birds is to respond only if the inputs arriving from both ears coincide in a precise manner (10–100 μs), while avoiding a response when the input comes from only one ear. (*A*) A flowchart of a simplistic algorithm to achieve this computation. Inputs to each ear are summed sublinearly, and the input from both ears is then summed linearly and compared with threshold. Thus only if there are inputs from both ears will the sum exceed the threshold, whereas if the input arrives only to one ear the output is not large enough. (*B*) Agmon-Snir et al. (1998) showed that dendrites of these neurons might implement a similar algorithm. Inputs from each ear arrive on one dendrite. Sublinear summation is achieved by the mutual shunting of the excitatory inputs, and the threshold is implemented via the spike-generation mechanism. (*C*) Neurons are known to be involved in much more sophisticated computations, such as face recognition (Kreiman et al. 2000). An algorithm to solve a face recognition task is one of the holy grails of computer science. At present, we do not know precisely how single neurons are involved in this computation. An essential first step is feature extraction from the image, which clearly involves a lot of network preprocessing before features are fed into the individual cortical neuron. The flowchart implements a three-layer model of dendritic processing (see Häusser & Mel 2003) to integrate the input. The way such a flowchart is mapped onto the real geometry of a cortical pyramidal neuron (*right panel*) remains unknown.

of dendrites provide computational functions in their own right, as discussed below.

Delay lines via dendritic filtering. In terms of signal propagation, dendrites behave like electrical cables with medium-quality insulation. As such, passive dendrites linearly filter the input signal as it spreads to the site of initiation, where it is compared with the threshold. This filtering tends to attenuate the dendritic signal as a function of the distance it travels and the frequency of the original signal. Thus a brief and sharp excitatory postsynaptic potential (EPSP) that originates in the dendrite will be transformed into a much smaller and broader signal when it arrives at the soma (**Figure 2**). As a consequence, the time-to-peak of a synaptic event, and thus the delay of the resulting output spikes, depends on the location of the synapse in the dendritic tree. Rall (1964) recognized that this property may be exploited to perform simple computations. First, for single inputs, by acting as a delay line, the dendrites can thus "label" particular inputs on distinct regions of the dendritic tree by the latency of the resulting output spikes. In fact, EPSPs with different somatic shape are likely to affect the somatic output spike trains in different ways (Fetz & Gustafsson 1983). Second, for multiple inputs, the time course of the somatic voltage response depends on the temporal order of activation of the dendritic synapses (in contrast to the scenario where they are all located on the soma) (Rall 1964; see below, and see also **Figure 6**).

Parallel processing and local computations. Synaptic inputs onto dendrites do not only inject current but also locally change the membrane conductance to certain ions. This leads to a nonlinear interaction between multiple inputs if they colocalize in time and space. When two excitatory inputs are active together at short distance, each depolarizes the membrane and reduces the driving force for the other input, and thus, theoretically, the response to the simultaneous activation is smaller than the sum of the individual responses (Rall et al. 1967). In this context, dendrites might be beneficial because they enable the spatial separation of inputs to minimize their interaction. In some cases, however, this possible sublinear summation may actually be advantageous (Agmon-Snir et al. 1998, section on coincidence detection in auditory neurons, p. 519) (see **Figure 1**). It also provides a mechanism for saturation of inputs, thus preventing overexcitation of the neuron by a group of synapses.

Nonlinear interactions are especially prominent between excitatory synapses and shunting inhibition. Shunting inhibition usually describes inhibition that changes the total conductance of the membrane but does not cause any voltage change when activated on its own. In this case it is more convenient to think of the inhibition as reducing the input resistance of the cell, effectively reducing the voltage response to excitatory current. This property of inhibition can be mathematically formulated as having a divisive effect on excitatory signals (Fatt & Katz 1953), providing a mechanism for simple arithmetic computation in dendrites (Blomfield 1974). Rall was the first to recognize that the effectiveness of this interaction has a strong spatial component. When excitatory and inhibitory inputs are widely separated from each other on different dendritic branches, then the inputs will tend to sum linearly at the soma. In contrast, when the excitatory and inhibitory inputs are located adjacent to each other, then the inhibition can produce a highly nonlinear "shunting" of the excitatory input (Rall 1964). An elegant recent experimental study by Liu (2004) demonstrated that such an inhibitory effect can be localized to a single dendritic branch. Theoretical work has shown that inhibition is also effective when it is located on the path between the excitatory input and the soma (Jack et al. 1975, Koch et al. 1983, Rall 1964). Thus, the relative location of synaptic inhibition versus excitation determines whether

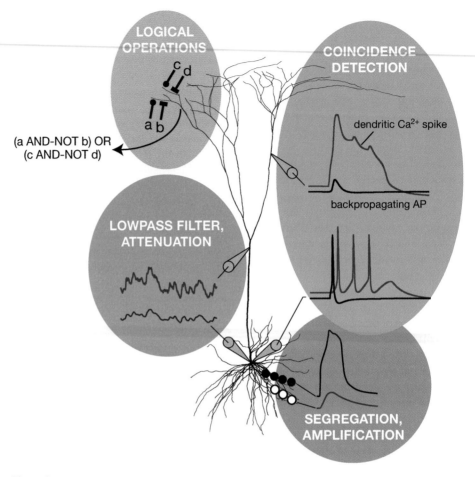

Figure 2

The dendritic computational toolkit. A schematic figure highlighting four key dendritic mechanisms, mapped onto a layer 5 pyramidal neuron morphology, which can allow dendrites to act as computational elements. These mechanisms can coexist in the same neuron and be active in parallel or in a hierarchical manner. *Bottom left*: Passive dendrites act as passive filters. A high-frequency fluctuating current injected in the dendritic pipette will evoke high-frequency and large-amplitude local voltage responses, but the response recorded by the somatic pipette will be attenuated and smoothed (low pass filtered).
Top left: Nonlinear interaction between excitation and shunting inhibition on small dendritic branches can implement logical operations. The branch marked by an arrow sums up the current from the two subtrees, such that its output would be a logical OR on their output. Each of the subtrees will in turn inject current if and only if the excitation AND-NOT the inhibition will be active. *Bottom right*: Dendrites can help reduce or amplify the mutual interaction between excitatory inputs. Excitatory inputs to the same branch tend to sum sublinearly, whereas inputs on different branches sum linearly. Thus mapping input to different branches can reduce this effect. In neurons with active dendrites, however, clusters of inputs active synchronously on the same branch can evoke a local dendritic spike, which leads to significant amplification of the input. Synapses onto a different branch (*open circles*) are only slightly influenced by this spike. *Top right*: In layer 5 cortical pyramidal neurons, as depicted here, coincidence detection between the apical and basal dendritic compartments is achieved by active dendritic mechanisms. A backpropagating action potential, which coincides with a distal synaptic input, will trigger a dendritic Ca^{2+} spike, which depolarizes the whole apical dendrite and drives a burst of spikes in the axon. See text for further details.

inhibition predominantly counteracts a specific set of (neighboring) excitatory synapses or whether it acts on the global set of excitatory synapses.

Although inhibition can act in a graded manner, it has been predicted that, in principle, synaptic inhibition may be able to veto an excitatory signal effectively, depending on the location and the strength of the inhibitory conductances (Jack et al. 1975, Koch et al. 1983, Rall 1964). For example, the result of the combined operation of a neighboring pair of excitatory and inhibitory inputs will cause somatic depolarization if and only if excitatory input AND NOT inhibitory input is active (**Figure 2**). This AND-NOT function is a Boolean logical operation, of the exact same kind implemented in modern computers and studied in mathematical computational theory. Whether dendrites really implement a network of Boolean gates is not clear, but this is exactly the kind of formalism required for a deep understanding of dendritic computation, namely a formal mathematical entity that would clarify the operations performed by dendrites. Koch et al. (1983) cleverly showed that logic operations can be linked to the less formal notions of computation used by physiologists to devise a model of a retinal ganglion neuron that has a directional selectivity to moving visual inputs (**Figures 2** and **6**; see below). Because the branch points in the dendritic tree can be seen as summing up the currents in individual branches, each tree can be seen as summing over many logical gates, and thus the whole dendrite can implement complex functions. Note that a key issue in the implementation of such a mechanism is the addressing of the right synapses to the right dendrite. In fact, as a general rule, as we see below, any computation that exploits local nonlinearity mechanisms is bound to require the addressing of the relevant synaptic inputs to the relevant locality in the dendrite (Mehta 2004, Poirazi & Mel 2001). It remains to be seen whether the power of dendritic computation can itself provide a constraint for targeting of synaptic inputs at the appropriate locations and whether such addressing indeed takes place as a basic phenomenon in the brain (Chklovskii et al. 2004).

Computations in Active Dendrites

Dendritic excitability as a feedback mechanism. Solely on the basis of anatomical observations, Cajal formulated the law of dynamic polarization (Cajal 1911), which states that in the nervous system information flows in one direction: from dendrites to soma to axon. In the past decade it has become clear that in many types of neurons the presence of excitable ionic currents in the dendrites supports dendritic action potentials that travel in the reverse direction, from the soma into the dendrites (Stuart et al. 1997). Computationally this "backpropagation" has major consequences because it implies that the neuron is no longer an open-loop system but has an internal mechanism of feedback. It is thus no longer the case that feedback is a property only of the network, but rather it is a salient property of each element of the network. Moreover, the feedback conveyed by the backpropagating action potential is highly sophisticated and has many important consequences for dendritic function, and also for synaptic plasticity (Magee & Johnston 1997, Linden 1999). For example, a single backpropagating action potential can activate slow dendritic voltage-gated ionic currents, which in turn flow back towards the spike initiation zone in the axon, often resulting in additional action potentials (see below). Thus, the somatic action potential can under favorable conditions trigger a burst due to its interaction with the dendrites (Carruth & Magee 1999; Williams & Stuart 1999). Modeling studies show that this interaction between soma and dendrites can be captured by a reduced, two-compartment model of the neuron and critically depends on the coupling coefficient between the two compartments. This coupling is governed by dendritic morphology and the distribution and properties of dendritic voltage-gated channels and synaptic activity (Doiron et al. 2002, Mainen & Sejnowski 1996, Pinsky & Rinzel 1994, Schaefer et al. 2003, Vetter et al. 2001). The firing patterns of neurons are thus potentially tunable simply by changing dendritic properties. The interplay between somatic spikes and dendritic response could be exploited computationally, e.g., as a slope detector (Kepecs et al. 2002), or for feature detection in sensory systems (Oswald et al. 2004).

Amplification of synaptic inputs. The fact that passive dendrites attenuate the synaptic input on its way to the soma raises a long-standing question: Why have so many distal synapses if their activity is not going to affect the output whatsoever? Investigators have proposed that other mechanisms are involved in synaptic integration, effectively endowing each synapse with equal "vote" and creating a "dendritic democracy." Resolving the importance of these different mechanisms in different neuronal types is important because the presence or absence of compensatory mechanisms leads to fundamentally different views of neuronal function (Häusser & Mel 2003). Here we briefly outline the various scenarios and the supporting experimental evidence. Four major mechanisms have been proposed: synaptic scaling, synaptic boosting, local dendritic spikes, and global dendritic spikes.

1. Synaptic scaling: In this scenario, the conductances of distal synapses are scaled according to their distance from the soma, so as to equalize their efficacies. First, indirect evidence for this mechanism of "dendritic democracy" was found in motoneurons (Iansek & Redman 1973) and has been more recently supported by studies in hippocampal CA1 pyramidal neurons (Magee 2000). However, this mechanism may not be general because other major types of neurons do not seem to follow this rule (Williams & Stuart 2002).

2. Subthreshold boosting: Inward voltage-dependent dendritic currents can amplify synaptic inputs on their way to the soma, thus compensating for their attenuation. Although it is clear that there exist dendritic currents to support this scenario (Cook & Johnston 1997, 1999; Migliore & Shepherd 2002), there is contradictory experimental evidence regarding whether such boosting plays an important role, and whether it stems from dendritic or somatic currents (Oviedo & Reyes 2002, Schwindt & Crill 1995, Stuart & Sakmann 1995).

3. Local dendritic spikes: A powerful mechanism for overcoming dendritic attenuation is the local dendritic spikes triggered by coactivation of synaptic inputs. The regenerative inward currents required for triggering such spikes can be provided by voltage-gated sodium channels, voltage-gated calcium channels, or synaptically activated N-methyl-D-aspartate (NMDA) receptor channels. Such spikes could be triggered by synaptic inputs or local dendritic current injections. The spatial extent of these dendritic spikes is highly variable, and so is their effect on the somatic voltage (Gasparini et al. 2004, Golding & Spruston 1998, Mel 1993, Polsky et al. 2004, Schiller et al. 1997, Softky 1994, Stuart et al. 1997, Williams & Stuart 2002). Theoretical studies have recently shown that such a mechanism could lead to a substantial increase in the computational power of the neuron (see below; Poirazi & Mel 2001; **Figure 5**).

4. Global dendritic spikes: Layer 5 cortical pyramidal neurons represent an extreme case where, owing to the length of the apical dendrite, many synapses are located so distally that, in the absence of any compensation mechanisms, they would have virtually no effect on the somatic voltage (Cauller & Connors 1994; Stuart & Spruston 1998). Not

only is there marked passive attenuation in these neurons, but active currents may further attenuate the signal (see below). Recently experiments have shown that these neurons exhibit a second spike-initiation zone near the main branch point of the apical dendrite (about 500–650 μm from the soma). Investigators have suggested that the distal apical dendritic tree can act as a separate synaptic integration region, having its own separate spike-initiation zone at this location. When the distal compartment crosses the threshold, a dendritic Ca^{2+} spike is initiated, resulting in a huge dendritic depolarization that drives the somatic region to initiate action potentials. In this way the distal compartment communicates with the soma (Larkum et al. 1999b, 2001; Schiller et al. 1997; Williams 2004; Yuste et al. 1994) (**Figure 4**). To some extent this mechanism is also likely to be operational in other types of pyramidal neuron, such as those in hippocampal CA1 (Golding et al. 2002).

Compressive dendritic nonlinearities. Dendrites express not only voltage-gated inward currents, but also they are rich in other classes of voltage-gated currents that counteract regenerative excitation and thus can be thought to be responsible for maintaining the balance of excitability in the dendritic tree. Some of these currents, such as A-type K^+ currents or hyperpolarization-activated inward I_h currents, are located at higher densities in the distal part of the dendrites (Hoffman et al. 1997, Lörincz et al. 2002, Magee 1998, Williams & Stuart 2000; reviewed by Migliore & Shepherd 2002). This arrangement of nonregenerative currents in the dendrites is puzzling because it enhances the attenuation experienced by synaptic inputs. However, in view of the rich complement of excitable currents in the dendrites such mechanisms are likely required to maintain dendritic stability. In addition to their global balancing effect, these currents can take part in more local interactions. One example for such an interaction has been shown for A-type K^+ currents. When a synaptic input is active on a dendritic branch, the depolarization of the branch inactivates A-type K^+ currents in this branch. This in turn facilitates the ability of backpropagating action potentials to invade this branch more easily, which may provide a gating mechanism for plasticity in dendritic branches (Hoffman et al. 1997, Magee et al. 1998).

Coincidence detection. The simplest nonlinear operation is multiplication. In case of binary variables, multiplication is identical to the logical AND operation (the result is 1 if and only if the two inputs are 1), which can also be described in terms of coincidence detection. We have already described how the interaction of excitation and inhibition could implement this operation, but the regenerative inward currents expressed in the dendritic membrane provide numerous alternative mechanisms. These forms of coincidence detection can operate on a highly local scale (down to the level of a few spines) or on the scale of the entire neuron.

Numerous experimental studies have provided evidence that active dendrites can generate local dendritic spikes given synaptic input that is sufficiently clustered in space and time (the latter being the essential requirement for a coincidence detector). Such spikes can be generated by any combination of the voltage-gated regenerative inward currents known to be present in the dendritic membrane. For example, the current driven by NMDA receptor activation is known to be highly voltage dependent. Recently investigators (Cai et al. 2004, Polsky et al. 2004, Schiller et al. 2000) (see **Figure 5**) showed that synchronous synaptic inputs onto the same dendritic branch of layer 5 pyramidal neurons depolarize the membrane and create a positive feedback loop such that the current through the NMDA receptor depolarizes the membrane and recruits more NMDA-mediated current (supported by activation of dendritic Na$^+$ and Ca^{2+} channels). This all-or-none phenomenon is termed an NMDA spike, and its spatial extent is limited to a short region of the dendrite by active and passive mechanisms. In CA1 pyramidal neurons, a similar local coincidence-detection mechanism exists, based primarily on a different voltage-dependent current (Na$^+$) (Ariav et al. 2003). These dendritic mechanisms provide the neuron with the ability to detect coincidences in neighboring synaptic inputs on a very fast timescale, previously thought to be restricted to auditory neurons. The same mechanism was previously suggested in models to explain the variability in output spike trains of cortical neurons (Softky 1994).

Dendritic mechanisms also exist for reporting coincident pre- and postsynaptic activity. At distal synapses on the apical dendrite of layer 5 pyramidal neurons, pairing postsynaptic action potentials and synaptic input can trigger highly nonlinear amplification of backpropagating dendritic action potentials via recruitment of voltage-gated Na$^+$ channels (Stuart & Häusser 2001) (**Figure 3**). A similar supralinear interaction has also been observed in CA1 pyramidal neurons when pairing backpropagating spikes and EPSPs, where the contribution of A-type K$^+$ channel inactivation is more prominent (Johnston et al. 2000, Magee & Johnston 1997). This form of coincidence detection exhibits a narrow time window (∼10 ms), similar to that required for induction of synaptic plasticity when pairing APs and EPSPs, and may thus act as a substrate for the induction of synaptic plasticity.

Finally, in layer 5 pyramidal neurons excitatory synaptic input to the distal apical tuft that coincides with backpropagation of the action potential results in a large dendritic Ca$^+$ spike, which in turn propagates toward the soma and drives the axon to fire a burst of action potentials (Larkum et al. 1999b, Schiller et al. 1997, Stuart & Häusser 2001) (**Figure 4**). This mechanism thus enables the detection of coincident activation of synaptic inputs to the two major compartments of the dendritic tree, and may thus be involved in reporting simultaneous activity across different cortical layers. This coincidence detection mechanism is potentially tunable, either by changing dendritic geometry or by modulating channel densities and properties (Vetter et al. 2001, Schaefer et al. 2003).

Dendritic subunits: neurons within a neuron. Nonlinear mechanisms in dendrites can vary widely in the spatial extent of the resulting electrical and chemical signals. Some events spread across the entire dendritic tree, whereas others remain very local (Häusser & Mel 2003). Focusing on the local mechanisms, Mel and colleagues (Mel 1993; Poirazi et al. 2003a; Poirazi & Mel 2001; Polsky et al. 2004) have developed a framework that breaks the dendrites into many tiny computational units. The basic nonlinearity in individual branchlets is based on the NMDA spike (Schiller et al. 2000, Schiller & Schiller 2001) and is modeled as a sigmoidal function (**Figure 5**). Each subunit thus integrates its inputs and passes them through a sigmodial nonlinearity function. This gives each piece of dendrite the computational power of a small unit similar to those conventionally used in neural networks. The output of each subunit is conveyed to the soma, in terms of passive dendritic integration. The picture that emerges from this analysis is of a two-layered neuronal network that resides within a single neuron. This analysis is supported by detailed modeling of single neurons, showing that the predictive power of the two-layer neural network description is very good (Poirazi et al. 2003b), and by experiments in layer 5 pyramidal neurons (Polsky et al. 2004). The attractiveness of this approach stems from the fact that two-layer neural networks are general-purpose computation machines, which have been extensively studied and can implement very powerful computations.

This analysis also poses important questions about the way the neuron learns to

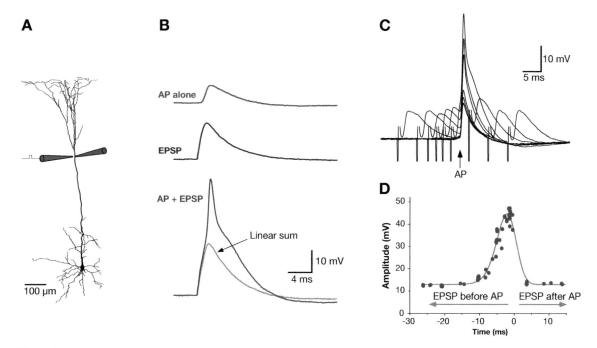

Figure 3

Coincidence detection of EPSPs and action potentials. (*A*) Schematic illustration of the recording configuration. A dendritic recording (*red pipette*) is made from the distal apical dendrite of a layer 5 pyramidal neuron, and synaptic input local to the recording site is activated with a stimulation electrode (*blue pipette*). (*B*) *Top trace*: backpropagating action potential initiated by somatic current injection (2.5 nA) recorded 720 μm from the soma. *Middle trace*: evoked EPSP recorded at the same dendritic location. *Bottom trace*: response to pairing the backpropagating action potential with the EPSP ("AP + EPSP"). For comparison, the linear sum of the action potential plus EPSP is also shown (*grey trace*). Stimulus artifacts are blanked for clarity. (*C*) Superimposed sweeps of evoked EPSPs at different times before and after action potentials initiated by somatic current injections (1.5 nA) at the time indicated by the arrow. Dendritic recording 480 μm from the soma. (*D*) Plot of action potential amplitude (measured at the time of action potential peak) versus the time difference between EPSP and somatic action potential onset (same cell as in *C*). The time of somatic action potential peak is defined as zero; negative time corresponds to when EPSPs were evoked before action potentials, and positive time corresponds to when EPSPs were evoked after action potentials. The smooth line is the fit with a skewed Gaussian. Modified from Stuart & Häusser (2001).

compute its input-output function. In this review we must assume that neurons are learning to compute what they compute. The leading theory for how this is achieved is Hebbian plasticity (Hebb 1949). But if dendrites are implementing these nonlinear subunits, Hebbian plasticity will not exploit the power of this model. Plasticity will help to drive learning within each individual subunit, but the number of inputs in each of these units is small and it is not clear how the relevant inputs will get there in the first place. Poirazi & Mel (2001) have proposed a learning algorithm by which synaptic connections are continuously remodelled until they hit the correct dendritic subunit. They show that such a learning algorithm implemented in a dendritic tree can be much more powerful than the Hebbian learning scheme. Although there exists conflicting evidence for ongoing remodeling of synaptic inputs in the adult brain (Grutzendler et al. 2002, Holtmaat et al. 2005, Mizrahi & Katz 2003, Trachtenberg et al. 2002), it is still an open question whether this algorithm indeed is implemented in real neural circuits. Other challenges for the model are its integration with the dendritic tree. All the model is currently taking from the dendritic tree is the ability to have independent subunits. However, it ignores global nonlinearities and modulations evident in the dendrites. Moreover, in a two-layer neural network, the coefficients from each unit in the first layer to the next are modifiable. Here, in comparison, they are fixed and determined by the dendritic tree. Finally, although the two-layer neural network model reliably predicts the steady-state firing rates of the pyramidal neurons, it neglects the temporal properties of spike firing that have been closely linked to dendritic excitability (e.g., Ariav et al. 2003, Larkum et al. 1999b).

Chemical computation. The expression of voltage-gated calcium channels in the dendritic membrane (Migliore & Shepherd 2002) immediately provides a biochemical readout of electrical excitability. In particular, dendritic calcium signals activated by backpropagating action potentials reliably encode the level of axonal spike firing in apical dendrites of pyramidal cells. This therefore provides a "frequency code" where firing rate is read out using a dendritic biochemical signal (Helmchen et al. 1996). This readout can also have a nonlinear frequency dependence if it involves activation of a dendritic calcium spike (Larkum et al. 1999a). The calcium signal can in turn activate voltage-gated potassium currents, thus acting as a feedback regulator of excitability, which changes the

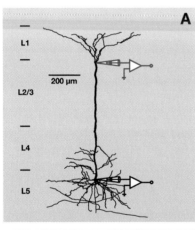
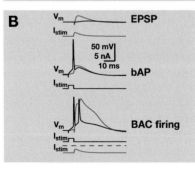
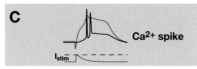
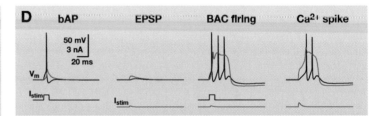
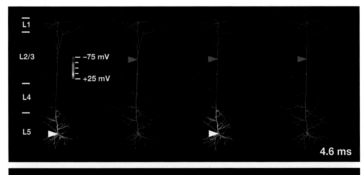
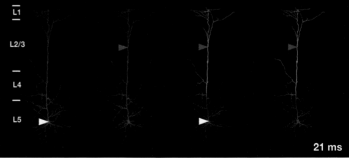

input-output gain of the neuron (Sobel & Tank 1994) (see below). A similar readout of dendritic excitability can be provided by intracellular dendritic Na^+ signals (Rose & Konnerth 2001), which may in turn regulate excitability via activation of Na^+-activated K^+ channels.

Whereas the various regenerative dendritic mechanisms for coincidence detection discussed above will also be read out via voltage-gated calcium channels to generate large dendritic calcium signals, the biochemical intracellular signaling pathways in dendrites may themselves contribute to coincidence detection in unique ways. For example, the IP3 receptor is cooperatively activated by both calcium and IP3, which allows for coincidence detection of calcium and IP3 delivered by different sources, such as action potentials and synaptic mGluR activation (Nakamura et al. 1999, Wang et al. 2000). This form of coincidence detection can be spatially segregated to particular types of dendrite (Nakamura et al. 2002). A further mechanism for coincidence detection and/or intracellular calcium amplification can be implemented via dendritic calcium-induced calcium release from stores (Emptage et al. 1999). Finally, the lowly calcium-buffering proteins localized in dendrites may themselves allow for a simple form of coincidence detection, generating a supralinear dendritic calcium signal by buffer saturation (Maeda et al. 1999). The large calcium signals generated by these different forms of coincidence detection can remain highly localized (Wang et al. 2000), or they can spread to other regions of the dendritic tree, assisted by further regenerative calcium release from stores (Larkum et al. 2003, Nakamura et al. 2002). Barlow proposed that such processing by intracellular networks can implement a second layer of computation

---

Figure 4

Coincidence detection across dendritic compartments. (*A*) Reconstruction of a layer 5 pyramidal neuron; the locations of recording pipettes (soma, *black*; dendrite, *red*) are depicted schematically. (*B*) Distal current injection of 1.1 nA in the shape of an EPSP (I_{stim}, *red*) evoked only weak somatic (*black*) depolarization (*upper panel*). Threshold current injection (5 ms) into the soma (*black*) produced an AP that propagated back into the apical dendritic arbor (backpropagating action potential, bAP, *red trace, middle panel*). Combination of somatic and dendritic current injection generates several somatic APs and a dendritic Ca^{2+} spike (backpropagating action potential–activated Ca^{2+} spike firing, BAC firing; *lower panel*). The dashed line indicates the current threshold for a dendritic Ca^{2+} spike alone. (*C*) A dendritic Ca^{2+} spike was evoked by 2 nA current injection into the apical dendrite alone. Thus, the bAP reduced the threshold for dendritic Ca^{2+} spike by 0.9 nA (45% coupling). (*D*) A model of channel density distributions and kinetics was constructed to reproduce BAC firing in reconstructed model neurons (Schaefer et al. 2003). The electrical response of the reconstructed model neurons to dendritic and somatic current injection was investigated using the same protocols as in the experiment (*A–C*). (*Upper panels*) bAP: Threshold somatic current injection evoked a bAP ($I_{stim} = 1.9$ nA). EPSP: Distal EPSP-like current injection was adjusted to BAC firing threshold, which was 0.6 nA. Only a small somatic depolarization can be detected ($\Delta V \leq 2.5$ mV). BAC firing: Pairing the bAP with the dendritic EPSP-like current injection resulted in a large and long-lasting dendritic depolarization. Ca^{2+} spike: Large distal EPSP-like current injection (1.7 nA) elicited a Ca^{2+} spike. Thus, the bAP reduced the threshold for dendritic Ca^{2+} spikes by 1.1 nA, which resulted in a coupling of 1.1 nA/1.7 nA = 65%. Voltages were measured at the positions indicated by triangles in lower panels; (*red*: dendritic recording/current injection; *black*: somatic recording/current injection). (*Lower panels*) Same as upper panels but showing membrane potential in the entire dendritic tree. Voltages are color coded as indicated in the upper left. The position of current injection is indicated by the red (dendritic) and white (somatic) arrowheads. At the time of AP initiation (4.6 ms after the beginning of the somatic current injection), depolarization due to the bAP has already spread into the apical dendrite (in the case of bAP and BAC firing). After 21 ms, the voltage deflection due to the bAP has decayed back to baseline. Note that the spread of depolarization is almost the same for a dendritically elicited Ca^{2+} spike and BAC firing. Modified from Larkum et al. (1999b) and Schaefer et al. (2003).

coupled to, but semi-independent from, the electrical signaling in the plasma membrane (Barlow 1996). Such a "two level" arrangement could have enormous computational power, which is only beginning to be explored.

EXAMPLES OF REAL-WORLD DENDRITIC COMPUTATION

In the previous section we described how the passive and active properties of the dendrites can endow them with computational features. The key question, of course, is whether the

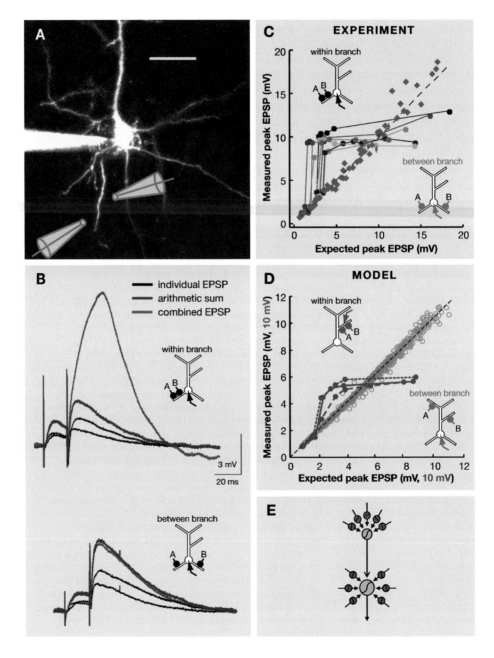

brain takes advantage of these building blocks to perform computations. It is extremely difficult to show directly that a particular computational strategy is both necessary and sufficient to explain the computational behavior of networks. However, a few favorable cases have provided strong circumstantial evidence for dendritic computation playing a key and possibly essential role in computations performed by a neural network.

Directional Selectivity

Perhaps the most extensively studied computation on the single-cell level is direction selectivity. Direction-selective neurons respond to image motion in a preferred (PREF) direction but not in the opposite NULL direction. They can be found in many species from fly eyes to mammalian cortex, and in all these cases a role for dendritic computation has been proposed. One of the first and most convincing experiments demonstrating dendritic involvement in directional selectivity was provided by Single & Borst (1998). Using imaging of dendritic calcium signals in tangential cells of the fly visual system in vivo, they showed that the input to each dendritic branch, and thus the resulting dendritic calcium signal, is already directional selective, but the dendritic filtering is required to maintain a coherent response free from spatial pattern properties of the visual scene to ensure a purely direction-selective output signal in the axon.

In this section we focus on work on the direction-selective retinal ganglion cells (DSGC) described by Barlow & Levick (Barlow et al. 1964, Barlow & Levick 1965) as a case study for dendritic computation. Rall (1964) provided the first model for how dendrites can implement a directionally selective

Figure 5

Dendritic multiplication in pyramidal cell dendritic branches. (*A*) Two stimulating electrodes were positioned near selected basal dendrites of a layer 5 pyramidal neuron filled with the calcium-sensitive dye Oregon Green BAPTA-1 (200 μM). A dendritic branch was visualized using a confocal microscope, and two stimulation electrodes were positioned in close proximity to the selected branch (*blue*). Scale bar, 75 μm. (*B*) Electrodes were activated first individually (*black traces*) and then simultaneously (*red traces*), and somatic EPSPs were recorded. Blue traces show the arithmetic sum of the two individual responses. Voltage traces are averages of four individual sweeps. Left traces show within-branch summation. The two electrodes were positioned near the same dendritic branch, separated by 20 μm (150 μm from the soma). Right traces show between-branch summation, where the two electrodes stimulated different branches and summation was linear. (*C*) Summary plot shows predicted versus actual combined responses in seven basal dendrites and one apical oblique dendrite (*pink curve*). Colored circles show sigmoidal modulation of within-branch summation (*blue and yellow*, without bicuculline to block GABAergic inhibition; *dark green trace*, with locally applied 10 μM bicuculline; *five remaining traces*, 1 μM bicuculline). Dashed line denotes exact linear summation. Green diamonds show between-branch summation experiments (12 branch pairs, 4 of them apical oblique dendrites). (*D*) Modeling data: Summation of single-pulse EPSPs in the apical oblique dendrites of a CA1 pyramidal cell model showed a similar overall pattern (Poirazi et al 2003a), including sigmoidally modulated within-branch summation (*red circles*) and linear between-branch summation (*open green circles*). Within-branch data for dendrites are attached to the apical trunk 92 μm (*short dashes*), 232 μm (*solid*), and 301 μm (*long dashes*) from the soma. Because of the uneven distances to the somatic recording electrode, recordings shown were made within the respective dendrite; for these data, axis values are scaled up 10, thus 0 mV, 20 mV, 40 mV, and so on. Modified from Polsky et al. 2004. (*E*) Schematic representation of a speculative neural network model based on the data shown in *A–C* and **Figure 4** (see also Häusser & Mel 2003). Blue branches represent the distal apical inputs, and red branches denote the basal inputs. Together, these inputs constitute the first layer of the network model, each performing supralinear summation of synaptic inputs as shown in *B* (*small circles with sigmoids*). The outputs of this first layer feed into two integration zones: one near the apical tuft (*top*) and one near the soma. These integration zones constitute the second layer of the network model (*large circles with sigmoids*).

unit (**Figure 6A**). The idea is very simple: Synaptic input generated on the distal part of the dendrite is delayed at the soma by the dendritic filtering compared with proximal input. If synaptic inputs are activated in a sequence starting from the distal location of the dendrite toward the soma (the centripetal direction), then the EPSPs in the soma will sum effectively and the resulting somatic voltage would be large. In contrast, activating the same inputs in the centrifugal direction would result in a much lesser degree of summation because the proximal EPSP will decay by the time the distally originated EPSP will arrive. Assuming that the voltage peak is translated into action potentials, the neuron will show directional selectivity. Although the mechanism proposed is clearly feasible, there

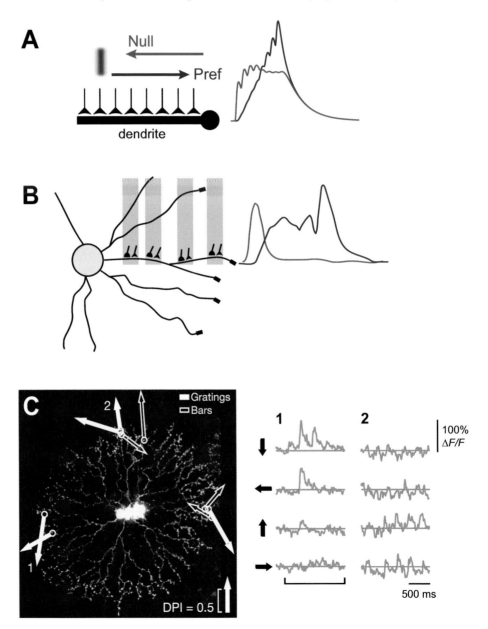

is currently little direct evidence to support this model in cases where directional selectivity has been found (Anderson et al. 1999, Euler et al. 2002).

An ongoing debate exists about identifying the earliest level of neurons that show directional selectivity. Koch et al. (1982) proposed that the nonlinear interaction between the excitation and inhibition can explain directional selectivity in retinal ganglion cells. The essential assumption was that there is an asymmetry in the spatial distribution of inhibitory and excitatory inputs to the cell such that the inhibition biased and shifted to the NULL direction. In this way when the inputs are sweeping through the receptive field in the PREF direction the excitation is acting before the inhibition and the integration of excitatory inputs cause the neuron to respond. If the input moves in the NULL direction, then the inhibition is "on the path" of the excitation and vetoes it, preventing the neuron from responding. Recently, Taylor et al. (Taylor & Vaney 2002) showed, using intracellular recording from direction-selective retinal ganglion cells (DSRGC), that indeed such asymmetry in the inhibition exists, but these results are debatable (see also Borg-Graham 2001, Fried et al. 2002, Taylor & Vaney 2000).

The other possibility is that the input to the DSRGC is already direction selective. Using similar assumptions to the model by Koch, Borg-Graham & Grzywacz (1992) (**Figure 6B**) showed that the directional selectivity could be computed in the individual dendritic branches of starburst amacrine cells (SBAC), which are presynaptic to the retina ganglion cells. These cells do not have appropriate axons; rather each dendrite has an "output" synapse at its distal end. By using two-photon optical imaging of Ca^{2+} concentrations in the dendrites of SBAC, Euler et al. (2002) showed that indeed the Ca^{2+} concentration at the tip of the dendrites of SBAC is direction selective (**Figure 6C**). However, in contrast to the model, the response is still selective in the presence of $GABA_A$ blockers. In summary, it seems that directional selectivity is indeed computed by individual dendrites of SBAC, but the mechanism by which it is computed is still not fully understood.

Coincidence Detection in Auditory Neurons

Another system in which the contribution of the dendrites to computation has been demonstrated is the sound localization system of chicks (Agmon-Snir et al. 1998). In this system a special type of neuron is responsible for computing the time difference between sounds arriving to the two ears. Each

Figure 6

Dendritic mechanisms for directional selectivity. (*A*) A linear model exploiting the filtering properties of passive dendritic cables. When the input advances to the right (the preferred direction), the first EPSPs are widened by dendritic filtering, which gives time for the later input to sum temporally and build a large voltage response. When the input moves in the null direction, the first large EPSP decays by the time the last EPSP arrives at the soma, and a smaller peak response is achieved. Note that this mechanism is not very robust because the difference in peak amplitude between the two scenarios is small, and the time integral of the voltage (corresponding to the total amount of charge) is identical. (*B*) A model of the starburst amacrine cell in the retina (Borg-Graham 1992). The input to each amacrine dendrite is composed of excitatory and inhibitory inputs that have symmetric receptive fields. Although the starburst cell is radially symmetric, the symmetry breaks with respect to the direction-selective circuit because the outputs to the directionally selective ganglion cells are on the distal tip of each dendrite. Furthermore, these outputs are formally direction selective, in the sense that the integral of the response depends on direction, because of the nonlinear interaction between excitation and inhibition, as described by Rall (1964) and Koch et al. (1982). (*C*) Imaging of internal Ca^{2+} concentration from dendrites of starburst amacrine cells shows that these dendrites are indeed direction selective (Euler & Denk 2002). The experiments are not consistent with the mechanism in *A*, but they do not completely agree with the model in *B* either, because blocking $GABA_A$ receptors retains some of the directional selectivity.

neuron responds only to a very precise time difference, which corresponds to a specific location in space. The neurons contain only two major dendrites, and each dendrite receives inputs only from one ear. The inputs are supposedly arranged in such a way that there is a constant delay between the inputs arriving from one ear and the inputs arriving from the second. Coincident inputs from both ears arriving to the two dendrites are summed up at the soma and cause the neuron to emit action potentials. However, when coincident spikes arrive from the same ear, they arrive at the same dendrite and thus their summation is sublinear, resulting in a subthreshold response (**Figure 1A**). Moreover, Rubel and colleagues (Smith & Rubel 1979) showed that there is an inverse relationship between the preferred frequency of the neurons and their dendritic length, supporting the hypothesis that the dendrites are directly contributing to the computation. This is in agreement with the model because for high-frequency inputs the dendritic filtering causes accidental spikes that are out of phase to summate and cross the threshold. Thus the advantage of the dendrites in the low-frequency range becomes a burden in the high-frequency range, and the auditory coincidence detection neuron is better off with shorter dendrites.

Temporal Integration Over Long Timescales

All the computations discussed above take place on relatively short timescales; the neuron responds almost instantaneously to the input. Computations over longer timescales, such as those required for working memory, are usually attributed to network phenomena or involving molecular dynamics. In this context, computation of time integration becomes a challenging problem. How does a system integrate transients and maintain the computed integral for a long period, far longer than its intrinsic time constant? One example of such a system is the oculomotor system in the goldfish, where neurons maintain stable firing rates corresponding to the position of the eye and switch between them rapidly when saccades occur. Although previous work has focused largely on network explanations for such forms of time integration, dendritic mechanisms (Brody et al. 2003) in single neurons may also contribute. In particular, one intrinsic mechanism that could be involved in such a phenomenon is bistability. This is a property of a dynamical system exhibiting more than one stable point. Once driven to approach a stable point, it stays there. Neurons expressing specific types of voltage-gated ionic currents, for example, can show bistability. Some indirect evidence for dendritic bistability exists (Booth & Rinzel 1995, Milojkovic et al. 2004). Recently, Loewenstein & Sompolinsky (2003) presented a model in which a specifically dendritic bistability enables the dendrite to become a time integrator. In this model, the dendritic concentration of Ca^{2+} is bistable. If at one end of the dendrite it is forced to be in the up state, and at the other edge to be in the down state, then a standing wave of Ca^{2+} concentration is created. The location of the wavefront represents the result of the integral such that incoming transient synaptic input repositions the location of the front. As long as no input arrives, the front will keep its position and the dendrite will preserve the "working memory." A related study, based on an earlier model (Rosen 1972), has been proposed on the basis of dendritic bistability involving voltage-sensitive conductances (Goldman et al. 2003).

Image Processing in Dendrites of Fly Neurons

A key component of visual information processing by the fly nervous system is convolution, whereby an image is smoothed to remove noise and improve extraction of salient features. This operation is thought to be performed by a group of neurons in the lobula plate in the third visual ganglion known as the horizontally sensitive tangential cells. These

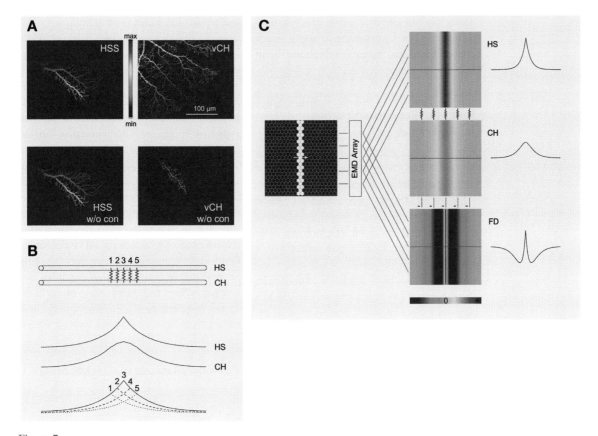

Figure 7

Image processing in visual interneurons of the fly. (*A*) *Top*: Spread of membrane potential in a model of neurons, which are horizontally sensitive of the southern area of the visual field (HSS) (*left*) and ventral centrifugal horizontal cells (vCH) model (*right*) after local current injection into HSS. *Bottom*: same as top panels, but HSS and vCH models were not connected to each other. Current was injected in HSS (*left*) and vCH (*right*). (*B*) Simplified model of HS and CH neurons. *Top*: Two cylinders (HS and CH) are connected by five linear conductances surrounding the location of current injection. *Middle panel*: in HS, the signal spreads with an exponential decay. The CH spread is broader. *Bottom panel*: The CH spread can be approximated by the sum of passive spread through each conductance. (*C*) Consequences of the CH cell dendritic image blurring for relative motion detection. An array of elementary motion detectors computes the image motion in a retinotopic way, feeding onto the dendrites of HS and figure-detection (FD) cells. This motion representation is blurred in the dendrites of the CH cell via dendro-dendritic connections between HS and CH cells. By conveying inhibitory dendro-dendritic input to FD cells, being subtracted from the retinotopic input, an enhancement of the motion edges is achieved. Modified from Cuntz et al. 2003.

neurons respond to visual motion in a directionally selective way and can be divided into two groups of neurons: the horizontal system (HS) cells and the centrifugal horizontal (CH) cells. Specific CH cells, the ventral CH (vCH) cells, are electrically coupled via dendritic gap junctions to HS cells. Retinotopic input to HS cells is already filtered by the electrotonic decay of membrane potential in HS cell dendrites. As shown in **Figure 7** (Cuntz et al. 2003), the coupling via gap junctions then imposes the filtered membrane potential of the HS cell onto the CH dendritic tree, where another round of low-pass filtering takes place. The end result is a blurred and de-noised version of the original image.

Thus, the biophysical properties of dendritic trees can implement non-trivial image processing operations in a simple and elegant manner.

Looming Sensitive Neurons in the Locust

The lobula giant movement detector (LGMD) is an identified neuron in the locust visual system whose output firing rate is most sensitive to objects approaching a collision course (looming visual stimuli), indicating a forthcoming collision (**Figure 8**). The timing of the peak firing rate comes with a fixed delay after the time at which a looming object reaches a fixed-threshold angular size, on the retina, independent of the object's approach speed or size (Gabbiani et al. 1999, 2004). A mathematical model supported by experimental results predicts that this behavior could be explained as a multiplication of two parameters of the approaching object, namely its angular size and speed of approach (Gabbiani et al. 2002). The LGMD neuron has a unique dendritic structure composed of a fan-like tree and two additional sub trees. The synaptic input is segregated such that excitatory, motion-sensitive input arrives on the major fan-like tree, and inhibitory size-sensitive inputs arrive in separated ON/OFF channels at the remaining two subtrees. The multiplication is thought to be implemented such that each of the relevant parameters is encoded logarithmically in one of these subtrees, and the dendritic sum of the two types of inputs results with the sum of logarithm (corresponding to the logarithm of the multiplication). The spiking mechanism on this combined input implements an approximate exponentiation that inverses the logarithm, and the result of the multiplication is thus encoded in the firing rate. The accessibility of the LGMD neuron to dendritic recordings and optical imaging in vivo makes it a promising candidate for understanding the biophysics of a high-level computation in dendrites in the near future.

Forward Masking of Cricket Songs

Omega neurons in female crickets respond to the male calling song with bursts of action potentials. The response of an omega to a particular sound can be dramatically attenuated if it is preceded by an identical but louder sound. This suppression is known as

---

Figure 8

Computation in dendrites of locust looming-sensitive neurons. (*A*) The LGMD neuron's dendritic tree consists of three distinct subfields (*A–C*). Subfield *A* receives motion-sensitive excitatory inputs, whereas subfields *B* and *C* receive phasic inhibition related to object size. As in many invertebrate neurons the soma lies outside the electrical signal propagation path, and spikes are generated at the point where the axon is thinnest. (*B*) Schematic illustration of the neuronal inputs received by the LGMD. Postsynaptic inhibitory regions of the LGMD are illustrated in red and excitatory ones in green. Green and red dots represent inhibitory and excitatory synapses, respectively. Yellow dots indicate cell bodies. *Bottom*: A solid object of size 2ℓ is approaching the animal on a direct collision course with a constant velocity v. The angle subtended by the object is represented by θ. (*C*) Top panel shows the time course of θ for the looming stimulus, and the middle panel shows individual spike trains of the LGMD neuron in response to repetitions of this stimulus. The blue line above the spike trains represents the average instantaneous firing rate, and its peak is marked by a star. *Bottom*: The relation between the peak firing time relative to collision as a function of $\ell/|v|$ is nearly linear. This is equivalent to the angular size subtended by the object being a fixed constant δ ms prior to the peak, independent of the stimulation. This angular size is typically in the range of $15°$–$35°$. Thus, LGMD's peak firing time acts as an angular threshold detector. (*D*) Top three traces are intracellular dendritic recordings in response to a looming stimulus, and bottom three traces show the response after picrotoxin injection to the lobula. Picrotoxin prolongs the responses, and the peak firing rate no longer predicts collisions. Adapted from Gabbiani et al. 2004.

forward masking and represents a simple form of gain control, allowing the female cricket to focus on the loudest male in the presence of multiple competing males. Sobel & Tank (1994) have shown that biochemical dendritic signaling underlies this simple computation. By imaging dendritic calcium signals in omega neurons, they showed that a loud simulated calling song triggers a large, long-lasting dendritic calcium transient associated with a potassium conductance that suppresses excitability. The time course of the calcium signal was tightly correlated with the time course of forward masking. To demonstrate

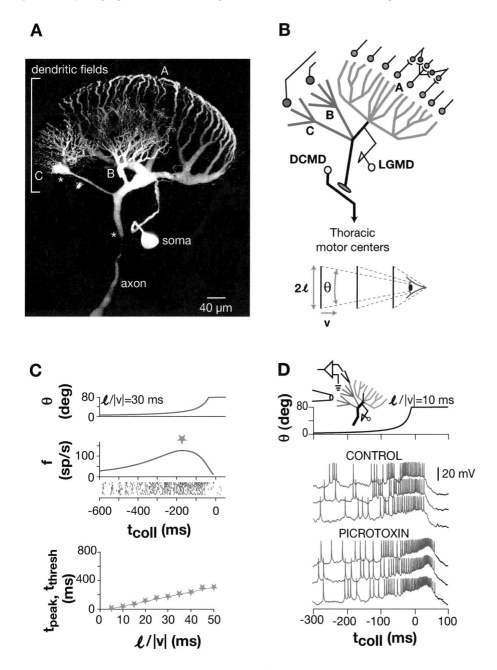

a causal relationship, Sobel & Tank showed that buffering dendritic calcium prevents the hyperpolarization and reduction in excitability. They also demonstrated that uncaging calcium in the dendrite produces a similar hyperpolarization and dampening of excitability to that generated by the sound. These findings demonstrate how dendritic mechanisms—conversion of action potentials into a calcium signal and then activation of a potassium conductance—can be directly linked to a computational task relevant to behavior.

Dendritic Mechanisms and Behavior

Two recent pioneering experimental studies have opened the door for exploring dendritic function of mammalian neurons in an entirely new framework. By using molecular techniques to manipulate dendritic ion channels in transgenic animals, Nolan et al. (2004) and Bernard et al. (2004) have demonstrated that it is possible to link the excitable properties of distal dendrites with network-level and behavioral phenomena. Bernard et al. (2004) demonstrated that in an animal model of temporal lobe epilepsy, the excitability of CA1 pyramidal cell dendrites is enhanced by downregulation of dendritic A-type K^+ channels by phosphorylation, combined with reduced expression of these channels. Although the evidence remains indirect, the enhanced dendritic excitability associated with this channelopathy may contribute to the observed reduction in seizure threshold in the hippocampus.

Nolan and colleagues (2004) generated a transgenic mouse with a forebrain-restricted deletion of the HCN1 gene, which encodes the hyperpolarization-activated cation current I_h. These mice exhibit enhanced performance in hippocampal-dependent learning and memory tasks. On the cellular and network level, the mice demonstrate enhanced theta oscillations and LTP. Because previous work has demonstrated that HCN1 channels are highly concentrated in the distal dendrites of CA1 pyramidal neurons (Magee 1998), the authors examined integration of proximal and distal synaptic input to CA1 pyramidal neurons. They demonstrated that distal perforant path synaptic inputs are selectively enhanced in the HCN1 knockout and that LTP of these inputs is also enhanced, whereas LTP at the more proximal Schaffer collateral input is unchanged. These results provide some of the best available evidence linking integration of distal dendritic synaptic input with behavior and point to the importance of independent regulation of excitability in subcompartments of the dendritic tree.

CHALLENGES FOR THE FUTURE: A WISH LIST FOR DENDRITIC COMPUTATION

The ultimate challenge for those interested in dendritic computation is to show that computation conveys a significant advantage in the operation of real neural circuits. This advantage is difficult to show directly, given that dendrites also have other properties not directly related to computation, such as their important structural role in determining brain wiring (Chklovskii 2004). Here we outline several of the key challenges faced by experimenters working at different levels to understand the contribution of dendrites to computation in the mammalian brain.

For Molecular Biologists: Designer Dendrites

To manipulate the computational properties of dendritic trees, we want to be able to manipulate dendritic shape and the spatial distribution and properties of voltage-gated channels using genetic tools. The molecular regulation of dendritic growth and shape has become a burgeoning field over the past decade, and many kinases have been identified that can be regulated to produce changes in dendritic form (Scott & Luo 2001). In parallel, our understanding of ion channel trafficking regulation and the local synthesis of ion channels in dendrites has also made substantial progress (Horton & Ehlers 2003, Misonou et al. 2004, Misonou & Trimmer 2004). Within the next decade it should be possible, using transgenic and viral techniques, to selectively modify single neurons or entire populations of neurons to generate dendritic subunits with defined shapes and electrical properties. Such "designer dendrites" can be used to identify the role of specific dendritic mechanisms, such as action potential backpropagation, for the function and formation of specific neural networks and ultimately can serve as a well-defined bridge between molecules and behavior.

Many dendritic mechanisms require that specific kinds of synaptic input are addressed to specific regions of the dendritic tree or furthermore require that synaptic inputs are highly spatially clustered. Although there is already considerable evidence from decades of anatomical work that precise spatial targeting of certain kinds of synaptic inputs is achieved in some cell types (Freund & Buzsaki 1996, Somogyi et al. 1998), identifying the molecular mechanisms responsible for such targeting will be very important for understanding how synaptic connectivity interacts with and defines dendritic computational mechanisms (Ango et al. 2004). It will be of great interest to harness these mechanisms to identify and manipulate the spatial arrangement of specific types of synaptic inputs to the dendritic tree. These mechanisms could be exploited first to label inputs selectively conveying different streams of information. This will allow us to test directly to what extent inputs carrying similar or divergent information are clustered on neighboring stretches of dendrite, or whether targeting is essentially random on the scale of microns to dozens of microns. Second, once patterns of spatial clustering have been identified, then by manipulating molecular mechanisms it may be possible to disrupt or redirect such clustering in a defined way to permit causal links to be made between the

HOW CAN WE PROVE THAT DENDRITES ARE INVOLVED IN COMPUTATION?

Proving that dendrites are both necessary and sufficient for a particular computation relevant to behavior is a very difficult challenge. Necessity is within reach for the relatively simple case where the computation is accomplished by a single identified neuron, as is the case for some invertebrate sensory neurons (see text) or when the underlying biophysical mechanism depends on a single channel type. Demonstrating sufficiency is much more difficult, particularly because dendritic computation is a process that is tightly interlinked with the proper functioning of the entire system. Nevertheless, we outline here a list of objectives that must be achieved to prove that dendrites are required for computation. These should not necessarily be addressed in a linear sequence; rather, it will be beneficial to attack these problems in parallel.

Identify the Computation:

Probing the contribution of dendrites to computation is possible only when the computation of the neuron bearing the dendrites is identified. This requires identifying a simple behavior that involves a recognizable kind of computation (e.g., filtering, convolution, pattern recognition) and tracing it to the neurons responsible.

Defining the Mechanism:

Use recordings (e.g., electrophysiological or imaging) from dendrites of these neurons in an accessible preparation (e.g., brain slices) to define the dendritic signals and biophysical mechanisms that may underlie the behavior.

Correlation in the Intact Preparation:

Use recordings from dendrites in an intact preparation to show strong correlations between dendritic signals linked with the identified computation and the behavior of the animal.

Manipulation to Define a Causal Link:

Manipulate a dendritic mechanism to determine if it is both necessary and sufficient to explain the computation. Selectively knock out the mechanism and demonstrate that the behavior is impaired. Activate or modify the dendritic mechanism to demonstrate that the behavior is modified in the expected direction.

Modeling the Computation:

Use modeling to define an algorithm that describes the computation, or sequence of computations, performed by the dendrites that can plausibly explain the behavior. Modeling of single neurons and neural networks can also be used to confirm that the computation can convey a significant benefit (which can help to establish sufficiency).

local nonlinear computations described above and behavior.

bAP: backpropagating action potential

CH: centrifugal horizontal

DSGC: direction-selective retinal ganglion cells

EPSP: excitatory postsynaptic potential

GABA$_A$: γ-aminobutyric acid type A

HS: horizontal system

LGMD: lobula giant movement detector

NMDA: N-methyl-D-aspartate

SBAC: starburst amacrine cells

VCH: ventral centrifugal horizontal

For Neurophysiologists: Putting Dendrites Back into the Network

Most of our understanding of dendritic function has come from studies in isolated preparations (brain slices and culture preparations). Although this approach has been very successful in defining the biophysical basis of dendritic excitability and identifying computational subunits in neurons, it is associated with two main problems. First, the baseline conditions in these preparations are often very different from those pertaining in the intact brain, where the presence of high levels of background synaptic input can fundamentally change dendritic processing (Destexhe et al. 2003, Williams 2004). Second, the technical limitations imposed by slice experiments, together with uncertainty about the spatiotemporal pattern of synaptic inputs to single neurons in the intact brain, have made it difficult to identify which of the many mechanisms studied in vitro are also operating in vivo and may actually be relevant for computation in the intact brain. Fortunately, it is now possible to investigate dendritic function directly in vivo. Both electrophysiological (Buzsaki & Kandel 1998, Kamondi et al. 1998, Larkum & Zhu 2002, Loewenstein et al. 2005) and optical (Helmchen et al. 1999, Svoboda et al. 1999, Waters et al. 2004) techniques now exist for recording electrical and chemical dendritic signaling in anesthetized and awake, head-restrained animals. The prospect of using two-photon imaging techniques to monitor dendritic signaling in freely moving animals (Helmchen et al. 2001) should permit correlations to be established between dendritic events associated with computation (e.g., dendritic spikes) and behavior. In combination with the molecular tools outlined above and new tools for noninvasively manipulating neuronal activity in defined populations (Fetcho & Bhatt 2004, Lima & Miesenböck 2005), it should be possible to identify the causal links between dendritic computation and behavior.

For the Theorist: Proving the Benefits of Dendritic Computation

Ultimately, understanding the role of dendrites in neural computation requires a theory. This theory must identify the benefits of having dendrites and reveal the basic principles used to provide these benefits. To make advances toward such a theory, efforts should be made in three major directions. First, we need to construct algorithms based on the dendritic toolkit and show how specific computations can be achieved, making predictions that are experimentally testable (e.g., Agmon-Snir et al. 1998). Second, given that realistic modeling of single neurons has reached a relatively mature phase, we need to take advantage of the capability of such models to simulate conditions that are very difficult to test experimentally. One important task is to explore how different components of the dendritic toolkit interact with each other (e.g., how do local interactions between excitation and inhibition influence local dendritic spikes?). It is also essential to use these models to see how realistic conditions, such as noise, neuromodulation, and adaptation, affect the computational properties of the dendrites. It is a major challenge to understand how stability of the algorithmic computation can be maintained in the face of these variables, such that it is resistant to them or such that these variables can be synergistically exploited. The third challenge is to put dendrites back into networks. This will be greatly aided by the construction of reduced models of dendritic neurons (e.g., Rall 1964, Pinsky & Rinzel 1994) which capture essential features of dendritic function that could be exploited for computation. The ultimate step will be to build artificial neural networks incorporating such reduced models of the single neuron to demonstrate to what degree dendritic algorithms enhance the performance of neural networks in well-defined tasks.

CONCLUSIONS

Although dendrites have been studied for decades, the field of dendritic computation is still in its infancy. This is partly because dendrites remain relatively inaccessible and have only recently begun to yield their secrets to the onslaught of multiple new experimental tools. However, the real challenge is a deeper one, faced by many areas of neuroscience (and biology in general): how to evaluate the importance of mechanisms on the molecular and cellular level for computation at the behavioral level. The ability not only to record electrical and chemical signals in the intact brain but also to manipulate the structure and function of dendrites using molecular tools will hopefully allow us to move from the descriptive level, correlating dendritic signals linked to computation with behavior, toward directly testing the causal nature of these links. Such experiments will provide a deeper understanding of how single neurons contribute to computation in the brain and should inspire the development of novel neural network architectures with the computational powers of real brains.

ACKNOWLEDGMENTS

We are grateful to Peter Dayan, Lyle Graham, Julian Jack, Bartlett Mel, Arnd Roth, and Idan Segev for many helpful discussions and for comments on the manuscript. We thank the HFSP, Gatsby Foundation, and Wellcome Trust for financial support.

LITERATURE CITED

Agmon-Snir H, Carr CE, Rinzel J. 1998. The role of dendrites in auditory coincidence detection. *Nature* 393:268–72

Anderson JC, Binzegger T, Kahana O, Martin KA, Segev I. 1999. Dendritic asymmetry cannot account for directional responses of neurons in visual cortex. *Nat. Neurosci.* 2:820–24

Ango F, di Cristo G, Higashiyama H, Bennett V, Wu P, Huang ZJ. 2004. Ankyrin-based subcellular gradient of neurofascin, an immunoglobulin family protein, directs GABAergic innervation at purkinje axon initial segment. *Cell* 119:257–72

Ariav G, Polsky A, Schiller J. 2003. Submillisecond precision of the input-output transformation function mediated by fast sodium dendritic spikes in basal dendrites of CA1 pyramidal neurons. *J. Neurosci.* 23:7750–58

Barlow HB. 1996. Intraneuronal information processing, directional selectivity and memory for spatio-temporal sequences. *Network* 7:251–59

Barlow HB, Hill RM, Levick WR. 1964. Retinal ganglion cells responding selectively to direction and speed of image motion in the rabbit. *J. Physiol.* 173:377–407

Barlow HB, Levick WR. 1965. The mechanism of directionally selective units in rabbit's retina. *J. Physiol.* 178:477–504

Bernard C, Anderson A, Becker A, Poolos NP, Beck H, Johnston D. 2004. Acquired dendritic channelopathy in temporal lobe epilepsy. *Science* 305:532–35

Blomfield S. 1974. Arithmetical operations performed by nerve cells. *Brain Res.* 69:115–24

Booth V, Rinzel J. 1995. A minimal, compartmental model for a dendritic origin of bistability of motoneuron firing patterns. *J. Comput. Neurosci.* 2:299–312

Borg-Graham LJ. 2001. The computation of directional selectivity in the retina occurs presynaptic to the ganglion cell. *Nat. Neurosci.* 4:176–83

Borg-Graham LJ, Grzywacz NM. 1992. A model of the directional selectivity circuit in retina: transformation by neuron singly and in concert. In *Single Neuron Computation*, ed. T McKenna, J Davis, SF Zornetzer, pp. 347–76. San Diego: Academic

Brody CD, Romo R, Kepecs A. 2003. Basic mechanisms for graded persistent activity: discrete attractors, continuous attractors, and dynamic representations. *Curr. Opin. Neurobiol.* 13:204–11

Buzsaki G, Kandel A. 1998. Somadendritic backpropagation of action potentials in cortical pyramidal cells of the awake rat. *J. Neurophysiol.* 79:1587–91

Cai X, Liang CW, Muralidharan S, Kao JP, Tang CM, Thompson SM. 2004. Unique roles of SK and Kv4.2 potassium channels in dendritic integration. *Neuron* 44:351–64

Cajal SR. 1911. *Histologie du Système Nerveux de l'Homme et des Vertébrés*. Paris: Maloine

Carruth M, Magee JC. 1999. Dendritic voltage-gated ion channels regulate the action potential firing mode of hippocampal CA1 pyramidal neurons. *J. Neurophysiol.* 82:1895–901

Cauller LJ, Connors BW. 1992. Functions of very distal dendrites: experimental and computational studies of layer I synapses on neocortical pyramidal cells. In *Single Neuron Computation*, ed. T McKenna, J Davis, SF Zornetzer, pp. 199–229. Boston: Academic

Chklovskii DB. 2004. Synaptic connectivity and neuronal morphology: two sides of the same coin. *Neuron* 43:609–17

Chklovskii DB, Mel BW, Svoboda K. 2004. Cortical rewiring and information storage. *Nature* 431:782–88

Cook EP, Johnston D. 1997. Active dendrites reduce location-dependent variability of synaptic input trains. *J. Neurophysiol.* 78:2116–28

Cook EP, Johnston D. 1999. Voltage-dependent properties of dendrites that eliminate location-dependent variability of synaptic input. *J. Neurophysiol.* 81:535–43

Cuntz H, Haag J, Borst A. 2003. Neural image processing by dendritic networks. *Proc. Natl. Acad. Sci. USA* 100:11082–85

Destexhe A, Rudolph M, Pare D. 2003. The high-conductance state of neocortical neurons in vivo. *Nat. Rev. Neurosci.* 4:739–51

Doiron B, Laing C, Longtin A, Maler L. 2002. Ghostbursting: a novel neuronal burst mechanism. *J. Comput. Neurosci.* 12:5–25

Emptage N, Bliss TV, Fine A. 1999. Single synaptic events evoke NMDA receptor-mediated release of calcium from internal stores in hippocampal dendritic spines. *Neuron* 22:115–24

Euler T, Denk W. 2001. Dendritic processing. *Curr. Opin. Neurobiol.* 11:415–22

Euler T, Detwiler PB, Denk W. 2002. Directionally selective calcium signals in dendrites of starburst amacrine cells. *Nature* 418:845–52

Fatt P, Katz B. 1953. The effect of inhibitory nerve impulses on a crustacean muscle fibre. *J. Physiol.* 121:374–89

Fetcho JR, Bhatt DH. 2004. Genes and photons: new avenues into the neuronal basis of behavior. *Curr. Opin. Neurobiol.* 14:707–14

Fetz EE, Gustafsson B. 1983. Relation between shapes of post-synaptic potentials and changes in firing probability of cat motoneurones. *J. Physiol.* 341:387–410

Freund TF, Buzsáki G. 1998. Interneurons of the hippocampus. *Hippocampus* 6:347–470

Fried SI, Münch TA, Werblin FS. 2002. Mechanisms and circuitry underlying directional selectivity in the retina. *Nature* 420:411–14

Gabbiani F, Krapp HG, Hatsopoulos N, Mo CH, Koch C, Laurent G. 2004. Multiplication and stimulus invariance in a looming-sensitive neuron. *J. Physiol. Paris* 98:19–34

Gabbiani F, Krapp HG, Koch C, Laurent G. 2002. Multiplicative computation in a visual neuron sensitive to looming. *Nature* 420:320–24

Gabbiani F, Krapp HG, Laurent G. 1999. Computation of object approach by a wide-field, motion-sensitive neuron. *J. Neurosci.* 19:1122–41

Gasparini S, Migliore M, Magee JC. 2004. On the initiation and propagation of dendritic spikes in CA1 pyramidal neurons. *J. Neurosci.* 24:11046–56

Golding NL, Spruston N. 1998. Dendritic sodium spikes are variable triggers of axonal action potentials in hippocampal CA1 pyramidal neurons. *Neuron* 21:1189–200

Golding NL, Staff NP, Spruston N. 2002. Dendritic spikes as a mechanism for cooperative long-term potentiation. *Nature* 418:326–31

Goldman MS, Levine JH, Major G, Tank DW, Seung HS. 2003. Robust persistent neural activity in a model integrator with multiple hysteretic dendrites per neuron. *Cereb. Cortex* 13:1185–95

Grutzendler J, Kasthuri N, Gan WB. 2002. Long-term dendritic spine stability in the adult cortex. *Nature* 420:812–16

Häusser M, Mel B. 2003. Dendrites: bug or feature? *Curr. Opin. Neurobiol.* 13:372–83

Häusser M, Spruston N, Stuart GJ. 2000. Diversity and dynamics of dendritic signaling. *Science* 290:739–44

Hebb D. 1949. *The Organization of Behavior*. New York: Wiley

Helmchen F, Fee MS, Tank DW, Denk W. 2001. A miniature head-mounted two-photon microscope. High-resolution brain imaging in freely moving animals. *Neuron* 31:903–12

Helmchen F, Imoto K, Sakmann B. 1996. Ca^{2+} buffering and action potential-evoked Ca^{2+} signaling in dendrites of pyramidal neurons. *Biophys. J.* 70:1069–81

Helmchen F, Svoboda K, Denk W, Tank DW. 1999. In vivo dendritic calcium dynamics in deep-layer cortical pyramidal neurons. *Nat. Neurosci.* 2:989–96

Hoffman DA, Magee JC, Colbert CM, Johnston D. 1997. K^+ channel regulation of signal propagation in dendrites of hippocampal pyramidal neurons. *Nature* 387:869–75

Holtmaat AJ, Trachtenberg JT, Wilbrecht L, Shepherd GM, Zhang X, et al. 2005. Transient and persistent dendritic spines in the neocortex in vivo. *Neuron* 45:279–91

Horton AC, Ehlers MD. 2003. Neuronal polarity and trafficking. *Neuron* 40:277–95

Iansek R, Redman SJ. 1973. The amplitude, time course and charge of unitary excitatory post-synaptic potentials evoked in spinal motoneurone dendrites. *J. Physiol.* 234:665–88

Jack JJB, Noble D, Tsien RY. 1975. *Electric Current Flow in Excitable Cells*. Oxford, UK: Oxford Univ. Press

Johnston D, Hoffman DA, Magee JC, Poolos NP, Watanabe S, et al. 2000. Dendritic potassium channels in hippocampal pyramidal neurons. *J. Physiol.* 525:75–81

Kamondi A, Acsady L, Buzsaki G. 1998. Dendritic spikes are enhanced by cooperative network activity in the intact hippocampus. *J. Neurosci.* 18:3919–28

Kepecs A, Wang XJ, Lisman J. 2002. Bursting neurons signal input slope. *J. Neurosci.* 22:9053–62

Koch C, Poggio T, Torre V. 1982. Retinal ganglion cells: a functional interpretation of dendritic morphology. *Phil. Trans. R. Soc. London* 298:227–64

Koch C, Poggio T, Torre V. 1983. Nonlinear interactions in a dendritic tree: localization, timing, and role in information processing. *Proc. Natl. Acad. Sci. USA* 80:2799–802

Koch C, Segev I. 2000. The role of single neurons in information processing. *Nat. Neurosci.* 3(Suppl.):1171–77

Kreiman G, Koch C, Fried I. 2000. Category-specific visual responses of single neurons in the human medial temporal lobe. *Nat. Neurosci.* 3(9):946–53

Larkum ME, Kaiser KM, Sakmann B. 1999a. Calcium electrogenesis in distal apical dendrites of layer 5 pyramidal cells at a critical frequency of back-propagating action potentials. *Proc. Natl. Acad. Sci. USA* 96:14600–4

Larkum ME, Watanabe S, Nakamura T, Lasser-Ross N, Ross WN. 2003. Synaptically activated Ca^{2+} waves in layer 2/3 and layer 5 rat neocortical pyramidal neurons. *J. Physiol.* 549:471–88

Larkum ME, Zhu JJ. 2002. Signaling of layer 1 and whisker-evoked Ca^{2+} and Na^+ action potentials in distal and terminal dendrites of rat neocortical pyramidal neurons in vitro and in vivo. *J. Neurosci.* 22:6991–7005

Larkum ME, Zhu JJ, Sakmann B. 1999b. A new cellular mechanism for coupling inputs arriving at different cortical layers. *Nature* 398:338–41

Larkum ME, Zhu JJ, Sakmann B. 2001. Dendritic mechanisms underlying the coupling of the dendritic with the axonal action potential initiation zone of adult rat layer 5 pyramidal neurons. *J. Physiol.* 533:447–66

Lima SQ, Miesenböck G. 2005. Remote control of fly behavior through genetically targeted photostimulation of neurons. *Cell*. In press

Linden D. 1999. The return of the spike: postsynaptic action potentials and the induction of LTP and LTD. *Neuron* 22:661–66

Liu G. 2004. Local structural balance and functional interaction of excitatory and inhibitory synapses in hippocampal dendrites. *Nat. Neurosci.* 7:373–79

Loewenstein Y, Sompolinsky H. 2003. Temporal integration by calcium dynamics in a model neuron. *Nat. Neurosci.* 6:961–67

Loewenstein Y, Mahon S, Chadderton P, Kitamura K, Sompolinsky H, et al. 2005. Bistability of cerebellar Purkinje cells modulated by sensory stimulation. *Nat. Neurosci.* 8:202–11

Lörincz A, Notomi T, Tamás G, Shigemoto R, Nusser Z. 2002. Polarized and compartment-dependent distribution of HCN1 in pyramidal cell dendrites. *Nat. Neurosci.* 5:1185–93

Maeda H, Ellis-Davies GC, Ito K, Miyashita Y, Kasai H. 1999. Supralinear Ca^{2+} signaling by cooperative and mobile Ca^{2+} buffering in Purkinje neurons. *Neuron* 24:989–1002

Magee J, Hoffman D, Colbert C, Johnston D. 1998. Electrical and calcium signaling in dendrites of hippocampal pyramidal neurons. *Annu. Rev. Physiol.* 60:327–46

Magee JC. 1998. Dendritic hyperpolarization-activated currents modify the integrative properties of hippocampal CA1 pyramidal neurons. *J. Neurosci.* 18:7613–24

Magee JC. 2000. Dendritic integration of excitatory synaptic inputs. *Nat. Rev. Neurosci.* 1:181–90

Magee JC, Johnston D. 1997. A synaptically controlled, associative signal for Hebbian plasticity in hippocampal neurons. *Science* 275:209–13

Mainen ZF. 1999. Functional plasticity at dendritic synapses. In *Dendrites*, ed. G Stuart, N Spruston, M Häusser, pp. 310–38. Oxford, UK: Oxford Univ. Press

Mainen ZF, Sejnowski TJ. 1996. Influence of dendritic structure on firing pattern in model neocortical neurons. *Nature* 382:363–66

McCulloch WS, Pitts WH. 1943. A logical calculus of the ideas immanent in nervous activity. *Bull. Math. Biophys.* 5:115–33

Mehta MR. 2004. Cooperative LTP can map memory sequences on dendritic branches. *Trends Neurosci.* 27:69–72

Mel BW. 1993. Synaptic integration in an excitable dendritic tree. *J. Neurophysiol.* 70:1086–101

Migliore M, Shepherd GM. 2002. Emerging rules for the distributions of active dendritic conductances. *Nat. Rev. Neurosci.* 3:362–70

Milojkovic BA, Radojicic MS, Goldman-Rakic PS, Antic SD. 2004. Burst generation in rat pyramidal neurones by regenerative potentials elicited in a restricted part of the basilar dendritic tree. *J. Physiol.* 558:193–211

Misonou H, Mohapatra DP, Park EW, Leung V, Zhen D, et al. 2004. Regulation of ion channel localization and phosphorylation by neuronal activity. *Nat. Neurosci.* 7:711–18

Misonou H, Trimmer JS. 2004. Determinants of voltage-gated potassium channel surface expression and localization in Mammalian neurons. *Crit. Rev. Biochem. Mol. Biol.* 39:125–45

Mizrahi A, Katz LC. 2003. Dendritic stability in the adult olfactory bulb. *Nat. Neurosci.* 6:1201–7

Nakamura T, Barbara JG, Nakamura K, Ross WN. 1999. Synergistic release of Ca^{2+} from IP3-sensitive stores evoked by synaptic activation of mGluRs paired with backpropagating action potentials. *Neuron* 24:727–37

Nakamura T, Lasser-Ross N, Nakamura K, Ross WN. 2002. Spatial segregation and interaction of calcium signalling mechanisms in rat hippocampal CA1 pyramidal neurons. *J. Physiol.* 543:465–80

Nolan MF, Malleret G, Dudman JT, Buhl DL, Santoro B, et al. 2004. A behavioral role for dendritic integration: HCN1 channels constrain spatial memory and plasticity at inputs to distal dendrites of CA1 pyramidal neurons. *Cell* 119:719–32

Oswald AM, Chacron MJ, Doiron B, Bastian J, Maler L. 2004. Parallel processing of sensory input by bursts and isolated spikes. *J. Neurosci.* 24:4351–62

Oviedo H, Reyes AD. 2002. Boosting of neuronal firing evoked with asynchronous and synchronous inputs to the dendrite. *Nat. Neurosci.* 5:261–66

Pinsky PF, Rinzel J. 1994. Intrinsic and network rhythmogenesis in a reduced Traub model for CA3 neurons. *J. Comput. Neurosci.* 1:39–60

Poirazi P, Brannon T, Mel BW. 2003a. Arithmetic of subthreshold synaptic summation in a model of a CA1 pyramidal cell. *Neuron* 37:977–87

Poirazi P, Brannon T, Mel BW. 2003b. Pyramidal neuron as a 2-layer neural network. *Neuron* 37:989–99

Poirazi P, Mel BW. 2001. Impact of active dendrites and structural plasticity on the memory capacity of neural tissue. *Neuron* 29:779–96

Polsky A, Mel BW, Schiller J. 2004. Computational subunits in thin dendrites of pyramidal cells. *Nat. Neurosci.* 7:621–27

Rall W. 1964. Theoretical significance of dendritic trees for neuronal input-output relations. In *Neural Theory and Modeling*, ed. R Reiss, pp. 73–97. Stanford, CA: Stanford Univ. Press

Rall W, Burke RE, Smith TG, Nelson PG, Frank K. 1967. Dendritic location of synapses and possible mechanisms for the monosynaptic EPSP in motoneurons. *J. Neurophysiol.* 30:1169–93

Rose CR, Konnerth A. 2001. NMDA receptor-mediated Na^+ signals in spines and dendrites. *J. Neurosci.* 21:4207–14

Rosen MJ. 1972. A theoretical neural integrator. *IEEE Trans. Biomed. Eng.* 19:362–67

Schaefer AT, Larkum ME, Sakmann B, Roth A. 2003. Coincidence detection in pyramidal neurons is tuned by their dendritic branching pattern. *J. Neurophysiol.* 89:3143–54

Schiller J, Major G, Koester HJ, Schiller Y. 2000. NMDA spikes in basal dendrites of cortical pyramidal neurons. *Nature* 404:285–89

Schiller J, Schiller Y. 2001. NMDA receptor-mediated dendritic spikes and coincident signal amplification. *Curr. Opin. Neurobiol.* 11:343–48

Schiller J, Schiller Y, Stuart G, Sakmann B. 1997. Calcium action potentials restricted to distal apical dendrites of rat neocortical pyramidal neurons. *J. Physiol.* 505:605–16

Schwindt PC, Crill WE. 1995. Amplification of synaptic current by persistent sodium conductance in apical dendrite of neocortical neurons. *J. Neurophysiol.* 74:2220–24

Scott EK, Luo L. 2001. How do dendrites take their shape? *Nat. Neurosci.* 4:359–65

Segev I, London M. 1999. A theoretical view of passive and active dendrites. In *Dendrites*, ed. G Stuart, N Spruston, M Häusser, pp. 205–30. Oxford, UK: Oxford Univ. Press

Segev I, London M. 2000. Untangling dendrites with quantitative models. *Science* 290:744–50

Single S, Borst A. 1998. Dendritic integration and its role in computing image velocity. *Science* 281:1848–50

Smith DJ, Rubel EW. 1979. Organization and development of brain stem auditory nuclei of the chicken: dendritic gradients in nucleus laminaris. *J. Comp. Neurol.* 186:213–39

Sobel E, Tank DW. 1994. In vivo Ca^{2+} dynamics in a cricket auditory neuron: an example of chemical computation. *Science* 263:823–26

Softky W. 1994. Sub-millisecond coincidence detection in active dendritic trees. *Neurosci.* 58:13–41

Somogyi P, Tamas G, Lujan R, Buhl EH. 1998. Salient features of synaptic organisation in the cerebral cortex. *Brain Res. Rev.* 26:113–35

Stuart G, Sakmann B. 1995. Amplification of EPSPs by axosomatic sodium channels in neocortical pyramidal neurons. *Neuron* 15:1065–76

Stuart G, Schiller J, Sakmann B. 1997. Action potential initiation and propagation in rat neocortical pyramidal neurons. *J. Physiol.* 505:617–32

Stuart G, Spruston N, Sakmann B, Häusser M. 1997. Action potential initiation and backpropagation in central neurons. *Trends Neurosci.* 20:125–31

Stuart GJ, Häusser M. 2001. Dendritic coincidence detection of EPSPs and action potentials. *Nat. Neurosci.* 4:63–71

Stuart GJ, Spruston N. 1998. Determinants of voltage attenuation in neocortical pyramidal neuron dendrites. *J. Neurosci.* 18:3501–10

Svoboda K, Helmchen F, Denk W, Tank DW. 1999. Spread of dendritic excitation in layer 2/3 pyramidal neurons in rat barrel cortex in vivo. *Nat. Neurosci.* 2:65–73

Taylor WR, Vaney DI. 2002. Diverse synaptic mechanisms generate direction selectivity in the rabbit retina. *J. Neurosci.* 22:7712–20

Trachtenberg JT, Chen BE, Knott GW, Feng G, Sanes JR, et al. 2002. Long-term in vivo imaging of experience-dependent synaptic plasticity in adult cortex. *Nature* 420:788–94

Vetter P, Roth A, Häusser M. 2001. Action potential propagation in dendrites depends on dendritic morphology. *J. Neurophysiol.* 83:3177–82

Wang SS, Denk W, Häusser M. 2000. Coincidence detection in single dendritic spines mediated by calcium release. *Nat. Neurosci.* 3:1266–73

Waters J, Helmchen F. 2004. Boosting of action potential backpropagation by neocortical network activity in vivo. *J. Neurosci.* 24:11127–36

Williams SR. 2004. Spatial compartmentalization and functional impact of conductance in pyramidal neurons. *Nat. Neurosci.* 7:961–67

Williams SR, Stuart GJ. 1999. Mechanisms and consequences of action potential burst firing in rat neocortical pyramidal neurons. *J. Physiol.* 521:467–82

Williams SR, Stuart GJ. 2000. Site independence of EPSP time course is mediated by dendritic I(h) in neocortical pyramidal neurons. *J. Neurophysiol.* 83:3177–82

Williams SR, Stuart GJ. 2002. Dependence of EPSP efficacy on synapse location in neocortical pyramidal neurons. *Science* 295:1907–10

Williams SR, Stuart GJ. 2003. Role of dendritic synapse location in the control of action potential output. *Trends Neurosci.* 26:147–54

Yuste R, Gutnick MJ, Saar D, Delaney KR, Tank DW. 1994. Ca^{2+} accumulations in dendrites of neocortical pyramidal neurons: an apical band and evidence for two functional compartments. *Neuron* 13:23–43

Optical Imaging and Control of Genetically Designated Neurons in Functioning Circuits

Gero Miesenböck[1] and Ioannis G. Kevrekidis[2]

[1]Department of Cell Biology, Yale University School of Medicine, New Haven, Connecticut 06520; email: gero.miesenboeck@yale.edu

[2]Departments of Chemical Engineering and Mathematics, and Program in Applied and Computational Mathematics, Princeton University, Princeton, New Jersey 08544; email: yannis@princeton.edu

Abstract

Proteins with engineered sensitivities to light are infiltrating the biological mechanisms by which neurons generate and detect electrochemical signals. Encoded in DNA and active only in genetically specified target cells, these proteins provide selective optical interfaces for observing and controlling signaling by defined groups of neurons in functioning circuits, in vitro and in vivo. Light-emitting sensors of neuronal activity (reporting calcium increase, neurotransmitter release, or membrane depolarization) have begun to reveal how information is represented by neuronal assemblies, and how these representations are transformed during the computations that inform behavior. Light-driven actuators control the electrical activities of central neurons in freely moving animals and establish causal connections between the activation of specific neurons and the expression of particular behaviors. Anchored within mathematical systems and control theory, the combination of finely resolved optical field sensing and finely resolved optical field actuation will open new dimensions for the analysis of the connectivity, dynamics, and plasticity of neuronal circuits, and perhaps even for replacing lost—or designing novel—functionalities.

Contents

- INTRODUCTION: OBSERVATION AND CONTROL 534
- A BIOLOGICAL APPROACH TO OBSERVATION AND CONTROL OF NEURONAL CIRCUITS 535
- SENSING 539
  - Sensing Membrane Potential 539
  - Sensing Calcium 542
  - Sensing Neurotransmitter Release . 544
- ACTUATION 546
  - Positive Actuation 546
  - Negative Actuation 550
- LINKING SENSING AND ACTUATION 552

INTRODUCTION: OBSERVATION AND CONTROL

The classical scientists were observers, because they were forced to be observers. The focus of classical science was celestial mechanics. When one studies the solar system and attempts to explain and predict its idiosyncrasies, it is apparent that neither human theory nor human observation is going to have any effect on the actual motion of the moon, the rising of Venus, or the occurrence of the next eclipse. Consequently, it is natural to think solely in terms of descriptive theories of the universe. [...] Coming closer to contemporary times, as man began to understand more of the fundamental forces of nature, we note a change in the scientist from the role of the observer to that of the doer. "Doing," of course, has always been the fundamental objective of science. We understand so that we can do something with that understanding. When we can do nothing with our understanding, we try to console ourselves with thoughts of *ars gratia artis*, "knowledge for knowledge's sake," and other picturesque phrases. But it should be clearly understood that this is a type of intellectual sour grapes. (Bellman 1968, pp. 11–12)

This passage, a motivation from hindsight of the thinking that led Richard Bellman to the invention of dynamic programming, a pillar of modern control theory, suggests that scientific fields shift emphasis from observation to control as they mature. Many branches of biology have already undergone such shifts; others, such as systems neuroscience, remain predominantly observational. Biochemistry and genetics, for example, each require reconstitution of function—that is, predictive control over experimental outcomes—to establish causality and mechanism. The purpose of such predictive control is heuristic as much as it is pragmatic: In contrast to Bellman, we not only understand in order to do; we also do in order to understand.

Observation and control require sensor and actuator interfaces, the universal mediators of interactions in natural and engineered systems. Sensors and actuators can be realized in a nearly endless variety of materials and mechanisms and assume a nearly endless variety of sizes and shapes. Cajal's law of dynamic polarization, for example, may be distilled to the notion that neurons form specialized sensor and actuator domains—dendrites and axons—through which they detect and influence the activities of other neurons. Organisms interact with each other and their environments through sensory and motor systems. The stability and operability of engineered devices, from chemical reactors to airplanes, depends on continuous sensing and actuation.

The properties of sensors and actuators quite naturally determine which aspects of a system's behavior can be observed and influenced, and to which extent. The observability and reachability theorems of control theory formalize these intuitive ideas (see Sidebar; Brockett 1970, Bélanger 1995). To illustrate these concepts, consider a microelectrode, which for the experimental neuroscientist has traditionally played the roles of both voltage or current sensor as well as that of current-passing actuator. (In the configuration typically used for intracellular investigations, sensor and actuator functions are linked in closed loop via feedback to establish voltage or current clamp. Feedback leads to a number of interesting possibilities, some of which are considered in the final section of this review. Here, we are concerned with the basic properties of the devices employed to sense and actuate neuronal activity.)

Used as a sensor, the electrode reports voltage or current fluctuations at a single recording site. The observable voltage or current waveforms are composites shaped by many variables that can neither be directly observed nor reliably inferred: the locations, time courses, and magnitudes of individual synaptic potentials or currents, for example, or the active conductances and passive cable properties that modulate them. Used as an actuator, the electrode can control membrane potential near the site of impalement but not necessarily in more remote locations, particularly if the cellular geometry is complex. The state in which the neuron's membrane is isopotential at every location—a control problem commonly known as the problem of "space clamp"—may thus not be reachable. The biological reasons that limit reachability are similar to those that limit observability, echoing a fundamental duality (see Sidebar).

Additional constraints emerge if one attempts to observe and control neuronal assemblies or circuits with electrodes rather than individual neurons. These constraints reflect practical limits on the number of cells that can be monitored or actuated simultaneously through physical contact or impalement and the difficulty of identifying and targeting functionally relevant circuit elements in intact neural tissue. It is probably no overstatement that the detailed dynamics of functioning circuits have generally remained unobservable and that arbitrary circuit states have not been experimentally reachable.

OBSERVABILITY AND REACHABILITY

Consider a system whose state-space representation consists of the differential equations

$$\frac{dx}{dt} = Ax(t) + Bu(t),$$
$$y(t) = Cx(t),$$
$$x(0) = x_0.$$

$x(t)$ is the system state (x_0 is the initial state), $u(t)$ is a control input, and $y(t)$ is a measured output; for our purposes these are finite real vectors, and the dimension of $x(t)$ is n. The matrices A, B, and C are the dynamics, actuator, and sensor matrices, respectively. We take them to be constant so that this is a linear, time-invariant (LTI) system. An LTI system is observable if the initial state $x(0) = x_0$ can be uniquely deduced from knowledge of the control input $u(t)$ and measured output $y(t)$ for all t between 0 and any $T > 0$. A necessary and sufficient condition for observability is that the observability matrix

$$O = \begin{bmatrix} C \\ CA \\ CA^2 \\ \vdots \\ CA^{n-1} \end{bmatrix}$$

has rank n.

An LTI system is reachable if for every state x_1 and every $T > 0$ there exists a control function $u(t)$, $0 < t \leq T$, which takes the system from the initial state $x(0) = x_0$ to the state x_1 at $t = T$. A necessary and sufficient condition for reachability is that the reachability matrix

$$R = \begin{bmatrix} B & AB & A^2B & \ldots & A^{n-1}B \end{bmatrix}$$

has rank n. The apparent similarity of the observability and reachability matrices O and R, expressed in the correspondences $A \sim A^T$ and $C \sim B^T$, indicates a fundamental duality between observability and reachability, and, more generally, between control and estimation. (The superscript T indicates the transpose of a matrix.)

A BIOLOGICAL APPROACH TO OBSERVATION AND CONTROL OF NEURONAL CIRCUITS

Recognition of these difficulties has motivated the development of an alternative, biologically grounded approach to sensing and actuation of neuronal function. Rather than relying on "hard" tools from physics and electrical engineering, this new, "soft" approach seeks to convert the biological mechanisms underlying neuronal excitability and communication into experimental mechanisms for observing and controlling activity. At its core lie sensor proteins that report changes in the physiological states of neurons or synapses through the emission of light (Miesenböck & Rothman 1997, Miyawaki et al. 1997, Siegel & Isacoff 1997) and actuator proteins that effect such changes in response to exogenous optical triggers (Zemelman et al. 2002). Encoded in DNA and placed under the control of suitable regulatory sequences determining their expression, sensor and actuator proteins can be localized biosynthetically to defined classes of neurons in intact tissue. The problem of identifying neurons for observation and control is thus solved organically: A biological magic wand that automatically highlights all members of a functionally relevant population of cells avoids the tedium, arbitrariness, and numerical limitations of experimenter-controlled impalement of individual neurons with electrodes or of assigning spikes recorded by extracellular devices to a handful of cells in close proximity.

The use of biological sensors and actuators stands in a long tradition of experimental strategies that rely on biological principles to cut through biological complexity. Progress in molecular biology, for example (Crick 1999), has been powered by the ability to recombine DNA fragments at precise junctions, read nucleotide sequences, synthesize proteins of defined amino acid sequences, and recognize specific biomolecules in a broad cellular context. Each of these tasks poses a problem that would frustrate classical organic chemistry: recognition of short, rare sequence motifs in long, monotonous polymers; precisely controlled stepwise assembly and disassembly of polymers from chemically similar building blocks; and molecular discrimination by small free energy differences. Harnessing evolved solutions—restriction enzymes, DNA polymerases, RNA-directed peptide synthesis by the ribosome, base pairing, and antibodies—provided the keys to areas that had previously been locked. The premise of the approaches we review is that an increasing reliance on biological principles will have a similarly transformative impact on systems neuroscience.

Because gene expression patterns provide the platform on which the biological strategies for observing and controlling neuronal function are built, the most fundamental limitations on observability and reachability necessarily derive from them. (As discussed below, the currently available sensor and actuator proteins also impose often considerable constraints on what can be observed and done, but many of these constraints seem technical rather than fundamental.) The fundamental limitations are twofold. The first, obvious problem is to find regulatory sequences or mechanisms that provide selective genetic access to the neurons of interest—or, more subtly, to classify neurons genetically. The faith that gene expression patterns have something meaningful to say about neural organization and function has spawned several large-scale efforts to map the sites in the mammalian central nervous system (CNS) at which individual genes are active (Reymond et al. 2002, Gong et al. 2003; **http://www.gensat.org**, **http://www.brainatlas.org**). These efforts lay the foundation for a future functional anatomy of the brain, but they cannot bring it forth themselves. To do so would require the solution of the inverse problem: not the collection of an inventory of the brain regions or cells in which individual genes are expressed, but a principled synthesis that identifies cell types with common functionalities through distinctive patterns of gene expression. Because it must be expected that such cell types are specified combinatorially rather than through the expression of single "signature" genes, the expression of sensor and actuator proteins will ultimately also have to be controlled combinatorially. The development of intersectional methods that draw on the activities of two (or possibly more) promoter elements to regulate expression (Awatramani et al. 2003, Branda & Dymecki 2004) is an important first step in this direction.

A second, deeper problem is that the object of interest in most neuroscientific experiments is not a population of genetically homogeneous neurons but instead a heterogeneous operational unit—a circuit—defined by connectivity or coactivity rather than by a common pattern of gene expression. Two of the most informative applications of genetically encoded sensors reported to date, for instance, have not visualized activity within a single class of neurons but instead have traced transformations of activity between two or more genetically distinct populations (**Figure 1**; Ng et al. 2002, Wang et al. 2003). Fortunate problem choice made these analyses possible. Both studies focused on the antennal lobe of *Drosophila melanogaster*, a circuit in the olfactory system whose connectivity is sufficiently well understood and invariant to permit comparisons between individuals. Such interindividual comparisons were necessary because different genotypes had to be produced to express the sensor proteins in different classes of neurons. Clear anatomical landmarks—the glomeruli of the antennal lobe—aided in the registration of images obtained from different individuals and allowed the subsequent synthesis of a coherent functional picture.

If well-defined connectivity, interindividual invariance, and identifiable landmarks are missing, circuitry analysis becomes forbiddingly difficult. Traditional methods of "circuit breaking"—that is, electrode searches for pairwise connections—become painfully inadequate when faced with complex circuit topologies. Perhaps the inadequacy of these "hard" approaches suggests, once again, a resort to the "soft" stratagems of biology? Could genetic programs be devised that trace functionally linked cellular assemblies, even if these assemblies lack a unifying genetic signature? Connectivity- or activity-based cellular automata might then explore and expose neural circuit architectures autonomously and in so doing also serve as vectors for the distribution of sensor and actuator proteins.

In a formal sense, cellular automata are programs that operate on a collection of abstract "cells" arranged in a grid of a certain topology (Ulam 1962, von Neumann 1966). The cells can exist in a minimum of two states—"on" and "off." These states are updated in discrete time steps via rules based on the states of neighboring cells. To see how a cellular automaton could be programmed to trace functional connectivity, think of the automaton's abstract cells as concrete neurons, and of the automaton's abstract grid as the actual network of synaptic connections. Assume that each cell possesses two states: one in which a certain gene or set of genes is "off," and another in which these genes are "on." Starting with all cells in the "off" state, the program is initialized by switching one cell or a small cluster of cells "on." This could be accomplished by illumination of the cell so that gene expression from a light-sensitive promoter turns on (Shimizu-Sato et al. 2002), by localized application of a chemical compound that induces gene expression, or by infection with a virus. The gene whose expression is activated specifies the rule that determines how the states of neighboring cells will be updated. If the gene encodes an agent that is transmitted through synaptic contacts and induces subsequent gene expression in pre- or postsynaptic cells, thereby switching these cells "on," the resulting cellular automaton traces anterograde or retrograde connectivity. Biological precedents for this type of behavior exist. Infection with certain herpes viruses, for example, causes the production of viral particles that spread—in anterograde or retrograde direction—to synaptically coupled cells and turn these cells into secondary sources of virus (Martin & Dolivo 1983, Ugolini et al. 1989, Card et al. 1990, Strack & Loewy 1990). Certain neurotoxins (Schwab & Thoenen 1976, Bizzini et al. 1977, Schwab et al. 1979, Coen et al. 1997, Maskos et al. 2002) and lectins

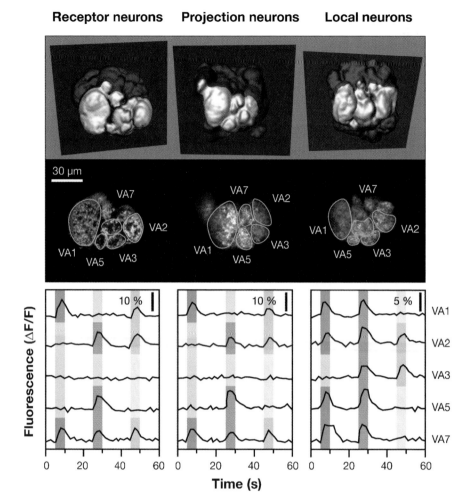

Figure 1

Optical imaging of odor-evoked synaptic activity in the antennal lobe of *Drosophila melanogaster*. Synapto-pHluorin, a genetically encoded sensor of neurotransmitter release (see also **Figure 4**), was expressed in three populations of neurons innervating the antennal lobe: olfactory receptor neurons (*left*), projection neurons (*middle*), and local inhibitory interneurons (*right*). Stacks of optical sections through the antennal lobes were collected by two-photon laser-scanning microscopy and assembled into three-dimensional anatomical models that reveal the glomerular structure of the lobes (*top*). Regardless of the identity of the neurons expressing synapto-pHluorin, identical sets of glomeruli—for example, the ventral-anterior set encompassing VA1, VA2, VA3, VA5, and VA7—are discernible in individual optical sections (*middle*). Time courses of fluorescence at the indicated focal planes reveal odor responses of synapto-pHluorin-positive receptor neurons, projection neurons, and local neurons within these glomeruli (*bottom*). Three test odors were used to stimulate odor responses; the presentations of these odors are marked by blue, red, and yellow vertical bars. To facilitate comparisons among different sets of neurons, color-encoded response matrices are superimposed on the fluorescence traces: Intense color denotes an odor-evoked synapto-pHluorin transient, and faded color denotes its absence. Note that odors are represented as combinations of active glomeruli, and that information is transmitted faithfully from receptor neurons to projection neurons, the first- and second-order neurons, respectively, of the circuit. Odors elicit broad but specific patterns of inhibitory activity in local neurons. Inhibitory cross talk among glomeruli may help extract higher-order odor features—such as coactivation of specific sets of glomeruli—from sensory input. Modified, with permission, from Ng et al. (2002).

(Gerfen et al. 1982, Ruda & Coulter 1982, Shipley 1985, Horowitz et al. 1999, Yoshihara et al. 1999, Braz et al. 2002) also possess the capacity to cross synapses. The domains mediating the synaptic transfer of these agents might be fused to engineered transactivator functions that induce their own expression in recipient cells. Such a mechanism therefore also specifies a connectivity-tracing rule.

Automata that delineate sets of coactive neurons could be built around promoters responding to activity-driven calcium influx (Greenberg et al. 1986, Morgan & Curran 1991, West et al. 2001, Wilson et al. 2002). The expression patterns of immediate early genes controlled by these promoters—most notably c-*fos*—provide coarse maps of activity (Morgan et al. 1987, Smeyne et al. 1992). To improve the spatial and temporal resolution of these maps, intersectional rules that incorporate multiple AND gates could be programmed. Gene expression might, for instance, be induced only in neurons that are electrically active, AND belong to a certain type, AND produce spikes within a given window of time. The first AND gate would be derived from an activity-dependent promoter element; the second, from a cell type–specific element; and the third, from a control element regulated, for example, by a fast-acting chemical compound. Administration and withdrawal of the compound would then open and close temporal windows during which active neurons can become genetically marked. These windows could be made to coincide with certain behavioral tasks or with periods of actuation of other neurons to identify coactive followers.

Regardless of the precise rule that specifies the nature of the cellular automaton, the genetic mechanisms that permit its self-reproduction will be able to control the expression of sensor and actuator proteins simultaneously. The automaton then not only acts to exhibit circuits, but also prepares these circuits for further analysis and control.

SENSING

Analog computation by neurons requires cellular machinery that detects and responds to information-carrying signals. This machinery provides the raw material for the construction of genetically encoded sensor proteins: Voltage-gated ion channels sense changes in membrane potential (Siegel & Isacoff 1997); calcium-responsive proteins detect changes in calcium concentration (Miyawaki et al. 1997, Romoser et al. 1997); synaptic vesicle proteins are linked to transmitter release (Miesenböck & Rothman 1997, Miesenböck et al. 1998); and the effectors of second messengers respond to changes in second messenger concentrations (Zaccolo et al. 2000). Because biological signals almost universally register as conformational changes or translocation events, permutation of a few design principles based on a handful of conformationally or environmentally sensitive fluorophores can generate a vast slew of sensors (for reviews, see van Roessel & Brand 2002, Zhang et al. 2002). Our focus here is on the three categories of probes that sense variables representing the principal inputs and outputs of neural computations: membrane potential, calcium concentration, and neurotransmitter release. These sensors also illustrate the original design principles (Miyawaki et al. 1997, Siegel & Isacoff 1997, Miesenböck et al. 1998, Baird et al. 1999).

Sensing Membrane Potential

Voltage-gated ion channels are electromechanical devices that undergo a sequence of voltage-driven conformational changes. Regardless of ionic selectivity or voltage dependence, they possess a canonical structure exhibiting four-fold symmetry (Hille 2001). Each of the four subunits or domains lines a sector of the central ion conduction pathway (helices S5 and S6 in the linear sequence) (Doyle et al. 1998), and each contains a peripheral voltage-sensing element (composed of helices S1 through S4). In response to a depolarizing voltage step, positive charge in helix S4 of the voltage-sensing element electrophoreses across the membrane (Bezanilla 2000); this movement pulls apart the S6 gate helices (Jiang et al. 2002) occluding the inner mouth of the channel and opens the permeation pathway. The flow of ions through the channel is quickly blocked by a cytoplasmic peptide plug (fast or N-type inactivation; Hoshi et al. 1990) and a subsequent conformational rearrangement that collapses the channel's selectivity filter (slow or C-type inactivation; Hoshi et al. 1991).

Each of the three principal steps in the activation sequence of a voltage-gated ion channel—voltage sensing, gating, and inactivation—has been coupled to a small fluorescence change (**Figure 2**). In sensors termed FlaSh (Siegel & Isacoff 1997) and SPARC (Ataka & Pieribone 2002), a green fluorescent protein (GFP) is inserted downstream of the S6 gate helices of nonconducting mutants of Shaker potassium and $Na_V 1.4$ sodium channels, respectively. The kinetics of SPARC are very fast, with a voltage-fluorescence relationship indicative of the motions of a single voltage sensor domain (**Figure 2**, *middle*; Ataka & Pieribone 2002). The much slower kinetics of FlaSh (**Figure 2**, *top*) match those of C-type inactivation (Hoshi et al. 1991, Siegel & Isacoff 1997), a more extensive but poorly understood rearrangement of the channel structure in which a collapsed extracellular selectivity filter is stabilized by interactions of the intracellular C-terminal domains (Loots & Isacoff 1998). Substitutions of different fluorescent proteins in FlaSh have unmasked additional kinetic components—mostly from channel gating—in the fluorescence signal (Guerrero et al. 2002). FlaSh thus reports a weighted sum of gating and inactivation processes; the weights of the individual contributions are determined by the nature of the fluorescent module. The third type of fluorescent membrane potential probe, termed VSFP1 (Sakai et al. 2001), consists of an isolated voltage-sensor domain (that is, a $K_V 2.1$ channel truncated after the S4 helix) to which a fluorescence resonance energy transfer (FRET) pair of cyan and yellow fluorescent proteins (Miyawaki et al. 1999) has been appended. A small FRET signal tracks the movements of the "gating" charge (**Figure 2**, *bottom*).

So far, no genetically encoded membrane potential sensor has reported an action potential in a neuron. In *Xenopus* oocytes, where most validating experiments have been performed (Siegel & Isacoff 1997, Ataka & Pieribone 2002), the amount of fluorescence appearing at the cell surface (F_0) is minute even after prolonged periods of protein expression, which suggests that folding, assembly, and/or transport problems hinder the efficient export of the sensors from the endoplasmic reticulum (ER) or that chromophore maturation is impaired. The small amount of fluorescence that does reach the cell surface may represent a trace fraction that folds, assembles, or matures correctly or that simply overwhelms the quality control system of the ER by mass action, owing to high-level expression from large doses of microinjected mRNA. If expression levels are lower, as will generally be the case in neurons, the surface fraction will disappear, and so will any voltage-dependent signals. The one ray of hope in these dire straits is that the small fractional fluorescence changes observed ($\Delta F/F_0 = 0.5$–5% per 100 mV; **Figure 2** and Siegel & Isacoff 1997, Sakai et al. 2001, Ataka & Pieribone 2002) probably underestimate the true amplitudes of the voltage responses because fluorescence from nonresponsive intracellular sensor pools would be expected to contaminate the signals. If the nonresponsive pools could be eliminated and high levels of surface expression achieved, reasonably robust voltage responses could result.

A simple and elegant alternative to electromechanical voltage probes are sensors that operate without moving parts. Rather than relying on voltage-driven conformational rearrangements, these solid-state sensors would report direct effects of the transmembrane electric field on the electric dipole moments of their chromophores. Fluorescence is a poor readout in these circumstances: Even the large electric fields of ~100,000 V/cm experienced by synthetic fluorophores within the plasma membrane cause only small fluorescence intensity changes (less than 5% per 100 mV) or electrochromic shifts (~3 nm per 100 mV) (Cohen et al. 1978, Loew et al. 1978, 1979). More sensitive to applied electric forces than fluorescence is the nonlinear optical phenomenon of second-harmonic generation (2HG) (Bouevitch et al. 1993, Ben-Oren et al. 1996), which arises from coherent scattering of a fundamental incident optical field into a frequency-doubled "harmonic" field (Bloembergen 1965). Because 2HG is electric dipole–forbidden in media with inversion symmetry, it is generally observed only in highly anisotropic structures or at interfaces at which this symmetry is broken (Shen 1989). The intercalation of lipophilic dyes into one leaflet of a biological membrane, for example, produces an oriented, asymmetrical dipole array in which the phased second-harmonic signals from individual chromophores can sum constructively (Bouevitch et al. 1993, Ben-Oren et al. 1996, Moreaux et al. 2000, Moreaux et al. 2001).

A first attempt to emulate such an arrangement with genetically encoded second-harmonic generators has failed (Roorda et al. 2004). Using two lipid anchors, a prenyl group at the C-terminus and a palmitoylation motif within an internal polypeptide loop, the cylinder-shaped GFP molecule was tied flush, with its axis membrane-parallel and its electronic transition dipole moment roughly membrane-orthogonal, to the cytoplasmic face of the plasma membrane. The resulting array of genetically encoded dipoles generated intense two-photon fluorescence but no detectable second-harmonic power, which suggests that the arrangement lacked the orientational anisotropy and/or packing density required for efficient 2HG: In contrast to fluorescence, whose intensity scales linearly with the surface density of fluorophores, harmonic power is a quadratic function of the

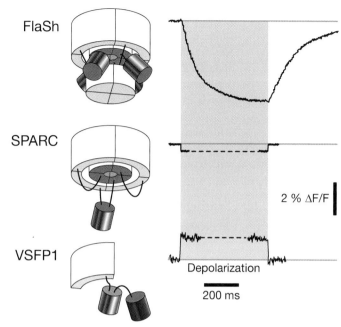

Figure 2

Sensing membrane potential. Genetically encoded membrane potential sensors are derivatives of voltage-gated ion channels, whose central pore-forming domains are surrounded by a peripheral ring of voltage-sensing modules. The schematic structures of these sensors (*left*) are viewed from an intracellular vantage point looking toward the cytoplasmic face of the plasma membrane. FlaSh (*top*) is based on the Shaker potassium channel of *Drosophila*, a non-covalent homotetramer. Tetramerization of Shaker subunits involves the assembly of a "hanging gondola" of T1 domains below the cytoplasmic opening of the ion permeation pathway. The hanging gondola occupies the space immediately below the inner mouth of the channel and thereby displaces the four cylinder-shaped GFP molecules toward the periphery. To reach their attachment sites near the pore, the polypeptide cables connecting the GFPs traverse windows in the suspension of the hanging gondola. SPARC (*middle*) is based on the $Na_V 1.4$ sodium channel from rat skeletal muscle. A single GFP molecule is inserted into the cytoplasmic loop connecting domains II and III of the covalently linked channel tetramer. VSFP1 (*bottom*) consists of an isolated voltage-sensing domain of a $K_V 2.1$ potassium channel, fused to a FRET pair of cyan and yellow fluorescent proteins. Although VSFP1 is depicted as a monomer, it might tetramerize because of the presence of an N-terminal T1 domain (*not shown*). Fluorescence responses (*right*) to steps from polarized to depolarized membrane voltages (–80 to 10 mV for FlaSh; –120 to –10 mV for SPARC; –80 to 20 mV for VSFP1) were recorded by voltage-clamp fluorimetry in *Xenopus* oocytes (FlaSh and SPARC) or HEK cells (VSFP1). Traces represent averages of 20 sweeps (FlaSh), 45 sweeps (SPARC), or single-pass recordings (VSFP1). Fluorescence data, with permission, from Siegel & Isacoff (1997), Ataka & Pieribone (2002), and Sakai et al. (2001).

surface density of radiating dipoles (Moreaux et al. 2001). As in the case of the electromechanical voltage sensors, however, there is a glimmer of hope. Because fluorescence is not a prerequisite for 2HG, polypeptides other than GFP, or even nonnative lipids synthesized by ectopically expressed enzymes, might be able to serve as solid-state voltage probes. These polypeptides or lipids might be saddled with less molecular bulk than GFP and easier to array, with semicrystalline precision, at or within the plasma membrane.

Sensing Calcium

If membrane potential changes are not directly observable, the classical alternative is to measure the changes in intracellular calcium concentration that precede or follow voltage changes as synaptic or voltage-gated calcium channels open (Tank et al. 1988, Regehr et al. 1989). Most genetically encoded calcium sensors rely on calmodulin (Miyawaki et al. 1997, Romoser et al. 1997), a dumbbell-shaped protein that binds two calcium ions each in its globular N- and C-terminal domains. The calcium-binding domains are connected by a linker that stiffens in the presence of calcium alone but melts, with steep calcium dependence reflecting cooperativity among the four binding sites, in the presence of a helical target peptide, which the calcium-saturated calmodulin embraces (Ikura et al. 1992).

Different readout mechanisms report these calcium- and target peptide–dependent conformational changes. The first sensors, termed cameleons, have employed FRET within a polypeptide concatenated from fluorescence donor protein, calmodulin, calmodulin target (the M13 peptide from myosin light chain kinase), and fluorescence acceptor protein (Miyawaki et al. 1997, 1999). Although cameleons have proven adequate in cell culture, zebrafish (Higashijima et al. 2003), and *Caenorhabditis elegans*, where spontaneous calcium elevations in muscle as well as neuronal transients evoked by intense electrical stimulation (Kerr et al. 2000) and touch (Suzuki et al. 2003) have been detected, they have often not yielded interpretable signals in other organisms.

Subsequent sensor generations have coupled calcium detection and fluorescence report in an unusual topology (**Figure 3**): To accommodate calmodulin, the cylindrical shell of a green or yellow fluorescent protein is cracked at a strategic location overlying the chromophore (Baird et al. 1999). The calmodulin graft is either stitched seamlessly into the polypeptide backbone of the fluorescent host (camgaroo; Baird et al. 1999, Griesbeck et al. 2001) or left dangling near the open cleft. An M13 target is placed across the cleft from calmodulin so that calcium-dependent engagement of the target will narrow or close the gap in the protein shell (pericam and G-CaMP; Nagai et al. 2001, Nakai et al. 2001). In each case, tight conformational coupling between calcium sensor and fluorophore allows for dynamic ranges that in purified proteins match those of small-molecule calcium indicators (i.e., fluorescence changes of about eightfold at saturating versus basal calcium concentrations), with effective dissociation constants of ~ 0.5–8 μM.

Dynamic ranges and calcium affinities tend to be lower in living cells, possibly because the calmodulin moieties of the sensors participate in nonproductive interactions with endogenous calmodulin targets. Consistent with this idea, some studies (Hasan et al. 2004), but not others (Pologruto et al. 2004), have found evidence that $\sim 50\%$ of the intracellular sensor population is immobile and unresponsive to calcium in vertebrate neurons. This immobile fraction may be eliminated, and larger dynamic ranges restored, by replacing calmodulin with a non-neuronal calcium-sensing moiety, such as a muscle-specific troponin C (Heim & Griesbeck 2004). Nevertheless, the signal intensities resulting from the expression of G-CaMP have proven amply sufficient for analyses of odor representations by first-, second-, and third-order neurons in the olfactory system of flies with single-cell resolution (J.W. Wang et al. 2003,

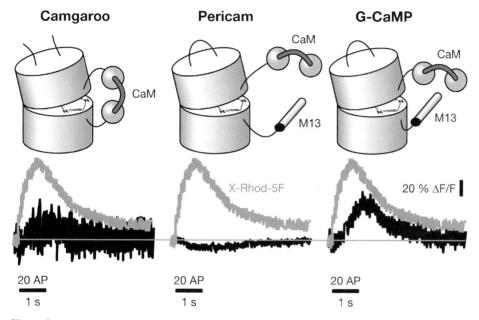

Figure 3

Sensing calcium. Genetically encoded calcium sensors incorporate fluorescent proteins with clefts in the cylindrical shells surrounding the chromophores (*top*). In camgaroo (*left*), the cleft is bridged by a calmodulin (CaM) insert; in pericam (*middle*) and G-CaMP (*right*), calmodulin and an M13 target peptide are situated across from each other, on opposite sides of the cleft. Although pericam and G-CaMP employ a very similar design, they show subtle differences in the kind of fluorescent protein used, the dimensions of the cleft in the protein shell, and the lengths and sequences of the polypeptide linkers that attach calmodulin and M13. Fluorescence responses of the calcium sensors to trains of 20 action potentials (AP) at 20 Hz were recorded by two-photon laser-scanning microscopy in cultured rat hippocampal slices (*bottom*). Single lines intersecting the apical dendrites of pyramidal neurons were scanned at 500 Hz. The neurons were held in whole-cell current clamp and filled through the patch pipettes with X-Rhod-5F, a synthetic calcium indicator dye. Fluorescence responses of camgaroo2 (*left*), inverse pericam (*middle*), and G-CaMP (*right*) are averages of 4–8 trials. Copies of the X-Rhod-5F signal (*grey traces*) are overlaid for comparison. Fluorescence data, with permission, from Pologruto et al. (2004).

Suh et al. 2004, Y. Wang et al. 2004). Because the fluorescence output of G-CaMP is a highly nonlinear function of calcium concentration and action potential number, however, not all responsive neurons may experience calcium transients large enough to exceed the threshold of detection (Pologruto et al. 2004 and below). Caution must therefore be exercised in ascribing physiological significance to the "sparseness" of the observable activity patterns; what is observable may be only the tip of an iceberg. In vertebrates, light-evoked responses of individual retinal ganglion cells have been detected in retinal whole mounts from transgenic mice expressing pericam and camgaroo; odor-evoked population responses of sensory afferents and granule cells have been recorded from the olfactory bulbs of these mice in vivo (Hasan et al. 2004).

Side-by-side comparisons of the impulse and frequency responses of camgaroo, pericam, and G-CaMP with those of the synthetic calcium dye X-Rhod-5F in cultured hippocampal brain slices, however, have revealed that the genetically encoded calcium sensors cannot yet serve as reliable surrogate reporters of electrical activity (**Figure 3**; Pologruto et al. 2004). Owing to cooperativity among the four calcium-binding sites in calmodulin, the fluorescence report is a highly nonlinear function of the number of action potentials. Small action-potential numbers and low-frequency activity are generally underreported because only trains of >5–33 action potentials at 20–50 Hz are able to push the signal-to-noise ratio beyond 2 (**Figure 3**). The frequencies of periodic stimuli at, beyond, and in all likelihood even below 20 Hz cannot be recovered from the fluorescence record (Pologruto et al. 2004).

Sensing Neurotransmitter Release

Synapses are the sites of apposition between the actuator and sensor domains through which neurons communicate. Reporters of neurotransmitter release offer a direct view of the function of the actuator compartment, i.e., the presynaptic terminal (Miesenböck et al. 1998). Calcium sensors located near postsynaptic transmitter- or voltage-gated calcium channels can provide a complementary view of how presynaptic actuation is sensed by postsynaptic targets (Ji et al. 2004).

To convert transmitter release into an optical signal, a switchable light emitter is anchored at the inner surface of synaptic vesicles through fusion with an integral membrane protein (**Figure 4**, *top*). The light emitter is a pH-sensitive variant of GFP, termed pHluorin, whose fluorescence is reversibly quenched by the high abundance of protons inside resting vesicles (Miesenböck et al. 1998). Vacuolar proton pumps maintain the interior of synaptic vesicles at a pH of ∼5.7, which corresponds to a roughly 50-fold higher proton concentration than that of the pH-neutral extracellular medium. Vesicle fusion establishes a proton conduit between the vesicle interior and the cell exterior that dissipates the pH gradient. As a result, the chromophores of pHluorins located in the membranes of fused vesicles are deprotonated, and the resulting 20-fold increase in their fluorescence signals vesicle release (**Figure 4**; Miesenböck et al. 1998, Sankaranarayanan et al. 2000). The fluorescence increase lasts until the vesicle membrane is recycled and the interior of the reformed vesicle is acidified.

Amplitude and waveform of the synapto-pHluorin signal due to vesicle exo- and endocytosis reflect the itinerary of the synaptic vesicle protein that carries the light emitter; this protein is, in general, the v-SNARE VAMP-2/synaptobrevin (Miesenböck et al. 1998). At hippocampal synapses in culture, ∼10%–20% of the total synapto-pHluorin pool is present at the cell surface under unstimulated conditions; this pool elevates background fluorescence and reduces the signal-to-noise ratio (Sankaranarayanan et al. 2000, Sankaranarayanan & Ryan 2000). If a synapse has a low basal release rate and the vesicular and surface synapto-pHluorin pools do not intermix, prestimulus photobleaching (Gandhi & Stevens 2003, Samuel et al. 2003) can selectively eliminate the fluorescent surface pool and make even single vesicle turnovers detectable (**Figure 4**, *top*; Gandhi & Stevens 2003). Without prestimulus photobleaching, ∼20 action potentials are required for signal-to-noise ratios exceeding 2 at single boutons (**Figure 4**, *bottom*; Sankaranarayanan & Ryan 2000). Assuming release probabilities in the range of 0.1–0.5 (Hessler et al. 1993, Rosenmund et al. 1993), 20 action potentials translate into 2–10 vesicle fusion events. In contrast to the highly nonlinear calcium sensors, synapto-pHluorins report presynaptic activity linearly because fluorescence quenching and dequenching involve a single protonation site.

Once a synaptic vesicle exocytoses, its synapto-pHluorin complement remains fluorescent for an extended, but variable, period of time that is terminated by vesicle retrieval and acidification (Miesenböck et al. 1998, Sankaranarayanan & Ryan 2000, Sankaranarayanan & Ryan 2001, Gandhi & Stevens 2003). Because individual vesicles add and subtract unitary quanta of fluorescence as they exocytose and recycle, the time course of fluorescence recorded from a synapse reflects the convolution of two physiological variables: the timing of individual release events

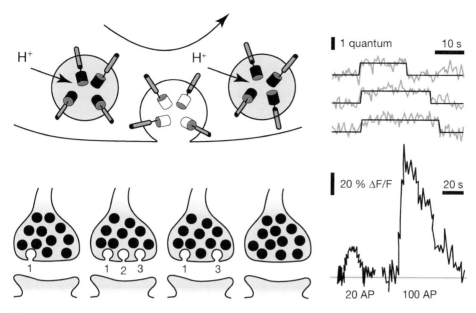

Figure 4

Sensing neurotransmitter release. Genetically encoded sensors of neurotransmitter release, termed synapto-pHluorins, are fusion proteins between a pH-sensitive GFP (pHluorin) and the synaptic vesicle transmembrane protein VAMP-2/synaptobrevin. In the context of individual synaptic vesicles (*top*), synapto-pHluorins located in resting vesicles (internal pH ∼5.7) are "off" because their chromophores tend to be in the protonated, non-fluorescent state. Upon exocytosis, the vesicle interior becomes continuous with the extracellular space, pH rises to ∼7.4, and the pHluorin complement of the fused vesicle switches "on." Fluorescence time courses of individual presynaptic terminals formed by cultured hippocampal neurons reveal stepwise up (+1 quantum) and down (−1 quantum) transitions following a single electrically evoked action potential. These transitions presumably correspond to the release and retrieval of individual synaptic vesicles. The time elapsing between up and down transitions is variable because of the stochastic nature of vesicle retrieval. Fluorescence data, with permission, from Gandhi & Stevens (2003). In the context of a presynaptic terminal (*bottom*), fluorescence changes reflect changes in the balance of vesicle exo- and endocytosis: Exocytosis adds fluorescence; endocytosis removes fluorescence. Because synaptic vesicle membrane is reinternalized after a variable delay following exocytosis (compare vesicles 1, 2, and 3), the synapto-pHluorin signal represents the convolution of each synaptic impulse with the time course of the associated vesicle turnover. Trains of 20 and 100 action potentials (AP) at 10 Hz cause transient fluorescence increases detectable in single-pass recordings from individual presynaptic terminals of cultured hippocampal neurons. Fluorescence relaxes back to baseline with time constants ranging from ∼4–50 s; the decay curves approximate the distribution of waiting times for endocytosis. Fluorescence data, with permission, from Sankaranarayanan & Ryan (2000).

and the distribution of waiting times for the retrieval process (**Figure 4**). Waiting times may last from a few hundred milliseconds to several dozens of seconds and accordingly limit the observability of fast events.

Synapto-pHluorins have revealed how a neuron in *C. elegans* encodes an error signal that partly drives the thermotactic behavior of the worm (Samuel et al. 2003); how the transmission of information between three populations of neurons in the fly's olfactory system is arranged to allow for the extraction of odor features (**Figure 1**; Ng et al. 2002); and how sensory afferents relay odor input to the olfactory bulbs of mice (Bozza et al. 2004). As might be expected for sensors of synaptic transmission, synapto-pHluorins have also found a niche in studies of synaptic plasticity. They have detected the recruitment of previously inactive synapses in the antennal lobes of flies by Pavlovian conditioning (Yu et al. 2004) and helped distinguish different modes of serotonin-triggered synaptic reorganization in *Aplysia* (Kim et al. 2003).

ACTUATION

Positive Actuation

With roots reaching back to Galvani, the electrode has been the chief experimental tool for controlling excitable cells. Its use as an actuator has opened windows on neurobiological phenomena as diverse as evoked neurotransmitter release (Katz 1969), the topographic organization of somatosensory and motor cortices (Penfield & Rasmussen 1950), decision making (Salzman et al. 1990, Salzman & Newsome 1994), and the specification of simple actions by cortical pyramidal cells (Brecht et al. 2004). Despite these accomplishments, electrodes suffer from a number of important shortcomings: They are serial devices that do not permit the members of a dispersed population of neurons to be addressed in parallel. They have no built-in ability to discriminate among different functional classes of neurons. They provide only crude spatial resolution, stimulate axons of passage, can cause significant tissue damage, and require mechanical stability, which makes them very cumbersome to use in unrestrained animals. And they require an a priori decision about which location to stimulate, impeding discovery.

Because moving light beams is easier than moving electrodes, the search for optical methods for stimulating neurons began with the advent of sufficiently powerful laser light sources. Early efforts (and a recent second coming in two-photon form; Hirase et al. 2002) focused light of high intensity directly on the neuronal plasma membrane (Fork 1971). Illuminated cells were depolarized to action potential threshold, possibly because the high-intensity beam created microscopic pores in the plasma membrane. The insertion of a fluorescent dye into the membrane increased the efficacy (but, unfortunately, also the toxicity) of direct photostimulation and permitted a first systematic optical search for the sources of synaptic input to an electrically recorded neuron in an invertebrate ganglion (Farber & Grinvald 1983). However, despite its ability to act "at a distance" rather than through physical contact, the stimulating light beam remained the functional equivalent of an electrode: It operated serially, exerting its effect at only one focal point at a time. The exact location of the stimulation site still had to be predetermined by the experimenter, on the basis of visible anatomy rather than a functional characteristic, such as a transmitter phenotype.

This situation remained fundamentally unchanged with the synthesis of the first caged neurotransmitters (Walker et al. 1986, Wilcox et al. 1990, Wieboldt et al. 1994), an important practical advance that has made photostimulation virtually synonymous with optical uncaging of excitatory transmitters, in particular, L-glutamate (Callaway & Katz 1993, Wieboldt et al. 1994). The principle of this approach is to render a neurotransmitter biologically inert by chemical modification with a photoremovable blocking group (Kaplan et al. 1978; Walker et al. 1986, 1988; Wilcox et al. 1990; Wieboldt et al. 1994); illumination then produces a concentration jump from the caged to the active transmitter. The photoreleased agonist binds to its cognate receptors (primarily AMPA receptors in the case of glutamate) and gates open their conduction pathways. Channel opening enables current flow that depolarizes the membrane (Callaway & Katz 1993). Because virtually all neurons express glutamate receptors, photolytic uncaging of glutamate is, for all practical purposes, a universally applicable stimulation method. The down side of the ubiquitous presence of glutamate receptors, however, is that the site of stimulation can often be localized only poorly: Activity is elicited in all neurons whose dendritic arbors cross the uncaging beam. In scattering neural tissue, this sphere of activation may extend over several hundred microns.

To avoid indiscriminate stimulation in such circumstances, uncaging reactions have been designed that show a second-order dependence on the concentration of incident photons; sufficient photon fluxes for uncaging are then reached only within the focus (Denk et al. 1990, Denk 1994, Pettit et al. 1997, Furuta et al. 1999, Matsuzaki et al. 2001). Restricting stimulation to a small focal volume, however, has also reduced the scope of the neurobiological problems amenable to optical probing: Photostimulation is now increasingly preoccupied with subcellular structures such as dendritic branches or spines (Wang et al. 2000; Matsuzaki et al. 2001, 2004). This is a sharp departure from the vision of the pioneers, who saw the potential of optics first and foremost in the analysis of circuits (Fork 1971, Farber & Grinvald 1983, Callaway & Katz 1993). Clearly, meaningful analyses of circuits or systems would require actuators that combine massive parallelism with a built-in selectivity for distinct functional classes of neurons.

Gene expression patterns can provide the cell-type selectivity necessary to solve the specificity problem (Zemelman et al. 2002). Imagine a situation in which neural tissue is illuminated broadly, but only a subset of all illuminated neurons can decode and transduce the diffusely broadcast optical signal into electrical activity. If the "receiver" of the optical signal is a protein encoded in DNA, the responsive subset of neurons can be restricted genetically to certain cell types or circuit elements. Localizing the susceptibility to stimulation is an exact inversion of the logic of all previously considered stimulation methods, which, regardless of whether electrical or optical, must localize the position of the stimulus; they therefore target anatomical locations rather than functionally circumscribed populations of neurons. Because sensitivity to light is built into each target neuron, advance knowledge of its spatial coordinates is unnecessary. Multiple neurons can be addressed simultaneously and precisely, in moving animals, without undesirable cross talk to neighboring cells that are functionally distinct (Lima & Miesenböck 2005).

Neurons employ two principal classes of transduction systems to generate electrical signals in response to external stimuli (Hille 2001). Metabotropic signaling systems consist of heptahelical receptors that communicate with their effectors through heterotrimeric G proteins. The activation of some of these effectors is coupled to changes in membrane potential. Ionotropic signaling systems effect changes in membrane potential directly, via chemically or physically gated conductances. Both metabotropic and ionotropic systems provide raw material for the construction of genetically encoded "phototriggers" (**Figure 5**; Zemelman et al. 2002, 2003). Although this material is typically borrowed from sensory systems, its use is not limited to experiments that shuffle or add sensory receptors at the periphery (Zuker et al. 1988, Troemel et al. 1997, Tobin et al. 2002, Smallwood et al. 2003). On the contrary, the power of the approach lies in its potential to "skip the peripheral processes" (Julesz 1971) and actuate central neurons directly (Zemelman et al. 2002, Lima & Miesenböck 2005).

Because the photoreceptors of vertebrate and invertebrate eyes are naturally equipped with genetically encoded receivers that allow them to respond to light, the search for optically controlled actuators that could be transplanted to nonphotoreceptor cells initially concentrated on them (Zemelman et al. 2002). A comparison of classical phototransduction mechanisms in vertebrates and invertebrates suggested that the invertebrate transduction machinery, which operates with depolarizing photocurrents (Hardie 1991, Ranganathan et al. 1991) and a bistable chromophore (Hillman et al. 1983), could serve as a portable light-controlled source of depolarizing current (**Figure 5**). Indeed, coexpression of three photoreceptor proteins of the fly, *Drosophila melanogaster*, is able to trigger light-evoked action potentials in hippocampal neurons (Zemelman et al. 2002).

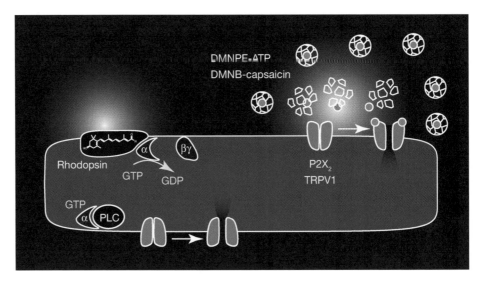

Figure 5

Metabotropic and ionotropic phototriggers. In metabotropic phototriggers such as chARGe (*left*), illuminated rhodopsin catalyzes nucleotide exchange on its cognate heterotrimeric G-protein, which as a result dissociates into a GTP-bound α subunit and a βγ heterodimer. The α subunit activates a phosphoinositol-specific phospholipase C (PLC). In a poorly understood mechanism, a product of PLC causes the opening of nonselective cation channels in the plasma membrane, which leads to membrane depolarization and the generation of action potentials. Ionotropic phototriggers (*right*) are ligand-gated sodium or calcium channels, such as the purinoceptor $P2X_2$ or the capsaicin receptor TRPV1. The channels are optically activated when an illuminating beam shatters the photolabile nitrobenzyl "cages" (DMNPE and DMNB) that sequester the active ligands, ATP and capsaicin.

The three actuator proteins are the blue-sensitive opsin, NinaE (O'Tousa et al. 1985, Zuker et al. 1985), arrestin-2 (Hyde et al. 1990, LeVine et al. 1990, Yamada et al. 1990), and the α subunit of the cognate heterotrimeric G protein (Lee et al. 1990). To allude to these components and their explosive properties when coexpressed, they are referred to collectively as "chARGe" (Zemelman et al. 2002). In addition to the three genetically encoded components, chARGe requires a fourth, the chromophore all-*trans* (or 11-*cis*) retinal. Functional rhodopsin is reconstituted from empty, unliganded NinaE in the membrane of chARGed cells by incubation with the chromophore. The metarhodopsin-like intermediate formed with the *trans* isomer is then photoconverted to rhodopsin (Zemelman et al. 2002).

Photostimulation of chARGed neurons is a strictly cell-autonomous process involving, in all likelihood, the activation of an endogenous TRP channel; it persists unattenuated when all excitatory synaptic communication between neurons is blocked (Zemelman et al. 2002). Alternating periods of light and darkness cause alternating episodes of electrical activity and quiescence. Action potentials, however, often appear and disappear with unpredictable lag periods and frequencies after the light stimulus is applied and removed (Zemelman et al. 2002). The tight coupling between stimulus timing and response timing and between stimulus intensity and response frequency that characterizes the native photoreceptor (Hardie 1991, Ranganathan et al. 1991) is thus relaxed in a chARGed neuron driven by only the minimal phototransduction machinery.

After the original description of chARGe, photocurrents evoked by two nonclassical opsins have been recorded in non-neuronal cells. These opsins do not require the coexpression of two auxiliary proteins (i.e., arrestin and $G_q\alpha$) and might thus simplify the genetic manipulations required to sensitize neurons to light; they remain, however, dependent on retinal. One of these atypical transducers is melanopsin (Melyan et al. 2005, Panda et al. 2005, Qiu et al. 2005), the nonimaging opsin expressed by a small population of intrinsically photosensitive retinal ganglion cells. Like NinaE in chARGe, melanopsin signals through $G_{q/11}$ and phospholipase C to a TRP channel; like NinaE, it possesses a built-in, possibly arrestin-dependent photoisomerase function that can regenerate the active chromophore. The other atypical transducer is the algal bacteriorhodopsin relative channelrhodopsin-2 (Nagel et al. 2003), which may unite the functions of retinal-dependent light sensor and passive ionic conductance within one and the same protein—a surprising but auspicious combination.

Phototriggers based on ionic conductances (Zemelman et al. 2003) promise to overcome the reachability problems of chARGe—that is, their inability to control spike times and spike frequencies accurately and to elicit firing rates above 7.5 Hz in pyramidal neurons (Zemelman et al. 2002). Instead of relying on a metabotropic cascade triggered by light, the excitatory stimulus is now transduced by directly gated, agonist-controlled ion channels. The agonists are supplied optically through photorelease from caged precursors (**Figure 5**). Ionotropic phototriggers thus bear some similarity to photostimulation with caged glutamate (Callaway & Katz 1993): In either case, a transmitter molecule is rendered biologically inert by chemical modification with a photoremovable blocking group; illumination produces a concentration jump from the caged to the active agonist; agonist binds to ionotropic receptors and gates open their conduction pathways; the resulting current leads to membrane depolarization. The key difference, however, is that the receptor on which the agonist acts is not a ubiquitously expressed, endogenous protein (such as the AMPA receptor gated by glutamate), but instead is a heterologous channel that is normally absent from the nervous system of interest. Agonist (the "key") can therefore spark action potentials only in those neurons that have been genetically programmed to express the cognate "lock" (Zemelman et al. 2003). This ensures cell-type specificity.

Three ionotropic phototrigger candidates from two protein families have been identified (Zemelman et al. 2003): the ATP-gated channel $P2X_2$ (Brake et al. 1994, Valera et al. 1994), as well as the capsaicin and menthol receptors, TRPV1 and TRPM8, respectively, which are both members of the superfamily of TRP channels (Caterina et al. 1997, McKemy et al. 2002, Peier et al. 2002). All three candidates are expressed primarily in pain-sensing neurons of the peripheral nervous system, thus minimizing the potential for cross talk in the CNS. [In vertebrates, but not in other organisms (Lima & Miesenböck 2005), the presence of endogenous purinoceptors at some central synapses limits the utility of $P2X_2$ as a selectively addressable actuator. We anticipate that the resulting controllability problems will be overcome with the help of orthogonal receptor-ligand pairs—mutant P2X receptors that lack sensitivity to ATP but can be gated by unnatural nucleotide analogs. The exceptionally simple architecture of P2X channels should facilitate the necessary engineering efforts.] Caged ligands currently exist for two of the three channels, making these the true and tested "photo"-triggers: $P2X_2$ is gated effectively by photolysis of P^3-[1-(4,5-dimethoxy-2-nitrophenyl)ethyl]-ATP (DMNPE-ATP) (Kaplan et al. 1978, Zemelman et al. 2003); and TRPV1 is gated by photolysis of 4,5-dimethoxy-2-nitrobenzyl-capsaicin (DMNB-capsaicin) (Zemelman et al. 2003). As expected for an ionotropic mechanism, spikes follow light pulses quickly, reliably, and predictably; responses terminate rapidly; spike frequencies can be tuned by varying the photon dose; and firing rates of ≥ 40 Hz are reachable.

The first application of a genetically encoded phototrigger in a systems setting was in the fruit fly, *Drosophila melanogaster* (Lima & Miesenböck 2005). Flies were engineered to express the phototrigger P2X$_2$ (which has no endogenous counterparts in flies); caged ATP was microinjected into the CNS. The freely moving animals responded to brief pulses of laser illumination with behaviors characteristic of the sites of phototrigger expression: When the phototrigger was located in one of two highly restricted, nonoverlapping sets of neurons in the giant fiber system that controls escape behaviors (such as jumping and the initiation of flight), illuminated flies opened and flapped their wings and took flight (**Figure 6**); when the phototrigger was expressed in dopaminergic neurons, illumination caused changes in locomotor activity and walking path selections (**Figure 7**). To ensure that these responses reflected the direct optical activation of central neuronal targets rather than confounding visual input, the flies were blinded with the help of a mutation that eliminates an essential phototransduction component. Blind animals rarely if ever flew spontaneously, suggesting that the initiation of flight is visually gated. Flight, however, could be restored reliably, even in the absence of vision, by photostimulating neurons in the giant fiber system directly (Lima & Miesenböck 2005). Artificial neural signals may thus be used to repair or bypass behavioral deficits.

Negative Actuation

In contrast to an intracellular electrode, which can pass depolarizing as well as hyperpolarizing current, genetically encoded phototriggers are able to exert positive control only: They can depolarize but not hyperpolarize. Restricting control inputs to be non-negative is a common type of control constraint; for example, it is not possible to apply a negative dose of insulin to elevate the blood glucose level. Rather, a second type of actuator—the administration of an insulin-antagonizing hormone, or of glucose itself—is required to balance positive and negative control. Genetic loss-of-function approaches play this role vis-à-vis triggers of activity.

Reflecting perhaps their historical origins in the lesion studies of clinical neurology, many genetic loss-of-function approaches still cause lasting defects rather than tunable actuation. The only form of control that can be exerted over these negative effectors is transcriptional; once induced, their impact is, for all practical purposes, permanent even if the targeted neurons survive the intervention. Clostridial neurotoxins, such as the light chain of tetanus toxin, for example, cleave SNARE proteins required for synaptic vesicle exocytosis and thereby block transmission (Schiavo et al. 1992, Sweeney et al. 1995). The induced block does not subside when transcription of the gene encoding the toxin is turned off, but only after all mRNA and toxin molecules have been degraded and a sufficient fraction of the proteolyzed SNAREs have been replaced with newly synthesized, functional proteins. The same limitation applies to surface-tethered toxins that *cis*-inactivate carriers of synaptic or action currents (Ibañez-Tallon et al. 2004) and to the overexpression of potassium channels—such as inward rectifiers (Johns et al. 1999), open rectifiers (Nitabach et al. 2002), or engineered voltage-gated channels (White et al. 2001)—that clamp the membrane potential at or below resting value and shunt excitatory synaptic currents.

Efforts to inhibit the function of specific neurons temporarily rather than permanently tap metabotropic or ionotropic signaling systems that couple to hyperpolarizing conductances—a variation on the familiar mechanisms of metabotropic and ionotropic phototriggers (**Figure 5**). In contrast to the phototriggers, however, the currently available negative actuators are regulated pharmacologically rather than optically, and temporal control over their activities is accordingly coarse. The metabotropic system consists of the fly allatostatin receptor and its peptide ligand, and in some instances it must also include

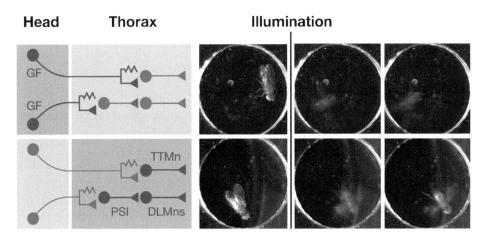

Figure 6

Genetically targeted photostimulation of the giant fiber system in *Drosophila melanogaster*. The giant fiber system is a reflex circuit responsible for escape behaviors such as jumping and flight. It originates with a pair of giant fiber (GF) neurons in the head, which project their axons to the thoracic ganglion, where they form mixed electrical and chemical synapses with the TTMn and PSI neurons. TTMn is the motor neuron innervating the tergotrochanteral (jump) muscle; PSI controls the dorsal longitudinal (flight) muscles indirectly via chemical synapses with the DLMn motor neurons. Flies expressing the genetically encoded phototrigger $P2X_2$ in either the GF neurons (*top*) or the thoracic group of neurons (TTMn, PSI, and DLMns, *middle*) respond to brief illumination with jumping and extended periods of high-frequency wing beating. (Actual flight is prohibited in the small experimental chamber.) The fact that different artificial "command impulses" elicit identical motor output suggests that rhythmic wing movement represents a periodic attractor in the dynamic landscape of the circuit (*bottom*). This periodic orbit, symbolized by the closed yellow path surrounding the lake, is reached by default from all initial states that lie within its "basin of attraction," just like water will spontaneously flow into the lake from nearby regions of the mountainous terrain. A variety of command impulses that lift the circuit from its initial state of quiescence (*right foreground*) into the basin of attraction of the periodic orbit can specify transitions from quiescence to flight activity robustly and economically. Modified, with permission, from Lima & Miesenböck (2005).

the two subunits of the G protein–coupled inwardly rectifying potassium (GIRK) channel that serves as the effector (Lechner et al. 2002). Ligand activation of the receptor reduces the input resistance and thereby raises the spike threshold of vertebrate neurons 13-fold; the effect reverses within several minutes after the allatostatin peptide is removed. The ionotropic silencing system is composed of two subunits of a glutamate-gated chloride

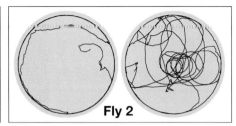

Figure 7

Genetically targeted photostimulation of dopaminergic neurons in *Drosophila melanogaster*. Flies expressing the genetically encoded phototrigger $P2X_2$ in dopaminergic neurons were analyzed in a 25-mm circular arena (*grey circles*). Movement trajectories of two individuals during two-minute observation periods before and after illumination are shown. Note the increase in locomotor activity, the tendency toward centripetal movement, and the display of characteristic dopamine-induced circling behavior following photostimulation. Modified, with permission, from Lima & Miesenböck (2005).

channel from *C. elegans* that can also be operated by the anthelminthic drug ivermectin (Slimko et al. 2002). To avoid paradoxical activation of the channel during excitatory glutamatergic transmission in heterologous systems, the channel's glutamate response may be attenuated about sixfold by a mutation in the ligand-binding domain (Li et al. 2002). Because the ivermectin effect takes many hours to subside, however, temporal control over this negative actuator approaches transcriptional timescales (Slimko et al. 2002). In fact, the slow return of neuronal excitability after ivermectin is removed may be due to the turnover of activated channels rather than the dissociation of ligand.

The only optically controlled negative actuator of neuronal activity—and, as such, the closest analog in temporal acuity but not in mechanism of the phototriggers discussed above—is a voltage-gated potassium channel that can be blocked and unblocked with light (Banghart et al. 2004). The blocking group, a tetraethylammonium (TEA) function, is held near the external mouth of the channel on a photoisomerizable azobenzene arm; this arm is disulfide-bonded to an engineered cysteine residue at the perimeter of the pore. Light-driven extension and retraction of the arm imposes and relieves the TEA block. Despite the appeal of its reversible photochemistry, the practical utility of the device is undermined by the inevitable loss of neuronal excitability that accompanies the expression of the underivatized channel (Banghart et al. 2004) and the reduction in excitability that lingers if the chemical derivatization reaction cannot be driven to completion.

LINKING SENSING AND ACTUATION

The ability to sense and the ability to actuate come together in feedback—actuation based on sensing. Most scientific exploration involves feedback at a fundamental level: Experiments are designed on the basis of the results of previous experiments. Real-time sensing, actuation, and computation—enabled through fast interfaces and algorithms—open the possibility of control with more tailored goals. One example is the design and tuning of feedback laws that keep a system at a setpoint, such as a neuron under voltage clamp. Every time a conductance opens that drives the membrane potential away from the setpoint, the deviation is sensed, and a force (actuation) is applied in the form of current to bring the neuron back to the prescribed state. What is an unstable state for the open-loop system—a neuron whose membrane potential would fluctuate and occasionally enter the explosively nonlinear regime of spiking—becomes a stable state by closing the feedback loop.

The application of analogous principles of real-time sensing, actuation, and computation to cell assemblies rather than single neurons—enabled through optical field sensing and actuation—is expected to generate powerful new strategies for exploring, understanding, and perhaps even designing the behavior of neural circuits. Although we recognize that many technical hurdles must be overcome before this possibility can be fully realized (most notably, the development of fast sensors), we hope that rapid progress in the ability to control physical and chemical processes foreshadows similar advances in biology. For most of the past century, sensing and actuation in the physical sciences were relatively coarse grained and slow. Sensors were few, had poor spatial resolution, and were often incapable of responding in real time. Actuators were even less developed, to the extent that systems—think of a chemical reaction—could be addressed only through changes in global variables such as temperature, pressure, or pH. The computations required to solve optimal control problems were difficult or impossible to perform online.

The past few decades have seen a vast change in sensing and real-time computation. Accurate measurements with progressively higher spatial and temporal resolution have yielded much richer data sets and shortened feedback delays. Novel actuators have become increasingly fine grained, and improvements in the spatial resolution of control have followed suit. Rather than through global variables, physicochemical processes can now be controlled via spatiotemporally resolved signals at microscopic scales: Laser beams address catalytic chemical reactions (Wolff et al. 2001) or affect the surface tension of fluid interfaces (Garnier et al. 2003); digital projectors guide and control the reaction fronts of photochemical processes (Sakurai et al. 2002); colloidal particle assemblies are optically manipulated through lithographic masks (Hayward et al. 2000) or individually with optical tweezers (Grier 2003).

The well-developed mathematical apparatus of optimal control and high-dimensional data analysis generalizes immediately from the physical to the life sciences. In nervous systems, as in other complex systems, macroscopic, coarse-grained behavior (such as the emergence of spatial or temporal order) results from the interactions of individual agents, which collectively possess an enormous number of degrees of freedom. Modern data-analysis techniques, going beyond principal component analysis, hold the promise of uncovering coarse-grained descriptions inherent in such high-dimensional data sets. Diffusion maps, along with the associated fast algorithms, may yield nontrivial macroscopic descriptors of circuit states (Belkin & Niyogi 2003, Coifman et al. 2005, Nadler et al. 2005), similar to the thermodynamic variables that encapsulate microscopic behavior in statistical mechanics. Feedback could be used adaptively to enrich the available data sets by enhancing data collection in regions of state space that the open-loop circuit would very seldom (or never) spontaneously visit. Finding data-driven, collective observables likely to represent the natural modes of a system is crucial in model identification and in effectively reducing microscopic dynamics to collective dynamics. Such an effective reduction will enhance understanding and facilitate the design of experiments.

Which types of problems, then, would benefit from a newfound authority over the function of neural circuits, granted by the incorporation of optical sensors and actuators into precisely defined groups of neurons? The first task on which fine-grained sensing and actuation can be brought to bear is system identification: exploring the connectivity and dynamics of naturally existing circuits. Here, a prescribed optical input is used to address a particular neuron, and the responses of this neuron, as well as those of other, synaptically coupled neurons, are sensed. The resulting point-to-point transfer

2HG: second-harmonic generation

DMNB: 4,5-dimethoxy-2-nitrobenzyl

DMNPE: P^3-1-(4,5-dimethoxy-2-nitrophenyl)ethyl

ER: endoplasmic reticulum

FRET: fluorescence resonance energy transfer

GFP: green fluorescent protein

TEA: tetraethylammonium

functions reveal circuit connectivity and relate inputs to observables and outputs. This problem, in a nutshell, motivated the development of conventional photostimulation techniques: the analysis of connections between individual neurons, regions, or layers (Fork 1971, Farber & Grinvald 1983, Callaway & Katz 1993, Dantzker & Callaway 2000, Sawatari & Callaway 2000, Brivanlou et al. 2004), and their developmental or experience-driven rearrangements (Dalva & Katz 1994, Shepherd et al. 2003). The ability to actuate specific cell types through genetically encoded phototriggers will refine the resolution of point-to-point mapping; the ability to control extended yet genetically defined populations of neurons will facilitate the exploration of dynamic circuit modes.

The inverse problem is also open to analysis: Which input causes a particular firing pattern—say, a given spike sequence in a neuron or circuit? Which input is required to recall a memory or elicit a specific behavior? A first, simple representative of this type of experiment was an optical search for different neuronal "command impulses" capable of triggering the oscillatory circuit that underlies wing movement in insect flight (**Figure 6**; Lima & Miesenböck 2005).

Neuronal responses are typically state-dependent—that is, they vary as functions of the neuron's current state and its recent activity history (Llinás & Sugimori 1980, Turrigiano et al. 1994, Marder et al. 1996, Egorov et al. 2002, Loewenstein et al. 2005), as well as the activities of synaptically connected neurons (Destexhe & Paré 1999, Hô & Destexhe 2000, Chance et al. 2002). Feedback could likely be used to bring a neuron or circuit to a prescribed state so that input-output measurements may be performed reproducibly and their state-dependent statistics quantified. The control action required to maintain a system in a given state provides a mirror image of the spontaneous evolution from that state—in the same way that the force exerted by a constraint quantifies how the system would evolve in its absence.

Control and measurement are thus intimately linked (see Sidebar). Analogies can be drawn again with recordings of the currents required to keep a neuron under voltage clamp and with umbrella sampling techniques in molecular dynamics, where computational feedback is used to prepare a molecular system consistent with a given macroscopic state, and the force exerted by this computational feedback probes the local free energy surface (Torrie & Valleau 1974, Ryckaert et al. 1977).

The ability to initialize a neuronal assembly at a given state and then let it go free, so that one can observe its spontaneous evolution from that state, might allow one to estimate a rate of change—a time derivative—that could be used to predict future (or deduce past) behavior. If the fundamental assumption of smoothness in time can be met, the mathematics involved are very simple—i.e., Taylor series—and at the same time very powerful. Consider the task of experimentally locating the attracting stationary states of a circuit, such as the persistent firing patterns thought to represent short-term memories (Fuster 1973, Goldman-Rakic 1995, Seung et al. 2000, Egorov et al. 2002). Computationally, such stationary states are found not through slow evolution of the system, but with the help of iterative fixed-point algorithms like Newton-Raphson: The time derivatives of the system in its current state, as well as in nearby states, are calculated, and this linearized, Jacobian information is then used to accelerate the approach to the ultimate stationary state. When a circuit can be initialized in nearby defined states and brief trajectories of its evolution from these states recorded, a Newton-Raphson-type search for stationary states can, in effect, be implemented experimentally. This type of search for stationary states might be called "equation-free experimentation" to reflect a direct analogy with the recently developed strategy of equation-free computation (Kevrekidis et al. 2003, 2004); in equation-free computation, short bursts of appropriately initialized microscopic simulations are paired with macroscopic time-steppers to accelerate system analysis and design; here, actual experiments take the place of the microscopic simulations.

Electrophysiology with only a small number of electrode actuators severely limits reachability. Optical field actuation will allow one to sample a much richer space of input structures and explore a correspondingly richer output space. Even the effective connectivity among circuit elements is no longer constrained by the biological hardware because connections can also be simulated. Spatiotemporally variegated actuation could, for example, couple the activities of neurons that are not physically linked with prescribed correlations in space and time or temporarily remove neurons from a circuit without physically severing their connections. Recorded activity histories and their statistics might be used to alter the strengths of virtual connections in accordance with specified learning rules. Conversely, persistent actuation might cause modifications of the topologies and strengths of actual connections, engineer new functionalities, and reveal principles governing biological plasticity.

If virtual connections can be realized through feedback loops, then one can envision simulating some of the external connections of an explanted neural tissue, such as a cortical or hippocampal slice. Because sensory interfaces have provided the sole portals through which distributed inputs could be supplied to neuronal circuits, sensory systems have received a perhaps disproportionate amount of neuroscientific attention, whereas the function of circuits at some synaptic distance from sensory surfaces has remained comparatively obscure. The inability to control such circuits in reduced, experimentally accessible preparations has forced scientists to observe spontaneous or pharmacologically evoked phenomena with uncertain significance. The combination of sensing and actuation may herald a change in the roles of these scientists from observers to doers.

ACKNOWLEDGMENTS

We thank Lawrence Cohen and Leslie Loew for advice on the electrochromic properties of voltage-sensitive dyes. The authors' work is supported by the NIH (G.M.), AFOSR Dynamics and Control (I.G.K.), and DARPA (I.G.K.).

LITERATURE CITED

Ataka K, Pieribone VA. 2002. A genetically targetable fluorescent probe of channel gating with rapid kinetics. *Biophys. J.* 82:509–16

Awatramani R, Soriano P, Rodriguez C, Mai JJ, Dymecki SM. 2003. Cryptic boundaries in roof plate and choroid plexus identified by intersectional gene activation. *Nat. Genet.* 35:70–75

Baird GS, Zacharias DA, Tsien RY. 1999. Circular permutation and receptor insertion within green fluorescent proteins. *Proc. Natl. Acad. Sci. USA* 96:11241–46

Banghart M, Borges K, Isacoff E, Trauner D, Kramer RH. 2004. Light-activated ion channels for remote control of neuronal firing. *Nat. Neurosci.* 7:1381–86

Belkin M, Niyogi P. 2003. Laplacian eigenmaps for dimensionality reduction and data representation. *Neural Comput.* 15:1373–96

Bélanger PR. 1995. *Control Engineering: A Modern Approach*. Fort Worth, TX: Saunders. 471 pp.

Bellman R. 1968. *Some Vistas of Modern Mathematics: Dynamic Programming, Invariant Embedding, and the Mathematical Biosciences*. Lexington: Univ. Ky. Press. 141 pp.

Ben-Oren I, Peleg G, Lewis A, Minke B, Loew L. 1996. Infrared nonlinear optical measurements of membrane potential in photoreceptor cells. *Biophys. J.* 71:1616–20

Bezanilla F. 2000. The voltage sensor in voltage-dependent ion channels. *Physiol. Rev.* 80:555–92

Bizzini B, Stoeckel K, Schwab M. 1977. An antigenic polypeptide fragment isolated from tetanus toxin: chemical characterization, binding to gangliosides and retrograde axonal transport in various neuron systems. *J. Neurochem.* 28:529–42

Bloembergen N. 1965. *Nonlinear Optics*. New York: Benjamin. 222 pp.

Bouevitch O, Lewis A, Pinevsky I, Wuskell JP, Loew LM. 1993. Probing membrane potential with nonlinear optics. *Biophys. J.* 65:672–79

Bozza T, McGann JP, Mombaerts P, Wachowiak M. 2004. In vivo imaging of neuronal activity by targeted expression of a genetically encoded probe in the mouse. *Neuron* 42:9–21

Brake AJ, Wagenbach MJ, Julius D. 1994. New structural motif for ligand-gated ion channels defined by an ionotropic ATP receptor. *Nature* 371:519–23

Branda CS, Dymecki SM. 2004. Talking about a revolution: the impact of site-specific recombinases on genetic analyses in mice. *Dev. Cell* 6:7–28

Braz JM, Rico B, Basbaum AI. 2002. Transneuronal tracing of diverse CNS circuits by Cre-mediated induction of wheat germ agglutinin in transgenic mice. *Proc. Natl. Acad. Sci. USA* 99:15148–53

Brecht M, Schneider M, Sakmann B, Margrie TW. 2004. Whisker movements evoked by stimulation of single pyramidal cells in rat motor cortex. *Nature* 427:704–10

Brivanlou IH, Dantzker JL, Stevens CF, Callaway EM. 2004. Topographic specificity of functional connections from hippocampal CA3 to CA1. *Proc. Natl. Acad. Sci. USA* 101:2560–65

Brockett RW. 1970. *Finite Dimensional Linear Systems*. New York: Wiley. 244 pp.

Callaway EM, Katz LC. 1993. Photostimulation using caged glutamate reveals functional circuitry in living brain slices. *Proc. Natl. Acad. Sci. USA* 90:7661–65

Card JP, Rinaman L, Schwaber JS, Miselis RR, Whealy ME, et al. 1990. Neurotropic properties of pseudorabies virus: uptake and transneuronal passage in the rat central nervous system. *J. Neurosci.* 10:1974–94

Caterina MJ, Schumacher MA, Tominaga M, Rosen TA, Levine JD, Julius D. 1997. The capsaicin receptor: a heat-activated ion channel in the pain pathway. *Nature* 389:816–24

Chance FS, Abbott LF, Reyes AD. 2002. Gain modulation from background synaptic input. *Neuron* 35:773–82

Coen L, Osta R, Maury M, Brulet P. 1997. Construction of hybrid proteins that migrate retrogradely and transynaptically into the central nervous system. *Proc. Natl. Acad. Sci. USA* 94:9400–5

Cohen LB, Salzberg BM, Grinvald A. 1978. Optical methods for monitoring neuron activity. *Annu. Rev. Neurosci.* 1:171–82

Coifman RR, Lafon S, Lee AB, Maggioni M, Nadler B, et al. 2005. Geometric diffusions as a tool for harmonic analysis and structure definition of data. Part I: diffusion maps. *Proc. Natl. Acad. Sci. USA*. In press

Crick F. 1999. The impact of molecular biology on neuroscience. *Philos. Trans. R. Soc. London B. Biol. Sci.* 354:2021–25

Dalva MB, Katz LC. 1994. Rearrangements of synaptic connections in visual cortex revealed by laser photostimulation. *Science* 265:255–58

Dantzker JL, Callaway EM. 2000. Laminar sources of synaptic input to cortical inhibitory interneurons and pyramidal neurons. *Nat. Neurosci.* 3:701–7

Denk W. 1994. Two-photon scanning photochemical microscopy: mapping ligand-gated ion channel distributions. *Proc. Natl. Acad. Sci. USA* 91:6629–33

Denk W, Strickler JH, Webb WW. 1990. Two-photon laser scanning fluorescence microscopy. *Science* 248:73–76

Destexhe A, Paré D. 1999. Impact of network activity on the integrative properties of neocortical pyramidal neurons in vivo. *J. Neurophysiol.* 81:1531–47

Doyle DA, Cabral JM, Pfuetzner RA, Kuo A, Gulbis JM, et al. 1998. The structure of the potassium channel: molecular basis of K^+ conduction and selectivity. *Science* 280:69–77

Egorov AV, Hamam BN, Fransen E, Hasselmo ME, Alonso AA. 2002. Graded persistent activity in entorhinal cortex neurons. *Nature* 420:173–78

Farber IC, Grinvald A. 1983. Identification of presynaptic neurons by laser photostimulation. *Science* 222:1025–27

Fork RL. 1971. Laser stimulation of nerve cells in Aplysia. *Science* 171:907–8

Furuta T, Wang SS, Dantzker JL, Dore TM, Bybee WJ, et al. 1999. Brominated 7-hydroxycoumarin-4-ylmethyls: photolabile protecting groups with biologically useful cross-sections for two photon photolysis. *Proc. Natl. Acad. Sci. USA* 96:1193–200

Fuster JM. 1973. Unit activity in prefrontal cortex during delayed-response performance: neuronal correlates of transient memory. *J. Neurophysiol.* 36:61–78

Gandhi SP, Stevens CF. 2003. Three modes of synaptic vesicular recycling revealed by single-vesicle imaging. *Nature* 423:607–13

Garnier N, Grigoriev RO, Schatz MF. 2003. Optical manipulation of microscale fluid flow. *Phys. Rev. Lett.* 91:54501-1–4

Gerfen CR, O'Leary DD, Cowan WM. 1982. A note on the transneuronal transport of wheat germ agglutinin-conjugated horseradish peroxidase in the avian and rodent visual systems. *Exp. Brain Res.* 48:443–48

Goldman-Rakic PS. 1995. Cellular basis of working memory. *Neuron* 14:477–85

Gong S, Zheng C, Doughty ML, Losos K, Didkovsky N, et al. 2003. A gene expression atlas of the central nervous system based on bacterial artificial chromosomes. *Nature* 425:917–25

Greenberg ME, Ziff EB, Greene LA. 1986. Stimulation of neuronal acetylcholine receptors induces rapid gene transcription. *Science* 234:80–83

Grier DG. 2003. A revolution in optical manipulation. *Nature* 424:810–16

Griesbeck O, Baird GS, Campbell RE, Zacharias DA, Tsien RY. 2001. Reducing the environmental sensitivity of yellow fluorescent protein. Mechanism and applications. *J. Biol. Chem.* 276:29188–94

Guerrero G, Siegel MS, Roska B, Loots E, Isacoff EY. 2002. Tuning FlaSh: redesign of the dynamics, voltage range, and color of the genetically encoded optical sensor of membrane potential. *Biophys. J.* 83:3607–18

Hardie RC. 1991. Whole-cell recordings of the light induced current in dissociated Drosophila photoreceptors: evidence for feedback by calcium permeating the light-sensitive channels. *Proc. R. Soc. London B.* 245:203–10

Hasan MT, Friedrich RW, Euler T, Larkum ME, Giese G, et al. 2004. Functional fluorescent Ca^{2+} indicator proteins in transgenic mice under TET control. *PLoS Biol.* 2:e163

Hayward RC, Saville DA, Aksay IA. 2000. Electrophoretic assembly of colloidal crystals with optically tunable micropatterns. *Nature* 404:56–59

Heim N, Griesbeck O. 2004. Genetically encoded indicators of cellular calcium dynamics based on troponin C and green fluorescent protein. *J. Biol. Chem.* 279:14280–86

Hessler NA, Shirke AM, Malinow R. 1993. The probability of transmitter release at a mammalian central synapse. *Nature* 366:569–72

Higashijima S, Masino MA, Mandel G, Fetcho JR. 2003. Imaging neuronal activity during zebrafish behavior with a genetically encoded calcium indicator. *J. Neurophysiol.* 90:3986–97

Hille B. 2001. *Ion Channels of Excitable Membranes*. Sunderland, MA: Sinauer. 814 pp.

Hillman P, Hochstein S, Minke B. 1983. Transduction in invertebrate photoreceptors: role of pigment bistability. *Physiol. Rev.* 63:668–772

Hirase H, Nikolenko V, Goldberg JH, Yuste R. 2002. Multiphoton stimulation of neurons. *J. Neurobiol.* 51:237–47

Hô N, Destexhe A. 2000. Synaptic background activity enhances the responsiveness of neocortical pyramidal neurons. *J. Neurophysiol.* 84:1488–96

Horowitz LF, Montmayeur JP, Echelard Y, Buck LB. 1999. A genetic approach to trace neural circuits. *Proc. Natl. Acad. Sci. USA* 96:3194–99

Hoshi T, Zagotta WN, Aldrich RW. 1990. Biophysical and molecular mechanisms of Shaker potassium channel inactivation. *Science* 250:533–38

Hoshi T, Zagotta WN, Aldrich RW. 1991. Two types of inactivation in Shaker K^+ channels: effects of alterations in the carboxy-terminal region. *Neuron* 7:547–56

Hyde DR, Mecklenburg KL, Pollock JA, Vihtelic TS, Benzer S. 1990. Twenty Drosophila visual system cDNA clones: one is a homolog of human arrestin. *Proc. Natl. Acad. Sci. USA* 87:1008–12

Ibañez-Tallon I, Wen H, Miwa JM, Xing J, Tekinay AB, et al. 2004. Tethering naturally occurring peptide toxins for cell-autonomous modulation of ion channels and receptors in vivo. *Neuron* 43:305–11

Ikura M, Clore GM, Gronenborn AM, Zhu G, Klee CB, Bax A. 1992. Solution structure of a calmodulin-target peptide complex by multidimensional NMR. *Science* 256:632–38

Ji G, Feldman ME, Deng KY, Greene KS, Wilson J, et al. 2004. Ca^{2+}-sensing transgenic mice: postsynaptic signaling in smooth muscle. *J. Biol. Chem.* 279:21461–68

Jiang Y, Lee A, Chen J, Cadene M, Chait BT, MacKinnon R. 2002. The open pore conformation of potassium channels. *Nature* 417:523–26

Johns DC, Marx R, Mains RE, O'Rourke B, Marban E. 1999. Inducible genetic suppression of neuronal excitability. *J. Neurosci.* 19:1691–97

Julesz B. 1971. *Foundations of Cyclopean Perception*. Chicago: Univ. Chicago Press. 406 pp.

Kaplan JH, Forbush B 3rd, Hoffman JF. 1978. Rapid photolytic release of adenosine 5′-triphosphate from a protected analogue: utilization by the Na:K pump of human red blood cell ghosts. *Biochemistry* 17:1929–35

Katz B. 1969. *The Release of Neural Transmitter Substances*. Liverpool, UK: Liverpool Univ. Press. 60 pp.

Kerr R, Lev-Ram V, Baird G, Vincent P, Tsien RY, Schafer WR. 2000. Optical imaging of calcium transients in neurons and pharyngeal muscle of *C. elegans*. *Neuron* 26:583–94

Kevrekidis IG, Gear CW, Hummer G. 2004. Equation-free: the computer-aided analysis of complex multiscale systems. *Am. Inst. Chem. Eng. J.* 50:1346–55

Kevrekidis IG, Gear CW, Hyman JM, Kevrekidis PG, Runborg O, Theodoropoulos K. 2003. Equation-free coarse-grained multiscale computation: enabling microscopic simulators to perform system-level tasks. *Comm. Math Sci.* 4:715–62

Kim JH, Udo H, Li HL, Youn TY, Chen M, et al. 2003. Presynaptic activation of silent synapses and growth of new synapses contribute to intermediate and long-term facilitation in Aplysia. *Neuron* 40:151–65

Lechner HA, Lein ES, Callaway EM. 2002. A genetic method for selective and quickly reversible silencing of mammalian neurons. *J. Neurosci.* 22:5287–90

Lee YJ, Dobbs MB, Verardi ML, Hyde DR. 1990. dgq: a Drosophila gene encoding a visual system-specific G alpha molecule. *Neuron* 5:889–98

LeVine H 3rd, Smith DP, Whitney M, Malicki DM, Dolph PJ, et al. 1990. Isolation of a novel visual-system-specific arrestin: an in vivo substrate for light-dependent phosphorylation. *Mech. Dev.* 33:19–25

Li P, Slimko EM, Lester HA. 2002. Selective elimination of glutamate activation and introduction of fluorescent proteins into a Caenorhabditis elegans chloride channel. *FEBS Lett.* 528:77–82

Lima SQ, Miesenböck G. 2005. Remote control of behavior through genetically targeted photostimulation of neurons. *Cell* 121:141–52

Llinás R, Sugimori M. 1980. Electrophysiological properties of in vitro Purkinje cell somata in mammalian cerebellar slices. *J. Physiol.* 305:171–95

Loew LM, Bonneville GW, Surow J. 1978. Charge shift optical probes of membrane potential. Theory. *Biochemistry* 17:4065–71

Loew LM, Scully S, Simpson L, Waggoner AS. 1979. Evidence for a charge-shift electrochromic mechanism in a probe of membrane potential. *Nature* 281:497–99

Loewenstein Y, Mahon S, Chadderton P, Kitamura K, Sompolinsky H, et al. 2005. Bistability of cerebellar Purkinje cells modulated by sensory stimulation. *Nat. Neurosci.* 8:202–11

Loots E, Isacoff EY. 1998. Protein rearrangements underlying slow inactivation of the Shaker K^+ channel. *J. Gen. Physiol.* 112:377–89

Marder E, Abbott LF, Turrigiano GG, Liu Z, Golowasch J. 1996. Memory from the dynamics of intrinsic membrane currents. *Proc. Natl. Acad. Sci. USA* 93:13481–86

Martin X, Dolivo M. 1983. Neuronal and transneuronal tracing in the trigeminal system of the rat using the herpes virus suis. *Brain Res.* 273:253–76

Maskos U, Kissa K, St Cloment C, Brulet P. 2002. Retrograde trans-synaptic transfer of green fluorescent protein allows the genetic mapping of neuronal circuits in transgenic mice. *Proc. Natl. Acad. Sci. USA* 99:10120–25

Matsuzaki M, Ellis-Davies GC, Nemoto T, Miyashita Y, Iino M, Kasai H. 2001. Dendritic spine geometry is critical for AMPA receptor expression in hippocampal CA1 pyramidal neurons. *Nat. Neurosci.* 4:1086–92

Matsuzaki M, Honkura N, Ellis-Davies GC, Kasai H. 2004. Structural basis of long-term potentiation in single dendritic spines. *Nature* 429:761–66

McKemy DD, Neuhausser WM, Julius D. 2002. Identification of a cold receptor reveals a general role for TRP channels in thermosensation. *Nature* 416:52–58

Melyan Z, Tarttelin EE, Bellingham J, Lucas RJ, Hankins MW. 2005. Addition of human melanopsin renders mammalian cells photoresponsive. *Nature* 433:741–45

Miesenböck G, De Angelis DA, Rothman JE. 1998. Visualizing secretion and synaptic transmission with pH-sensitive green fluorescent proteins. *Nature* 394:192–95

Miesenböck G, Rothman JE. 1997. Patterns of synaptic activity in neural networks recorded by light emission from synaptolucins. *Proc. Natl. Acad. Sci. USA* 94:3402–7

Miyawaki A, Llopis J, Heim R, McCaffery JM, Adams JA, et al. 1997. Fluorescent indicators for Ca^{2+} based on green fluorescent proteins and calmodulin. *Nature* 388:882–87

Miyawaki A, Griesbeck O, Heim R, Tsien RY. 1999. Dynamic and quantitative Ca^{2+} measurements using improved cameleons. *Proc. Natl. Acad. Sci. USA* 96:2135–40

Moreaux L, Sandre O, Charpak S, Blanchard-Desce M, Mertz J. 2001. Coherent scattering in multi-harmonic light microscopy. *Biophys. J.* 80:1568–74

Moreaux L, Sandre O, Mertz J. 2000. Membrane imaging by second-harmonic generation microscopy. *J. Opt. Soc. Am. B.* 17:1685–94

Morgan JI, Cohen DR, Hempstead JL, Curran T. 1987. Mapping patterns of c-fos expression in the central nervous system after seizure. *Science* 237:192–97

Morgan JI, Curran T. 1991. Stimulus-transcription coupling in the nervous system: involvement of the inducible proto-oncogenes fos and jun. *Annu. Rev. Neurosci.* 14:421–51

Nadler B, Lafon S, Coifman RR, Kevrekidis IG. 2005. Diffusion maps, spectral clustering and the reaction coordinates of dynamical systems. *Appl. Comp. Harm. Anal.* In press

Nagai T, Sawano A, Park ES, Miyawaki A. 2001. Circularly permuted green fluorescent proteins engineered to sense Ca^{2+}. *Proc. Natl. Acad. Sci. USA* 98:3197–202

Nagel G, Szellas T, Huhn W, Kateriya S, Adeishvili N, et al. 2003. Channelrhodopsin-2, a directly light-gated cation-selective membrane channel. *Proc. Natl. Acad. Sci. USA* 100:13940–45

Nakai J, Ohkura M, Imoto K. 2001. A high signal-to-noise Ca^{2+} probe composed of a single green fluorescent protein. *Nat. Biotechnol.* 19:137–41

Ng M, Roorda RD, Lima SQ, Zemelman BV, Morcillo P, Miesenböck G. 2002. Transmission of olfactory information between three populations of neurons in the antennal lobe of the fly. *Neuron* 36:463–74

Nitabach MN, Blau J, Holmes TC. 2002. Electrical silencing of Drosophila pacemaker neurons stops the free-running circadian clock. *Cell* 109:485–95

O'Tousa JE, Baehr W, Martin RL, Hirsh J, Pak WL, Appleburg ML. 1985. The Drosophila ninaE gene encodes an opsin. *Cell* 40:839–50

Panda S, Nayak SK, Campo B, Walker JR, Hogenesch JB, Jegla T. 2005. Illumination of the melanopsin signaling pathway. *Science* 307:600–4

Peier AM, Moqrich A, Hergarden AC, Reeve AJ, Andersson DA, et al. 2002. A TRP channel that senses cold stimuli and menthol. *Cell* 108:705–15

Penfield W, Rasmussen T. 1950. *The Cerebral Cortex of Man. A Study of Localization of Function.* New York: Macmillan. 248 pp.

Pettit DL, Wang SS, Gee KR, Augustine GJ. 1997. Chemical two-photon uncaging: a novel approach to mapping glutamate receptors. *Neuron* 19:465–71

Pologruto TA, Yasuda R, Svoboda K. 2004. Monitoring neural activity and $[Ca^{2+}]$ with genetically encoded Ca^{2+} indicators. *J. Neurosci.* 24:9572–79

Qiu X, Kumbalasiri T, Carlson SM, Wong KY, Krishna V, et al. 2005. Induction of photosensitivity by heterologous expression of melanopsin. *Nature* 433:745–49

Ranganathan R, Harris GL, Stevens CF, Zuker CS. 1991. A Drosophila mutant defective in extracellular calcium-dependent photoreceptor deactivation and rapid desensitization. *Nature* 354:230–32

Regehr WG, Connor JA, Tank DW. 1989. Optical imaging of calcium accumulation in hippocampal pyramidal cells during synaptic activation. *Nature* 341:533–36

Reymond A, Marigo V, Yaylaoglu MB, Leoni A, Ucla C, et al. 2002. Human chromosome 21 gene expression atlas in the mouse. *Nature* 420:582–86

Romoser VA, Hinkle PM, Persechini A. 1997. Detection in living cells of Ca^{2+}-dependent changes in the fluorescence emission of an indicator composed of two green fluorescent protein variants linked by a calmodulin-binding sequence. A new class of fluorescent indicators. *J. Biol. Chem.* 272:13270–74

Roorda RD, Hohl TM, Toledo-Crow R, Miesenböck G. 2004. Video-rate nonlinear microscopy of neuronal membrane dynamics with genetically encoded probes. *J. Neurophysiol.* 92:609–21

Rosenmund C, Clements JD, Westbrook GL. 1993. Nonuniform probability of glutamate release at a hippocampal synapse. *Science* 262:754–57

Ruda M, Coulter JD. 1982. Axonal and transneuronal transport of wheat germ agglutinin demonstrated by immunocytochemistry. *Brain Res.* 249:237–46

Ryckaert JP, Ciccotti G, Berendsen HJC. 1977. Numerical integration of Cartesian equations of motion of a system with constraints—molecular dynamics of n-alkanes. *J. Comput. Phys.* 23:327–41

Sakai R, Repunte-Canonigo V, Raj CD, Knöpfel T. 2001. Design and characterization of a DNA-encoded, voltage-sensitive fluorescent protein. *Eur. J. Neurosci.* 13:2314–18

Sakurai T, Mihaliuk E, Chirila F, Showalter K. 2002. Design and control of wave propagation patterns in excitable media. *Science* 296:2009–12

Salzman CD, Britten KH, Newsome WT. 1990. Cortical microstimulation influences perceptual judgements of motion direction. *Nature* 346:174–77

Salzman CD, Newsome WT. 1994. Neural mechanisms for forming a perceptual decision. *Science* 264:231–37

Samuel AD, Silva RA, Murthy VN. 2003. Synaptic activity of the AFD neuron in Caenorhabditis elegans correlates with thermotactic memory. *J. Neurosci.* 23:373–76

Sankaranarayanan S, De Angelis D, Rothman JE, Ryan TA. 2000. The use of pHluorins for optical measurements of presynaptic activity. *Biophys. J.* 79:2199–208

Sankaranarayanan S, Ryan TA. 2000. Real-time measurements of vesicle-SNARE recycling in synapses of the central nervous system. *Nat. Cell Biol.* 2:197–204

Sankaranarayanan S, Ryan TA. 2001. Calcium accelerates endocytosis of vSNAREs at hippocampal synapses. *Nat. Neurosci.* 4:129–36

Sawatari A, Callaway EM. 2000. Diversity and cell type specificity of local excitatory connections to neurons in layer 3B of monkey primary visual cortex. *Neuron* 25:459–71

Schiavo G, Benfenati F, Poulain B, Rossetto O, Polverino de Laureto P, et al. 1992. Tetanus and botulinum-B neurotoxins block neurotransmitter release by proteolytic cleavage of synaptobrevin. *Nature* 359:832–35

Schwab ME, Suda K, Thoenen H. 1979. Selective retrograde transsynaptic transfer of a protein, tetanus toxin, subsequent to its retrograde axonal transport. *J. Cell Biol.* 82:798–810

Schwab ME, Thoenen H. 1976. Electron microscopic evidence for a transsynaptic migration of tetanus toxin in spinal cord motoneurons: an autoradiographic and morphometric study. *Brain Res.* 105:213–27

Seung HS, Lee DD, Reis BY, Tank DW. 2000. Stability of the memory of eye position in a recurrent network of conductance-based model neurons. *Neuron* 26:259–71

Shen YR. 1989. Surface properties probed by second-harmonic and sum-frequency generation. *Nature* 337:519–25

Shepherd GM, Pologruto TA, Svoboda K. 2003. Circuit analysis of experience-dependent plasticity in the developing rat barrel cortex. *Neuron* 38:277–89

Shimizu-Sato S, Huq E, Tepperman JM, Quail PH. 2002. A light-switchable gene promoter system. *Nat. Biotechnol.* 20:1041–44

Shipley MT. 1985. Transport of molecules from nose to brain: transneuronal anterograde and retrograde labeling in the rat olfactory system by wheat germ agglutinin-horseradish peroxidase applied to the nasal epithelium. *Brain Res. Bull.* 15:129–42

Siegel MS, Isacoff EY. 1997. A genetically encoded optical probe of membrane voltage. *Neuron* 19:735–41

Slimko EM, McKinney S, Anderson DJ, Davidson N, Lester HA. 2002. Selective electrical silencing of mammalian neurons in vitro by the use of invertebrate ligand-gated chloride channels. *J. Neurosci.* 22:7373–79

Smallwood PM, Olveczky BP, Williams GL, Jacobs GH, Reese BE, et al. 2003. Genetically engineered mice with an additional class of cone photoreceptors: implications for the evolution of color vision. *Proc. Natl. Acad. Sci. USA* 100:11706–11

Smeyne RJ, Schilling K, Robertson L, Luk D, Oberdick J, et al. 1992. fos-lacZ transgenic mice: mapping sites of gene induction in the central nervous system. *Neuron* 8:13–23

Strack AM, Loewy AD. 1990. Pseudorabies virus: a highly specific transneuronal cell body marker in the sympathetic nervous system. *J. Neurosci.* 10:2139–47

Suh GS, Wong AM, Hergarden AC, Wang JW, Simon AF, et al. 2004. A single population of olfactory sensory neurons mediates an innate avoidance behaviour in Drosophila. *Nature* 431:854–59

Suzuki H, Kerr R, Bianchi L, Frokjaer-Jensen C, Slone D, et al. 2003. In vivo imaging of *C. elegans* mechanosensory neurons demonstrates a specific role for the MEC-4 channel in the process of gentle touch sensation. *Neuron* 39:1005–17

Sweeney ST, Broadie K, Keane J, Niemann H, O'Kane CJ. 1995. Targeted expression of tetanus toxin light chain in Drosophila specifically eliminates synaptic transmission and causes behavioral defects. *Neuron* 14:341–51

Tank DW, Sugimori M, Connor JA, Llinás RR. 1988. Spatially resolved calcium dynamics of mammalian Purkinje cells in cerebellar slice. *Science* 242:773–77

Tobin DM, Madsen DM, Kahn-Kirby A, Peckol EL, Moulder G, et al. 2002. Combinatorial expression of TRPV channel proteins defines their sensory functions and subcellular localization in *C. elegans* neurons. *Neuron* 35:307–18

Torrie GM, Valleau JP. 1974. Monte-Carlo free-energy estimates using non-Boltzmann sampling—application to subcritical Lennard-Jones fluid. *Chem. Phys. Lett.* 28:578–81

Troemel ER, Kimmel BE, Bargmann CI. 1997. Reprogramming chemotaxis responses: sensory neurons define olfactory preferences in *C. elegans*. *Cell* 91:161–69

Turrigiano G, Abbott LF, Marder E. 1994. Activity-dependent changes in the intrinsic properties of cultured neurons. *Science* 264:974–77

Ugolini G, Kuypers HG, Strick PL. 1989. Transneuronal transfer of herpes virus from peripheral nerves to cortex and brainstem. *Science* 243:89–91

Ulam SM. 1962. On some mathematical problems connected with patterns of growth of figures. In *Proceedings of Symposia in Applied Mathematics*, pp. 215–24. Providence: Am. Math. Soc. Vol. 14

Valera S, Hussy N, Evans RJ, Adami N, North RA, et al. 1994. A new class of ligand-gated ion channel defined by P2x receptor for extracellular ATP. *Nature* 371:516–19

von Neumann J. 1966. *Theory of Self-Reproducing Automata*. Urbana: Univ. Ill. Press. 388 pp.

van Roessel P, Brand AH. 2002. Imaging into the future: visualizing gene expression and protein interactions with fluorescent proteins. *Nat. Cell Biol.* 4:E15–20

Walker JW, McCray JA, Hess GP. 1986. Photolabile protecting groups for an acetylcholine receptor ligand. Synthesis and photochemistry of a new class of o-nitrobenzyl derivatives and their effects on receptor function. *Biochemistry* 25:1799–805

Walker JW, Reid GP, McCray JA, Trentham DR. 1988. Photolabile 1-(2-nitrophenyl)ethyl phosphate esters of adenine nucleotide analogs. Synthesis and mechanism of photolysis. *J. Am. Chem. Soc.* 110:7170–77

Wang JW, Wong AM, Flores J, Vosshall LB, Axel R. 2003. Two-photon calcium imaging reveals an odor-evoked map of activity in the fly brain. *Cell* 112:271–82

Wang SS, Khiroug L, Augustine GJ. 2000. Quantification of spread of cerebellar long-term depression with chemical two-photon uncaging of glutamate. *Proc. Natl. Acad. Sci. USA* 97:8635–40

Wang Y, Guo HF, Pologruto TA, Hannan F, Hakker I, et al. 2004. Stereotyped odor-evoked activity in the mushroom body of Drosophila revealed by green fluorescent protein-based Ca2+ imaging. *J. Neurosci.* 24:6507–14

West AE, Chen WG, Dalva MB, Dolmetsch RE, Kornhauser JM, et al. 2001. Calcium regulation of neuronal gene expression. *Proc. Natl. Acad. Sci. USA* 98:11024–31

White BH, Osterwalder TP, Yoon KS, Joiner WJ, Whim MD, et al. 2001. Targeted attenuation of electrical activity in Drosophila using a genetically modified K(+) channel. *Neuron* 31:699–711

Wieboldt R, Gee KR, Niu L, Ramesh D, Carpenter BK, Hess GP. 1994. Photolabile precursors of glutamate: synthesis, photochemical properties, and activation of glutamate receptors on a microsecond time scale. *Proc. Natl. Acad. Sci. USA* 91:8752–56

Wilcox M, Viola RW, Johnson KW, Billington AP, Carpenter BK, et al. 1990. Synthesis of photolabile "precursors" of amino acid neurotransmitters. *J. Org. Chem.* 55:1585–89

Wilson Y, Nag N, Davern P, Oldfield BJ, McKinley MJ, et al. 2002. Visualization of functionally activated circuitry in the brain. *Proc. Natl. Acad. Sci. USA* 99:3252–57

Wolff J, Papathanasiou AG, Kevrekidis IG, Rotermund HH, Ertl G. 2001. Spatiotemporal addressing of surface activity. *Science* 294:134–37

Yamada T, Takeuchi Y, Komori N, Kobayashi H, Sakai Y, et al. 1990. A 49-kilodalton phosphoprotein in the Drosophila photoreceptor is an arrestin homolog. *Science* 248:483–86

Yoshihara Y, Mizuno T, Nakahira M, Kawasaki M, Watanabe Y, et al. 1999. A genetic approach to visualization of multisynaptic neural pathways using plant lectin transgene. *Neuron* 22:33–41

Yu D, Ponomarev A, Davis RL. 2004. Altered representation of the spatial code for odors after olfactory classical conditioning; memory trace formation by synaptic recruitment. *Neuron* 42:437–49

Zaccolo M, De Giorgi F, Cho CY, Feng L, Knapp T, et al. 2000. A genetically encoded, fluorescent indicator for cyclic AMP in living cells. *Nat. Cell Biol.* 2:25–29

Zemelman BV, Lee GA, Ng M, Miesenböck G. 2002. Selective photostimulation of genetically chARGed neurons. *Neuron* 33:15–22

Zemelman BV, Nesnas N, Lee GA, Miesenböck G. 2003. Photochemical gating of heterologous ion channels: remote control over genetically designated populations of neurons. *Proc. Natl. Acad. Sci. USA* 100:1352–57

Zhang J, Campbell RE, Ting AY, Tsien RY. 2002. Creating new fluorescent probes for cell biology. *Nat. Rev. Mol. Cell Biol.* 3:906–18

Zuker CS, Cowman AF, Rubin GM. 1985. Isolation and structure of a rhodopsin gene from D. melanogaster. *Cell* 40:851–58

Zuker CS, Mismer D, Hardy R, Rubin GM. 1988. Ectopic expression of a minor Drosophila opsin in the major photoreceptor cell class: distinguishing the role of primary receptor and cellular context. *Cell* 53:475–82

Subject Index

A

Actin, 26–46
    actin remodeling, 30, 32–33, 43–46
        actin binding proteins and, 30
    axonal and dendritic growth and, 25–26, 30, 33, 41–46
    synapse structure and, 32–35, 43–46
    synaptogenesis and, 30–32
Actin binding proteins (ABPs), 30, 33, 46
    actin remodeling and, 30, 33, 43–46
Actin cytoskeleton, 25–46
    actin dynamics and, 29–30
    morphological plasticity of synapses and, 25–46
Active zone, 251–63
Adaptive gain model of locus coeruleus-norepinephrin function, 403, 417–39
    exploration versus exploitation and, 420–23
    learning and, 431–32
    optimization of task performance and, 403, 418–20, 433–34
    plasticity and, 431–32
    top-down evaluative information from frontal cortex and, 423, 431
    See also Locus coeruleus
Adult neurogenesis in mammalian CNS, 223–42
    antidepressants and, 239
    depression and, 239
    endogenous neurogenesis in vivo, 226–29, 236–39
        analysis of, 226–29
        modulation of, 236–39
    ischemic brain injury and, 239
    nerve guidance and, 234–35, 242
    neural stem cell and, 223, 231–42
    neural progenitors, 223, 229
    neuronal fate, 231–34, 242
    proliferation, 231–34, 236
    specification, 231, 231–34, 236
    neurodenerative diseases and, 240–41
    neurogenesis in vitro and ex vitro analysis of, 229–30
    neuronal fate and, 231–34, 242
    neuronal migration and, 234–35, 242
    radiation injury and, 240
    seizures and 239–40
    stress and, 239
    synaptogenesis and, 235–36, 242
    synaptic plasticity and, 235–36
Aggregation
    C. elegans and, 477–81
        neuroendocrine regulation of, 481
Alzheimer's disease
    axon degeneration in, 140
    genetic influences on brain structure and, 11–12
Amygdala
    autism and deficit in, 112–13
Anterior cingulate cortex (ACC), 403, 423–31
    locus coeruleus and
        task utility and, 403, 423–31
Antidepressants
    adult neurogenesis and, 239
Aperture problem
    motion perception and, 157, 169–73
Apoptosis
    sympathetic nervous system development and, 202–4
        BDNF and, 202–4

NGF-TrkA signaling and, 203–4
    p75 and, 202–4
Artemin, 194–95
Asperger syndrome, 110
Attention
    autism and, 109–19
    locus coeruleus and, 406
    modulation
        middle temporal visual area and, 176–77
Autism, 109–19
    amygdala deficit in, 112–13
    analytical brain and, 114
    empathizing-systemizing theory of, 109–11
    executive dysfunction theory of, 111
    Fragile X syndrome and, 119
    medial frontal cortex and, 113
    neural connectivity and, 116–18
    phenotype, 118–19
    social brain and, 112–13
    weak central coherence in, 111–12
Axon branching
    retinotopic maps and, 327–48
Axon degeneration, 139–40, 142–49
    See also Axon elimination
Axon elimination, 127–49
    degeneration in neurological diseases, 139–40, 142–49
        dying back degeneration, 139–40
        Wallerian degeneration, 139–40, 143–49
    developmental pruning of connections, 127–39
        area-specific subcortical projection of layer 5 of neocortex, 133–36
        delayed interstitial branching and, 133–36
        eye-specific retinal connections and, 131–33
        neuronal remodeling in insects, 139
        retinotopic maps and, 133, 135, 137–38
        species-unique projection patterns and, 138
        synapse elimination at neuromuscular junction and, 130–31
        topographic maps and, 136–38
    mechanisms of axon degeneration, 142–48
        Drosophila mushroom body and, 142–43, 146–47
        ubiquitin-proteasome system (UPS) and, 143–47
        Wallerian degeneration and, 143–49
    mechanisms of axon retraction, 140–42
        cytoskeleton and, 140–41
        intracellular signaling pathways and, 141–42
        RhoGTPases, 141–42

    Semaphorins and, 142
Axon guidance
    retinotopic maps and, 327–48
Axon growth
    sympathetic, 191–200
        GDNF and, 194
        growth inhibition, 198–99
        growth mechanisms, 199
        initiation of, 193
        proximal axon extension and, 193–97
        target innervation and, 197
Axon-target interactions
    GENSTAT screen and, 100

B

BAC transgenic expression vectors, 100–1
Behavior
    neuronal substrates of complex behaviors in *C. elegans*, 451–89
Behavior regulation
    neuromodulatory systems and, 403–39
        historical perspective, 405–6
Behavioral performance
    optimization of
        locus coeruleus and, 403–4, 406–39
Blindness
    cross-modal plasticity and, 387–96
    occipital cortex activation and, 387–96
Bone derived neurotrophic factor, 202–4, 207
Bone morphogenetic protein, 204–7
Brain-derived neurotrophic factor, 202–4
    p75 and
        apoptosis and, 202–4
Brain development and circuitry
    gene expression screens in CNS and, 89, 97–102
        in situ hybridization, 98–99
        reporter gene assays, 98–100
        transcriptional profiling via microarray, 101–2
    gene function screens in CNS and, 89–97
        ENU mutagenesis, 92–95
        gene targeting, 96–97
        gene trap screening, 95–96
        spontaneous mutant mice, 90–92
Brain injury
    plasticity and, 384–87
Brain map, 1–17
    defined, 5
Brain mapping, 1–17
    brain structure and, 1–17

cortical pattern matching and, 6–9
functional versus structural, 5
genetic influences on brain structure and, 1–2, 9–17
registration and, 6
statistical, 8–9
Brain morphology
See Brain structure
Brain structure
environmental influences on, 16
genetics of, 1–17
candidate genes and brain function, 11
specific genes and, 11–12
heritability and
cortical gray matter distribution and, 9–17
MRI studies and, 10–11
intelligence and, 15–17

C

Caenorhabditis elegans, 451–89
aggregation and aerotaxis, 477–81
neuroendocrine regulation of, 481
calcium oscillations
defecation motor program (DMP) and, 472–76
rhythmic behavior and, 471–77
chemotaxis toward water-soluble chemicals, 464–65
foraging, 451, 462
area-restricted search, 462–63
food-induced slowing response, 463–64
learning in, 451, 486–89
glutamatergic signaling and, 486–87
locomotion, 454–71
mating, 451, 481–86
gene products and, 484–86
neuronal control of, 483–84
neuronal substrates of complex behaviors in, 451–89
nociception, 457–62
olfaction, 465–68
plasticity, 451, 486–89
sensory processing in, 451
thermosensation, 468–71
Calcium oscillations
defecation motor program (DMP) and, 472–76
rhythmic behavior and, 471–77
Cell-adhesion molecules
target recognition and, 256–57
vertebrate synaptogenesis and, 256–60
neuroligin, 258, 260
synaptic cell-adhesion molecule (SynCAM), 258–60
Center-surround antagonism, 157, 160, 165–68
Chemoaffinity hypothesis, 330
Chemotaxis toward water-soluble chemicals
C. elegans and, 464–65
Cholinergic switch
sympathetic nervous system development and, 191, 195, 208–11
Circadian rhythms
calcium oscillations in *C. elegans*
defecation motor program (DMP) and, 472–76
plasticity and, 467
rhythmic behavior and, 471–77
Cognition
locus coeruleus-norepinephrine and dopamine systems and, 432, 438–39
psychiatric disorders and, 432, 438–39
Color vision
cytochrome oxidase and, 303–19
V1 and V2 circuitry and, 303–19
See also Visual system
Computation
analog, 539
dendritic computation, 503–27
chemical computations and, 505, 516–17
dendritic voltage-dependent channels and, 505, 509–16, 520
directional selectivity, 517–19
forward masking of cricket songs and, 523
image processing in fly neurons and, 520–21
lobula giant movement detector (LGMD) in locusts and, 521–23
passive dendrites and nonlinear interactions between synaptic inputs, 505–9
sound localization and, 519–20
transgenic animals and, 523–24
working memory and, 520
Cortical pattern matching, 6–9
Cortical stimulation
plasticity and, 377–96
Cross-modal plasticity, 387–96
Cytochrome oxidase, 303–19
color vision and, 303–19
histochemistry
V1 organization and, 303–10
V2 organization and, 312–15

magno, parvo, and konio geniculate signals to
V2 and, 303–19
Cytoskeleton, 25–46, 251–66
  actin, 25–46
    actin dynamics and, 29–30
    morphological plasticity of synapses and,
25–46
    mechanics of vertebrate synaptogenesis and,
251–66
    postsynaptic organization of, 29
    presynaptic organization of, 28–29
  signaling
    axon retraction and, 127, 140–41

D

Decision making, 403, 407–8, 411–13, 417–39
  locus coeuruleus-norepinephrine system and,
403, 407–8, 411–13, 417–39
  mathematical model for decision processes,
417–18
    sequential probability ratio test (SPRT),
418
Defecation motor program (DMP) of *C. elegans*,
472–76
  genetic analysis of, 473–74
Dementia
  heritability and brain structure and, 11–12
Dendrites
  dendritic computation, 503–27
    working memory and, 520
  dendritic computational toolkit, 504–24
    chemical computations and, 505, 516–17
    dendritic voltage-dependent channels and,
505, 509–16, 520
    passive dendrites and nonlinear interactions
between synaptic inputs, 505–9
  dendritogenesis
    bone morphogenetic protein and, 204–7
    nerve growth factor and, 204–7
    target-derived signals and, 204–7
  directional selectivity, 517–19
    directional-selective retinal ganglion cells
and, 517–19
  forward masking of cricket songs and, 523
  image processing in fly neurons and, 520–21
  lobula giant movement detector (LGMD) in
locusts and, 521–23
  sound localization and, 519–20
  transgenic animals and, 523–24
Dendritic spine, 25–26, 30, 33, 41–46
  actin binding proteins and, 30, 33, 41–46

  long-term plasticity and, 43–45
  short-term plasticity and, 41–43
Depression
  adult neurogenesis and, 239
Directional-selective retinal ganglion cells
  dendritic computation and, 517–19
DJ-1
  Parkinson's disease and 58, 67–68, 73
Dorsal paired medial (DPM) neurons
  memory formation in *Drosophila* and, 280,
282–83
Drosophila melanogaster
  odor-evoked sympathetic activity in, 537–38
  olfactory learning and memory in, 275–97
    attenal lobe and, 283–84
    conditioned stimulus pathway and, 278–84
    conserved behavioral functions in mammals
and, 295–97
    dorsal paired medial (DPM) neurons and,
280, 282–83
    genes and gene products required for,
277–78, 287–95
    mushroom body-mediated, 275–76, 284–95
Dystonia, 383–84

E

Empathy
  empathizing-systemizing theory of autism,
109–11
  neural basis of, 112–13
ENU mutagenesis screens, 92–95
  dominant and semi-dominant genome-wide,
93–94
  genotype-driven genetics and, 95
  recessive genome-wide, 94–95
Ephs and ephrins, 327–48
  retinotopic maps and
    bifunctional and bidirectional actions of,
338–48
    EphBs and eprins Bs and dorsoventral
mapping, 339
    multiple actions and models, 342–43
EprhinB
  vertebrate synaptogenesis and, 257–58
Extrastriate cortex, 157–59
  early exploration of, 158–59

F

Feedback, 552–55
Fibroblast growth factor
  vertebrate synaptogenesis and, 255–56, 258

Flynn Effect, 14
Fragile X syndrome, 119
Frontal cortex
    locus coeruleus regulation and, 403, 423–30
        anterior cingulate cortex (ACC), 403, 423–31
        orbitofrontal cortex (OFC), 403, 423–31
Frontal lobe
    autism and, 109, 113–18

G

g
    Spearman's, 3–4, 16
Gene expression screens in CNS, 89–103
    in situ hybridization, 98–99
    proteomics, 102
    reporter gene assays, 98–100
        axon-target interactions and, 100
        tangential migratory patterns and, 100–1
        transcriptional profiling via microarray, 101–2
Gene function screens in CNS, 89–97
    ENU mutagenesis, 92–95
    gene targeting, 96–97
    gene trap screening, 95–96
    spontaneous mutant mice, 90–92

H

Heritability, 1–17
Huntington's disease
    axon degeneration in, 140

I

In situ hybridization, 98–99
Intelligence, 1–17
    brain structure and, 15–17
    definition, 3
    environmental influences on, 2, 14–15
    fluid and crystallized, 5
    genetics of, 1–17
    heritability of, 12–17
    Raven's Progressive Matrices, 4
    social, 112–13
        neural basis of empathy and, 112–13
    Spearman's g, 3–4, 16
    tests of, 3–5
        Binet-Simon tests, 3
        scholastic aptitude tests (SAT), 3
        Stanford-Binet Intelligence Tests, 3
        Wechsler Adult Intelligence Scale (WAIS), 3, 16
        Wechsler Intelligence Scale for Children (WISC), 3, 16
    unitary versus multiple, 4–5
Ion channels
    dendritic voltage-dependent channels and, 505, 509–16, 520
    dendritic computation and, 505, 509–16, 520
    sensing intracellular calcium concentration, 542–44
    sensing membrane potential of voltage-gated, 539–42
IQ
    See Intelligence
Ischemic brain injury
    adult neurogenesis and, 239

L

Lateral geniculate nucleus (LGN), 303–19
    color, form, and motion in physiology of, 310–12
    intracortical circuitry in, 307–10
Learning, 451, 486–89
    C. elegans and
        glutamatergic signaling and, 486–87
Learning and memory
    actin and long-term structural changes at synapses, 45–46
    long-term depression (LTD), 25, 43–45
        stable actin remodeling and, 43–45
    long-term potentiation (LTP), 25, 43–45
        stable actin remodeling and, 43–45
    olfactory learning and memory formation in *Drosophila*, 275–95
        attenal lobe and, 283–84
        conditioned stimulus pathway and, 278–84
        conserved behavioral functions in mammals of, 295–97
        dorsal paired medial (DPM) neurons and, 282–83
        genes and gene products required for, 277–78, 287–95
        mushroom body-mediated, 275–76, 284–95
        unconditioned stimulus information and, 279–84
Learning mutants, 275, 287–95
    olfactory memory formation in *Drosophila* and, 275, 287–95
    crammer and, 294
    CREB2 and, 288, 290

Notch and, 290–91
pumilio and, 294–95
rsh[1] and, 291–93
rutabaga and, 287–88
staufen and, 294
Lobula giant movement detector (LGMD) in locusts
 dendritic computation and, 521–23
Locus coeruleus
 adaptive gain model of function of, 403, 417–39
  exploration versus exploitation and, 420–23
  learning and, 431–32
  plasticity and, 431–32
  top-down evaluative information from frontal cortex and, 423, 431
 frontal cortex and regulation of, 423–30
  anterior cingulate cortex (ACC), 403, 423–31
  orbitofrontal cortex (OFC), 403, 423–31
 human electrophysiological findings and, 436–37
 phasic activity mode of, 403, 407–8, 410–14, 418–20, 423–39
  decision making and, 403, 407–8, 411–13, 417–39
  optimization of task performance and, 403, 418–20, 433–34
  selectivity and plasticity of, 431–32
 tonic activity mode of, 403, 407–8, 413–14, 423–39
  adaptive adjustments of gain, 420–23
  anterior cingulate cortex (ACC) and, 403, 423–31
  baseline activity increase and, 413–14
  orbitofrontal cortex (OFC) and, 403, 423–31
  task-related utility and, 403, 423–31
Locus coeruleus-norepinephrine function, 403–39
 adaptive gain model of function, 403, 417–39
  exploration versus exploitation and, 420–23
  learning and, 431–32
  optimization of task performance and, 403, 418–20, 433–34
  plasticity and, 431–32
  top-down evaluative information from frontal cortex and, 423, 431
 dopamine systems and
  cognition and, 432, 438–39
 empirical and modeling studies, 408–17
  physiological mechanisms and behavioral effects, 415–17
  regulating neural processing and behavior, 408–10
  two modes of operation, 410–14
 optimization of behavioral performance and, 403–39
Long-term depression (LTD), 25, 43–45
 stable actin remodeling and, 43–45
Long-term potentiation (LTP), 25, 43–45
 stable actin remodeling and, 43–45

M

Membrane trafficking, 251, 260–63
 postsynaptic assembly, 261–63
 presynaptic assembly, 260–63
Memory
 sustained neural network activity and, 357, 363–65
 working memory
  dendritic computation and, 520
Memory formation
 in *Drosophila*, 275–97
 See also Learning and memory; Olfactory learning and memory formation in *Drosophila*
Middle temporal visual area (MT or V5), 157–80
 acceleration coding in, 179
 attention modulation and, 176–77
 center-surround antagonism and, 157, 160, 165–68
 cortical feedforward inputs to, 161
 depth perception and, 174–75
 functional organization of, 162–64
 magnocellular inputs to, 157, 160
 motion perception and, 157, 164–80
  aperture problem, 157, 169–73
  Bayesian model of, 179
  computation of velocity, 168–73
  motion opponency, 173
  null direction, 168
  pooling and parsing, 173–74
  preferred direction, 159, 178
  smooth pursuit eye movements and, 176
  structure-from-motion, 157, 174–75
  vector-average hypothesis of, 178–79
  winner-take-all hypothesis of, 178–79
 single neuron sensitivity, 177–78
Mitochondrial complex-I
 Parkinson's disease and, 55, 68–73
Morphological plasticity in synapses, 25–46
 actin cytoskeleton and stable, 43–46
 actin cytoskeleton and transient, 41–43

Motion perception, 157, 164–80
  aperture problem, 157, 169–73
  Bayesian model of, 179
  computation of velocity, 168–73
  motion opponency, 173
  null direction, 168
  pooling and parsing, 173–74
  preferred direction, 159, 178
  smooth pursuit eye movements and, 176
  structure-from-motion, 157, 174–75
  vector-average hypothesis of, 178–79
  winner-take-all hypothesis of, 178–79
Motor system plasticity, 377–87, 395–96
  risk of unwanted change and
    faulty practice, 383–84
    focal hand dystonia in musicians and, 383–84
Mushroom bodies
  memory formation in *Drosophila* and, 275–76, 284–95
    atypical PKCs and, 293–94
    crammer and, 294
    CREB2 and, 288, 290
    genes and gene products and, 287–95
    Notch and, 290–91
    pumilio and, 294–95
    rsh^1 and, 291–93
    rutabaga and, 287–88
    staufen and, 294

N

Nerve guidance, 234–35, 242
Network connectivity
  autism and, 109, 116–18
Neural networks, 357–72, 503–27
  chaotic spiking activity, 357, 359–63, 366–68
    background spiking, 357–59
    balance of excitation and inhibition and, 359, 366–68
  dendritic computation and, 503–27
  firing rate networks, 360
  integrate-and-fire networks, 360–62
  oscillatory network activity, 357–59, 362, 365–66
  signal propagation, 357, 368–72
    avalanche model, 368–70
    propagation of firing rates, 368, 372
    synfire chains, 368–72
  sustained responses to transient stimuli, 357–59, 361, 363–65
    memory and, 357, 363–65

Neural stem cell, 223–42
  adult
    neural progenitors, 223, 229
    neuronal fate, 231–34
    proliferation, 231–34, 236
    specification, 231, 231–34, 236
Neurodegeneration
Neurodegenerative disease
  axon degeneration in, 127, 139–40, 143–49
    dying back degeneration, 139–40
    Wallerian degeneration, 139–40, 143–49
  Parkinson's disease
    molecular pathogenesis of, 55–74
Neurogenesis
  adult, mammalian, 223–42
  antidepressants and, 239
  depression and, 239
  endogenous neurogenesis in vivo, 226–29, 236–39
    analysis of, 226–29
    modulation of, 236–39
  ischemic brain injury and, 239
  nerve guidance and, 234–35, 242
  neural stem cell and, 223–42
    neural progenitors, 223, 229
    neuronal fate, 231–34
    proliferation, 231–34, 236
    specification, 231, 231–34, 236
  neurodenerative diseases and, 240–41
  neurogenesis in vitro and ex vitro
    analysis of, 229–30
  neuronal fate and, 242
  neuronal migration and, 234–35, 242
  radiation injury and, 240
  seizures and 239–40
  stress and, 239
  synaptogenesis and, 235–36, 242
  synaptic plasticity and, 235–36
Neuromodulation
  locus coeruleus-norepinephrine system and, 403–39
    adaptive gain model of function of, 403, 417–39
    historical perspective, 405–6
    performance optimization and, 403–39
  plasticity of human brain cortex and, 377–96
Neuromuscular junction (NMJ)
  synapse elimination at, 130–31, 149
Neuronal activity-regulated pentraxin (Narp)
  synaptogenesis and, 257–59
Neuronal circuits
  actuation, 546–55

Subject Index 571

negative,
positive, 546–50
sensing and, 552–55
observation and control of, 534–55
actuation, 546–55
biological approach to, 535–39
feedback and, 552–55
reachability and, 535, 555
sensing intracellular calcium concentration, 542–44
sensing membrane potential of voltage-gated ion channels, 539–42
sensing neurotransmitter release, 544–46
Neuronal fate
adult regeneration and, 231–34, 242
Neuronal remodeling
insect metamorphosis and, 139
Neurotrophins, 191–211
sympathetic nervous system development and, 191–211
BDNF, 199, 202–4, 207, 210
CNTF, 208
GDNF, 194, 209
NGF, 197–211
NT-3, 195–98, 210
Nitrosative stress
Parkinson's disease and, 55

O

Observation and control, 534–55
biological approach to, 535–39
feedback and, 552–55
reachability and, 535
sensing intracellular calcium concentration, 542–44
sensing membrane potential of voltage-gated ion channels, 539–42
sensing neurotransmitter release, 544–46
Occipital cortex activation
blindness and, 387–96
cross-modal plasticity and, 387–96
Olfaction
C. elegans and, 465–68
Olfactory learning and memory formation in *Drosophila*, 275–97
attenal lobe and, 283–84
conditioned stimulus pathway and, 278–84
dorsal paired medial (DPM) neurons and, 280, 282–83
genes and gene products required for, 277–78, 287–95

conserved behavioral functions in mammals of, 295–97
mushroom body-mediated, 275–76, 284–95
atypical PKCs and, 293–94
crammer and, 294
CREB2 and, 288, 290
Notch and, 290–91
pumilio and, 294–95
rsh^1 and, 291–93
rutabaga and, 287–88
staufen and, 294
unconditioned stimulus information and, 279–84
Optic tectum (OT)
retinotopic maps and, 327–48
Optical imaging
neuronal circuits and, 533–55
actuation and, 546–55
feedback and, 552–55
odor-evoked synaptic activity in *Drosophila*, 537–38
reachability and, 535, 555
sensing intracellular calcium concentration, 542–44
sensing membrane potential of voltage-gated ion channels, 539–42
sensing neurotransmitter release, 544–46
Optimization of behavioral performance, 403, 418–20, 433–34
locus coeruleus and task performance, 403, 418–20, 433–34
anterior cingulate cortex (ACC), 403, 423–31
obitofrontal cortex and, 403, 423–31
Orbitofrontal cortex (OFC), 403, 423–31
locus coeruleus and
task-related utility and, 403, 423–31
Oxidative stress
Parkinson's disease and, 55, 61, 68–73

P

p75, 191, 202–4
apoptosis and, 202–4
BDNF and, 202–4
Parkin
Parkinson's disease and, 55, 58, 61–65, 73
Parkinson's disease, 55–74, 140
axon degeneration in, 140
heritability of, 56–57
molecular pathogenesis of, 55
DJ-1, and, 58, 67–68

genetic contribution to, 57–68
  mitochondrial complex-I and, 55, 68–73
  nitrosative stress and, 55
  oxidative stress and, 55, 61, 68–73
  parkin and, 55, 58, 61–65, 73
  PINK1, 58, 66–67, 73
  α-synuclein and, 55, 57–61, 73
  ubiquitin-proteasome system and, 55, 65, 71–72
 nitrosative stress and, 55
 oxidative stress and, 55, 61, 68–73
 PARK1- and PARK4-linked, 57–61
 PARK2-linked, 57, 61–65
 PARK5-linked, 57, 65–66
 PARK6-linked, 57, 66–67
 PARK7-linked, 57
 PARK8-linked, 57, 72–74
 PARK9-linked, 57
 PARK10-linked, 57
 PARK11-linked, 57
PINK1
 Parkinson's disease pathogenesis and, 58, 66–67, 73
Plasticity
 brain injury and, 384–87
 cross-modal, 387–96
  occipital cortex activation and blindness, 387–96
 as intrinsic property and ongoing state of the nervous system, 378–79, 396
 locus coeruleus and, 431–32
 mapping between brain activity and behavior, 382–84
  pathology and, 384
 morphological changes in synapse, 25–46
  actin cytoskeleton and, 25–46
 motor system, 377–87, 395–96
 risk of unwanted change and
  faulty practice, 383–84
  focal hand dystonia in musicians and, 383–84
 structural
  adult neurogenesis and, 223
 synaptic, 25, 35, 41–46
  actin cytoskeleton and, 25, 35, 41–46
  activity-dependent, 41–46
  adult neurogenesis and, 235–36
  C. elegans and, 451, 486–89
  learning and memory and, 45–46
  long-term, 43–45
  short-term, 41–43
 two steps of, 379–82, 396

  visually deprived visual system and, 387–96
Philosophy of science
  observation and control, 534–35
Postsynaptic density (PSD)
  vertebrate synaptogenesis and, 251–53, 259–61, 263
Prion disease
  axon degeneration in, 140
Progenitor
  adult neurogenesis and, 223, 229
Proteomics, 102
Psychiatric disorders
  locus coeruleus-norepinephrine and dopamine systems and, 432, 438–39

R

Radiation injury
  adult neurogenesis and, 240
Raven's Progressive Matrices, 4
Regeneration
  adult neurogenesis in mammalian CNS and, 223–42
  See also Adult neurogenesis in mammalian CNS
Remodeling
  actin, 30, 32–33, 43–46
  neuronal, 139
  synaptic, 30, 30, 32–33, 45–46
Retinal ganglion cells (RGC), 131–33, 137–38, 327–48, 517–19
  axon elimination to develop eye-specific pattern connections, 131–33
  directional-selective
   dendritic computation and, 517–19
  retinotopic maps and, 327–48
Retinotopic maps, 137–38, 327–48
  axon elimination and, 133, 135, 137–38
  axon extension and target overshoot, 333–35
  axon guidance and, 327–48
  Ephs and Ephrins, 327–48
   bidirectional signaling and, 338–48
   bifunctional action for ephrin-As, 337–38
   EphBs and ephrin-Bs and dorsoventral mapping, 339
   multiple actions and models, 342–43
  lateral-medial mapping by directed growth of interstitial branches, 338
  mechanisms for anterior-posterior (AP) branch specificity, 337–38
  opposing AP gradients of branch inhibitors and, 338–39

patterned activity and axon repellents, 344–46
retinal ganglion cells and, 328–29, 331
species differences, 343–44
RGM, 329, 331
RhoGTPases, 127, 141–42
axon retraction and, 141–42

S

Schizophrenia
genetic influences on brain structure and, 11–13
Semaphorins
mechanisms of axon retraction and, 142
topographical guidance and, 329
Sequential probability ratio test (SPRT), 418
Signal propagation, 357, 368–72
avalanche model, 368–70
propagation of firing rates, 368, 372
synfire chains, 369–72
Social brain, 112–13
Social cognition
autism and, 109, 112–13
Sound localization
dendritic computation and, 519–20
Stanford-Binet Intelligence Tests, 3
Stem cells, 223, 231–42
neural
neural progenitors, 223, 229
neuronal fate, 231–34
proliferation, 231–34, 236
specification, 231, 231–34, 236
Stereopsis
axon elimination and, 131–32
Stress
adult neurogenesis and, 239
Striate cortex
See Primary visual cortex
Stroke
cortical plasticity following, 385–87
Superior colliculus (SC), 328–30, 332
Sympathetic nervous system development, 191–211
artemin and, 194–95
axon growth and, 191–200
GDNF and, 194
growth inhibition, 198–99
growth mechanisms, 199
initiation of, 193
proximal axon extension and, 193–97
target innervation and, 197
dendrite formation and, 191, 204–11

bone morphogenetic protein and, 204–7
nerve growth factor and, 204–7
target-derived signals and, 204–7
diversification and, 208–11
cholinergic switch and, 195, 208–11
CNTF and, 208
neurotrophins and, 191–211
BDNF, 199, 202–4, 207, 210
CNTF, 208
GDNF, 194, 209
NGF, 197–211
NT-3, 195–98, 210
sympathetic neuron survival, 200–2
NGF signaling and, 201–2
synaptogenesis and, 207–8, 210–11
bone derived neurotrophic factor and, 207
nerve growth factor and, 207
target-derived prodeath factors and, 202–4
BDNF and, 202–4
NGF-TrkA signaling and, 203–4
p75 and, 202–4
Synapses, 25–46
actin cytoskeleton and, 25–46
elimination
axon pruning and, 130–31
vertebrate synaptogenesis and, 264–65
formation and maintenance of
actin and, 30–34, 32–33, 45–46
learning and memory and, 45–46
morphological changes in
actin cytoskeleton and stable, 43–46
actin cytoskeleton and transient, 41–43
remodeling of
actin and, 30, 32–33, 43–46
Synaptic cell-adhesion molecule (SynCAM)
vertebrate synaptogenesis and
target recognition in, 258–60
Synaptic plasticity, 25, 35, 41–46, 451, 486–89
activity-dependent, 41–46
learning in *C. elegans* and, 451, 486–89
glutamatergic signaling and, 486–87
learning and memory and, 45–46
long-term
actin remodeling and, 43–45
short-term
actin dynamics and, 41–43
Synaptic vesicle cycle, 25–46
actin cytoskeleton and, 25–46
reserve synaptic vesicle pool and, 36
synaptic vesicle delivery to active zone and, 36, 38

synaptic vesicle endocytosis and, 40
synaptic vesicle exocytosis and, 38–40
Synaptogenesis, 30–46, 207–8, 210–11, 251–66
   actin and, 30–32
   active zone and, 251–63
   activity-dependent regulation of, 264–66
      ubiquitin and, 265–66
   bone derived neurotrophic factor and, 207
   cell-adhesion molecules and, 256–60
      neuroligin, 258, 260
      synaptic cell-adhesion molecule (SynCAM), 258–60
   EprhinB and, 257–58
   fibroblast growth factor and, 255–56, 258
   maturation, 263
   mechanisms of vertebrate, 251–66
   membrane trafficking and, 251, 260
      postsynaptic assembly, 261–63
      presynaptic assembly, 260–63
   nerve growth factor and, 207
   neuronal activity-regulated pentraxin (Narp) and, 257–59
   postsynaptic density and (PSD), 251–53, 259–61, 263
   target recognition and, 254–57
   vertebrate, 251–66
   Wnt and, 255–56, 258
Synfire chains, 368–72

T

Target recognition
   synapse specificity and
      cell-adhesion molecules and, 256–57
      diffusible target-derived factors and, 255–56
Task disengagement
   locus coeruleus and, 403, 423–31
Thermosensation
   C. elegans and, 468–71
Topographic maps, 136–38, 327–48
   axon elimination and, 136–38
   graded topographical guidance molecules, 330–31
Transgenic animals
   dendritic computation and, 523–24

U

Ubiquitin-proteasome system (UPS)
   mechanisms of axon degeneration and, 143–47
   Parkinson's disease and, 55, 65, 71–72
Utility, 403, 423–31
   locus coeruleus and task performance, 403, 423–31
      anterior cingulate cortex (ACC), 403, 423–31
      obitofrontal cortex (OCC) and, 403, 423–31
   task disengagement and, 403, 423–31

V

Visual system
   axon elimination and, 127, 131–32
   color vision, 303–19
   connections between V1 and V2, 312–15
      feedback, 315
      feedforward, 312–15
   cytochrome oxidase (CO), 303–19
      histochemistry of and V1 organization, 307–10
      histochemistry of and V2 organization, 312–15
      magno, parvo, and konio geniculate signals to V2 and, 303–19
   development of
      axon elimination and, 127, 131–32
   directional-selective retinal ganglion cells
      dendritic computation and, 517–19
   eye-specific retinal connections, 131–33
   integration of color, form, and motion, 303–19
   lateral geniculate nucleus (LGN), 303–19
      color, form, and motion in physiology of, 310–12
      intracortical circuitry in, 307–10
   middle temporal visual area (MT or V5), 157–80
      acceleration coding in, 179
      attention modulation and, 176–77
      center-surround antagonism and, 157, 160, 165–68
      cortical feedforward inputs to, 161
      depth perception and, 174–75
      functional organization of, 162–64
      magnocellular inputs to, 157, 160
      single neuron sensitivity, 177–78
   primary visual cortex (V1), 303–19
   retinotopic mapping, 133, 135, 137–38
   retinotopic maps and, 327–48
   secondary visual cortex (V2), 303–19
      intracortical circuitry in, 316–17

stripe physiology of, 317–18
visual cortical hierarchy
   revisited, 318–19
   tripartite model, 303–5, 310, 312
visually deprived visual system
   cross-modal plasticity and, 387–96
   See also Middle temporal visual area; Primary visual cortex; Secondary visual cortex
Vomoronasal organ (VMO), 329

W

Wallerian degeneration, 127, 139–40, 143–49
   axon degeneration and, 127, 139–40, 143–49
   mechanisms, 143–49
Wechsler Adult Intelligence Scale (WAIS), 3, 16
Wechsler Intelligence Scale for Children (WISC), 3, 16
Williams syndrome
   genetic deletion and, 13
Wnt
   vertebrate synaptogenesis and, 255–56, 258

Cumulative Indexes

Contributing Authors, Volumes 19–28

A

Abarbanel HDI, 24:263–97
Abbott LF, 28:357–76
Abraham WC, 19:437–62
Albright TD, 25:339–79
Allada R, 24:1091–119
Amedi A, 28:377–401
Andersen RA, 20:299–326; 25:189–220
Angleson JK, 22:1–10
Arnold AP, 20:455–77
Aston-Jones G, 28:403–50

B

Bacon JP, 27:29–51
Bal T, 20:185–215
Baltuch G, 22:219–40
Barbe MF, 20:1–24
Barde Y-A, 19:289–317
Bargmann CI, 21:279–308
Baron-Cohen S, 28:109–26
Barres BA, 23:579–612
Basbaum AI, 23:777–811
Bate M, 19:545–75
Baylor DA, 24:779–805
Bear MF, 19:437–62
Belmonte MK, 28:109–26
Bernard C, 25:491–505
Best PJ, 24:459–86
Betz WJ, 22:1–10
Bi G-Q, 24:139–66
Bibbig A, 27:247–78
Bigbee AJ, 27:145–67

Blagburn JM, 27:29–51
Boehning D, 26:105–31
Bohn LM, 27:107–44
Bonhoeffer T, 24:1071–89
Bonini NM, 26:627–56
Bookheimer S, 25:151–88
Born RT, 28:157–89
Bottjer SW, 20:455–77
Bouchard TJ Jr, 21:1–24
Boussaoud D, 20:25–42
Bowers BJ, 24:845–67
Boyden ES, 27:581–609
Bradley DC, 20:299–326; 28:157–89
Broadie K, 19:545–75
Bruijn LI, 27:723–49
Buck LB, 19:517–44
Buhl EH, 27:247–78
Buneo CA, 25:189–220
Buonomano D, 21:149–86
Buonomano DV, 27:307–40
Burns ME, 24:779–805

C

Callaerts P, 20:479–528
Callaway EM, 21:47–74
Caminiti R, 20:25–42
Cannon SC, 19:141–64
Caron MG, 27:107–44
Caterina MJ, 24:487–517
Caughey B, 26:267–98
Chelazzi L, 27:611–47
Chemelli RM, 24:429–58

Chen C, 20:157–84
Chen K, 22:197–217
Chiba A, 19:545–75
Chklovskii DB, 27:369–92
Choi DW, 21:347–75
Christie BR, 19:165–86
Clark RE, 27:279–306
Cleveland DW, 19:187–217; 27:723–49
Cloutier J-F, 26:509–63
Cochilla AJ, 22:1–10
Cohen JD, 24:167–202; 28:403–50
Colbert CM, 19:165–86
Colby CL, 22:319–50
Collinge J, 24:519–50
Conlon P, 25:491–505
Connors BW, 27:393–418
Corey DP, 20:563–90
Cowan WM, 23:343–91; 24:551–600; 25:1–50
Craig A, 28:251–74
Craig AD, 26:1–30
Craighero L, 27:169–92
Cumming BG, 24:203–38
Curran T, 24:1005–39

D

Dacey DM, 23:743–75
Darnell RB, 24:239–62
Dasen JS, 24:327–55
David S, 26:411–40
Davis RL, 28:275–302

Dawson TM, 28:57–84
Dawson VL, 28:57–84
Dayan P, 26:381–10
Deadwyler SA, 20:217–44
DeAngelis GC, 24:203–38
de Bono M, 28:451–501
De Camilli P, 26:701–28
deCharms RC, 23:613–47
de Leon RD, 27:145–67
DeLuca NA, 19:265–87
DePaulo JR, 20:351–69
DiAntonio A, 27:223–46
Dickinson A, 23:473–500
Dijkhuizen P, 25:127–49
Dillon C, 28:25–54
Donoghue JP, 23:393–415
Douglas RJ, 27:419–51
Doupe AJ, 22:567–631
Dreyfuss G, 20:269–98
Dubnau J, 21:407–44
Dunwiddie TV, 24:31–55

E

Eagleson KL, 20:1–24
Eatock RA, 23:285–314
Edgerton VR, 27:145–67
Edwards RH, 20:125–56
Emery P, 24:1091–119
Emeson RB, 19:27–52

F

Feldman JL, 26:239–66
Fernald RD, 27:697–722
Ferster D, 23:441–71
Fink DJ, 19:265–87
Fischbach GD, 20:425–54
Fischbeck KH, 19:79–107
Fischer U, 20:269–98
Fishell G, 25:471–90
Fisher SE, 26:57–80
Fitch RH, 20:327–49
Flanagan JG, 21:309–45
Fortini ME, 26:627–56
Francis NJ, 22:541–66
Frankland PW, 21:127–48
Fregni F, 28:377–401
Friedrich RW, 24:263–97
Friston K, 25:221–50
Fritzsch B, 25:51–101

Fuchs AF, 24:981–1004
Fukuchi-Shimogori T, 26:355–80

G

Gaiano N, 25:471–90
Gainetdinov RR, 27:107–44
Garbers DL, 23:417–39
García-Añoveros J, 20:563–90
Garner CC, 28:251–74
Gegenfurtner KR, 26:181–206
Gehring WJ, 20:479–528
Geppert M, 21:75–95
Ghosh A, 25:127–49
Gibson AD, 23:417–39
Gingrich JR, 21:377–405
Ginty DD, 26:509–63; 28:191–222
Glass JD, 19:1–26
Glebova NO, 28:191–222
Glimcher PW, 26:133–79
Glorioso JC, 19:265–87
Goda Y, 28:25–54
Goedert M, 24:1121–59
Goins WF, 19:265–87
Goldberg JL, 23:579–612
Goldberg ME, 22:319–50
Goldstein LSB, 23:39–72
González-Scarano F, 22:219–40
Goodman CS, 19:341–77
Goodwin AW, 27:53–77
Gordon JA, 27:193–222
Gottlieb DI, 25:381–407
Greenberg ME, 19:463–89
Greenspan RJ, 27:79–105
Griffin JW, 21:187–226
Grill-Spector K, 27:649–77
Grimwood PD, 23:649–711
Grove EA, 26:355–80
Gudermann T, 20:395–423

H

Halder G, 20:479–528
Hampson RE, 20:217–44
Harter DH, 23:343–91
Hatten ME, 22:511–39; 28:89–108

Hatton GI, 20:371–93
Häusser M, 28:503–32
He Z, 27:341–68
Heintz N, 28:89–108
Hemmati-Brivanlou A, 20:43–60
Hen R, 27:193–222
Hendry SHC, 23:127–53
Hensch TK, 27:549–79
Herrup K, 20:61–90
Hicke L, 27:223–46
Hickey WF, 25:537–62
Hildebrand JG, 20:591–629
Hlavin ML, 21:97–125
Ho TW, 21:187–226
Hobert O, 26:207–38
Honarpour N, 23:73–87
Horton JC, 28:303–26
Howe CL, 24:1217–81
Huang EJ, 24:677–736
Huber AB, 26:509–63
Hyman SE, 25:1–50

I

Ikeda S, 26:657–700
Insel TR, 27:697–722
Ip NY, 19:491–515

J

Jamison KR, 20:351–69
Jan L, 23:531–56
Jan LY, 20:91–123
Jan Y-N, 20:91–123; 23:531–56
Jessell TM, 22:261–94
Johnson PB, 20:25–42
Johnson RT, 19:1–26
Johnston D, 19:165–86
Jones EG, 22:49–103; 23:1–37
Joyner AL, 24:869–96
Julius D, 24:487–517

K

Kamiguchi H, 21:97–125
Kandel ER, 23:343–91
Kaplan JM, 21:279–308
Karschin A, 23:89–125

Kastner S, 23:315–41
Katoh A, 27:581–609
Katz LC, 22:295–318
Kauer JS, 24:963–79
Keshishian H, 19:545–75
Kevrekidis IG, 28:533–63
Kida S, 21:127–48
King DP, 23:713–42
Kiper DC, 26:181–206
Knowlton BJ, 25:563–93
Kogan JH, 21:127–48
Koh JY, 21:347–75
Kolodkin AL, 26:509–63
Konishi M, 26:31–55
Kopan R, 26:565–97
Kopnisky KL, 25:1–50
Koprivica V, 27:341–68
Korsmeyer SJ, 20:245–67
Koulakov AA, 27:369–92
Kuemerle B, 20:61–90
Kuhl PK, 22:567–631

L

Lacroix S, 26:411–40
Lai CSL, 26:57–80
Landis SC, 22:541–66
Lansbury PT Jr, 26:267–98
Laurent G, 24:263–97
LeBeau FEN, 27:247–78
LeDoux JE, 23:155–84
Lee KJ, 22:261–94
Lee MK, 19:187–217
Lee VM-Y, 24:1121–59
Lefkowitz RJ, 27:107–44
Lemaire P, 27:453–85
Lemke G, 24:87–105
Lemmon V, 21:97–125
Lester HA, 23:89–125
Levitt P, 20:1–24; 25:409–32
Lewin GR, 19:289–317
Lewis DA, 25:409–32
Lichtman JW, 22:389–442
Liu A, 24:869–96
Liu Y, 20:125–56
Lo DC, 22:295–318
Logothetis NK, 19:577–621
London M, 28:503–32
Long MA, 27:393–418
Lu B, 23:531–56
Luo L, 28:127–56

M

MacDermott AB, 22:443–85
MacKinnon DF, 20:351–69
Magee JC, 19:165–86
Malach R, 27:649–77
Malenka RC, 23:185–215; 25:103–26
Malinow R, 25:103–26
Marder E, 21:25–45
Maren S, 24:897–931
Maricq AV, 28:451–501
Marín O, 26:441–83
Martin KAC, 27:419–51
Martin R, 25:491–505
Martin SJ, 23:649–711
Martinez S, 21:445–77
Masino SA, 24:31–55
Masland RH, 23:249–84
Mason P, 24:737–77
Matthews G, 19:219–33
Mauk MD, 27:307–40
Mayeux R, 26:81–104
McAllister AK, 22:295–318
McCormick DA, 20:185–215
McEwen BS, 22:105–22
McGaugh JL, 27:1–28
McGue M, 21:1–24
McInnes RR, 26:657–700
McKhann GM, 21:187–226
McLaughlin T, 28:327–55
McNamara JO, 22:175–95
Meaney MJ, 24:1161–92
Meinertzhagen IA, 27:453–85
Melton D, 20:43–60
Menzel R, 19:379–404
Merabet LB, 28:377–401
Merry DE, 20:245–67
Merzenich MM, 21:149–86
Michael WM, 20:269–98
Miesenböck G, 28:533–63
Mignot E, 25:283–313
Miller EK, 24:167–202
Miller KD, 23:441–71
Miller S, 20:327–49
Miller TM, 27:723–49
Minai A, 24:459–86
Ming G-l, 28:223–50
Mitchell GS, 26:239–66
Mobley WC, 24:1217–81
Mogil JS, 23:777–811
Mombaerts P, 22:487–509

Monaco AP, 26:57–80
Moore DJ, 28:57–84
Morris RGM, 23:649–711
Mueller BK, 22:351–88
Müller U, 19:379–404
Murthy VN, 26:701–28
Musunuru K, 24:239–62

N

Nakielny S, 20:269–98
Nattie EE, 26:239–66
Newsome WT, 21:227–77
Nicola SM, 23:185–215
Nijhawan D, 23:73–87
Nishina PM, 26:657–700
Noebels JL, 26:599–625

O

O'Donnell WT, 25:315–38
Okamura Y, 27:453–85
Oksenberg JR, 25:491–505
Olanow CW, 22:123–44
O'Leary DDM, 28:127–56; 327–55
Olshausen BA, 24:1193–1215
Olson CR, 26:331–54
Orr HT, 23:217–47

P

Pacione LR, 26:657–700
Packard MG, 25:563–93
Parker AJ, 21:227–77
Pascual-Leone A, 28:377–401
Paulson HL, 19:79–107
Pfaff SL, 25:251–81
Polleux F, 25:127–49
Poo M-M, 24:139–66
Pouget A, 26:381–410
Premont RT, 27:107–44
Price DL, 21:479–505
Puelles L, 21:445–77

Q

Quinn WG, 24:1283–309

R

Rabinovich MI, 24:263–97
Radcliffe RA, 24:845–67
Raisman G, 20:529–62
Rajan K, 28:357–376
Raviola E, 23:249–84
Raymond JL, 27:581–609
Read HL, 23:501–29
Reichardt LF, 24:677–736
Reid RC, 23:127–53
Reyes A, 24:653–75
Reynolds JH, 27:611–47
Rice DS, 24:1005–39
Ridd M, 22:197–217
Rizzolatti G, 27:169–92
Robinson FR, 24:981–1004
Roder J, 21:377–405
Role LW, 22:443–85
Romo R, 24:107–37
Rosbash M, 24:1091–119
Rosen KM, 20:425–54
Rosenfeld MG, 24:327–55
Roses AD, 19:53–77
Ross ME, 24:1041–70
Rougon G, 26:207–38
Roy RR, 27:145–67
Rubel EW, 25:51–101
Rubenstein JLR, 21:445–77; 26:441–83
Rubin LL, 22:11–28

S

Sala C, 24:1–29
Salinas E, 24:107–37
Sanes JN, 23:393–415
Sanes JR, 22:389–442
Saper CB, 25:433–69
Schall JD, 22:241–59
Scheiffele P, 26:485–508
Schöneberg T, 20:395–423
Schreiner CE, 23:501–29
Schultz G, 20:395–423
Schultz W, 23:473–500
Schuman EM, 24:299–325
Schwartz AB, 27:487–507
Scott MP, 24:385–428
Searle JR, 23:557–78
Segal RA, 19:463–89; 26:299–330
Selkoe DJ, 26:565–97
Sheinberg DL, 19:577–621
Sheng M, 24:1–29
Shepherd GM, 20:591–629
Shih JC, 22:197–217
Shimamura K, 21:445–77
Shirasaki R, 25:251–81
Shooter EM, 24:601–29
Siegelbaum SA, 19:235–63; 22:443–85
Silva AJ, 21:127–48
Simerly RB, 25:507–36
Simoncelli EP, 24:1193–215
Simpson L, 19:27–52
Sincich LC, 28:303–26
Sinton CM, 24:429–58
Sisodia SS, 21:479–505
Snyder LH, 20:299–326
Snyder SH, 26:105–31
Sofroniew MV, 24:1217–81
Song H, 28:223–50
Squire LR, 27:279–306
Srinivasan MV, 27:679–96
Staddon JM, 22:11–28
Stark CEL, 27:279–306
Steinman L, 25:491–505
Steward O, 24:299–325
Stoner GR, 25:339–79
Stopfer M, 24:263–97
Strittmatter WJ, 19:53–77
Südhof TC, 21:75–95; 24:933–62; 27:509–47
Surmeier DJ, 23:185–215
Sutter ML, 23:501–29
Szego MJ, 26:657–700

T

Taheri S, 25:283–313
Takahashi JS, 23:713–42; 24:1091–119
Tallal P, 20:327–49
Tanaka K, 19:109–39
Tanji J, 24:631–51
Tatton WG, 22:123–44
Tennissen AM, 24:807–43
Thompson KG, 22:241–59
Thompson PM, 28:1–23
Tillakaratne NJK, 27:145–67
Toga AW, 28:1–23
Tonegawa S, 20:157–84
Traub RD, 27:247–78
Trojanowski JQ, 24:1121–59
Tully T, 21:407–44

U

Ungerleider LG, 23:315–41

V

Vanderhaeghen P, 21:309–45
Vogels TP, 28:357–76
Volkovskii A, 24:263–97

W

Waddell S, 24:1283–309
Waites CL, 28:251–74
Walsh CA, 24:1041–70
Wandell BA, 22:145–73
Wang X, 23:73–87
Warren ST, 25:315–38
Watts AG, 24:357–84
Wechsler-Reya R, 24:385–428
Wehner JM, 24:845–67
West AB, 28:57–84
Wheat HE, 27:53–77
White AM, 24:459–86
White FJ, 19:405–36
White J, 24:963–79
Whitford KL, 25:127–49
Whitney KD, 22:175–95
Whittington MA, 27:247–78
Williams KC, 25:537–62
Willie JT, 24:429–58
Wise RA, 19:319–40
Wise SP, 20:25–42
Wolpaw JR, 24:807–43
Wong ROL, 22:29–47

X

Xing J, 20:299–326

Y

Yamasaki M, 21:97–125
Yanagisawa M, 24:429–58

Yancopoulos GD, 19:491–515
Yang Z, 23:39–72
Yu L, 23:777–811
Yuste R, 24:1071–89

Z

Zador A, 23:613–47
Zagotta WN, 19:235–63
Zeitzer JM, 25:283–313

Zeki S, 24:57–86
Zemel RS, 26:381–410
Zhang S, 27:679–96
Zoghbi HY, 23:217–47

Chapter Titles, Volumes 19–28

Affect and Emotion

Neurobiology of Pavlovian Fear Conditioning	S Maren	24:897–931
An Integrative Theory of Locus Coeruleus-Norepinephrine Function: Adaptive Gain and Optimal Performance	G Aston-Jones, JD Cohen	28:403–50

Attention

Novel Neural Modulators	D Boehning, SH Snyder	26:105–31
Attentional Modulation of Visual Processing	JH Reynolds, L Chelazzi	27:611–47

Auditory System

Adaptation in Hair Cells	RA Eatock	23:285–314
Modular Organization of Frequency Integration in Primary Auditory Cortex	CE Schreiner, HL Read, ML Sutter	23:501–29
Auditory System Development: Primary Auditory Neurons and Their Targets	EW Rubel, B Fritzsch	25:51–101
Coding of Auditory Space	M Konishi	26:31–55

Autonomic Nervous System

The Central Autonomic Nervous System: Conscious Visceral Perception and Autonomic Pattern Generation	CB Saper	25:433–69

Basal Ganglia

Dopaminergic Modulation of Neuronal Excitability in the Striatum	SM Nicola, DJ Surmeier, RC Malenka	23:185–215
Learning and Memory Functions of the Basal Ganglia	MG Packard, BJ Knowlton	25:563–93

Behavioral Neuroscience

Emotion Circuits in the Brain	JE LeDoux	23:155–84
Quantitative Genetics and Mouse Behavior	JM Wehner, RA Radcliffe, BJ Bowers	24:845–67
The Neurobiology of Visual-Saccadic Decision Making	PW Glimcher	26:133–79
Genetic Approaches to the Study of Anxiety	JA Gordon, R Hen	27:193–222
How the Brain Processes Social Information: Searching for the Social Brain	TR Insel, RD Fernald	27:697–722
Neuronal Substrates of Complex Behaviors in *C. elegans*	M de Bono, AV Maricq	28:451–501

Cerebellum

The Role of the Cerebellum in Voluntary Eye Movements	FR Robinson, AF Fuchs	24:981–1004
Cerebellum-Dependent Learning: The Role of Multiple Plasticity Mechanisms	ES Boyden, A Katoh, JL Raymond	27:581–609

Cerebral Cortex

Inferotemporal Cortex and Object Vision	K Tanaka	19:109–39
Patterning and Specification of the Cerebral Cortex	P Levitt, MF Barbe, KL Eagleson	20:1–24
Sleep and Arousal: Thalamocortical Mechanisms	DA McCormick, T Bal	20:185–215
Multimodal Representation of Space in the Posterior Parietal Cortex and Its Use in Planning Movements	RA Andersen, LH Snyder, DC Bradley, J Xing	20:299–326
Local Circuits in Primary Visual Cortex of the Macaque Monkey	EM Callaway	21:47–74
Cortical and Subcortical Contributions to Activity-Dependent Plasticity in Primate Somatosensory Cortex	EG Jones	23:1–37

Touch and Go: Decision-Making Mechanisms in Somatosensation	R Romo, E Salinas	24:107–37
An Integrative Theory of Prefrontal Cortex Function	EK Miller, JD Cohen	24:167–202
Sequential Organization of Multiple Movements: Involvement of Cortical Motor Areas	J Tanji	24:631–51
Molecular Control of Cortical Dendrite Development	KL Whitford, P Dijkhuizen, F Polleux, A Ghosh	25:127–49
Intentional Maps in Posterior Parietal Cortex	RA Andersen, CA Buneo	25:189–20
Neuronal Circuits of the Neocortex	RJ Douglas, KAC Martin	27:419–51

Circadian and other Rhythms

Molecular Genetics of Circadian Rhythms in Mammals	DP King, JS Takahashi	23:713–42

Clinical Neuroscience

Human Immunodeficiency Virus and the Brain	JD Glass, RT Johnson	19:1–26
Apolipoprotein E and Alzheimer's Disease	WJ Strittmatter, AD Roses	19:53–77
Trinucleotide Repeats in Neurogenetic Disorders	HL Paulson, KH Fischbeck	19:79–107
Sodium Channel Defects in Myotonia and Periodic Paralysis	SC Cannon	19:141–64
Addictive Drugs and Brain Stimulation Reward	RA Wise	19:319–40
The Role of Vesicular Transport Proteins in Synaptic Transmission and Neural Degeneration	Y Liu, RH Edwards	20:125–56
Genetics of Manic Depressive Illness	DF MacKinnon, KR Jamison, JR DePaulo	20:351–69
Human Autoimmune Neuropathies	TW Ho, GM McKhann, JW Griffin	21:187–226
Zinc and Brain Injury	DW Choi, JY Koh	21:347–75
Mutant Genes in Familial Alzheimer's Disease and Transgenic Models	DL Price, SS Sisodia	21:479–505
Etiology and Pathogenesis of Parkinson's Disease	CW Olanow, WG Tatton	22:123–44
Microglia as Mediators of Inflammatory and Degenerative Diseases	F González-Scarano, G Baltuch	22:219–40
Glutamine Repeats and Neurodegeneration	HY Zoghbi, HT Orr	23:217–47

The Emergence of Modern Neuroscience: Some Implications for Neurology and Psychiatry	WM Cowan, DH Harter, ER Kandel	23:343–91
Paraneoplastic Neurologic Disease Antigens: RNA-Binding Proteins and Signaling Proteins in Neuronal Degeneration	K Musunuru, RB Darnell	24:239–62
The Developmental Biology of Brain Tumors	R Wechsler-Reya, MP Scott	24:385–428
Prion Diseases of Humans and Animals: Their Causes and Molecular Basis	J Collinge	24:519–50
Neurodegenerative Tauopathies	VM-Y Lee, M Goedert, JQ Trojanowski	24:1121–59
Nerve Growth Factor Signaling, Neuroprotection, and Neural Repair	MV Sofroniew, CL Howe, WC Mobley	24:1217–81
The Human Genome Project and Its Impact on Psychiatry	WM Cowan, KL Kopnisky, SE Hyman	25:1–50
Beyond Phrenology: What Can Neuroimaging Tell Us About Distributed Circuitry?	K Friston	25:221–50
A Decade of Molecular Studies of Fragile X Syndrome	WT O'Donnell, ST Warren	25:315–38
Schizophrenia as a Disorder of Neurodevelopment	DA Lewis, P Levitt	25:409–32
Multiple Sclerosis: Deeper Understanding of Its Pathogenesis Reveals New Targets for Therapy	L Steinman, R Martin, C Bernard, P Conlon, JR Oksenberg	25:491–505
Epidemiology of Neurodegeneration	R Mayeux	26:81–104
Molecular Approaches to Spinal Cord Repair	S David, S Lacroix	26:411–40
The Biology of Epilepsy Genes	JL Noebels	26:599–625
Unraveling the Mechanisms Involved in Motor Neuron Degeneration in ALS	LI Bruijn, TM Miller, DW Cleveland	27:723–49
Molecular Pathophysiology of Parkinson's Disease	DJ Moore, AB West, VL Dawson, TM Dawson	28:57–84
Autism: A Window Onto the Development of the Social and the Analytic Brain	S Baron-Cohen, MK Belmonte	28:109–26

Computational Approaches

From Biophysics to Models of Network Function	E Marder	21:25–45
Odor Encoding as an Active, Dynamical Process: Experiments, Computation and Theory	G Laurent, M Stopfer, RW Friedrich, MI Rabinovich, A Volkovskii, HDI Abarbanel	24:263–97
Natural Image Statistics and Neural Representation	EP Simoncelli, BA Olshausen	24:1193–1215
Inference and Computation with Population Codes	A Pouget, P Dayan, R Zemel	26:381–410
Maps in the Brain: What Can We Learn from Them?	DB Chklovskii, AA Koulakov	27:369–92
Neural Network Dynamics	TP Vogels, K Rajan, LF Abbott	28:357–76
Dendritic Computation	M London, M Häusser	28:503–32

Cytoskeleton and Axonal Transport

Neuronal Intermediate Filaments	MK Lee, DW Cleveland	19:187–217
Neurodegenerative Tauopathies	VM-Y Lee, M Goedert, JQ Trojanowski	24:1121–59

Degeneration and Repair

The Nogo Signaling Pathway for Regeneration Block	Z He, V Koprivica	27:341–68
The Actin Cytoskeleton: Integrating Form and Function at the Synapse	C Dillon, Y Goda	28:25–54
Axon Retraction and Degeneration in Development and Disease	L Luo, DDM O'Leary	28:127–56
Growth and Survival Signals Controlling Sympathetic Nervous System Development	NO Glebova, DD Ginty	28:191–222

Developmental Neurobiology

Mechanisms and Molecules that Control Growth Cone Guidance	CS Goodman	19:341–77
The Drosophila Neuromuscular Junction: A Model System for Studying Synaptic Development and Function	H Keshishian, K Broadie, A Chiba, M Bate	19:545–75

Vertebrate Neural Induction	A Hemmati-Brivanlou, D Melton	20:43–60
Pax-6 in Development and Evolution	P Callaerts, G Halder, WJ Gehring	20:479–528
Adhesion Molecules and Inherited Diseases of the Human Nervous System	ML Hlavin, H Kamiguchi, M Yamasaki, V Lemmon	21:97–125
Regionalization of the Prosencephalic Neural Plate	JLR Rubenstein, K Shimamura, S Martinez, L Puelles	21:445–77
Retinal Waves and Visual System Development	ROL Wong	22:29–47
The Specification of Dorsal Cell Fates in the Vertebrate Central Nervous System	KJ Lee, TM Jessell	22:261–94
Growth Cone Guidance: First Steps Towards a Deeper Understanding	BK Mueller	22:351–88
Development of the Vertebrate Neuromuscular Junction	JR Sanes, JW Lichtman	22:389–442
Central Nervous System Neuronal Migration	ME Hatten	22:511–39
Cellular and Molecular Determinants of Sympathetic Neuron Development	NJ Francis, SC Landis	22:541–66
Control of Cell Divisions in the Nervous System: Symmetry and Asymmetry	B Lu, L Jan, Y-N Jan	23:531–56
The Relationship Between Neuronal Survival and Regeneration	JL Goldberg, BA Barres	23:579–612
Glial Control of Neuronal Development	G Lemke	24:87–105
Signaling and Transcriptional Mechanisms in Pituitary Development	MG Rosenfeld, JS Dasen	24:327–55
The Developmental Biology of Brain Tumors	R Wechsler-Reya, MP Scott	24:385–428
Early Days of the Nerve Growth Factor Proteins	EM Shooter	24:601–29
Neurotrophins: Roles in Neuronal Development and Function	EJ Huang, LF Reichardt	24:677–736
Early Anterior/Posterior Patterning of the Midbrain and Cerebellum	A Liu, AL Joyner	24:869–96
Role of the Reelin Signaling Pathway in Central Nervous System Development	DS Rice, T Curran	24:1005–39
Human Brain Malformations and Their Lessons for Neuronal Migration	ME Ross, CA Walsh	24:1041–70
Maternal Care, Gene Expression, and the Transmission of Individual Differences in Stress Reactivity Across Generations	MJ Meaney	24:1161–92

Nerve Growth Factor Signaling, Neuroprotection, and Neural Repair	MV Sofroniew, CL Howe, WC Mobley	24:1217–81
Molecular Control of Cortical Dendrite Development	KL Whitford, P Dijkhuizen, F Polleux, A Ghosh	25:127–49
Transcriptional Codes and the Control of Neuronal Identity	R Shirasaki, SL Pfaff	25:251–81
Large-Scale Sources of Neural Stem Cells	DI Gottlieb	25:381–407
The Role of Notch in Promoting Glial and Neural Stem Cell Fates	N Gaiano, G Fishell	25:471–90
Generating the Cerebral Cortical Area Map	EA Grove, T Fukuchi-Shimogori	26:355–80
Cell Migration in the Forebrain	O Marín, JLR Rubenstein	26:441–83
Cell-Cell Signaling During Synapse Formation in the CNS	P Scheiffele	26:485–508
Signaling at the Growth Cone: Ligand-Receptor Complexes and the Control of Axon Growth and Guidance	AB Huber, AL Kolodkin, DD Ginty, J-F Cloutier	26:509–63
Critical Period Regulation	TK Hensch	27:549–79
Molecular Gradients and Development of Retinotopic Maps	T McLaughlin, DDM O'Leary	28:327–55

Glia, Schwann Cells, and Extracellular Matrix

Glial Control of Neuronal Development	G Lemke,	24:87–105
The Role of Notch in Promoting Glial and Neural Stem Cell Fates	N Gaiano, G Fishell	25:471–90
Central Nervous System Damage, Monocytes and Macrophages, and Neurological Disorders in AIDS	KC Williams, WF Hickey	25:537–62

Higher Cognitive Functions

Monoamine Oxidase: From Genes to Behavior	JC Shih, K Chen, M Ridd	22:197–217
Mechanisms of Visual Attention in the Human Cortex	S Kastner, LG Ungerleider	23:315–41
Neuronal Coding of Prediction Errors	W Schultz, A Dickinson	23:473–500
Consciousness	JR Searle	23:557–78

The Mirror-Neuron System	G Rizzolatti, L Craighero	27:169–92

Hippocampus

Spatial Processing in the Brain: The Activity of Hippocampal Place Cells	PJ Best, AM White, A Minai	24:459–86
Selectivity in Neurotrophin Signaling: Theme and Variations	RA Segal	26:299–330

History of Neuroscience

Spatial Processing in the Brain: The Activity of Hippocampal Place Cells	PJ Best, AM White, A Minai	24:459–86
Viktor Hamburger and Rita Levi-Montalcini: The Path to the Discovery of Nerve Growth Factor	WM Cowan	24:551–600
Early Days of the Nerve Growth Factor Proteins	EM Shooter	24:601–29
Flies, Genes, and Learning	S Waddell, WG Quinn	24:1283–309
The Human Genome Project and Its Impact on Psychiatry	WM Cowan, KL Kopnisky, SE Hyman	25:1–50

Ion Channels

Structure and Function of Cyclic Nucleotide-Gated Channels	WN Zagotta, SA Siegelbaum	19:235–63
Cloned Potassium Channels from Eukaryotes and Prokaryotes	LY Jan, YN Jan	20:91–123
Gain of Function Mutants: Ion Channels and G Protein-Coupled Receptors	HA Lester, A Karschin	23:89–125

Language

Birdsong and Human Speech: Common Themes and Mechanisms	AJ Doupe, PK Kuhl	22:567–631
Functional MRI of Language: New Approaches to Understanding the Cortical Organization of Semantic Processing	S Bookheimer	25:151–88
Deciphering the Genetic Basis of Speech and Language Disorders	SE Fisher, CS Lai, AP Monaco	26:57–80

Learning and Memory

Learning and Memory in Honeybees: From Behavior to Neural Substrates	R Menzel, U Müller	19:379–404

Neurobiology of Speech Perception	RH Fitch, S Miller, P Tallal	20:327–49
CREB and Memory	AJ Silva, JH Kogan, PW Frankland, S Kida	21:127–48
Gene Discovery in Drosophila: New Insights for Learning and Memory	J Dubnau, T Tully	21:407–44
Synaptic Plasticity and Memory: An Evaluation of the Hypothesis	SJ Martin, PD Grimwood, RGM Morris	23:649–711
An Integrative Theory of Prefrontal Cortex Function	EK Miller, JD Cohen	24:167–202
Spatial Processing in the Brain: The Activity of Hippocampal Place Cells	PJ Best, AM White, A Minai	24:459–86
Neurobiology of Pavlovian Fear Conditioning	SA Maren	24:897–931
Learning and Memory Functions of the Basal Ganglia	MG Packard, BJ Knowlton	25:563–93
The Amygdala Modulates the Consolidation of Memories of Emotionally Arousing Experiences	JL McGaugh	27:1–28
The Medial Temporal Lobe	LR Squire, CEL Stark, RE Clark	27:279–306
Olfactory Memory Formation in *Drosophila*: From Molecular to Systems Neuroscience	RL Davis	28:275–302

Miscellaneous

An Urge to Explain the Incomprehensible: Geoffrey Harris and the Discovery Control of the Pituitary Gland	G Raisman	20:529–62
The Cell Biology of the Blood-Brain Barrier	LL Rubin, JM Staddon	22:11–28
Making Brain Connections: Neuroanatomy and the Work of TPS Powell, 1923–1966	EG Jones	22:49–103
The Human Genome Project and Its Impact on Psychiatry	WM Cowan, KL Kopnisky, SE Hyman	25:1–50
Wired for Reproduction: Organization and Development of Sexually Dimorphic Circuits in the Mammalian Forebrain	RB Simerly	25:507–36
Electrical Synapses in the Mammalian Brain	BW Connors, MA Long	27:393–418

Molecular Neuroscience

RNA Editing	L Simpson, RB Emeson	19:27–52

Physiology of the Neurotrophins	GR Lewin, Y-A Barde	19:289–317
Intracellular Signaling Pathways Activated by Neurotrophic Factors	RA Segal, ME Greenberg	19:463–89
The Neurotrophins and CNTF: Two Families of Collaborative Neurotrophic Factors	NY Ip, GD Yancopoulos	19:491–515
RNA Transport	S Nakielny, U Fischer, WM Michael, G Dreyfuss	20:269–98
ARIA: A Neuromuscular Junction Neuregulin	GD Fischbach, KM Rosen	20:425–54
Rab3 and Synaptotagmin: The Yin and Yang of Synaptic Membrane Fusion	M Geppert, TC Südhof	21:75–95
Inducible Gene Expression in the Nervous System of Transgenic Mice	JR Gingrich, J Roder	21:377–405
Microtubule-Based Transport Systems in Neurons: The Roles of Kinesins and Dyneins	LSB Goldstein, Z Yang	23:39–72
Apoptosis in Neural Development and Disease	D Nijhawan, N Honarpour, X Wang	23:73–87
PDZ Domains and the Organization of Supramolecular Complexes	C Sala, MH Sheng	24:1–29
Protein Synthesis at Synaptic Sites on Dendrites	O Steward, E Schuman	24:299–325
Signaling and Transcriptional Mechanisms in Pituitary Development	MG Rosenfeld, JS Dasen	24:327–55
Prion Diseases of Humans and Animals: Their Causes and Molecular Basis	J Collinge	24:519–50
Early Days of the Nerve Growth Factor Proteins	EM Shooter	24:601–29
Neurotrophins: Roles in Neuronal Development and Function	EJ Huang, LF Reichardt	24:677–736
α-Latrotoxin and its Receptors: Neurexins and CIRL/Latrophilins	TC Südhof	24:933–62
AMPA Receptor Trafficking and Synaptic Plasticity	R Malinow, RC Malenka	25:103–26
Transcriptional Codes and the Control of Neuronal Identity	R Shirasaki, SL Pfaff	25:251–81
New Insights into the Diversity and Function of Neuronal Immunoglobulin Superfamily Molecules	G Rougon, O Hobert	26:207–38

Protofibrils, Pores, Fibrils, and Neurodegeneration: Separating the Responsible Protein Aggregates from the Innocent Bystanders	B Caughey, PT Lansbury Jr.	26:267–98
Cell-Cell Signaling During Synapse Formation in the CNS	P Scheiffele	26:485–508
Notch and Presenilin: Regulated Intramembrane Proteolysis Links Development and Degeneration	DJ Selkoe, R Kopan	26:565–97
The Biology of Epilepsy Genes	JL Noebels	26:599–625

Motor Systems

Premotor and Parietal Cortex: Corticocortical Connectivity and Combinatorial Computations	SP Wise, D Boussaoud, PB Johnson, R Caminiti	20:25–42
Plasticity and Primary Motor Cortex	JN Sanes, JP Donoghue	23:393–415
Neuropeptides and the Integration of Motor Responses to Dehydration	AG Watts	24:357–84
Sequential Organization of Multiple Movements: Involvement of Cortical Motor Areas	J Tanji	24:631–51
The Role of the Cerebellum in Voluntary Eye Movements	FR Robinson, AF Fuchs	24:981–1004
Cortical Neural Prosthetics	AB Schwartz	27:487–507

Neural Networks

The Significance of Neuronal Ensemble Codes During Behavior and Cognition	SA Deadwyler, RE Hampson	20:217–44
Neural Representation and the Cortical Code	RC deCharms, A Zador	23:613–47
Breathing: Rhythmicity, Plasticity, Chemosensitivity	JL Feldman, GS Mitchell, EE Nattie	26:239–66
Cellular Mechanisms of Neuronal Population Oscillations in the Hippocampus In Vitro	RD Traub, A Bibbig, FEN LeBeau, EH Buhl, MA Whittington	27:247–78
The Neural Basis of Temporal Processing	MD Mauk, DV Buonomano	27:307–40

Neuroethology

Developmental Plasticity in Neural Circuits for a Learned Behavior	SW Bottjer, AP Arnold	20:455–77

Neurogenetics

The Compartmentalization of the Cerebellum	K Herrup, B Kuemerle	20:61–90
Bcl-2 Gene Family in the Nervous System	DE Merry, SJ Korsmeyer	20:245–67
Genetic and Environmental Influences on Human Behavioral Differences	M McGue, TJ Bouchard Jr.	21:1–24
Signal Transduction in the Caenorhabditis elegans Nervous System	CI Bargmann, JM Kaplan	21:279–308
Quantitative Genetics and Mouse Behavior	JM Wehner, RA Radcliffe, BJ Bowers	24:845–67
Human Brain Malformations and their Lessons for Neuronal Migration	ME Ross, CA Walsh	24:1041–70
Flies, Genes, and Learning	S Waddell, WG Quinn	24:1283–1309
Selectivity in Neurotrophin Signaling: Theme and Variations	RA Segal	26:299–330
Human Neurodegenerative Disease Modeling Using Drosophila	NM Bonini, ME Fortini	26:627–56
E Pluribus Unum, Ex Uno Plura: Quantitative and Single-Gene Perspectives on the Study of Behavior	RJ Greenspan	27:79–105
The Neurobiology of the Ascidian Tadpole Larva: Recent Developments in an Ancient Chordate	IA Meinertzhagen, P Lemaire, Y Okamura	27:453–85
Genetics of Brain Structure and Intelligence	AW Toga, PM Thompson	28:1–23

Neuronal Membranes

Influence of Dendritic Conductances on the Input-Output Properties of Neurons	A Reyes	24:653–75

Neuronal Plasticity

Active Properties of Neuronal Dendrites	D Johnston, JC Magee, CM Colbert, BR Christie	19:165–86

Molecular Genetic Analysis of Synaptic Plasticity, Activity-Dependent Neural Development, Learning, and Memory in the Mammalian Brain	C Chen, S Tonegawa	20:157–84
Function-Related Plasticity in Hypothalamus	GI Hatton	20:371–93
Cortical Plasticity: From Synapses to Maps	D Buonomano, MM Merzenich	21:149–86
Stress and Hippocampal Plasticity	BS McEwen	22:105–22
Neurotrophins and Synaptic Plasticity	AK McAllister, LC Katz, DC Lo	22:295–318
Synaptic Modification by Correlated Activity: Hebb's Postulate Revisited	G-Q Bi, M-M Poo	24:139–66
Activity-Dependent Spinal Cord Plasticity in Health and Disease	JR Wolpaw, AM Tennissen	24:807–43
Morphological Changes in Dendritic Spines Associated with Long-Term Synaptic Plasticity	R Yuste, T Bonhoeffer	24:1071–89

Neuropeptides

Odor Encoding as an Active, Dynamical Process: Experiments, Computation and Theory	G Laurent, M Stopfer, RW Friedrich, MI Rabinovich, A Volkovskii, HDI Abarbanel	24:263–97
Neuropeptides and the Integration of Motor Responses to Dehydration	AG Watts	24:357–84
To Eat or Sleep? The Role of Orexin in Coordination of Feeding and Arousal	JT Willie, RM Chemelli, CM Sinton, M Yanagisawa	24:429–58
Imaging and Coding in the Olfactory System	JS Kauer, J White	24:963–79

Neuroscience Techniques

Gene Transfer to Neurons Using Herpes Simplex Virus-Based Vectors	DJ Fink, NA DeLuca, WF Goins, JC Glorioso	19:265–87
Imaging and Coding in the Olfactory System	JS Kauer, J White	24:963–79
Large-Scale Genomic Approaches to Brain Development and Circuitry	ME Hatten, N Heintz	28:89–108
Optical Imaging and Control of Genetically Designated Neurons in Functioning Circuits	G Miesenböck, IG Kevrekidis	28:533–63

Neurotrophins

Early Days of the Nerve Growth Factor Proteins	EM Shooter	24:601–29
Neurotrophins: Roles in Neuronal Development and Function	EJ Huang, LF Reichardt	24:677–736

Olfaction/Taste

Information Coding in the Vertebrate Olfactory System	LB Buck	19:517–44
Mechanisms of Olfactory Discrimination: Converging Evidence for Common Principles Across Phyla	JG Hildebrand, GM Shepherd	20:591–629
Molecular Biology of Odorant Receptors in Vertebrates	P Mombaerts	22:487–509
Guanylyl Cyclases as a Family of Putative Odorant Receptors	AD Gibson, DL Garbers	23:417–39
Odor Encoding as an Active, Dynamical Process: Experiments, Computation and Theory	G Laurent, M Stopfer, RW Friedrich, MI Rabinovich, A Volkovskii, HDI Abarbanel	24:263–97
Imaging and Coding in the Olfactory System	JS Kauer, J White	24:963–79

Pain

Pain Genes?: Natural Variation and Transgenic Mutants	JS Mogil, L Yu, AI Basbaum	23:777–811
The Vanilloid Receptor: A Molecular Gateway to the Pain Pathway	MJ Caterina, D Julius	24:487–517
Contributions of the Medullary Raphe and Ventromedial Reticular Region to Pain Modulation and Other Homeostatic Functions	P Mason	24:737–77
Pain Mechanisms: Labeled Lines Versus Convergence in Central Processing	AD Craig	26:1–30

Plasticity

Plasticity of the Spinal Neural Circuitry After Injury	VR Edgerton, NJK Tillakaratne, AJ Bigbee, RD de Leon, RR Roy	27:145–67
Adult Neurogenesis in the Mammalian Central Nervous System	G-l Ming, H Song	28:223–50

The Plastic Human Brain Cortex	A Pascual-Leone, A Amedi, F Fregni, LB Merabet	28:377–401

Receptors and Receptor Subtypes

Functional and Structural Complexity of Signal Transduction via G-Protein-Coupled Receptors	T Gudermann, T Schöeneberg, G Schultz	20:395–423
The Ephrins and Eph Receptors in Neural Development	JG Flanagan, P Vanderhaeghen	21:309–45
The Vanilloid Receptor: A Molecular Gateway to the Pain Pathway	MJ Caterina, D Julius	24:487–517
Role of the Reelin Signaling Pathway in Central Nervous System Development	DS Rice, T Curran	24:1005–39
AMPA Receptor Trafficking and Synaptic Plasticity	R Malinow, RC Malenka	25:103–26

Sleep and Sleep Disorders

To Eat or Sleep? The Role of Orexin in Coordination of Feeding and Arousal	JT Willie, RM Chemelli, CM Sinton, M Yanagisawa	24:429–58
The Role of Hypocretins (Orexins) in Sleep Regulation and Narcolepsy	S Taheri, JM Zeitzer, E Mignot	25:283–313

Somatosensory System

Touch and Go: Decision-Making Mechanisms in Somatosensation	R Romo, E Salinas	24:107–37
Sensory Signals in Neural Populations Underlying Tactile Perception and Manipulation	AW Goodwin, HE Wheat	27:53–77

Synapses/Synaptic Transmission

Neurotransmitter Release	G Matthews	19:219–33
Synaptic Regulation of Mesocorticolimbic Dopamine Neurons	FJ White	19:405–36
Long-Term Depression in Hippocampus	MF Bear, WC Abraham	19:437–62

Monitoring Secretory Membrane with FM1-43 Fluorescence	AJ Cochilla, JK Angleson, WJ Betz	22:1–10
Autoimmunity and Neurological Disease: Antibody Modulation of Synaptic Transmission	KD Whitney, JO McNamara	22:175–95
Presynaptic Ionotropic Receptors and the Control of Transmitter Release	AB MacDermott, LW Role, SA Siegelbaum	22:443–85
PDZ Domains and the Organization of Supramolecular Complexes	C Sala, MH Sheng	24:1–29
The Role and Regulation of Adenosine in the Central Nervous System	TV Dunwiddie, SA Masino	24:31–55
Synaptic Modification by Correlated Activity: Hebb's Postulate Revisited	G-Q Bi, M-M Poo	24:139–66
Influence of Dendritic Conductances on the Input-Output Properties of Neurons	A Reyes	24:653–75
α-Latrotoxin and its Receptors: Neurexins and CIRL/Latrophilins	TC Südhof	24:933–62
Cell Biology of the Presynaptic Terminal	VN Murthy, P De Camilli	26:701–28
Control of Central Synaptic Specificity in Insect Sensory Neurons	JM Blagburn, JP Bacon	27:29–51
Desensitization of G Protein–Coupled Receptors and Neuronal Functions	RR Gainetdinov, RT Premont, LM Bohn, RJ Lefkowitz, MG Caron	27:107–44
Ubiquitin-Dependent Regulation of the Synapse	A DiAntonio, L Hicke	27:223–46
The Synaptic Vesicle Cycle	TC Südhof	27:509–47
Mechanisms of Vertebrate Synaptogenesis	CL Waites, A Craig, CC Garner	28:251–74

Vision

The Molecules of Mechanosensation	J García-Añoveros, DP Corey	20:563–90
Localization and Globalization in Conscious Vision	S Zeki	24:57–86
The Physiology of Stereopsis	BG Cumming, GC DeAngelis	24:203–38
Activation, Deactivation, and Adaptation in Vertebrate Photoreceptor Cells	ME Burns, DA Baylor	24:779–805

Natural Image Statistics and Neural Representation	EP Simoncelli, BA Olshausen	24:1193–1215
Contextual Influences on Visual Processing	TD Albright, GR Stoner	25:339–79
The Circuitry of V1 and V2: Integration of Color, Form, and Motion	LC Sincich, JC Horton	28:303–26

Visual System

Visual Object Recognition	NK Logothetis, DL Sheinberg	19:577–621
Sense and the Single Neuron: Probing the Physiology of Perception	AJ Parker, WT Newsome	21:227–77
Computational Neuroimaging of Human Visual Cortex	BA Wandell	22:145–73
Neural Selection and Control of Visually Guided Eye Movements	JD Schall, KG Thompson	22:241–59
Space and Attention in Parietal Cortex	CL Colby, ME Goldberg	22:319–50
The Koniocellular Pathway in Primate Vision	SHC Hendry, RC Reid	23:127–53
Confronting Complexity: Strategies for Understanding the Microcircuitry of the Retina	RH Masland, E Raviola	23:249–84
Neural Mechanisms of Orientation Selectivity in the Visual Cortex	D Ferster, KD Miller	23:441–71
Parallel Pathways for Spectral Coding in Primate Retina	DM Dacey	23:743–75
The Role and Regulation of Adenosine in the Central Nervous System	TV Dunwiddie, SA Masino	24:31–55
Localization and Globalization in Conscious Vision	S Zeki	24:57–86
The Physiology of Stereopsis	BG Cumming, GC DeAngelis	24:203–38
Activation, Deactivation, and Adaptation in Vertebrate Photoreceptor Cells	ME Burns, DA Baylor	24:779–805
Natural Image Statistics and Neural Representation	EP Simoncelli, BA Olshausen	24:1193–1215
Color Vision	KR Gegenfurtner, DC Kiper	26:181–206
Brain Representation of Object-Centered Space in Monkeys and Humans	CR Olson	26:331–54

Progress Toward Understanding the Genetic and Biochemical Mechanisms of Inherited Photoreceptor Degenerations	LR Pacione, MJ Szego, S Ikeda, PM Nishina, RR McInnes	26:657–700
The Human Visual Cortex	K Grill-Spector, R Malach	27:649–77
Visual Motor Computations in Insects	MV Srinivasan, S Zhang	27:679–96
Structure and Function of Visual Area MT	RT Born, DC Bradley	28:157–89

Annual Reviews — Your Starting Point for Research Online
http://arjournals.annualreviews.org

- Over 900 Annual Reviews volumes—more than 25,000 critical, authoritative review articles in 31 disciplines spanning the Biomedical, Physical, and Social sciences— available online, including all Annual Reviews back volumes, dating to 1932
- Current individual subscriptions include seamless online access to full-text articles, PDFs, Reviews in Advance (as much as 6 months ahead of print publication), bibliographies, and other supplementary material in the current volume and the prior 4 years' volumes
- All articles are fully supplemented, searchable, and downloadable — see http://neuro.annualreviews.org
- Access links to the reviewed references (when available online)
- Site features include customized alerting services, citation tracking, and saved searches

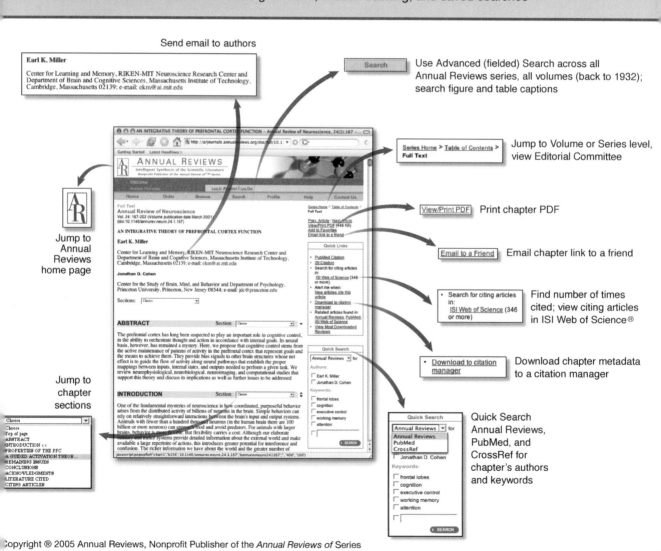

Copyright ® 2005 Annual Reviews, Nonprofit Publisher of the *Annual Reviews of* Series